COMPREHENSIVE ANALYTICAL CHEMISTRY

Elsevier
Radarweg 29, PO Box 211, 1000 AE Amsterdam, The Netherlands
The Boulevard, Langford Lane, Kidlington, Oxford OX5 1GB, UK

First edition 2006

Notice
No responsibility is assumed by the publisher for any injury and/or damage to persons
or property as a matter of products liability, negligence or otherwise, or from any use
or operation of any methods, products, instructions or ideas contained in the material
herein. Because of rapid advances in the medical sciences, in particular, independent
verification of diagnoses and drug dosages should be made

Library of Congress Cataloging-in-Publication Data
A catalog record for this book is available from the Library of Congress

British Library Cataloguing in Publication Data
A catalogue record for this book is available from the British Library

ISBN-13: 978-0-444-52259-7
ISBN-10: 0-444-52259-X
ISSN: 0166-526X

For information on all Elsevier publications
visit our website at books.elsevier.com

Printed and bound in The Netherlands

06 07 08 09 10   10 9 8 7 6 5 4 3 2 1

COMPREHENSIVE ANALYTICAL CHEMISTRY

ADVISORY BOARD

Wilson & Wilson's

COMPREHENSIVE ANALYTICAL CHEMISTRY

Edited by

D. BARCELÓ

Research Professor
Department of Environmental Chemistry
IIQAB-CSIC
Jordi Girona 18-26
08034 Barcelona
Spain

Wilson & Wilson's

COMPREHENSIVE ANALYTICAL CHEMISTRY

VOLUME 47

MODERN INSTRUMENTAL ANALYSIS

Edited by

S. AHUJA

Ahuja Consulting, Inc., 1061 Rutledge Court, NW, Calabash, NC 28467, USA

N. JESPERSEN

Chemistry Department, St. John's University, 8000 Utopia Parkway, Jamaica, NY 1143920, USA

ELSEVIER

AMSTERDAM – BOSTON – HEIDELBERG – LONDON – NEW YORK – OXFORD – PARIS
SAN DIEGO – SAN FRANCISCO – SINGAPORE – SYDNEY – TOKYO

CONTRIBUTORS TO VOLUME 47

Satinder Ahuja
  Ahuja Consulting, Inc., 1061 Rutledge Court, Calabash, NC 284675, USA
Harry G. Brittain
  Center for Pharmaceutical Physics, 10 Charles Road, Milford, NJ 08848, USA
Richard D. Bruce
  Johnson & Johnson Pharmaceutical Research and Development, PO Box 776, Welsh and McKean Roads, Spring House, PA 19477, USA
David J. Burinsky
  GlaxoSmitKline, Five Moore Dr., PO Box 13398, Research Triangle Park, NC 27709, USA
Emil W. Ciurczak
  77 Park Road, Golden Bridge, NY 10526, USA
Diane M. Diehl
  Waters Corp., 34 Maple St., Midford, MA 01757, USA
Sue M. Ford
  Toxicology Program, St. John's University, 8000 Utopia Parkway, Jamaica, NY 11439, USA
Pamela M. Grillini
  Pfizer Corp., PO Box 4077, Eastern Point Rd., Groton, CT 06340, USA
Abul Hussam
  Department of Chemistry and Biochemistry, George Mason University, Fairfax, VA 22030, USA
Neil Jespersen
  St. John's University, Chemistry Department, 8000 Utopia Parkway, Jamaica, NY 1143920, USA
M. Jimidar
  Global Analytical Development, Johnson & Johnson Pharmaceutical Research and Development, A Division of Janssen Pharmaceutica n.v., Turnhoutseweg 30, B-2340 Beerse, Belgium
C.H. Lochmüller
  Department of Chemistry, Duke University, 2819 McDowell Rd., Durham, NC 27705-5604, USA
Linda Lohr
  Pfizer Inc., Global Research & Development, Eastern Point Rd., Groton, CT 06340, USA

Brian L. Marquez
  Analytical Research & Development Department, Pfizer Global Research & Development, Eastern Point Rd., Groton, CT 06340, USA
Gary Martin
  Schering-Plough Corp., Chemical and Physical Sciences, 556 Morris Avenue, MS S7-E2, Summit, NJ 07901
Mary Ellen P. McNally
  DuPont Crop Protection, E.I. duPont deNemours and Co., INC, Stine Haskell Research Center, 1090 Elkton Road, S315/2224, Newark, DE 19711-3507
Yu Qin
  Department of Chemistry, Renmin University of China, Beijing 100872, China
Douglas Raynie
  Department of Chemistry and Biochemistry, South Dakota State University, Brookings, SD 57007, USA
Terence H. Risby
  Division of Toxicological Sciences, Department of Environmental Health Sciences, Bloomberg School of Public Health, The Johns Hopkins University, Baltimore, MD, USA
Thomas R. Sharp
  Analytical Research & Development Department, Pfizer Global Research & Development, Eastern Point Rd., Groton, CT 06340, USA
Nicholas H. Snow
  Department of Chemistry and Biochemistry, Seton Hall University, 4000 South Orange Ave., South Orange, NJ 07079, USA
Martin Telting-Diaz
  Department of Chemistry, Brooklyn College of The City University of New York, Brooklyn, NY 11210-2889, USA
Alan H. Ullman
  Proctor and Gamble Co., 6100 Center Hill Avenue, Cincinnati, OH 45224, USA
Enju Wang
  Chemistry Department, St. John's University, 8000 Utopia Parkway, Jamaica, NY 11439, USA

WILSON AND WILSON'S

COMPREHENSIVE ANALYTICAL CHEMISTRY

VOLUMES IN THE SERIES

Volumes in the series

Volumes in the series

Contents

Chapter 1. Overview
  Satinder Ahuja

Contents

Contents

Contents

Chapter 8. *Atomic spectroscopy*
Terence H. Risby

Chapter 9. *Emission methods*
C.H. Lochmüller

Contents

Contents

Contents

Chapter 13. *Thin-layer chromatography*
Pamela M. Grillini

Chapter 14. *Gas chromatography*
Nicholas H. Snow

Contents

Chapter 15. High-pressure liquid chromatography
Satinder Ahuja

Contents

Contents

Contents

Chapter 19. Hyphenated methods

Thomas R. Sharp and Brian L. Marquez

Contents

Contents

Contents

Chapter 21. *Problem solving and guidelines for method selection*
Alan H. Ullman and Douglas E. Raynie

Contents

range from a high school diploma to a Ph.D.

Preface

Modern analytical chemistry frequently requires precise analytical measurements, at very low concentrations, with a variety of instruments. A high-resolution separation has to be generally performed with a selective chromatographic method prior to quantification. Therefore the knowledge of instrumentation used in chemical analysis today is of paramount importance to solve current problems and assure future progress in various fields of scientific endeavor. These include chemistry, biochemistry, pharmaceutical chemistry, medicinal chemistry, biotechnology, nanotechnology, archaeology, anthropology, environmental sciences, and a variety of other scientific disciplines. The instruments may be operated by a variety of people in industry, government, or academic fields, whose educational backgrounds can range from a high school diploma to a Ph.D. with postdoctoral experience. Increased knowledge relating to the principles of the instrumentation and separation methodologies allows optimal usage of instrumentation with more meaningful data generation that can be interpreted reliably.

This book covers the fundamentals of instrumentation as well as the applications, to lead to better utilization of instrumentation by all scientists who plan to work in diverse scientific laboratories. It should serve as an educational tool as well as a first reference book for the practicing instrumental analyst.

This text has been broadly classified into six sections:

1. Overview, Sampling, Evaluation of Physical Properties, and Thermal Analysis (Chapters 1–4)
2. Spectroscopic Methods (Chapters 5–11)
3. Chromatographic Methods (Chapters 12–16)
4. Electrophoretic and Electrochemical Methods (Chapters 17–18)
5. Hyphenated Methods, Unique Detectors (Chapters 19–20)
6. Problem Solving and Guidelines for Method Selection (Chapter 21)

Each of these sections has one or more chapters covering important aspects of the titled subject. Every chapter is a distinct entity that provides the relevant information deemed necessary by the author. All chapters contain important information about applications that illustrate the use of the methods. They also include an appropriate set of review questions.

This book is planned to enable the reader to understand the proper utilization of instrumentation and various methodologies that are based on them. It provides an understanding of how various methods may be combined to obtain a comprehensive picture of the nature of sample in question. The reader will also gain a better understanding of how to select the best method for solving a given problem.

Aimed as the text for a special course or graduate students, it provides the necessary background information for various laboratory personnel. The text should be very useful for managers in academia, industrial, and government laboratories interested in improving their scientific knowledge and background.

September 1, 2005

Satinder Ahuja
Neil Jespersen

Series editor's preface

Modern instrumental analysis has a long history in the field of analytical chemistry, and that makes it difficult to prepare a book like this one. The main reason is the continuous improvement in the instrumentation applied to the analytical field. It is almost impossible to keep track of the latest developments in this area. For example, at PITTCON, the largest world exhibition in analytical chemistry, every year several new analytical instruments more sophisticated and sensitive than the previous versions are presented for this market.

Modern Instrumental Analysis is written to reflect the popularity of the various analytical instruments used in several different fields of science. The chapters are designed to not only give the reader the understanding of the basics of each technique but also to give ideas on how to apply each technique in these different fields. The book contains 21 chapters and covers sampling; spectroscopic methods such as near infrared, atomic and emission methods, nuclear magnetic resonance and mass spectrometric methods; separation methods with all chromatographic techniques; electrochemical methods and hyphenated methods. The last chapter of the book, a complementary chapter, is very useful and is practically oriented to problem-solving, giving guidelines for method selection.

Considering all the chapters indicated above, the book is suitable for a wide audience, from students at the graduate level to experienced researchers and laboratory personnel in academia, industry and government. It is a good introductory book from which one can then go on to more specialized books such as the ones regularly published in the Comprehensive Analytical Chemistry series. Such a general book describing modern analytical methods has been needed since the start of this series and it is obvious that after a certain amount of time, a maximum of 10 years, an update of the book will be needed. Finally, I would like to specially thank the two editors and all the contributing authors of this book for their time and efforts in preparing this excellent and useful book on modern instrumental analysis.

D. Barceló
Depart. Environmental Chemistry, IIQAB-CSIC
Barcelona, Spain

Chapter 1

Overview

Satinder Ahuja

1.1   INTRODUCTION

Modern analytical chemistry generally requires precise analytical measurements at very low concentrations, with a variety of instruments. Frequently, high-resolution separations have to be achieved with selective chromatographic methods prior to analytical determinations [1,2]. Therefore, the knowledge of instrumentation used in chemical analysis today is of paramount importance to assure future progress in various fields of scientific endeavor. This includes various disciplines of chemistry such as biochemistry, pharmaceutical chemistry, medicinal chemistry, biotechnology, and environmental sciences. Instruments can be operated by a variety of people in industry, government, or academic fields, with a wide range of educational backgrounds. The optimal usage of instrumentation with more meaningful data generation that can be interpreted reliably is possible only with the improved knowledge of the principles of the instrumentations used for measurement as well as those utilized to achieve various separations. This book covers the fundamentals of instrumentation as well as the applications that should lead to better utilization of instrumentation by all scientists who plan to work in diverse scientific laboratories. It should serve as an educational tool as well as a first reference book for the practicing instrumental analyst.

This text has been broadly classified into the following areas:

General information (sampling/sample preparation and basic properties)
Spectroscopic methods
Chromatographic methods
Electrophoretic and electrochemical methods

Comprehensive Analytical Chemistry 47
S. Ahuja and N. Jespersen (Eds)
Volume 47 ISSN: 0166-526X DOI: 10.1016/S0166-526X(06)47001-X

Combination of chromatographic and spectroscopic methods, and
unique detectors
Problem solving and guidelines for method selection.

To provide a concise overview of this book, the contents of various
chapters are highlighted below. Also included here is an additional information that will help to round out the background information to
facilitate the performance of modern instrumental analysis.

1.2   SAMPLING AND SAMPLE PREPARATION

Prior to performance of any analysis, it is important to obtain a representative sample. Difficulties of obtaining such a sample and how it is
possible to do the best job in this area are discussed in Chapter 2.
Another important consideration is to prepare samples for the required
analytical determination(s). Sample preparation is frequently a major
time-consuming step in most analyses because it is rarely possible to
analyze a neat sample. The sample preparation generally entails a
significant number of steps prior to analysis with appropriate instrumental techniques, as described in this book. This chapter presents
various methods used for collecting and preparing samples for analysis
with modern time-saving techniques. It describes briefly the theory and
practice of each technique and focuses mainly on the analysis of organic
compounds of interest (analytes) in a variety of matrices, such as the
environment, foods, and pharmaceuticals.

1.3   EVALUATIONS OF BASIC PHYSICAL PROPERTIES

Some of the simplest techniques such as determinations of melting and
boiling points, viscosity, density, specific gravity, or refractive index can
serve as a quick quality control tool (Chapter 3). Time is often an important consideration in practical chemical analysis. Raw materials
awaiting delivery in tanker trucks or tractor-trailers often must be
approved before they are offloaded. Observation and measurement of
basic physical properties are the oldest known means of assessing
chemical purity and establishing identity. Their utility comes from the
fact that the vast majority of chemical compounds have unique values
for their melting points, boiling points, density, and refractive index.
Most of these properties change dramatically when impurities are
present.

1.4 THERMAL METHODS

Thermal analysis measurements are conducted for the purpose of evaluating the physical and chemical changes that may take place in a sample as a result of thermally induced reactions (Chapter 4). This requires interpretation of the events observed in a thermogram in terms of plausible thermal reaction processes. The reactions normally monitored can be endothermic (melting, boiling, sublimation, vaporization, desolvation, solid–solid phase transitions, chemical degradation, etc.) or exothermic (crystallization, oxidative decomposition, etc.) in nature. Thermal methods have found extensive use in the past as part of a program of preformulation studies, since carefully planned work can be used to indicate the existence of possible drug–excipient interactions in a prototype formulation. It is important to note that thermal methods of analysis can be used to evaluate compound purity, polymorphism, solvation, degradation, drug–excipient compatibility, and a wide variety of other thermally related characteristics.

1.5 GENERAL PRINCIPLES OF SPECTROSCOPY AND SPECTROSCOPIC ANALYSIS

The range of wavelengths that the human eye can detect varies slightly from individual to individual (Chapter 5). Generally, the wavelength region from 350 to 700 nm is defined as the visible region of the spectrum. Ultraviolet radiation is commonly defined as those wavelengths from 200 to 350 nm. Technically, the near-infrared (NIR) region starts immediately after the visible region at 750 nm and ranges up to 2,500 nm. The classical infrared region extends from 2500 nm (2.5 μm) to 50,000 nm (50 μm). The energies of infrared radiation range from 48 kJ/mol at 2500 nm to 2.4 kJ/mol at 50,000 nm. These low energies are not sufficient to cause electron transitions, but they are sufficient to cause vibrational changes within molecules. This is why infrared spectroscopy is often called vibrational spectroscopy. The principles involved in these spectroscopic techniques are discussed in this chapter.

1.6 NEAR-INFRARED SPECTROSCOPY

As mentioned above, the NIR portion of the electromagnetic spectrum is located between the visible and mid-range infrared sections, roughly 750–2500 nm or 13,333–4000 cm^{-1}. It mainly consists of overtones and combinations of the bands found in the mid-range infrared region ($4000–200\ cm^{-1}$). In general, NIR papers did not begin to appear in earnest until the 1970s, when commercial instruments became easily available owing to the work of the US Department of Agriculture (USDA). Some of these developments are discussed under instrumentation in Chapter 6. After the success of the USDA, food producers, chemical producers, polymer manufacturers, gasoline producers, etc. picked up the ball and ran with it. The last to become involved, mainly for regulatory reasons, are the pharmaceutical and biochemical industries.

1.7   X-RAY DIFFRACTION AND FLUORESCENCE

All properly designed investigations into the solid-state properties of pharmaceutical compounds begin with understanding of the structural aspects involved. There is no doubt that the primary tool for the study of solid-state crystallography is X-ray diffraction. To properly comprehend the power of this technique, it is first necessary to examine the processes associated with the ability of a crystalline solid to act as a diffraction grating for the electromagnetic radiation of appropriate wavelength. Chapter 7 separately discusses the practice of X-ray diffraction as applied to the characterization of single crystals and to the study of powdered crystalline solids.

Heavy elements ordinarily yield considerably more intense X-ray fluorescence (XRF) spectrum, because of their superior fluorescent yield bands than the light elements. This feature can be exploited to determine the concentration of inorganic species in a sample or the concentration of a compound that contains a heavy element in some matrix. The background emission detected in an XRF spectrum is usually due to scattering of the source radiation. Since the scattering intensity from a sample is inversely proportional to the atomic number of the scattering atom, it follows that background effects are more pronounced for samples consisting largely of second-row elements (i.e., organic molecules of pharmaceutical interest). Hence, background correction routines play a major role in transforming raw XRF spectra into spectra suitable for quantitative analysis.

1.8   ATOMIC SPECTROSCOPY

Elemental analysis at the trace or ultratrace level can be performed by a number of analytical techniques; however, atomic spectroscopy remains the most popular approach. Atomic spectroscopy can be subdivided into three fields: atomic emission spectroscopy (AES), atomic absorption spectroscopy (AAS), and atomic fluorescence spectroscopy (AFS) that differ by the mode of excitation and the method of measurement of the atom concentrations. The selection of the atomic spectroscopic technique to be used for a particular application should be based on the desired result, since each technique involves different measurement approaches. AES excites ground-state atoms (atoms) and then quantifies the concentrations of excited-state atoms by monitoring their special deactivation. AAS measures the concentrations of ground-state atoms by quantifying the absorption of spectral radiation that corresponds to allowed transitions from the ground to excited states. AFS determines the concentrations of ground-state atoms by quantifying the radiative deactivation of atoms that have been excited by the absorption of discrete spectral radiation. All of these approaches are discussed in Chapter 8. It should be noted that AAS with a flame as the atom reservoir and AES with an inductively coupled plasma have been used successfully to speciate various ultratrace elements.

1.9 EMISSION SPECTROSCOPIC MEASUREMENTS

The term emission refers to the release of light from a substance exposed to an energy source of sufficient strength (Chapter 9). The most commonly observed line emission arises from excitation of atomic electrons to higher level by an energy source. The energy emitted by excited atoms of this kind occurs at wavelengths corresponding to the energy level difference. Since these levels are characteristic for the element, the emission wavelength can be characteristic for that element. For example, sodium and potassium produce different line spectra. Emission methods can offer some distinct advantages over absorption methods in the determination of trace amounts of material. Furthermore, it may be possible to detect even single atoms.

1.10 NUCLEAR MAGNETIC RESONANCE SPECTROSCOPY

Nuclear magnetic resonance (NMR) spectroscopy is used to study the behavior of the nuclei in a molecule when subjected to an externally applied magnetic field. Nuclei spin about the axis of the externally applied magnetic field and consequently possess an angular momentum. The group of nuclei most commonly exploited in the structural

S. Ahuja

characterization of small molecules by NMR methods are the spin 1/2 nuclei, which include ^{1}H, ^{13}C, ^{19}F, and ^{31}P. NMR is amenable to a broad range of applications. It has found wide utility in the pharmaceutical, medical, and petrochemical industries as well as across the polymer, materials science, cellulose, pigment, and catalysis fields. The vast diversity of NMR applications may be due to its profound ability to probe both chemical and physical properties of molecules, including chemical structure and molecular dynamics. Furthermore, it can be applied to liquids, solids, or gases (Chapter 10).

1.11 MASS SPECTROMETRY

Mass spectrometry is arguably one of the most versatile analytical measurement tools available to scientists today, finding application in virtually every discipline of chemistry (i.e., organic, inorganic, physical, and analytical) as well as biology, medicine, and materials science (Chapter 11). The technique provides both qualitative and quantitative information about organic and inorganic materials, including elemental composition, molecular structure, and the composition of mixtures. The combination of the technique with the powerful separation capabilities of gas chromatography (GC-MS) and liquid chromatography (LC-MS) led to the development of new kinds of mass analyzers and the introduction of revolutionary new ionization techniques. These new ionization techniques (introduced largely in the past 15 years) are primarily responsible for the explosive growth and proliferation of mass spectrometry as an essential tool for biologists and biochemists. The information derived from a mass spectrum is often combined with that obtained from techniques such as infrared spectroscopy and NMR spectroscopy to generate structural assignments for organic molecules. The attributes of mass spectrometry that make it a versatile and valuable analytical technique are its sensitivity (e.g., recently, a detection limit of approximately 500 molecules, or 800 yoctomoles has been reported) and its specificity in detecting or identifying unknown compounds.

1.12 THEORY OF SEPARATIONS

Separation models are generally based on a mixture of empirical and chemical kinetics and thermodynamics. There have been many attempts to relate known physical and chemical properties to observation

6

and to develop a "predictive theory." The goal has been to improve our understanding of the underlying phenomena and provide an efficient approach to method development (see Chapter 12). The latter part of the goal is of greater interest to the vast majority of people who make use of separation methods for either obtaining relatively pure materials or in isolating components of a chemical mixture for quantitative measurements.

1.13   THIN-LAYER CHROMATOGRAPHY

Thin-layer chromatography (TLC) is one of the most popular and widely used separation techniques because it is easy to use and offers adequate sensitivity and speed of separations. Furthermore, multiple samples can be run simultaneously. It can be used for separation, isolation, identification, and quantification of components in a sample. The equipment used for performance of TLC, including applications of TLC in drug discovery process, is covered in Chapter 13. This technique has been successfully used in biochemical, environmental, food, pharmacological, and toxicological analyses.

1.14   GAS CHROMATOGRAPHY

The term gas chromatography (GC) refers to a family of separation techniques involving two distinct phases: a stationary phase, which may be a solid or a liquid; and a moving phase, which must be a gas that moves in a definite direction. GC remains one of the most important tools for analysts in a variety of disciplines. As a routine analytical technique, GC provides unparalleled separation power, combined with high sensitivity and ease of use. With only a few hours of training, an analyst can be effectively operating a gas chromatograph, while, as with other instrumental methods, to fully understand GC may require years of work. Chapter 14 provides an introduction and overview of GC, with a focus on helping analysts and laboratory managers to decide when GC is appropriate for their analytical problem, and assists in using GC to solve various problems.

1.15   HIGH-PRESSURE LIQUID CHROMATOGRAPHY

The phenomenal growth in chromatography is largely due to the introduction of the technique called high-pressure liquid chromatography, which is frequently called high-performance liquid chromatography (both are abbreviated as HPLC; see discussion in Chapter 15 as to which term is more appropriate). It allows separations of a large variety of compounds by offering some major improvements over the classical column chromatography, TLC, GC; and it presents some significant advantages over more recent techniques such as supercritical fluid chromatography (SFC), capillary electrophoresis (CE), and electrokinetic chromatography.

1.16   SUPERCRITICAL FLUID CHROMATOGRAPHY

Supercritical fluid chromatography employs supercritical fluid instead of gas or liquid to achieve separations. Supercritical fluids generally exist at conditions above atmospheric pressure and at an elevated temperature. As a fluid, the supercritical state generally exhibits properties that are intermediate to the properties of either a gas or a liqiud. Chapter 16 discusses various advantages of SFC over GC and HPLC and also provides some interesting applications.

1.17   ELECTROMIGRATION METHODS

The contribution of electromigration methods to analytical chemistry and biological/pharmaceutical science has been very significant in the last several decades. Electrophoresis is a separation technique that is based on the differential migration of charged compounds in a semi-conductive medium under the influence of an electric field (Chapter 17). It is the first method of choice for the analyses of proteins, amino acids, and DNA fragments. Electrophoresis in a capillary has evolved the technique into a high-performance instrumental method. The applications are widespread and include small organic-, inorganic-, charged-, neutral-compounds, and pharmaceuticals. Currently, CE is considered an established tool that analytical chemists use to solve many analytical problems. The major application areas still are in the field of DNA sequencing and protein analysis, as well as low-molecular weight compounds (pharmaceuticals). The Human Genome Project was completed many years earlier than initially planned, because of contributions of CE. The CE technology has grown to a certain maturity, which has allowed development and application of robust analytical methods. This technique is already described as general monographs in European Pharmacopoeia and the USP. Even though both electromigration and chromatographic methods evolved separately over many decades, they have converged into a single method called capillary electrochromato-graphy. This technique has already shown a lot of promise.

1.18 ELECTROCHEMICAL METHODS

The discussion of electrochemical methods has been divided into two areas, potentiometry and voltammetry.

1.18.1 Potentiometry

In potentiometry, the voltage difference between two electrodes is measured while the electric current between the electrodes is maintained under a nearly zero-current condition (Chapter 18a). In the most common forms of potentiometry, the potential of a so-called indicator electrode varies, depending on the concentration of the analyte while the potential arising from a second reference electrode is ideally a constant. Most widely used potentiometric methods utilize an ion-selective electrode membrane whose electrical potential to a given measuring ion, either in solution or in the gas phase, provides an analytical response that is highly specific. A multiplicity of ion-selective electrode designs ranging from centimeter-long probes to miniaturized microm-eter self-contained solid-state chemical sensor arrays constitute the basis of modern potentiometric measurements that have become in-creasingly important in biomedical, industrial, and environmental ap-plication fields. The modern potentiometric methods are useful for a multitude of diverse applications such as batch determination of medi-cally relevant electrolytes (or polyions in undiluted whole blood), in vivo real time assessment of blood gases and electrolytes with non-thrombogenic implantable catheters, multi-ion monitoring in indus-trial process control via micrometer solid-state chemical-field effect transistors (CHEMFETs), and the determination of heavy metals in environmental samples.

1.18.2 Voltammetry

This is a dynamic electrochemical technique, which can be used to study electron transfer reactions with solid electrodes. A voltammo-gram is the electrical current response that is due to applied exci-tation potential. Chapter 18b describes the origin of the current in steady-state voltammetry, chronoamperometry, cyclic voltammetry, and square wave voltammetry and other pulse voltammetric techniques.

The employment of these techniques in the study of redox reactions provides very useful applications.

1.19   HYPHENATED METHODS

The combination of separation techniques with spectroscopy has provided powerful tools. For example, UV and various other spectroscopic detectors have been commonly used in HPLC (or LC), SFC, and CE (see Chapters 15–17). To fully utilize the power of combined chromatography-mass spectrometry techniques, it is necessary to understand the information and nuances provided by mass spectrometry in its various forms (Chapter 19). In principle, these novel systems have solved many of the problems associated with structure elucidation of low-level impurities. It must be emphasized that the tried-and-true method of off-line isolation and subsequent LC-MS is still the most reasonable approach in some cases. In general, some compounds behave poorly, spectroscopically speaking, when solvated in particular solvent/solvent–buffer combinations. The manifestation of this poor behavior can be multiple stable conformations and/or equilibration isomers in the intermediate slow-exchange paradigm, resulting in very broad peaks on the NMR time scale. Therefore, it can be envisioned that the flexibility of having an isolated sample in hand can be advantageous. There are additional issues that can arise, such as solubility changes, the need for variable temperature experiments, the need to indirectly observe heteroatoms other than ^{13}C (triple resonance cryogenic probe required), as well as many others. It must be stated that LC-NMR and its various derivatives are ideally suited for looking at simple regiochemical issues in relative complex systems. In some cases, full structure elucidation of unknown compounds can be completed. Although this will become more routine with some of the recent advances such as peak trapping in combination with cryogenic flow-probes. There are many elegant examples of successful applications of LC-NMR and LC-NMR/MS to very complex systems. The principle advantage is that in most cases one can collect NMR and MS data on an identical sample, thus eliminating the possibility of isolation-induced decomposition.

1.20   UNIQUE DETECTORS

Special detectors have been developed to deal with various situations. A few interesting examples are discussed below. These include optical sensors (optrodes), biosensors, bioactivity detectors, and drug detectors. Some of the new developments in the area of lab on chip have also been summarized.

1.20.1 Optical sensors

Chemical sensors are simple devices based on a specific chemical recognition mechanism that enables the direct determination of either the activities or concentrations of selected ions and electrically neutral species, without pretreatment of the sample. Clearly, eliminating the need for sample pretreatment is the most intriguing advantage of chemical sensors over other analytical methods. Optical chemical sensors (optodes or optrodes) use a chemical sensing element and optical transduction for the signal processing; thus these sensors offer further advantages such as freedom from electrical noise and ease of miniaturization, as well as the possibility of remote sensing. Since its introduction in 1975, this field has experienced rapid growth and has resulted in optodes for a multitude of analytes (see Chapter 20a).

Optical biosensors can be designed when a selective and fast bioreaction produces chemical species that can be determined by an optical sensor. Like the electrochemical sensors, enzymatic reactions that produce oxygen, ammonia, hydrogen peroxide, and protons can be utilized to fabricate optical sensors.

1.20.2 Bioactivity detectors

Routine monitoring of food and water for the presence of pathogens, toxins, and spoilage-causing microbes is a major concern for public health departments and the food industry. In the case of disease-causing contamination, the identification of the organism is critical to trace its source. The Environmental Protection Agency monitors drinking water, ambient water, and wastewater for the presence of organisms such as nonpathogenic coliform bacteria, which are indicators of pollution. Identifying the specific organism can aid in locating the source of the pollution by determining whether the organism is from humans, livestock, or wildlife. Pharmaceutical companies monitor water-for-injection for bacterial toxins called pyrogens, which are not removed by filter-sterilization methods. Chapter 20b discusses the need for methods to quickly detect biothreat agents, which may be introduced as bacteria, spores, viruses, rickettsiae, or toxins, such as the botulism toxin or ricin.

1.20.3 Drug detectors

Drug detection technologies are used in various law enforcement scenarios, which present challenges in terms of instrumentation development. Not only "must be detected" drugs vary, but the amounts to be detected range from microgram quantities left as incidental contamination from drug activity to kilograms being transported as hidden caches. The locations of hidden drugs create difficulties in detection as well. Customs and border agents use drug-detection technology to intercept drugs smuggled into the country in bulk cargo or carried by individuals. Correctional facilities search for drugs being smuggled into the prison by mail (e.g., small amounts under stamps) or by visitors, and furthermore need to monitor drug possession inside the prison. Law enforcement representatives may use detectors in schools to find caches of dealers. Other users of drug detection technology include aviation and marine carriers, postal and courier services, private industry, and the military. The task is to find the relatively rare presence of drugs in a population of individuals or items, most of which will be negative. Consequently, for drug detection purposes, the ability to find illicit contraband is of greater importance than accurate quantitation. Chapter 20c describes various instruments that have been developed for drug detection.

Microchip Arrays: An interest in developing a lab on chip has led to some remarkable advancements in instrumentation [3]. In the early 1990s, microfabrication techniques borrowed from the electronic industry to create fluid pathways in materials such as glass and silicon [4]. This has led to instrumentation such as NanoLC-MS [5]. The main focus of these methods is analysis of biologically sourced material for which existing analytical methods are cumbersome and expensive. The major advantages of these methods are that a sample size 1000–10,000 times smaller than with the conventional system can be used, high-throughput analysis is possible, and a complete standardization of analytical protocol can be achieved. A large number of sequence analyses have to be performed in order to build a statistically relevant database of sequence variation versus phenotype for a given gene or set of genes [6]. For example, patterns of mutation in certain genes confer resistance to HIV for various antiretrovirals. Sequence analyses of many HIV samples in conjunction with phenotype data can enable researchers to explore and design better therapeuticals. Similarly, the relationship of mutation in cancer-associated genes, such as p53, to disease severity can be addressed by massive sequence analyses.

DNA probe array systems are likely to be very useful for these analyses. Active DNA microchip arrays with several hundred test sites are being developed for evolving genetic diseases and cancer diagnostics. These approaches can lead to high-throughput screening (HTS) in drug discovery. The Human Genome Project has been a major driving force in the development of suitable instruments for genome analysis. The companies that can identify genes that are useful for drug discovery are likely to reap the harvest in terms of new therapeutic agents and therapeutic approaches and commercial success in the future.

1.21 PROBLEM SOLVING AND GUIDELINES FOR METHOD SELECTION

Analytical chemistry helps answer very important questions such as "What is in this sample, and how much is there?" It is an integral part of how we perceive and understand the world and universe around us, so it has both very practical and also philosophical and ethical implications. Chapter 21 deals with the analytical approach rather than orienting primarily with techniques, even when techniques are discussed. It is expected that this will further develop critical-thinking and problem-solving skills and, at the same time, gain more insight into an understanding and appreciation of the physical world surrounding us. This chapter will enable the reader to learn the type of thinking needed to solve real-world problems, including how to select the "best" technique for the problem.

REFERENCES

1   S. Ahuja, *Trace and Ultratrace Analysis by HPLC*, Wiley, New York, 1992.
2   S. Ahuja, *Chromatography and Separation Science*, Academic, San Diego, 2003.
3   S. Ahuja, *Handbook of Bioseparations*, Academic, San Diego, 2000.
4   *Am. Lab.*, November. (1998) 22.
5   *PharmaGenomics*, July/August, 2004.
6   T. Kreiner, *Am. Lab.*, March, (1996) 39.

Chapter 2

Sampling and sample preparation

Satinder Ahuja and Diane Diehl

2.1 INTRODUCTION

Sample preparation (SP) is frequently necessary for most samples and still remains one of the major time-consuming steps in most analyses. It can be as simple as dissolution of a desired weight of sample in a solvent prior to analysis, utilizing appropriate instrumental techniques described in this book. However, it can entail a number of procedures, which are listed below, to remove the undesirable materials [1] (also see Section 2.3 below):

1. Removal of insoluble components by filtration or centrifugation
2. Precipitation of undesirable components followed by filtration or centrifugation
3. Liquid–solid extraction followed by filtration or centrifugation
4. Liquid–liquid extraction
5. Ultrafiltration
6. Ion-pair extraction
7. Derivatization
8. Complex formation
9. Freeze-drying
10. Solid-phase extraction
11. Ion exchange
12. Use of specialized columns
13. Preconcentration
14. Precolumn switching
15. Miscellaneous cleanup procedures

The application of robotics to enhance laboratory productivity has met with some success in laboratories with high-volume repetitive routine assays. However, the requirements for extensive installation, development, validation, and continuing maintenance effort have slowed

Comprehensive Analytical Chemistry 47
S. Ahuja and N. Jespersen (Eds)
Volume 47 ISSN: 0166-526X DOI: 10.1016/S0166-526X(06)47002-1

the growth of robotics in pharmaceutical laboratories [2]. Many robots in pharmaceutical R&D labs have been abandoned because they were not sufficiently flexible or reliable to automate the varying applications to warrant the continuing upkeep. The ideal dream of full automation, where one simply inputs samples (tablets, capsules, or other pharmaceutical products) and receives viable data with absolute reliability and without frequent troubleshooting, is generally not realized by laboratory robotics.

This chapter presents an overview of the various methods for collecting and preparing samples for analysis, from classical to more modern techniques. We provide below a general overview, outlining some of the theory and practice of each technique. Our focus is mainly on the analysis of organic compounds of interest (analytes) in a variety of matrices, such as environment, food, and pharmaceuticals. For further reading, the analysts are referred to a more detailed discussion of these techniques in various textbooks and key references [3–7].

2.2 SAMPLING

Successful quantitative analysis starts with the sample collection and ends with the interpretation of the data collected on them. Depending on the nature of the matrix, various sample collection techniques are employed. The sample should be representative of the bulk material. For example, with environmental samples, which include soil, water, and air, the first important consideration of representative sampling is right at the point of collection because you need to consider intrinsic heterogeneity of most materials [8]. Distribution of minerals in earth, dispersion of leachates from toxic waste into our aquifers, discharge of damaging chemicals into streams and rivers, and the buildup of air pollutants in urban areas all present complex sampling challenges and require sample-specific strategies to obtain clear and unbiased overviews. These considerations become exceedingly important whenever you are performing analysis of trace and ultratrace components [1,9,10].

Let us consider a difficult example of sampling of mined materials (sand, dolomite, and limestone) used for the manufacture of glass. The materials are delivered in railroad cars; each may contain up to 2 tons of material and are susceptible to vertical separation on the basis of size and density of particles because of vibrations during transportation [8]. To obtain a representative sample, sampling is done with a hollow conical stainless steel spear, on a vertical basis, which collects approximately 500 g each time. The sample is collected at six different locations per railroad car. This means a total of 3 kg sample is collected per railroad car, so you can imagine the magnitude of sample size you may have at hand if you multiply this by the number of railroad cars. Enormous difficulty will be encountered to select a representative sample if you were to analyze 0.5 g portion(s) of the bulk sample. Clearly, it would be necessary to homogenize the sample thoroughly if the analysis of the composite was the objective (see discussion below).

2.2.1 Sample size reduction

Particle size reduction is necessary to assure that various materials in the sample have roughly the same particle size [11]. Even after the particle size has been made reasonably uniform, the sample size has to be reduced to a manageable amount that can be analyzed, e.g., 0.5 g from a bulk sample of several kilograms in the example discussed above. Some of the approaches that have been used follow: coning and quartering, proportional divider, and straight-line sampler (for details see Ref. [8]).

It is important to divide the sample without discrimination with respect to size, density, or any other physical characteristics of the particles.

2.2.2 Sampling error

The relationship between the sampling error and particle size can be seen in Fig. 2.1. Various lines in that show the error relating to the amount sampled in each case. It is clear that a larger sample size reduces error significantly; however, it is not logical to work with very large quantities. The sampling error for the normal sample for analysis, which is likely to be anywhere between 0.1% and 1% of the whole sample, can be seen in Fig. 2.1. The sampling error can be minimized by grinding the particles down to fine size, assuming there are no issues relating to the stability of the analytes.

To minimize sampling errors in the pharmaceutical industry, an effective quality assurance approach utilizes sampling at various stages of the manufacturing process to provide a better indication of the true content of a product. Let us review an example of sampling tablets (drug product) for potency analysis; i.e., amount of the active ingredient present in each tablet. A batch of tablets can contain as many as a million or more tablets made with a high-speed tablet-punching

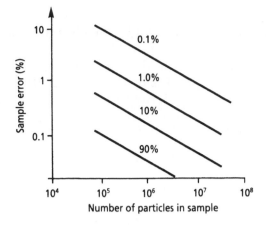

Fig. 2.1. Plot that shows the relationship between sample error, the number of particles in a sample, and different percentages of original material sampled [8].

machine. Since a relatively small number of tablets are finally tested to pass or fail a batch, several testing steps are built in to reduce variations that are due to sampling uncertainty and thus assure quality of the batch. Prior to manufacturing a batch of tablets, the active pharmaceutical ingredient (API) is mixed carefully with the inert ingredients (excipients) to produce a granulation that ensures that the API is uniformly dispersed and that it would become more readily available to reach the bloodstream when administered to patients. During the development phase, tests are conducted at various intervals in the mixing process to assure optimum mixing time. During production, the granulation is tested to assure that reliable mixing has been accomplished for each batch before tablets are manufactured. The tablets are manufactured with a tablet-punching machine, where an individual weighed portion is punched into an individual tablet. The weight of the API and the weight of the tablet can influence the accuracy of this process. Tablets are sampled at various portions of the run to assure that the tablet-punching machine is indeed producing uniform tablets in terms of dosage of API. The final batch needs to be sampled to represent the whole batch. This sample is drawn from various portions of the whole batch to assure fair representation of the batch. An analytical chemist then selects 20 tablets (a composite sample), in a random manner, from the bulk sample. The composite sample is powdered to produce a uniform powder from which a portion is drawn for the analysis. Duplicate analyses are generally carried out to arrive at the potency of the batch.

TABLE 2.1

Variation of dosage of Pertofrane capsules and Tofranil tablets[a]

| | % Tablets or capsules within indicated dosage | | | | | | | |
| | ±2% | | ±5% | | ±10% | | ±15% | |
Dosage (mg)	Tablets	Capsules	Tablets	Capsules	Tablets	Capsules	Tablets	Capsules
10	27.7	32.5	57.4	65.0	90.3	96.7	99.4	100.0
25	38.1	49.7	79.6	67.8	99.2	94.1	100.0	100.0
50	45.2	42.0	82.9	88.7	99.0	100.0	100.0	100.0

[a]Adapted from Ref. [12].

To assure tablet uniformity for example, the United States Pharmacopoeia requires that a sample size of 30 tablets be tested as follows: Test 10 tablets: the batch passes if no tablet is outside of plus or minus 15% of specified limits; if one tablet out of 10 is outside 15% of specified limits, then 20 more tablets are tested; and no more than 2 tablets can be outside the plus or minus 25% of the specified limits.

Table 2.1 demonstrates how excellent dosage uniformity can be achieved [12] for Tofranil tablets (10, 25, and 50 mg) and Pertofrane capsules (10, 25, and 50 mg). Of the 730 individual tablets analyzed, over 90% were within 10% of indicated dosage, and all except one tablet (Tofranil 10 mg) were within 15% of the indicated dosage. And all were within the 25% limits. Of the 590 capsules analyzed, all were within 15% of the indicated dosage.

This example demonstrates how various testing steps can assure that representative results can be obtained, even with a relatively small sample size, especially when steps have been taken to assure that the whole batch is reasonably uniform.

It is important to remember that the confidence in our final results depends both on the size of our sample, which adequately represents the whole batch, and steps that have been taken to assure uniformity of the whole batch. In addition, the magnitude of interfering effects should be considered.

One should never forget that all sampling investigations are subject to experimental error. Careful consideration of sampling can keep this error to a minimum. To assure valid sampling, major consideration should be given to the following [13]:

1. Determination of the population or the bulk from which the sample is drawn.

2. Procurement of a valid gross sample.
3. Reduction of the gross sample to a suitable sample for analysis.

It is desirable to reduce the analytical uncertainty to one-third or less of sampling uncertainty [14]. Interested readers should consider sampling theory [15].

2.3 SAMPLE PREPARATION

A survey on trends of sample preparation (SP) revealed that pharmaceutical and environmental analysts are more likely to employ sample preparation techniques [16]. Figure 2.2 shows various sample preparation steps that may be employed in sample preparation. Most analysts use 1 to 4 steps for sample preparation; however, some can use more than 7. Interestingly, a large number of analysts have to perform analyses of components in the parts-per-million range or below.

As discussed above, after the sample has been collected, a decision has to be taken as to whether the entire sample or a portion of the sample needs to be analyzed. For low-level analytes, the entire sample is often utilized. When the concentration of the analyte is suspected to be well above a detection limit, then a portion of the collected sample can be used. The sample must be homogenous so that the data generated from the subsample are representative of the entire sample. For example, a soil sample may need to be sieved, large pieces ground, and then mixed for the sample to be homogenous.

The next crucial portion of the process is the sample preparation, which allows the analyte to be detected at a level that is appropriate for detection by the instrument. The sample must be prepared in such a way that the matrix does not interfere with the detection and measurement of the analytes. Often, this requires complete separation of the sample matrix from the analyte(s) of interest. In other cases, the analytes can be measured in situ or first derivatized and then measured in situ. Both the nature of the analyte and the matrix dictate the choice of sample preparation (see Ref. [1]).

Table 2.2 shows a comparison of various extraction methods for solid samples [17]. It appears that one can take anywhere from 0.1 to 24 h for the extraction process. Microwave-assisted sample preparation requires minimal time; however, if cost is a consideration, Soxhlet extraction is least costly but requires the longest sample preparation time.

Sampling and sample preparation

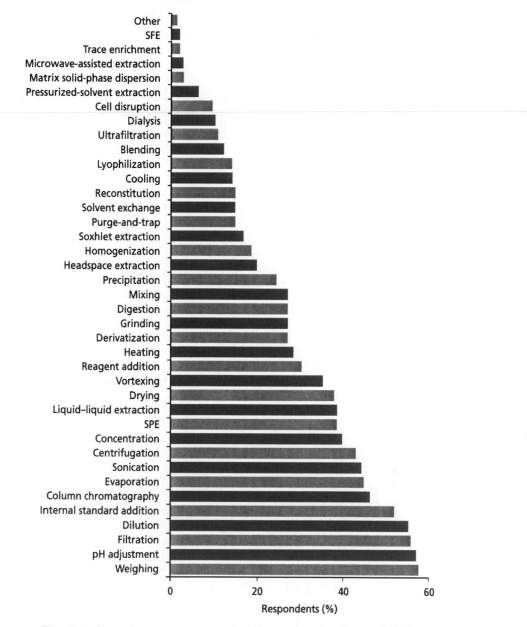

Fig. 2.2. Sample preparation procedures commonly used [16].

TABLE 2.2

Comparison of extraction methods for solid sample preparation[a]

Method	Sample size (g)	Solvent volume (ml)	Temperature (°C)	Typical operating pressure (psi)	Time[b] (h)	Automation level	Number of samples[c]	Cost[d]
Sonication	20–50	100–300	Ambient–40	Atmospheric	0.5–1.0	None	1 (serial), many (batch)	Low
Traditional Soxhlet	10–20	200–500	40–100	Atmospheric	12–24	None	1 (serial)	Very low
Modern Soxhlet	10–20	50–100	40–100	Atmospheric	1–4	Mostly	6 (batch)	Moderate
SFE	5–10	10–20[e]	50–150	2000–4000	0.5–1.0	Fully	44 (serial)	High
Pressurized fluid	1–30	10–45	50–200	1500–2000	0.2–0.3	Fully	24 (serial), 6 (batch)	High
Closed-vessel microwave-assisted	2–5	30	100–200	1500–2000	0.1–0.2	Mostly	12 (batch)	Moderate
Open-vessel microwave-assisted	2–10	20–30	Ambient	Atmospheric	0.1–0.2	Mostly	6 (batch)	Moderate

[a] Adapted from Ref. [17].
[b] Total processing time per sample from weighing to collection.
[c] Maximum number of samples that commercial instruments can handle; serial means one sample processed at a time, and batch means multiple samples at a time.
[d] Very low = less than $1000; low = less than $10,000; moderate = $10,000–20,000; high = more than 20,000.
[e] Solvent volume when organic modifier is used to affect polarity.

2.3.1 Types of samples

Our focus in this chapter is on the analysis of organic analytes in sample matrices that are organic and/or inorganic in nature. These organic analytes can be subclassified into volatile, semivolatile, or nonvolatile. The matrix can be gas (or volatile samples), solid, or liquid. Both the anticipated concentration of the analyte and the type of sample dictate the instrumentation that can be used, as well as the sample preparation technique required.

2.3.1.1 Volatile samples

Volatile samples are usually analyzed by gas chromatography. There are many different sample pretreatment methods for gases. Let us review these techniques:

Grab sampling

A gaseous sample is pulled into an empty container such as a metal canister. In the laboratory, the sample is often chilled to isolate the volatile compounds. The container may then simply be rinsed with a solvent to capture these compounds. The solvent can then be directly injected into a suitable instrument for analysis, such as a gas chromatograph (GC).

Solid-phase trapping

A gaseous sample is passed through a solid material, such as silica gel or polyurethane foam (PUF), in a tube. A glass fiber filter is often put in front of the solid support to capture particle-phase constituents, while the vapor-phase compounds are captured on the solid support. This is used for semivolatile analytes, such as polycyclic aromatic hydrocarbons and pesticides. The solid support is then usually extracted in the lab with a solvent (see techniques described later in this chapter), and then the techniques used for liquid samples are followed.

Liquid trapping

In this case, the gas is bubbled through a liquid that has an affinity for the analyte of interest. After a certain period of time, the liquid is then analyzed, often directly by injection into a GC.

Head-space sampling

The solid or liquid sample is placed into a glass vial, which is heated, with enough empty space in the vial above the sample to allow the analyte to reach equilibrium between the gas and solid (or liquid) phase. The gas phase is sampled and analyzed, usually by GC. This type of sampling is generally used for analyzing trace amounts of volatile samples.

Purge and trap

This technique is similar to head-space sampling in which the sample is placed in a heated vial. However, the head-space vapors are continually removed by a flow of inert gas. This gas is then trapped either on a solid support or is cooled and then thermally desorbed into the GC. This procedure is used preferably over head-space sampling if the analytes are in low concentration or have unfavorable partition coefficients. An interesting application of this technique is in the area of food quality control—specifically in the wine industry [18]. 2,4,6-Trichloroanisole (TCA) is the analyte responsible for cork taint in bottles of wine. In this research, samples of wine and cork were extracted with pentane, rotary evaporated, and reconstituted in an aqueous solution of sodium chloride and sulfuric acid, and then submitted to purge and trap, with gas chromatographic analysis. Another method that utilizes purge and trap is the analysis of volatile organic compounds in drinking water [19]. Compounds that contain multiple halogens such as dichlorobromomethane are potentially harmful to humans. In this research, helium was passed through water samples to strip out the analytes, which were trapped in a capillary trap and thermally desorbed in a purge and trap system.

2.3.1.2 Solid samples

Solid samples often require more creativity in preparation. The approach of dissolve, dilute, and shoot (inject into a chromatograph) is quite common for the analysis of drug substances. In a drug product (tablets or capsules) where API can be easily solubilized, the process can be reduced to dissolve, filter, and shoot as well. However, for drug products, a more elaborate process of grind, extract, dilute, and filter is generally employed to extract the API from the tablets (a solid matrix composed of inactive ingredients). As high accuracy and precision are mandatory in assays for drug products, volumetric flasks (25–500 ml) are used to attain very precise and sufficient volumes for solubilizing the API in the case of the tablets. This approach is straightforward for immediate-release dosage forms but can be challenging for controlled-release products and formulations of low-solubility APIs [2].

2.3.1.2.1 Pretreatment steps

Before many of the techniques for processing solid samples can be employed, the solid sample must be homogenized. There are many physical techniques to accomplish this goal. First, grinding, such as with a mortar and pestle or with a mechanical grinder puts the sample into a fine particulate state, making the extraction techniques more effective and more efficient. However, if thermally labile analytes are present, care must be taken to avoid heating induced by friction. For softer samples, ball milling is the preferred grinding method. In this case, the sample is placed in a container; porcelain, steel, or some other inert material is added to the container, and the entire device is shaken or rotated. For very sticky samples, liquid nitrogen may be used to force the sample into a solid brittle state.

An additional step must be taken for samples that are wet, such as a soil sample or plant tissue. The amount of water present in a sample can affect the total reported concentration. Therefore, samples are often analyzed for water content before analysis. In this case, samples are weighed, dried (usually in an oven), and then reweighed. Care must be taken if the analytes are volatile or if the sample may decompose under heating.

More elaborate sample preparation is often needed for complex sample matrices, e.g., lotions and creams. Many newer SP technologies such as solid-phase extraction (SPE), supercritical fluid extraction (SFE), pressurized fluid extraction, accelerated solvent extraction (ASE), and robotics are frequently utilized (see Ref. [2]). Dosage forms such as suppositories, lotions, or creams containing a preponderance of hydrophobic matrices might require more elaborate SP and sample cleanup, such as SPE or liquid–liquid extraction.

For samples that are more complex, some type of liquid-extraction process to remove the analytes of interest may be required. Many of these techniques have been around for over 100 years. These techniques are widely accepted by various global regulatory agencies, such as the EPA in the USA, and therefore have not changed much over time. Some improvements to reduce the amount of organic solvents used have been made and are detailed below:

Solid–liquid extraction

This is one of the simplest extraction methods. The sample is placed in a vessel, and the solvent is added to dissolve the sample components. The container is shaken for a given period of time, and then the insoluble components are removed by filtering or centrifugation. This process works well if the analytes are readily soluble in the solvent and not tightly bound to the matrix. The sample is often heated to boil/reflux to speed up the process if thermal lability is not a problem.

Solution–solvent extraction

The analytes of interest are solubilized in aqueous solvent at the desired pH and extracted with an organic solvent (see Section 2.3.1.3

for more details on liquid–liquid extraction). Let us review a particularly challenging sample preparation technique for water-soluble vitamins from multivitamin tablets, which entails diversity of analytes of varied hydrophobicities and pK_a. Water-soluble vitamins (WSVs) include ascorbic acid (vitamin C), niacin, niacinamide, pyridoxine (vitamin B-6), thiamine (vitamin B-1), folic acid, riboflavin (vitamin B-2), and others [20]. While most WSVs are highly water soluble, riboflavin is quite hydrophobic and insoluble in water. Folic acid is acidic, while pyridoxine and thiamine are basic. In addition, ascorbic acid is light sensitive and easily oxidized. The SP for fat-soluble vitamins (FSVs) A, E, and D presents a different set of challenges. The extraction strategy employs a two-step approach, using mixed solvents of different polarity and acidity as follows. A two-phase liquid–liquid extraction system is used to isolate the FSV into hexane, while retaining other analytes and excipients in the aqueous layer. A capped centrifuge test tube rather than a separatory funnel is used to minimize extraction time and solvent volumes. The final hexane solution is injected directly into the HPLC system with a mobile phase consisting of 85% methanol/water. Chromatographic anomalies from injecting a stronger solvent are avoided by reducing the injection volume to 5 µl.

Sonication

Sonication helps improve solid–liquid extractions. Usually a finely ground sample is covered with solvent and placed in an ultrasonic bath. The ultrasonic action facilitates dissolution, and the heating aids the extraction. There are many EPA methods for solids such as soils and sludges that use sonication for extraction. The type of solvent used is determined by the nature of the analytes. This technique is still in widespread use because of its simplicity and good extraction efficiency. For example, in research to determine the amount of pesticide in air after application to rice paddy systems, air samples collected on PUF were extracted by sonication, using acetone as the solvent. The extraction recoveries were between 92% and 103% [21].

Soxhlet extraction

This extraction method remains the most widely used method of extraction for solid samples. The sample is placed in a thimble, which sits inside the Soxhlet apparatus. A constant refluxing of the solvent causes the liquid to drip on top of the sample and leach through the thimble (taking analyte with it). The design of the system is such that once the liquid reaches a certain level, a siphoning mechanism returns the liquid to the boiling vessel. The liquid continues boiling and refluxing for 18–24 h. While this is a slow process, once the setup is complete, the operator can walk away and return to it at the end of the operation.

One of the requirements of this approach is that the analytes must be stable at the boiling point of the solvent, since the analytes collect in the flask. The solvent must show high solubility for the analyte and none for the sample matrix. Since this is one of the oldest methods of sample preparation, there are hundreds of published methods for all kinds of analytes in as many matrices. For example, XAD-2 resin (styrene-divinylbenzene) that was used to collect air samples to monitor current usage of pesticides in Iowa was Soxhlet-extracted for 24 h with hexane/acetone [22]. This is common in environmental sample analysis, and one will see rows and rows of these systems in environmental laboratories. One disadvantage of Soxhlet extraction is the amount of solvent consumed, though modern approaches to solid sample extraction have tried to reduce the amount of solvent used.

In recent years, several modifications to Soxhlet extractions have been developed. Two such methods are automated focused microwave-assisted Soxhlet extraction (FMASE) [23,24] and ultrasound-assisted Soxhlet extraction [25]. In one experiment with the FMASE setup [23], the sample cartridge is irradiated by microwaves, allowing for faster extraction. Soil samples contaminated with polychlorinated biphenyls were extracted by FMASE and conventional Soxhlet extraction. The FMASE method produced data that are as accurate and precise as the conventional method. The benefits of this newer method are the time-savings (70 min versus 24 h) and the recycling of 75% of the organic extraction solvent. Similarly, utilizing ultrasound-assisted Soxhlet extraction [25] for determining the total fat content in seeds, such as sunflower and soybeans, resulted in data comparable to conventional Soxhlet extraction, with RSDs less than 1.5%, and with extraction times of 90 min versus 12 h.

Accelerated solvent extraction (ASE)

In this technique, organic solvents at high temperature under pressure are used to extract analytes. The process consists of placing the sample in a stainless-steel vessel, introducing the solvent, pressurizing the vessels, heating of the sample cell under constant pressure, extraction, transferring the extract to a sealed vial with a fresh wash of the solid sample, nitrogen purge of the cell, and then loading the next sample. With commercial systems, this process is all automated. Another advantage of this system is the reduction in both extraction times (10–20 min versus 18–24 h) and solvent usage (20 ml versus 150–500 ml) from a traditional Soxhlet extraction. The US EPA has certified an ASE extraction method for the extraction of polychlorinated biphenyls, PAHs, organochloride and organophosphorus pesticides from soil, Method 3545 [26].

Supercritical fluid extraction (SFE)

The popularity of this extraction method ebbs and flows as the years go by. SFE is typically used to extract nonpolar to moderately polar analytes from solid samples, especially in the environmental, food safety, and polymer sciences. The sample is placed in a special vessel and a supercritical gas such as CO_2 is passed through the sample. The extracted analyte is then collected in solvent or on a sorbent. The advantages of this technique include better diffusivity and low viscosity of supercritical fluids, which allow more selective extractions. One recent application of SFE is the extraction of pesticide residues from honey [27]. In this research, liquid–liquid extraction with hexane/acetone was termed the conventional method. Honey was lyophilized and then mixed with acetone and acetonitrile in the SFE cell. Parameters such as temperature, pressure, and extraction time were optimized. The researchers found that SFE resulted in better precision (less than 6% RSD), less solvent consumption, less sample handling, and a faster extraction than the liquid–liquid method [27].

SFE may also be coupled to GC and HPLC systems [28] for a simple on-line extraction and analysis system. Lycopene is determined in food products such as tomatoes and tomato products, using SFE coupled to HPLC with an HPLC column used for trapping and analysis. The method is short, requires small sample amounts, and has good linearity and sensitivity. Because the entire system is closed, there is little chance for the lycopene to degrade.

Microwave-assisted extraction (MAE)

Most average homes contain a microwave oven. Researchers have taken that same technology and applied it to sample extraction. Using microwave technology with acidic solutions has become a commonplace replacement for traditional acid digests. The sample is placed in a closed, chemical-resistant vessel and heated in a microwave. This process is much faster than hot-plate techniques and has become widely accepted by such agencies as the US EPA.

Combining microwave heating with solvent extraction (MAE) has gained status in the scientific community. Compounds absorb microwave energy approximately in proportion to their dielectric constants: the higher the constant, the higher the level of absorption of the microwave energy. There are two ways of running MAE: using a microwave-absorbing solvent with a high dielectric constant or use of a

non-microwave absorbing solvent of low dielectric constant. With the
solvent that absorbs microwave energy, the solvent and sample are in a
closed vessel. This causes the solvent to heat to above its boiling point
and forces a quick extraction under moderate pressure. In the non-
microwave absorbing approach, the sample and solvent are in an open
vessel. In this case, the sample itself absorbs the energy and heats up.
The analytes are released from the hot sample into the cool surrounding liquid.

The advantages of MAE are short extraction times (10 min), extraction of many samples at one time (up to 14, depending on the system),
and less organic solvent consumed. In one recent study [29], MAE was
used to extract paclitaxel from Iranian yew trees. The needles of the
tree were air-dried and ground. The needles were covered with methanol–water and placed in the MAE apparatus. Extractions took
9–16 min. The extracts were filtered and analyzed by HPLC. Further
optimization of the method resulted in less than 10% RSDs for precision and greater than 87% recovery. The overall benefits of the MAE
method are reduced extraction times (15–20 min versus 17 h), minimal
sample handling, and 75–80% reduction in solvent consumption [29].

2.3.1.3   Liquid samples

Liquid samples, other than those that are inherently liquid, can arise
from the solid sample extraction techniques described above. As mentioned previously, sometimes a simple dilute-and-shoot approach can be
utilized, i.e., add solvent to the sample and then inject directly into the
instrument. Other times, evaporation of residual liquid can be utilized—then the sample is either directly injected, or if the sample is
evaporated to dryness, a new solvent can be added. Often, however, the
residual matrix causes interference and the following techniques can be
employed for further sample cleanup.

Microdialysis

A semipermeable membrane is placed between two liquids, and the
analytes transfer from one liquid to the other. This technique is used
for investigating extracellular chemical events as well as for removing
large proteins from biological samples prior to HPLC analysis.

Lyophilization

Aqueous samples are frozen, and water is removed by sublimation
under vacuum; this procedure is especially good for nonvolatile analytes.

Filtration

Liquid is passed through a filter paper or membrane to remove suspended solids.

Centrifugation

A liquid sample with solids is spun at high speed to force the solids to the bottom of the vessel. The liquid is decanted and often further processed.

Liquid–liquid extraction (LLE)

Liquid–liquid extraction is a form of solvent extraction in which the solvents produce two immiscible liquid phases. The separation of analytes from the liquid matrix occurs when the analyte partitions from the matrix–liquid phase to the other. The partition of analytes between the two phases is based on their solubilities when equilibrium is reached. Usually, one of the phases is aqueous and the other is an immiscible organic solvent. Large, bulky hydrophobic molecules like to partition into an organic solvent, while polar and/or ionic compounds prefer the aqueous phase.

A review of several classic equilibrium equations is in order. The Nernst distribution law states that a neutral species will distribute between two immiscible solvents with a constant ratio of concentrations.

$$K_D = C_o/C_{aq}$$

where K_D is the distribution constant, C_o is the concentration of the analyte in the organic phase, and C_{aq} is the concentration of the analyte in the aqueous phase. Another equation that is more appropriate is the fraction of analyte extracted, E,

$$E = C_o V_o/(C_o V_o + C_{aq} V_{aq}) = K_D V/(1 + K_D V)$$

where V_o is the volume of the organic layer, V_{aq} the volume of aqueous phase, and V the phase ratio V_o/V_{aq}.

Extraction is an equilibrium process, and therefore a finite amount of solute might be in both phases, necessitating other processing steps or manipulation of the chemical equilibria.

The organic solvent should preferably be insoluble in water, should be of high volatility for easy removal, and should have a great affinity for the analytes. There are several mechanisms for increasing K_D such as changing the organic solvent so the analyte more readily partitions into that layer, manipulation of the charge state to make the analyte uncharged and more soluble in the organic layer, or "salting out" the analyte from the aqueous layer by adding an inert, neutral salt to the aqueous layer.

The actual extraction is usually carried out in a separatory funnel, with tens of milliliters of solvents. Most extractions require two to three extractions, resulting in large volumes of solvent that must be removed, often by rotary evaporation. Even with the large amounts of solvents generated, LLE is perceived as easy to do and has been used for the extraction of acyclovir in human serum [30] and amphetamine in urine [31].

One advance in the area of LLE is the use of solid supports that facilitate the partitioning of the analyte(s) of interest. LLE extraction methods involving nonpolar matrices often suffer from the formation of emulsions, and using the solid support is a possible solution. In one study, polychlorinated biphenyls, dioxins, and furans were extracted from the lipid fraction of human blood plasma [32], using diatomaceous earth as the solid support. Long glass columns (30 cm) were packed with several layers of Chem-Elut (a Varian product) and sodium chloride. The plasma samples were diluted with water and ethanol and passed over the columns. A mixture of isopropanol and hexane (2:3) was passed over the column and the LLE was performed. It can be concluded that the LLE with the solid support is easier to perform and can be applied to other lipid heavy matrices such as milk [32].

2.4   SOLID-PHASE EXTRACTION

One of the most widely used techniques across all disciplines is solid-phase extraction (SPE). A simple approach to SPE is shown in Fig. 2.3. These columns provide rapid extraction, minimize emulsion problems, and eliminate most sample handling.

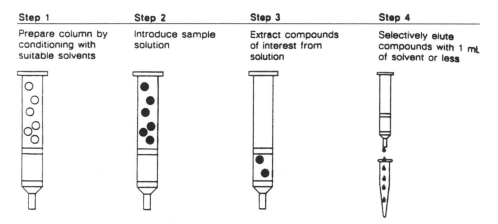

Step 1	Step 2	Step 3	Step 4
Prepare column by conditioning with suitable solvents	Introduce sample solution	Extract compounds of interest from solution	Selectively elute compounds with 1 mL of solvent or less

Fig. 2.3. Rapid extraction with solid-phase columns.

The basic technique of removing undesired components from a solution with the help of solids has been used for sample cleanup for a long time. The recoveries from a quick cleanup for waste solvents based on a sample filtration through a Florisil and sodium sulfate column are given in Table 2.3. It can be seen from the data in this table that a large number of pesticides are retained by Florisil. These can be recovered for appropriate analysis.

More sophisticated modifications have been developed relatively recently to remove various matrix effects from the sample (see Biological Matrices in Section 2.4). For example, a solid sorbent is sandwiched between two frits in a barrel-type device. The particles are often irregularly shaped particles of 20–40 μm in diameter, which allows for better flow through the device. The same mechanisms that apply to HPLC also apply to SPE. However, because of the large particle diameters, the efficiency is much lower than in HPLC.

TABLE 2.3

Florisil filtration efficiency[a]

Compound	Sample concentration (ppm)	% Recovery[b]	Standard deviation
Aldrin	2	93	9.5
BHC	2	86	6.1
Lindane	2	92	13.5
Chlordane	2	79	4.7
DDD	2	69	16.0
DDE	2	92	11.0
DDT	2	89	7.2
Dieldrin	2	88	3.5
Endosulfan	2	91	7.6
Endrin	2	98	7.6
Heptachlor	2	97	2.9
Toxaphene	3	90	16.5
Arochlor 1016	20	93	2.0
Arochlor 1221	20	95	6.4
Arochlor 1232	20	100	8.2
Arochlor 1242	20	93	8.3
Arochlor 1248	10	95	9.1
Arochlor 1254	10	86	9.7
Arochlor 1260	10	87	12.2

[a]Adapted from Pedersen, B.A. and Higgins, G.M., LCGC, 6 (1988) 1016.
[b]Average recovery from triplicate analysis.

SPE is an elegant technique because of the variety of ways the devices can be used. Over half of the uses of SPE are to remove any unwanted matrix from the analytes of interest—in other words, separate the analytes from everything else in the sample. The interferences either pass through the sorbent unretained or become adsorbed to the solid support. This technique has been used in just about every application area from bioanalytical analyses in the pharmaceutical industry to removal of dyes from food samples to the removal of humic acids from soil samples. The second major benefit of SPE is the concentration of the analytes in trace analysis. For example, several liters of river water can be passed over a few grams of sorbent—the analytes adsorb to the sorbent, to be later eluted with a small volume of solution. Other uses include solvent switching or phase exchange, in which the sample is switched from one type of solvent to another, often from aqueous to organic; and solid-phase derivatization in which the SPE sorbent is coated with a special reagent, such as dinitrophenylhydrazine (DNPH). DNPH devices contain acidified reagent DNPH coated on a silica sorbent. Air samples are collected, and these SPE devices are used for quantitation of aldehydes and ketones by reaction to form the hydrazone derivative. DNPH-silica is specified in several EPA procedures for the analysis of carbonyl compounds in air.

Besides the barrel configurations, an explosion of other formats has become available over the years. One type of device includes a disk or membrane that contains a small amount of particulate material. This format has the advantage that large amounts of sample can be passed over the disk in much less time than in larger bed masses. However, large holdup volumes can be observed. To facilitate automated, high-throughput SPE, 96-well and 384-well plates have become commonplace in the pharmaceutical industry. Typically, 2–60 mg of sorbent are contained in each well—making fast SPE of biological fluids such as urine and plasma possible. Other advances in the format include a microelution plate design in which small volumes (25 µl) of solvent can be used to elute a sample, aiding in sample concentration and elimination of the need to evaporate and reconstitute.

The first reversed-phase SPE sorbents were based on silica gel particles, similar to the particles used in HPLC. A number of phases are available ranging from C8 to C18 to anion- and cation-exchange functionalities. Recent advances in particle technology have included polymeric materials that combine the benefits of a water-wettable particle to retain polar analytes with a reversed-phase, hydrophobic moiety to retain large bulky molecules. These polymers are available with many types of ion-exchange moieties, allowing for selective retention of acids and bases.

The practical application of SPE is relatively straightforward. The sorbent is first washed with organic solvent to both remove any residual components remaining from manufacture (although today's SPE phases are extensively cleaned) and to allow for proper wetting of the chromatographic surface. This is then followed by equilibration with an aqueous solvent. The sample is then loaded onto the sorbent. Usually, vacuum is applied to facilitate flow through the cartridge— depending on the nature of the sample preparation step, the analyte either passes through the cartridge while the matrix components are retained, or the analyte is retained on the sorbent. Subsequent wash steps are then used to "lock on" the analyte to the material. For example, if a basic analyte is passed through a cation-exchange sorbent, an acid wash is used to ensure that the base is in its charged state so it can "stick" to the cartridge. Further wash steps, often organic solvents, are then used to remove the unwanted analytes from the sorbent. A final elution step of an organic solvent, sometimes with an acidic or basic modifier, is used to elute the analyte of interest from the sorbent.

Method development in SPE begins with an analysis of the sample matrix as well as the sample analytes. One of the most efficient mechanisms of SPE is to use mixed-mode ion exchange sorbents. If the analytes are bases, cation exchange is the most selective SPE mode. If the analytes are acids, anion exhange is the most efficient. However, it is sometimes easier to capture the matrix components and allow the analytes to pass through the sorbent unretained.

Solid-phase microextraction (SPME)

In the 1990s, Pawliszyn [3] developed a rapid, simple, and solvent-free extraction technique termed solid-phase microextraction. In this technique, a fused-silica fiber is coated with a polymer that allows for fast mass transfer—both in the adsorption and desorption of analytes. SPME coupled with GC/MS has been used to detect explosive residues in seawater and sediments from Hawaii [33]. Various fibers coated with carbowax/divinylbenzene, polydimethylsiloxane/divinylbenzene, and polyacrylate are used. The SPME devices are simply immersed into the water samples. The sediment samples are first sonicated with acetonitrile, evaporated, and reconstituted in water, and then sampled by SPME. The device is then inserted into the injection port of the GC/MS system and the analytes thermally desorbed from the fiber. Various parameters such as desorption temperature, sample adsorption time, and polymer coating have been evaluated.

SPME has also been used in the area of head-space sampling of volatile compounds such as flavors and fragrances [34,35]. In these experiments, the analytes are adsorbed onto the polymer-coated fiber that is in contact with the head space in the vial. The fiber is then inserted into the injection port of a GC and thermally desorbed.

Biological matrices

Protein precipitation is used routinely in bioanalytical laboratories in the pharmaceutical industry. Plasma is mixed with an excess (3 to 5 times) of organic solvent (typically acetonitrile or methanol) or an acid such as formic acid. The proteins precipitate out of solution, the sample is centrifuged, and the supernatant is analyzed. While this method is relatively fast and simple, the extract contains salts and lipids that can interfere with subsequent analyses. Often, a second technique such as SPE is used for further cleanup. Table 2.4 exhibits various samples that

TABLE 2.4

Samples analyzed by solid-phase extraction[a]

Sample	Matrix	Solid-phase column	Detectability (ng/ml)
3-Methoxy-4-hydroxyphenyl glycol	Plasma	Alumina	1–10
Oxytetracycline	Fish tissues	C-8	5–10[b]
Doxefazepam	Plasma	C-18	0.1[c]
Cortisol	Urine	C-18	
Basic drugs	Plasma	CN	Therapeutic levels
Δ^9-Tetrahydrocannabinol	Plasma	C-18	2, 100 pg
Ibuprofen	Plasma	C-2	1.3[c]
Ranitidine	Plasma, other fluids	CN	2
Chlorpromazine and metabolites	Plasma	C-8	
Cyclotrimethylenetrinitramine	Biological fluids	C-18	
Sotalol	Plasma, urine	C-8	10
Cyclosporin A	Serum, urine	Cyanopropyl	
Cyclosporine	Blood	C-18	10
Growth factors	Urine	C-1	200–1400-fold enrichment
Carbamazine and metabolites	Plasma	C-18	50

[a]Adapted from Ref. [10].
[b]Value is in ng/g.
[c]Value is in µm/ml.

can be analyzed by SPE from different matrices. The type of samples includes urine, plasma, blood, various biological fluids, and fish tissue. The detectability levels are in ng/ml range.

Three important concerns drive sample preparation from biological matrices [36]:

- removing interferences,
- concentrating analyte(s) of interest, and
- improving analytical system performance.

Innovative SPE technology employing microelution plate design is shown in Fig. 2.4. Tip design of the well plate provides a smaller surface

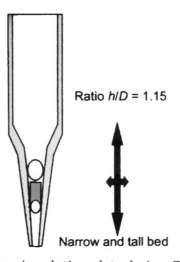

Ratio *h*/*D* = 1.15

Narrow and tall bed

Fig. 2.4. Schematic of the microelution plate design. Tip design of the 96-well plate affords a smaller surface area and larger depth, more like an HPLC column, and permits good flow through the plate and low holdup volume on the order of nanoliters [36].

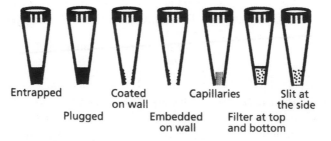

Entrapped   Coated on wall   Capillaries   Slit at the side

Plugged   Embedded on wall   Filter at top and bottom

Fig. 2.5. Types of micropipette tips [37].

TABLE 2.5

Common strategies to accelerate sample preparation[a]

Strategy	Aim
Specific detection	Avoid response of unwanted or increased response of target analyte(s)
Large volume injection	Miniaturization of sample preparation
Automation	Bulk processing
Combined analysis	Efficiency improvement

[a]Adapted from Ref. [38].

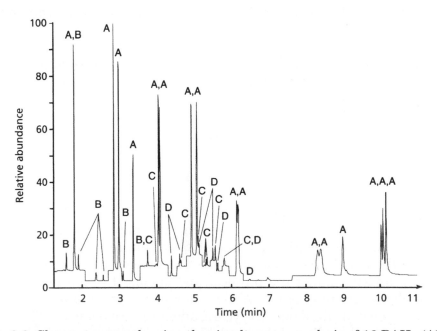

Fig. 2.6. Chromatogram showing the simultaneous analysis of 16 PAHs (A), 7 chlorobenzenes (B), 10 organochlorine pesticides (C), and 7 PCBs (D). Column: RapidMS (10 ml × 530 µm i.d., 0.12 µm df, Varian, The Netherlands). GC oven: 50°C (1 min) to 190°C (3 min) at 30°C/min to 280°C at 40°C/min. MS: time-selected SIM mode with dual ion confirmation. Analyte concentration levels were between 50 and 175 ppb [38].

area and a larger depth similar to an HPLC column and permits good flow through the plate with low holdup volume. This design allows utilization of volumes as low as 25 ml. Higher sensitivity has been achieved for analysis of amitriptyline-spiked plasma samples at 0.1 ng/ml than with protein-precipitated samples.

The micropipette tip containing solid phases is a relatively new sample preparation technique that permits handling of microliter to submicroliter amounts of liquid samples, using the techniques of SPE, dialysis, and enzyme digestion. Various phases (reversed-phase, affinity, size-exclusion, etc.) are packed, embedded, or coated on the walls of pipette, permitting liquid samples to be transferred without undue pressure drop or plugging (Fig. 2.5).

2.5 ENHANCED SAMPLE THROUGHPUT

Some common throughput enhancement strategies for environmental samples are given in Table 2.5 (for details, see Ref. [38]).

The problems relating to increased contamination levels and/or insufficient sensitivity may be overcome by using matrix–solid phase dispersion, MS detection in selected ion monitoring (SIM) mode, and/or large volume injection. An example of combined analysis that utilizes specific detection is shown in Fig. 2.6. It entails simultaneous analyses of PAHs, PCBs, chlorobenzene, and organochlorine pesticides in soil.

2.6 SUMMARY AND CONCLUSIONS

A variety of methods for the sample collection and preparation of gases, solids, and liquids have been covered in this chapter. Each technique discussed involves separating the analytes of interest from the components of the matrix. Selection of the best sample preparation technique involves understanding the physical and chemical properties of both the analytes and the matrix. While many of the techniques discussed here have been utilized over the years, advances are being continuously made. A careful reading of the available methodologies is recommended prior to selecting a suitable methodology.

REFERENCES

1   S. Ahuja, *Trace and Ultratrace Analysis by HPLC*, Wiley, New York, 1992.
2   S. Ahuja and M.W. Dong, *Handbook of Pharmaceutical Analysis by HPLC*, Elsevier, Amsterdam, The Netherlands, 2005.
3   J. Pawliszyn, *Sampling and Sample Preparation for Field and Laboratory Samples*, Elsevier, Amsterdam, The Netherlands, 2002.
4   S. Mitra, *Sample Preparation Techniques in Analytical Chemistry*, Wiley, New York, 2003.

5   T.C. Lo, M.H.I. Baird and C. Hanson, *Handbook of Solvent Extraction*, Krieger Publishing, Melbourne, FL, 1991.

6   D.A. Wells, *High Throughput Bioanalytical Sample Preparation: Method and Automation Strategies*, Elsevier, The Netherlands, 2003.

7   R.E. Majors, *LCGC*, 13 (1995) 742.

8   R.F. Cross, *LCGC*, 18 (2000) 468.

9   S. Ahuja, *Ultratrace Analysis of Pharmaceuticals and Other Compounds of Interest*, Wiley, New York, 1986.

10 S. Ahuja, *Trace and Residue Analysis in Kirk-Othmer Encyclopedia of Chemical Technology*, Wiley, New York, 1997.

11 R.E. Majors, *LCGC*, 16 (1998) 438.

12 S. Ahuja, Study on Dosage Variations on Individual Capsules and Tablets of Desipramine and Imipramine Hydrochloride. Presented at A.Ph.A. Meeting, Miami, FL, May, 1968; *J. Pharm. Sci.* 57(11) (1968) 1979–1982.

13 B. Kratochvil and J.K. Taylor, *Anal. Chem.*, 53 (1981) 924A.

14 W.J. Youden, *J. Assoc. Off. Anal. Chem.*, 50 (1967) 1007.

15 P.M. Gy, *LCGC*, 12 (1994) 808.

16 R.E. Majors, *LCGC*, 20 (2002) 1098.

17 R.E. Majors, *LCGC*, 17 (1999) S8.

18 N. Campillo, N. Aguinaga, P. Vinas, I. Lopez-Garcia and M. Hernandez-Cordoba, *J. Chromatogr. A*, 1061 (2004) 85–91.

19 N. Campillo, P. Vinas, I. Lopez-Garcia, N. Aguinaga and M. Hernandez-Cordoba, *J. Chromatogr. A*, 1035 (2004) 1–8.

20 M.W. Dong, G.D. Miller and R.K. Paul, *J. Chromatogr.*, 14 (1996) 794.

21 F. Ferrari, D.G. Karpouzas, M. Trevisan and E Capri, *Environ. Sci. Technol.*, 39 (2005) 2968–2975.

22 A. Peck and K. Hornbuckle, *Environ. Sci. Technol.*, 39 (2005) 2952–2959.

23 J. Luque-Garcia and M. Luque de Castro, *J. Chromatogr. A*, 998 (2003) 21–29.

24 S. Morales-Munoz, J. Luque-Garcia and M. Luque de Castro, *J. Chromatogr. A*, 1026 (2004) 41–46.

25 J. Luque-Garcia and M. Luque de Castro, *J. Chromatogr. A*, 1034 (2004) 237–242.

26 K. Wenzel, B. Vrana, A. Hubert and G. Schuurmann, *Anal. Chem.*, 76(18) (2004) 5503–5509.

27 S. Rissato, M. Galhiane, F. Knoll and B. Apon, *J. Chromatogr. A*, 1048 (2004) 153–159.

28 J. Pol, T. Hyotylainen, O. Ranta-Aho and M. Riekkola, *J. Chromatogr. A*, 1052 (2004) 25–31.

29 M. Talebi, A. Ghassempour, Z. Talebpour, A. Rassouli and L. Dolatyari, *J. Sep. Sci.*, 27 (2004) 1130–1136.

30 G. Bahrami, S. Mirzaeei and A. Kiani, *J. Chromatogr. B*, 816 (2005) 327–331.

31 S. Wang, T. Wang and Y. Giang, *J. Chromatogr. B*, 816 (2005) 131–143.

32 H. Wingfors, M. Hansson, O. Papke, S. Bergek, C.A. Nilsson and P. Hag-
lund, *Chemosphere*, 58 (2005) 311–320.
33 F. Monteil-Rivera, C. Beaulieu and J. Jawari, *J. Chromatogr. A*, 1066
(2005) 177–187.
34 C. Bicchi, C. Cordero and P. Rubiolo, *J. Chromatogr. Sci.*, 42 (2004)
402–409.
35 M. Rosario-Ramirez, M. Estevez, D. Morcuende and R.J. Cava, *Agric.
Food Chem.*, 52 (2004) 7637–7643.
36 Z. Lu, C. Mallet, D.M. Diehl and Y. Cheng, *LCGC*, 22(Supplement) (2004).
37 A. Shukla and R.E. Majors, *LCGC*, 23 (2005) 646.
38 J. Vercammen, K. Francois, M. Wolf, Vermeulen, and K. Welvaert, *LCGC*,
23 (2005) 594.

Further Reading

P. Gy, *Sampling for Analytical Purposes*, Wiley, England, 1998.
C.J. Koester and A. Moulik, *Anal. Chem.*, 77 (2005) 3737–3754.
S. Mitra (Ed.), *Sample Preparation Techniques in Analytical Chemistry*, Wiley,
New Jersey, 2003.
S.C. Moldoveanu, *J Chromatogr. Sci.*, 42 (2004) 1–14.
F. Settle (Ed.), *Handbook of Instrumental Techniques for Analytical Chemis-
try*, Prentice-Hall, New Jersey, 1997.

REVIEW QUESTIONS

1. Develop a method for extracting pesticides from soil samples, and
   explain the reasoning for selecting that method.
2. Develop a method for measuring the amounts of a new pharmaceutical active ingredient in plasma.
3. Explain new techniques for improving Soxhlet extraction.
4. Explain microwave-assisted extraction.

Chapter 3

Evaluation of basic physical properties

Neil Jespersen

3.1   INTRODUCTION

Some of the simplest techniques and instruments are valuable tools for chemical analysis. This chapter is designed to remind students that simple, rapid methods are advantageous in many situations. These methods are often used for quality control purposes. The methods discussed here are melting and boiling points, viscosity, density or specific gravity and refractive index.

Time is often an important consideration in practical chemical analysis. Raw materials awaiting delivery in tanker trucks or tractor-trailers often must be approved before they are offloaded. Batches of product prepared by compounders on the factory floor must be analyzed and approved before the packaging process can begin. Analysis is often performed again on packaged products before they are sent to the warehouse. To maintain a smoothly running manufacturing system, the analytical chemist must have rapid, informative methods to assay a wide variety of samples. Also important in quality control is the establishment of standards that are agreed upon by all involved in the manufacturing process. Lax enforcement of standards brings into question the utility of the analysis and threatens product quality.

Observation and measurement of physical properties are the oldest known means of assessing chemical purity and establishing identity. Their utility comes from the fact that the vast majority of chemical compounds have unique values for their melting points, boiling points, density and refractive index. In addition, viscosity and conductance are rapid methods for overall properties of substances. The *Handbook of Chemistry and Physics* has approximately 100 tables listing the refractive index, density, specific gravity, viscosity and conductivity of aqueous solutions of common inorganic and organic solutes at varying concentrations. Most of these properties also change dramatically if

Comprehensive Analytical Chemistry 47
S. Ahuja and N. Jespersen (Eds)
Volume 47 ISSN: 0166-526X DOI: 10.1016/S0166-526X(06)47003-3

impurities are present, thus making them ideal for quality control criteria. Additional, rapid methods such as Fourier transform infrared spectrometer (FT-IR) and rapid gas and liquid chromatographic techniques are also used in quality control, and are discussed elsewhere.

3.2 THEORY

3.2.1 Melting and boiling points

Melting occurs when a crystal structure is heated to the point when the kinetic energy of the molecules is sufficient to overcome the lattice energy of a crystal. The lower the lattice energy, the lower is the melting point of a substance. Boiling occurs when the temperature of a liquid is increased to the point when the vapor pressure of the liquid is the same as the ambient atmospheric pressure. Both of these properties are summarized in the heating curve as shown in Fig. 3.1. Melting is the transition from the solid phase to the liquid phase. In particular, the melting process occurs at a fixed temperature when the liquid and solid phases are both present in the sample. Similarly, boiling occurs at

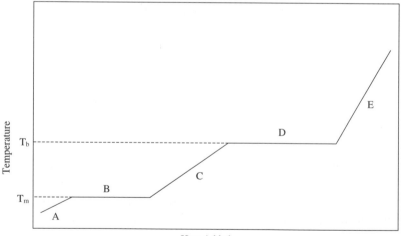

Fig. 3.1. Heating curve for a pure substance at constant pressure. Segment A represents heating the solid; segment B the liquid and solid in equilibrium as melting occurs; segment C the heating of the liquid; segment D vaporizing the liquid where liquid and vapor are in equilibrium and segment E heating of the vapor. T_m and T_b represent the melting and boiling points, respectively.

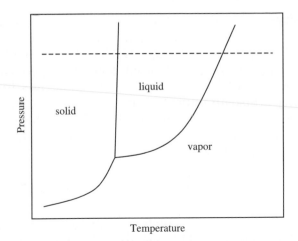

Fig. 3.2. A stylized phase diagram for a simple pure substance. The dashed line represents 1 atm pressure and the intersection with the solid–liquid equilibrium line represents the normal boiling point and the intersection with the liquid–vapor equilibrium line represents the normal boiling point.

a fixed temperature as long as the liquid and vapor phases are in equilibrium.

Figure 3.1 illustrates that the melting temperature varies little with pressure. Boiling points on the other hand can change more significantly with pressure changes. As a result, we define the normal boiling point (and also the normal melting point) as those measured when the atmospheric pressure is 1.00 atm.

The melting point and boiling points will vary with external atmospheric pressure as shown in the phase diagram in Fig. 3.2. The effect of impurities on melting and boiling points is understandable when we see that an added impurity has the effect of decreasing the triple point of the substance along the solid–gas equilibrium line in Fig. 3.2. The result is always a decrease in the melting point and an increase in the boiling point of a substance. For a small group of substances, the quantitative relationship of the freezing point depression and the boiling point elevation is known and has been used to determine molecular masses using

$$\Delta T = km = k \left(\frac{g_{\text{solute}}/MM_{\text{solute}}}{kg_{\text{solvent}}} \right)$$

where MM_{solute} represents the molecular mass of a non-dissociating solute.

3.2.2   Viscosity

Viscosity measurement or rheology. Viscosity is an important property of many fluids, particularly consumer products. Pancake syrup that does not appear "thick" or a "runny" shampoo has little consumer appeal. Viscosity is also an important measurement for liquid polymers. It is a measure of polymer chain length and branching. On the other hand, viscosity does not have the sensitivity to detect small amounts of impurities in pure substance.

There are two variables used in the description of fluid flow: shear stress and shear strain. Stress is measured in units of Pascals and the strain is dimensionless.

A Newtonian fluid is one in which the ratio of shear stress to the rate of shear strain is constant . This parameter is the viscosity η. That is,

$$\eta = \frac{\text{shear stress}}{\text{shear strain}}$$

The unit of viscosity is the poise $(g\,cm^{-1}\,s^{-1})$. Kinematic viscosity v is often used and is defined as

$$v = \eta/\rho$$

where ρ is the density of the fluid. The unit of kinematic viscosity is the Stokes $(cm^2\,s^{-1})$, many consumer products are specified in terms of their kinematic viscosity in centistokes (cst). The viscosity of water ranges from 1.79 centipoise (cP) at 0°C to 0.28 cP at 100°C. Olive oil has a viscosity of 84 cP and glycerin has a viscosity of 1490 cP, both at 20°C.

3.2.3   Density

Density is the mass per unit volume.

$$\rho = \text{mass}/\text{volume}$$

Density is also dependent on temperature and tabulated values of density are valid only at the specified temperature. A related but more versatile is the specific gravity. This is the ratio of the density of a substance to the density of water at the same temperature.

$$\text{Specific gravity} = \rho_{\text{sample}}\Big/\rho_{\text{water}}$$

The density of a substance in any desired units (e.g. g/cm^3 or Kg/quart) can be obtained by multiplying the specific gravity by the density of water in those units (i.e. e.g. g/cm^3 or Kg/quart respectively).

3.2.4 Refractive index

Refractive index is a measure of the velocity of light in air divided by the velocity of light in the compound of interest. Accurate measurement of refractive indices requires careful temperature control. The refractive index is sensitive to the purity of the substance.

3.2.5 Conductivity

Conductance of a solution is a measure of its ionic composition. When potentials are applied to a pair of electrodes, electrical charge can be carried through solutions by the ions and redox processes at the electrode surfaces. Direct currents will result in concentration polarization at the electrodes and may result in a significant change in the composition of the solution if allowed to exist for a significant amount of time. Conductance measurements are therefore made using alternating currents to avoid the polarization effects and reduce the effect of redox processes if they are reversible.

Charge conductance depends upon the ions involved. The ability of an ion to carry charge through a solution is dependent upon its transport number, also known as the transference number. This indicates that some ions will be more effective in conducting charge than others. Hydrogen ions and hydroxide ions are almost 10 times as effective at transporting charge as other ions. Table 3.1 tabulates these transference numbers. As a result, conductance is not useful for specific analyses, but is useful for assessing the overall nature of a solution. Conductance has its major use in assessing the ionic composition of water and monitoring of deionization or distillation processes. Consumer products may have characteristics that can be evaluated by conductivity also.

3.3   INSTRUMENTATION

3.3.1 Melting and boiling points

Melting point instruments consist of a resistance heater to increase temperature, a sample holder and a temperature-measuring device. In its simplest form, an oil bath can be heated with a Bunsen burner, while a capillary tube with a few milligrams of sample is attached to a thermometer with the sample next to the mercury bulb.

Heating is most often achieved using electrical resistive devices to heat a relatively massive metal block. This ensures that the temperature

TABLE 3.1

Examples of some ionic conductivity data. Sum of anionic and cationic conductivities may be used as estimates of salt conductivity

Cation	Ionic conductivity $10^{-4}\, m^2\, S\, mol^{-1}$	Anion	Ionic conductivity $10^{-4}\, m^2\, S\, mol^{-1}$
Ag^+	61.9	Br^-	78.1
$1/3Al^{3+}$	61	Cl^-	69.3
$1/2Cu^{2+}$	53.6	$1/2CO_3^{2-}$	69.3
$1/2Fe^{2+}$	54	ClO_4^-	64.6
$1/3Fe^{3+}$	68	F^-	55.4
H^+	349	$1/2HPO_4^{2-}$	33
K^+	73.48	$H_2PO_4^-$	33
$1/3La^{3+}$	69.7	$1/3PO_4^{3-}$	69.0
NH_4^+	73.5	MnO_4^-	61.3
$1/2Pb^{2+}$	71	NO_3^-	71.42
$1/2UO_2^{2+}$	32	OH^-	198
$1/2Zn^{2+}$	52.8	SO_4^{2-}	80.0

measured at one point on the block by the temperature sensor is the same as that experienced by the sample. Lower heating rates also help assure that the sensor and the sample are at the same temperature.

Visual observation is the historical method for observing melting points. Some instruments use photoelectric methods to detect changes that are interpreted as the onset of melting. Instruments that digitally record images of the melting process along with automatically determining the melting point are available. The latter provides a check on the automated process and also a permanent record.

Temperature measurement was historically done with a mercury thermometer. Modern instruments have electronic temperature sensors that can be coupled with digital temperature readouts. Digital temperature monitoring also allows the operator to record the observed melting point with the press of a keypad button. Data can be stored within the instrument or transmitted to a computer or laboratory information management system (LIMS).

3.3.2 Accuracy and precision

The accuracy and precision of melting point determinations depend on several factors. Calibration of the temperature sensor is the first and foremost requirement for accurate results. Checks of the calibration can be made using standard reference materials.

The second factor that affects the accuracy and precision of the results is the uniformity of the temperature throughout the system. In particular, the temperature sensor and the sample must be at the same temperature. In systems where the temperature is changing, it is difficult to assure that all components are at the same temperature.

Heat must flow from the heater to the sensor and sample and then into the sensor and sample. In many instruments the electrical resistance heater heats a relatively large metal block that acts as a heat reservoir, and then heat from the metal block is transferred to the temperature sensor and sample. Using a thermometer and a sample in a capillary tube in this example, there are considerations of heat flow through the glass bulb of the thermometer and the capillary tube. Only after traveling through the glass of the thermometer and capillary tube will the mercury and sample receive any of the heat that started at the heater. Each of these segments has a thermal gradient where the highest temperature will always be the heater and the lowest temperature will be the thermometer or the sample. Unfortunately, the thermal gradients in the thermometer and sample will rarely be identical and an error will result.

Thermal gradients cause the error in melting points and a decrease in the thermal gradient will decrease the error. There are several ways to decrease thermal gradients. First, if the thermal conductivity of the system is increased then thermal gradients will decrease. Metals have high thermal conductivity while the glass in the system has a low thermal conductivity. Second, lowering the heating rate will decrease thermal gradients. A heating rate of $2°C\,min^{-1}$ will have a much lower thermal gradient than one of $10°C\,min^{-1}$.

A final contribution to the accuracy and precision of the melting point is the ability of the instrument operator to determine when the sample melts and at what temperature, preferably simultaneously. This contribution to the error is also related to the heating rate.

3.3.3 Attributes of a good melting point determination

For optimum results, a melting point determination should have as many of the following characteristics as possible. The temperature sensor should be electronic, small and metallic in nature to quickly and accurately reflect the temperature of the system. The electronic nature of the sensor will allow the temperature to be recorded without physically reading a scale or meter. The sample should be in a thin wall capillary tube, and a minimum of sample should be used. The heating rate should be as slow as economically possible. A video recording of the sample and the temperature sensor signal may be valuable for review and archival purposes.

The need for highly accurate melting point determinations is rare. For quality control or routine purification assessment experiments, much less is required. Impurities equivalent to less than 1% wt/wt will result in melting point changes that can be readily observed in many compounds.

3.3.4   The melting process

Pure substances undergoing a phase change do so at a well-defined temperature that we call the melting point. Melting points are not significantly affected by atmospheric pressure. Ideally, the melting point should be measured in a system that is large enough to place a temperature sensor into a container where the solid and liquid phases co-exist. When observed in a melting-point apparatus, all substances require a finite amount of time to absorb enough heat to melt. This results in an *apparent* melting point range. If the heating rate is decreased, the melting point range will decrease accordingly and become zero when the heating rate approaches zero. To record the start and end of the melting process as a melting point range is without any firm theoretical foundation.

Impure substances have melting points that are very dependent upon the amount of impurity present. For a few substances this is quantified as the molal freezing point depression constant. The result is that melting points can be a very useful indicator of purification efforts. As long as each purification step in a process results in a higher melting point, the substance has been made more pure. This same concept allows the quality control chemist to have a very sensitive method for detecting impurities that is lower than anticipated.

Melting points of impure substances or mixtures often do have distinct melting point ranges. The reason for this is that when the mixture starts melting, the liquid form is enriched in the lower melting component. The remaining solid is enriched (slightly purified) in the higher melting component. The result is a true temperature range for the melting process. Some melting point instruments are shown in Figs. 3.3–3.5

3.3.5 Boiling points

Boiling points can be determined in some of the instruments that are used to determine melting points. The concepts are entirely analogous. As shown in the phase diagram, non-volatile impurities will increase the boiling points of substances. This makes the boiling point as effective as melting points for rapid evaluation of purity in research, production or quality control.

The characteristic of a boiling point is that the vapor pressure of the liquid is equal to the atmospheric pressure. Therefore, while atmospheric pressure has no effect on melting points, it can have a significant effect on boiling points. "Normal" boiling points are those measured with an atmospheric pressure of 760 Torr or 1.00 atm of pressure.

That boiling produces bubbles of vapor creates an additional problem for performing the experiment. If a bubble of gas forms at the bottom of a capillary tube, its expansion and rise to the top of the capillary will expel the rest of the liquid. This is due to the fact that the surface tension of most liquids combined with the narrow bore of the capillary will not allow fluid to drain around the bubble as it rises. The solution is that a larger sample tube and sample is required for the experiment. One advantage of the boiling point experiment is that the thermal conductivity of a liquid is higher than its solid because of the mobility of the molecules.

3.3.6 Viscosity (rheology)

The technical term for the study of and the measurement of viscosity is rheology. Viscosity measurements can be made in a variety of ways. The simplest method is to determine the time it takes for a metal ball to fall through a specified distance of a liquid in a graduated cylinder. The American Society of Testing Materials (ASTM) specifies several methods for the determination of viscosity of specific substances using this method. When the radius of the ball is small compared to the radius of the graduated cylinder, the viscosity will be

$$\eta = \tfrac{2}{9} r^2 g \frac{(\rho_{ball} - \rho_{liquid})}{kv}$$

where r is the radius of the ball, g the acceleration of gravity, ρ_{ball} and ρ_{liquid} are the densities of the ball and liquid, respectively, and k is a

439-025

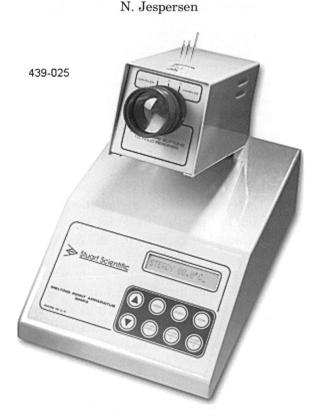

Fig. 3.3. A Jencons Scientific Inc. melting point instrument.

constant that corrects the velocity for the fact that the cylinder is not infinitely wide.

A method similar to the falling-ball method is the bubble method. A vial is filled with liquid, leaving sufficient room for a bubble that can equal the diameter of the vial. The vial is inverted and the time required for the bubble to pass two predetermined marks is determined. The viscosity will be directly proportional to the time required for the bubble to rise.

Another sample method uses a cup with a hole in the bottom. The cup is dipped into the liquid (often a paint or similar fluid) and removed. The time required for all of the liquid to drain from the cup is a measure of the viscosity. Cups calibrated for a variety of viscosity ranges are available.

A special glass device called a kinematic viscometer tube is shown in Fig. 3.6. To use the device, the liquid is added to the tube to fill the large bulb on the right. The tube is submerged in a thermostatted bath so

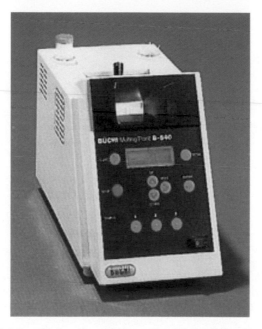

Fig. 3.4. Büchi Melting Point Apparatus. Büchi Melting Point Models B-540 and B-545 can determine melting points on routine and non-routine substances in research and quality control labs of pharmaceutical and chemical companies. Both models have a temperature range from ambient to 400°C and a selection of nine heating speeds. Heating (from 50 to 350°C) takes approximately 7 min and cooling (from 350 to 50°C) takes approximately 10 min. The units can simultaneously determine melting points of three samples or two boiling points.

that all three bulbs are immersed. After the system is equilibrated (10–30 min), suction is used to pull the liquid up to fill both of the small bulbs. When the suction is released, the liquid will flow back into the large bulb. Timing is started when the liquid level reaches the mark between the two small bulbs and it is stopped when it reaches the other mark. These tubes are usually precalibrated at a specified operating temperature so that

$$\eta = kt$$

The viscosity of a liquid can also be determined by measuring the torque needed to rotate a cylinder in the liquid. Brookfield viscometers and rheometers fall into this class of instrument (Fig. 3.7). The viscometer measures the torque produced when a spindle is rotated at constant velocity in a liquid. The Rheometer produces a constant torque

Fig. 3.5. OptiMelt MPA100—Automated Melting Point System. This instrument has automatic and visual determination of melting points. There is PID-controlled temperature ramping and a digital movie is taken of each melt.

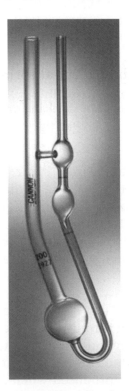

Fig. 3.6. Kinematic viscometer tube from Fisher Scientific.

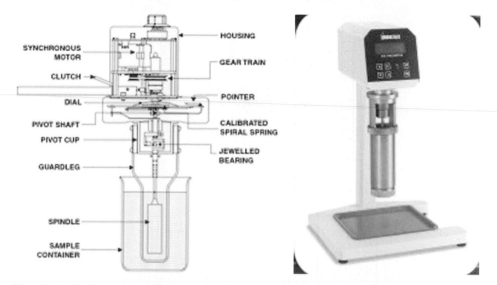

Fig. 3.7. Hydrometer (left) with large weighted bulb and calibrated scale. Pycnometer (right) is filled with fluid and when thermometer is inserted the liquid overflows into the small bulb.

and relates that to the rate of rotation. In the viscometer a constant rpm motor through a calibrated spiral spring drives the spindle. The greater the viscosity, the more the spring is displaced on an analog scale in some models. In digital models a rotary variable displacement transducer detects the position of the spring.

3.3.7 Density

Measurement of density is often a rapid and easily mastered technique. Different procedures are used for liquids, solids and gases.

3.3.7.1 Liquids

The easiest measure of density is obtained using a hydrometer. The hydrometer is a weighted glass vessel that will float in liquids having a precalibrated density range. The hydrometer shown in Fig. 3.8 has a large bulb with a narrow stem. Inside the stem is a calibrated scale. When the hydrometer is immersed in a liquid a slight spin is applied to the stem to keep the hydrometer centered in the graduated cylinder. When equilibrated in a liquid it will displace a volume of liquid equal to its mass and then float on the liquid. The depth to which the hydrometer sinks is inversely proportional to the density. The density of the

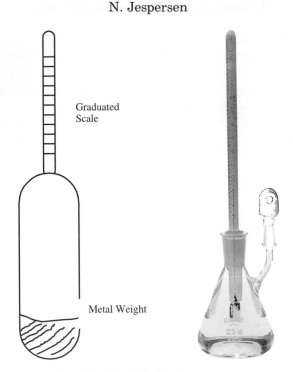

Graduated
Scale

Metal Weight

Fig. 3.8. Brookfield viscometers.

liquid can be read from the calibrated scale on the stem. A more time-consuming process is to calibrate the volume of a small glass vessel called a pycnometer with thermostatted distilled water by weighing the pycnometer before and after filling with water. The mass of water can be converted into the volume using known density tables. Using this volume, the mass of an unknown liquid is determined and the density is calculated as

density = mass/volume

3.3.7.2 Solids

Densities of solids are determined by determining the volume of the solid, either by measurement of the solid's dimensions or by displacement of a liquid and then determining the mass on a laboratory balance. The density can also be determined by weighing the object and then weighing it again when suspended in a liquid of known density.

Additionally, there is interest in determining the density of porous objects, powders or very irregular crystals. It is often difficult to obtain an accurate measure of the volume of these substances. One method used to determine the volume of these substances is to allow an inert gas to expand from a chamber with a known pressure and volume ($P_{initial}V_{initial}$) into an evacuated chamber of known volume, V_{known}. Measuring the final pressure of the expanded gas allows the calculation of the final volume since $P_{initial}V_{initial} = P_{final}V_{final}$. The difference between V_{final} and V_{known} is the volume of the powder or porous object.

3.3.8 Refractive index

The refractive index is another very rapid analytical method for determining purity and identity of a substance. It is uniquely useful for quality control monitoring of raw materials and finished products. Many students encounter a refractometer first as a "universal" detector for high-performance liquid chromatography.

Refractometers were first produced for scientific work in the late 1800s, and continue to be made mainly by the Abbe Corporation.

Theory: The initial understanding of the refraction of light dates back to Maxwell's study of electromagnetic radiation. Ernst Abbe invented the first commercial refractometer in 1889 and many refractometers still use essentially the same design.

Light is refracted, or bent, as it passes between media with different refractive indices. Snell's Law describes refraction mathematically as

$$\eta_1 \sin \theta_1 = \eta_2 \sin \theta_2$$

Refractive indices of some common materials are shown in Table 3.2.

The angles are measured from the normal to the interface between the two media (Fig. 3.9).

When light is moving from a medium with a large refractive index to one with a smaller refractive index, the phenomenon of total internal reflection can occur when the refracted angle is 90° or more. In a

TABLE 3.2

Refractive indices of some common materials

Medium	Index
Air (STP)	1.00029
Water (20°C)	1.33
Crown glass	1.52
Flint glass	1.65

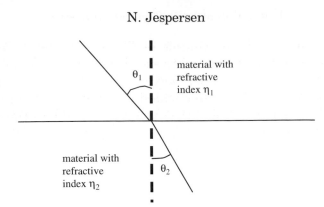

Fig. 3.9. Illustration of the refraction of light. The top medium has a refractive index that is smaller than the one at the bottom, $\eta_1 < \eta_2$, resulting in $\theta_1 > \theta_2$.

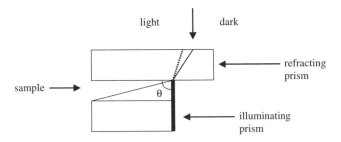

Fig. 3.10. Schematic diagram of a refractometer. The illuminating prism provides light to the sample ranging in angles from $0°$ (from the normal) to θ.

refractometer there are two prisms, the illuminating prism and the refracting prism. The illuminating prism has low refractive index and the refracting prism has a high refractive index that generally represents the limits of refractive index that the instrument can determine. The sample is placed between the two prisms. In addition, the surface of the incident prism in contact with the sample is frosted to produce a diffuse light source. In Fig. 3.10 the furthest point on the refracting prism that can be illuminated is shown. A light ray from the far end of the illuminating prism will have the largest angle from the normal and have the largest refracted angle. Therefore, the spot where the rays are traced to the top of the refracting prism indicates the furthest that the refracting prism will be illuminated. The dashed line shows how a sample with a different refractive index will result in a different position of the demarcation line between the dark and light portions of the refracting prism. White light will produce an indistinct line between the dark and light portions because different wavelengths of light will be refracted by different amounts (i.e., it is dispersed). Use of special Amici prisms can reduce this dispersion and result in a distinct line between the light and dark regions. Use of a single wavelength such as the sodium D line at 588 nm produces a sharp demarcation line.

In practice the viewing optics are moved until the light/dark line is centered in the eyepiece. The position of the eyepiece is coupled to the refractive index scale that is read independently.

3.3.9   Operation

Use of a refractometer is extremely simple. The illuminating and refracting prisms are held together with a hinge and a latch. When the latch is opened, the prism faces can be cleaned with an appropriate solvent. When the latch is closed there is a small opening that will accept a few drops of sample. In a manual instrument the eyepiece of the instrument is moved until the screen is half-light and half-dark. If the demarcation line is not sharp, the lens is rotated to sharpen the line. Finally, a scale is read with a second small magnifier, or on some instruments a button is pressed that illuminates an internal scale using the same eyepiece.

Automatic instruments are usually reflection-type instruments that measure the deflection of a beam of light as it passes from one medium, into the sample and then is reflected back to a detector. The angle at which the light beam exits the medium is related to the refractive index of the sample. Automated instruments are calibrated with standard substances of precisely known refractive index prior to use.

3.3.10   Analysis using refractometry

Table 3.3 simple organic liquids, which illustrates a variety of refractive indices that substances have and the fact that the refractive index is a useful property for identifying a substance.

The refractive index also varies with the amount of substance in a mixture. Most often, refractive index is used to assess the concentration of sugar in wine, soft drinks, cough medicines and other preparations having relatively high concentrations of sucrose. Refractive index is also used to determine the concentration of alcohol in fermented products. For sucrose solutions the refractive index varies from 1.3330 (pure water) to 1.5033 when the solution contains 85% sucrose. This is an increase of approximately 0.0002 in the refractive index for each 0.1%

TABLE 3.3

Refractive indices of some common materials. Refractive indices to 5 decimal places are commonly measured

Compound	Formula	n^D
Aniline	$C_6H_5NH_2$	1.5863
Acetic acid	CH_3COOH	1.3716
Benzaldehyde	C_6H_5CHO	1.5463
Benzene	C_6H_6	1.5011
Hexanoic acid	$CH_3(CH_2)_4COOH$	1.4163
Cyclohexane	C_6H_{12}	1.4266
Hexane	C_6H_{14}	1.3751
Carbon tetrachloride	CCl_4	1.4601
Nicotine	$C_{10}H_{14}N_2$	1.5282
Phenol	C_6H_5OH	1.5408
Tetradecane	$C_{14}H_{30}$	1.429
Trimethylamine	$(CH_3)_3N$	1.3631
o-Xylene	$(CH_3)_2C_6H_4$	1.5055
m-Xylene	$(CH_3)_2C_6H_4$	1.4972
p-Xylene	$(CH_3)_2C_6H_4$	1.4958

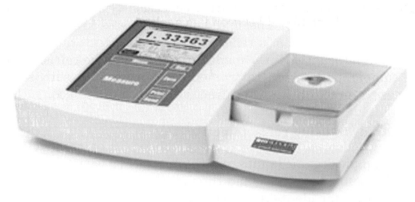

Fig. 3.11. A modern refractometer. Sample is placed in the depression on the right side of the instrument.

change in sucrose concentration. Similarly, ethanol varies from 1.3330 to 1.3616 at 50% ethanol and changes by 0.0003–0.0004 for each 0.5% change at low concentrations (Figs. 3.11 and 3.12). The more commonly used tables of refractive index and concentration are found in handbooks such as the *Handbook of Chemistry and Physics*.

Fig. 3.12. An older Abbe refractometer. Thermostatted water enters and exits through the tubes on the front. The front housing also opens to allow sample to be placed between two analyzer plates.

Use of refractive index for quantitative measurements requires careful temperature control. Change of temperature by 1°C can result in a decrease in refractive index of 0.0002–0.0004 for aqueous solutions. Solutions of ethanol can see a decrease in refractive index of as much as $0.0008°C^{-1}$.

3.3.11 Conductivity

The conductance of a solution is the inverse of its resistance, and conductance has units of $ohms^{-1}$ or mohs. The higher the conductance of a solution, the lower is its electrical resistance. A conductivity meter and conductivity cell are used to determine the effective resistance of a solution. The conductivity cell consists of a pair of platinized platinum electrodes with an area of approximately $1.0\,cm^2$ with spacers designed to hold the electrodes rigidly parallel and at a fixed distance from each other. The cell can be standardized with solutions of known conductivity to obtain the cell constant, k so that the instrument response R

can be converted directly into conductivity

conductivity (mhos) $= k\,R$

Classical conductivity meters are forms of the Wheatstone bridge. The Wheatstone bridge is a null device. It produces an "off-balance" potential when the four resistances making up the bridge do not satisfy the null condition in the following equation.

$R_{cell}R_1 = R_{var}R_2$

Where two arms of the bridge are $R_1 = R_2$ and the conductivity cell is R_{cell} and $R_{variable}$ is adjustable to balance the bridge. When balanced, or nulled so that there is no off-balance potential $R_{cell} = R_{variable}$. The null condition could be observed electronically or by sound since the bridge was usually powered by 1000 Hz signals of approximately 1 V.

TABLE 3.4

Examples of the conductivity of some common substances

Examples of conductivity values	Conductivity at 25°C
Water	$\mu S\,cm^{-1}$
Ultra-pure water	0.055
Distilled water	0.5–5
Rain water	20–100
Mineral water	50–200
River water	250–800
Tap water	100–1500
Surface water	30–7000
Waste water	700–7000
Brackish water	1000–8000
Ocean water	40000–55000
Chemicals	$mS\,cm^{-1}$
KCl 0.01 M	1.41
$MgSO_4$	5.81
KCl 0.1 M	12.9
H_2SO_4	82.6
KCl 1.0 M	112
NaOH 50%	150
NaOH 5%	223
NaCl saturated	251
HCl 10%	700
HNO_3 31% (highest known)	865

Modern conductivity meters use a variety of electrode surfaces, carbon for instance. Two- and four-pole electrode systems are also used. The latter are designed to minimize polarization effects. The instrument itself applies a constant current to the conductance cell so that the applied voltage is always 200 mV. The instrument monitors the resulting current and voltage, taking the I/V ratio that is equal to $1/R$ or the conductance. In addition, the instrument can have a variety of conductivity ranges. Along with increasing conductivity, the applied frequency is increased to help minimize polarization effects. Capacitance correction is also made to increase the accuracy of the instrument.

Table 3.4 lists some solutions and their conductivities that can be used for calibration purposes.

REVIEW QUESTIONS

1. A spherical object has a diameter of 2.32 cm and a mass of 24.3 g. What is its density?
2. A cylindrical object has a length of 3.45 cm and a diameter of 1.6 cm. With a mass of 35.6 g, what is the density of the object?
3. A pycnometer weighs 23.4562 g when empty. When filled with water at 23.2°C it weighs 37.2362 g. The pycnometer is filled with an unknown liquid and it has a mass of 35.8739 g. What is the density of this substance? What is the specific gravity of this substance?
4. A pycnometer filled with an organic liquid weighs 45.2316 g. Filled with water at 26.4°C the pycnometer weighs 48.2386 g. The empty pycnometer weighs 21.2323 g. What is the density of the unknown liquid? What is its specific gravity?
5. Argon in a 500 ml chamber at 1.342 atm pressure is expanded into an identical chamber that contains 3.65 g of a powder substance. The final pressure is 0.772 atm. What is the volume of the solid? What is the density of the solid?
6. Helium in a 200 ml chamber has a pressure of 1534 Torr. When expanded into a chamber twice its size and containing 23.45 g of sample the final pressure is 607 Torr. What is the volume and density of the sample?
7. Why do concentrated sugar solutions have a high viscosity?
8. Is the viscosity of gasoline expected to be high or low? Explain your answer.

9. What classes of substances are expected to have sharp melting points? Which ones will have extended melting point ranges? Why?
10. A mixed melting point is often used to confirm the identity of an organic substance. How is this done and why does it work?
11. A solid has a melting point of 80 and a boiling point of 218°C. Use the tables in the *Handbook of Chemistry and Physics* to suggest the identity of the compound.
12. A solid has a melting point of 121 +/− 2°C and a boiling point of 250 +/− 3°C. What is the possible identity of this compound? Use the tables in the *Handbook of Chemistry and Physics*.
13. Why are sugar and alcohol content in foods commonly determined by refractive index determinations?
14. If there were more than one possible answer to question 11 or 12, what additional experiment would be best to decide the identity of the compound? Select from the methods in this chapter.
15. Suggest a reason why the specific conductance of hydronium and hydroxide ions is so much larger than the specific conductance of all other ions.
16. Suggest an industrial use for conductance measurements.

Chapter 4

Thermal analysis

Harry G. Brittain and Richard D. Bruce

4.1   INTRODUCTION

Thermal methods of analysis can be defined as those techniques in which a property of the substance under study is determined as a function of an externally applied and programmed temperature. Dollimore [1] has listed three conditions that define the usual practice of thermal analysis:

1. The physical property and the sample temperature should be measured continuously.
2. Both the property and the temperature should be recorded automatically.
3. The temperature of the sample should be altered at a predetermined rate.

Measurements of thermal analysis are conducted for the purpose of evaluating the physical and chemical changes, which may take place as a result of thermally induced reactions in the sample. This requires that the operator subsequently interpret the events observed in a thermogram in terms of plausible thermal reaction processes. The reactions normally monitored can be *endothermic* (melting, boiling, sublimation, vaporization, desolvation, solid–solid phase transitions, chemical degradation, etc.) or *exothermic* (crystallization, oxidative decomposition, etc.) in nature.

Thermal methods have found extensive use in the past as part of a program of preformulation studies, since carefully planned work can be used to indicate the existence of possible drug–excipient interactions in a prototype formulation [2]. It should be noted, however, that the use of differential scanning calorimetry (DSC) for such work is less in vogue than it used to be. Nevertheless, in appropriately designed applications, thermal methods of analysis can be used to evaluate compound purity,

Comprehensive Analytical Chemistry 47
S. Ahuja and N. Jespersen (Eds)
Volume 47 ISSN: 0166-526X DOI: 10.1016/S0166-526X(06)47004-5

polymorphism, solvation, degradation, drug–excipient compatibility, and a wide variety of other thermally related characteristics. Several reviews are available regarding the scope of such investigations [2–6].

4.2   DETERMINATION OF MELTING POINT

The melting point of a substance is defined as the temperature at which the solid phase exists in equilibrium with its liquid phase. This property is of great value as a characterization tool since its measurement requires relatively little material, only simple instrumentation is needed for its determination (see Chapter 3), and the information can be used for compound identification or in an estimation of purity. For instance, melting points can be used to distinguish among the geometrical isomers of a given compound, since these will normally melt at non-equivalent temperatures. It is a general rule that pure substances will exhibit sharp melting points, while impure materials (or mixtures) will melt over a broad range of temperature.

When a substance undergoes a melting phase transition, the high degree of molecular arrangement existing in the solid becomes replaced by the disordered character of the liquid phase. In terms of the kinetic molecular approach, the melting point represents the temperature at which the attractive forces holding the solid together are overcome by the disruptive forces of thermal motion. The transition is accompanied by an abrupt increase in entropy and often an increase in volume.

The temperature of melting is usually not strongly affected by external pressure, but the pressure dependence can be expressed by the Clausius–Clapeyron equation:

$$\frac{\mathrm{d}T}{\mathrm{d}p} = \frac{T(V_\mathrm{L} - V_\mathrm{S})}{\Delta H} \tag{4.1}$$

where p is the external pressure, T is the absolute temperature, V_L and V_S are the molar volumes of liquid and solid, respectively, and ΔH is the molar heat of fusion. For most substances, the solid phase has a larger density than the liquid phase, making the term $(V_\mathrm{L} - V_\mathrm{S})$ positive, and thus an increase in applied pressure usually raises the melting point. Water is one of the few substances that exhibits a negative value for $(V_\mathrm{L} - V_\mathrm{S})$, and therefore one finds a decrease in melting point upon an increase in pressure. This property, of course, is the basis for the winter sport of ice-skating.

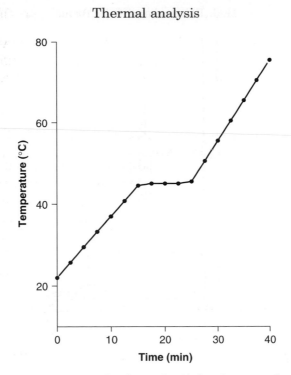

Thermal analysis

Fig. 4.1. Melting point curve of a hypothetical substance, having a melting point around 45°C.

If a solid is heated at a constant rate and its temperature monitored during the process, the melting curve as illustrated in Fig. 4.1 is obtained. Below the melting point, the added heat merely increases the temperature of the material in a manner defined by the heat capacity of the solid. At the melting point, all heat introduced into the system is used to convert the solid phase into the liquid phase, and therefore no increase in system temperature can take place as long as solid and liquid remain in equilibrium with each other. At the equilibrium condition, the system effectively exhibits an infinite heat capacity. Once all solid is converted to liquid, the temperature of the system again increases, but now in a manner determined by the heat capacity of the liquid phase.

Measurements of melting curves can be used to obtain very accurate evaluations of the melting point of a compound when slow heating rates are used. The phase transition can also be monitored visually, with the operator marking the onset and completion of the melting process. This is most appropriately performed in conjunction with optical microscopy, thus yielding the combined method of thermomicroscopy or hot-stage microscopy [7].

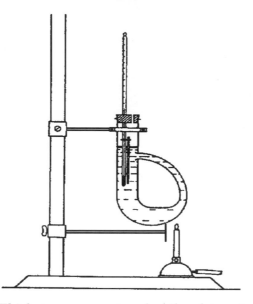

Fig. 4.2. Typical Thiele-type apparatus for the determination of melting points.

A thorough discussion of apparatus suitable for the determination of melting points has been provided by Skau [8]. One of the most common methods involves placing the analyte in a capillary tube, which is immersed in a batch whose temperature is progressively raised by an outside heating force. The Thiele arrangement (which is illustrated in Fig. 4.2) is often used in this approach. The analyst observes the onset and completion of the melting process, and notes the temperatures of the ranges with the aid of the system thermometer. The thermometer should always be calibrated by observing the melting points of pure standard compounds, such as those listed in Table 4.1.

For pharmaceutical purposes, the melting range or temperature of a solid is defined as those points of temperature within which the solid coalesces or is completely melted. The general method for this methodology is given in the United States Pharmacopeia as a general test [9].

The determination of melting point as a research tool has long been supplanted by superior technology, although synthetic chemists still routinely obtain melting point data during performance of chemical synthesis. However, at one time such measurements provided essential information regarding the structure of chemical compounds, and their careful determination was a hallmark of such work. For instance, Malkin and coworkers conducted a long series of X-ray diffraction and

TABLE 4.1

Corrected melting points of compounds suitable as reference materials in the calibration of thermometers

Melting point (°C)	Material
0	Ice
53	p-Dichlorobenzene
90	m-Dinitrobenzene
114	Acetanilide
122	Benzoic acid
132	Urea
157	Salicylic acid
187	Hippuric acid
200	Isatin
216	Anthracene
238	Carbanilide
257	Oxanilide
286	Anthraquinone
332	N, N-diacetylbenzidine

melting point analyses of every possible glycerol ester, and used this work to study the polymorphism in the system. Three distinct structural phases were detected for triglycerides [10], diglycerides [11], and monoglycerides [12], and categorized largely by their melting points. In all cases, solidification of the melt yielded the metastable α-phase, which could be thermally transformed into the β′-phase, and this form could eventually be transformed into the thermodynamically stable β-phase. As shown in Fig. 4.3, each form was characterized by a characteristic melting point, and these were found to be a function of the number of carbon atoms in the aliphatic side chains. In general, the melting points of the α-forms lay along a smooth line, while the melting points of the β′- and β-forms followed a zig-zag dependence with the number of carbons.

As it turns out, there are pharmaceutical implications associated with the polymorphism of glycerol esters, since phase transformation reactions caused by the melting and solidification of these compounds during formulation can have profound effects on the quality of products. For instance, during the development of an oil-in-water cream formulation, syneresis of the aqueous phase was observed upon using certain sources of glyceryl monostearate [13]. Primarily through the use of variable temperature X-ray diffraction, it was learned that

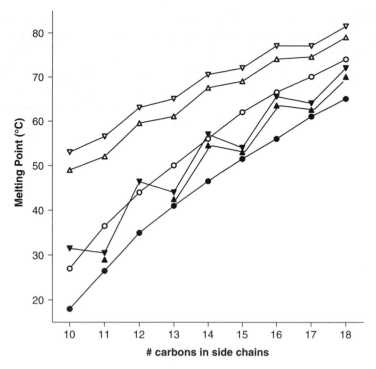

Fig. 4.3. Dependence of the melting points of various polymorphs of monoglycerides and triglycerides upon the number of carbon atoms in the aliphatic sidechains. Data are shown for the α-forms of monoglycerides (O) and triglycerides (ℓ), the β′-forms of monoglycerides (△) and triglycerides (▲), and the β-forms of monoglycerides (∇) and triglycerides (▼).

this material would undergo changes in phase composition upon melting and congealing. The ability of glyceryl monostearate to break the oil-in-water emulsion was directly related to the composition of the raw material and in the degree of its expansion (or lack thereof) during the congealing process. Knowledge of the melting behavior of this excipient, as influenced by its source and origin, proved essential to the transferability of the formulation in question.

4.3 DIFFERENTIAL THERMAL ANALYSIS

4.3.1 Background

Differential thermal analysis (DTA) consists of the monitoring of the difference in temperature existing between a solid sample and a reference as a function of temperature. Differences in temperature between the sample and reference are observed when a process takes place that requires a finite heat of reaction. Typical solid-state changes of this type would be phase transformations, structural conversions, decomposition reactions, and desolvation of solvatomorphs. These processes require either the input or the release of energy in the form of heat, which in turn translates into events that affect the temperature of the sample relative to a non-reactive reference.

Although a number of attempts have been made to use DTA as a quantitative tool, such applications are not trivial. However, the technique has been used to study the kinetics associated with inter-phase reactions [14], and as a means to study enthalpies of fusion and formation [15].

However, for most studies, DTA has been mostly used in a qualitative sense as a means to determine the characteristic tempera-tures of thermally induced reactions. Owing to the experimental conditions used for its measurement, the technique is most useful for the characterization of materials that evolve corrosive gases during the heating process. The technique has been found to be highly useful as a means for compound identification based on the melting point considerations, and has been successfully used in the study of mixtures.

4.3.2 Methodology

Methodology appropriate for the measuring of DTA profiles has been extensively reviewed [16–18]. A schematic diagram illustrating the essential aspects of the DTA technique is shown in Fig. 4.4. Both the sample and the reference materials are contained within the same furnace, whose temperature program is externally controlled. The outputs of the sensing thermocouples are amplified, electronically subtracted, and finally shown on a suitable display device. If the observed ΔH is positive (endothermic reaction), the temperature of the sample will lag behind that of the reference. If the ΔH is negative (exothermic reaction), the temperature of the sample will exceed that of the reference. One of the great advantages associated with DTA analysis is that the analysis can usually be performed in such a manner that corrosive gases evolved from the sample do not damage expensive portions of the thermal cell assembly.

Wendlandt has provided an extensive compilation of conditions and requirements that influence the shape of DTA thermograms [18]. These can be divided into instrumental factors (furnace atmosphere, furnace geometry, sample holder material and geometry, thermocouple details,

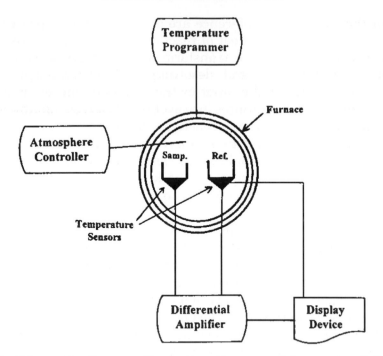

Fig. 4.4. Schematic diagram illustrating the essential aspects of the DTA technique.

heating rate, and thermocouple location in sample) and sample characteristics (particle size, thermal conductivity, heat capacity, packing density, swelling or shrinkage of sample, mass of sample taken, and degree of crystallinity). A sufficient number of these factors are under the control of the operator, thus permitting selectivity in the methods of data collection. The ability to correlate an experimental DTA thermogram with a theoretical interpretation is profoundly affected by the details of heat transfer between the sample and the calorimeter [19].

The calibration of DTA systems is dependent on the use of appropriate reference materials, rather than on the application of electrical heating methods. The temperature calibration is normally accomplished with the thermogram being obtained at the heating rate normally used for analysis [20], and the temperatures known for the thermal events used to set temperatures for the empirically observed features. Recommended reference materials that span melting ranges of pharmaceutical interest include benzoic acid (melting point 122.4°C), indium (156.4°C), and tin (231.9°C).

4.3.3 Applications

Historically, one of the most important uses of DTA analysis has been in the study of interactions between compounds. In an early study, the formation of 1:2 association complexes between lauryl or myristyl alcohols with sodium lauryl or sodium myristyl sulfates have been established [21]. In a lesson to all who use methods of thermal analysis for such work, the results were confirmed using X-ray diffraction and infrared absorption spectroscopic characterizations of the products.

The use of DTA analysis as a means to deduce the compatibility between a drug substance and its excipients in a formulation proved to be a natural application of the technique [22]. For instance, Jacobson and Reier used DTA analysis to study the interaction between various penicillins and stearic acid [23]. For instance, the addition of 5% stearic acid to sodium oxacillin monohydrate completely obliterated the thermal events associated with the antibiotic. It seems that the effect of lubricants on formulation performance was as problematic then as it is now, and DTA served as a useful method in the evaluation of possible incompatibilities. Since that time, many workers employed DTA analysis in the study of drug–excipient interactions, although the DTA method has been largely replaced by DSC technology.

Proceeding along a parallel track, Guillory and coworkers used DTA analysis to study complexation phenomena [2]. Through the performance of carefully designed studies, they were able to prove the existence of association complexes and deduced the stoichiometries of these. In this particular work, phase diagrams were developed for 2:1 deoxycholic acid/menadione, 1:1 quinine/phenobarbital, 2:1 theophylline/phenobarbital, 1:1 caffeine/phenobarbital, and 1:1 atropine/phenobarbital. The method was also used to prove that no complexes were formed between phenobarbital and aspirin, phenacetin, diphenylhydantoin, and acetaminophen.

In its heyday, DTA analysis was very useful for the study of compound polymorphism and in the characterization of solvate species of drug compounds. It was used to deduce the ability of polymorphs to undergo thermal interconversion, providing information that could be used to deduce whether the system in question was monotropic or enantiotropic in nature. For instance, the enthalpies of fusion and transition were measured for different polymorphs of sulfathiazole and methylprednisolone [24]. The DTA thermograms shown in Fig. 4.5 demonstrate that Form-I is metastable with respect to Form-II, even

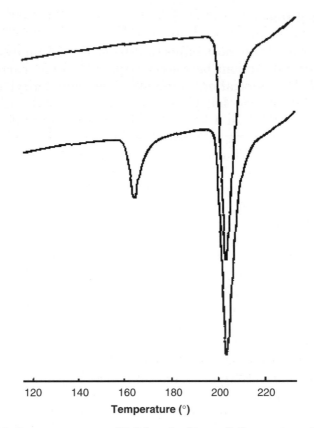

Fig. 4.5. DTA thermograms sulfathiazole, Form-I (lower trace) and Form-II (upper trace). Adapted from Ref. [24].

though the enthalpies of fusion of the two forms were almost equal. However, the enthalpy of transition was found to be significant.

Form-I of chloroquine diphosphate melts at 216°C, while Form-II melts at 196°C [25]. The DTA thermogram of Form-I consists of a simple endotherm, while the thermogram of Form-II is complicated. The first endotherm at 196°C is associated with the melting of Form-II, but this is immediately followed by an exothermic transition corresponding to the crystallization of Form-I. This species is then observed to melt at 216°C, establishing it as the thermodynamically more stable form at the elevated temperature.

DTA analysis proved to be a powerful aid in a detailed study that fully explained the polymorphism and solvates associated with several sulfonamides [26]. For instance, three solvate species and four true polymorphs were identified in the specific instance of sulfabenzamide.

Quantitative analysis of the DTA thermograms was used to calculate the enthalpy of fusion for each form, with this information then being used to identify the order of relative stability. Some of these species were found to undergo phase conversions during the heating process, but others were noted to be completely stable with respect to all temperatures up to the melting point.

It is always possible that the mechanical effects associated with the processing of materials can result in a change in the physical state of the drug entity [27], and DTA analysis has proven to be a valuable aid in this work. For instance, the temperature used in the drying of spray-dried phenylbutazone has been shown to determine the polymorphic form of the compound [28]. A lower melting form was obtained at reduced temperatures (30–40°C), while a higher melting material was obtained when the material was spray-dried at 100–120°C. This difference in crystal structure would be of great importance in the use of spray-dried phenylbutazone since the dried particles exhibited substantially different crystal morphologies.

The reduction of particle size by grinding can also result in significant alterations in structural properties, and DTA analysis has been successfully used to follow these in appropriate instances. In one study, methisazone was found to convert from one polymorph to another upon micronization, and the phase transformation could be followed through a study of the thermal properties of materials ground for different times [29]. In another study, it was found that extensive grinding of cephalexin monohydrate would effectively dehydrate the material [30]. This physical change was tracked most easily through the DTA thermograms, since the dehydration endotherm characteristic of the monohydrate species became less prominent as a function of the grinding time. It was also concluded that grinding decreased the stability of cephalexin, since the temperature for the exothermic decomposition was observed to decrease with an increase in the grinding time.

4.4   DIFFERENTIAL SCANNING CALORIMETRY

4.4.1   Background

In many respects, the practice of DSC is similar to the practice of DTA, and analogous information about the same types of thermally induced reactions can be obtained. However, the nature of the DSC experiment makes it considerably easier to conduct quantitative analyses, and this aspect has ensured that DSC has become the most widely used method of thermal analysis. The relevance of the DSC technique as a tool for pharmaceutical scientists has been amply documented in numerous reviews [3–6,31–32], and a general chapter on DSC is documented in the United States Pharmacopeia [33].

In the DSC method, the sample and the reference are maintained at the same temperature and the heat flow required to keep the equality in temperature is measured. DSC plots are therefore obtained as the differential rate of heating (in units of W/s, cal/s, or J/s) against temperature. The area under a DSC peak is directly proportional to the heat absorbed or evolved by the thermal event, and integration of these peak areas yields the heat of reaction (in units of cal/s g or J/s g).

When a compound is observed to melt without decomposition, DSC analysis can be used to determine the absolute purity [34]. If the impurities are soluble in the melt of the major component, the van't Hoff equation applies:

$$T_s = T_o - \{R(T_o)^2 X_i\}/\{F\Delta H_f\} \tag{4.2}$$

where T_s is the sample temperature, T_o is the melting point of the pure major component, X_i is the mole fraction of the impurity, F is the fraction of solid melted, and ΔH_f is the enthalpy of fusion of the pure component. A plot of T_s against $1/F$ should yield a straight line, whose slope is proportional to X_i. This method can therefore be used to evaluate the absolute purity of a given compound without reference to a standard, with purities being obtained in terms of mole percent. The method is limited to reasonably pure compounds that melt without decomposition. The assumptions justifying Eq. (4.2) fail when the compound purity is below approximately 97 mol%, and the method cannot be used in such instances. The DSC purity method has been critically reviewed, with the advantages and limitations of the technique being carefully explored [35–37].

4.4.2 Methodology

Two types of DSC measurement are possible, which are usually identified as power-compensation DSC and heat-flux DSC, and the details of each configuration have been fully described [1,14]. In power-compensated DSC, the sample and the reference materials are kept at the same temperature by the use of individualized heating elements, and the observable parameter recorded is the difference in power inputs to the two heaters. In heat-flux DSC, one simply monitors the heat differential between the sample and reference materials, with the methodology not being terribly different from that used for DTA. Schematic diagrams of the two modes of DSC measurement are illustrated in Fig. 4.6.

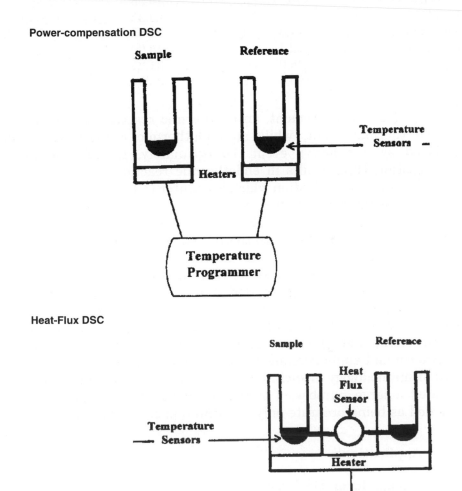

Fig. 4.6. Schematic diagrams of the power-compensation and heat-flux modes of DSC measurement.

TABLE 4.2

Melting temperatures and enthalpies of fusion for compounds suitable as reference materials in DSC

Material	Melting point (°C)	Enthalpy of fusion (kJ/mol)
Naphthalene	80.2	19.05
Benzil	94.8	23.35
Acetamide	114.3	21.65
Benzoic acid	122.3	18.09
Diphenylacetic acid	148.0	31.27
Indium	156.6	3.252

In the DTA measurement, an exothermic reaction is plotted as a positive thermal event, while an endothermic reaction is usually displayed as a negative event. Unfortunately, the use of power-compensation DSC results in endothermic reactions being displayed as positive events, a situation which is counter to IUPAC recommendations [38]. When the heat-flux method is used to detect the thermal phenomena, the signs of the DSC events concur with those obtained using DTA, and also agree with the IUPAC recommendations.

The calibration of DSC instruments is normally accomplished through the use of compounds having accurately known transition temperatures and heats of fusion, and a list of appropriate DSC standards is provided in Table 4.2. Once a DSC system is properly calibrated, it is easy to obtain melting point and enthalpy of fusion data for any compound upon integration of its empirically determined endotherm and application of the calibration parameters. The current state of methodology is such, however, that unless a determination is repeated a large number of times, the deduced enthalpies must be regarded as being accurate only to within approximately 5%.

4.4.3 Applications

In its simplest form, DSC is often thought of as nothing more than a glorified melting point apparatus. This is so because many pure compounds yield straightforward results consisting of nothing more than one event, the melting of a crystalline phase into a liquid phase. For example, acetaminophen was found to have an onset temperature of 170.0°C and a peak of 170.9°C, with an enthalpy of fusion equal to 116.9 J/g (see Fig. 4.7). In displaying DSC plots, it is important to indicate the ordinate scale to be interpreted as the endothermic or

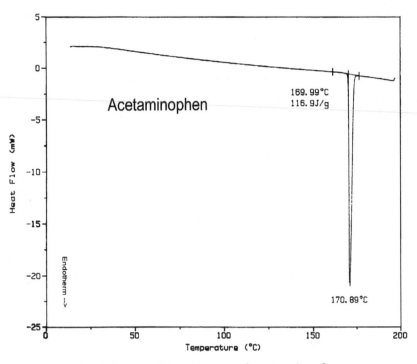

Fig. 4.7. DSC thermogram of acetaminophen.

exothermic response of the measuring instrument, due to differences in calorimetric cell arrangements between different equipment manufacturers.

Figure 4.8 shows the comparison of three lots of loperamide hydrochloride, each obtained from a different supplier. The displayed thermograms represent normal behavior for this material, and while the figure shows the uniqueness of each source, the variations were within acceptable limits. Owing to the decomposition that followed on the end of the melting endotherm, specific heats of fusion were not calculated in this case.

Much more information can be obtained from the DSC experiment than simply an observation of the transition from a solid to a liquid phase. A plot of heat flow against temperature is a true depiction of the continuity of the *heat capacity at constant pressure* (C_p). If the entire temperature range of a given process is known, the physical state of a material will reflect the usefulness of that material at any temperature point on the plot. For polyethylene terephthalate (see Fig. 4.9), a step-shaped transition is interpreted as a change in C_p resulting from a

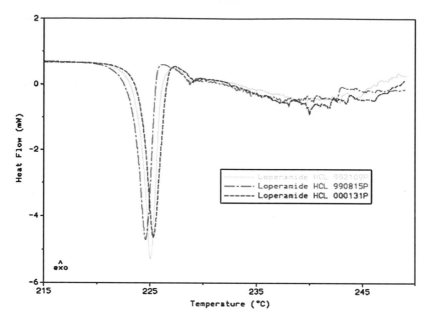

Fig. 4.8. DSC thermograms of three lots of loperamide hydrochloride obtained from different suppliers.

transition from an amorphous, rigid ("glassy") state to an amorphous, non-rigid ("plastic") state. The temperature at the inflection point in such a transition is called the *glass transition temperature* (T_g).

Exothermic events, such as crystallization processes (or recrystallization processes) are characterized by their *enthalpies of crystallization* (ΔH_c). This is depicted as the integrated area bounded by the interpolated baseline and the intersections with the curve. The onset is calculated as the intersection between the baseline and a tangent line drawn on the front slope of the curve. Endothermic events, such as the melting transition in Fig. 4.9, are characterized by their *enthalpies of fusion* (ΔH_f), and are integrated in a similar manner as an exothermic event. The result is expressed as an enthalpy value (ΔH) with units of J/g and is the physical expression of the crystal lattice energy needed to break down the unit cell forming the crystal.

Single scan DSC information is the most commonly used instrumental mode, but multiple scans performed on the same sample can be used to obtain additional information about the characteristics of a material, or of the reproducibility associated with a given process.

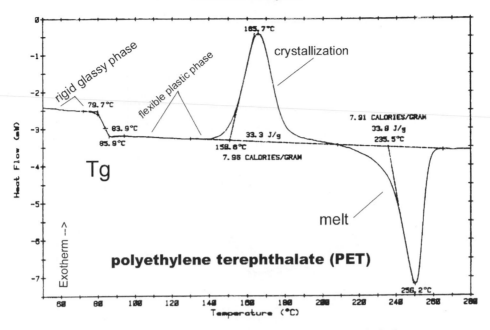

Fig. 4.9. DSC thermogram of polyethylene terephthalate.

4.4.3.1 Cyclical differential scanning calorimetry

An example of the supercooling phenomenon was found when comparing seven lots of microcrystalline wax that were obtained from the same supplier. While all of the lots exhibited similar endothermic events at approximately 65 and 88°C, there were some minor variations observed when the samples were reheated. These observations are typical for waxes and represent differing degrees of crystalline cure. Since these materials are families of several different wax analogs, their melting behavior is not sharply defined. The results are appropriately illustrated as an overlay plot of the heating and cooling curves for each lot (see Fig. 4.10). The results for all seven lots were within acceptable ranges for this material and did not differentiate one lot from another. As a result, the lots were treated as being equivalent materials for their intended formulation purposes.

The three overlay plots shown in Fig. 4.11 illustrate another example of how process variations can lead to unexpected changes in material behavior. Consider the instance of a spray congealing process where a mixture of one or more waxes with a melting point below that of the

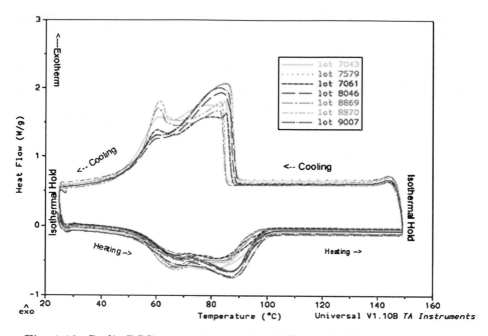

Fig. 4.10. Cyclic DSC comparisons of seven lots of microcrystalline wax.

API (acetaminophen in this case) is heated above the wax melting point, but below that of the drug component. This process results in a suspension of the drug particles in the molten wax. When a nozzle is used to spray the atomized suspension onto cooled air, the molten droplets of suspension congeal, thus allowing the liquid components to encapsulate the solid API and resulting in packets of API particles encased in solidified wax.

In normal processing the mixture is maintained at $80\pm5°C$ prior to spraying. An aliquot of this mixture was removed from the melting pot, allowed to cool, and a fragment examined by DSC (labeled as "melt prior to spraying (2)"). A portion of the melting pot charge was spray-congealed and the product collected until a clog developed in the lines requiring an increase in hating to soften the blockage. A sample of the normal product analyzed by DSC produced the result shown in the overlay as "spray-congealed product (1)". Meanwhile, during the clog-removal step, the remaining mixture in the melting pot was exposed to a temperature excursion between approximately 100 and 110°C. Within moments of arriving at this higher temperature, the suspension in the melting pot spontaneously formed a precipitate while a clear molten liquid phase developed above the collapsed solid. When the melting pot

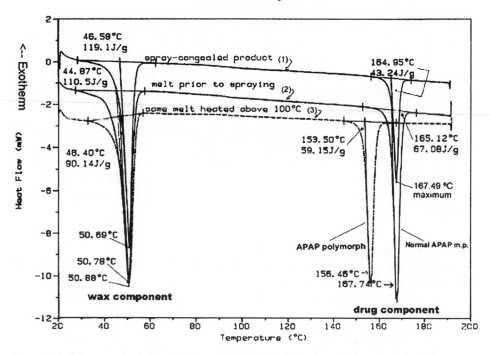

Fig. 4.11. Unexpected formation of a metastable phase of acetaminophen as a result of exposure to the molten wax formulation during spray-congeal processing. *Note*: The three curves have been manually offset on the *Y*-axis from the normal zero milliwatt baselines in order to display the relative *X*-axis (temperature) differences between the three samples.

mixture was quickly separated on a filter, the precipitate at the bottom of the pot had formed a crystalline material. The surprising change was noted when a sample of this new solid phase was checked by DSC. Labeled as "same melt heated above 100°C (3)", this material produced an endothermic response in the wax component region that was identical to the two other samples.

But the drug component region of the thermogram showed endothermic melting event shifted approximately 10°C lower in temperature, and the complete absence of the normal endotherm. This finding stood in sharp contrast to the other two samples, which showed normal melting endotherms in both the wax and drug component regions of the thermogram. Later experiments confirmed the occurrence of this metastable form of acetaminophen that was not observed with pure acetaminophen.

In fact, Fig. 4.7 shows that no other endothermic or exothermic transition is observed when acetaminophen is heated to its normal melting point region. The occurrence of the new crystalline form is characterized by a similar ΔH_f value of 59.15 J/g (see curve (3) of Fig. 4.11), when compared to ΔH_f value of 67.08 J/g for the spray-congealed acetaminophen product (see curve (2) of Fig. 4.11). Only the drug alone yielded a higher ΔH_f value (116.9 J/g, Fig. 4.7), since for the pure drug substance there would be no other material capable of perturbing the crystal formation of the material. The molten wax system therefore provided a unique environment in which acetamino-phen molecules were able to orient themselves into a different unit cell having a lower melting point and a lower enthalpy of fusion, probably due to the generation of alternative nucleation possibilities associated with the presence of the molten wax.

4.4.3.2   Utility in studies of polymorphism

Polymorphism is the ability of the same chemical substance to exist in different crystalline structures that have the same empirical composition [39,40]. It is now well established that DSC is one of the core technologies used to study the phenomenon. Polymorphic systems are often distinguished on the basis of the type of interconversion between the different forms, being classified as either enantiotropic or monotropic in nature.

When a solid system undergoing a thermal change in phase exhibits a reversible transition point at some temperature below the melting points of either of the polymorphic forms of the solid, the system is described as exhibiting *enantiotropic polymorphism*, or *enantiotropy*. On the other hand, when a solid system undergoing thermal change is characterized by the existence of only one stable form over the entire temperature range, then the system is said to display *monotropic polymorphism*, or *monotropy*.

An example of monotropic behavior consists of the system formed by anhydrous ibuprofen lysinate [41,42]. Figure 4.12 shows the DSC thermogram of this compound over the temperature range of 20–200°C, where two different endothermic transitions were noted for the substance (one at 63.7°C and the other at 180.1°C). A second cyclical DSC scan from 25 to 75°C demonstrated that the 64°C endotherm, generated on heating, had a complementary 62°C exotherm, formed on cooling (see Fig. 4.13). The superimposable character of the traces in the thermograms demonstrates that both these processes were reversible, and indicates that the observed transition is associated with an enantiotropic phase interconversion [41]. X-ray powder (XRPD) diffraction patterns acquired at room temperature, 70°C, and

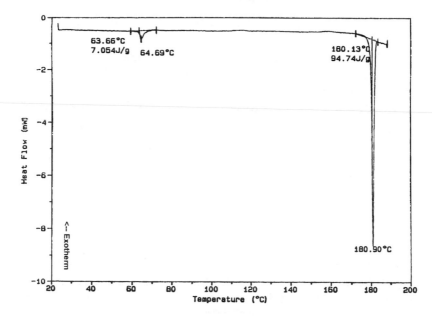

Fig. 4.12. DSC thermogram of non-solvated ibuprofen lysinate, illustrating the enantiotropic conversion of the metastable phase to the more stable phase (64°C endotherm) and subsequent melting of the stable form (181°C endotherm).

on sample cooled back to room temperature confirmed the same cyclical behavior was observed for the system.

The polymorphs of tristearin are monotropically related, as evidenced in the DSC thermograms shown in Figs. 4.14 and 4.15, showing a ripening effect between a kinetically formed lower melting form and a conversion to a higher melting stable form. The monotropic nature of the tristearin solid system is initially confirmed by the fact that the form melting at 58.2°C exhibits a single endothermic transition, when the sample is removed from a bottle stored at room temperature for longer than 3 days (see the top overlay plot in Fig. 4.14).

However, if that same DSC pan is immediately quench-cooled (i.e., cooled from 120 to –20°C in less than 5 min) and its DSC thermogram obtained immediately, three linked thermal events are observed (see the bottom overlay plot in Fig. 4.14). A melting endotherm at 46.6°C transitions into a recrystallization exotherm at 48.7°C, only to transition again into a second larger endotherm at 57.3°C. When the same process was observed using hot-stage light microscopy, the system was observed to begin melting, but it never fully achieved a one-phase

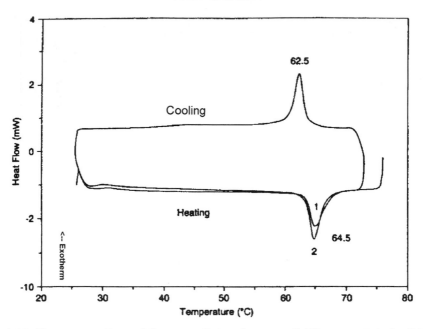

Fig. 4.13. Demonstration of the enantiotropic reversibility associated with the phase conversion between the non-solvated polymorphs of ibuprofen lysinate.

liquid state until the temperature exceeded 65°C. Without the DSC as a guide, the visual observation would have been interpreted as a broad single melt over a wide temperature range.

In another experiment, the newly melted material from the second rescan (bottom trace, Fig. 4.14) was slowly allowed to cool from 120 to 20°C over a two-day time period. As shown in Fig. 4.16, the DSC thermogram of this sample showed the same three-part pattern, but the ratio of lower melting metastable form to the higher melting stable form was greatly shifted in favor of the thermodynamically stable form.

It is worth noting that a monotropic polymorphic system offers the potential of annealing the substance to achieve the preferred form of the thermodynamically stable phase. The use of the most stable form is ordinarily preferred to avoid the inexorable tendency of a metastable system to move toward the thermodynamic form. This is especially important especially if someone elects to use a metastable phase of an excipient as part of a tablet coating, since physical changes in the properties of the coating can take place after it has been made. Use of the most stable form avoids any solid–solid transition that could

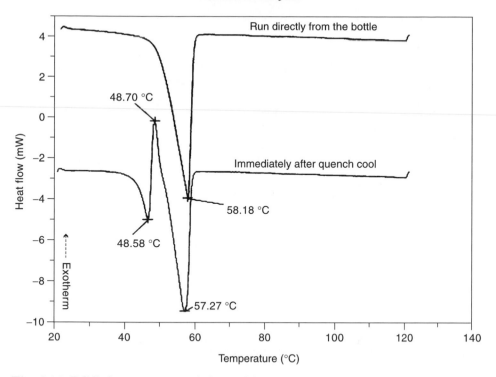

Fig. 4.14. DSC thermograms of the stable (upper trace) and metastable (lower trace) forms of tristearin.

negatively impact the quality (release rate, surface roughness, particle flowability, etc.) of a coating.

4.4.3.3 *Characterization of phase transformations associated with compression*

Often the character of materials in a mixture undergoes solid-state rearrangements due to the pressures associated with compaction, which may or may not be polymorphic in nature. Consider the pre-compression powder blend, whose DSC thermogram is shown in Fig. 4.16, and which features the presence of four endothermic transitions. In the post-compression, ground tablet sample whose DSC thermogram is shown in Fig. 4.17, the endotherms having maxima at 86.5 and 106°C remain relatively constant (maxima at 85.3 and 104.2°C). On the other hand, the third endotherm in the pre-compression thermogram shows considerable attrition in the post-compression sample, and an additional endotherm (not previously observed in the pre-compression sample)

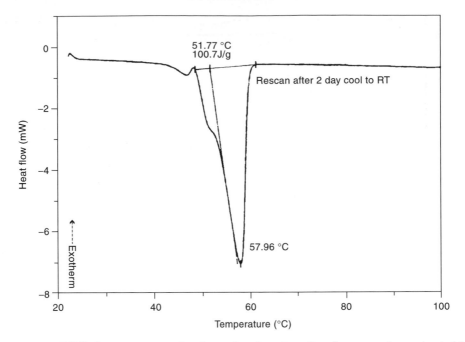

Fig. 4.15. DSC thermogram of tristearin showing the decrease in metastable phase content as a result of a two-day annealing process.

appears with a maximum at 186.8°C. These changes in thermal profile were traced to a pressure-induced polymorphic transition of one of the excipient ingredients in the formulation.

Another example of pressure-induced polymorphism is seen in the case of amiloride hydrochloride, where ball-milling Form-B causes a solid-state phase transformation into Form-A [43]. These workers deduced the phase relationship between two different pressure-induced polymorphs of the dihydrate, as well as the alternative route to one of those dihydrate forms that used the anhydrous form as the source material and effected the phase transformation through storage at high degrees of relative humidity storage.

4.4.3.4   The value of measurements of glass transition temperatures
C.M. Neag has provided a concise example of how measurements of glass transition temperatures by DSC can help determine the comparative property differences of a group of related materials [44]:

> Probably the best understood and most commonly used property of polymers, glass transition temperatures are important in virtually

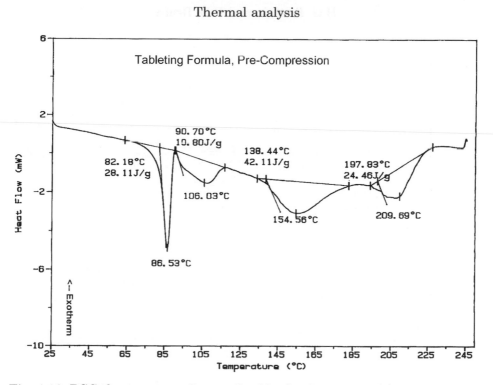

Fig. 4.16. DSC thermogram of a powder blend prior to its compression into a tablet.

every phase of a coating's development and manufacture. The T_g marks a polymer's transition from an amorphous glass to a rubbery solid and defines the limits of processability for most polymers ... the T_g is most commonly assigned to the extrapolated onset of the transition .... Close examination of the DSC heat flow curves gives outstanding clues about the character of the polymers being analyzed. Compared to a typical T_g, the transitions in {Figure 16} are very broad—covering some 40 to 50°C—and quite shallow, falling less than 0.1 cal/s/g from beginning to end. The character of the glass transition region in a typical DSC is quite different. The temperature range of this region is usually no more than about 25°C wide and usually drops more than 0.5 cal/s/g over the T_g range. The differences in these particular T_g's probably stem from the combined effects of monomer sequence distribution [45] and end group effects related to the relatively low molecular weight [46] of these copolymers. The polymers used in this experiment were all low-molecular-weight tetramers (number average molecular weight < 5000) composed of

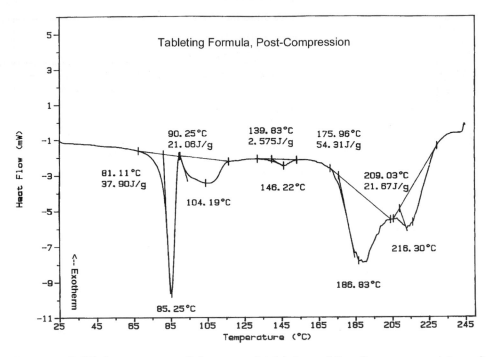

Fig. 4.17. DSC thermogram of the ground tablet resulting from compression of the powder blend of Fig. 16.

various combinations of methylated and butylated acrylics. Van Krevlan [47] provides a more comprehensive overview of polymer properties that could have an influence on the assignment of the glass transition temperature.

4.4.3.5 Modeling freeze/thaw cycles in stability samples

Drug formulations can be exposed to variable temperatures during storage. While this is not usually a desirable case, anticipating the effect of freeze/thaw cycles on a drug substance or a formulation may avoid costly reformulation owing to problems discovered at a later time in the product scale-up process. In the case chosen for illustration, the bulk formulation is stored frozen, thawing only a small portion for intravenous use a short period before administration to the patient.

Figure 4.18 shows the results of a cyclic DSC evaluation of a sample of the aqueous IV formulation. The sample was analyzed in an open aluminum pan, being cooled and then heated (under nitrogen purge) through a series of three and a half freeze/thaw cycles at a temperature ramp of 10°C per min over the range of –50 to +25°C. This range was

Thermal analysis

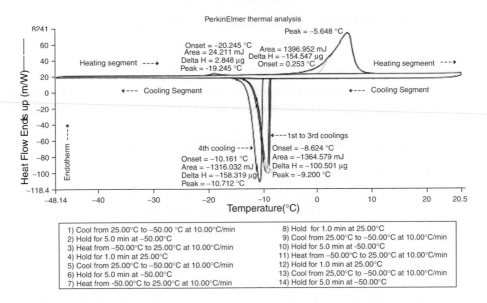

Fig. 4.18. DSC freeze/thaw cycles of an aqueous intravenous formulation.

TABLE 4.3

Summary of temperature peak maxima measured for various lots of an intravenous product across the full cycle range of DSC freeze/thaw cycles

Lot number	Initial freezing maximum (°C)	Second freezing maximum (°C)	Third freezing maximum (°C)	Fourth freezing maximum (°C)	First endotherm maximum (°C)	Second endotherm (melt) maximum (°C)
A	−9.203	−9.203	−9.203	−10.712	−19.245	5.648
B	−13.982	−14.799	−15.308	−15.807	−18.913	6.975
C	−14.241	−15.527	−15.527	−15.751	−18.780	6.317
D	−11.005	−11.421	−12.384	−12.678	−18.751	6.314
E	−14.666	−15.018	−15.089	−15.089	−18.676	7.305
F	−9.454	−11.229	−11.949	−11.980	−18.768	6.313

chosen to contain the complete freeze/thaw behavior range of the sample, and the specific method segments are shown at the bottom of the resulting thermogram of Fig. 4.18. All six samples were run in an identical manner but only one was chosen to illustrate the technique, and the results of all six lots are shown in Table 4.3.

The sample thermogram by the presence of an endotherm associated with a melting transition, and characterized by onset and peak temperatures of 0.29 and 5.65°C, respectively, and an enthalpy of

89

fusion equal to 164.55 J/g. Upon cooling, a reciprocal event occurs as an exotherm due to crystallization, which was characterized onset and peak temperatures of approximately −8 and −9°C, respectively, and an enthalpy of crystallization approximately 160 J/g. It is worth noting that the recrystallization exotherm is not a mirror image of the melting endotherm, but that both events are nearly of the same magnitude in the enthalpy values.

This suppression of the freezing point relative to the melting point of the sample is indicative of the phenomenon of supercooling. This occurs when a liquid does not condense to a solid crystalline state, either due to short-lived kinetics invoked in a flash cooling situation or because of steric hindrance of the individual molecules from forming a crystal lattice at the thermodynamically optimum temperature. All of the samples showed superimposable heating profiles, while there were variations in the cooling cycles that were consistent neither from sample to sample nor within sample heat/cooling loops. As mentioned above, it is likely that kinetic or steric factors involved in the condensation of each sample led to such variations. It is significant that the characteristic melting point temperature remained a constant, showing that the solid phase was eventually obtained at the beginning of the experiment (regardless of cycle) produced identical melting endotherms. Results from the cyclic DSC experiments also showed that no apparent thermal degradation changes took place in any of the samples over the range of −50 to +25°C.

Supercooling has been observed in an extreme form in molten ibuprofen if the molten solid is allowed to cool from the melting point to room temperature without vibration in a smooth-lined container [48]. For instance, undisturbed rac-ibuprofen can exist as an oil phase for several hours to a few days. If disturbed, however, an exothermic recrystallization proceeds and bulk crystalline material rapidly grows vertically out of the oil phase so energetically that the system emits an audible cracking sound.

4.4.3.6 Determination of the freezing point of a 2% aqueous solution of dielaidoylphosphatidylcholine

Dielaidoylphosphatidylcholine (DEPC) has been used as a membrane model to study the interactions of bioactive membrane-penetrating agents such as melittin (a bee venom peptide), which is composed of a hydrophobic region including hydrophobic amino acids and a positively charged region including basic amino acids. When liposomes of phosphatidylcholine were prepared in the presence of melittin, reductions in the phase transition enthalpies were observed [49]. In attempting to define the molecular mechanism of action of bioactive membrane-penetrating agents and how they induce structural perturbations in phospholipid multilayers, a potentially helpful model involving DEPC helped confirm the important role played by the phospholipid bilayers in the association of invasive agents with cell membranes.

When pure phospholipids are suspended in excess water, they form multilamellar liposomes consisting of concentric spheres of lipid bilayers interspersed with water. It is possible to obtain the complete main phase transition for the DEPC solution without having to go below 0°C. It is critical that these solutions not be cooled below 0°C as the water component can freeze and destroy the liposome structures. The effect of changing the cooling rate on the main phase transition for the DEPC solution was determined. Figure 4.19 displays the results obtained by cooling the DEPC solution at rates of 0.5, 1.0, and 2.0°C per minute. As the cooling rate is increased, the temperature of the observed peak significantly decreases. This data may be used to assess the time-dependency, or kinetics, of the main phase transition for the DEPC solution [50].

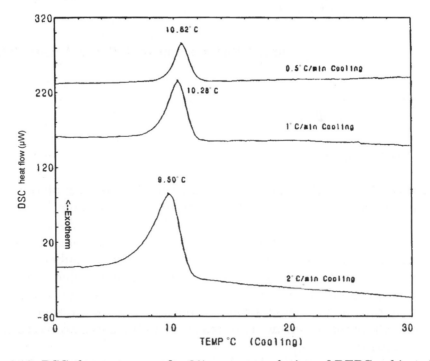

Fig. 4.19. DSC thermograms of a 2% aqueous solution of DEPC subjected to various heating rates.

4.5 THERMOGRAVIMETRY

4.5.1 Background

Thermogravimetry (TG) is a measure of the thermally induced weight loss of a material as a function of the applied temperature [51]. TG analysis is restricted to studies, which involve either a mass gain or loss, and is most commonly used to study desolvation processes and compound decomposition. TG analysis is a very useful method for the quantitative determination of the total volatile content of a solid, and can be used as an adjunct to Karl Fischer titrations for the determination of moisture.

TG analysis also represents a powerful adjunct to DTA or DSC analysis, since a combination of either method with a TG determination can be used in the assignment of observed thermal events. Desolvation processes or decomposition reactions must be accompanied by weight changes, and can be thusly identified by a TG weight loss over the same temperature range. On the other hand, solid–liquid or solid–solid phase transformations are not accompanied by any loss of sample mass and would not register in a TG thermogram.

When a solid is capable of decomposing by means of several discrete, sequential reactions, the magnitude of each step can be separately evaluated. TG analysis of compound decomposition can also be used to compare the stability of similar compounds. The higher the decomposition temperature of a given compound, the more positive would be the ΔG value and the greater would be its stability.

4.5.2 Methodology

Measurement of TG consists of the continual recording of the mass of the sample as it is heated in a furnace, and a schematic diagram of a TG apparatus is given in Fig. 4.20. The weighing device used in most devices is a microbalance, which permits the characterization of milligram quantities of sample. The balance chamber itself is constructed so that the atmosphere may be controlled, which is normally accomplished by means of a flowing gas stream. The furnace must be capable of being totally programmable in a reproducible fashion, whose inside surfaces are resistant to the gases evolved during the TG study.

It is most essential in TG design that the temperature readout pertain to that of the sample, and not to that of the furnace. To achieve

Thermal analysis

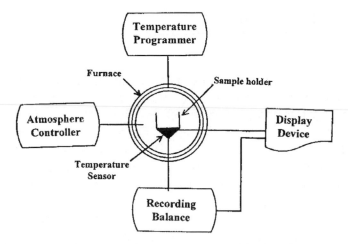

Fig. 4.20. Schematic diagram of apparatus suitable for the measurement of thermogravimetry.

this end, the thermocouple or resistance thermometer must be mounted as close to the sample pan as possible, in direct contact if this can be achieved.

Although the TG methodology is conceptually simple, the accuracy and precision associated with the results are dependent on both instrumental and sample factors [52]. The furnace heating rate used for the determination will greatly affect the transition temperatures, while the atmosphere within the furnace can influence the nature of the thermal reactions. The sample itself can play a role in governing the quality of data obtained, with factors such as sample size, nature of evolved gases, particle size, heats of reaction, sample packing, and thermal conductivity all influencing the observed thermogram.

4.5.3 Applications

4.5.3.1 Determination of the solvation state of a compound
TG can be used as a rapid method to determine the solvation state of a compound. For example, Fig. 4.21 contains the weight loss profiles for a compound having a molecular weight of 270.23, and which is capable of being isolated as an anhydrate crystal form, or as the monohydrate and dihydrate solvatomorphs. Evaluation of the thermograms indicates effectively no temperature-induced weight loss for the anhydrate substance, as would be anticipated. The theoretical weight loss for the monohydrate solvatomorph was calculated to be 6.25%, which agrees

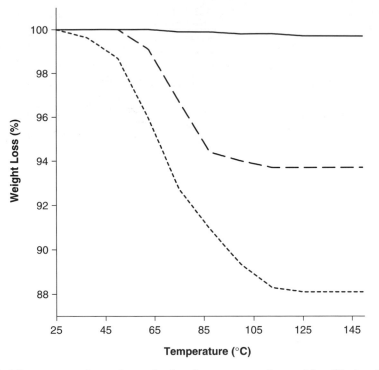

Fig. 4.21. Thermogravimetric analysis of a compound capable of being isolated as an anhydrate crystal form (solid trace), and as the monohydrate (dashed trace) and dihydrate (dotted trace) solvatomorphs.

well with the experimentally determined value of 6.3%, and therefore confirms existence of the monohydrate. The theoretical weight loss for the dihydrate solvatomorph was calculated to be 11.76%, which agrees well with the experimentally determined value of 11.9%.

4.5.3.2   Use of thermogravimetry to facilitate interpretation of differential scanning calorimetry thermograms

TG is a powerful adjunct to DSC studies, and are routinely obtained during evaluations of the thermal behavior of a drug substance or excipient component of a formulation. Since TG analysis is restricted to studies involving either a gain or a loss in sample mass (such as desolvation decomposition reactions), it can be used to clearly distinguish thermal events not involving loss of mass (such as phase transitions).

For example, an overlay of the DSC and TG thermograms for an active pharmaceutical ingredient is presented in Fig. 4.22. The TG thermogram shows a total weight loss of 5.21% over the range of 25.1–140.0°C, which is associated with the loss of water and/or solvent. Over this same temperature range, a small initial rise in the baseline and a broad endotherm (peak temperature of 129.5°C) are observed in the DSC thermogram. Above 140°C, a small endotherm is observed at a peak temperature of 168.1°C.

The TG thermogram indicates that slightly more than half of the total weight loss (approximately 3.1%) occurred over the range of 25–100°C. This loss of weight corresponded in the DSC thermogram to a slight rise in the baseline. The remainder of the weight loss (approximately 2.1%) corresponded to the endotherm at 129.5°C. It was determined that the enthalpy of the endotherm was greater than the enthalpy associated with the small rise in the baseline preceding 100°C. Because the enthalpy of the endotherm corresponded to only 40% of the total weight loss, it was conjectured that something more than the loss of water/solvent was contributing to the endothermic transition. Through the use of XRPD, it was later shown that the sample consisted of a mixture of crystalline and amorphous substances, suggesting that the DSC endotherm was a result of a solid-state transition, possibly that of a crystalline to amorphous transition.

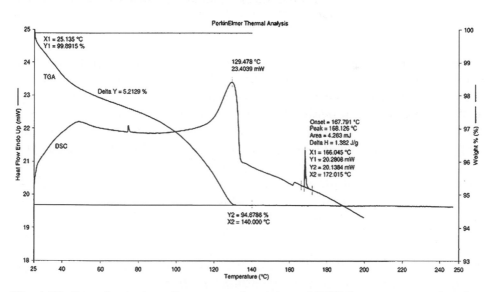

Fig. 4.22. Complementary thermogravimetric and DSC thermograms, showing loss of a volatile component and solid–solid conversion of some of the sample from the amorphous to the crystalline phase.

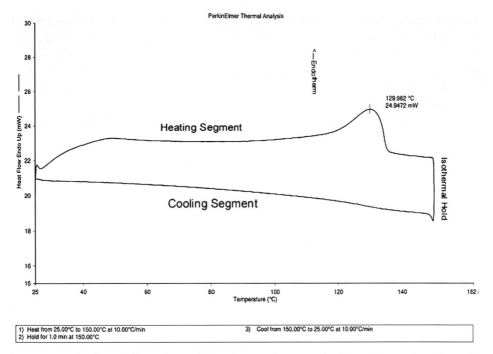

Fig. 4.23. Cyclic DSC studies of the drug substance in Fig. 22 proving that the compound does not degrade at temperatures near or below 150°C.

This conclusion was confirmed by XRPD analysis of an aliquot of the material that had been placed in a convection oven at 150°C for 5 min. Prior to this, cyclic DSC was used to simulate the conditions that the sample would be subjected to prior to being analyzed by XRPD. The results of this DSC experiment showed that the solid-state identity of the sample that had been heated to 150°C and cooled did not change, since no additional rise in baseline was noted during the 150°C isothermal step nor were additional exothermic peaks observed in the cooling cycle (see Fig. 4.23). Comparison of the diffraction patterns of the drug substance before and after heating showed that heating the sample to 150°C caused peak broadening, a reduction in peak intensity, and an increase in baseline curvature. This was taken as confirmation of a phase conversion (partial in this case) from a crystalline to an amorphous state.

4.5.3.3   Estimating the isothermal lifetime of pharmaceutical coatings using thermogravimetric decomposition kinetics
When formulating with crystalline drugs that have received a Biopharmaceutical classification system (BCS) ranking as being either

type-II or type-IV, modifying the solid phase by using crystal-lattice-disrupting excipients [53–56] (Gelucire® 44/14) will result in formation of amorphous solid solutions, or at least crystalline or semi-crystalline phases, that exhibit lower crystal lattice energies and thus yield enhanced solubility and dissolution rates. Dissolution is likely to be the rate-determining step in type-II drugs, while type-IV drugs are generally problem compounds for which in vitro dissolution results may not be reliable or predictive.

Determining the decomposition rate and expected thermal lifetime of formulation components at the elevated temperatures of formulation processes is essential for avoiding thermal decomposition of the formulation components during processing. Such processes include melt-extrusion casting, spray-congealing, and hot-melt fluid-bed coating or enrobing of drug substances.

TA instruments has developed automated thermogravimetric analysis and related kinetic programs that enable a rapid determination of decomposition rates to be made. The following excerpt from a TA application brief [57] explains the method:

Thermogravimetric Analysis provides a method for accelerating the lifetime testing of polymers waxes and other materials so that short-term experiments can be used to predict in-use lifetime. A series of such tests, performed at different oven temperatures, creates a semi-logarithmic plot of lifetime versus the reciprocal of failure temperature. The method assumes first order kinetics and uses extrapolation to estimate the long lifetimes encountered at normal use temperature …. Many polymers are known to decompose with first order kinetics. For those that do not, the earliest stages of decomposition can be approximated well with first order cal kinetics … In the TGA approach, the material is heated at several different rates through its decomposition region. From the resultant thermal curves, the temperatures for a constant decomposition level are determined. The kinetic activation energy is then determined from a plot of the logarithm of the heating rate versus the reciprocal of the temperature of constant decomposition level. This activation energy may then be used to calculate estimated lifetime it a given temperature or the maximum operating temperature for a given estimated lifetime. This TGA approach requires a minimum of three different heating profiles per material. However, even with the associated calculations, the total time to evaluate a material is less than one day. With an automated TGA ... the actual operator time is even lower with overnight valuation being possible.

In a hot-melt patent [58] involving a coating applied as a hot-melt spray into a fluidized bed of acetaminophen, the molten coating consisted of an 88:12 w/w% ratio of Carnauba Wax to Polyaldo® 10-1-S (polyglycerol esters of fatty acids, specifically, decaglyceryl mono-stearate [59]). The drug delivery system of the present invention is preferably prepared by the following steps. The pharmaceutically active ingredient is placed in a fluidized bed. Melted wax and emulsifier (along with other ingredients) are stirred together. The emulsifier/wax mixture is then added to the fluidized bed. The type of fluidized bed is not critical, as top spray, Wurster and rotor type, fluidized beds may be employed in the present invention. The fluidized bed should provide an air stream of at least about 40°C above the melting temperature of the emulsifier/wax mixture. An atomization air temperature of about 125°C is adequate for most systems. The melted coating material is delivered into the fluidized bed under pressure through a nozzle to create droplets of the emulsifier/wax mixture. The addition of the emulsifier/wax system is then applied to the surface of the pharma-ceutically active ingredient. Another advantage of the present inven-tion is that no solvents, either water or non-aqueous, are required in order to prepare the drug delivery system.

Regarding the above formulation and process, the thermal stability was determined by estimating the isothermal lifetime from thermo-gravimetric data. Process scientists for scale-up equipment design and temperature specifications sought answers to the following four basic questions:

1. Are the formulation components thermally sensitive?
2. What is excessive temperature for this process?
3. What are the lifetimes of materials in this process?
4. What is the predicted thermal storage lifetime of the coating?

A number of assumptions needed to be made in order to implement the thermogravimetric decomposition kinetics method. To minimize entrapped internal volatile components as the TG heating progresses, powders or ground specimens with high-surface areas are preferable for use. Sample size should be held to 3 ± 1 mg and lightly pressed flat to minimize layering effects in mass loss during TGA heating. Variations in sample particle size distribution can be controlled without the loss of volatile components by using a nitrogen-blanketed environment (glove box) to grind each sample in a micro mill under liquid nitrogen, followed by passing the resultant specimen through a 150-mesh sieve (150 µm screen openings).

In addition, a rational relationship must be established between the TG results and the process being modeled. The method applies to well-defined decomposition profiles characterized by smooth, continuous mass change and a single maximum rate. Plots of log (heating rate) against the reciprocal temperature (i.e., $1/K$) must be linear and approximately parallel. Self-heating or diffusion of volatiles can become rate-determining conditions at high heating rates, and such conditions would invalidate the TG kinetic model. The value of calculated activation energy is independent of reaction order, an assumption that holds for early stages of decomposition. Finally, use of a half-life value check is required, where a sample held for 1 h at the 60-min half-life temperature should lose approximately 50% of its mass. Experimental results that do not come close indicate a poor fit for the predictive model, and results that do not pass this value check should not be considered valid.

Figure 4.24 displays a single TG thermogram of a 10°C per min scan of the hot-melt coating formulation showing percent weight loss (or percent decomposition conversion) points that will form the data set required to build the decomposition kinetic model. The model uses at least three different heating rates of three aliquots of the sample. Figure 4.24 presents the overlaid weight loss curves for the hot-melt coating mixture at four different heating rates (scanned at 1, 2, 5, and 10°C per min).

The first step in the data analysis process is to choose the level of decomposition. A selection level early in the decomposition is desired since the mechanism is more likely to be related to the process of the actual failure onset point of the material (i.e., thermal decomposition). The analyst must be cautious to use former experience with the construction of the model construction of the method so as not to select a level too early and cross material failure with the measurement of some volatilization that is not involved in the failure mechanism. A value of 5% decomposition level (sometimes called "conversion") is a commonly chosen value. This is the case in the example in Fig. 4.25, and all other calculations from the following plots were based on this level.

Figure 4.26 further illustrates why the 5% weight loss level was a good choice for this system. This level provides the best compromise between avoiding simple moisture loss if a lower level selects intersections too early in the TG experiments, or mixed-mechanistic

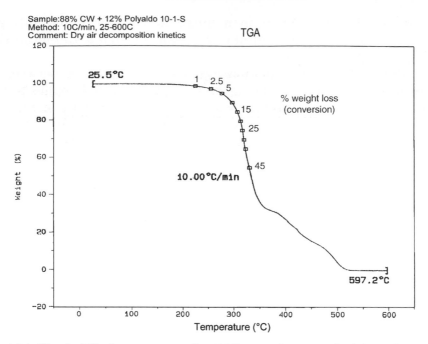

Fig. 4.24. Single TG thermogram of a 10°C per min scan of a hot-melt coating formulation, showing the percent weight loss conversion points for building the kinetic model.

decompositions that tend to develop at higher loss levels later in the TG scan. Using the selected value of conversion, the temperature (in degrees Kelvin) at that conversion level is measured for each thermal curve. A plot of the logarithm of the heating rate versus the corresponding reciprocal temperature at constant conversion is prepared, which should produce a straight line. Further, as mentioned above, all the related plots at other levels should also be both linear and have slopes nearly parallel to each other. If the particular specimen decomposition mechanism were the same at all conversion levels, the lines would all have the same slope. However, this is not the case for the example being provided. The lines for the low conversion cases were quite different from those of 5% and higher conversion, so 5% conversion was judged to be the optimal point of constant conversion for the model.

To calculate the activation energy (E_{act}), Flynn and Wall [60] adapted the classic Arrhenius equation into Eq. (4.3) to reflect the

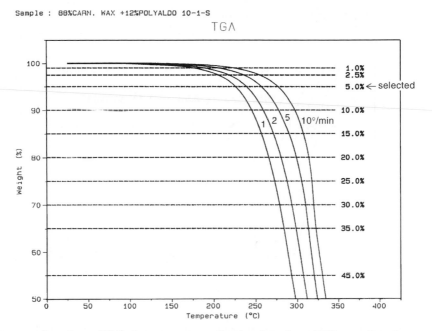

Fig. 4.25. Overlay of TG thermograms obtained at four different heating rates, showing conversion iso-percentile levels.

parameters of the TG multiplexed method:

$$E = \frac{-R}{b}\left[\frac{d\log\beta}{d(1/T)}\right] \tag{4.3}$$

where E is the activation energy (units of J/mol), R is the universal gas constant (8.314 J/mol K), T is the absolute temperature at constant conversion, β is the heating rate (units of °C per min), and b is a constant (equal to 0.457). The value of the bracketed term in the above equation is the slope of the lines plotted in Fig. 4.26. The value for the constant b is derived from a lookup table in reference [60], and varies depending upon the value of E/RT. Thus, an iterative process must be used where E is first estimated, a corresponding value for b is chosen, and then a new value for E is calculated. This process is continued until E no longer changes with successive iterations. For the given decomposition reaction for the values of E/RT between 29 and 46, the value for b is within $\pm 1\%$ of 0.457, thus this value is chosen for the first iteration.

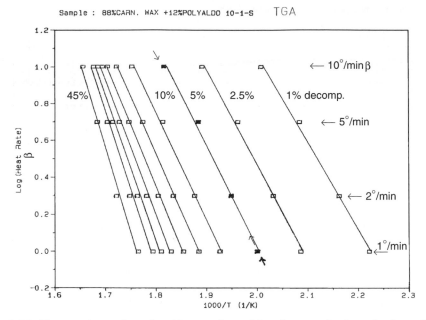

Fig. 4.26. Comparison plots for 10 preselected levels, graphed as the logarithm of heating rates versus the corresponding reciprocal temperatures at constant conversion.

Toop [61] has postulated a relationship between the activation energy and the estimated lifetime:

$$\ln t_f = \frac{E}{RT_f} + \ln\left[\frac{E}{\beta R} \bullet P(X_f)\right] \tag{4.4}$$

where t_f is the estimated time to failure (units of minutes), T_f is the failure temperature (in degrees Kelvin), $P(X_f)$ is a function whose values depend on E at the failure temperature, T_c is the temperature for 5% loss at β (degrees K), and the other symbols have their meaning as before.

To calculate the estimated time to failure, the value for T_c at the constant conversion point is first selected for a slow heating rate. This value is then used along with the activation energy (E) to calculate the quantity E/RT. This value is then used to select a value for log $P(X_f)$ from the numerical integration table given in Toop's paper, and the numerical value for $P(X_f)$ can then be calculated by taking the antilogarithm. Selection of a value for *failure temperature* (T_f) (or *operation temperature* in the case of the hot-melt coating process) permits the calculation of t_f from Toop's postulated Eq. (4.4).

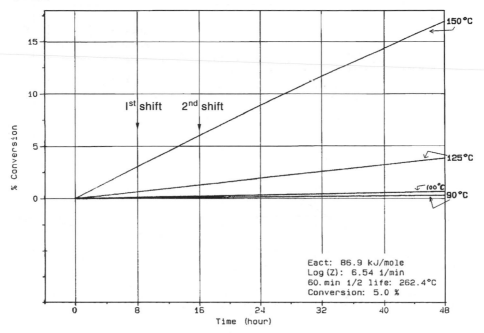

Comment: Plot for Hot Melt Processing in a dry air atmosphere.

Fig. 4.27. Estimated time (in hours) to thermal decomposition failure of hot-melt coating held at various constant temperatures.

Rearrangement of Eq. (4.4) yields a relation that may be used to calculate the maximum use temperature (T_f) for a given lifetime (t_f):

$$T_f = \frac{E/R}{\ln t_f} + \ln\left[\frac{E}{\beta R} \bullet P(X_f)\right] \qquad (4.5)$$

This equation may be used to create a plot, similar to those in Figs. 4.27 and 4.28, in which the logarithm of the estimated lifetime is plotted against the reciprocal of the failure temperature. From plots of this type, the dramatic increases in estimated lifetimes for small decreases in temperature can be more easily visualized.

The value of the method can be seen by reconsidering responses to the four basic questions of the TG decomposition kinetic model. The first question concerned whether the formulation components were thermally sensitive, and at what operational hold times the constant temperature decomposition was under 2%. From Fig. 4.27, the model predicted that an operating temperature in the range of 90–100°C

103

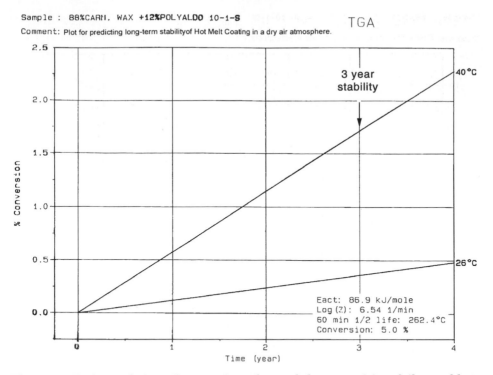

Sample : 88%CARN. WAX +12%POLYALDO 10-1-S

TGA

Comment: Plot for predicting long-term stabilityof Hot Melt Coating in a dry air atmosphere.

3 year
stability

40 °C

26 °C

Eact: 86.9 kJ/mole
Log (Z): 6.54 1/min
60 min 1/2 life: 262.4°C
Conversion: 5.0 %

% Conversion

Time (year)

Fig. 4.28. Estimated time (in years) to thermal decomposition failure of hot-melt coating held at various constant temperatures.

would be an appropriate holding temperature. The kinetic model successfully predicted that the hot-melt coating formulation remained thermally stable for 48 h in dry air and required no oxidation protection.

The second question asked what would be an excessive temperature for this process. It was recommended that process hot spots (i.e., zones higher than 100°C) should be avoided. This requirement was met by keeping the heating lines, the walls of the melting pot, and the spray head thermally jacketed to maintain the appropriate internal soak temperature. As a result, the model presented a potential for hot spots at the skin surfaces of the lines and equipment walls. This needed to be investigated for its decomposition potential, and in fact, after several batches were processed, the flexible heat-traced lines had to be discarded because of a buildup of a blacked residue on the inner tubing walls. The kinetic model predicted how many batches could be run before this necessary replacement maintenance was required.

The third question concerned the lifetimes of materials in the process. Figure 4.27 showed that the coating mixture could be held in the melting pot for at least two consecutive 8-h work shifts in the production plant without significant decomposition, if the coating batch was maintained at or below 100°C.

The final question concerned the predicted thermal storage lifetime of the coating, which would probably be a good measure of stability lifetime of the hot-melt-coated particle. As shown in Fig. 4.28, the coating mixture was predicted to be thermally stable in dry air at 40°C for more than four years.

4.6 ALTERNATE METHODS OF THERMAL ANALYSIS

Most workers in the pharmaceutical field identify thermal analysis with the melting point, DTA, DSC, and TG methods just described. Growing in interest are other techniques available for the characterization of solid materials, each of which can be particularly useful to deduce certain types of information. Although it is beyond the scope of this chapter to delve into each type of methodology in great detail, it is worth providing short summaries of these. As in all thermal analysis techniques, the observed parameter of interest is obtained as a function of temperature, while the sample is heated at an accurately controlled rate.

4.6.1 Modulated differential scanning calorimetry (MDSC)

In MDSC, the sample is exposed to a linear heating rate that has a superimposed sinusoidal oscillation, which provides the three signals of time, modulated temperature, and modulated heat flow. The total heat flow is obtained from averaging the modulated heat flow (equivalent to normal DSC), while the sample heat capacity is calculated from the ratio of the modulated heat flow amplitude and the modulated heating rate by a Fourier transform. The reversing heat flow is calculated by multiplication of the heat capacity with the negative heating rate, and the non-reversing heat flow is calculated as the difference between the total and reversing heat flows.

MDSC is particularly useful for the study of reversible (related to the heat capacity) thermal reactions, and is less useful for non-reversing (kinetically controlled) reactions. Examples of reversible thermal events include glass transitions, heat capacity, melting, and enantiotropic phase transitions. Examples of non-reversible events include vaporization, decomposition, cold crystallization, relaxation phenomena, and mono-tropic phase transitions. The ability of MDSC to differentiate between reversible and non-reversible thermal events can yield improved separation of overlapping transitions.

The technique appears to be particularly useful in the characteriza-tion of glass transition phenomena. The utility of MDSC in the study of glass transitions can lead to methods for determination of the amorphous content in a substance [62,63].

4.6.2 Evolved gas analysis (EGA)

In this technique, both the amount and composition of the volatile component are measured as a function of temperature. The composi-tion of the evolved gases can be determined using gas chromatography, mass spectrometry, or infrared spectroscopy.

4.6.3 Thermo-mechanical Analysis (TMA)

The deformation of the analyte, under the influence of an externally applied mechanical stress, is followed as a function of temperature. When the deformation of the sample is followed in the absence of an external load, the technique is identified as *thermodilatometry*.

4.6.4 Thermoptometry

This category refers to a variety of techniques in which some optical property of the sample is followed during the heating procedure. Observable quantities could be the absorption of light at some wavelength (*thermospectrophotometry*), the emission of radiant energy (*thermoluminescence*), changes in the solid refractive index (*thermo-refractometry*), or changes in the microscopic particle characteristics (*thermomicroscopy*). The latter state is often referred to as *hot-stage microscopy*.

4.6.5 Dielectric analysis

As applied to thermal analysis, dielectric analysis consists of the measurement of the capacitance (the ability to store electric charge) and conductance (the ability to transmit electrical charge) as functions of applied temperature. The measurements are ordinarily conducted over a range of frequencies to obtain full characterization of the system.

The information deduced from such work pertains to mobility within the sample, and has been extremely useful in the study of polymers.

REFERENCES

1   D. Dollimore, Thermoanalytical Instrumentation and Applications. In: G.W. Ewing (Ed.), *Analytical Instrumentation Handbook*, 2nd ed., Marcel Dekker, New York, 1997, pp. 947–1005, Chapter 17.
2   J.K. Guillory, S.C. Hwang and J.L. Lach, *J. Pharm. Sci.*, 58 (1969) 301.
3   D. Ghiron, *J. Pharm. Biomed. Anal.*, 4 (1986) 755.
4   D. Ghiron-Forest, C. Goldbronn and P. Piechon, *J. Pharm. Biomed. Anal.*, 7 (1989) 1421.
5   I. Townsend, *J. Therm. Anal.*, 37 (1991) 2031.
6   A.F. Barnes, M.J. Hardy and T.J. Lever, *J. Therm. Anal.*, 40 (1993) 499.
7   M. Kuhnert-Brandstätter, *Thermomicroscopy in the Analysis of Pharmaceuticals*, Pergamon Press, Oxford, 1971.
8   E.L. Skau and J.C. Arthur. Determination of melting and freezing temperatures. In: A. Weissberger and B.W. Rossiter (Eds.), *Physical Methods of Chemistry*, vol. I, part V, Wiley-Interscience, New York, 1971, pp. 137–171, Chapter III.
9   Melting range or temperature. General Test <741>, *United States Pharmacopeia*, vol. 27, United States Pharmacopeial Convention, Rockville, MD, 2004, pp. 2324–2325.
10 C.E. Clarkson and T. Malkin, *J. Chem. Soc.*, (1934) 666–671; ibid., (1948) 985–987.
11 T. Malkin and M.R. El-Shurbagy, *J. Chem. Soc.* (1936) 1628–1634.
12 T. Malkin, M.R. El-Shurbagy and M.L. Meara, *J. Chem. Soc.* (1937) 1409–1413.
13 R. O'Laughlin, C. Sachs, H.G. Brittain, E. Cohen, P. Timmins and S. Varia, *J. Soc. Cosmet. Chem.*, 40 (1989) 215–229.
14 D.M. Speros and R.L. Woodhouse, *J. Phys. Chem*, 67 (1963) 2164–2168 ibid., 72 (1968) 2846–2851
15 I. Kotula and A. Rabczuk, *Thermochim. Acta*, 126 (1988) 61–66 ibid., 126 (1988) 67–73 ibid., 126 (1988) 75–80
16 W. Smykatz-Kloss, *Differential Thermal Analysis*, Springer, Berlin, 1974.
17 M.I. Pope and M.D. Judd, *Differential Thermal Analysis*, Heyden, London, 1977.
18 W.W. Wendlandt, *Thermal Methods of Analysis*, Wiley-Interscience, New York, 1964.
19 G.M. Lukaszewski, *Lab. Pract.*, 15 (1966) 664.
20 M.J. Richardson and P. Burrington, *J. Therm. Anal.*, 6 (1974) 345.
21 H. Kung and E.D. Goddard, *J. Phys. Chem.*, 67 (1963) 1965–1969.

22 J.K. Guillory, S.C. Hwang and J.L. Lach, *J. Pharm. Sci.*, 58 (1969) 301–308.

23 H. Jacobson and G. Reier, *J. Pharm. Sci.*, 58 (1969) 631–633.

24 P. Van Aerde, J.P. Remon, D. DeRudder, R. Van Severen and P. Braeckman, *J. Pharm. Pharmacol.*, 36 (1984) 190–191.

25 S.S. Yang and J.K. Guillory, *J. Pharm. Sci.*, 61 (1972) 26–40.

26 H.G. Brittain, *J. Pharm. Sci.*, 91 (2002) 1573–1580.

27 Y. Matsuda, S. Kawaguchi, H. Kobayshi and J. Nishijo, *J. Pharm. Sci.*, 73 (1984) 173–179.

28 K.C. Lee and J.A. Hersey, *J. Pharm. Pharmacol.*, 29 (1977) 249–250.

29 M. Otsuka and N. Kaneniwa, *Chem. Pharm. Bull.*, 32 (1984) 1071–1079.

30 M. Otsuka and N. Kaneniwa, *Chem. Pharm. Bull*, 31 (1983) 4489–4495 ibid., 32 (1984) 1071–1079

31 J.L. Ford and P. Timmins, *Pharmaceutical Thermal Analysis*, Ellis Horwood, Ltd., Chichester, 1989.

32 D. Giron, *Acta Pharm. Jugosl.*, 40 (1990) 95.

33 Thermal analysis, General chapter <891>, *United States Pharmacopeia*, 28th ed., United States Pharmacopeial Convention, Rockville, MD, 2005, pp. 2501–2503.

34 R.L. Blaine and C.K. Schoff, *Purity Determinations by Thermal Methods*, ASTM Press, Philadelphia, 1984.

35 F.F. Joy, J.D. Bonn and A.J. Barnard, *Thermochim. Acta*, 2 (1971) 57.

36 E.F. Palermo and J. Chiu, *Thermochim. Acta*, 14 (1976) 1.

37 A.A. van Dooren and B.W. Muller, *Int. J. Pharm.*, 20 (1984) 217.

38 R.C. Mackenzie, *Pure Appl. Chem.*, 57 (1985) 1737.

39 H.G. Brittain, *Polymorphism in Pharmaceutical Solids*, Marcel Dekker, New York, 1999.

40 J. Bernstein, *Polymorphism in Molecular Crystals*, Clarendon Press, Oxford, 2002.

41 J.A. McCauley, *ALCHE Symposium Series: Particle Design Via Crystallization*, 87(284) (1990) 58–63.

42 R. Pfeiffer, Impact of Solid State factors on Drug Development, Polymorphs and Solvates of Drugs. Purdue University School of Pharmacy Short Course (1990).

43 M.J. Jozwiakowski, S.O. Williams and R.D. Hathaway, *Int. J. Pharm.*, 91 (1993) 195–207.

44 C.M. Neag, Coatings characterization by thermal analysis. In: J.V. Koleske (Ed.), Paint and coating Testing Manual. *Gardner-sward Handbook*, 14th ed., ASTM, West Conshohocken, PA, 1995, Chapter 75, *ASTM Man. 17* (1995) 843–845.

45 N.W. Johnston, *J. Macromol. Sci., Rev. Macromol. Chem.*, C14 (1976) 215–250.

46 J.M.G. Cowie and P.M. Toporowski, *Eur. Poly. J.*, 4 (1968) 621.

47 D.W. van Krevlan, *Properties of Polymers*, 3rd ed., Elsevier, New York, 1990, p. 49.
48 R.D. Bruce, J.D. Higgins, S.A. Martellucci and T. Gilmor, Ibuprofen. In: H.G. Brittain (Ed.), *Analytical Profiles of Drug Substances and Excipients*, vol. 27, Academic Press, San Diego, 2001, pp. 265–300, Chapter 6.
49 Y. Higashino, *J. Biochem. (Tokyo)*, 130 (2001) 393–397.
50 W.J. Sichina, *Seiko Instruments Application Brief DSC-7*, pp. 2–3.
51 C.J. Keattch and D. Dollimore, *Introduction to Thermogravimetry*, 2rd ed., Heyden, London, 1975.
52 C. Duval, *Inorganic Thermogravimetric Analysis*, 2nd ed., Elsevier, Amsterdam, 1963.
53 R.D. Bruce, J.D. Higgins III, S.A. Martellucci and Stephen A., Sterol esters in tableted solid dosage forms, United States Patent 6,376,481, April 23, 2002.
54 B. Burruano, R.D. Bruce and M.R. Hoy, Method for producing water dispersible sterol formulations, United States Patent 6,110,502, August 29, 2000.
55 R.D. Bruce, B. Burruano, M.R. Hoy and J.D. Higgins III, Method for producing water dispersible sterol formulations, United States Patent 6,054,144, April 25, 2000.
56 R.D. Bruce, B. Burruano, M.R. Hoy and N.R. Paquette, Method for producing dispersible sterol and stanol compounds, United States Patent 6,242,001, June 5, 2001.
57 Estimation of Polymer Lifetime by TGA Decomposition Kinetics, *TA Instruments Thermal Analysis Application Brief TA-125*, pp. 1–4.
58 J. Reo and W.M. Johnson, Tastemasked Pharmaceutical System, United States Patent 5,891,476, April 6, 1999 (see Example 4).
59 CAS-Number 79777-30-3, Lonza Product Data Sheet: US Polyaldo 10-1-S.pdf. URL: http://www.lonza.com/group/en/products_services/products/catalog/groups.ParSys.5500.File0.tmp?path=product_search_files/db_files/US_Polyaldo%2010-1-S.pdf
60 J.H. Flynn, *Polym. Lett.*, B4 (1966) 323.
61 D.J. Toop, *IEEE Trans. Elec. Ins.*, El-6 (1971) 2.
62 R. Saklatvala, P.G. Royall and D.Q.M. Craig, *Int. J. Pharm.*, 192 (1999) 55–62.
63 S. Guinot and F. Leveiller, *Int. J. Pharm.*, 192 (1999) 63–75.

REVIEW QUESTIONS

1. How does melting point of a substance relates to its purity?
2. Provide two application of DTA.
3. What is DSC?
4. Describe various applications of DSC.

Chapter 5

General principles of spectroscopy and spectroscopic analysis

Neil Jespersen

5.1 INTRODUCTION

Many of the laws of optics were discovered or rediscovered in the period called the Renaissance. Isaac Newton studied the properties of prisms and their ability to separate white light into what we now call the visible spectrum and also prepared lenses to use in telescopes. Laws of optics such as the law of reflection,

$$\sin \theta_{incident} = \sin \theta_{reflected} \tag{5.1}$$

and Snell's Law of refraction,

$$\eta_{incident} \sin \theta_{incident} = \eta_{refracted} \sin \theta_{refracted} \tag{5.2}$$

where η is the refractive index defined as the ratio of the speed of light in a vacuum to the speed of light in the given medium, date from this period.

In more modern times, infrared, "beyond red," radiation was discovered by William Herschel in 1800. He found that the temperature recorded by a thermometer increased from the violet to the red in the visible region. Going beyond the red he found the temperature continued to increase instead of decreasing if the light ended at the end of the visible spectrum. Recognition of the utility of the infrared spectral region for chemical analysis is credited to W.W. Coblentz and it was not until the mid-1900s that infrared spectroscopy became an established technique. Apparently, noting Herschel's discovery of infrared radiation, Johann Wilhelm Ritter discovered ultraviolet radiation in 1801 by noting that silver chloride was reduced to silver metal when exposed to violet visible light and was even more efficiently reduced by radiation beyond the violet end of the visible spectrum.

Comprehensive Analytical Chemistry 47
S. Ahuja and N. Jespersen (Eds)
Volume 47 ISSN: 0166-526X DOI: 10.1016/S0166-526X(06)47005-7

James Clerk Maxwell predicted the existence of electromagnetic waves in 1864 and developed the classical sine (or cosine) wave description of the perpendicular electric and magnetic components of these waves. The existence of these waves was demonstrated by Heinrich Hertz 3 years later.

Diffraction of light by transmission and reflection gratings was used to demonstrate the existence of light waves and led to the development of the diffraction equation.

$$n\lambda = d(\sin \theta_{incident} \pm \sin \theta_{reflected}) \qquad (5.3)$$

where n represents the order of diffraction and d stands for the spacing between grooves in a diffraction grating.

Finally, in the early 20th century Albert Einstein explained the photoelectric effect based on quantized packets of electromagnetic radiation called photons. These quickly led to the familiar relationships of the energy of a photon,

$$E = h\nu = h\frac{c}{\lambda} \qquad (5.4)$$

and the particle–wave duality expressed by DeBroglie in 1938

$$\lambda = \frac{h}{mv} \qquad (5.5)$$

Use of spectroscopy for chemical analysis most probably can be related to alchemists' use of flame tests for the qualitative determination of elemental composition. Comparison of colors of solutions, colorimetry, also emerged at that time. Development of light sources, dispersing devices, optics, detectors, modern transducers and digital technology has led to continuing improvement of spectroscopic instrumentation. Modern instrumentation for spectroscopic analysis in the ultraviolet, visible and infrared spectral regions is based on sophisticated principles and engineering, yet is often simple enough to produce accurate results with a minimum of training.

5.1.1 Spectral regions

This chapter covers ultraviolet, visible and infrared spectroscopies, the most commonly used range of wavelengths employed by chemists today. The range of wavelengths that the human eye can detect varies slightly from individual to individual. Generally, the wavelength region from 350 to 700 nm is defined as the visible region of the spectrum. The energy of a mole of photons ranges from 170 to 340 kJ mol^{-1} and may be compared to the approximate bond energy for a C–C bond of $350\,kJ\,mol^{-1}$ and a C–H bond of $412\,kJ\,mol^{-1}$. This amount of energy is sufficient to cause electronic transitions within molecules and in some instances can cause ionization and bond breaking.

Ultraviolet radiation is commonly defined as the wavelengths from 200 to 350 nm. Around 200 nm oxygen absorbs strongly, part of the process to produce ozone in the upper atmosphere, and makes measurements difficult. One solution to the problem is to evacuate the instrument, giving rise to the terminology that wavelengths from 200 to 100 nm are in the "vacuum UV" region. The energies of photons in UV region range from 340 to $595\,kJ\,mol^{-1}$. These energies are sufficiently high to cause ionization and bond breaking. As a result, electromagnetic radiation starting at 300 nm or down is often called ionizing radiation.

Technically, the infrared region starts immediately after the visible region at 700 nm. From 700 to 2500 nm is the near infrared, NIR, region and its use is discussed in Chapter 6. The classical infrared region extends from 2500 ($2.5\,\mu m$) to 50,000 nm ($50\,\mu m$). Infrared spectroscopists often use wavenumbers to describe the infrared spectral region. A wavenumber is the reciprocal of the wavelength when the wavelength is expressed in centimeters and has the symbol, ν. As a result 2500 nm is $4000\,cm^{-1}$ and 50,000 nm is $200\,cm^{-1}$. Multiplication of ν by the speed of light, $3 \times 10^{10}\,cm\,s^{-1}$, gives the frequency that is directly proportional to the energy. The energies of infrared radiation range from $48\,kJ\,mol^{-1}$ at 2500 nm to $2.4\,kJ\,mol^{-1}$ at 50,000 nm. These low energies are not sufficient to cause electron transitions but they are sufficient to cause vibrational changes within molecules. Infrared spectroscopy is often called vibrational spectroscopy.

5.1.2   Spectra

A spectrum is a plot of some measure of the electromagnetic radiation absorbed by a sample versus the wavelength or energy of the electromagnetic radiation. For example, it is common practice to plot the absorbance versus wavelength for spectra in the ultraviolet and visible spectral regions as shown below (Fig. 5.1).

Historically, in the infrared region spectra have been represented as percent transmittance versus wavenumber as shown in Fig. 5.2.

Infrared spectra plotted as absorbance versus wavelength are becoming more common especially with instruments that are computer controlled and can make the change with a few commands.

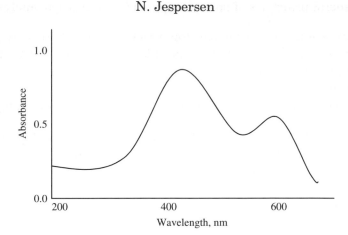

Fig. 5.1. Typical format, absorbance versus wavelength, of ultraviolet and visible spectra. Peaks in these regions tend to be broad.

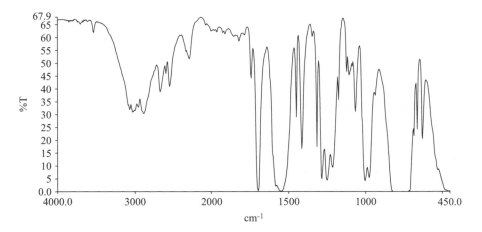

Fig. 5.2. Typical format, percentage transmittance (%T) versus wavenumber, for infrared spectra. This is a spectrum of benzoic acid.

5.1.3 Origin of the spectra

All spectra are due to the absorbance of electromagnetic radiation energy by a sample. Except for thermal (kinetic) energy, all other energy states of matter are quantized. Quantized transitions imply precise energy levels that would give rise to line spectra with virtually no linewidth. Most spectral peaks have a definite width that can be explained in several ways. First, the spectral line-width can be related to the lifetime of the excited state using the Heisenberg uncertainty principle. The width of a peak in terms of ΔE is $h/2\pi\tau$, where τ is the lifetime of the excited state,

$$\Delta E = \frac{h}{2\pi\tau} \tag{5.6}$$

The second contribution to the line-width is Doppler broadening. While the transition energy ΔE may be constant, the frequency and therefore the energy of radiation increases if the molecule is approaching the source and decreases if the molecule is receding from the source. In terms of energy

$$\Delta E = 2E_0 \left(\frac{2kT\ln 2}{mc^2}\right)^{1/2} \tag{5.7}$$

Collisional line broadening can be written in a form similar to Eq. (5.6)

$$\Delta E = \frac{h}{2\pi\tau_c} \tag{5.8}$$

the difference being that τ_c represents the time between collisions and in Eq. (5.6) it represents the excited state lifetime.

Finally, the width of peaks in most spectra is due to the fact that the peak actually represents an "envelope" that describes the outline of a group of closely spaced, unresolved, peaks. In the infrared region the rotational energy levels are superimposed on the vibrational energy levels giving rise to many closely spaced transitions that are generally not resolved. In the visible and ultraviolet regions the vibrational and rotational energy levels are superimposed on the electronic transitions giving rise to very wide absorbance bands.

Rotational transitions from one state to another (e.g., J_0–J_1) require the least energy and these transitions usually occur in the microwave region of the spectrum. The energy of microwaves ranges from 156 to 325 J. Microwave spectra tend to have very sharp peaks.

Vibrational transitions (e.g., v_0–v_1) require more energy than rotational transitions and this amount of energy is generally found in the infrared region of the spectrum. Infrared spectra have sharp peaks with some width to them.

Each atom within a molecule has three degrees of freedom for its motion in three-dimensional space. If there are N atoms within a molecule there are $3N$ degrees of freedom. However, the molecule as a whole has to move as a unit and the x, y, z transitional motion of the entire molecule reduces the degrees of freedom by three. The molecule also has rotational degrees of freedom. For a non-linear molecule this rotation has three degrees of freedom reducing the number of degrees of freedom to $3N-6$. A linear molecule can be rotated around its axis with no change and only two significant rotations. Therefore, a linear molecule has $3N-5$ degrees of freedom. This calculation indicates the maximum number of transitions a molecule can have.

The number of peaks actually observed in an infrared spectrum is often less than the maximum because some of the vibrations are energetically identical or degenerate. A real molecule will often have two or more vibrations that may differ only by their orientation in space. These will have exactly the same energy and result in one absorption peak. In addition to the degeneracy of vibrational modes, there is also the requirement that a vibration result in a change in the dipole moment of the molecule needs to be observed.

The number of peaks in an IR spectrum may increase due to overtones. Normally, the vibrational level is allowed to change by ± 1. If the vibrational energy level changes by ± 2 or more (a "forbidden" transition), an overtone results. It is also possible for two normal mode vibrations to combine into a third.

To illustrate the above concept, we consider the possible and observed peaks for H_2S and CS_2. H_2S is a non-linear molecule and is expected to have $3N-6 = 3$ spectroscopic peaks. The diagram below shows the three possible vibrations as a symmetrical stretch, and asymmetric stretch and a motion called scissoring (Fig. 5.3).

For CS_2, it is a linear molecule and should have $3N-5 = 4$ vibrational modes. There is a symmetrical stretch, an asymmetrical stretch and a bending in-plane and a bending out-of-plane modes of vibration.

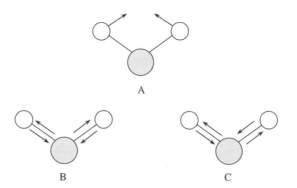

Fig. 5.3. The three vibrational modes of H2S. (A) Represents the scissoring motion, (B) is the symmetrical stretch and (C) is the asymmetrical stretch.

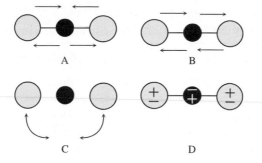

Fig. 5.4. Vibrational modes for CS2. (A) Represents the symmetrical stretch, (B) is the asymmetrical stretch, (C and D) are the out-of-plane and in-plane bending motions. The + and – symbols indicate motion toward and away from the viewer, respectively.

The two bending motions are identical; they differ only by the direction in space of the motion. Therefore they degenerate and appear as only one peak (Fig. 5.4).

Without derivation, we may consider the bonds between atoms as a spring connecting two atoms in a harmonic oscillator. The energy difference between two vibrational energy levels is

$$E_v = \left(v + \frac{1}{2}\right)\left(\frac{h}{2\pi}\right)\left(\frac{k}{\mu}\right)^{1/2} \tag{5.9}$$

where v is the vibrational quantum number, h represents Planck's constant, k is the Hooke's Law restoring force constant and μ is the reduced mass

$$\text{Reduced mass} = \frac{m_1 m_2}{m_1 + m_2} \tag{5.10}$$

The approximate energy change for vibrational transitions may be calculated using the approximate force constants and subtracting $E_{v=1} - E_{v=0} = \Delta E$ (Table 5.1).

The effect of isotopic substitution on the position of a peak can also be estimated using this relationship.

For ultraviolet and visible spectroscopy the transitions are between electronic energy levels. Certain groupings of atoms, particularly in organic molecules have electrons that can be excited from one energy level to another with the energies found in the ultraviolet and visible regions of the spectrum. These groupings are called chromophores and usually account for most of the absorption of energy by a molecule.

TABLE 5.1

Force constants for bond types useful for estimation vibrational frequencies

Type of bond	Force constant, k (mdyne Å^{-1})
C–H	5
C–F	6
N–H	6.3
O–H	7.7
C–Cl	3.5
C–C	10
C=C	12
C≡C	15.6
C≡N	17.7

Source: From Mann, C.K., Vickers, T.J. and Gulick, W.M., *Instrumental Analysis*, Harper & Row, New York, 1974, p. 483.

Electrons in an organic compound are usually σ (sigma bond electrons), n (non-bonding electrons) and π (pi bond electrons). These electrons may be excited to their corresponding antibonding levels, σ^* σ, π^* π. The non-bonding electrons may be excited to the σ^* or π^* levels. Of these transitions, only the π^* π has a large molar absorptivity (ca. $10,000 \, \text{l} \, \text{mol}^{-1} \, \text{cm}^{-1}$) along with a low enough energy to occur in the ultraviolet or visible regions.

Independent double bonds (i.e., without conjugation) have molar absorptivities that are approximately multiples of the molar absorptivity for one double bond. For instance, 1,4-hexadiene has twice the molar absorptivity of 1-hexene but absorbs at the same wavelength.

For conjugated, non-aromatic substances, both the molar absorptivity and wavelength of maximum absorption increase. An example of this is the comparison of 1-hexene that absorbs at 177 nm with a molar absorptivity of 12,000 while 1,3,5-hexatriene absorbs at 268 nm and has a molar absorptivity of 42,500.

Aromatic compounds have very high molar absorptivities that usually lie in the vacuum ultraviolet region and are not useful for routine analysis. Modest absorption peaks are found between 200 and 300 nm. Substituted benzene compounds show dramatic effects from electron-withdrawing substituents. These substituents are known as auxochromes since they do not absorb electromagnetic radiation but they have a significant effect on the main chromophore. For example, phenol and aniline have molar absorptivities that are six times the molar absorptivity of benzene or toluene at similar wavelengths.

Charge-transfer spectra represent one of the most important classes of spectra for analytical chemistry since the molar absorptivities tend to be very large. Charge-transfer can occur in substances, usually complexes that have one moiety that can be an electron donor and another that can be an electron acceptor. Both the donor and acceptor must have a small difference in their energy levels so that the electron can be readily transferred from the donor to the acceptor orbitals and back again. One example is the well-known, deep-red color of the iron (III) thiocyanate ion. The process appears to be

$$(Fe^{3+}SCN^-)^{2+} + hv \rightarrow (Fe^{2+}SCN)^{2+} \tag{5.11}$$

an electron from the thiocyanate is excited to an orbital of iron, effectively reducing it to iron (II) and the thiocyanate radical. The electron rapidly returns to the thiocyanate to repeat the process.

5.2   SPECTROSCOPIC ANALYSIS

5.2.1   Qualitative relationships

Infrared spectra differ markedly from the typical ultraviolet or visible spectrum. Infrared spectra are marked by many relatively sharp peaks and the spectra for different compounds are quite different. This makes infrared spectroscopy ideal for qualitative analysis of organic compounds.

For qualitative analysis the infrared spectrum is divided roughly into half. The region from 4000 to 2500 cm^{-1} is the group region and 2500 to 200 cm^{-1} is the fingerprint region. In the group region, there are fairly well-defined ranges at which different functional groups absorb. For example, the nitrile group (–C≡N) has a sharp line at 2260–2240 cm^{-1} and the –OH group has a large broad peak at 3000 cm^{-1}. A brief table of some functional groups is given below (Table 5.2).

The fingerprint region is an area that has many peaks and it allows us to distinguish between different substances that may have the same functional groups. All alcohols will have a large, broad peak at 3000 cm^{-1}, however, each alcohol will have a distinctively different number and position of peaks in the fingerprint region.

Significant tabulations of spectra are available in hardcopy or in electronic databases. In addition, compilations of common absorption bands based on functional group or vibrational mode are also available.

The ultraviolet and visible spectra are usually comprised of a few broad peaks at most. This is due to the fact that electronic transitions

TABLE 5.2

Absorption band ranges for some typical functional groups

Functional group	Frequency range (cm^{-1})
Alcohol –OH	O–H stretch 3550–3200 broad
Ketone –C=O	C=O stretch 1870–1540 strong
Aldehyde –C=O	C=O stretch 1740–1720
Acids –COOH (dimers)	C=O stretch 1700–1750 strong
O–H stretch 3300–2500 broad	
Amines, primary –NH$_2$	N–H stretch 3500–3400 doublet, weak
Amino acids	NH$_3^+$ stretch 3100–2600 strong
Esters	C=O stretch 1750–1735
C–O stretch 1300–1000	
Ethers C–O–C	C–O–C asym. str. 1170–1050 strong
Halogens –CH$_2$X	CH$_2$ wag 1333–1110

Source: Specific absorption frequencies vary from compound to compound.

dominate these spectral regions. Table 5.3 lists some of the electronic transitions that can occur and their approximate wavelength ranges and molar absorptivities.

Qualitatively, absorbance in the ultraviolet region of the spectrum may be taken to indicate one or more unsaturated bonds present in an organic compound. Other functional groups can also absorb in the UV and visible regions. The portion of a molecule that absorbs the electromagnetic radiation is called a chromophore. Fully saturated compounds only absorb in the vacuum UV region. Unsaturated bonds absorb electromagnetic radiation as a π–π^* transition. The energy difference is small between these two states and the molar absorptivities are relatively high.

Some salts, particularly of transition metals, are highly colored and absorb in the UV and visible regions. Salts and complexes that have the highest molar absorptivities tend to absorb electromagnetic radiation by a charge-transfer process. In the charge-transfer process, an electron is promoted from one part of a complex to another causing one part of the complex to be oxidized and the other to be reduced as in Eq. (5.11).

5.2.2 Quantitative relationships

Spectroscopic measurements for the UV, visible and infrared regions are most conveniently and reliably made by determining the absorbance of a

TABLE 5.3

Examples of some electronic transitions in the ultraviolet region

Electronic transition	Maximum wavelength (nm)	Maximum molar absorptivity $(\mathrm{lmol}^{-1}\,\mathrm{cm}^{-1})$	Example
$\sigma \to \sigma^*$	135	–	Ethane
$n \to \sigma^*$	173	200	Methyl chloride
$n \to \sigma^*$	259	400	Methyl iodide
$\pi \to \pi^*$	165	11,000	Ethylene
$\pi \to \pi^*$	217	21,000	1,3-Butadiene
$\pi \to \pi^*$	188	900	Acetone
$n \to \pi^*$	290	17	Acetaldehyde
$n \to \pi^*$	204	41	Acetic acid
Aromatic $\pi \to \pi^*$	180	60,000	Benzene
Aromatic $\pi \to \pi^*$	200	8000	Benzene
Aromatic $\pi \to \pi^*$	255	215	Benzene
Aromatic $\pi \to \pi^*$	210	6200	Toluene
Aromatic $\pi \to \pi^*$	270	1450	Toluene

Source: From Silverstein, R.M., Bassler, G.C. and Morrill, T.C. *Spectrometric Identification of Organic Compounds*, 4th ed., John Wiley and Sons, New York, 1981, p. 312.

solution. The absorbance tends to be a robust measure that is reproducible and only slightly affected by common variables of temperature and trace impurities. The absorbance of a system is determined by measuring the intensity of light at a given wavelength, I_0 and then measuring the intensity with a sample in the same beam of light, I. The ratio of these intensities is the transmittance, T.

$$T = \frac{I}{I_0} \tag{5.12}$$

When using a single-beam spectrometer, I_0 is measured when a reagent blank is used to "zero" the absorbance scale. The value of I is then measured when the sample is inserted into the spectrometer. On the other hand, when using a double-beam instrument both the reagent blank, I_0, and the sample, I, are measured continuously and the appropriate ratio is determined electronically.

Most infrared measurements are transmittance values plotted as a spectrum. To convert data from an infrared spectrum to usable absorbance values involves the following steps: First, the peak of interest is located and a tangent is drawn from one shoulder to the other to create a baseline. Then a vertical line is constructed from the peak to

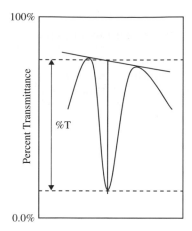

Fig. 5.5. Illustration of the method to determine the percent transmittance of a peak in an infrared spectrum.

the baseline. The percent transmittance of the peak is the difference in transmittance from the baseline intersection to the peak. This procedure is illustrated in Fig. 5.5.

The absorbance is defined as

$$A = -\log T = (\%T/100) \tag{5.13}$$

The Beer–Lambert Law relates the absorbance to concentration in two alternate forms depending on the units used for the concentration:

$$A = a\,b\,c \quad \text{or} \quad A = \varepsilon\,b\,c \tag{5.14}$$

Modern terminology defines A as the absorbance, a as the absorptivity, b as the optical path length and c as the concentration. In the second equation ε represents the molar absorptivity. Table 5.4 compares these terms.

The Beer–Lambert Law assumes that the electromagnetic radiation being absorbed is monochromatic. In practical instruments it is not a single wavelength but a band of wavelengths that enter the sample. The middle of this band of wavelengths is called the nominal wavelength. It can be shown that as long as a, or ε, is relatively constant over the band of wavelengths, the absorbances of each wavelength can be added to obtain the total absorbance that obeys the Beer–Lambert Law,

$$A_{\text{total}} = A_{\lambda_1} + A_{\lambda_2} + A_{\lambda_3} + \cdots + \tag{5.15}$$

If the absorptivities or molar absorptivities are not approximately equal, the linear relationships of the Beer–Lambert Law will not hold.

TABLE 5.4

Terminology and units used for the Beer–Lambert law

	A	a or ε	b	c
$A = abc$	Absorbance dimensionless	Absorptivity $(l\,g^{-1}\,cm)$	Optical path length (cm)	Concentration $(g\,l^{-1})$
$A = \varepsilon bc$	Absorbance dimensionless	Molar absorptivity $l\,mol^{-1}\,cm$	Optical path length (cm)	Concentration $mol\,l^{-1}$

In practical situations the absorbance of a sample is determined by making two measurements, the first to determine I_0 and the second to determine I. The determination of I_0 is used to cancel a large number of experimental factors that could affect the result. When measuring I_0 the sample container must closely match the unknown container in all ways except for the analyte content. The cuvettes should be a matched pair if a double beam instrument is used and the same cuvette can be used for both the blank and sample with a single beam instrument. The blank solution filling the cuvette should be identical to the solvent that the sample is dissolved in, except for the sample itself. If done correctly, the least-squares line for the calibration graph will come very close to the 0,0 point on the graph.

5.2.3 Single component analysis

The simplest spectroscopic analysis is of one compound that absorbs electromagnetic radiation strongly at a wavelength where no other substance absorbs. In this case a series of standard solutions, that have absorbances between zero and 1.0 are prepared. Each of the standards is measured and a plot of absorbance versus concentration is drawn. A spreadsheet program can be used to record the data and generate the graph along with a least squares line that has a slope of εb. The absorbances of unknown solutions are then determined. The absorbances of the unknown solutions must fall between the highest and lowest absorbances of the standard solutions. Unknowns with absorbances that are too high must be diluted and those with low absorbances must be concentrated. The graph may be used to determine the concentration by drawing a horizontal line from the absorbance of the unknown to the least squares line and then a vertical line to the concentration axis. This method has a drawback that it may not be possible to determine the concentration to more than two significant figures. Alternatively, the equation for the least squares line may be used to solve for the unknown concentration. The drawback of using the equation is that too many significant figures may be kept and that unknown absorbances far out-side the range of the standards may be inappropriately used.

Example. 10.0 ml of an aqueous solution containing the Mn^{2+} ion is reacted with KIO_4 to produce the permanganate ion. The final mixture is diluted to 100 ml. A stock solution of 100 ppm Mn^{2+} is produced by dissolving 0.100 g of manganese metal and diluting to 1.00 l. Standard solutions are prepared by pipeting 1.00, 3.00, 5.00, 7.00 and 10.00 ml of the stock solution into separate flasks and reacting with KIO_4 in the same manner as the unknown. The absorbances of the standards were determined, in a 1.00 cm cuvette, to be 0.075, 0.238, 0.359, 0.533 and 0.745, respectively. The absorbance of the unknown was determined to be 0.443. What is the concentration of the unknown and the molar absorptivity of the MnO_4^- ion under these conditions.

Solution. Calculate the concentration of the standard solutions as:

$$(ppm\ stock)\ (volume\ stock) = (ppm\ std)\ (volume\ std)$$
$$(100\ ppm)\ (1.00\ ml) = ppm\ std\ (100\ ml\ std)$$
$$ppm\ std = 1.00$$

repeat process for remaining standard solutions

Enter data into a spreadsheet and obtain a graph of absorbance versus concentration of the standards. Obtain the least-squares line and its equation (Fig. 5.6).

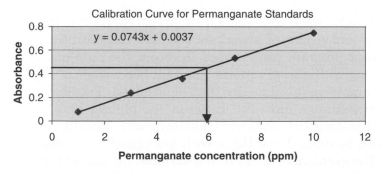

Fig. 5.6. Calibration curve for permanganate standards. Line is a least-squares linear regression for the data. Graphical interpolation is illustrated for an unknown with an absorbance of 0.443.

To determine the concentration using the graphical method draw a horizontal line from $A = 0.443$ to the least-squares line. Then construct a vertical line from that point to the concentration axis and read the value of the unknown. In this case it appears to be slightly less than 6.0 ppm. Using the least-squares equation we get

$$0.443 = 0.0743x + 0.0037 \text{ and the value of } x = 5.91 \text{ ppm}$$

To calculate the absorptivity and molar absorptivity, we see that 0.0743 is the slope of the line and the slope is ab. Since $b = 1.00$ cm, $a = 0.743 \text{ cm}^{-1} \text{ppm}^{-1}$.

We know that $1\,\text{ppm} = 1\,\text{mg}\,\text{l}^{-1}$ therefore $a = 0.743\,\text{l}\,\text{cm}^{-1}\,\text{mg}^{-1}$ inserting 10^{-3} for the prefix m results in $a = 743\,\text{l}\,\text{cm}^{-1}\,\text{g}^{-1}$. To convert a into ε we divide the mass by the molar mass of permanganate ($MnO_4^- = 118.93\,\text{g}\,\text{mol}^{-1}$) to get $\varepsilon = 88,363\,\text{l}\,\text{mol}^{-1}\,\text{cm}^{-1}$.

5.2.4 Mixture analysis

If a mixture of two or more substances is absorbing electromagnetic radiation at the same nominal wavelength, their absorbances will be additive,

$$A_{\text{total}} = A_1 + A_2 + \cdots + \tag{5.16}$$

If each of the substances in a mixture has different spectra, it will be possible to determine the concentration of each component. In a two-component mixture measurement of the absorbance at two (appropriately chosen) different wavelengths will provide two simultaneous equations that can be easily solved for the concentration of each substance.

$$A_{\lambda_1} = \varepsilon_{a\lambda_1} bc_a + \varepsilon_{b\lambda_1} bc_b$$
$$A_{\lambda_2} = \varepsilon_{a\lambda_2} bc_a + \varepsilon_{b\lambda_2} bc_b$$

Since the slope of the calibration curve is εb, we need to construct four calibration curves, two at the first wavelength for compounds a and b and two at the second wavelength for compounds a and b. Once the four slopes are determined along with the absorbance of the unknown at the two wavelengths, we have two equations in two unknowns that can be solved algebraically or with simple matrix methods.

Example. A mixture of two compounds, X and Y, needs to be analyzed. The ε_{max} for compound X is at 632 nm, while ε_{max} for compound Y is at 447 nm. Standards are prepared for both X and Y and the calibration curves give the following results:

$\varepsilon_{X632}\, b = 8879$

$\varepsilon_{Y632}\, b = 2210$

$\varepsilon_{X447}\, b = 3480$

$\varepsilon_{Y447}\, b = 6690$

The absorbances at the two wavelengths were $A_{632} = 0.771$ and $A_{447} = 0.815$. What are the concentrations of the compounds X and Y? Substitute the given data into two equations for the total absorbance at 632 and 447 nm to get

$0.771 = 8879c_x + 2210c_y$

$0.815 = 3480c_x + 6690c_y$

multiply the bottom equation by 8879/3480 to get

$2.079 = 8879c_x + 17069c_y$

subtract the first equation from the new second one to get

2.079	$=$	$8879c_x$	$+17069c_y$
-0.771	$=$	$-8879c_x$	$-2210c_y$
1.308	$=$	$0.00c_x$	$+14859c_y$

then

$c_y = 8.80 \times 10^{-5}\,\mathrm{mol\,l^{-1}}$

$c_x = (0.771 - 0.194)/8879 = 6.50 \times 10^{-5}\,\mathrm{mol\,l^{-1}}$

5.2.5   Method of standard additions

In certain circumstances the matrix, defined as everything except the analyte, contributes significantly to the absorbance of a sample and is also highly variable. One method that can be used to improve results is the method of standard additions. The basic idea is to add standard to the analyte so that the standard is subjected to the same matrix effects as the analyte. This method assumes that the system obeys the Beer–Lambert Law.

General principles of spectroscopy and spectroscopic analysis

As an example, consider a sample that contains the dichromate ion. The absorbance of the unknown may be readily determined at 425 nm, which is close to the maximum for the dichromate ion. For this unknown the absorbance measured will be 0.525. Now, we will add 5.00 ml of a 3.00 mM solution of $K_2Cr_2O_7$ to 25.0 ml of the unknown and remeasure the absorbance and find it to be 0.485. The calculations are

Original unknown : $A_{unk} = \varepsilon b c_{unk}$

Unknown with added standard $A_{unk+std} = \varepsilon b c_{unk+std}$
The ratio of these two equations is

$$\frac{A_{unk}}{A_{unk+std}} = \frac{\varepsilon b c_{unk}}{\varepsilon b c_{unk+std}} = \frac{c_{unk}}{c_{unk+std}}$$

For the denominator we can use the dilution equation

$$c_{unk}v_{unk} + c_{std}v_{std} = c_{unk+std}v_{unk+std} = c_{unk+std}(v_{unk} + v_{std})$$

$$c_{unk+std} = \frac{c_{unk}v_{unk} + c_{std}v_{std}}{v_{unk} + v_{std}}$$

$$\frac{A_{unk}}{A_{unk+std}} = \frac{c_{unk}}{\frac{c_{unk}v_{unk}+c_{std}v_{std}}{v_{unk}+v_{std}}}$$

Looking at this equation we have measured the two absorbances, we know the volumes of the unknown and standard and we know the concentration of the standard. Only one unknown, the concentration of the unknown is left to calculate.

$$\frac{0.525}{0.485} = \frac{c_{unk}}{\frac{c_{unk}(25.0 \text{ ml})+(3.00 \text{ mM})(5.00 \text{ ml})}{(25.0 \text{ ml}+5.00 \text{ ml})}}$$

$c_{unk} = 5.53$ mM dichromate

An alternate method that lends itself to analysis using a database is to add varying amounts of standard to a fixed volume of unknown in separate volumetric flasks. The flasks are all filled to the mark, mixed well and measured. As an example let us take a solution containing an unknown amount of Cu^{2+}; 25.0 ml of the unknown is pipetted into each of five 50 ml volumetric flasks. Added into these flasks are 0.0 ml, 3.0 ml 5.0 ml, 7.0 ml and 10.0 ml of a 0.600 ppm solution of Cu^{2+}. Each flask is then filled to the mark with 1.0 M ammonia solution to develop the color. The measured absorbances were 0.326, 0.418, 0.475, 0.545 and 0.635, respectively. The concentrations of each of the standards in the

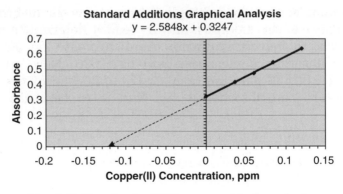

Fig. 5.7. Standard additions graphical analysis.

flasks are 0.0, 0.036, 0.060, 0.084 and 0.12 ppm, respectively. We can write five equations and solve any pair of them

$A_1 = 0.326 = \varepsilon bc_{\text{unk}}$

$A_2 = 0.418 = \varepsilon bc_{\text{unk}} + \varepsilon b(0.036 \text{ ppm})$

$A_3 = 0.475 = \varepsilon bc_{\text{unk}} + \varepsilon b(0.060 \text{ ppm})$

$A_4 = 0.545 = \varepsilon bc_{\text{unk}} + \varepsilon b(0.084 \text{ ppm})$

$A_5 = 0.635 = \varepsilon bc_{\text{unk}} + \varepsilon b(0.12 \text{ ppm})$

Dividing the second equation by the first equation yields

$$\frac{0.418}{0.326} = \frac{\varepsilon bc_{\text{unk}} + \varepsilon b(0.036 \text{ ppm})}{\varepsilon bc_{\text{unk}}}$$

$$1.282 = 1.0 + \frac{0.036 \text{ ppm}}{c_{\text{unk}}}$$

$0.282c_{\text{unk}} = 0.036 \text{ ppm}$

$c_{\text{unk}} = 0.128 \text{ ppm } Cu^{2+}$

We can also plot the data in this problem and extrapolate to the x-axis intercept that will be $-c_{\text{unk}}$. On the graph below the concentration is approximately 0.125 ppm (Fig. 5.7).

5.2.6 Photometric error

The basic measurements of absorbance spectroscopy are actually I_0 and I that determine the transmittance. The uncertainty in the measurement of the transmittance can be evaluated as a relative uncertainty in the concentration. The Beer-Lambert Law can be rewritten as

$$c = \frac{-1}{\varepsilon b} \log T \qquad (5.17)$$

converting to natural logarithms results in

$$c = \frac{-0.434}{\varepsilon b} \ln T \qquad (5.18)$$

Take the partial derivative of this equation yields

$$\delta c = \frac{-0.434}{\varepsilon b T} \delta T \qquad (5.19)$$

Divide this equation by the first equation to obtain

$$\frac{\delta c}{c} = \frac{-0.434}{\log T} \frac{\delta T}{T} \qquad (5.20)$$

This expression is interpreted as the relative uncertainty in the concentration, as related to the relative uncertainty of the transmittance measurements, $\delta T / T$. The graph below illustrates the effect of a 1% uncertainty in transmission measurements on the percent relative uncertainty in the concentration (Fig. 5.8).

The minimum uncertainty (ca. 3%) of photometric error ranges from approximately 20 to 60% transmittance or an absorbance range of 0.2–0.7, a 5% relative error in concentration has a photometric range of 0.1–1.0.

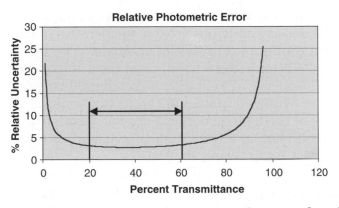

Fig. 5.8. Relative photometric error in concentration as a function of %T. Optimum range of transmittance is shown as 20–60%T for approximately 3% error for a 1% error in T.

5.3 INSTRUMENTATION

All spectrometers have the following basic units: a source of electromagnetic radiation, a dispersion device, sample holder, optical devices for collimating and focusing, a detection device and a data readout or storage system. There are also a variety of ways in which these parts are assembled into the entire spectrometer.

5.3.1 Sources of electromagnetic radiation

High-intensity radiation in the visible region of the spectrum is obtained from a simple tungsten light bulb. This bulb is essentially a black-body emitter and the relative intensity of the wavelengths of light emitted depends on the temperature of the tungsten wire as shown below.

Radiation in the infrared region of the spectrum is obtained from heated ceramic devices such as the Nernst glower or Globar. The Globar is made of silicon carbide and is heated to approximately 800–1500°C to emit black-body radiation in the infrared region of the spectrum. Coils of nichrome wire also emit infrared radiation when electrically heated.

Sources of electromagnetic radiation for the ultraviolet region of the spectrum are high-pressure mercury or xenon lamps or low-pressure deuterium or hydrogen discharge lamps. The mercury and xenon discharge lamps contain a gas that conducts electricity when a high voltage is applied to its electrodes. In the process the gas is excited and emits photons characteristic of the element when returning to the ground state. If the pressure in the tube is low, a characteristic atomic spectrum will be obtained. However, at higher pressures line broadening occurs and a wide distribution of wavelengths will be emitted. The hydrogen and deuterium lamps are low temperature and power lamps that provide a continuous ultraviolet spectrum. The hydrogen or deuterium molecule is excited electrically and then it dissociates to release the energy. The energy of excitation is distributed between the kinetic energies of the hydrogen atom and the photon emitted. Since the kinetic energies of the hydrogen atoms are not quantized, the energy of the photon is also not quantized, resulting in a broad range of energies in the ultraviolet region being emitted.

5.3.2 Optical components

Within the typical spectrometer there is need to collimate electromagnetic radiation into parallel light rays, light needs to be redirected and it also needs to be focused. All of these operations are done using optical devices of lenses, and mirrors. The thin lens equation

$$\frac{1}{\text{source}} + \frac{1}{\text{focal point}} = \frac{1}{\text{focal length}} \tag{5.21}$$

describes the positioning of a light source and a lens so that the angular emission of a light source can be converted into a collimated beam. When a light source is placed at the focal length from the lens, a collimated beam that has its focal point at an extremely large distance will be produced. A collimated light beam will be focused on a point equal to the focal length to reverse the process. Lenses, and prisms, disperse light because the refractive index of the lens is different from the refractive index of air. If the refractive index of the lens was the same for all wavelengths, then all wavelengths of light in an instrument would be perfectly collimated or focused. However, the refractive index often depends on the wavelength of light (that is why a prism works) and the focal point is wavelength-dependent. The change in refractive index with wavelength is called the dispersion, and substances with large dispersions are valued for preparing prisms. The reflection of a mirror does not depend upon the refractive index, particularly front-coated aluminum mirrors. The result is that parabolic mirrors can achieve the same collimating and focusing functions as lenses. They are superior because they minimize the aberrations due to refractive index effects and also do not decrease the light intensity as much as light passing through a lens. Wherever possible, modern instruments replace lenses with parabolic mirrors.

Dispersion devices. Dispersion of light was first achieved using a glass prism. It was discovered that the prism worked because different wavelengths of light had different refractive indices in glass. The result was that each wavelength was "bent" at a different angle when emerging from the prism, producing the separation of white light into the rainbow of colors. This dispersion was not linear and instrument design was very difficult using prisms. Prisms also had the same disadvantage as lenses in that some light was absorbed passing through the prism, decreasing the overall light intensity. Because of their lack of use in modern instruments, further discussion of prisms is omitted.

Reflection gratings greatly decreased the problems formerly associated with prisms. In a reflection grating light is dispersed linearly from one end of the spectral region to the other. Gratings being reflective devices also minimize losses due to absorption of light.

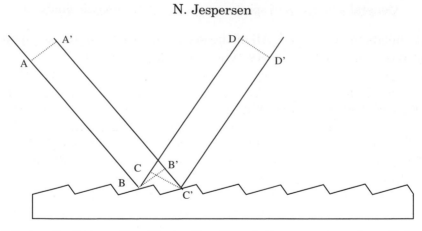

Fig. 5.9. A reflection grating illustrating the relative geometry of two light rays that may result in constructive or destructive reinforcement.

Figure 5.9 shows a reflection grating with rays illustrating just one set of angles for the light source and light output. In this diagram the line segments $\overline{AB} = \overline{A'B'}$ and $\overline{CD} = \overline{C'D'}$ and the line segments $\overline{BC}$ and $\overline{B'C'}$ may or may not be the same length. Geometrically, $\overline{BC} = d \sin \theta_{\text{incident}}$ and $\overline{B'C'} = d \sin \theta_{\text{reflected}}$ the difference between the two segments must be an integer number of wavelengths for constructive reinforcement yielding

$$n\lambda = d(\sin \theta_{\text{incident}} \pm \sin \theta_{\text{reflected}}) \tag{5.22}$$

The incident and reflection angles are measured from the normal (perpendicular) to the grating.

In addition to the important incident and reflected angles for the grating, the blaze angle or the angle of the major reflective surface is important because it helps to concentrate the reflected light in the first order and also determines the usable wavelength range of the grating. The blaze angle, β, is where the angles of incidence and reflection are identical and all wavelengths reinforce and is the angle that the major reflective surface makes with the grating.

$$n\lambda = 2d \sin \beta \tag{5.23}$$

Physically a blaze of white light is observed in the visible region. A grating will typically be useful from approximately one-third of the blaze wavelength to three times the blaze wavelength in the first order.

Gratings also diffract light in the second, third and higher orders. If a grating reinforces light at 600 nm in the first order, it will also reinforce light at 300 nm in the second order and 200 nm in the third order. Gratings are usually paired with simple optical filters to remove unwanted light. In the above case, ordinary borosilicate glass will absorb virtually all of the ultraviolet radiation from the visible 600 nm light.

Modern gratings may be etched on concave surfaces so that they will serve a dual purpose of diffracting light and also focusing the radiation. This decreases the number of parts in a spectrometer and also decreases losses in intensity by having fewer optical parts.

Creating the master grating that commercial replica gratings are duplicated from is a painstaking task with many possibilities for imperfections. Imperfections in the grating may cause interferences such as stray radiation and unfocused images. Advances in laser technology enable precise layouts of reflection gratings by using the interference patterns of intersecting laser beams. The resulting pattern can be used to sensitize a photoresist that can then be dissolved with an organic solvent and then the exposed surface can be etched to produce the grating. Holographic patterns produce extremely high quality master gratings and the replica gratings are of equal high quality. Another advantage of holographic grating production is that the technology is not limited to flat surfaces. Excellent concave gratings can be formed with the advantages mentioned previously.

As with prisms, there are other devices that have been historically used for dispersing or filtering electromagnetic radiation. These include interference filters and absorption filters. Both of these are used for monochromatic instruments or experiments and find little use compared to more versatile instruments. The interested reader is referred to earlier versions of instrumental analysis texts.

5.3.3 Detectors

Detectors for each region of the spectrum differ because of the unique properties of either the radiation itself or the source of the electromagnetic radiation. Light sources produce plentiful amounts of photons in the visible region and the photon energy is sufficient so that a simple phototube or phototransistor will generate enough electron flow to measure. In the ultraviolet region of the spectrum, the available light sources produce relatively few photons when compared to the visible light sources. Therefore, measurement of ultraviolet photons uses a special arrangement of a phototube called a photomultiplier to obtain a measurable electrical current. It is not difficult to generate sufficient photons in the infrared region but the photons produced are of such low energy that devices to rapidly measure infrared radiation have just been recently developed.

N. Jespersen

Phototubes are vacuum tubes with a large anode coated with a photoemissive substance such as cadmium sulfide. A positive voltage of approximately 90 V on the cathode attracts electrons dislodged by photons
from the cadmium sulfide. The current measured by an ammeter is proportional to the number of photons entering the phototube. Figure 5.10
represents a schematic diagram of a phototube measuring circuit.

A phototransistor or photodiode may also be used to detect visible light.
Both devices have p–n junctions. In the photodiode the photon ejects an
electron from the p semiconductor to the n semiconductor. The electron
cannot cross back across the p–n junction and must travel through the
circuitry, an ammeter to return to the p material. In a phototransistor,
usually an npn type, the base (p-type semiconductor) is enlarged and
photosensitive. Photons dislodge electrons that act as if a potential was
applied to the base. This results in an amplified flow of electrons proportional to the number of photons striking the base (Fig. 5.11).

For detection of ultraviolet photons the preferred device is the photomultiplier tube. This device has an electron emissive anode (called a
dynode) that photons strike and eject electrons. However, the electrons
are attracted toward a second electron emissive surface where each
electron generated at the first dynode will eject several electrons toward
a third dynode. Approximate 10–12 dynodes are arranged as shown in
Fig. 5.12. Each dynode is biased so that it is 90 V more positive than the
previous dynode so that the ejected electrons are attracted from one
dynode to the next.

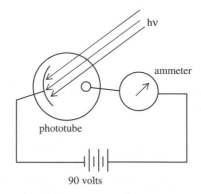

Fig. 5.10. Schematic diagram of a simple phototube circuit. Photons (hv) strike
the CdS-coated anode and electrons are ejected and attracted toward the positive cathode and return through the circuit. The ammeter monitors the flow
of electrons that are proportional to the intensity of the photons.

134

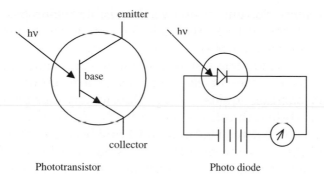

Phototransistor   Photo diode

Fig. 5.11. Schematic diagrams of a phototransistor and a photodiode.

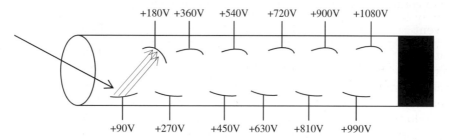

Fig. 5.12. A schematic diagram of a photomultiplier tube.

Infrared radiation has a very low energy and cannot eject electrons from most common photoemissive surfaces. The initial infrared sensors were temperature-sensing devices. Thermocouples and thermistors are forms of bolometers used for detecting infrared radiation.

5.3.4 Transmitting surfaces

Each of the three spectral regions has different requirements for materials that can be used for sample containers, lenses and prisms if used. Visible light is transmitted readily through borosilicate glass and this glass also has good dispersing properties if lenses and prisms are to be used. Disposable plastic cuvettes are also available for spectroscopy in the visible region. There are also a large number of solvents that are useful in the visible region. These include water, liquid alkanes, benzene, toluene, halogenated hydrocarbons, acetonitrile, ethers and esters.

In the ultraviolet region of the spectrum quartz optical materials are required. There are some plastic materials that may also be used in the

135

ultraviolet region. Solvents usable in the visible region also may not be appropriate for the ultraviolet. Table 5.5 lists some common solvents and their cutoff wavelengths.

The infrared region has a strong, obvious absorbance peak for the -OH group. The surfaces of glass and water are eliminated as useful optical materials. For liquids, sodium chloride crystals are used to construct fixed path-length cells as shown in Fig. 5.13. The path length of the cells is usually between 0.1 mM and 1 mM because most samples are pure compounds. Other transmitting materials used in the infrared are listed, with their usable wavelength range in the table below.

TABLE 5.5

Some common solvents and their UV cutoff wavelengths

Solvent	Lower λ limit (nm)
Water	180
Hexane	195
Cyclohexane	200
Ethanol	210
Diethyl ether	215
Chloroform	245
Carbon tertachloride	260
Dimethylsulfoxide	268
Toluene	284
Acetone	330
Methyl isobutyl ketone	334

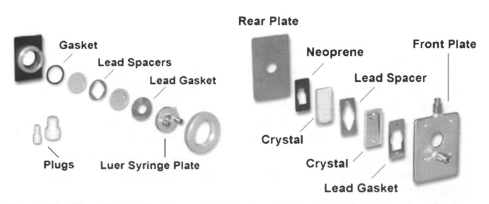

Fig. 5.13. Diagrams of commercial fixed path length infrared cells. Lead, or Teflon® spacers provide the cell thickness and also seal the cell from leakage.

5.3.5   Sample handling

Typically, the ultraviolet and visible regions are used for quantitative analysis. By far, the majority of the samples are liquids and are dilute solutions that will achieve the lowest photometric error (between 0.2 and 0.8 absorbance units). A knowledge of the molar absorptivity allows calculation of the usable concentration range for analysis, and samples can be diluted appropriately. Cuvettes having optical path lengths of 1.0, 10.0 and even 100.0 mm are available. Gas samples can also be examined in the ultraviolet and visible regions. Best results are obtained if a reagent blank containing all of the solution components except the analyte is used.

Infrared analysis is usually used as a qualitative method to identify substances. Liquids are usually analyzed as pure substances in cells with very small optical path lengths of 0.1–1.0 mm. Usable spectra can be obtained by placing a drop of relatively non-volatile sample between two sodium chloride plates, allowing them to be held together by capillary action.

It is often necessary to determine the optical path length of salt cells since they are subject to wear and erosion from moisture. To determine the optical path length, b, a spectrum is obtained on the empty cell. Reflections from the internal walls of the cell create an interference pattern that looks like a series of waves in the spectrum. Using as many well-formed waves as possible, the start and ending frequencies (in cm^{-1}) are determined along with the total number of waves. The optical path length is then calculated from the following relationship:

$$b = \frac{\text{number of waves}}{2(\text{wavenumber}_2 - \text{wavenumber}_1)}$$

where the wavenumbers have units of cm^{-1}.

Recently, polyethylene and Teflon® mesh sample holders have been used. A drop of sample is placed on the mesh and spread to a relatively uniform thickness for analysis. These holders can often be rinsed and reused. A very convenient alternative to liquid sample holders is the technique called attenuated total reflection or ATR. The ATR cell is a crystal of gallium arsenide, GaAs; and the infrared radiation enters one end of the trapezoidal crystal. With the angles adjusted to obtain total internal reflection, all of the IR radiation passes through the crystal and exits the other end as shown in Fig. 5.14.

However, as the IR waves strike the surfaces to be reflected, part of the wave emerges from the crystal. This can be absorbed by a sample on the other side of the crystal. The GaAs crystal is unaffected by water

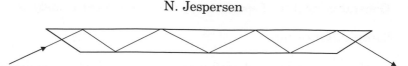

Fig. 5.14. Schematic diagram of an ATR gallium arsenide crystal and the total internal reflection of a light ray. The sample is placed on top of the crystal and interacts with the evanescent wave producing the spectrum.

and aqueous samples can be studied. Solutions can be analyzed providing the solvent does not absorb in the infrared region of interest or if a reference cell can be used to cancel the absorbance of the solvent.

Solid samples can be analyzed in the IR by preparing a solution in a suitable solvent or by preparing a KBr pellet containing the sample. A KBr pellet is prepared by mixing approximately 0.5% sample with very pure and dry KBr (e.g., 1 mg sample and 200 mg KBr). The sample and KBr are ground together to a very fine powder and transferred to a high-pressure press. At approximately 2000 psi the mixture fuses into a solid pellet that can be mounted and scanned in the spectrometer. Presence of water will cloud the pellet and very dry KBr is required, and some presses have the ability to remove water by vacuum while the pellet is being fused. Gaseous samples are readily, and often, analyzed by infrared spectroscopy. Gas cells with optical path lengths of 10 cm fit most IR spectrometers. Additional path length may be had by arranging mirrors in a gas cell to allow the radiation to pass through the cell several times before exiting.

5.4 PUTTING THE PARTS TOGETHER

5.4.1 Spectrometers for the visible region

Simple spectrometers that cover the region from 350 to 1000 nm are available for modest cost and are useful for routine analysis. These spectrometers are usually single beam instruments that are set up according to the block diagram in Fig. 5.15, and Fig. 5.16 illustrates the actual configuration of a commercial instrument.

A single beam instrument requires that a reagent blank be used to determine I_0 (set 0.0 absorbance or 100% T) at each wavelength before measuring I for the sample (that electronically is transmitted to the meter as absorbance or %T). Inserting the reagent blank, zeroing the instrument and then inserting the sample and manually reading the absorbance is time consuming. One solution, made possible by digital electronics, is to measure I_0 for the reagent blank at all desired wavelengths at one time and store the data in memory and then insert the sample and measure I at

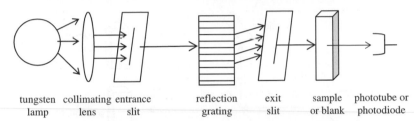

| tungsten | collimating | entrance | | reflection | exit | sample | phototube or |
| lamp | lens | slit | | grating | slit | or blank | photodiode |

Fig. 5.15. Block diagram of a single beam visible spectrometer.

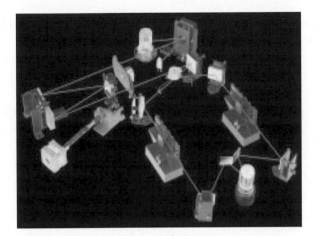

Fig. 5.16. A schematic diagram of a single beam instrument, the Genesis 2000 produced by Thermo Spectra. Layout of parts for a UV–Vis spectrometer (with permission of Thermo inc.).

the same wavelengths. The spectrum is then calculated and displayed on a computer screen. This approach requires very stable electronics to assure that there are minimal system changes between measurement of I_0 and I. The older Beckman DU-7 and many of the FT-IR instruments operate in this manner. While separate recording of the reagent blank and sample intensities is one solution, double beam instruments surmount both problems mentioned above.

A double beam instrument splits the electromagnetic radiation into two separate beams, one for the reagent blank, and the other for the sample. There are two ways to do this. The first method uses a mirror that is half silvered and half transparent. As shown in Fig. 5.17 this results in a continuous beam of light for both the sample and reagent blank.

After passing through the sample and reagent blank, the two beams can be monitored at separate detectors and then combined electronically to obtain the ratio of I/I_0.

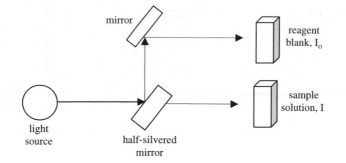

Fig. 5.17. Essentials of a double beam instrument in space. This is characterized by two continuous beams of light.

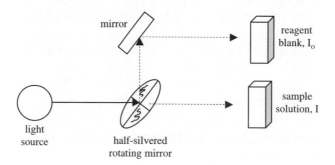

Fig. 5.18. Schematic diagram of a double-beam instrument in space.

The second type of double-beam instrument is one where the light source is divided into two beams by a rotating sector mirror that alternately reflects and transmits the light. This results in a chopped beam of light that alternately passes through the reagent blank and the sample as shown in Fig. 5.18.

The double-beam in-space spectrometer has alternating "segments" of light impinging on the sample and the reagent blank. These beams can be recombined and focused on a single detector. The result will be a square-wave type of signal as shown in Fig. 5.19.

The square wave produced by the double-beam in space spectrometer is preferred since there is only one detector and the signal is a square-wave that is essentially an alternating current. Alternating currents are much easier to manipulate electronically. In particular they can be easily amplified and noise that is either direct current noise or high-frequency noise can be filtered from the signal.

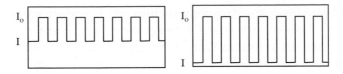

Fig. 5.19. Representations of the output of the detector for a double-beam in-space spectrometer. The first panel shows a system that has a relatively large I/I_0 ratio while panel 2 has a smaller I/I_0 ratio and a larger absorbance.

5.4.2 Rapid spectroscopy

Spectroscopic instruments are used for quantitative and qualitative analyses in stand-alone situations for the most part. However, there are situations where spectroscopic measurements are used as detectors for other instruments, in particular high-performance liquid chromatography (HPLC) discussed in Chapter 15. In addition it is recognized that the signal-to-noise ratio can be increased by repeatedly adding spectra so that the signal increases with each measurement, N, and the noise only increases as the square root of N, $N^{1/2}$. Two types of instrumentation were developed to meet these needs. First is the photodiode array (PDA) spectrometer for the ultraviolet and visible regions of the spectrum and the second is the Fourier transform infrared spectrometer, FT-IR, for the infrared region.

Diode array spectrometers are also known as PDA spectrometers designed to measure the desired band of wavelengths at the same time. This is achieved by placing a linear array of photodiodes in the path of the dispersed beam of UV–Vis radiation. The radiation has passed through the sample, often an HPLC flow cell, prior to dispersion. The grating disperses the radiation so that it is linearly, in terms of wavelength, dispersed at the focal plane. If each of the diodes is of a certain width, then each diode will intercept a given band of radiation. The size of the band of radiation observed is related to the resolution of the instrument. Considering the range from 200 to 700 nm, it would take 500 photodiodes to achieve one nanometer resolution. The speed of a diode array instrument depends on the speed at which a computer can access, sample, measure and discharge the voltage developed on each diode. The number of diodes sampled also has an effect on the rate at which spectra can be obtained. Current photodiode instruments can obtain spectra with resolutions of 1–3 nm (256–1024 diodes/ 200–1100 nm range) at a rate of up to 2500 spectra sec^{-1}. This makes the PDA ideal as a versatile detector for HPLC applications.

Fourier transform spectroscopy technology is widely used in infrared spectroscopy. A spectrum that formerly required 15 min to obtain on a continuous wave instrument can be obtained in a few seconds on an FT-IR. This greatly increases research and analytical productivity. In addition to increased productivity, the FT-IR instrument can use a concept called Fleggetts Advantage where the entire spectrum is determined in the same time it takes a continuous wave (CW) device to measure a small fraction of the spectrum. Therefore many spectra can be obtained in the same time as one CW spectrum. If these spectra are summed, the signal-to-noise ratio, S/N can be greatly increased. Finally, because of the inherent computer-based nature of the FT-IR system, databases of infrared spectra are easily searched for matching or similar compounds.

The core of the FT-IR is the Michaelson interferometer (Fig. 5.20) and the mathematical relationship between the frequency and time domains of a spectrum that is called the Fourier transform. The Michaelson interferometer is diagrammed below. A beam of infrared radiation from a source as described previously is collimated and directed through the sample to a 50% transmitting mirror, beamsplitter. The split beams reflect off two plane mirrors directly back to the beamsplitter where they recombine. One of the plane mirrors is fixed but the other moves so that the paths that the two light beams travel are not equal. The difference in the distance from the 50% mirror to the moving and fixed mirrors is called the retardation, δ. The difference in distance that one light beam travels compared to another is therefore 2δ. As with a reflection grating, if $2\delta = n\lambda$ then that set of wavelengths (this includes the $n = 2$, 3, ... overtones) will be reinforced while all others will be attenuated to some extent. Those wavelengths where 2δ is $(n+0.5)\lambda$ will be 180° out of phase and completely attenuated.

The Fourier transform allows the mathematical conversion between the time domain and the frequency domain of a spectrum. The names for these domains refer to the x-axis of their conventional graphical representations. Figure 5.21 illustrates how these are important in terms of conventional spectra. Importantly, the time and the frequency domains contain exactly the same information.

For the two waves represented in Fig. 5.21 the power at any time t is

$$P(t) = k_A \cos(2\pi v_A t) + k_B \cos(2\pi v_B t) \tag{5.24}$$

or the algebraic sum of A and B as shown in C of Fig. 5.21.

Frequencies in the infrared region are between 10^{13} and 10^{14} Hz, they are approximately three orders of magnitude greater in the ultraviolet

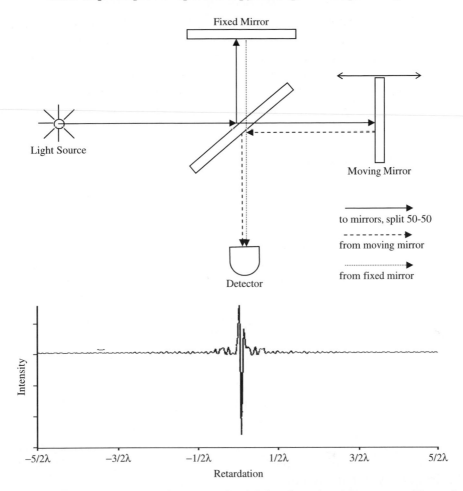

Fig. 5.20. (Top) Schematic diagram of a Michaelson interferometer. Retardation determines difference in optical path between fixed mirror and moving mirror. When retardation, δ, is $\lambda/2$ light with a wavelength equal to λ will be reinforced. (Bottom) Interference pattern from the Michaelson interferometer. Major peak where $\delta = 0$ is where all wavelengths are reinforced.

and visible regions. There are no detectors available that are fast enough to measure waves at this frequency. Signals that vary in the range of 100–10,000 Hz can be accurately measured with modern electronics. The Michaelson interferometer not only produces an interferogram, but the interferogram has a fundamental frequency that can approximately be a factor of 10^{10} lower because of a process called modulation. Modulation is the process of changing a high frequency to a lower one or a low frequency to a higher one, which was developed along with the broadcast

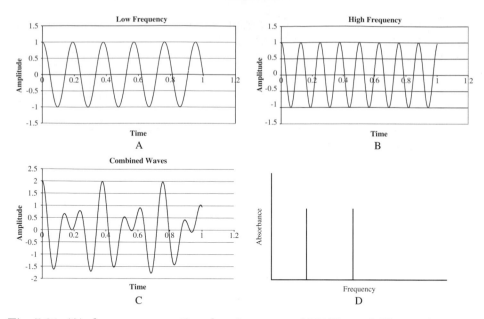

Fig. 5.21. (A) shows a conventional cosine wave of 500 Hz and (B) a cosine wave of 333 Hz, (C) the algebraic sum of (A) and (B), (D) represents the cosine wave in (A) in the frequency domain, (D) the representation of (C) in the frequency domain.

industry to propagate voice and video information at television and radio frequencies. In FT-IR the signal is modulated by the interferometer to a measurable frequency. In order to avoid aliasing, it is necessary to sample an alternating current signal at frequency that is greater than twice the frequency of the signal.

To understand the process, consider the instrument itself. The mirror moves at a constant velocity v_M and the time it takes for the mirror to move $\lambda/2$ is τ then $v_M\tau = \lambda/2$ of mirror movement or λ in total difference traveled by the two light beams. Therefore, τ is the time for one wavelength or $1/\tau$ the frequency of the *light striking the detector*. From this

$$f(\text{s}^{-1}) = \frac{1}{\tau(\text{s})} = \frac{v_M(\text{cm s}^{-1})}{\lambda(\text{cm})/2} = 2v_M(\text{cm s}^{-1})\bar{v}(\text{cm}^{-1}) \qquad (5.25)$$

If the velocity of the mirror is $0.1\,\text{cm s}^{-1}$ and the wavelength is $8.0\,\mu\text{m}$ (the center of the IR region) the frequency of the modulated IR radiation is

$$\text{Frequency} = \frac{2(0.1\ \text{cm s}^{-1})}{(8.0 \times 10^{-4}\ \text{cm})} = 250\ \text{s}^{-1} \qquad (5.26)$$

A similar calculation in the visible region at 400 nm results in a modulated frequency of $125,000\,\mathrm{s}^{-1}$ that is difficult to measure.

Resolution is the ability to distinguish two closely spaced spectral peaks. If the infrared resolution is

$$\Delta \bar{v} = \bar{v}_2 - \bar{v}_1 \tag{5.27}$$

Equation 5.24 will be at a maximum, at zero retardation and reach its maximum again when the waves are in phase when $v_A t - v_B t$ is unity or

$$1 = \delta \bar{v}_2 - \delta \bar{v}_1 \tag{5.28}$$

From this

$$\Delta \bar{v} = \bar{v}_2 - \bar{v}_1 = \frac{1}{\delta} \tag{5.29}$$

so that the resolution is approximately equal to the reciprocal of the total retardation. The retardation, δ, is equal to twice the total mirror movement. If the mirror moves a total of 1.0 cm the $\delta = 2.0$ cm and the resolution will be $0.5\,\mathrm{cm}^{-1}$.

5.5   STATISTICS FOR SPECTROSCOPY

5.5.1   Signal averaging

As known from statistics the standard deviation is

$$s = \sqrt{\frac{\sum (x_i - \bar{x})^2}{N - 1}} \tag{5.30}$$

And the uncertainty or confidence limit will be

$$u - \bar{x} = \frac{ts}{\sqrt{N}} \tag{5.31}$$

where u is the true mean and is the measured mean. As the number of measurements, N, increases, the uncertainty about the value of the mean will decrease with the square root of N. Similarly, the signal-to-noise ratio will increase as the number of measurements increases because the signal is additive with the number of measurements and the noise increases as the square root of N.

$$\frac{\text{Signal}}{\text{Noise}} = \left(\frac{\text{Signal}}{\text{Noise}}\right)\left(\frac{N}{\sqrt{N}}\right) \tag{5.32}$$

145

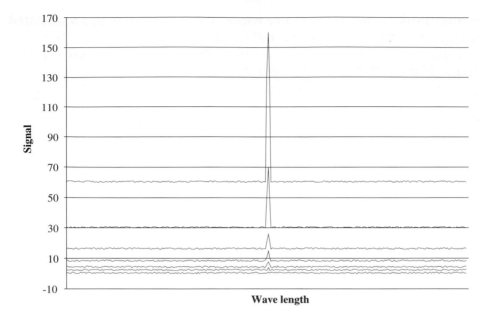

Fig. 5.22. Illustration of the increase in signal to noise ratio when repetitive scans are added. Bottom trace represents a S/N ratio of 1.0. Successive traces represent summation of 4, 16, 50, 100, 1600 and 10,000 repetitive scans.

For a spectrum the same principles occur. Repetitive spectra that are added to each other show the signal emerging from the noise as shown in Fig. 5.22.

5.5.2   Statistics for a calibration curve

Calibration curves are plots of concentration (x_i) versus some response of the instrument (y_i). Concentration values are assumed to be the most precise and all of the error is associated with the response measurement. With those definitions, we write the equation for the least-squares line as $y = mx + b$ and, omitting the derivation, find that

$$m = \frac{n \sum (x_i y_i) - \sum x_i \sum y_i}{n \sum (x_i^2) - \left(\sum x_i\right)^2} \tag{5.33}$$

$$b = \frac{n \sum (x_i^2) \sum y_i - \sum x_i y_i \sum x_i}{n \sum (x_i^2) - \left(\sum x_i\right)^2} \tag{5.34}$$

Solving the two equations above is easiest using a spreadsheet starting with one column of x_i values and another with y_i values. Another column

146

can be constructed to obtain x_iy_i values and two more columns to obtain the squares of the x_i and y_i values. Finally, each of these columns can be summed. The appropriate values can be used to determine the slope m and the intercept b. Once m and b have been calculated another column can be added to the spreadsheet to determine the vertical deviation, d_i, of each y_i value from the least-squares line and the square of the deviation d_i^2

$$d_i = (y_i - y) = y_i - (mx_i + b) \tag{5.35}$$

The standard deviation of the y values is calculated as

$$s_y = \sqrt{\frac{\sum d_i^2}{n - 2}} \tag{5.36}$$

From this the standard deviations of the slope, s_m, and intercept, s_b, are

$$s_m = \sqrt{\frac{s_y^2 n}{n \sum (x_i^2) - \left(\sum x_i \right)^2}} \tag{5.37}$$

$$s_b = \sqrt{\frac{s_y^2 \sum x_i^2}{n \sum (x_i^2) - \left(\sum x_i \right)^2}} \tag{5.38}$$

Preparation of a calibration curve has been described. From the fit of the least-squares line we can estimate the uncertainty of the results. Using similar equations we can determine the standard deviation of the calibration line (similar to the standard deviation of a group of replicate analyses) as

$$s_{line} = \sqrt{\frac{n \sum (y_i^2) - \left(\sum y_i \right)^2 - m^2 \left(n \sum (x_i^2) - \left(\sum x_i \right)^2 \right)}{n(n - 2)}}$$

$$s_{sample} = \frac{s_{line}}{m} \sqrt{\left(\frac{1}{M} + \frac{1}{n} + \frac{n \left(\bar{y}_{sample} - \bar{y}_{cal} \right)^2}{m^2 \left(n \sum (x_i^2) - \left(\sum x_i \right)^2 \right)} \right)}$$

The standard deviation for an analytical result will be

$$S_{sample} = \frac{S_{line}}{m} \sqrt{\left(\frac{1}{M} + \frac{1}{n} + \frac{n\left(\bar{y}_{sample} - \bar{y}_{cal}\right)^2}{m^2\left(n\sum(x_i^2) - \left(\sum x_i\right)^2\right)} \right)}$$

where M replicates analysis of the sample that have a mean of $\bar{y}_{sample}$ and $\bar{y}_{cal}$ is the mean of the n samples used to construct the calibration curve.

5.5.3   Signal-to-noise ratio

The signal-to-noise ratio is an important parameter that allows us to evaluate the quality of an instrument and spectra. The determination of the signal-to-noise ratio requires a measurement of some quantity of the signal and a measurement of the noise. Measurement of the signal is generally a straightforward difference between a reagent blank and the sample or spectral baseline and peak amplitude. Noise is more difficult. Many instruments appear to have little or no noise. Experiments must be made to expand the scale sufficiently to observe the random fluctuations called noise. Once noise can be observed, its best measure is the root mean square of the noise (rms noise). The term rms means that we have taken the square root of the mean of the squares of a representative number of measurements of the noise. This is not an easy task. Assuming the noise is a sine wave we can estimate

$$N_{rms} = 0.707\left(\frac{N_{p-p}}{2}\right) \tag{5.39}$$

We can measure the peak-to-peak noise, N_{p-p}, rather easily and calculate the equivalent in rms noise.

5.5.4   Limit of detection

Detection limit is the minimum amount of signal that can be observed with some certainty that there is a signal at all. This requires that the signal be at least three times the rms noise of the experiment.

5.5.5   Limit of quantitation

It is recognized that the detection limit is at the extreme of the instrument's capabilities. As such it is very difficult to quantify a signal that is on the verge of non-existence. For quantitative measurements, most analysts take a value that is 10–20 times as large as the limit of detection as the lower limit for quantitation. If an instrument can detect 15 ppb of a herbicide, the lowest level, it could be used to quantitate that same herbicide is approximately 0.15 ppm.

5.5.6   Sensitivity

The sensitivity of a method is a measure of its ability to distinguish one concentration from another. If a particular instrument could be used to determine the concentration of a heavy metal such as lead and could reliably distinguish a 25 ppb solution from a 30 ppb solution, it would be more sensitive than an instrument that could barely tell the difference between a 25 ppb solution and a 50 ppb solution. The best quantitative measure of the sensitivity of an instrument and/or an analytical method is to determine the slope of the calibration curve. The greater the slope, the more sensitive the instrument and/or method.

REVIEW QUESTIONS

1. Determine the energy in $kJ\,mol^{-1}$ for electromagnetic radiation with the following wavelengths:
   a.   225 nm
   b.   3650 nm
   c.   450 mm
   d.   750 nm
   e.   15 µm

2. Determine the energy in $kJ\,mol^{-1}$ for electromagnetic radiation with the following wavelengths:
   a.   180 nm
   b.   6.0 µm
   c.   12.0 µm
   d.   645 nm
   e.   300 nm

3. Compare the results in problem(s) 1 and/or 2 to the average bond energies for C–C, C=C, C–H and C–Cl bonds. Which can be considered ionizing radiation?

4. Based on minimizing the photometric error, what range of absorbances is optimal for absorbance spectroscopy? What is the relative dynamic range of absorbance measurements?

5. Optimal results will be obtained for what range of concentrations if the molar absorptivity ($l\,mol^{-1}\,cm$) of the analyte is (assume a 1.0 cm optical path length)
   a. 150
   b. 1.2×10^5
   c. 655
   d. 1025
   e. 25

6. Optimal results will be obtained for what range of concentrations if the molar absorptivity ($l\,mol^{-1}\,cm$) of the analyte is: (assume a 1.0 mm optical path length)
   a. 6.2×10^4
   b. 15
   c. 575
   d. 2500
   e. 125

7. The figure below represents the noise of a spectrometer detector. Estimate the peak-to-peak noise, and the rms noise of this detector. If an analyte produces a signal of 6.3 pA, will it be above or below the limit of detection?

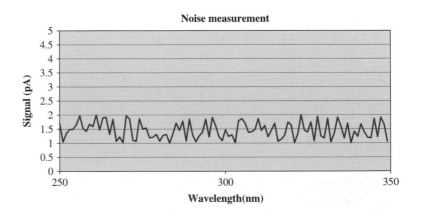

8. The figure below illustrates the visible absorbance spectrum of substance A, what is the appropriate analytical wavelength for determining the concentration of A?

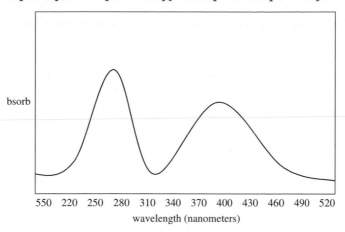

9. The figure below illustrates the visible absorbance spectrum of substance B, what is the appropriate analytical wavelength for determining the concentration of B?

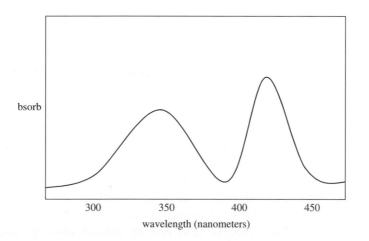

10. A mixture of two substances A and B, shows absorbance peaks at 465 and 720 nm, respectively. The slope of a calibration plot for substance A is 18,350 at 465 nm and 884 at 720 nm. For substance B the slope at 465 nm is 1024 and at 720 nm it is 12,240. What is the concentration of A and B in a sample that has an absorbance of 0.566 at 465 nm and an absorbance of 0.728 at 720 nm?

11. A mixture of two substances X and Y, show absorbance peaks at 365 and 620 nm, respectively. The slope of a calibration plot for substance X is 14,350 at 365 nm and 804 at 620 nm. For substance

Y the slope at 365 nm is 1154 and at 620 nm it is 17,240. What is the concentration of X and Y in a sample that has an absorbance of 0.566 at 365 nm and an absorbance of 0.628 at 720 nm?

12. A 5.00 ml sample containing iron is mixed with hydroxylamine hydrochloride to reduce the iron (III) to iron (II). The solution is then mixed with an excess of phenanthroline and the absorbance is measured and found to be 0.448. A second 5.00 ml solution of the same unknown is mixed with 1.00 ml of 2.0×10^{-4} M Fe^{2+} and is then treated the same way as the original sample. The absorbance is found to be 0.525. What is the concentration of the iron in the sample?

13. A 10.0 ml solution containing proteins is reacted with biuret solution and the absorbance is measured as 0.356. Another 10.0 ml sample of the same protein solution is mixed with 5.0 ml of a protein solution known to contain $1.6 \, \mu g \, ml^{-1}$ of protein. The mixture is reacted in the same way as the original unknown and the absorbance is found to be 0.562. What is the concentration of the protein in the sample?

14. Use a spreadsheet such as EXCEL to obtain a graph of absorbance versus concentration for the following data.

Sample	Absorbance	Concentration $(mol \, l^{-1})$
blank	0.00	0.00
1	0.238	0.000200
2	0.455	0.000400
3	0.665	0.000600
4	0.878	0.000800

Also determine the slope and intercept of the least-squares line for this set of data. Determine the concentration and standard deviation of an analyte that has an absorbance of 0.335.

15. Use a spreadsheet such as EXCEL to obtain a graph of absorbance versus concentration for the following data.

Sample	Absorbance	Concentration $(mol \, l^{-1})$
blank	0.00	0.00
1	0.165	0.000200
2	0.321	0.000400
3	0.505	0.000600
4	0.687	0.000800

Also determine the slope and intercept of the least-squares line for this set of data. Determine the concentration and standard deviation of an analyte that has an absorbance of 0.335.

16. What wavelength will be observed at a detector that is placed at 23.5° when the light source strikes the reflection grating at an angle of 45.3° and the grating has 1250 lines per mm. Assume that the order of diffraction is first order. What is the second order wavelength observed?

17. What wavelength will be observed at a detector that is placed at 43.5° when the light source strikes the reflection grating at an angle of 38.3° and the grating has 3250 lines per mm. Assume that the order of diffraction is first order. What is the third order wavelength observed?

18. What is the useful wavelength range for a grating with 10,000 lines per mm and a blaze angle of 41.6°? What spectral region is this?

19. What is the useful wavelength range of a grating with 1560 lines per mM that has 2500 lines per mm. What spectral region is this?

20. How many lines per mm will be needed for a grating that will be able to resolve spectral lines that are 10 nm apart?

21. How many lines per mm are needed to have a resolution of $2.0 \, \text{cm}^{-1}$ in the infrared?

22. What is the distance that a mirror in a Michaelson interferometer must move to have a resolution of $1.0 \, \text{cm}^{-1}$ in the infrared?

23. What distance must a mirror move in order to have a resolution of 1.0 nm in the visible region of the spectrum?

24. Fill in the blank spaces in the following table, where needed units are given.

	A	ε	b	c	T
1	0.450		1.00 cm	$3.2 \times 10^{-4} \, \text{M}$	
2	0.223	2840	dm	$7.8 \times 10^{-5} \, \text{M}$	
3		12,100	10.0 cm	$4.2 \times 10^{-6} \, \text{M}$	
4		546	0.25 cm		34.2%
5	0.665		1.00 mm	$7.6 \times 10^{-5} \, \text{M}$	

25. Fill in the blank spaces in the following table, where needed units are given.

	A	a	b	c	T
1	0.650		1.00 cm	7.7×10^{-4} ppm	
2		$2{,}7361\,\mathrm{g}^{-1}$ cm	1.00 cm	ppb	22.3%
3	0.239		1.0 mM	0.00634 wt%	
4		$471\,\mathrm{g}^{-1}$ cm	10.0 cm	ppm	0.652
5	0.849	$7{,}845\mathrm{lg}^{-1}$ cm	mm	2.35 ppb	

26. Six 25 ml volumetric flasks are filled with 10 ml of the analyte and then 1, 2, 3, 4, 5 and 6 ml of a standard solution containing $6.5 \times 10^{-3}\,\mathrm{mol\,l^{-1}}$ of the same analyte. 5.00 ml of color-developing reagent is added to each flask and enough distilled water is added to bring each flask to exactly 25.0 ml. The absorbances of the five solutions were 0.236, 0.339, 0.425, 0.548, 0.630 and 0.745, respectively. Use a spreadsheet to obtain a graph of the data and extrapolate the data to obtain the information needed to determine the initial concentration of the analyte. From the data, estimate the uncertainty of the result.

27. It is observed that the infrared spectrum obtained with a continuous wave infrared spectrometer has increasing resolution as the scan speed is decreased. Explain this observation.

28. Explain how changing the solvent polarity can be used in certain circumstances to determine the nature of the transition causing an observed absorbance.

29. Explain what types of quantized absorbances are expected in the ultraviolet, visible and infrared spectral regions.

30. Give plausible reasons why Fourier transform techniques are used for the infrared region but not the visible and ultraviolet spectral regions.

31. If the C=O stretch is found at $1856\,\mathrm{cm}^{-1}$ what wavenumber would we expect the same stretch to occur at if the oxygen atom was the $^{18}\mathrm{O}$ isotope?

32. The nitrile stretch frequency is $2354\,\mathrm{cm}^{-1}$. What is the wavenumber of the same stretch if the nitrogen isotope has a mass of 16 rather than 14?

33. The optical path length of an infrared cell can be determined using the method shown in the text. Determine the optical path lengths for the following sodium chloride cells.

   a. Fifteen interference fringes are observed between 11 and 6 μm.

   b. Twenty-two interference fringes are observed between $2500\,\mathrm{cm}^{-1}$ and $1000\,\mathrm{cm}^{-1}$.

     c.   Twenty-six interference fringes are observed between 14.6 and $8.2\,\mu m$.

     d.   Sixteen interference fringes are observed between 2085 and $855\,cm^{-1}$.

34.   Provide plausible reasons why interference fringes are not observed for the typical 1.0 cm quartz cuvette used in the ultraviolet region.

Chapter 6

Near-infrared spectroscopy

Emil W. Ciurczak

6.1   INTRODUCTION

The near-infrared portion of the electromagnetic spectrum is located between the visible and mid-range infrared (MIR) sections, roughly 750–2500 nm or 13,333–4000 cm^{-1} (see Fig. 6.1). It consists (mainly) of overtones and combinations of the bands found in the mid-range infrared region (4000–200 cm^{-1}). The region was discovered by Sir William Herschel in 1800 [1]. Sir William was attempting to discover the color of light that carried the heat of sunlight. He used a glass prism to split the colors from white light and arranged a series of thermometers, wrapped in dark cloth, such that they would each be exposed to a different colored light.

Not much happened in the visible region, but as he allowed the thermometer to be located next to the red band, he noticed a dramatic

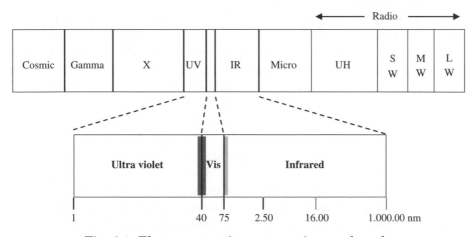

Fig. 6.1. Electromagnetic spectrum in wavelength.

Comprehensive Analytical Chemistry 47
S. Ahuja and N. Jespersen (Eds)
Volume 47 ISSN: 0166-526X DOI: 10.1016/S0166-526X(06)47006-9

increase in temperature. He (correctly) assumed there was a type of light, which, while invisible to the naked eye, was capable of carrying heat. Since it was "beyond the red" portion, he named it "infrared." The truth be told, mid-range infrared radiation does not penetrate glass (re: your car in summertime), so what he discovered was the *near*-infrared (NIR). The significance of the discovery was that it was the first evidence of "light" (now called electromagnetic radiation) outside the visible region.

In 1881, Abney and Festing [2], using newly developed photographic plates, recorded the spectra of organic liquids in the $1-2\mu$ range. Inspired by this work, W. W. Coblentz built a rock salt spectrometer with a sensitive thermopile connected to a mirror galvanometer [3]. While it took a day to produce a single spectrum, he managed to produce several hundred spectra of organic compounds, publishing his results in a series of papers in 1905. The regions of the spectrum related to groups, such as –OH, became apparent, although, he discovered that no two compounds had the same spectrum.

While good, commercial instruments were not generally available, research was being performed in the near-infrared (NIR). One of the first quantitative measurements was at Mount Wilson observatory in 1912; F. E. Fowler measured the moisture in the atmosphere [4]. Later, in 1938, Ellis and Bath [5] measured the amount of water in gelatin. During the early 1940s, Barchewitz [6] performed analyses of fuels, and Barr and Harp [7] published the spectra of vegetable oils. Later in the 1940s, Harry Willis of ICI characterized polymers and used NIR to measure the thickness of polymer films. WW II emphasized the use of mid-range IR for synthetic rubber, pushing the instrument manufacturers to commercialize IR spectrometers.

In general, NIR papers did not begin in earnest until the 1970s, when commercial instruments became easily available because of the work of the US Department of Agriculture (USDA)[8–12]. Some of these developments will be discussed in Section 6.3. After the success of the USDA, food producers, chemical producers, polymer manufacturers, gasoline producers, etc. picked up the ball and ran with it. The last to become involved, mainly for regulatory reasons, are the pharmaceutical and biochemical industries.

In number of labs, the NIR is a rapid, non-destructive test. It is used (everywhere) for water determination. For the petroleum industry, it is routinely used for octane and betaine values, to determine levels of additives, and as a test for unsaturation. The polymer companies, in addition to identification, use NIR for molecular weight, cross-linking, iodine value (unsaturation) block copolymer ratios, and numerous physical attributes. Thus, it is logical that textiles are also analyzed using NIR: sizing, coatings, heat-treatments, dyes, and blend levels (cotton, Dacron, polyester, etc.) are all measured.

In agriculture and food, NIR has been a powerful tool for decades. All shipments of grain leaving US ports are analyzed for moisture, fat, protein, and starch via NIR. Processed foods are also a prime venue for NIR: percent of fat in cheese spread, hardness of wheat, and freshness of meats are just some of the applications in food.

In pharmaceuticals, NIR is used for, of course, moisture, polymorphic (drug) forms, percent crystallinity, isomer purity, tablet/capsule assay, coating levels, evaluation of dissolution times, and numerous process tests. It is a rapid means for the Food and Drug Administration to check for counterfeit drugs, and for the Drug Enforcement Agency to ascertain what type of materials are impounded in "drug raids."

6.2 BASIC THEORY

In short, near-infrared spectra arise from the same source as mid-range (or "normal") infrared spectroscopy: vibrations, stretches, and rotations of atoms about a chemical bond. In a classical model of the vibrations between two atoms, Hooke's Law was used to provide a basis for the math. This equation gave the lowest or base energies that arise from a harmonic (diatomic) oscillator, namely:

$$v = 1/2\pi(k/\mu)^{1/2} \tag{6.1}$$

where v is the vibrational frequency; k the classical force constant and μ reduced mass of the two atoms.

This gives a reasonable approximation of the fundamental vibrational frequency of a simple diatomic molecule. Indeed, it is quite close to the average value of a two-atom stretching frequency within a polyatomic molecule. Since NIR is based upon the hydrogen-X bands within a molecule, this simplified equation would lead to reduced masses for CH, OH, and NH or 0.85, 0.89, and 0.87, respectively. It would seem, based upon these values, that there would be no differentiation among moieties. However, with actual electron donating and withdrawing properties of adjacent atoms, hydrogen bonding, and van der Waal's forces actually changing these values, spectra certainly do exist.

Since we recognize that the allowed energy levels for molecular vibrations are not a continuum, but have distinct values, the manner in which we calculate them is slightly more complex.

The energy levels are described by quantum theory and may be found by solving the time-independent Schroedinger equation by using the vibrational Hamiltonian for a diatomic molecule [13,14].

$$\frac{-h^2\delta^2\psi(X)}{2m\delta(X)} + V(X)\psi(X) = E\psi(X) \tag{6.2}$$

Solving this equation gives complicated values for the ground and excited states, as well. Using a simplified version of the equation, more "usable" levels may be discerned (here, the "echoes" of Hooke's Law are seen)

$$E_v = (v + 1/2)h/2\pi(k/\mu)^{1/2} \qquad (v = 0, 1, 2 \ldots) \tag{6.3}$$

Rewriting this equation, substituting the quantum term hv, the equation becomes

$$E_v = (v + 1/2)hv \qquad (v = 0, 1, 2 \ldots) \tag{6.4}$$

Polyatomic molecules, with the many layers of vibrational levels, can be treated, to a first approximation, as a series of diatomic, independent, and harmonic oscillators. This general equation may be expressed as

$$E(v_1, v_2, v_3, \ldots) = \sum_{i=1}^{3N-6} (v_i + 1/2)hv \quad (v_1, v_2, v_3, \ldots = 0, 1, 3 \ldots) \tag{6.5}$$

In a case where the transition of an energy state is from 0 to 1 in any one of the vibrational states ($v_1, v_2, v_3, \ldots$), the transition is considered as *fundamental* and is allowed by selection rules. When a transition is from the ground state to $v_i = 2, 3, \ldots$, and all others are zero, it is known as an *overtone*. Transitions from the ground state to a state for which $v_i = 1$ and $v_j = 1$ simultaneously are known as *combination bands*. Other combinations, such as $v_i = 1$, $v_j = 1$, $v_k = 1$, or $v_i = 2$, $v_j = 1$, etc., are also possible. In the strictest form, overtones and combinations are not allowed, however they do appear (weaker than fundamentals) due to anharmonicity or Fermi resonance.

In practice, the harmonic oscillator has limits. In the "ideal" case, the two atoms can approach and recede with no change in the attractive force and without any repulsive force between electron clouds. In reality, the two atoms will dissociate when far enough apart, and will be repulsed by van der Waal's forces as they come closer. The net effect is the varying attraction between the two in the bond. When using a quantum model, the energy levels would be evenly spaced, making the overtones forbidden.

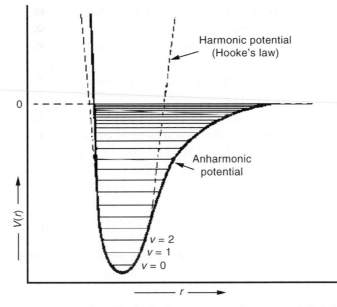

Fig. 6.2. Energy curve for Hooke's law versus Quantum Model of harmonic oscillator.

In the actual (working) model (see Fig. 6.2), the levels become closer as the vibrational number increases. It is this "anharmonicity" that allows the overtones to exist. The equation describing this phenomenon is

$$E_v = (v + 1/2)h\omega_e - (v + 1/2)^2\omega_e X_e + \text{higher terms} \qquad (6.6)$$

where $\omega_e = (1/2\pi)(K_e/\mu_e)^{1/2}$ is a vibrational frequency; $\omega_e X_e$ the anharmonicity constant (this is usually between 1 and 5%); K_e the harmonic force constant ($K = {\sim}5 \times 10^5$ dyn/cm for single bonds, ${\sim}10 \times 10^5$ dyn/cm for double bonds, and ${\sim}15 \times 10^5$ dyn/cm for triple bonds); and μ_e the reduced mass of the two atoms.

Using these factors and a fundamental vibration at 3500 nm, the first overtone would be

$$n = 3500/2 + (3500 \times [0.01, 0.02, \ldots]) \qquad (6.7)$$

This equation will give values from 1785 to 1925 nm. In reality, the first overtone would likely be at 3500/2± several nanometers, usually to a longer wavelength. The anharmonicity gives rise to varying distances between overtones. As a consequence, two overlapping peaks may be separated at a higher overtone.

E.W. Ciurczak

To make the spectrum more complicated, *Combination Bands* also exist. These are simply two or more bands that are physically adjacent or nearby on a molecule that add or subtract their energies (in the frequency domain) to produce a new or separate band. For example, the molecule SO_2, according to the formula for allowed bands,

$$\#bands = 3N - 6 \tag{6.8}$$

should have three absorption bands, where N are the number of atoms. Those are the symmetric stretch (at $1151\,cm^{-1}$), the asymmetric stretch ($1361\,cm^{-1}$), and the O–S–O bend ($519\,cm^{-1}$). These are allowed by Group Theory. However, four other bands appear in the SO_2 spectrum: 606, 1871, 2305, and $2499\,cm^{-1}$.

These may be explained in the following manner. One band is explained as the anharmonic overtone of the symmetric stretch at $1151\,cm^{-1}$, occurring at $2305\,cm^{-1}$, with the $3\,cm^{-1}$ difference attributed to the anharmonic constant. The other three bands may be explained as combination bands.

Since two bands may combine as $v_a - v_b$ or $v_a + v_b$ to create a new band. Using these concepts, the band assignments for SO_2 may be seen as [15],

v (cm^{-1})	Assignment
519	v_2
606	$v_1 - v_2$
1151	v_1
1361	v_3
1871	$v_2 + v_3$
2305	$2v_1$
2499	$v_1 + v_3$

Any unknown (isolated) band may be deduced from first principles; unfortunately, there is considerable overlap in the NIR region, but this is where Chemometrics will be discussed. An idealized spectrum of combinations and overtones is seen in Fig. 6.3.

Another potential source of peaks in the NIR is called *Fermi resonance*. This is where an overtone or combination band interacts strongly with a fundamental band. The math is covered in any good theoretical spectroscopy text, but, in short, the two different-sized, closely located peaks tend to normalize in size and move away from one another. This leads to difficulties in "first principle" identification of peaks within complex spectra.

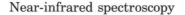

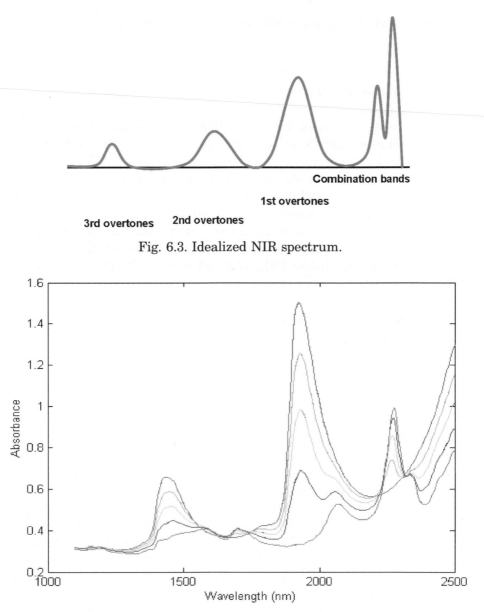

Fig. 6.3. Idealized NIR spectrum.

Fig. 6.4. Spectra of water–methanol mixtures.

Figure 6.4 shows the spectra of a series of water/methanol mixtures. In reality, NIR is used for often complex mixtures relying on chemometrics in lieu of actual spectral interpretation. Thus, while based on "real" spectroscopic principles, NIR is seldom about spectroscopy.

The skills needed to master NIR are physics, statistics, chemometrics, and optics.

6.3   INSTRUMENTATION

Since NIR was developed by the USDA for food products, the first (commercial) mode developed was diffuse reflection. The earliest work was performed on instruments which had, in essence, NIR as an "extra." The UV/Vis hardware (e.g., Cary model 10) had an additional detector and could be used through the NIR. This one fact explains why so much literature uses nanometers for units instead of wave numbers.

Another reason for nm instead of cm^{-1} is that mid-IR spectroscopists use the energy (wave numbers) to do spectral interpretation. With the massive overlapping in the NIR, coupled with hydrogen bonding, the NIR spectra are not easily "interpreted," so nanometers tend to remind us of that fact.

The first NIR instruments were, in reality, developed for the UV and Vis regions of the spectrum. They were made by seven companies: Beckman, Cary (now owned by Varian), Coleman, Perkin-Elmer, Shimadzu, Unicam, and Zeiss. Based on the work of Karl Norris and coworkers in the USDA, the Illinois Department of Agriculture solicited bids from companies to produce a "pure" NIR instrument, capable of measuring protein, oil, and moisture in soybeans.

The first commercial unit was produced by Dickey-John. It contained a tungsten-halogen lamp, six interference filters, and uncooled lead sulfide (PbS) detectors, using a 0–45° geometry. That is, the light struck the sample straight on and the light was collected at 45° to the normal.

The samples had dry matter over 85% and were ground to pass through a 1 mm screen and then packed in a quartz-windowed cup. The unit was demonstrated at the 1971 Illinois State Fair. After the success of this instrument, Neotec (later Pacific Scientific, then NIR Systems, then Perstorp, then FOSS) built a rotating (tilting) filter instrument. Both instruments were dedicated, analog systems, neither of which was considered "user-friendly."

In the middle of the 1970s, Technicon Instruments had Dickey-John produce a filter instrument for them, named the InfraAlyzer. The first, numbered 2.5, featured dust proof optics and internal temperature control. This improved stability and ruggedness made it practical for consumers to operate. Technicon also introduced the gold plated integrating sphere.

Since the 1980s, numerous instruments with varied construction have been introduced. In most spectroscopic techniques there are just a few technologies involved. In mid-range IR, there remain a number of grating type monochromators used, but in the whole, interferometers rule. The so-called "FT-(Fourier Transform) IRs" are the standard for the mid-range IR. For UV and Visible, gratings (either scanning or fixed with diode arrays) are the norm. However, with NIR, a plethora of wavelength selections are available.

In addition to interference filters, NIR manufacturers use holographic gratings (moving and fixed), interferometers, polarization interferometers, diode arrays, acoustic-optic tunable filters, as well as some specialty types.

While most other techniques use a limited amount of detectors (e.g., silica for visible, photomultipliers for UV) and MIR has a small number, NIR uses many types of semiconductors for detectors. The original PbS detectors are still one of the largest used in NIR, however, indium gallium arsenide (InGaAs), indium arsenide (InAs), indium antimonide (InSb), and lead selenide (PbSe) are among the semiconductor combinations used, both cooled and ambient.

Most samples analyzed by NIR are "as is" samples, typical of agriculture, now food, polymers, and pharmaceuticals. Because of this a large number of sample presentation techniques have been developed: cups, dipping fiber optics, spinning cups, flow-through cells, paired fiber probes, cuvettes, windows in production systems, and non-contact systems. In fact, much of the engineering goes into sample presentation. This places a lot of the system's success or failure on features having little to do with spectroscopy.

The size, speed, noise levels, precision, accuracy, and cost vary among the instruments. Higher cost does not necessarily mean better performance. Unlike "typical" spectroscopy, where the sample is reduced and, often, diluted in a non-interfering matrix, samples in NIR are read "as is" and one size instrument does NOT fit all. The application will determine the instrument (an idea that escapes many instrument salespeople).

As a general statement, the one thing that all NIR spectrometers, built in the past 25 years, have in common is that they are all single beam. This means that periodic wavelength, noise, and linearity checks must be made. In typical chemical or instrumental analyses (i.e. titrations or HPLC), a standard is run in parallel with an unknown and the result calculated from the response of the two. This is defined as an analysis or assay.

In NIR, a series of samples are scanned and then analyzed by a referee method. An equation is generated and used for future unknowns. This equation is used after the instrument is checked for compliance with initial performance criteria (at the time of the equation calibration). No standard is available for process or "natural" samples. The value(s) is gleaned from chemometric principles. This is defined as a prediction.

Thus, while a well-maintained instrument is important for any chemical/physical measurement, in NIR, without a concurrent standard for comparison, it is critical that the instrument be continuously calibrated and maintained. Since the major manufacturers of equipment have worked with the pharmaceutical industry, this has been formalized into what is called IQ/OQ/PQ, or Instrument Qualification, Operational Qualification, and Performance Qualification. The first is routinely performed (at first) by the manufacturer in the lab/process location, the second *in situ* by the user with help from the manufacturer, and the third is product/use dependent. These formal tests apply to all instruments in any industry.

6.4 MATH TREATMENTS

Since most quantitative applications are on mixtures of materials, complex mathematical treatments have been developed. The most common programs are Multiple Linear Regression (MLR), Partial Least Squares (PLS), and Principal Component Analyses (PCA). While these are described in detail in another chapter, they will be described briefly here.

MLR is based on classical least squares regression. Since "known" samples of things like wheat cannot be prepared, some changes, demanded by statistics, must be made. In a Beer's law plot, common in calibration of UV and other solution-based tests, the equation for a straight line

$$Y = mX + b \tag{6.9}$$

represents the line where Y is the absorbance generated by the corresponding concentration, X. Since we are considering a "true" Beer's law plot, zero concentration has zero absorbance, so that the "b" or intercept is always zero. In this case, the better known or least error-prone values, by convention, plotted along the X-axis, are the concentrations of materials. This is simply because a balance and volumetric glassware are used to make the solutions and, in this case, the absorbance is less accurate and is plotted on the Y-axis.

In most NIR measurements, the sample is not a simple solution, but either a complex solution or a complex solid. In this case, the material (or property) being assessed is measured by a referee method, not weighed into a flask from a balance. As a consequence, the more reliable (accurate) measurement is the absorbance. Thus, the "Inverse Beer's Law" equation becomes

$$A = \xi bc \qquad (6.10)$$

where A is the absorbance measured at a specific wavelength; ξ the molar absorptivity (at that wavelength in $L\,mol^{-1}cm^{-1}$); b the path length (in cm); c the concentration (in $mol\,L^{-1}$); and but is now written as

$$C = b_1 A + b_0 \qquad (6.11)$$

where A is the absorbance measured at a specific wavelength; C the concentration; b_1 the constant that incorporates ξ and b; and b_0 the intercept of the calibration line.

Since there are few clear absorbance peaks in a complex mixture, because of the overlap several wavelengths are often needed to generate a linear, descriptive equation. The equation then takes on the form of

$$C = b_1 A_1 + b_2 A_2 + \cdots + b_n A_n + b_0 \qquad (6.12)$$

where C is the concentration; $b_1, \ldots, b_n$ are constants for wavelengths 1 through n; and $A_1, \ldots, A_n$ are absorbance at wavelengths 1 through n.

If an MLR equation needs more than five wavelengths, it is often better to apply one of the other multivariate algorithms mentioned above. Since the information which an analyst seeks is spread throughout the sample, methods such as PLS and PCA use much or all of the NIR spectra to determine the information sought.

One example of MLR is seen in Figs. 6.5 and 6.6. Figure 6.5 shows the second derivative spectra of various combinations of two polymorphic forms of a crystalline drug substance. Since the chemistry is identical, only hydrogen bonding differences affect the spectra. The wavelength where the calibration is made is highlighted. The resulting calibration curve (two wavelength MLR equation) is seen in Fig. 6.6.

The differences of consequence will be highlighted here. Both show the variances within a set of spectra and attempt to define it. How they are developed and applied is slightly different, however.

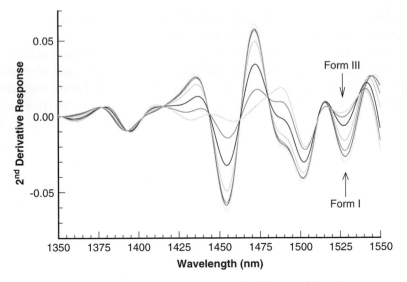

Fig. 6.5. Second derivative spectra of polymorphic mixtures.

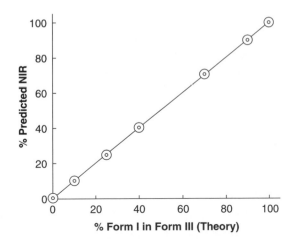

Fig. 6.6. Calibration curve for polymorph determination.

PCA is based only on the variances among spectra. No content information is used to generate the preliminary factors. In a series of mixtures of water and methanol (shown in Fig. 6.3), for instance, the first Principal Component (see Fig. 6.7) shows the positive and negative "lobes" representing the shifting of water in a positive direction and methanol in a negative direction. This is based solely on the change in

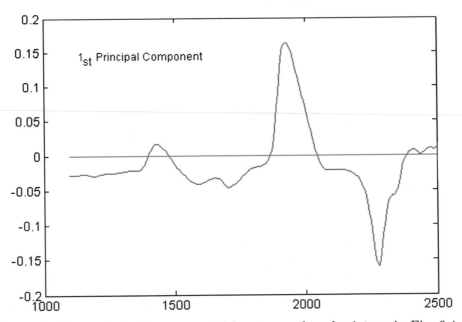

Fig. 6.7. First principal component of water–methanol mixture in Fig. 6.4.

spectra; no assay values have been given. Since the introduction of PLS, PCA has been used almost entirely in qualitative analyses, where numbers may not be available or relevant (which is the "better" lactose for production use?). In a purely qualitative use, PCA is used to show both differences and similarities among groups of materials. Figure 6.8 shows two groups of sample tablets, "clustered" purely on physical differences.

More definitive information may also be gleaned from PCA. In Fig. 6.9, three principal component scores are graphed to show how the amount of roasting in coffee beans may be ascertained.

In PLS, the variance is generated using the quantities generated by the referee analytical method. Therefore, the factors in a PLS analysis (especially the first) resemble the spectrum of the active ingredient (assuming a quantitative analysis of a constituent). When measuring a physical attribute (hardness, polymorphic form, elasticity), the PLS factor may not resemble any of the materials present in the mixture.

The first PLS factor represents the manner in which the spectra change with respect to the analytical values attached. That is (for normal spectra, not derivatives) as the correlation between change in absorbance and constituent value is greatest, there is a large "peak" or

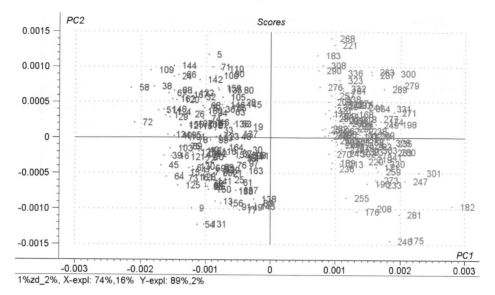

Fig. 6.8. Two-dimensional graph of PC scores of two groups of pharmaceutical tablets.

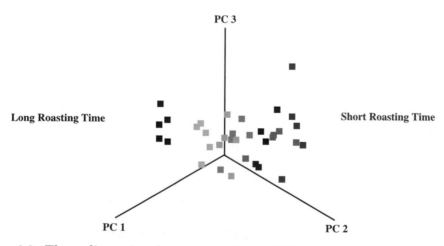

Fig. 6.9. Three-dimensional representation of PC scores for roasted coffee beans, roasted for varying amounts of time.

upswing in the factor (in the positive direction). A negative correlation brings about a negative swing. [This factor may even be used to uncover wavelengths for MLR equations.] Since this relationship exists, PLS is primarily a quantitative algorithm.

170

6.5 APPLICATIONS

The largest bodies of references for NIR applications are in the fields of agriculture and food. Following closely behind are chemicals, petrochemicals, and polymers. Only recently has the pharmaceutical industry recognized the potential of NIR. Because of its ability to make rapid, non-destructive, and non-invasive measurements, NIR is widely used in process analyses.

In agricultural applications, the most commonly analyzed constituents are water, protein, starch, sugars, and fiber [16–20]. Such physical or chemical functions such as hardness of wheat, minerals, and food values have no actual relation to chemicals seen in the NIR. These are usually done by inferential spectroscopy. That is, the effect of minerals or the relationship of the spectra to *in vitro* reactions is used *in lieu* of chemical analyses to NIR active constituents. Considering that all shipments of grain from the US since the 1980s have been cleared by NIR, it can be argued that this is a critical application of the technique.

The same functions used in agriculture can be applied to processed foods. In baked goods, wheat gluten, various additives, starch damage, and water absorption are just some of the parameters measured [21–24]. Dairy products are also important and often analyzed by NIR. Moisture, fat, protein, lactose, lactic acid, and ash are common analytes in the dairy industry [25–28].

Other food/agricultural applications are in the beverage and fabrics/wool industries. Wood fibers are easily analyzed for lignin, wool and cotton for ability to accept dyes, and beverages, both soft and hard, may be analyzed for contents.

Polymers may be analyzed from the synthesis stage (reaction monitoring) through blending and actual fabrication of plastic end products [29–35]. The molecular weight, degree of polymerization and cross-linking, hydroxyl values, acid number, and saponification values are just some of the values monitored (non-destructively, in real time). A major polymer company in Delaware is said to have as many as 1200 units throughout its plants.

The most recent converts are in the health care industry. Pharmaceutical and biological applications have become myriad since the early 1980s. The first widespread application was for the identification/qualification of incoming raw materials. Since then, applications have appeared for moisture (bound and free), blend uniformity of powders, tablet and capsule assays, counterfeiting, polymorphism, degree of crystallinity, hardness (of tablets), dissolution prediction, isomerism, as

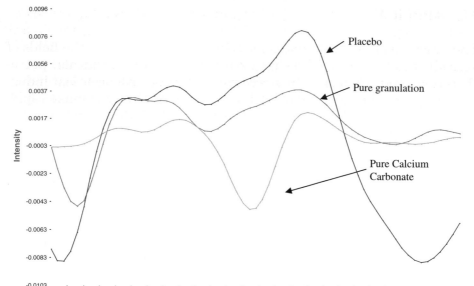

Fig. 6.10. Overlay of spectra of typical pharmaceutical ingredients for a tablet mixture

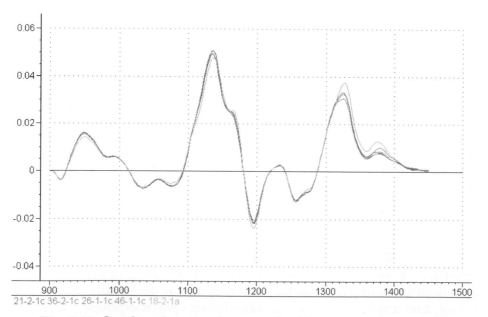

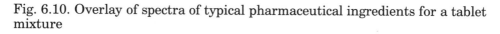

Fig. 6.11. Overlay of spectra for several tablets on a conveyor belt

well as synthesis monitoring [36–44]. Figure 6.10 shows the overlay of several components of a tablet mixture. The regions of interest are easily discerned for analysis.

In the process setting, NIR is fast enough to capture information at amazing speeds. Figure. 6.11 shows the spectra of tablets, generated as

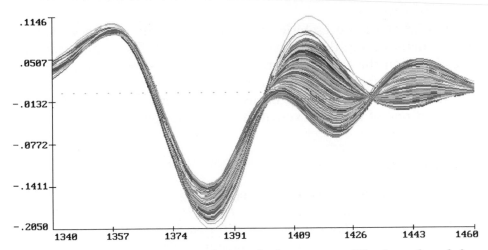

Fig. 6.12. Spectra from a granulation drying process. The bound and free water are seen at ~1420 and ~1440 nm, respectively.

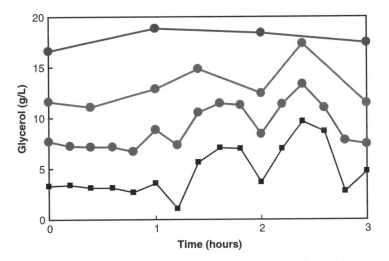

Fig. 6.13. Comparison of rates of sampling for a bioprocess (fermentation); the ability of NIR to measure in real time is compared with discrete sampling techniques.

173

they pass through the beam at conveyor-belt speeds. Medical and biological applications include skin and tissue monitoring, blood analysis, and fermentation of biologics [45–50]. Figure 6.12 shows the advantage of NIR in rapid sampling and analysis of glycerol in a fermentation process. The sampling rate for a bioprocess may be greatly increased with a NIR probe, as seen in Fig. 6.13, where manual and instrumental rates are compared.

Some of the more esoteric applications include currency counterfeiting, determining the quality of bowling alley floors, and watching crops ripen on the vine. In essence, any materials containing hydrogen are candidates for NIR analyses. It is fast, sensitive, rugged, accurate, and non-destructive. It may be used in contact or from a distance, making it one of the more versatile analytical applications today.

REFERENCES

1 W. Herschel, *Philos. Trans.*, 90 (1800) 225–283.
2 W. Abney and E.R. Festing, *Philos. Trans. Roy. Soc.*, 172 (1881) 887–918.
3 W.W. Coblentz, Investigations of Infrared Spectra Part 1. Publication No. 35, Carnagie Institute of Washington (1905).
4 F.E. Fowler, *Astrophys. J*, 35 (1912) 149–162.
5 J. Ellis and J. Bath, *J. Chem. Phys.*, 6 (1938) 732.
6 P. Barchewitz, *J. Chem. Phys.*, 45 (1943) 40.
7 I. Barr and W. Harp, *Phys. Rev.*, 63 (1942) 457.
8 I. Ben-Gera and K.H. Norris, *Israel J. Agric. Res.*, 18 (1968) 125.
9 I. Ben-Gera and K.H. Norris, *J. Food Sci.*, 33 (1968) 64.
10 K.H. Norris, R.F. Barnes, J.E. Moore and J.S. Shenk, *J. Anim. Sci.*, 43 (1976) 889.
11 F.E. Barton II and D. Burdick, *J. Agric. Food Chem.*, 27 (1979) 1248.
12 W.H. Butler and K.H. Norris, *Arch. Biochem. Biophys.*, 87 (1960) 31.
13 J.D. Ingle Jr. and S.R. Crouch, *Spectrochemical Analysis*, Prentice-Hall, Englewood Cliffs, NJ, 1988.
14 F.A. Cotton, *Chemical Applications of Group Theory*, 3rd ed., Wiley, New York, 1990.
15 R.S. Drago, *Physical Methods in Chemistry*, W.B. Saunders Co., Philadelphia, 1977 Chapter 6.
16 D.E. Honigs, G.M. Hieftje and T.B. Hirschfeld, *Appl. Spectrosc.*, 38 (1984) 844.
17 S.M. Abrams, J.S. Shenk, M.O. Westerhaus and F.E. Barton II, *J. Dairy Sci.*, 70 (1987) 806.
18 J.S. Shenk, I. Landa, M.R. Hoover and M.O. Westerhaus, *Crop Sci.*, 21 (1981) 355.

19   R.A. Isaac and W. Dorsheimer, Am. Lab., April, 1982.

20   T.H. Blosser and J.B. Reeves III, J. Dairy Sci., 69(Abstr. Suppl.) (1986) 136.

21   P.C. Williams, B.N. Thompson and D. Whetsel, Cereal Foods World, 26(5) (1981) 234.

22   B.G. Osborne, T. Fearn, A.R. Miller and S. Douglas, J. Sci. Food Agric., 35 (1984) 99.

23   B.G. Osborne, G.M. Barrett, S.P. Cauvain and T. Fearn, J. Sci. Food Agric., 35 (1984) 940.

24   K. Suzuki, C.E. McDonald and B.L. D'Appolonia, Cereal Chem., 63(4) (1986) 320.

25   A.M.C. Davies and A. Grant, Int. J. Food Sci. Technol., 22 (1987) 191.

26   T.C.A. McGann, Irish J. Food Sci. Technol., 2 (1978) 141.

27   D.A. Biggs, J. Assoc. Off. Anal. Chem., 55 (1972) 488.

28   J.D.S. Goulden, J. Dairy Sci., 24 (1957) 242.

29   A.M.C. Davies, A. Grant, G.M. Gavrel and R.V. Steeper, Analyst, 110 (1985) 643.

30   L.G. Weyer, J. Appl. Polym. Sci., 31 (1986) 2417.

31   S. Ghosh and J. Rodgers, Textile Res. J., 55(9) (1985) 234.

32   C. Tosi, Makromol. Chem., 112 (1968) 303.

33   T. Takeuchi, S. Tsuge and Y. Sugimura, Anal. Chem., 41(1) (1969) 184.

34   A. Giammarise, Anal. Letts., 2(3) (1969) 117.

35   W.H. Grieve and D.D. Doepken, Polym. Eng. Ser., April, 19–23 (1968).

36   E.W. Ciurczak, Pharm. Tech., 15(9) (1991) 141.

37   E.W. Ciurczak, 7th Ann. Symp. NIRA,, Technicon, Tarrytown, NY, 1984.

38   D.J. Wargo and J.K. Drennen, J. Pharm. Biomed. Anal., 14(11) (1996) 1414.

39   C. Tso, G.E. Ritchie, L. Gehrlein and E.W. Ciurczak, J. NIRS, 9 (2001) 165–184.

40   P.K. Aldrich, R.F. Mushinsky, M.M. Andino and C.L. Evans, Appl. Spectrosc., 48 (1994) 1272.

41   R. Gimet and T. Luong, J. Pharm. Biomed. Anal., 5 (1987) 205.

42   B.R. Buchanan, E.W. Ciurczak, A. Grunke and D.E. Honigs, Spectroscopy, 3(9) (1988) 54.

43   G. Ritchie, R. Roller, E.W. Ciurczak, H. Mark, C. Tso and S. MacDonald, J. Pharm. Biomed. Anal., 28(2) (2002) 251–260.

44   G. Ritchie, R. Roller, E.W. Ciurczak, H. Mark, C. Tso and S. MacDonald, J. Pharm. Biomed. Anal., 29(1–2) (2002) 159–171.

45   F.F. Jobsis, Science, 198 (1977) 1264.

46   D.M. Mancini, L. Bolinger, H. Li, K. Kendrik, B. Chance and J.R. Wilson, J. Appl. Physiol., 77(6) (1994) 2740.

47   V.V. Kupriyanov, R.A. Shaw, B. Xiang, H. Mantsch and R. Deslauriers, J. Mol. Cell. Cardiol., 29 (1997) 2431.

48   T. Wolf, U. Lindauer, H. Obrig, J. Drier, J. Back, A. Villringer and O. Dirnagl, J. Cereb. Blood Flow Metab., 16 (1996) 1100–1107.

49 R.A. Shaw, S. Kotowich, H.H. Mantsch and M. Leroux, *Clin. Biochem.*, 29(1) (1996) 11.
50 J.W. Hall, B. McNeil, M.J. Rollins, I. Draper, B.G. Thompson and G. Macaloney, *Appl. Spectrosc.*, 50(1) (1886) 102–108.

Further Reading

D.A. Burns and E.W. Ciurczak (Eds.), *Handbook of Near-Infrared Analysis*, 2nd ed., Marcel Dekker, New York, 2001.
H.L. Mark, *Principles and Practice of Spectroscopic Calibration*, Wiley, New York, 1991.
J. Workman and A. Springsteen (Eds.), *Applied Spectroscopy; A Compact Reference for Practitioners*, Academic Press, New York, 1998.
P. Williams and K. Norris (Eds.), *Near-Infrared Technology in the Agriculture and Food Industries*, 2nd ed, Amer. Assoc. Cereal Chem., St. Paul, MN, 2003.
E.R. Malinowski, *Factor Analysis in Chemistry*, 3rd ed., Wiley-Interscience, New York, 2002.
K.I. Hildrum, T. Isaksson, T. Naes and A. Tandburg (Eds.), *Near-Infrared Spectroscopy; Bridging the Gap between Data Analysis and NIR Applications*, Ellis Horwood, New York, 1992.

REVIEW QUESTIONS

1. When and how was near-infrared discovered?
2. What are some of the unique features of NIR?
3. What is the basis of absorption of light in the NIR region?
4. How many types of NIR instruments are in use today?
5. Why are Chemometrics necessary for most NIR measurements?
6. Why would NIR be more affected by hydrogen bonding than mid-range infrared?
7. Name three industrial applications where NIR would have an advantage over "classical" analyses.
8. Why would you guess that NIR instruments are so varied when compared with other spectroscopic methods?

Chapter 7

X-ray diffraction and x-ray fluorescence

Harry G. Brittain

7.1   X-RAY DIFFRACTION

7.1.1   Introduction

All properly designed investigations on the solid-state properties of pharmaceutical compounds begin with the understanding of the structural aspects involved. There is no doubt that the primary tool for the study of solid-state crystallography is x-ray diffraction. To properly comprehend the power of this technique, it is first necessary to examine the processes associated with the ability of a crystalline solid to act as a diffraction grating for the electromagnetic radiation of appropriate wavelength. Thereafter, we will separately discuss the practice of x-ray diffraction as applied to the characterization of single crystals and to the study of powdered crystalline solids.

7.1.1.1   Historical background

X-ray crystallography has its origins in the studies performed to discover the nature of the radiation emitted by cathode ray tubes, known at the time as "Roentgen rays" or "x-radiation." At the end of the 19th century, it was not clear whether these rays were corpuscular or electromagnetic in nature. Since it was known that they moved in straight lines, casted sharp shadows, were capable of crossing a vacuum, acted on a photographic plate, excited substances to fluoresce, and could ionize gases, it appeared that they had the characteristics of light. But at the same time, the mirrors, prisms, and lenses that operated on ordinary light had no effect on these rays, they could not be diffracted by ordinary gratings, and neither birefringence nor polarization could be induced in such beams by passage through biaxial crystals. These latter effects suggested the existence of a corpuscular nature.

Comprehensive Analytical Chemistry 47
S. Ahuja and N. Jespersen (Eds)
Volume 47 ISSN: 0166-526X DOI: 10.1016/S0166-526X(06)47007-0

The nature of the x-radiation ultimately became understood through studies of their diffraction, although the experimental techniques differed from those of the classical diffraction studies. The existence of optical diffraction effects had been known since the 17th century, when it was shown that shadows of objects are larger than they ought to be if light traveled past the bodies in straight lines undetected by the bodies themselves. During the 19th century, the complete theory of the diffraction grating was worked out, where it was established that in order to produce diffraction spectra from an ordinary grating, the spacings of the lines on the grating had to be of the same size magnitude as the wavelength of the incident light. Laue argued that if x-rays consisted of electromagnetic radiation having a very short wavelength, it should be possible to diffract the rays once a grating having sufficiently small spacings could be found.

The theory of crystals developed by Bravais [1] suggested that the atoms in a solid were in a regular array, and the estimates of atomic size showed that the distances between the sheets of atoms were probably about the same as the existing estimates for the wavelengths of the x-radiation. Laue suggested the experiment of passing an x-ray beam ray through a thin slice of crystalline zinc blende, since this material had an understood structure. Friedrich and Knipping tried the experiment, and found a pattern of regularly placed diffracted spots arranged around the central undeflected x-ray beam, thus showing that the rays were diffracted in a regular manner by the atoms of the crystal [2]. In 1913, Bragg reported the first x-ray diffraction determination of a crystal structure, deducing the structures of KCl, NaCl, KBr, and KI [3]. Eventually, the complicated explanation advanced by Laue to explain the phenomenon was simplified by Bragg, who introduced the concept of "reflexion," as he originally termed diffraction [4].

Bragg established that the diffraction angles were governed by the spacings between atomic planes within a crystal. He also reported that the intensities of diffracted rays were determined by the types of atoms present in the solid, and their arrangement within a crystalline material. Atoms with higher atomic numbers contain larger numbers of electrons, and consequently scatter x-rays more strongly than atoms characterized by lower atomic numbers. This difference in scattering power leads to marked differences in the intensities of diffracted rays, and such data provide information on the distribution of atoms within the crystalline solid.

7.1.1.2 Description of crystalline solids

Since a number of extensive discussions of crystallography are available [5–9], it is necessary to only provide the briefest outline of the nature of crystalline solids. An ideal crystal is constructed by an infinite regular spatial repetition of identical structural units. For the organic molecules of pharmaceutical interest, the simplest structural unit will contain one or more molecules. The structure of crystals is ordinarily explained in terms of a periodic *lattice*, or a three-dimensional grid of lines connecting points in a given structure. For organic molecules, a group of atoms is attached to a lattice point. It is important to note that the points in a lattice may be connected in various ways to form a finite number of different lattice structures. The crystal structure is formed only when a fundamental unit is attached identically to each lattice point, and extended along each crystal axis through translational repetition. Whether all of the lattice planes can be identified in a particular crystal will depend greatly upon the nature of the system, and the intermolecular interactions among the molecules incorporated in the structure. An example of a crystal lattice, where each repetitive unit contains two atoms, is shown in Fig. 7.1.

The points on a lattice are defined by three fundamental translation vectors, **a**, **b**, and **c**, such that the atomic arrangement looks the same in every respect when viewed from any point **r** as it does when viewed at point **r′**:

$$\mathbf{r'} = \mathbf{r} + n_1\mathbf{a} + n_2\mathbf{b} + n_3\mathbf{c} \tag{7.1}$$

where n_1, n_2, and n_3 are arbitrary integers. The lattice and translation vectors are said to be *primitive* if any two points from which an identical atomic arrangement is obtained through the satisfaction of Eq. (7.1) with a suitable choice of the n_1, n_2, and n_3 integers. It is common practice to define the primitive translation vectors to define the axes of the crystal, although other non-primitive crystal axes can be used for the sake of convenience. A lattice translation operation is defined as the displacement within a lattice, with the vector describing the operation being given by:

$$\mathbf{T} = n_1\mathbf{a} + n_2\mathbf{b} + n_3\mathbf{c} \tag{7.2}$$

The crystal axes, **a**, **b**, and **c**, form three adjacent edges of a parallelepiped. The smallest parallelepiped built upon the three unit translations is known as the *unit cell*. Although the unit cell is an imaginary construct, it has an actual shape and definite volume. The crystal

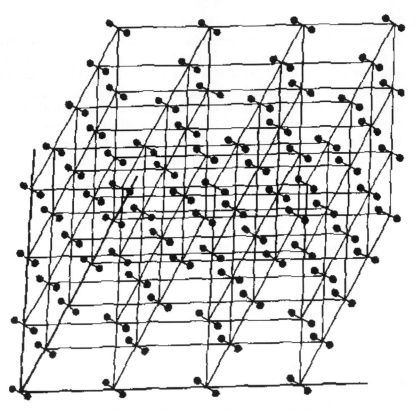

Fig. 7.1. Families of planes passing through lattice points.

structure is generated through the periodic repetition, by the three unit translations, of matter contained within the volume of the unit cell. A unit cell does not necessarily have a definite absolute origin or position, but does have the definite orientation and shape defined by the translation vectors. A cell will fill all space under the action of suitable crystal translation operations, and will occupy the minimum volume permissible.

The unit cell is defined by the lengths (a, b, and c) of the crystal axes, and by the angles (α, β, and γ) between these. The usual convention is that α defines the angle between the b- and c-axes, β the angle between the a- and c-axes, and γ the angle between the a- and b-axes. There are seven fundamental types of primitive unit cell (whose characteristics are provided in Table 7.1), and these unit cell characteristics define the seven *crystal classes*. If the size of the unit cell is known (i.e., α, β, γ, a, b, and c have been determined), then the unit cell volume (V) may be used

TABLE 7.1

The seven crystal classes, defined from their fundamental unit cells

System	Relationship between cell edges	Relationship between cell angles
Cubic	$a = b = c$	$\alpha = \beta = \gamma = 90°$
Tetragonal	$a = b \neq c$	$\alpha = \beta = \gamma = 90°$
Orthorhombic	$a \neq b \neq c$	$\alpha = \beta = \gamma = 90°$
Monoclinic	$a \neq b \neq c$	$\alpha = \gamma = 90°$ $\beta \neq 90°$
Triclinic	$a \neq b \neq c$	$\alpha \neq \beta \neq \gamma \neq 90°$
Hexagonal	$a = b \neq c$	$\alpha = \beta = 90°$ $\gamma = 120°$
Trigonal	$a = b = c$	$\alpha = \beta = 90°$ $\gamma \neq 90°$

to calculate the density (ρ) of the crystal:

$$\rho = \frac{ZM}{VA} \qquad (7.3)$$

where M is the molecular weight of the substance in the crystal, A the Avogadro's number, and Z the number of molecular units in the unit cell. When ρ is measured for a crystal (generally by flotation) for which the elemental composition of the molecule involved is known, then Z may be calculated. The value of Z is of great importance because it can provide molecular symmetry information from a consideration of the symmetry properties of the lattice.

The points of a lattice can be considered as lying in various sets of parallel planes. These, of course, are not limited to lying along a single Cartesian direction, but can instead be situated along any combination of axis directions permitted by the structure of the lattice. Consider a given set of parallel planes, which cuts across the a-, b-, and c-axes at different points along each axis. If the a-axis is divided into **h** units, the b-axis into **k** units, and c-axis into **l** units, then h k l are the *Miller indices* of that set of planes, and the set of planes is identified by its Miller indices as (h k l). If a set of planes happens to be parallel to one axis, then its corresponding Miller index is 0. For instance, a plane that lies completely in the ab-plane is denoted as the (001) plane, a plane that lies completely in the ac-plane is denoted as the (010) plane, a plane that lies completely in the bc-plane is denoted as the (100) plane, and the plane that equally intersects the a-, b-, and c-axes is denoted as

181

the (1 1 1) plane. The values of h, k, and l are independent quantities, and are defined only by their spatial arrangements within the unit cell.

7.1.1.3 Solid-state symmetry

Crystal lattices can be depicted not only by the lattice translation defined in Eq. (7.2), but also by the performance of various point symmetry operations. A *symmetry operation* is defined as an operation that moves the system into a new configuration that is equivalent to and indistinguishable from the original one. A *symmetry element* is a point, line, or plane with respect to which a symmetry operation is performed. The complete ensemble of symmetry operations that define the spatial properties of a molecule or its crystal are referred to as its *group*. In addition to the fundamental symmetry operations associated with molecular species that define the *point group* of the molecule, there are additional symmetry operations necessary to define the *space group* of its crystal. These will only be briefly outlined here, but additional information on molecular symmetry [10] and solid-state symmetry [11] is available.

Considering molecular symmetry first, the simplest operation is the *identity*, which leaves the system unchanged and hence in an orientation identical to that of the original. The second symmetry operation is that of *reflection* through a plane, and is denoted by the symbol σ. The effect of reflection is to change the sign of the coordinates perpendicular to the plane, while leaving unchanged the coordinates parallel to the plane. The third symmetry operation is that of inversion through a point, and is denoted by the symbol $\mathbf{i}$. In Cartesian coordinates, the effect of inversion is to change the sign of all three coordinates that define a lattice point in space. The fourth type of symmetry operation is the *proper rotation*, which represents the simple rotation about an axis that passes through a lattice point. Only rotations by angles of 2π (360°), $2\pi/2$ (180°), $2\pi/3$ (120°), $2\pi/4$ (90°), and $2\pi/6$ (60°) radians are permissible, these being denoted as one-fold (symbol $\mathbf{C_1}$), two-fold (symbol $\mathbf{C_2}$), three-fold (symbol $\mathbf{C_3}$), four-fold (symbol $\mathbf{C_4}$), and six-fold (symbol $\mathbf{C_6}$) proper rotation axes. The final type of symmetry operation is the *improper rotation*, which represents the combination of a proper rotation axis followed by the performance of a reflection operation. Once again, only improper rotations by angles of 2π (360°), $2\pi/2$ (180°), $2\pi/3$ (120°), $2\pi/4$ (90°), and $2\pi/6$ (60°) radians are permissible, and are denoted as one-fold (symbol $\mathbf{S_1}$), two-fold (symbol $\mathbf{S_2}$), three-fold (symbol $\mathbf{S_3}$), four-fold (symbol $\mathbf{S_4}$), and six-fold (symbol $\mathbf{S_6}$) proper rotation axes.

There are two additional symmetry operators required in the definition of space groups. These operators involve translational motion, for now we are moving molecules into one another along three dimensions. One of these is referred to as the *screw axis*, and the other is the *glide plane*. The screw operation rotates the contents of the cell by 180°, and then translates everything along its axis by one-half of the length of the parallel unit cell edge. The glide operation reflects everything in a plane, and then translates by one-half of the unit cell along one direction in the same plane.

The symmetry of a crystal is ultimately summed up in its crystallographic *space group*, which is the entire set of symmetry operations that define the periodic structure of the crystal. Bravais showed that when the full range of lattice symmetry operations was taken into account, one could define a maximum of 230 distinct varieties of crystal symmetry, or space groups. The space groups are further classified as being either symmorphic or non-symmorphic. A symmorphic space group is the one that is entirely specified by symmetry operations acting at a common point, and which do not involve one of the glide or screw translations. The symmorphic space groups are obtained by combining the 32 point groups with the 14 Bravais lattices, yielding a total of 73 space groups out of the total of 230. The non-symmorphic space groups are specified by at least one operation that involves a non-primitive translation, and one finds that there are a total of 157 non-symmorphic space groups.

7.1.1.4 *Diffraction of electromagnetic radiation by crystalline solids*

Laue explained the reflection of x-rays by crystals using a model where every atom of the crystal bathed in the beam of x-rays represents a secondary radiating source in which the wavelength and phase remain unchanged. Then, for a three-dimensional regular array, these secondary waves interfere with one another in such a way that, except for certain calculable directions, destructive interference occurs. In the special directions, however, there is constructive interference and strong scattered x-ray beams can be observed.

Bragg provided a much simpler treatment of the scattering phenomenon. He assumed that the atoms of a crystal are regularly arranged in space, and that they can be regarded as lying in parallel sheets separated by a definite and defined distance. Then he showed that scattering centers arranged in a plane act like a mirror to x-rays incident on them, so that constructive interference would occur for the direction of specular reflection. As can be envisioned from Fig. 7.1, an infinite number of sets of parallel planes can be passed through the points of a space lattice. If we now consider one set of planes, defined by a Miller index of $(h\ k\ l)$, each plane of this set produces a specular reflectance of the incident beam. If the incident x-rays are monochromatic (having a wavelength equal to λ), then for an arbitrary glancing angle of θ, the reflections from successive planes are out of phase with one another. This yields destructive interference in the scattered beams. However, by varying θ, a set of values for θ can be found so that the path difference between x-rays reflected by successive planes will be an integral number $(\mathbf{n})$ of wavelengths, and then constructive interference will occur.

Bragg's model is illustrated in Fig. 7.2, where the horizontal lines represent successive planes of the set. The spacing of these planes (denoted as d) is the perpendicular distance between them. The path difference is $\{(BC)+(CD)\}$, and for constructive interference this path length must equal $(\mathbf{n}\lambda)$. But since $\{(BC)+(CD)\}$ must be equal to $(2d\sin\theta)$, one deduces an expression universally known as Bragg's law:

$$2d\ \sin\ \theta = \mathbf{n}\lambda \tag{7.4}$$

Unlike the case of diffraction of light by a ruled grating, the diffraction of x-rays by a crystalline solid leads to the observation that constructive interference (i.e., reflection) occurs only at the critical Bragg angles. When reflection does occur, it is stated that the plane in question is reflecting in the $\mathbf{n}$th order, or that one observes $\mathbf{n}$th order diffraction for that particular crystal plane. Therefore, one will observe an x-ray scattering response for every plane defined by a unique Miller index of $(h\ k\ l)$.

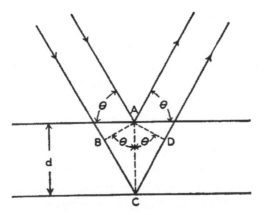

Fig. 7.2. Diffraction of electromagnetic radiation by planes of atoms in a crystalline solid.

Consider a crystal characterized by some unit cell that contains Z atoms per unit cell. Taking each of the Z atoms in succession as the origin of a lattice, one can envision the crystal as composed of Z identical interpenetrating lattices, with their displacements relative to each other being decided by the distribution of the Z atoms within the unit cell. All these lattices will be characterized by identical Bragg angles, but the reflections from the Z-independent lattices will interfere with one another. This interference will determine the intensity of the reflection for that plane, but the Bragg angle will be independent of the atomic distribution within the cell, and will depend only on the details of the external cell geometry.

The analysis of x-ray diffraction data is divided into three parts. The first of these is the geometrical analysis, where one measures the exact spatial distribution of x-ray reflections and uses these to compute the size and shape of a unit cell. The second phase entails a study of the intensities of the various reflections, using this information to determine the atomic distribution within the unit cell. Finally, one looks at the x-ray diagram to deduce qualitative information about the quality of the crystal or the degree of order within the solid. This latter analysis may permit the adoption of certain assumptions that may aid in the solving of the crystalline structure.

7.1.2 Single-crystal x-ray diffraction

The analysis of x-ray diffraction data consists of two main aspects. The first of these is a geometrical analysis, where the size and shape of a unit cell is computed from the exact spatial distribution of x-ray reflections. The second examination yields the atomic distribution within the unit cell from a study of the intensities of the various reflections. Throughout the process, one uses the x-ray diagram in order to deduce qualitative information about the quality of the crystal or the degree of order within the solid. Readers interested in learning more details of the process should consult any one of a number of standard texts for additional information [6–8,12–15].

At the end of the process, one obtains a large amount of information regarding the solid-state structure of a compound. Its three-dimensional structure and molecular conformation become known, as do the patterns of molecular packings that enable the assembly of the crystal. In addition, one obtains complete summaries of bond angles and bond lengths for the molecules in the crystal as well as detailed atomic coordinates for all of the atoms present in the solid. One generally finds crystallography papers divided as being either inorganic [16] or organic [17] in focus, but the studies of the crystallography of organic compounds are of most interest to the pharmaceutical community. After all, drugs are merely organic compounds that have a pharmacological function.

From the analysis of the x-ray diffraction of a single crystal, one obtains information on the three-dimensional conformation of the compound, a complete summary of the bond angles and bond lengths of the compound in question, and detailed information regarding how the molecules assemble to yield the complete crystal.

7.1.2.1 Case study: theophylline anhydrate and monohydrate

The type of structural information that can be obtained from the study of the x-ray diffraction of single crystals will be illustrated through an exposition of studies conducted on the anhydrate and hydrate phases of theophylline (3,7-dihydro-1,3-dimethyl-1H-purine-2,6-dione). The unit cell parameters defining the two phases are listed in Table 7.2, while the structure of this compound and a suitable atomic numbering system is located in Fig. 7.3.

As discussed above, one obtains a large amount of structural information upon completion of the data analysis, which will be illustrated using the example of theophylline monohydrate [20]. Table 7.3 contains a summary of intramolecular distances involving the non-hydrogen atoms in the structure, while Table 7.4 shows the intramolecular

TABLE 7.2

Crystallographic parameters defining the unit cells of theophylline anhydrate and theophylline monohydrate phase

	Anhydrate [18]	Monohydrate [19]
Crystal class	Orthorhombic	Monoclinic
Space group	$Pna2_1$	$P2_1/n$
Unit cell lengths	$a = 24.612 \text{ Å}$	$a = 4.468 \text{ Å}$
	$b = 3.8302 \text{ Å}$	$b = 15.355 \text{ Å}$
	$c = 8.5010 \text{ Å}$	$c = 13.121 \text{ Å}$
Unit cell angles	$\alpha = 90°$	$\alpha = 90°$
	$\beta = 90°$	$\beta = 97.792°$
	$\gamma = 90°$	$\gamma = 90°$
Molecules in unit cell	4	4
Cell volume	801.38 Å^3	891.9 Å^3
Density	1.493 g/cm^3	1.476 g/ml

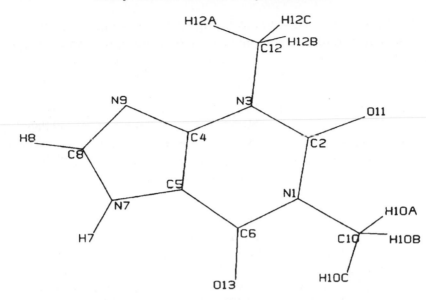

Fig. 7.3. Molecular structure and suitable atomic numbering system for theophylline.

TABLE 7.3

Intramolecular distances in theophylline monohydrate involving non-hydrogen atoms [20]

Atom(1)	Atom(2)	Distance (Å)[a]
O_{11}	C_2	1.219 (8)
O_{13}	C_6	1.233 (8)
N_1	C_2	1.420 (9)
N_1	C_6	1.399 (8)
N_1	C_{10}	1.486 (8)
N_3	C_2	1.379 (8)
N_3	C_4	1.375 (8)
N_3	C_{12}	1.469 (8)
N_7	C_5	1.407 (8)
N_7	C_8	1.331 (9)
N_9	C_4	1.365 (8)
N_9	C_8	1.345 (9)
C_4	C_5	1.350 (9)
C_5	C_6	1.421 (9)

[a]The estimated error in the last digit is given in parentheses.

TABLE 7.4

Intramolecular distances in theophylline monohydrate involving hydrogen atoms [20]

Atom(1)	Atom(2)	Distance (Å)
O_{14}	H_{14A}	0.833
O_{14}	H_{14B}	0.985
N_7	H_7	0.877
C_8	H_8	0.986
C_{10}	H_{10A}	0.926
C_{10}	H_{10B}	0.967
C_{10}	H_{10C}	0.950
C_{12}	H_{12A}	0.954
C_{12}	H_{12B}	0.973
C_{12}	H_{12C}	0.932

distances involving hydrogen atoms. Table 7.5 contains a summary of bond angles involving the non-hydrogen atoms, and Table 7.6 the bond angles involving hydrogen atoms.

The compilation of bond length and angle information, all of which must agree with known chemical principles, enables one to deduce the molecular conformation of the molecule in question. Such illustrations are usually provided in the form of ORTEP drawings, where atomic sizes are depicted in the form of ellipsoids at the 50% probability level. The molecular conformation of theophylline in the anhydrate and monohydrate phase is shown in Fig. 7.4. As would be expected from a rigid and planar molecule, these conformations do not differ significantly.

Of far greater use to structural scientists are the visualizations that illustrate how the various molecules assemble to constitute the unit cell, and the relationships of the molecules in them. This process is illustrated in Figs. 7.5–7.7 for our study of theophylline monohydrate, which first show the details of the unit cell and proceed on to provide additional information about the molecular packing motifs in the crystal. As one progresses through the depictions of Figs. 7.5–7.7, the role of the water molecules in the structure becomes clear. Theophylline molecules are effectively bound into dimeric units by the bridging water molecules, which ultimately form an infinite chain of water molecules hydrogen bonded to layers of theophylline molecules [19].

The molecular packing in the monohydrate phase is fundamentally different from that of the anhydrate phase, as shown in Fig. 7.8 [18].

TABLE 7.5

Intramolecular bond angles in theophylline monohydrate involving non-hydrogen atoms [20]

Atom(1)	Atom(2)	Atom(3)	Bond angle (degrees)[a]
C_2	N_1	C_6	127.6 (6)
C_2	N_1	C_{10}	115.4 (6)
C_6	N_1	C_{10}	116.9 (6)
C_2	N_3	C_4	119.6 (6)
C_2	N_3	C_{12}	118.6 (6)
C_4	N_3	C_{12}	121.7 (6)
C_5	N_7	C_8	105.9 (6)
C_4	N_9	C_8	102.2 (6)
O_{11}	C_2	N_1	121.2 (7)
O_{11}	C_2	N_3	122.7 (7)
N_1	C_2	N_3	116.1 (7)
N_3	C_4	N_9	125.1 (6)
N_3	C_4	C_5	121.6 (6)
N_9	C_4	C_5	113.4 (7)
N_7	C_5	C_4	104.5 (6)
N_7	C_5	C_6	130.6 (7)
C_4	C_5	C_6	124.9 (7)
O_{13}	C_6	N_1	121.7 (7)
O_{13}	C_6	C_5	128.1 (6)
N_1	C_6	C_5	110.2 (7)
N_7	C_8	N_9	114.1 (6)

[a]The estimated error in the last digit is given in parentheses.

Here, the theophylline molecules form hydrogen-bonded networks composed of one N–H$\cdots$N hydrogen bond and two bifurcated C–H$\cdots$O hydrogen bonds.

7.1.2.2 *Crystallographic studies of polymorphism*

It is now accepted that the majority of organic compounds are capable of being crystallized into different crystal forms [21–23]. One defines *poly-morphism* as signifying the situation where a given compound crystallizes into more than one structure, and where the two crystals would yield exactly the same elemental analysis. *Solvatomorphism* is defined as the situation where a given compound crystallizes in more than one structure, and where the two crystals yield differing elemental analyses owing to the inclusion of solvent molecules in the crystal structure. Theophylline, discussed above in detail, is a solvatomorphic system.

H.G. Brittain

TABLE 7.6

Intramolecular bond angles in theophylline monohydrate involving hydrogen atoms [20]

Atom(1)	Atom(2)	Atom(3)	Bond angle (degrees)
H_{14A}	O_{14}	H_{10B}	104.53
C_5	N_7	H_7	125.5
C_8	N_7	H_7	129.5
N_7	C_8	H_8	123.4
N_9	C_8	H_8	122.4
N_1	C_{10}	H_{10A}	109.95
N_1	C_{10}	H_{10B}	107.73
N_1	C_{10}	H_{10C}	109.25
H_{10A}	C_{10}	H_{10B}	110.10
H_{10A}	C_{10}	H_{10C}	111.65
H_{10B}	C_{10}	H_{10C}	108.07
N_3	C_{12}	H_{12A}	109.58
N_3	C_{12}	H_{12B}	108.63
N_3	C_{12}	H_{12C}	111.62
H_{12A}	C_{12}	H_{12B}	107.20
H_{12A}	C_{12}	H_{12C}	110.67
H_{12B}	C_{12}	H_{12C}	109.02

When considering the structures of organic molecules, one finds that different modifications can arise in two main distinguishable ways. Should the molecule be constrained to exist as a rigid grouping of atoms, these may be stacked in different motifs to occupy the points of different lattices. This type of polymorphism is then attributable to packing phenomena, and so is termed packing polymorphism. On the other hand, if the molecule in question is not rigidly constructed and can exist in distinct conformational states, then it can happen that each of these conformationally distinct modifications may crystallize in its own lattice structure. This latter behavior has been termed conformational polymorphism [24].

During the very first series of studies using single-crystal x-ray crystallography to determine the structures of organic molecules, Robertson reported the structure of resorcinol (1,3-dihydroxybenzene) [25]. This crystalline material corresponded to that ordinarily obtained at room temperature, and was later termed the α-form. Shortly thereafter, it was found that the α-form underwent a transformation into a denser crystalline modification (denoted as the β-form) when heated at about 74°C,

190

Anhydrate phase

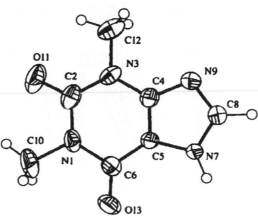

Monohydrate phase

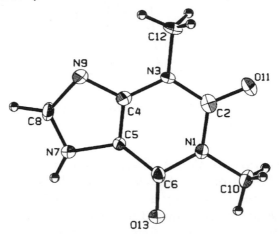

Fig. 7.4. Conformation of the theophylline molecule in the anhydrate phase [18], used with permission, and in the monohydrate phase [20].

and that the structure of this newer form was completely different [26]. The salient crystallographic properties of these two forms are summarized in Table 7.7, and the crystal structures of the α- and β-forms (viewed down the c-axis, or (0 0 1) crystal plane) are found in Fig. 7.9.

By its nature, resorcinol is locked into a single conformation, and it is immediately evident from a comparison of the structures in Fig. 7.9 that each form is characterized by a different motif of hydrogen bonding. In particular, the α-form features a relative open architecture that is maintained by a spiraling array of hydrogen bonding that ascends

191

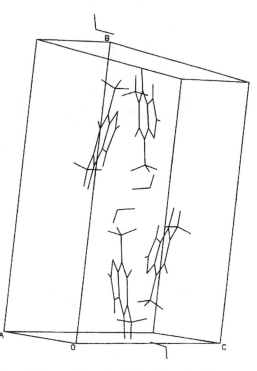

Fig. 7.5. Composition of the unit cell of the theophylline monohydrate crystal phase [20].

through the various planes of the crystal. In the view illustrated (defined by the *ab*-plane), the apparent closed tetramic grouping of hydroxyl groups is actually a slice through the ascending spiral. The effect of the thermally induced phase transformation is to collapse the open arrangement of the α-form by a more compact and parallel arrangement of the molecules in the β-form. This structural change causes an increase in crystal density on passing from the α-form $(1.278\,\text{g/cm}^3)$ to the β-form $(1.327\,\text{g/cm}^3)$. In fact, the molecular packing existing in the β-form was described as being more typical of hydrocarbons than of a hydroxylic compound [26].

Probucol (4,4'-[(1-methylethylidene)bis(thio)]-bis-[2,6-bis(1,1-dimethylethyl)phenol]) is a cholesterol-lowering drug that has been reported to exist in two forms [27]. Form II has been found to exhibit a lower melting point onset relative to Form I, and samples of Form II spontaneously transform to Form I upon long-term storage. The structures of these two polymorphic forms have been reported, and a summary of the crystallographic data obtained in this work is provided in

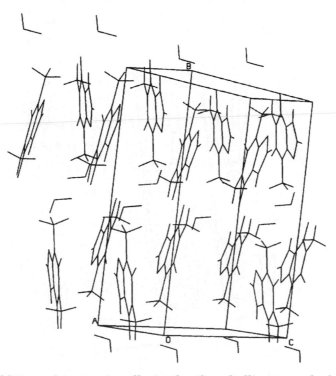

Fig. 7.6. Addition of two unit cells in the theophylline monohydrate crystal phase [20].

Table 7.8. In addition, detailed views of the crystal structures are given in Fig. 7.10.

The conformations of the probucol molecule in the two forms were found to be quite different. In Form II, the C–S–C–S–C chain is extended, and the molecular symmetry approximates C_{2v}. This molecular symmetry is lost in the structure of Form I, where now the torsional angles around the two C–S bonds deviate significantly from 180°. Steric crowding of the phenolic groups by the t-butyl groups was evident from deviations from trigonal geometry at two phenolic carbons in both forms. Using a computational model, the authors found that the energy of Form II was 26.4 kJ/mol higher than the energy of Form I, indicating the less-symmetrical conformer to be more stable. The crystal density of Form I was found to be approximately 5% higher than that of Form II, indicating that the conformational state of the probucol molecules in Form I yielded more efficient space filling.

Owing to the presence of the functional groups in drug substance molecules that promote their efficacious action, the crystallization

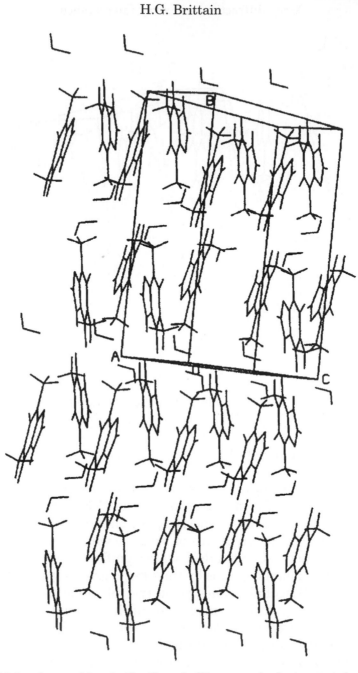

Fig. 7.7. Molecular packing in the theophylline monohydrate crystal phase [20].

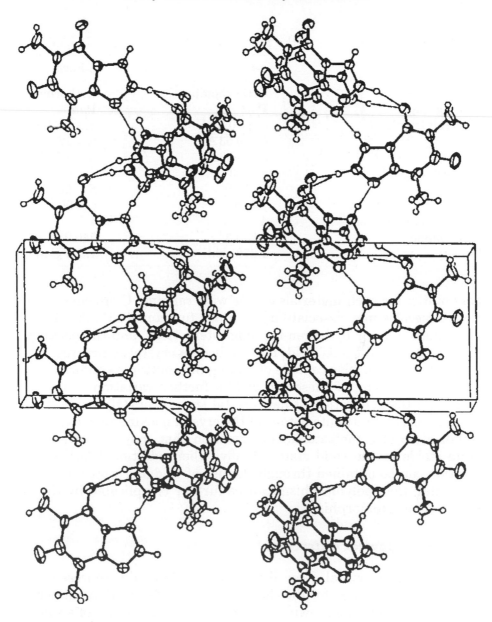

Fig. 7.8. Molecular packing in the theophylline anhydrate crystal phase. Adapted from Ref. [18].

TABLE 7.7

Crystallographic properties of the two polymorphs of resorcinol [24,25]

	α-Form	β-Form
Crystal class	Orthorhombic	Orthorhombic
Space group	**P**na	**P**na
Unit cell lengths	$a = 10.53\,\text{Å}$	$a = 7.91\,\text{Å}$
	$b = 9.53\,\text{Å}$	$b = 12.57\,\text{Å}$
	$c = 5.66\,\text{Å}$	$c = 5.50\,\text{Å}$
Unit cell angles	$\alpha = 90°$	$\alpha = 90°$
	$\beta = 90°$	$\beta = 90°$
	$\gamma = 90°$	$\gamma = 90°$
Molecules in unit cell	4	4
Cell volume	$567.0\,\text{Å}^3$	$546.9\,\text{Å}^3$
Density	$1.278\,\text{g/ml}$	$1.327\,\text{g/ml}$

possibilities for such materials can be wide ranging. Conformationally, rigid molecules may associate into various fundamental units through alterations in their hydrogen-bonding interactions, and the packing of these different unit cells yields non-equivalent crystal structures. When a drug molecule is also capable of folding into multiple conformational states, the packing of these can yield a further variation in unit cell types and the generation of new crystal structures. Often, the lattice interactions and intimate details of the crystal packing may lead to the stabilization of a metastable conformation that may nevertheless be quite stable in the solid state. Finally, when additional lattice stabilization can be obtained through the use of bridging water or solvate molecules, then polymorphism can be further complicated by the existence of solvatomorphism.

7.1.3 X-ray powder diffraction

Although single-crystal x-ray diffraction undoubtedly represents the most powerful method for the characterization of crystalline materials, it does suffer from the drawback of requiring the existence of a suitable single crystal. Very early in the history of x-ray diffraction studies, it was recognized that the scattering of x-ray radiation by powdered crystalline solids could be used to obtain structural information, leading to the practice of x-ray powder diffraction (XRPD).

The XRPD technique has become exceedingly important to pharmaceutical scientists, since it represents the easiest and fastest method

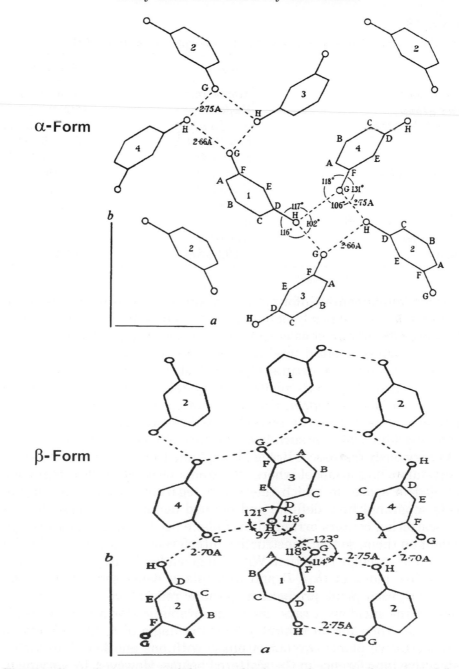

Fig. 7.9. Crystal structures of the α- and β-forms of resorcinol, as viewed down the c-axis (0 0 1 plane). The figure is adapted from data given in Refs. [24] and [25].

TABLE 7.8

Crystallographic properties of the two polymorphs of probucol [27]

	Form I	Form II
Crystal class	Monoclinic	Monoclinic
Space group	$P2_1/c$	$P2_1/n$
Unit cell lengths	$a = 16.972$ Å	$a = 11.226$ Å
	$b = 10.534$ Å	$b = 15.981$ Å
	$c = 19.03$ Å	$c = 18.800$ Å
Unit cell angles	$\alpha = 90°$	$\alpha = 90°$
	$\beta = 113.66°$	$\beta = 104.04°$
	$\gamma = 90°$	$\gamma = 90°$
Molecules in unit cell	4	4
Cell volume	3116.0 Å^3	3272.0 Å^3
Density	1.102 g/ml	1.049 g/ml

to obtain fundamental information on the structure of a crystalline substance in its ordinarily obtained form. Since the majority of drug substances are obtained as crystalline powders, the powder pattern of these substances is often used as a readily obtainable fingerprint for determination of its structural type. In fact, it is only by pure coincidence that two compounds might form crystals for which the ensemble of molecular planes happened to be identical in all space. One such example is provided by the respective trihydrate phases of ampicillin and amoxicillin [28], but such instances are uncommon.

As previously discussed, Bragg [4] explained the diffraction of x-rays by crystals using a model where the atoms of a crystal are regularly arranged in space, in which they were regarded as lying in parallel sheets separated by a definite and defined distance. He then showed that scattering centers arranged in a plane act like a mirror to x-rays incident on them, so that constructive interference would occurs for the direction of specular reflection. Within a given family of planes, defined by a Miller index of $(h\ k\ l)$ and each plane being separated by the distance $\mathbf{d}$, each plane produces a specular reflectance of the incident beam. If the incident x-rays are monochromatic (having wavelength equal to λ), then for an arbitrary glancing angle of θ, the reflections from successive planes are out of phase with one another. This yields destructive interference in the scattered beams. However, by varying θ, a set of values for θ can be found so that the path difference between x-rays reflected by successive planes will be an integral number ($\mathbf{n}$) of wavelengths, and then constructive interference will occurs.

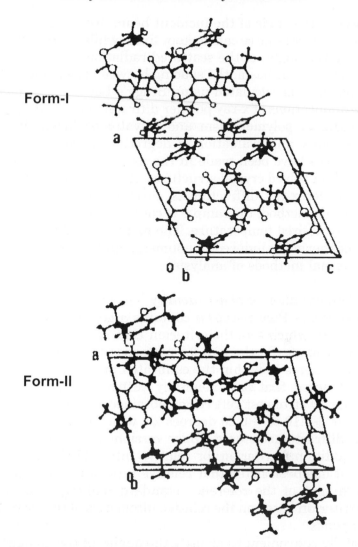

Fig. 7.10. Alternate representation of crystal structures of the α- and β-forms of resorcinol, as viewed down the c-axis (0 0 1 plane). The figure is adapted from data given in Refs. [24] and [25].

The mathematical relationship governing the mechanics of diffraction was discussed earlier as Bragg's law.

To measure a powder pattern, a randomly oriented powdered sample is prepared so as to expose all possible planes of a crystalline powder. The scattering angle, θ, is measured for each family of crystal planes by slowly rotating the sample and measuring the angle of diffracted x-rays with respect to the angle of the incident beam. Alternatively, the angle between sample and source can be kept fixed, while moving the detector to determine the angles of the scattered radiation. Knowing the wavelength of the incident beam, the spacing between the planes (identified as the d-spacings) is calculated using Bragg's Law.

Typical applications of x-ray powder diffraction methodology include the evaluation of polymorphism and solvatomorphism, the study of phase transitions, and evaluation of degrees of crystallinity. More recently, advances have been made in the use of powder diffraction as a means to obtain solved crystal structures. A very useful complement to ordinary powder x-ray diffraction is variable temperature x-ray diffraction. In this method, the sample is contained on a stage that can be heated to any desired temperature. The method is extremely useful for the study of thermally induced phenomena, and can be a vital complement to thermal methods of analysis.

7.1.3.1 *Determination of phase identity*

The United States Pharmacopeia contains a general chapter on x-ray diffraction [29], which sets the criterion that identity is established if the scattering angles in the powder patterns of the sample and reference standard agree to within the calibrated precision of the diffractometer. It is noted that it is generally sufficient that the scattering angles of the 10 strongest reflections obtained for an analyte agree to within either ± 0.10 or ± 0.20 degrees 2θ, whichever is more appropriate for the diffractometer used. Older versions of the general test contained an additional criterion for relative intensities of the scattering peaks, but it has been noted that relative intensities may vary considerably from that of the reference standard making it impossible to enforce a criterion based on the relative intensities of the corresponding scattering peaks.

It is usually convenient to identify the angles of the 10 most intense scattering peaks in a powder pattern, and to then list the accepted tolerance ranges of these based on the diffractometer used for the determinations. Such a representation has been developed for data obtained for racemic mandelic acid, and for its separated (S)-enantiomer, using a diffractometer system whose precision was known to be ± 0.15 degrees 2θ [30]. The criteria are shown in Table 7.9, the form of which enables the ready identification of a mandelic acid sample as being either racemic or enantiomerically pure. It should be noted that the powder pattern of the separated (R)-enantiomer would necessarily have to be identical to that of the separated (S)-enantiomer.

TABLE 7.9

XRPD identity test criteria for mandelic acid

Lower limit of acceptability for scattering peak (degrees 2θ)	Accepted angle for scattering peak (degrees 2θ)	Upper limit of acceptability for scattering peak (degrees 2θ)
Racemic phase		
11.271	11.421	11.571
16.264	16.414	16.564
18.607	18.757	18.907
21.715	21.865	22.015
22.173	22.323	22.473
23.345	23.495	23.645
25.352	25.502	25.652
27.879	28.029	28.179
33.177	33.327	33.477
41.857	42.007	42.157
Enantiomerically pure (S)-phase		
6.381	6.531	6.681
6.686	6.836	6.986
7.298	7.448	7.598
8.011	8.161	8.311
20.798	20.948	21.098
23.345	23.495	23.645
24.109	24.259	24.409
26.605	26.755	26.905
30.018	30.168	30.318
36.284	36.434	36.584

Useful tabulations of the XRPD patterns of a number of compounds have been published by Koundourellis and coworkers, including 12 diuretics [31], 12 vasodilators [32], and 12 other commonly used drug substances [33]. These compilations contain listings of scattering angles, d-spacings, and relative intensities suitable for the development of acceptance criteria.

Without a doubt, one of the most important uses of XRPD in pharmaceuticals is derived from application as the primary determinant of polymorphic or solvatomorphic identity [34]. Since these effects are due to purely crystallographic phenomena, it is self-evident that x-ray diffraction techniques would represent the primary method of determination. Owing to its ease of data acquisition, XRPD is particularly useful as a screening technique for batch characterization, following the same general criteria already described. It is prudent, however, to confirm the results of an XRPD study through the use of a confirmatory technique, such as polarizing light microscopy, differential scanning calorimetry, solid-state vibrational spectroscopy, or solid-state nuclear magnetic resonance.

The literature abounds with countless examples that illustrate how powder diffraction has been used to distinguish between the members of a polymorphic system. It is absolutely safe to state that one could not publish the results of a phase characterization study without the inclusion of XRPD data. For example, Fig. 7.11 shows the clearly distinguishable XRPD powder patterns of two anhydrous forms of a new chemical entity. These are easily distinguishable on the overall basis of their powder patterns, and one could place the identification on a more quantitative basis through the development of criteria similar to those developed for the mandelic acid system.

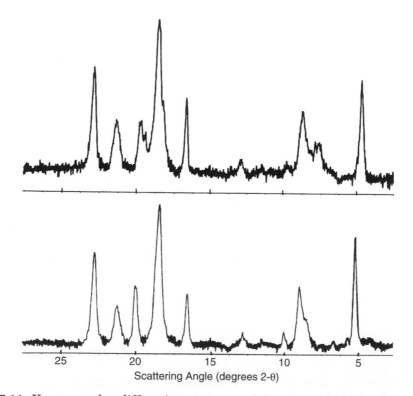

Fig. 7.11. X-ray powder diffraction patterns of the two anhydrous forms of a new chemical entity.

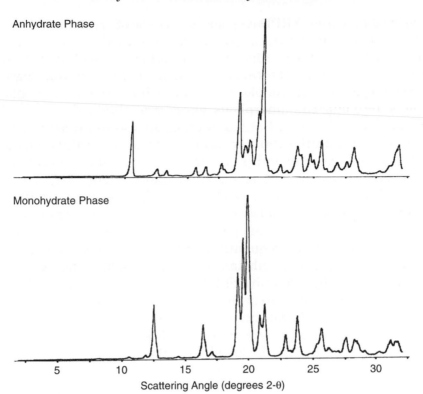

Fig. 7.12. X-ray powder diffraction patterns of the anhydrate and monohydrate phases of lactose.

XRPD can be similarly used to differentiate between the members of a solvatomorphic system. For instance, Fig. 7.12 shows the powder patterns obtained for the anhydrate and monohydrate phases of lactose. The existence of structural similarities in the two forms are suggested in that the main scattering peaks of each form are clustered near 20 degrees 2θ, but the two solvatomorphs are easily differentiated from an inspection of the patterns. The differentiation could also be rendered more quantitative through the development of a table similar to that developed for the mandelic acid system.

7.1.3.2 Degree of crystallinity

When reference samples of the pure amorphous and pure crystalline phases of a substance are available, calibration samples of known degrees of crystallinity can be prepared by the mixing of these. Establishment of a calibration curve (XRPD response vs. degree of crystallinity) permits the evaluation of unknown samples to be performed.

In a classic study that illustrates the principles of the method, an XRPD procedure was described for the estimation of the degree of crystallinity in digoxin samples [35]. Crystalline product was obtained commercially, and the amorphous phase was obtained through ball-milling of this substance. Calibration mixtures were prepared as a variety of blends from the 100% crystalline and 0% crystalline materials, and acceptable linearity and precision was obtained in the calibration curve of XRPD intensity vs. actual crystallinity. Figure 7.13 shows the powder pattern of an approximately 40% crystalline material, illustrating how these workers separated out the scattering contributions from the amorphous and crystalline phases.

Other studies have used quantitative XRPD to evaluate the degree of crystallinity in bulk drug substances, such as calcium gluceptate [36], imipenum [37], cefditoren pivoxil [38], and a Lumaxis analog [39].

When the excipient matrix in a formulation is largely amorphous, similar XRPD methods can be used to determine the amount of crystalline drug substance present in a drug product. The principles were established in a study that included a number of solid dose forms [40],

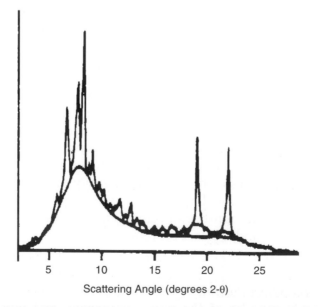

Scattering Angle (degrees 2-θ)

Fig. 7.13. X-ray powder diffraction pattern of digoxin, whose degree of crystallinity was reduced to 40% by ball-milling. The figure was adapted from data contained in Ref. [32].

and also in another study involving a determination of the amount of crystalline acetaminophen in some solid dispersions [41]. The same principles have been used to characterize the crystalline content within lyophilized solids [42], where the matrix is again an amorphous material.

7.1.3.3 Phase composition of mixtures

Once the characteristic XRPD patterns of one or more analytes have been established, it is usually possible to develop quantitative methods of analysis. The methodology is based on the premise that each component will contribute to the overall scattering by an amount that is proportional to its weight fraction in a mixture, and that the powder pattern of each analyte contains one or more peaks whose scattering angle is unique to that analyte. Quantitative analysis requires the use of reference standards that contribute known scattering peaks at appropriate scattering intensities. This can be achieved through the use of metal sample holders where, some of the metal scattering peaks are used as external standards. The author has had great success in this regard using aluminum sample holders, using the scattering peaks at 38.472 and 44.738 degrees 2θ to calibrate both intensity and scanning rate. Others have used internal standards, mixing materials such as elemental silicon or lithium fluoride into the sample matrix.

Although simple intensity correction techniques can be used to develop very adequate XRPD methods of quantitative analysis, the introduction of more sophisticated data acquisition and handling techniques can greatly improve the quality of the developed method. For instance, improvement of the powder pattern quality through the use of the Rietveld method has been used to evaluate mixtures of two anhydrous polymorphs of carbamazepine and the dihydrate solvatomorph [43]. The method of whole pattern analysis developed by Rietveld [44] has found widespread use in crystal structure refinement and in the quantitative analysis of complex mixtures. Using this approach, the detection of analyte species was possible even when their concentration was less than 1% in the sample matrix. It was reported that good quantitation of analytes could be obtained in complex mixtures even without the requirement of calibration curves.

The use of parallel beam optics as a means for determining the polymorphic composition in powder compacts has been discussed [45]. In this study, compressed mixtures of known polymorphic composition were analyzed in transmission mode, and the data were processed using profile-fitting software. The advantage of using transmission, rather than the reflectance, is that the results were not sensitive to the geometrical details of the compact surfaces and that spurious effects associated with preferential orientation were minimized.

The effects of preferred orientation in XRPD analysis can be highly significant, and are most often encountered when working with systems characterized by plate-like or tabular crystal morphologies. As discussed earlier, a viable XRPD sample is the one that presents equal numbers of all scattering planes to the incident x-ray beam. Any orientation effect that minimizes the degree of scattering from certain crystal planes will strongly affect the observed intensities, and this will in turn strongly affect the quantitation. The three polymorphs of mannitol are obtained as needles, and thus represented a good system for evaluating the effect of preferential orientation on quantitative XRPD [46]. Through the use of small particle sizes and sample rotation, the preferential orientation effects were held to a minimum, and allowed discrimination of the polymorphs at around the 1% level.

As a first example, consider ranitidine hydrochloride, which is known to crystallize into two anhydrous polymorphs. The XRPD patterns of the two forms are shown in Fig. 7.14, where it is evident that the two structures are significantly different [47]. For a variety of reasons, most workers are extremely interested in determining the quantity of Form II in a bulk Form I sample, and it is clear from the figure that the scattering peaks around 20 and 23.5 degrees 2θ would be particularly useful for this purpose. The analysis of Form II in bulk Form I represents an extremely favorable situation, since the most intense scattering peaks of the analyte are observed at scattering angles where the host matrix happens not to yield diffraction. In our preliminary studies, it was estimated that a limit of detection of 0.25% w/w Form II in bulk Form I could be achieved. On the other hand, if there was a reason to determine the level of Form I in a bulk Form II sample, then the scattering peaks around 9 and 25 degrees 2θ would suffice. For this latter instance, a preliminary estimate of limit of detection of 0.50% w/w Form I in bulk Form II was deduced [47].

The complications associated with preferential orientation effects were addressed in detail during studies of a benzoic acid/benzil system [48]. The use of various sample-packing methods was considered (vacuum free-fall, front-faced packing vs. rear-faced packing, etc.), but the best reduction in the effect was achieved by using materials having small particle sizes that were produced by milling. Through the use of sieving and milling, excellent linearity in diffraction peak area as a function of analyte concentration was attained. The authors deduced a

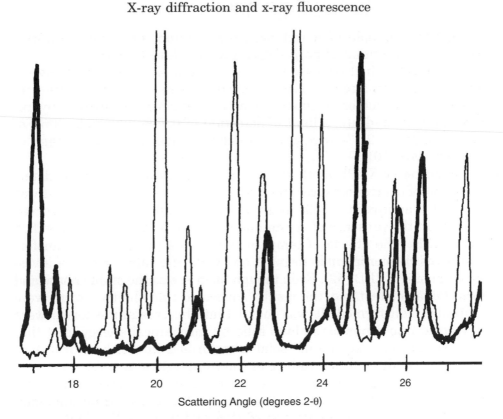

Fig. 7.14. X-ray powder diffraction patterns of ranitidine hydrochloride, Form I (thick trace) and Form II (thin trace) [47].

protocol for the development of a quantitative XRPD method that consisted of six main steps:

1. Calculation of the mass absorption coefficient of the drug substance
2. Selection of appropriate diffraction peaks for quantification
3. Evaluation of the loading technique for adequate sample size
4. Determination of whether preferred orientation effects can be eliminated through control of the sample particle size
5. Determination of appropriate milling conditions to obtain reproducibility in peak areas
6. Generation of calibration curves from physical mixtures.

After completion of these steps, one should then have a quantitative XRPD method that is suitable for analysis of real samples.

207

Another example of a well-designed method for the quantitative XRPD determination of polymorphs was developed for the phase analysis of prazosin hydrochloride [49]. As an example of an XRPD method for the determination of solvatomorphs, during the quantitation of cefepine dihydrochloride dihydrate in bulk samples of cefepine dihydrochloride monohydrate, a limit of detection of 0.75% w/w and a limit of quantitation of 2.5% w/w were associated with a working range of 2.5–15% w/w [50].

Quantitative XRPD methods have also been developed for the determination of drug substances in excipient matrices. For instance, it was found that approximately 2% of selegilin hydrochloride can be observed reliably in the presence of crystalline mannitol or amorphous modified starch [51]. The effects of temperature on the polymorphic transformation of chlorpropamide forms A and C during tableting were investigated using XRPD [52]. Even though form A was the stable phase and form C was metastable, the results suggested that the two forms were mutually transformed. It was found that the crystalline forms were converted into a non-crystalline solid by mechanical energy, and that the resulting non-crystalline solid transformed into either form A or form C depending on the nature of the compression process.

Since the separated enantiomers of a dissymmetric compound must crystallize in a different space group than does the racemic mixture, it should not be unanticipated that quantitative XRPD would be useful in the determination of enantiomeric composition. For instance, the differing XRPD characteristics of (S)-$(+)$-ibuprofen relative to the (RS)-racemate have been exploited to develop a sound method for the determination of the enantiomeric purity of ibuprofen samples [53].

7.1.3.4 Hot-stage x-ray diffraction

The performance of XRPD on a hot stage enables one to obtain powder patterns at elevated temperatures, and permits one to deduce structural assignments for thermally induced phase transitions. Determination of the origin of thermal events taking place during the conduct of differential thermal analysis or differential scanning calorimetry is not always straight-forward, and the use of supplementary XRPD technology can be extremely valuable. By conducting XRPD studies on a heatable stage, one can bring the system to positions where a Differential Scanning Calorimetry (DSC) thermogram indicates the existence of an interesting point of thermal equilibrium.

One of the uses for thermodiffractometry that immediately comes to mind concerns the desolvation of solvatomorphs. For instance, after the dehydration of a hydrate phase, one may obtain either a crystalline anhydrate phase or an amorphous phase. The XRPD pattern of a dc-hydrated hydrate will clearly indicate the difference. In addition, should one encounter an equivalence in powder patterns between the hydrate phase and its dehydrated form, this would indicate the existence of channel-type water (as opposed to genuine lattice water) [54].

An XRPD system equipped with a heatable sample holder has been described, which permitted highly defined heating up to 250°C [55]. The system was used to study the phase transformation of phenanthrene, and the dehydration of caffeine hydrate. An analysis scheme was developed for the data that permitted one to extract activation parameters for these solid-state reactions from a single non-isothermal study run at a constant heating rate.

In another study, thermodiffractometry was used to study phase transformations in mannitol and paracetamol, as well as the desolvation of lactose monohydrate and the dioxane solvatomorph of paracetamol [56]. The authors noted that in order to obtain the best data, the heating cycle must be sufficiently slow to permit the thermally induced reactions to reach completion. At the same time, the use of overly long cycle times can yield sample decomposition. In addition, the sample conditions are bound to differ relative to the conditions used for a differential scanning calorimetry analysis, so one should expect some differences in thermal profiles when comparing data from analogous studies.

The commercially available form of Aspartame is hemihydrate Form II, which transforms into hemihydrate Form I when milled, and a 2.5-hydrate species is also known [57,58]. XRPD has been used to study the desolvation and ultimate decomposition of the various hydrates. When heated to 150°C, both hemihydrate forms dehydrate into the same anhydrous phase, which then cyclizes into 3-(carboxymethyl)-6-benzyl-2, 5-dioxopiperazine if heated to 200°C. The 2.5-hydrate was shown to dehydrate into hemihydrate Form II when heated to 70°C, and this product was then shown to undergo the same decomposition sequence as directly crystallized hemihydrate Form II.

7.1.3.5 *XRPD as a stability-indicating assay method*
When the phase identity, or degree of crystallinity (or lack thereof), of a drug substance is important to its performance in a drug product, XRPD can serve as a vital stability-indicating assay method. There is no doubt that XRPD can be validated to the status of any other stability-indicating assay, and that one can use the usual criteria of method validation to establish the performance parameters of the method. This aspect would

be especially important when either a metastable or an amorphous form of the drug substance has been chosen for development. One may conduct such work either on samples that have been stored at various conditions and pulled at designated time points, or on substances that are maintained isothermally and the XRPD is periodically measured.

For example, amorphous clarithromycin was prepared by grind and spray-drying processes, and XRPD was used to follow changes in crystallinity upon exposure to elevated temperature and relative humidity [59]. Exposure of either substance to a 40°C/82% RH environment for seven days led to the formation of the crystalline form, but the spray-dried material yielded more crystalline product than did the ground material. This finding, when supported with thermal analysis studies, led to the conclusion that the amorphous substances produced by the different processing methods were not equivalent.

7.2   X-RAY FLUORESCENCE

7.2.1   Introduction

Although the use of quantum theory leads to a much greater understanding of x-ray fluorescence (XRF) phenomena, the theory developed by Bohr in the early 1900s proved sufficient to understand the nature of the processes involved. Bohr had been able to obtain a solution to the simultaneous equations for the electron angular momentum, the balance of coulombic attraction and centrifugal force, and Planck's equation, to derive a formula analogous to the empirical Balmer–Rydberg–Ritz equation that had been used to interpret the atomic spectrum of hydrogen. The following discussion is abstracted from several leading Refs. [60–63].

> From Bohr's postulates, it is possible to derive the energies of the possible stationary states responsible for the radiation that is absorbed or emitted by an atom consisting of a single electron and nucleus. The specification of these states permits one to then compute the frequency of the associated electromagnetic radiation. To begin, one assumes the charge on the nucleus to be Z times the fundamental electronic charge, e, and that Coulomb's law provides the attractive force, F, between the nucleus and electron:

$$F = Ze^2/r^2 \qquad (7.5)$$

where r is the radius of the electron orbit. This force must be exactly balanced by the centrifugal force on the rotating electron:

$$F = mr\omega^2 \tag{7.6}$$

where ω is the angular velocity of the electron, and m is its mass. Since equations (7.5) and (7.6) must be equal, we find that:

$$mr\omega^2 = Ze^2/r^2 \tag{7.7}$$

The angular momentum, L, of a circulating particle is defined as its linear momentum multiplied by the radius of its circular motion. Bohr's second postulate requires that the angular momentum of a rotating electron be an integral multiple of $\hbar$:

$$L = n\hbar = mr^2\,\omega \tag{7.8}$$

where n is an integer having the values of 1, 2, 3, etc. Solving equations (7.7) and (7.8) for the orbital radius yields:

$$r = \frac{n^2\hbar^2}{me^2Z} \tag{7.9}$$

One now finds that the orbital radius us restricted to certain values, termed orbits, having values equal to $\hbar^2/me^2Z$, $4\hbar^2/me^2Z$, $9\hbar^2/me^2Z$, etc. For the smallest allowed orbit of a hydrogen atom (defined by $Z = 1$ and $n = 1$), one finds that:

$$r_0 \equiv a_0 = \hbar^2/me^2 \tag{7.10}$$

and that $a_0 = 5.29 \times 10^{-11}$ m (0.529Å).

The Bohr theory also permits a calculation of the total energy, E, of the atom from Hamilton's equation:

$$E = T + V \tag{7.11}$$

where the kinetic energy, T, exists by virtue of the rotation motion of the electron:

$$T = 1/2mr^2\omega^2 = 1/2Ze^2/r \tag{7.12}$$

and the potential energy, V, exists by virtue of the position of the electron relative to the nucleus:

$$V = -Ze^2/r \qquad (7.13)$$

Substituting equations (7.12) and (7.13) into (7.11), one finds that the total energy is given by:

$$E = -1/2Ze^2/r \qquad (7.14)$$

and if the Bohr equation for r (7.9) is substituted into equation (7.14), one finds:

$$E = \frac{-mZ^2e^4}{2n^2\hbar^2} \qquad (7.15)$$

It is therefore concluded that only certain stationary states are possible, and that the state having the lowest total energy is defined by $n = 1$.

A transition between two states requires the absorption or emission of energy equal to the energy difference between the states. By the postulate of Bohr, this difference in energy must be quantized, so that

$$\Delta E = E_1 - E_2 = h\nu \qquad (7.16)$$

$$= \frac{me^4}{2n^2h}\left[\frac{1}{n_1^2} - \frac{1}{n_2^2}\right] \qquad (7.17)$$

7.2.2 Principles of x-ray fluorescence spectroscopy

As illustrated in Fig. 7.15, the electromagnetic radiation measured in an XRF experiment is the result of one or more valence electrons filling the vacancy created by an initial photoionization where a core electron was ejected upon absorption of x-ray photons. The quantity of radiation from a certain level will be dependent on the relative efficiency of the radiationless and radiative deactivation processes, with this relative efficiency being denoted at the fluorescent yield. The fluorescent yield is defined as the number of x-ray photons emitted within a given series divided by the number of vacancies formed in the associated level within the same time period.

To apply the Bohr energy level equation Eq. (7.17), one identifies the levels having $n = 1$ as being from the K shell, levels having $n = 2$ are

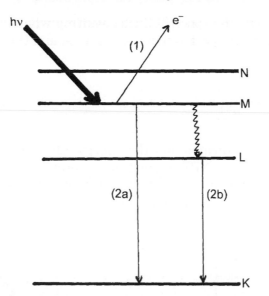

Fig. 7.15. Photophysics associated with x-ray photoelectron spectroscopy and x-ray fluorescence. As illustrated, in the XPS experiment one monitors the energy of the electron ejected from the M shell upon photoionization (process 1). In the XRF experiment, one monitors the fluorescence emitted from either the M shell after photoionization (process 2a), or from the L shell after photo ionization and radiationless decay (process 2b).

from the L shell, levels having $n = 3$ are from the M shell, and so on. When an electron of the L shell is transferred into a vacancy in the K shell, the energy released in the process equals the energy difference between the states:

$$\Delta E_{K_\alpha} = \frac{Z^2 m e^4}{2n^2 \hbar^2} \left[\frac{1}{1^2} - \frac{1}{2^2} \right] \tag{7.18}$$

This particular transition results in the emission of x-ray radiation known as the K_α line. For transfer from the M shell into the K shell, the energy of the K_β line is given by:

$$\Delta E_{K_\beta} = \frac{Z^2 m e^4}{2n^2 \hbar^2} \left[\frac{1}{1^2} - \frac{1}{3^2} \right] \tag{7.19}$$

Since both transitions described in Eqs. (7.18) and (7.19) terminate in the K shell, they are known as belonging to the spectrum of the

213

K series. Similarly, the spectral lines resulting when electrons fall into the L shell define the spectrum of the L series, and the energies of these lines are given by:

$$\Delta E_{\mathrm{L}} = \frac{me^4}{2n^2\hbar^2}\left[\frac{1}{2^2} - \frac{1}{n^2}\right] \tag{7.20}$$

Equation (7.20) indicates that the energy difference between the two states involved in an XRF transition is proportional to the atomic number of the element in question, a fact realized some time ago by Mosley and illustrated in Fig. 7.16. The simple equations given above do

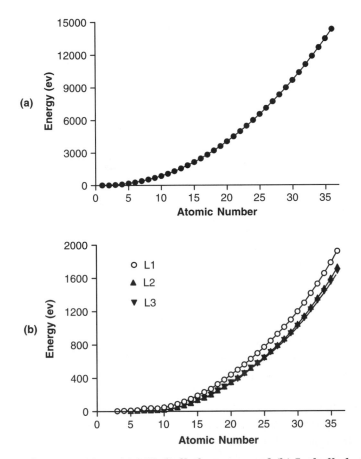

Fig. 7.16. Binding energies of (a) K shell electrons and (b) L shell electrons as a function of atomic number. The graphs were constructed from data tabulated in Ref. [56].

not account for the different orbitals that exist for each principal quantum number, and the inclusion of these levels into the spectral considerations explain the fine structure observed within almost all XRF bands. For instance, the L shell (defined by $n = 2$) contains the 2S and 2P levels, the transfer of electrons from these levels down to a vacancy in the K shell (defined by $n = 1$) results in the emission of x-ray radiation denoted as the $K_{\alpha 1}$ and the $K_{\alpha 2}$ lines.

A detailed summary of the characteristic XRF lines for the elements is available [64], as are more detailed discussions of the phenomena of x-ray absorption and XRF [65–68].

7.2.3   Experimental details

There are numerous types of instrumentation available for the measurement of XRF, but most of these are based either on wavelength dispersive methodology (typically referred to as WDX) or on the energy dispersive technique (typically known as EDX). For a detailed comparison of the two approaches for XRF measurement, the reader is referred to an excellent discussion by Jenkins [69].

The first XRF spectrometers employed the *wavelength dispersive* methodology, which is schematically illustrated in the upper half of Fig. 7.17. The x-rays emanating from the source are passed through a suitable filter or filters to remove any undesired wavelengths, and collimated into a beam that is used to irradiate the sample. For instance, one typically uses a thin layer of elemental nickel to isolate the K_{α} lines of a copper x-ray source from contamination by the K_{β} lines, since the K-edge absorption of nickel will serve to pass the K_{α} radiation but not the K_{β} radiation.

The various atoms making up the sample emit their characteristic XRF at an angle of θ relative to the excitation beam, and the resulting x-rays are discriminated by a monochromator that uses a single crystal as a diffraction grating. The diffraction off the crystal must obey Bragg's law, and this fact makes it clear that no single crystal can serve as an efficient diffraction grating for all wavelengths of x-rays. As a result, WDX instruments typically employ a variety of different crystals that the user can select to optimize the XRF region of greatest interest.

The detector can be either a gas-filled tube detector or a scintillation detector, which is systematically swept over the sample, and which measures the x-ray intensity as a function of the 2θ scattering angle. Through suitable calibration, each 2θ angle is converted into a wavelength value for display. The major drawback associated with WDX

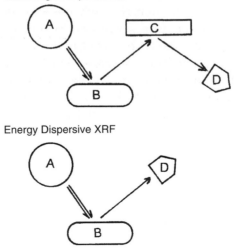

Fig. 7.17. Basic components of the apparatus used for the measurement of X-ray fluorescence by the wavelength and energy dispersive methods.WDX: X-rays from the source (A) are allowed to impinge on the sample (B); the resulting XRF is discriminated by the crystal (C), and finally measured by the detector (D). EDX: X-rays from the source (A) are allowed to impinge on the sample (B), and the resulting XRF is measured by the detector (D).

spectrometers is the reality derived from the Bragg scattering law that one cannot measure all wavelengths in an XRF spectrum in a single scan. Hence, one might be required to run multiple spectra if the range of elements to be studied is wide in terms of atomic number.

The *energy dispersive* methodology was developed as a means to permit an analyst to acquire the entire XRF spectrum simultaneously, therefore eliminating the requirement to acquire data in parcels. A simple schematic of an EDX spectrometer is illustrated in the lower half of Fig. 7.17. As with the WDX instrumentation, x-rays emanating from the source are filtered and collimated into a beam that is used to irradiate the sample. The XRF emitted by the sample is caused to fall onto a semiconductor diode detector, and one obtains the spectrum through the use of multi-channel analysis of the detector output. Although the instrumentation of the EDX method is simpler, it does not provide the same degree of wavelength resolution associated with WDX spectrometers.

One of the problems faced in XRF spectroscopy is the fact that the absolute sensitivity of an element decreases with atomic number, and this decrease is mostly considerable for light elements. For quite some time, XRF spectrometers could not detect elements having atomic numbers less than 12, which did not permit the analysis of any second row elements such as carbon, nitrogen, oxygen, and fluorine. Fortunately, advances in detector design have been made, and now through the use of ultra-thin beryllium windows analysts can now obtain reliable results on the light elements. When EDX is combined with scanning electron microscopy, one can even use windowless detectors to maximize the response of the light elements [70].

7.2.4   Qualitative XRF analysis

By virtue of their capacity to observe the entire range of x-ray fluorescence and speed of data acquisition, EDX spectrometers are admirably suited for qualitative analysis work. Equally well suited would be the WDX spectrometers capable of observing XRF over a wide range of scattering angles. As discussed earlier, the K_α and K_β XRF of a given element is directly related to its atomic number, and extensive tables are available that provide accurate energies for the fluorescence. Such information can be stored in the analysis computer, and used in a peak-match mode to easily identify the elemental origins of the fluorescing atoms within the sample. The elemental identification is facilitated by the fact that XRF originates from core electron levels that are only weakly affected by the details of chemical bonding. It must be remembered, however, that the selectivity associated with XRF analysis ends with elemental speciation, and that the technique is generally unable to distinguish between the same element contained in different chemical compounds.

X-ray fluorescence spectra of most elements consist largely of bands associated with the K, L, and M series. The XRF associated with the K series is dominated by the α_1/α_2 doublet, although some very weak β XRF can often be observed at higher energies. The L-series XRF will consist of three main groups of lines, associated with the α, β, and γ structure. Ordinarily, the α_1 line will be the strongest, and the β_1 line will be the next most intense. The peaks within the M series are observed only for the heavier elements, and consist of an unresolved α_1/α_2 doublet as the strongest feature, followed by a band of peaks associated with the β XRF.

For instance, the $K_{\alpha 1}$ and $K_{\alpha 2}$ lines of elemental copper are observed at wavelengths of 1.540 Å and 1.544 Å, respectively, and K_β lines will be observed at 1.392 Å and 1.381 Å [71]. The unresolved $L_{\alpha 1}$ and $L_{\alpha 2}$ lines are observed at a wavelength of 13.330, and L_β lines are observed at

13.053 Å and 12.094 Å. The presence of copper in a sample would be indicated by the presence of XRF peaks detected either at these wavelengths, or at their corresponding energies.

XRF qualitative analysis entails the identification of each line in the measured spectrum. The analysis begins with the assumption that the most intense lines will be due to either K_α or L_α emission, and these are used to match the observed lines with a given element. Once the most intense α lines are assigned, one then goes on to assign the β lines, and these assignments serve to confirm those made from the analysis of the α lines. Of course, this task has been made quite easy through the use of computer systems that store all the information and execute the peak matching analysis. For example, the EDX analysis of a fumarate salt of a drug substance was unexpectedly found to exhibit definite responses for chlorine, indicating that the sample actually consisted of mixed fumarate and hydrochloride salts [72].

In another qualitative study, EDX analysis was used to study the nature of the precipitate occasionally formed in Zn–insulin solutions [73]. Identification of the EDX peaks obtained for the crystalline precipitates enabled the deduction that the solid consisted of a Zn–insulin complex, and a rough analysis of the peak intensities indicated that the composition of the precipitate was comparable to that existing in the starting materials. The combination of the EDX technique with scanning electron microscopy enabled the analyses to be conducted on relatively few numbers of extremely small particles.

7.2.5 Quantitative XRF analysis

Owing to their superior fluorescent yield, heavy elements ordinarily yield considerably more intense XRF bands than the light elements. This feature can be exploited to determine the concentration of inorganic species in a sample, or the concentration of a compound that contains a heavy element in some matrix. Many potential XRF applications have never been developed owing to the rise of atomic spectroscopic methods, particularly inductively coupled plasma atomic emission spectrometry [74]. Nevertheless, under the right set of circumstances, XRF analysis can be profitably employed.

A number of experimental considerations must be addressed in order to use XRF as a quantitative tool, and these have been discussed at length [75,76]. The effects on the usual analytical performance parameters (accuracy, precision, linearity, limits of detection and quantitation, and ruggedness) associated with instrument are usually minimal.

However, the effects associated with sample handling, preparation, and presentation cannot be ignored, as they have the potential to exert a major influence over the quality of the analysis.

The background emission detected in an XRF spectrum is usually due to scattering of the source radiation. Since the scattering intensity from a sample is inversely proportional to the atomic number of the scattering atom, it follows that background effects are more pronounced for samples consisting largely of second row elements (i.e., organic molecules of pharmaceutical interest). Hence, background correction routines play a major role in transforming raw XRF spectra into spectra suitable for quantitative analysis.

The performance of a quantitative XRF analysis requires that one correct for the inter-element interactions that are associated with the matrix of the sample. For homogeneous samples, matrix effects can serve to enhance or diminish the response of a given element. The intensity retardation effects usually originate from competition among scattering units in the sample for the incident x-ray radiation, or from the attenuation of emitted XRF by its passage through the sample itself. Enhancement effects arise when emitted XRF is re-absorbed by other atoms in the sample, causing those atoms to exhibit a higher XRF intensity than would be associated with the incident x-ray radiation itself.

Since matrix and background effects are notoriously difficult to identify and control, quantitative XRF analysis usually entails the use of standards. One can write an expression relating the concentration of an analyte in a sample with the XRF intensity of one of its emission lines:

$$C_i = K I_i M B \tag{7.21}$$

where C_i is the concentration of the analyte exhibiting the intensity response I_i, K a constant composed of instrumental factors, M a constant composed of matrix factors, and B a constant composed of background factors. While it is conceivable that one could derive an equation describing how K would be calculated, it is clear that calculation of the M or B constants would be extraordinarily difficult.

However, when an internal standard containing an element of known concentration C_S is added to the sample, its concentration would be derived from its observed intensity I_S by:

$$C_S = K I_S M B \tag{7.22}$$

Since the analyte and standard are equally distributed in the analyzed sample, one can assume that the M and B constants would be the same for both species. In that case, dividing Eq. (7.21) by Eq. (7.22) and rearranging yields a relation enabling the calculation of the amount of analyte in the sample:

$$C_i = (I_i/I_S)C_S \qquad\qquad (7.23)$$

This method of using an internal standard to deal with matrix and background effects has, of course, been used for ages in analytical chemistry. Equally useful would be another analytical method, namely that of standard additions. In this latter approach, the sample is spiked with successively increasing amounts of the pure analyte, and the intercept in the response–concentration curve is used to calculate the amount of analyte in the original sample.

Total reflection x-ray fluorescence (TXRF) has become very popular for the conduct of microanalysis and trace elemental analysis [77–79]. TXRF relies on scatter properties near and below the Bragg angle to reduce background interference, and to improve limits of detection that can amount to an order of magnitude or moreover more traditional XRF measurements. As illustrated in Fig. 7.18, if x-rays are directed at a smooth surface at a very small angle, virtually all of the radiation will be reflected at an equally small angle. However, a few x-rays will excite atoms immediately at the surface, and those atoms will emit their characteristic radiation in all directions. One obtains very clean

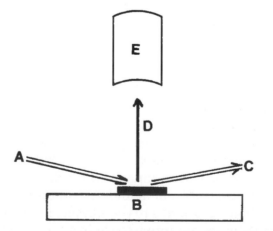

Fig. 7.18. Instrumental arrangement for the measurement of total reflection x-ray fluorescence. The x-rays from the source (A) are allowed to impinge on the sample mounted on a reflector plate (B). Most of the incident radiation bounces off the sample (C), but some results in the production of XRF (D), which is measured by the detector (E).

analytical signals when using the TXRF mode because there is essentially no backscatter radiation impinging on the detector.

TRXF was used to determine the trace elements in samples of lecithin, insulin, procaine, and tryptophan in an attempt to develop elemental fingerprints that could be used to determine the origin of the sample [80]. It was reported that through the use of matrix-independent sample preparation and an internal standard, one could use TXRF to facilitate characterization of the samples without the need for extensive pretreatment. In another work, a study was made of the capability of TXRF for the determination of trace elements in pharmaceutical substances with and without preconcentration [81].

Trace amounts of bromine in sodium diclofenac, sodium {2-[(2, 6-dichlorophenyl)amino] phenyl}acetate, have been determined using XRF [82], since the drug substance should not contain more than 100 ppm of organic bromine remaining after the completion of the chemical synthesis. Pellets containing the analyte were compressed over a boric acid support, which yielded stable samples for analysis, and selected XRF spectra obtained in this study are shown in Fig. 7.19. It was found that samples from the Far East contained over 4000 ppm of organic bromine, various samples from Europe contained about 500 ppm, while samples from an Italian source contained less than 10 ppm of organic bromine.

TXRF was used to characterize high-viscosity polymer dispersions [83], with special attention being paid to the different drying techniques and their effect on the uniformity of the deposited films. TXRF was also used as a means to classify different polymers on the basis of their incoherently scattered peaks [84]. Dispersive XRF has been used to assess the level of aluminum in antacid tablets [85].

Probably the most effective use of XRF and TXRF continues to be in the analysis of samples of biological origin. For instance, TXRF has been used without a significant amount of sample preparation to determine the metal cofactors in enzyme complexes [86]. The protein content in a number of enzymes has been deduced through a TXRF of the sulfur content of the component methionine and cysteine [87]. It was found that for enzymes with low molecular weights and minor amounts of buffer components that a reliable determination of sulfur was possible. In other works, TXRF was used to determine trace elements in serum and homogenized brain samples [88], selenium and other trace elements in serum and urine [89], lead in whole human blood [90], and the Zn/Cu ratio in serum as a means to aid cancer diagnosis [91].

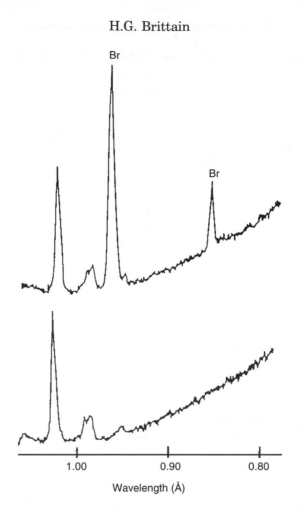

Fig. 7.19. XRF spectra of pure diclofenac (lower trace), and diclofenac containing organic bromine (upper trace). The bromine K_α peak is the marked peak at 1.04 Å, while the bromine K_β peak is the marked peak located at 9.93 Å. The figure was adapted from data in Ref. [82].

REFERENCES

1 A. Bravais, *Etudes Crystallographiques*, Gautier-Villars, Paris, 1866.
2 W. Friedrich, P. Knipping and M. Laue, *Sitzb. Kais: Akad. Wiss. Munchen* (1912) 303–322 *Chem. Abs.*, 7 (1912) 2009–2010.
3 W.H. Bragg and W.L. Bragg, *Proc. Roy. Soc. (London)*, A88 (1913) 428–438.
4 W.H. Bragg and W.L. Bragg, *X-Rays and Crystal Structure*, G. Bell & Sons, London, 1918.

5 W.L. Bragg, *The Crystalline State: A General Survey*, G. Bell and Sons, London, 1955.

6 M.J. Buerger, *X-Ray Crystallography*, Wiley, New York, 1942.

7 M.M. Woolfson, *An Introduction to X-Ray Crystallography*, Cambridge University Press, Cambridge, 1970.

8 J.P. Glusker and K.N. Trueblood, *Crystal Structure Analysis*, Oxford University Press, New York, 1972.

9 A.F. Wells, *Structural Inorganic Chemistry*, 5th ed., Clarendon Press, Oxford, 1984.

10 F.A. Cotton, *Chemical Applications of Group Theory*, 2nd ed., Wiley-Interscience, New York, 1971.

11 G. Burns and A.M. Glazer, *Space Groups for Solid State Scientists*, Academic Press, New York, 1978.

12 J.M. Bijovoet, N.H. Kolkmeyer and C.H. MacGillavry, *X-Ray Analysis of Crystals*, Butterworths, London, 1951.

13 A. Guinier, *X-Ray Crystallographic Technology*, Hilger and Watts, London, 1952.

14 B.D. Cullity, *Elements of X-Ray Diffraction*, Addison-Wesley Publishing Co., Reading, MA, 1956.

15 J.J. Rousseau, *Basic Crystallography*, Wiley, Chichester, 1998.

16 A.F. Wells, *Structural Inorganic Chemistry*, 5th ed., Clarendon Press, Oxford, 1984.

17 A.I. Kitaigorodskii, *Organic Chemical Crystallography*, Consultants Bureau, New York, 1961.

18 Y. Ebisuzaki, P.D. Boyle and J.A. Smith, *Acta Cryst.*, C53 (1997) 777–779.

19 C. Sun, D. Zhou, D.J.W. Grant and V.G. Young, *Acta Cryst.*, E58 (2002) o368–o370.

20 H.G. Brittain, G. Owoo and J.P. Jasinski, unpublished results.

21 H.G. Brittain, *Polymorphism in Pharmaceutical Solids*, Marcel Dekker, New York, 1999.

22 S.R. Byrn, R.R. Pfeiffer and J.G. Stowell, *Solid-State Chemistry of Drugs*, 2nd ed., SSCI Inc., West Lafayette, IN, 1999.

23 J. Bernstein, *Polymorphism in Molecular Crystals*, Clarendon Press, Oxford, UK, 2002.

24 J. Bernstein, Conformational polymorphism. In: G.R. Desiraju (Ed.), *Organic Solid State Chemistry*, Elsevier, Amsterdam, 1987, pp. 471–518 Chapter 13.

25 J.M. Robertson, *Proc. Roy. Soc. (London)*, A157 (1936) 79–99.

26 J.M. Robertson and A.R. Ubbelohde, *Proc. Roy. Soc. (London)*, A167 (1938) 122–135.

27 J.J. Gerber, M.R. Caira and A.P. Lötter, *J. Cryst. Spect. Res.*, 23 (1993) 863–869.

28 M.O. Boles, R.J. Girven and P.A.C. Gane, *Acta Cryst.*, B34 (1978) 461–466.

29 X-Ray Diffraction, General test <941>, *United States Pharmacopoeia 26*, The United States Pharmacopoeial Convention, Rockville, MD, 2003, pp. 2233–2234.

30 H.G. Brittain, Mandelic acid. In: H.G. Brittain (Ed.), *Analytical Profiles of Drug Substances and Excipients*, Vol. 29, Academic Press, San Diego, 2002, pp. 193–197 Chapter 6.

31 J.E. Koundourellis, C.K. Markopolou, F.A. Underwood and B. Chapman, *J. Chem. Eng. Data*, 37 (1992) 187–191.

32 J.E. Koundourellis, E.T. Malliou, R.A.L. Sullivan and B. Chapman, *J. Chem. Eng. Data*, 44 (1999) 656–660.

33 J.E. Koundourellis, E.T. Malliou, R.A.L. Sullivan and B. Chapman, *J. Chem. Eng. Data*, 45 (2000) 1001–1006.

34 H.G. Brittain, Methods for the characterization of polymorphs and solvates. In: H.G. Brittain (Ed.), *Polymorphism in Pharmaceutical Solids*, Marcel Dekker, New York, 1999, pp. 227–278 Chapter 6.

35 D.B. Black and E.G. Lovering, *J. Pharm. Pharmacol.*, 29 (1977) 684–687.

36 R. Suryanarayanan and A.G. Mitchell, *Int. J. Pharm.*, 24 (1985) 1–17.

37 L.S. Crocker and J.A. McCauley, *J. Pharm. Sci.*, 84 (1995) 226–227.

38 M. Ohta, Y. Tozuka, T. Oguchi and K. Yamamoto, *Chem. Pharm. Bull.*, 47 (1999) 1638–1640.

39 Z.G. Li, R.L. Harlow, C.M. Forris, R.E. Olson, T.M. Sielecki, J. Liu, R.D. Vickery and M.B. Maurin, *J. Pharm. Sci.*, 89 (2000) 1237–1242.

40 N.V. Phadnis, R.K. Cavatur and R. Suryanarayanan, *J. Pharm. Biomed. Anal.*, 15 (1997) 929–943.

41 M.M. de Villiers, D.E. Wurster, J.G. Van der Watt and A. Ketkar, *Int. J. Pharm.*, 163 (1998) 219–224.

42 S.-D. Clas, R. Faizer, R.E. O'Connor and E.B. Vadas, *Int. J. Pharm.*, 121 (1995) 73–79.

43 S.S. Iyengar, N.V. Phadnis and R. Suryanarayanan, *Powder Diffraction*, 16 (2001) 20–24.

44 H.M. Rietveld, *J. Appl. Crystallogr.*, 2 (1969) 65–71.

45 W. Cao, S. Bates, G.E. Peck, P.L.D. Wildfong, Z. Qiu and K.R. Morris, *J. Pharm. Biomed. Anal.*, 30 (2002) 1111–1119.

46 S.N. Campell Roberts, A.C. Williams, I.M. Grimsey and S.W. Booth, *J. Pharm. Biomed. Anal.*, 28 (2002) 1149–1159.

47 H.G. Brittain, unpublished results.

48 W.C. Kidd, P. Varlashkin and C.Y. Li, *Powder Diffraction*, 8 (1993) 180–187.

49 V.P. Tanninen and J. Yliruusi, *Int. J. Pharm.*, 81 (1992) 169–177.

50 D.E. Bugay, A.W. Newman and W.P. Findlay, *J. Pharm. Biomed. Anal.*, 15 (1996) 49–61.

51 J. Pirttimaki, V.-P. Lehto and E. Laine, *Drug Dev. Indust. Pharm.*, 19 (1993) 2561–2577.

52 M. Otsuka and Y. Matsuda, *Drug Dev. Indust. Pharm.*, 19 (1993) 2241–2269.

53 N.V. Phadnis and R. Suryanarayanan, *Pharm. Res.*, 14 (1997) 1176–1180.

54 G.A. Stephenson, E.G. Groleau, R.L. Kleemann, W. Xu and D.R. Rigsbee, *J. Pharm. Sci.*, 87 (1999) 536–542.

55 M. Epple and H.K. Cammenga, *Ber. Bunsenges. Phys. Chem.*, 96 (1992) 1774–1778.

56 P. Conflant and A.-M. Guyot-Hermann, *Eur. J. Pharm. Biopharm.*, 40 (1994) 388–392.

57 S.S. Leung and D.J.W. Grant, *J. Pharm. Sci.*, 86 (1997) 64–71.

58 S. Rastogi, M. Zakrzewski and R. Suryanarayanan, *Pharm. Res.*, 18 (2001) 267–273.

59 E. Yonemochi, S. Kitahara, S. Maeda, S. Yamamura, T. Oguchi and K. Yamamoto, *Eur. J. Pharm. Sci.*, 7 (1999) 331–338.

60 J.C. Davis, *Advanced Physical Chemistry*, Ronald Press, New York, 1965.

61 H. Eyring, J. Walter and G.E. Kimball, *Quantum Chemistry*, Wiley, New York, 1944.

62 W. Kauzmann, *Quantum Chemistry*, Academic Press, New York, 1957.

63 L. Pauling and E.B. Wilson, *Introduction to Quantum Mechanics*, McGraw-Hill, New York, 1935.

64 Y. Cauchois and C. Senemand, *Wavelengths of X-ray Emission Lines and Absorption Edges*, Volume 18 of International Tables of Selected Constants, Pergamon Press, Oxford, 1978.

65 H.K. Herglotz and L.S. Birks, *X-Ray Spectrometry*, Marcel Dekker, New York, 1978.

66 D.J. Fabian, *Soft X-Ray Band Spectra*, Academic Press, London, 1968.

67 R. Jenkins, *An Introduction to X-Ray Spectrometry*, Heyden, London, 1970.

68 R.E. Van Griken and A.A. Markowicz, *Handbook of X-Ray Spectrometry*, Marcel Dekker, New York, 1992.

69 R. Jenkins, Comparison of wavelength and energy dispersive spectrometers. In: *X-Ray Fluorescence Spectrometry*, 2nd ed., Wiley-Interscience, New York, 1999, pp. 111–121 Chapter 7.

70 A.W. Newman and H.G. Brittain, Optical and electron microscopies. In: *Physical Characterization of Pharmaceutical Solids*, Marcel Dekker, New York, 1995, pp. 127–156 Chapter 5.

71 R.O. Müller, *Spectrochemical Analysis by X-Ray Fluorescence*, Plenum, New York, 1972.

72 J.C. Berridge, *J. Pharm. Biomed. Anal.*, 14 (1995) 7.

73 P.J. Salemink, H.J.W. Elzerman, J. Th. Stenfert and T.C.J. Gribnau, *J. Pharm. Biomed. Anal.*, 7 (1989) 1261.

74 T. Wang, X. Jia and J. Wu, *J. Pharm. Biomed. Anal.*, 33 (2003) 639.

75 E.P. Betrin, *Principles and Practice of X-Ray Spectrometric Analysis*, 2nd ed., Plenum, New York, 1975.

76 R. Tertian and F. Claisse, *Principles of Quantitative X-Ray Fluorescence Analysis*, Heyden, London, 1982.
77 A. Prange, *Spectrochim. Acta*, B44 (1989) 437.
78 H. Aiginger, *Spectrochim. Acta*, B46 (1991) 1313.
79 P. Wobrauschek, *J. Anal. Atom. Spectromet.*, 13 (1998) 333.
80 M. Wagner, P. Rostam-Khani, A. Wittershagen, C. Rittmeyer, B.O. Kolbesen and H. Hoffmann, *Spectrochim. Acta*, B52 (1997) 961.
81 A. Kelki-Levai, I. Varga, K. Zih-Perenyi and A. Laszity, *Spectrochim. Acta*, B54 (1999) 827.
82 M.A.P. Da Re, F. Lucchini, F. Parisi and A. Salvi, *J. Pharm. Biomed. Anal.*, 8 (1990) 975.
83 C. Vasquez, *Spectrochim. Acta*, B59 (2004) 1215.
84 S. Boeykens and C. Vasquez, *Spectrochim. Acta*, B59 (2004) 1189.
85 C.A. Georgiades, *J. Assoc. Off. Anal. Chem.*, 73 (1990) 385.
86 A. Wittershagen, P. Rostam-Khani, O. Klimmek, R. Gross, V. Zickermann, I. Zickermann, S. Gemeinhardt, A. Kroger, B. Ludwig and B.O. Kolbesen, *Spectrochim. Acta*, B52 (1997) 1033.
87 M. Mertens, C. Rittmeyer and B.O. Kolbesen, *Spectrochim. Acta*, B56 (2001) 2157.
88 L.M. Marco, E.D. Greaves and J. Alvardo, *Spectrochim. Acta*, B54 (1999) 1469.
89 G. Bellisola, F. Pasti, M. Valdes and A. Torboli, *Spectrochim. Acta*, B54 (1999) 1481.
90 R.E. Ayalaa, E.M. Alvarez and P. Wobrauschek, *Spectrochim. Acta*, B46 (1991) 1429.
91 L.M. Marco, E. Jimenez, E.A. Hernandez, A. Rojas and E.D. Greaves, *Spectrochim. Acta*, B56 (2001) 2195.

REVIEW QUESTIONS

1. What is Bragg's Law?
2. What valuable information is provided by single-crystal x-ray diffraction?
3. Provide some of the applications of x-ray powder diffraction.
4. Provide some of the applications of x-ray fluorescence analysis.

Chapter 8

Atomic spectroscopy

Terence H. Risby

8.1 INTRODUCTION

Elemental analysis at the trace or ultratrace level can be performed by a number of analytical techniques and the most popular are based upon atomic spectroscopy. Atomic spectroscopy is subdivided into three fields, atomic emission spectroscopy (AES), atomic absorption spectroscopy (AAS), and atomic fluorescence spectroscopy (AFS) that differ by the mode of excitation and the method of measurement of the atom concentrations. The selection of the atomic spectroscopic technique to be used for a particular application should be based on the desired result since each technique involves different measurement approaches. AES excites ground state atoms (atoms) and then quantifies the concentrations of excited state atoms (atoms*) by monitoring their radiative deactivation. AAS measures the concentrations of ground state atoms by quantifying the absorption of spectral radiation that corresponds to allowed transitions from the ground to excited states. AFS determines the concentrations of ground state atoms by quantifying the radiative deactivation of atoms that have been excited by the absorption of discrete spectral radiation. The following schema summarizes these three analytical methods.

Basis of analytical measurement
AES measures a photon emitted when an excited atom deactives to the ground state

$$[ATOM] \underset{EXCITATION}{\overset{\Delta E}{\rightarrow}} [\textbf{\textit{ATOM}}^{*}] \underset{DEACTIVATION}{\overset{h\nu}{\rightarrow}} [ATOM]$$

Comprehensive Analytical Chemistry 47
S. Ahuja and N. Jespersen (Eds)
Volume 47 ISSN: 0166-526X DOI: 10.1016/S0166-526X(06)47008-2

AAS measures a photon absorbed when a ground state atom is excited

$$[\textbf{ATOM}] \quad \overset{h\nu}{\underset{EXCITATION}{\rightarrow}} \quad [ATOM^*] \quad \overset{h\nu}{\underset{DEACTIVATION}{\rightarrow}} \quad [ATOM]$$

AFS measures a photon emitted when an excited atom deactives to the ground state

$$[ATOM] \quad \overset{h\nu}{\underset{EXCITATION}{\rightarrow}} \quad [\textbf{ATOM}^*] \quad \overset{h\nu}{\underset{DEACTIVATION}{\rightarrow}} \quad [ATOM]$$

Brief history. Analytical atomic spectroscopy has taken more than 200 years to become the most widely used method for elemental analysis. Thomas Melville was the first to describe the principles of flame AES in 1752, but it took another 100 years before Kirchoff and Bunsen (1860) proposed the potential analytical relationship between ground or excited state atoms and the absorption or emission of discrete spectral radiation. All of these pioneering studies were performed by introducing solutions of metals by various means into alcohol flames or flames supported on Bunsen burners. However, the analytical utility of these early studies were limited by reproducibility of the analytical signal and this limitation was not solved until 1929 when Lundegardh introduced new designs for burners, nebulizers, gas control devices, and detection systems. These instrumental advances were used in most of the early flame photometers. Although Kirchoff and Bunsen had introduced the concept of atomic absorption in their original studies, it was not until 1955 that Walsh and his collaborators developed analytical AAS. This advance was due to their development of the sealed hollow cathode lamp as a spectral source of radiation that avoided the need for high-resolution monochromators to select and resolve the absorption lines. Eight years later Alkemade (1963) and Winefordner (1964) independently introduced the idea of analytical AFS. Finally in the 1960s and 1970s researchers (L'vov, Greenfield, Fassel, and West) introduced nonflame atomizers in order to minimize the spectral and chemical interferences that often occur in flames.

8.2 THEORY

8.2.1 Atomic emission spectroscopy

AES quantifies discrete radiation that is emitted by an excited atom when it deactivates to the ground state. This energy of excitation is provided by thermal, chemical, or electrical means. If the atom reservoir is in thermodynamic equilibrium then the Boltzmann distribution law gives the concentrations of atoms in the excited and ground states:

$$N_j/N_o = (g_j/g_o)e^{-E_j/KT}$$

where N_j and N_o are the number densities of atoms in the excited (jth state) and ground states, g_j and g_o the statistical weights of these states, E_j the energy difference between the jth and ground states, K the Boltzmann constant; and T the temperature (K) of the atom reservoir. The Boltzmann distribution law can only be used if excitation is produced by thermal collisions; the dominant process of excitation in flames. This equation is not valid to explain excitation caused either by chemical reactions in flames or by energetic collisions with excited species (electrons, ions, metastable atoms) that occur in electrical discharges or plasmas.

The concentration of atoms in the excited state is measured by monitoring their spectral deactivation to the ground state. The radiant power of this mechanism of deactivation is given by:

$$P = (hv_o/4\Pi)(g_j/g_o)A_{j\to o}l[M]e^{-E_j/KT}$$

where P is the flux of radiant energy per unit of solid angle and per unit surface area in a direction perpendicular to the flame surface. $A_{j\to o}$ is the transition probability per unit time of the transition from the jth to the ground state, and l is the thickness of atom reservoir along the axis of observation from which the emitted photons are monitored. [M] is the concentration of metal atoms and hv_o is the energy of the emitted photon. This equation demonstrates that the radiant power of the spectral deactivation is directly proportional to the concentration of the atoms and this linear relationship is followed providing that no interferences such as self-absorption occur. Self-absorption is the absorption of radiation by ground state atoms and this interference increases with the concentration of atoms. Also, this equation shows that small variations in temperature will produce larger variations in the radiant energy.

8.2.2 Atomic absorption spectroscopy

AAS measures the discrete radiation absorbed when ground state atoms are excited to higher energy levels by the absorption of a photon of energy. The radiant power of the absorbed radiation is related to the absorption coefficient of the ground state atoms using the

Beer Lambert equation:

$$I_{(\lambda)} = I_{o(\lambda)} 10^{-K_{(\lambda)}b}$$

where $I_{o(\lambda)}$ is the radiant power of the incident radiation of wavelength λ, $I_{(\lambda)}$ the radiant power of the transmitted radiation at wavelength λ, $K_{(\lambda)}$ the absorption coefficient of the ground state atom at wavelength λ, and b the path length.

This equation can be expressed in terms of absorbance ($A_{(\lambda)}$) where:

$$A_{(\lambda)} = \log\left(I_{(\lambda)}/I_{o(\lambda)}\right) = K_{(\lambda)}b$$

The usual method of excitation of the ground state atoms is to use an elemental spectral source (often a hollow cathode lamp) that emits the atomic spectra of the analyte element. If the width of the emission line from the spectral source is negligible compared to the absorption line of the ground state atoms, and if it is assumed that the absorption profile is determined by Doppler broadening, then the absorption coefficient integrated over the absorption-line profile can be approximated by the absorption coefficient at the absorption peak maximum (K_{max}). The relationship between K_{max} and the number density of ground state atoms is given by the following equation:

$$K_{\mathrm{max}} = \left(2\lambda^2/\lambda_{\mathrm{D}}\right)\left(\ln 2/\Pi\right)^{0.5}\left(\Pi e^2/mc^2\right)N_o f$$

where λ_{D} is the Doppler width of the line; λ the wavelength of the absorption maxima; e and m the charge and mass of an electron, respectively; c the velocity of light; and f the oscillator strength (average number of electrons per atom that can be excited by the incident radiation (λ)). Therefore, the absorbance is directly proportional to the concentration of atoms, provided that the absorption profile is dominated by Doppler broadening.

8.2.3   Atomic fluorescence spectroscopy

AFS quantifies the discrete radiation emitted by excited state atoms that have been excited by radiation from a spectral source. There are a number of mechanisms that are responsible for the atomic fluorescence signal: resonance fluorescence, step-wise fluorescence, direct-line fluorescence, and sensitized fluorescence. Generally, the lowest resonance transition ($1\rightarrow 0$) is used for AFS. If a line source is used for excitation and if the atomic vapor is dilute, then the radiant power of the atomic fluorescence signal (I_f) can be related to the concentration of ground state atoms by the following equation:

$$I_f = \frac{(e^2 \Omega_f L \lambda^2 f \delta \Phi I_L \Omega_A N_o)}{(6\Pi mc^2 \Delta \lambda_D)\left(\frac{2ln2}{\Pi}\right)^{0.5}}$$

where $\Omega_f/4\pi$ and $\Omega_A/4\pi$ are the solid angles of fluorescence and excitation that are measured by the instrument, or are incident upon the atom reservoir, respectively; L the length of the atom reservoir in the analytical direction; Φ the atomic fluorescence quantum efficiency; I_L the integrated radiant power for the incident beam per unit area; ∂ a correction factor that accounts for the relative line widths of the source and absorption profiles; and $\Delta \lambda_D$ the Doppler half width of the fluorescence profile.

On the basis of this equation it can be seen that the radiant power of atomic fluorescence signal is directly proportional to the concentration of the ground state atoms and to the radiant power of the exciting radiation. Therefore, increasing the intensity of the incident beam will improve the sensitivity of the technique.

8.3   INSTRUMENTATION

The following block schemas show the essential instrumental features of the various atomic spectroscopy techniques. Clearly, there are many similarities between these techniques. The subsequent discussions will describe the instrumental components of these techniques.

Atomic emission spectroscopy

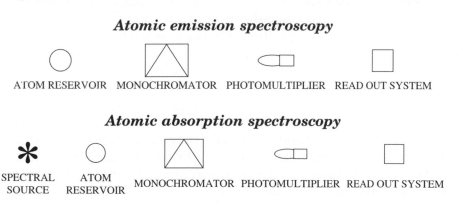

ATOM RESERVOIR   MONOCHROMATOR   PHOTOMULTIPLIER   READ OUT SYSTEM

Atomic absorption spectroscopy

SPECTRAL   ATOM   MONOCHROMATOR   PHOTOMULTIPLIER   READ OUT SYSTEM
SOURCE   RESERVOIR

T.H. Risby

Atomic fluorescence spectroscopy

ATOM RESERVOIR   MONOCHROMATOR   PHOTOMULTIPLIER   READ OUT SYSTEM

SPECTRAL SOURCE

8.3.1 Atom reservoirs

The production of atoms is a common critical requirement of all these techniques and there are various devices that can be used to generate the atoms in analytically useful spectroscopic states. The following schema summarizes the processes that occur during atomization:

Atomization

$$[SOLUTION] \xrightarrow[NEBULIZATION]{} [AEROSOL] \xrightarrow[DESOLVATION]{} [SOLID] \xrightarrow[VOLATILIZATION]{} [VAPOR] \xrightarrow[DISSOCIATION]{} [ATOM]$$

8.3.1.1 Flame atomizers

Flames have been traditionally the most popular atom reservoirs for all atomic spectroscopic techniques since they provide the most convenient way to generate atoms. Typically, solutions are aspirated into the oxidizer gas of a premixed hydrocarbon flame *via* a pneumatic nebulizer. Direct nebulizers pass the entire liquid aerosol that is generated into the flame, whereas indirect nebulizers pass only liquid aerosol particles of a given size and size distribution. During passage through the flame, the aerosol particles are desolvated, dissociated, and atomized. The efficiencies and reproducibilities of these processes will define the limit of detection that can be obtained and therefore considerable effort has been expended in developing efficient and quantitative nebulizers and atom reservoirs. The size and distribution of the aerosol particles will play a major role in the atomization efficiency since if the aerosol droplets are too large they may have insufficient residence time in the flame to be completely atomized or if the aerosol droplets are too small they can be desolvated in the nebulizer and be lost by collisions with the walls. Therefore, for a given composition of flame gases and burner there will be an optimum size and size distribution for the aerosol particles. The residence time available in the flame for atomization of the sample is dependent upon the flow rates of the fuel and oxidizer, and these flow rates are determined by the burning velocities of the particular flame gas mixture. Most atomic spectrometers employ indirect pneumatic nebulizers to generate a liquid aerosol of optimum size and size distribution and for these devices only about 10% of the aspirated sample reaches the flame.

The flame properties will also affect the atomization efficiency. The most popular flames used in analytical atomic spectroscopy are air–acetylene or nitrous oxide–acetylene. The former has a flame temperature of approximately 2300°C and the latter 2800°C. The increased atomization efficiency of the nitrous oxide–acetylene flame is not solely the result of the increase in flame temperature since this flame contains significant concentrations of excited cyanogen radicals (CN*) whose presence in the flame is exhibited by the emission of molecular bands in the region 650 nm (known as the red feather). The nitrous oxide–acetylene flame is recommended for those elements that form stable molecular species, such as refractory oxides, which can persist through the flame without atomization (such as aluminum, barium, beryllium, calcium, scandium, silicon, tantalum, titanium, uranium, vanadium, tungsten, zirconium, the lanthanides, and the rare earths). High concentrations of the reactive excited cyanogen radicals will reduce refractory oxides to atoms. The cooler air–acetylene flame is preferred for a different group of elements that have low ionization potentials (such as lithium, sodium, potassium, rubidium, and cesium). The hotter flame could ionize these elements with the result that the atom concentrations of the analyte species are reduced. There are a number of disadvantages to the use of flames as atom reservoirs and the most notable are the quantity of sample required for analysis, the brief residence time spent by the atom in the analytical zone of the flame, and the chemical environment within the flame. As a result, a number of nonflame atom reservoirs have been developed that generate atoms by electrical energy in controlled inert environments.

8.3.1.2   Nonflame atomizers
Nonflame atom reservoirs have been developed for specific atomic spectrometric techniques. Electrothermal atomizers (carbon rods, carbon furnaces, or tantalum ribbons) have been developed for AAS or AFS since they require the generation of ground state atoms, whereas

atmospheric pressure inductively coupled argon plasmas have been used to produce excited atoms for AES.

8.3.1.3 Electrothermal atomizers

The use of resistively heated carbon or tantalum rods, tubes, or filaments to generate atoms by thermal energy has increased the sensitivities of AAS and AFS by factors of 40–4000. The majority of this improvement in sensitivity is the result of increased residence time that the ground state atoms spend in the analytical zone, with minor contributions due to the reductions in chemical or spectral interferences from the flame. Some of the materials of construction of these atomizers can also play a role in the atomization processes since for example hot carbon can increase the reduction of the analyte species. The mode of sample introduction for these atomizers is to introduce a discrete aliquot of sample into the atomizer with a microsyringe (sample sizes are approximately 5 µl). The analyte is then dried, ashed, and atomized successively by resistive heating of the atomizer with a time-controlled ramp of low-voltage, high-current electricity. As a result of this mode of sample introduction, the analytical response is a transient pulse as opposed to a steady-state signal that is produced when a sample is nebulized continuously into a flame. Theoretically, the repeatability of the response obtained by electrothermal atomization should be lower than the flame since the peak signal for the former is dependent upon the precision with which the volume of the sample can be introduced. However, the electrothermal atomizers can be carefully controlled so that optimum temperatures can be obtained to dry, ash, and atomize the sample, which is not possible with flames. Also automatic syringes (usually injecting 50 µl) can be used to introduce the sample into the non-flame atomizer. Electrothermal atomizers can ash the sample *in situ* to destroy the sample matrix, which is a major advantage since flame atomizers require samples to be preashed. This advantage can also present difficulties unless precautions are taken to correct for absorption due to molecular species by background correction. The absorption profiles of molecular species are broad compared to the absorption profiles of atomic species. Currently, most instrument manufacturers use graphite furnace technology, which is based upon the research by L'vov from the early 1970s. The major advantage of the graphite furnace is that the atom vapor is maintained in the analyzer cell for significant periods of time allowing quantification to be performed. Also it is easy to generate reproducible temperature–time ramps to dry, ash, and atomize the elements of interest.

8.3.1.4 *Inductively coupled plasmas*

Greenfield and Fassel independently proposed the use of atmospheric pressure inductively coupled argon plasmas as atom reservoirs for atomic spectroscopy. This proposition has revolutionized the field of AES since this atom reservoir produces large concentrations of excited atoms for most elements in the periodic table. The plasma torch is produced by inductively coupling a high level (1.5 kW) of radio frequency energy (around 30 MHz) to a flowing stream of argon. This energy causes the argon to be ionized and various energetic species are produced, such as metastable argon atoms, excited argon atoms, argon ions and energetic electrons. These species will collide with the analyte to produce excited atoms and ions *via* various fragmentation and excitation mechanisms. The plasma torch has sufficient energy to excite most elements and will also populate multiple energy levels for a given element with the result that this atom reservoir is useful for multielement analysis. The population of different energy states for a given element allows more sensitive lines to be used for trace levels of analytes and less sensitive lines for higher concentrations of analyte atoms. This capability enables wide concentrations of samples to be determined without the problem of nonlinearity of response *versus* concentration. The analyte solutions are nebulized into the argon support using an ultrasonic nebulizer (3 MHz) and desolvation system. The desolvation system is necessary since this nebulizer is more efficient than pneumatic nebulizers and without predesolvation the plasma is cooled significantly. Inductively coupled plasmas have also been used successfully as the ionization source for elemental analysis by mass spectroscopy replacing arc and spark sources.

8.3.2 **Spectral sources**

8.3.2.1 *Continuous sources*

High-pressure electrical discharges were the first sources used in atomic spectroscopy. These sources consist of a sealed tube filled with a gas containing two electrodes. A voltage is applied between the electrodes and at a given voltage an electrical discharge is initiated. Electrons are accelerated by the potential difference between the electrodes and collide with the filler gas to produce excited molecules, atoms, and ions. At low gas pressures, the predominant output from these lamps is atomic line spectra characteristic of the filler gas, but as the pressure is increased the spectral output is broadened and a continuous spectra are produced. Hydrogen, deuterium, and xenon are the most widely used gases.

8.3.2.2 Line sources
Hollow cathode lamps

The introduction of sealed hollow cathode lamps by Walsh and his collaborators was the single event that revolutionized the field of atomic spectroscopy in the mid-1950s. These devices consist of a hollow cylindrical cathode manufactured from the element of interest. This cathode and an anode are sealed inside an optically transparent envelope with a quartz front window that is inline with the cathode. The lamp is filled with a low pressure (1–3 torr) of an inert gas (usually neon or argon). The hollow cathode lamp operates by producing inert gas discharge that sputters or vaporizes the element of interest from the cathode. These atomic species are subsequently excited by collision with inert gas ions, energetic inert gas atoms, or electrons to produce excited atoms that deactivate by the emission of characteristic photons. Cathodes can even be manufactured from nonelectrically conducting materials by the judicious choice of alloys. Once the discharge is struck a stable glow discharge is produced and the hollow cathode lamps can be operated with the minimum current. This operating procedure maintains a stable discharge and ensures that the spectral line output is not broadened. Hollow cathode lamps are available for most of the elements in the periodic table and multielement sources have been made using cathodes manufactured from mixtures of elements.

Electrodeless discharge lamps

In the 1960s and 1970s, Rains, West, Dagnall, and Kirkbright developed electrodeless discharge tubes as intense line sources for AAS and AFS. These lamps are easy to manufacture in the laboratory and consisted of sealed quartz tubes containing the element of interest or its halide, and low pressure (1–3 torr) of an inert gas usually argon. Energy in the microwave region (2.45 GHz) is supplied to the tube by placing it in a resonant cavity. The discharge is initiated by supplying electrons with a Tesla coil and intense atomic spectra of the element are obtained. Typically, these lamps produce higher intensity atomic spectra than the corresponding hollow cathode lamp however often the spectral outputs of these lamps are less stable.

8.3.3 Monochromators

The requirements for wavelength dispersion are very different for AES as compared to the spectral requirements of AAS and AFS. For AES it is essential to monitor only the radiation that results from the desired atomic transition and high-resolution monochromators (dispersion > 1.6 nm/mm of slit width) are required. The usual mode of operation is to monitor the intensity at a selected wavelength and then repeat the determination at a different wavelength. However, inductively coupled plasma-AAS with its ability to perform concurrent multi-element analysis requires a different design of monochromator. The most popular design involves monitoring the first-order spectrum of a concave grating spectrometer by placing suitable photosensitive devices on the Rowland circle. These photosensitive devices can be a photodiode array in which each element in the array can be monitored separately or else multiple fixed exit slits with photomultipliers placed at each slit. This latter arrangement is often called a quantometer or polychromator. Obviously there are a limited number of wavelengths that can be measured with this type of spectrometer with photomultipliers although advances in photodetection devices are reducing this limitation. The wavelength dispersion requirements for AASs and AFS are much less demanding than AES since the hollow cathode lamp is already producing radiation that is characteristic of the element under investigation. Therefore, the monochromator has only to separate the emission line of interest from other nonabsorbing lines. Theoretically, interference filters could replace these monochromators.

8.3.4 Read-out systems

Photomultipliers are generally used to convert the spectral radiation to an electrical current and often phase-sensitive lock-in amplifiers are used to amplify the resulting current. AES and AFS require similar read-out systems because both methods are measuring small signals. The difficulty associated with both these methods is the separation of the signal for the atomic transition of interest from the background radiation emitted by excited molecular species produced in the atom reservoir. AFS phase locks the amplifier detection circuit to the modulation frequency of the spectral source. Modulation of the source is also used in AAS.

8.4 GENERAL CONSIDERATIONS

8.4.1 Atomic emission spectroscopy

AES quantifies the deactivation of excited atoms. Atom reservoirs will also produce excited molecules that could interfere with the subsequent analysis since emission from excited molecular species is broad compared to the emission from excited atomic species. Multielement AES can identify and quantify concurrently all the elements contained in a sample, which is a major advantage compared to the single-element techniques, atom absorption, and AFS. Until the introduction of inductively coupled plasmas as atom reservoirs for AES, the number of elements that could be determined was limited by the available flame energy. Only those elements with low excitation energies could be determined unless arcs or sparks atom reservoirs were used. These latter types of atom reservoirs are not suitable for trace and ultratrace analysis as a result of extensive molecular spectral interferences. The current awareness of the importance of complex antagonistic and/or synergistic interactions between elements has increased the interest in rapid multielement analyses and inductively coupled plasma-AES is ideally suited for this application. Internal standards are generally added to the sample to aid identification and the selection of internal standards (such as gallium and yttrium) is based on the composition of the matrix. These elements enable chemical or spectral interferences to be subtracted. Theoretically, the inductively coupled plasma source should have sufficient energy to atomize all the elements in the sample and sample pretreatment should be minimal.

8.4.2 Atomic absorption spectroscopy

The instrumental requirement and cost of atomic absorption spectrometers are considerably less than those for multiwavelength atomic emission spectrometers. AAS quantifies the concentration of the element on the basis of the absorption of radiant energy by ground state atoms and the analytical response is based on the difference between the incident radiation and the transmitted radiation, i.e., the difference between two large signals. Therefore, it is imperative to use a spectral source with a very stable spectral output unless a double-beam spectrometer is used. Generally, the radiation from the spectral source is electronically modulated so that it can be selectively amplified with a lock-in amplifier. This mode of detection discriminates against the continuous background radiation from other species present in the atom reservoir. The usual sources of radiation for AAS are hollow cathode lamps that are available for most elements. The major limitation of AAS is the need to use a different hollow cathode lamp for each element since the spectral stability of multielement hollow cathode lamps is often poorer than single-element lamps. No spectral interferences from other elements are observed with AAS although spectral interferences can be produced by molecular species. Molecular spectral interferences can be minimized by the use of background correction with a hydrogen or deuterium continuum source or on the basis of the Zeeman effect, which occurs when energy levels are split by placing atoms in an external magnetic field. Molecular interferences are more significant with electrothermal atomizers than with flames. Since the magnitude of the absorbance is proportional to the path length, atom reservoirs have been designed so that the maximum concentration of ground state atoms are in the incident beam which is exactly opposite to the designs for atomic emission to AFS. The latter sources should be thin in the direction of measurement to avoid self-absorption. AAS can be used to quantify selectively any element that can be produced in the ground state, provided that a suitable spectral source is available. These spectral sources also allow facile optimization of the monochromator to the absorption maximum. The analytical response is displayed as a percent transmission or else as the logarithm of the percent transmission by the use of logarithmic amplifiers.

8.4.3 Atomic fluorescence spectroscopy

The instrumental requirements of AFS are the same as that of AAS with the exception that the incident radiation is at right angles to the analytical measurement direction. The atomic fluorescence signal is amplified with a phase-sensitive amplifier that is locked into the modulated incident radiation. AFS is more sensitive than AAS since the limit of detection is defined as the minimum detectable signal as opposed to minimum difference that can be measured between two large signals. Theoretically, the excitation source can be a continuum source, since only the radiation that has the energy that corresponds to the electronic transition will be absorbed and therefore the atom reservoir is acting as a high-resolution monochromator. However, practically most continuum sources do not have sufficient intensity at the wavelength of interest to produce analytically useful atomic fluorescence signals. AFS will quantify selectively any ground state atom that can be excited with incident radiation. The sensitivity of AFS is superior to the other atomic spectrometric techniques for a number of elements (Ag, Cu, Cd, Ni, Sb, Se, Te, Tl, and Zn). AFS is virtually free from spectral interferences although light scattering of the incident radiation can occur when samples with high solid contents are analyzed. This interference only occurs if resonance fluorescence is studied.

8.5 ANALYTICAL METHOD

Elemental analysis can be performed at ultratrace levels with any atomic spectrometric technique and the final selection is based on the identity and the number of elements to be determined. The initial step that is common to all analyses by atomic spectroscopy is the generation of a homogeneous solution.

8.5.1 Sample preparation

All reagents and solvents that are used to prepare the sample for analysis should be ultrapure to prevent contamination of the sample with impurities. Plastic ware should be avoided since these materials may contain ultratrace elements that can be leached into the analyte solutions. Chemically cleaned glassware is recommended for all sample preparation procedures. Liquid samples can be analyzed directly or after dilution when the concentrations are too high. Remember, all analytical errors are multiplied by dilution factors; therefore, using atomic spectroscopy to determine high concentrations of elements may be less accurate than classical gravimetric methods.

The preparation of aqueous solutions from solids is usually performed after the sample has been ground to a powder of uniform size. Sometimes, samples can be only sparingly soluble in water and therefore organic solvents may be used to dissolve the sample. Organic solvents can increase the sensitivities of atomic spectrometric analyses as a result of increases in the efficiencies of the nebulization of the analyte solutions. When organic solvents are used to dissolve samples nonselective ligands should be added to complex ionic species that would otherwise be insoluble in the organic solvent.

Although atomic spectroscopy can be element selective, large excesses of organic compounds present in the sample matrix may cause spectral interferences that will limit the sensitivities of the analyses. These interferences are caused by stable molecular species that are produced during atomization and various physical and chemical methods have been developed to remove them prior to the analysis. Physical methods use thermal (muffle furnace) or electrical (electrical discharges) energy in the presence of oxygen to completely oxidize the organic matrix to carbon dioxide and the elements of interest remain in an oxide ash. This inorganic residue is dissolved with acid solvents. Electrical discharges are preferred over the muffle furnaces when volatile elements (mercury, cadmium, lead, zinc, and metaloids) are present in the sample since the probability of the loss of these elements is increased with high temperatures and long ashing times. Chemical digestion with strong acids or strong bases in the presence of oxidizing agents are more commonly used since the final product can be diluted and analyzed directly. Careful selection of the acid used in the digestion should be made since explosive products, such as metal perchlorates or insoluble salts can be produced during the digestion. A different approach to the destruction of the organic matrix is to use solvent extraction coupled with selective complexing agents. This approach enables the elements of interest to be separated from the bulk sample and various extraction schemes have been proposed. The major advantage of this approach is that a concentration step is introduced during this sample preparation step.

8.5.2 Selection of an analytical procedure

Generally, extensive prior information is known about the sample in terms of the elemental composition and in these cases methods of analysis can be selected that will provide the desired result. However, if this information is not available or else a more general survey of ultratrace elements is required, then AES with an inductively coupled plasma source is the only atomic spectrometric technique that can provide these data at ultratrace levels.

The selection of a technique to determine the concentration of a given element is often based on the availability of the instrumentation and the personal preferences of the analytical chemist. As a general rule, AAS is preferred when quantifications of only a few elements are required since it is easy to operate and is relatively inexpensive. A comparison of the detection limits that can be obtained by atomic spectroscopy with various atom reservoirs is contained in Table 8.1. These data show the advantages of individual techniques and also the improvements in detection limits that can be obtained with different atom reservoirs.

Certain volatile elements must be analyzed by special analytical procedures as irreproducible losses may occur during sample preparation and atomization. Arsenic, antimony, selenium, and tellurium are determined *via* the generation of their covalent hydrides by reaction with sodium borohydride. The resulting volatile hydrides are trapped in a liquid nitrogen trap and then passed into an electrically heated silica tube. This tube thermally decomposes these compounds into atoms that can be quantified by AAS. Mercury is determined *via* the cold-vapor

TABLE 8.1

Detection limits for atomic spectroscopic analysis (µg/l)

Element	Atomic emission spectroscopy		Atomic absorption spectroscopy		Atomic fluorescence spectroscopy
	Flame	Inductively coupled plasma	Flame	Electrothermal atomizer	Flame
Ag		1	1.5	0.005	0.1
Al	200	1	50	0.1	2
As	50,000	2	150	0.03[a]	100
B	30,000	1	1000	20	
Ba	30	0.05	15	0.4	8
Be	1000	0.09	1.5	0.008	
Bi	40,000	1	30	0.05	3
Ca	5	0.05	2	0.01	20
Ce	10,000	1.5			
Cd	6000	0.1	0.8	0.002	8
Co	1000	0.2	7	0.15	1000
Cr	1000	0.2	5	0.004	1
Cs	5		15		
Cu	100	0.4	1.5	0.01	1
Dy	100	0.5	50		
Er	300	0.5	60		
Eu	3	0.5	30		
Fe	700	0.1	5	0.06	30
Ga	70	1.5	70		0.9
Gd	2000	1	2000		
Ge	600	1	300		15,000
Hf	75,000	1	300		
Hg	40,000	1	300	0.009[b]	100
In	30	1	50		100
Ir	100,000	1	900		
K	3	1	5		
La	1000	0.4	3000		
Li	0.003	0.3	1	0.06	
Lu	200	0.2	1000		
Mg	200	0.04	0.5	0.004	0.2
Mn	100	0.1	1.5	0.005	0.4
Mo	30	0.5	50	0.03	12
Na	1	0.5	0.5	0.005	100,000
Nb	1000	1.5	1500		
Nd	1000	2	1500		
Ni	600	0.5	10	0.07	2
P		4		130	
Pb	3000	1	10	0.05	30
Pd	1000	2	30	0.09	500
Rb	2	5	5	0.03	

continued

TABLE 8.1 (*continued*)

Element	Atomic emission spectroscopy		Atomic absorption spectroscopy		Atomic fluorescence spectroscopy
	Flame	Inductively coupled plasma	Flame	Electrothermal atomizer	Flame
Re	1000	0.5	750		
Rh	300	5	6		
Ru	300	1	100	1.0	
Sb	20,000	2	45	0.05	
Sc	70	0.2	30		
Se		4	100	0.03[a]	150
Si	5000	10	90	1	
Sn	600	2	150	0.1	0.3
Ta	18,000	1	1500		
Tb	1000	1	1000	50	
Te	20,000	2	30	0.03[a]	50
Th	150,000	3	100		
Ti	500	0.4	100	0.35	5
Tl	90	2	20	0.1	4
Tm	200	0.6	20		
U	10,000	11	12000		
V	300	0.5	100	0.1	50
W	4000	1	1500		
Y	300	0.2	100		
Yb	50	0.1	40		
Zn	50,000	0.2	2	0.02	0.04
Zr	50,000	0.5	500		

[a]Analysis *via* hydride.
[b]Analysis *via* cold-vapor technique.

technique with AAS. The cold-vapor technique employs chemical reducing agents to generate mercury atoms that are flushed into an absorption cell. Another method to generate mercury atoms uses amalgamation followed by the release of the mercury vapor by heating the amalgam. All of these procedures provide excellent sensitivities for volatile elements since the atoms can be directed into absorption cells that will optimize the analytical responses.

8.5.3 Quantification

The quantification of ultratrace elements by atomic spectroscopy should be performed on the basis of the addition of a series of known concentrations of the element(s) to the sample and quantifications are based on the method of standard additions. This method is preferred over comparisons of the analytical responses with standard calibration curves since the latter approach is fraught with potential problems. These standard additions should be made prior to any digestion or solvent extraction steps in order to correct any loss of sample during the sample preparation protocols. Additionally, this method will show whether the matrix is producing interferences with the selected analytical procedure.

8.5.4 Speciation

On the basis of the preceding discussion, it should be obvious that ultratrace elemental analysis can be performed without any major problems by atomic spectroscopy. A major disadvantage with elemental analysis is that it does not provide information on element speciation. Speciation has major significance since it can define whether the element can become bioavailable. For example, complexed iron will be metabolized more readily than unbound iron and the measure of total iron in the sample will not discriminate between the available and nonavailable forms. There are many other similar examples and analytical procedures that must be developed which will enable elemental speciation to be performed. Liquid chromatographic procedures (either ion-exchange, ion-pair, liquid–solid, or liquid–liquid chromatography) are the best methods to speciate samples since they can separate solutes on the basis of a number of parameters. Chromatographic separation can be used as part of the sample preparation step and the column effluent can be monitored with atomic spectroscopy. This mode of operation combines the excellent separation characteristics with the element selectivity of atomic spectroscopy. AAS with a flame as the atom reservoir or AES with an inductively coupled plasma have been used successfully to speciate various ultratrace elements.

8.6 CONCLUSIONS

Atomic spectroscopy is an excellent method of analysis for trace or ultratrace levels of many elements in the periodic table. The major disadvantage of all atomic spectroscopic methods is that they provide no information on the oxidation state of the element or its speciation. This disadvantage can be redressed by the use of selective reagents coupled with solvent extraction or by separation with various chromatographic methods. Another potential limitation is that most atomic spectroscopic methods require the sample be in solution. Instrumental advances in atomic spectroscopy during the last decade have been in two areas: design and operation of inductively coupled plasmas and the use of microprocessors for data display. Commercially available equipment has become more stable and user friendly.

REVIEW QUESTIONS

1. Why are spectral interferences less important in atomic absorption spectroscopy and atomic fluorescence spectroscopy than atomic emission spectroscopy?
2. Fluctuations in the coupling of radio frequency energy occur in inductively coupled plasmas. Which atomic spectroscopic technique will be more affected by these fluctuations? Explain your answer.
3. Which atomic spectroscopic technique will benefit from the availability of a tunable laser that provides radiation in the visible and ultraviolet region of the electromagnetic spectrum?
4. If you were asked to suggest improvements to instrumentation for atomic spectroscopy what areas would you propose? What is the basis for your selection?
5. What is the difference between the shape of the burner supporting a flame in atomic emission, atomic absorption, and atomic fluorescence spectroscopy? What is the theoretical basis for these differences?

Further Reading

Textbooks

J.W. Robinson, *Atomic Spectroscopy*, 2nd ed., Revised and Expanded, Marcel Dekker, NY, 1996.

J. Sneddon, Sample Introduction in Atomic Spectroscopy (Analytical Spectroscopy Library), Elsevier, New York, 1990.

R.M. Harrison and S. Raposomanikos, *Environmental Analysis using Chromatography Interfaced with Atomic Spectroscopy,* Ellis Horwood Series in Analytical Chemistry, Ellis Horwood Ltd., Chichester, 1992.

J. Sneddon and T.L. Thiem, *Lasers in Atomic Spectroscopy*, Wiley, New York, 1996.

M. Cullen, *Atomic Spectroscopy in Elemental Analysis*, CRC Press, Boca Raton, 2003.

Journals

List of specialized journals that present methods based upon atomic spectros-
copy.
Analytical Chemistry. Particularly the biennial reviews of atomic spectroscopy
that are published in the even years.
Analytica Chimica Acta
Talanta
Analyst
Journal of Analytical Atomic Spectrometry
Applied Spectroscopy
Spectrochimica Acta Part B

Chapter 9

Emission methods

C.H. Lochmüller

9.1   INTRODUCTION

The term emission here refers to the release of light from a substance
exposed to an energy source of sufficient strength. If you leave a poker
in a fire for a while and take it out, the fire exposed end will seem to
glow a reddish orange or even yellow. These colors are the *visible*
emission but even the warm handle of the poker emits light in the
longer wavelength *infrared*. Figure 9.1 shows the visible part of continuous and discrete or line emission spectra. The continuous spectrum
is rainbow-like in its transition from blue to red. Of more analytical
interest is emission spectrum showing discrete *lines*.

The most commonly observed line emission arises from excitation of
atomic electrons to higher level by an energy source. The energy emitted
by excited atoms of this kind occurs at wavelengths corresponding to the
energy level difference. Since these levels are characteristic for the
element involved, the emission wavelength can be characteristic for the
element involved. Sodium and potassium produce different line spectra.

Excitation can be achieved using *heat* as in a flame, an *electrical field*
or *light*. Different instrumental methods use different excitation

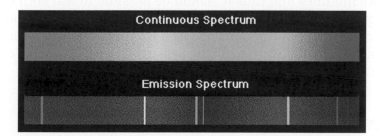

Fig. 9.1. Emission can be continuous across all wavelengths or at discrete
wavelengths (shorter wavelength on right and longer on the left).

Comprehensive Analytical Chemistry 47
S. Ahuja and N. Jespersen (Eds)
Volume 47 ISSN: 0166-526X DOI: 10.1016/S0166-526X(06)47009-4

sources ranging from the flame of a combustible gas to a beam of electrons or even x-rays. Many elemental composition determination methods are reasonably enough based on atomic emission spectroscopy or, more properly, spectrometry. The hardware and data collection strategies are different in form but share general principles. Emission spectrometry is not limited to atoms, of course, and molecular emission is a very powerful approach for both qualitative and quantitative determination as well.

Emission methods can offer some distinct advantages over absorption of light methods in the determination of trace amounts of material. It is possible to detect single photons and for that matter single atoms. Absorbance methods involve measuring the reduction of light intensity in a constant high level of light energy source. By recalling Beer's Law:

$$A = abc$$

where A is absorbance, c the molar concentration, b the path length and a the molar absorptivity, the problem is revealed. An absorbance of value 1 indicates a change in observed light of a factor of 10. Clearly at very low c, the measurement involves a very small change in the intensity or power of the light passing through path b. In emission there is no light if there is no emission and the challenge is to see the appearance of light in ideally total darkness. Put in human terms, on a dark night it is possible to see a burning match at a distance of almost 2 km but in bright sunlight it is a much bigger challenge to detect a match go out at the same distance.

Molecular emission is referred to as luminescence or fluorescence and sometimes phosphorescence. While atomic emission is generally instantaneous on a time scale that is sub-picoseconds, molecular emission can involve excited states with finite, lifetimes on the order of nanoseconds to seconds. Similar molecules can have quite different excited state lifetimes and thus it should be possible to use both emission wavelength and emission apparent lifetime to characterize molecules. The instrumental requirements will be different from measurements of emission, only in detail but not in principles, shared by all emission techniques.

9.2 ATOMIC EMISSION SPECTROMETRY

Perhaps, the most classical method involving atomic emission is also the simplest. If one takes a clean platinum wire fitted with a wooden handle and places the wire in the flame of a Bunsen burner the wire will heat and glow red. Dip the wire in a solution of sodium chloride and repeat the exposure to the flame. The flame glows yellow because of the flame-excited emission of sodium (there are actually two lines quite close together). Clean the wire in water and repeat with a solution of potassium chloride and observe a red flame instead. A mixture of sodium and potassium gives a yellow flame. What is needed is a method of wavelength selection and the simplest is a filter which will pass potassium's emission but not sodium. A sheet of "cobalt blue" glass will achieve this and the yellow flame will appear red if viewed through this filter.

This method is used to determine sodium and potassium in food, water and blood serum. The flame can be hydrogen/oxygen, methane/oxygen or methane/air fueled. Wavelength selection can be by filter, prism Fig. 9.2 or grating and by either one or two detectors.

In this case, sodium emission is monitored at a wavelength of 589.6 nm and potassium at a wavelength of 769.9 nm. The intensity of emission is calibrated with appropriate standards for the samples to be analyzed. In this way it is possible to automatically determine $\sim$100 values of sodium and potassium for 100 samples/h using modern clinical instruments. Limits of detection are sub-ppm and for serum values $\sim$140 mg/m the range of reproducibility is on the order of 2–3%.

It would be possible to write an entire book on the topic of emission spectrometry instrumentation devoted only to solution samples. There has been a literal mountain of research devoted to better thermal sources—gas flame, gas plasma and shrouded flames for often as a fluid sample in the form of an aerosol which is dried in the flame and the atoms in the "salt" are then excited. Clearly, the flow rate into a nebulizer that forms the aerosol must be constant, the droplet size consistent and more.

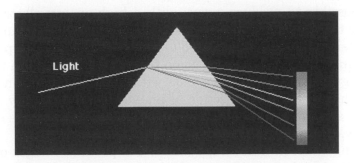

Fig. 9.2. Light from the emission source is broken into line components with a prism. A grating would provide the same result.

9.3   X-RAY EMISSION

X-ray emission is often called x-ray fluorescence (XRF) because the excitation source is a beam of x-rays of higher energy/shorter wavelength (see Chapter 7). X-ray emission has line structure and different elements produce characteristic x-rays of different wavelength although the spectroscopic axis is presented in energy terms rather than frequency of wavelength *per se*. This is likely a result of the historic origins of this measurement area in physics.

The production of x-rays in XRF requires a primary x-ray excitation source from an x-ray tube or a radioactive source. This x-ray energy strikes a sample, the x-ray photon can either be absorbed by the atom or be scattered through the sample material. The process in which an x-ray is absorbed by the atom is the origin of the spectra seen by the experiment's operator. Transferring all of its energy to an innermost electron is called the *photoelectric effect* and in this process, assuming the primary x-ray photon had sufficient energy, electrons are ejected from the inner shells of the sample atoms creating vacancies. These vacancies create an unstable condition for the atom and then return to a stable condition, electrons from the outer shells are transferred to the inner shells. The result is the emission a characteristic x-ray whose energy is the difference between the two binding energies of the corresponding shells. Each element has a unique set of energy levels and, as a result, each element produces x-rays of a unique energy values. With care, this permits elemental determination to be done in a nondestructive manner. In most cases, it is the innermost K and L shells that are involved in XRF detection. A typical x-ray spectrum from an irradiated sample will display multiple peaks of different intensities reflecting the distribution of elements and their relative mole fraction (Fig. 9.3).

It is possible to produce x-rays with a high-energy electron beam. There are two kinds of x-ray processes involved. One can be understood using classical physics and is called *brehmstrahlung*. In this process we come to understand that x-rays come from the energy given up by the high-energy electrons as they collide with matter. The second process involves the same K-shell ejection of an electron followed by return to the stable state of the atom with the emission of an x-ray as described previously.

Scanning electron microscopy (SEM) can produce images of surface and objects at high magnification. If the scanning of the electron beam is also coupled to detection of x-rays from the electron impact on the

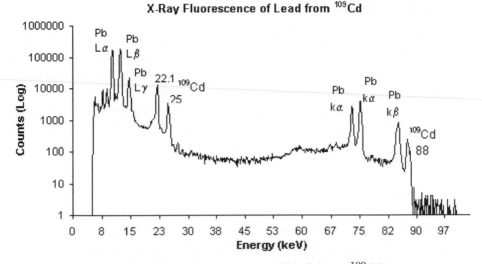

Fig. 9.3. X-ray fluorescence of lead from ^{109}Cd.

sample, it is possible to produce elemental maps of materials in this way. Many modern SEM instruments include this ability.

Lochmuller, Walter, Galbraith and coworkers realized that if one could use a source of excitation far removed in wavelength from K- and L-shell wavelengths, less background would be present in the collected spectra arising from scattered excitation source x-rays. Using high-purity hydrogen gas, one can produce hydrogen atoms that can be converted into free protons. These protons can then be accelerated in an electrical field on the order of MeV and if the geometry is correct a beam of high-energy protons can be made to bombard a sample. They named this technique PIXE for proton-induced, x-ray emission and applied it to chemical and biological sample materials. It is possible to quantitatively estimate—for example heavy metals—in tissue at the sub-ppb level. Gutknecht and Walter actually showed that it was possible to use PIXE to map heavy metal distribution in lyophilized human lung tissue and demonstrated that lead inhaled by tobacco smokers is concentrated in certain features in the lung rather than being uniformly distributed.

A great deal of discussion about elemental determination methods focuses on minor, trace, ultra-trace levels of analyte presence in relatively large volume samples. There is another area equally challenging and that involves elemental determination at major and minor concentration in very small volume/low mass samples. Lochmuller and Galbraith used PIXE to study the metal content of carbonic anhydrase (the enzyme that regulates bicarbonate-carbon dioxide in the blood-stream of mammals). In this case, researchers had removed the native Zn atom in bovine carbonic anhydrase and replaced it with Co. The goal was to study the effect on catalytic activity of different transition metals. The synthesis was done by exposing the metal-free form of carbonic anhydrase (the apo-form) to a solution of cobalt ions. Using PIXE, Lochmuller and Galbraith showed that the resulting protein contained Co, Ni, Cu and even Cr. The reason for the distribution of metals lies in the synthetic route and the relative stability constants of the products.

$$Me^{+++}apo\text{-}CA \leftrightarrow MeCA$$

Given the reaction and the very high stability constants involved, the production of cobalt carbonic anhydrase would require a solution not of "ACS-grade" cobalt nitrate but a 99.999999999999 … 999% pure cobalt nitrate solution. What happened in the lab synthesis was that trace metals in the ACS-grade salt were selectively bound to the apo-carbonic anhydrase because their stability constant advantage was orders of magnitude over that of cobalt. The sample used to discover this was sub-milligram in mass.

9.3.1 Detectors for x-rays

The engineering details of how one obtains x-ray emission spectra are beyond the scope of this chapter. The principle of detection is related to the same interaction with matter model used to explain the origin of x-rays. In the case of detection, x-rays impinge on a semi-conductor and expend their energy. The result is a current from electrons promoted to what is called the conduction band of the semi-conductor because of the energy transfer. The number of electrons is proportional to the energy transferred by a given x-ray photon and the total current over time is related to the number of atoms emitting the x-rays. In recent times Si-PIN photodiodes have been adapted for use as x-ray detectors.

9.4 MOLECULAR LUMINESCENCE/FLUORESCENCE

Molecular fluorescence is a powerful tool for analysis and has many applications in chemistry, biological chemistry and in the health sciences. The schematic instrumental geometry is shown in Fig. 9.4.

Light from the broadband source is "filtered" by a wavelength selector and then passed into the sample container. If fluorescence occurs,

252

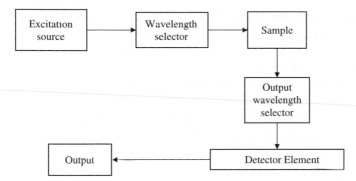

Fig. 9.4. Schematic of the components of a fluorescence detector.

then the light passes through the output wavelength selector to a detector and appears as output. Scanning the output detector will produce a spectrum.

The intensity of the light emitted is proportional to the number of molecules undergoing fluorescence. It is possible to quantify the concentration of the fluorescent molecules as a result. The output is given as $Output = kC$ where Output is not at the excitation wavelength. The C is concentration and k is a constant of proportionality. The k accounts for the power of the source, the efficiency of the fluorescence process (photons out for photons in) and the path length of light from the source in the sample and of the emitted light leaving the source toward the detector. The most common geometry is for the emission to be observed at right angles to the path of the excitation light. Note that the output should be zero if there is no fluorescence. Just as clearly, if the molar absorptivity of the sample at the excitation wavelength is zero, the output is also zero.

The dynamic range of the fluorescence experiment is related to a number of factors but it can be orders of magnitude. It is possible, for example, to determine quinine in water from nanomolar to millimolar concentration by direct measurement. Quinine fluorescence is familiar to most people that have noticed the blue glow of quinine tonic water in sunlight.

Scattering of excitation light in the direction of the detector and of the emitted light can influence the precision of the measurement. At very high concentration of analyte, emission may not be seen because most of the exciting light is absorbed near the wall of the sample cell. The fact is that most molecules do not fluoresce and cannot be determined using this method. This can be an advantage if the analyte of interest is fluorescent and is in a solution of otherwise non-fluorescent molecules.

Expected fluorescence yield can be complicated by a variety of quenching processes. Some of these include:

1. *Collisional quenching.* Molecules in the excited state collide with other molecules and transfer the excited-state energy in a spectrally dark fashion. Excitation did occur but fluorescence was not observed. Oxygen quenching is common example of solution impurity collisional quenching and oxygen is a particularly efficient quenching agent. Fortunately, purging or sparging the solution with nitrogen or argon will extract oxygen and eliminate this source of emission intensity loss.

2. *Energy transfer quenching.* If an impurity is present whose first excited singlet state is below that of the excited state of the analyte then energy can be transferred to the impurity and fluorescence is not seen. This does not have to involve collision and non-radiation transfer can occur. Aromatic molecules are particularly a source of this interference. Removal is an option but sometimes dilution is a solution if the desired fluorescence can still be measured at lower concentration.

3. *Concentration quenching.* In this case, quenching occurs because of the formation of an association between excited state and ground state, which if homo-molecular is called the excimer formation. The subsequent processes can be radiationless or the complex can emit at much longer wavelength and effectively not be detected.

Quenching can actually be used to create analytical methods. An example is that of fluorescent antibody assay. Here, a fluor is bound to an antibody but the binding constant is less than that for the antibody–antigen pair for which the antibody was generated. Binding quenches the fluorescence of the fluor. When a solution containing the native antigen is mixed with the fluor complex, the fluor is released and can fluoresce normally. There are numerous examples of this approach.

A clever adaptation of a special kind of fluorescence where the energy source is a chemical reaction is called chemiluminescence. The common firefly uses such a reaction to generate light in the green-yellow end of the visible light spectrum. The reaction involves an enzyme luciferinase on the substrate luciferin with ATP as the energy source. The luciferin and luciferinase can be obtained by harvesting these bugs. The analytical method introduced by DuPont involves the detection of bacteria in milk utilizing the fact that a living bacterial culture synthesizes ATP. By mixing the enzyme, substrate and milk in a buffer and measuring the light emitted is a direct measure of bacterial content. Absence of ATP effectively prevents the reaction and the process is quenched.

There are other examples of how quenching can be used in answering questions of accessibility to and of distance between chemically bound fluorophores. The first of these is best illustrated by the question of solution accessibility to fluorescent amino acid elements in a folded protein. Quenching of the fluor when in a crevice of an enzyme, can be studied using quenching agents of varying sizes dissolved in the contacting buffer solution. Fluors on the surface of a globular protein are readily accessible to heavy-metal ions, such as lead and are readily quenched. Those in the interior of the protein or in crevices where large ions cannot reach remain unquenched.

Lochmüller and coworkers used the formation of excimer species to answer a distance between site question related to the organization and distribution of molecules bound to the surface of silica xerogels such as those used for chromatography bound phases. Pyrene is a flat, poly aromatic molecule whose excited state is more pi-acidic than the ground state. An excited state of pyrene that can approach a ground state pyrene within $\sim 7\,\text{Å}$ will form an excimer $Pyr^* + Pyr \leftrightarrow (Pyr)_2^*$. Monomer pyrene emits at a wavelength shorter than the excimer and so isolated versus near-neighbor estimates can be made. In order to do this quantitatively, these researchers turned to measure lifetime because the monomer and excimer are known to have different lifetimes in solution. This is also a way to introduce the concept of excited state lifetime.

The excited state of a molecule can last for some time or there can be an immediate return to the ground state. One useful way to think of this phenomenon is as a time-dependent statistical one. Most people are familiar with the Gaussian distribution used in describing errors in measurement. There is no time dependence implied in that distribution. A time-dependent statistical argument is more related to "If I wait long enough it will happen!" view of a process. Fluorescence decay is not the only chemically important, time-dependent process, of course. Other examples are chemical reactions and radioactive decay.

C.H. Lochmüller

9.4.1 Statistical aspects of fluorescence decay

The following argument was used first by E. V. Schweidler in 1905 to describe radioactive decay but it applies to all similar kinetic processes. The fundamental assumption is that the probability p of an event occurring over a time interval dt is independent of past history of a molecule; it depends only on the length of time represented by dt and for sufficiently short intervals is just proportional to dt. Thus, $p = k\,dt$ where k is a constant of proportionality characteristic of the process being awaited. In fluorescence decay it is characteristic of the kind of molecule in chemical terms.

The probability of a molecule not returning to the ground state over the interval dt is simply $(1-p) = 1-k\,dt$. A molecule "surviving" dt then the probability of no return over the next interval is again $1-dt$. Using the law for compounding probability, the chance of surviving both is $(1-kt)^2$. Clearly for n such intervals the probability of remaining in the excited state is $(1-kt)^n$. The total time waited is $n\,dt = t$ and we then have $(1-kt/n)^n$. If the interval is made indefinitely small and n grows to infinity, we approach the limit in $(1-kt/n)^n$. This is identical to the expression $e^x = \lim (1+x/n)^n$ as n becomes infinite.

If now we consider a large number of molecules N_0, the fraction still in the excited state after time t would be $N/N_0 = e^{-kt}$ where N is the number unchanged at time t. This exponential law is familiar to chemists and biological scientists as the "first-order rate law" and by analogy fluorescence decay is a first-order process—plots of fluorescence intensity after an excitation event are exponential and each type of molecule has its own characteristic average lifetime.

How can lifetime measurements be used to answer the question of how many molecules of all present have nearest neighbors that can form excimer when the reactive groups on a surface are covalently bound to a chain terminating in a pyrene molecule? If we excite pyrene molecules using a pulse of light and observe the intensity of all light emitted from the sample, we will collect an exponential decay curve of light as a function of time. If there are two distinct lifetimes from two species, we will observe a biphasic curve. This curve is the sum of two exponential decays. We can fit this model to the data and derive the lifetimes and the relative fraction of each process. The result of the experiment done by Lochmuller and coworkers was a demonstration that there are some 10% of all the reacted sites that have no nearest neighbors within the distance needed to form excimer and that the surface is not uniform. Some think of this type of experiment as the use of a spectroscopic ruler.

Further reading

C.H. Lochmüller, J. Galbraith and R. Walter, Trace metal analysis in water by proton-induced x-ray emission analysis of ion-exchange membranes, *Anal. Chem.*, 46 (1974) 440.

C.H. Lochmüller, J. Galbraith, R. Willis and R. Walter, Metal-ion distribution in metallo-proteins by proton-induced x-ray emission analysis, *Anal. Biochem.*, 57 (1974) 618.

C.H. Lochmüller, A.S. Colborn, M.L. Hunnicutt and J.M. Harris, Bound pyrene excimer photophysics and the organization and distribution of reaction sites on silica, *J. Am. Chem. Soc.*, 106 (1984) 4077.

C.H. Lochmüller and S.S. Saavedra, Conformational changes in a soil fulvic acid measured by time-dependent fluorescence depolarization, *Anal. Chem.*, 58 (1986) 1978.

C.H. Lochmüller and S.S. Saavedra, Intrinsic florescence characteristics of apomyoglobin adsorbed to microparticulate silica, *Langmuir*, 3 (1987) 433.

C.H. Lochmüller and S.S. Saavedra, Interconversion of conformation of apomyoglobin adsorbed on hydrophobic silica gel, *J. Am. Chem. Soc.*, 109 (1987) 1244–1245.

REVIEW QUESTIONS

1. What is the most important analytical aspect of optical emission methods? Why are they in any sense better than optical absorption methods?

2. Emission lifetimes span the range of nano-seconds to hours. How wide a range is this compared to the time-span of recorded history on Earth? On the scale of useful chemical synthesis kinetics?

3. If flame emission is based on excitation of atoms formed by combustion in the flame, why does flame emission work well for sodium, potassium, cesium and some of the transition metals but not vanadium, molybdenum or the lanthanides?

4. Could heavy metal content be determined using a fluorescence method? Can you think of how to make a method, which could distinguish lead from cadmium?

5. How could photochemistry influence fluorescence emission?

Chapter 10

Nuclear magnetic resonance spectroscopy

Linda Lohr, Brian Marquez and Gary Martin

10.1   INTRODUCTION

Nuclear magnetic resonance (NMR) is amenable to a broad range of applications. It has found wide utility in the pharmaceutical, medical and petrochemical industries as well as across the polymer, materials science, cellulose, pigment, and catalysis fields to name just as a few examples. The vast diversity of NMR applications may be in part due to its profound ability to probe both chemical and physical properties including chemical structure as well as molecular dynamics. This gives NMR the potential to have a great breadth of impact compared with other analytical techniques. Furthermore, it can be applied to liquids, solids or gases. In some ways, it is a "universal detector" in that it detects all irradiated nuclei in a sample regardless of the source. Signals appear from all components in a mixture, proportional to their concentration. NMR is therefore a natural compliment to separation techniques such as chromatography, which provide a high degree of component selectivity in a mixture. NMR is also a logical compliment to mass spectrometry, since it can provide critical structural information. Compared to other solid-state techniques, NMR is exquisitely sensitive to small changes in local electronic environments, such as discerning individual polymorphs in a crystalline mixture.

Beyond the qualitative molecular information afforded by NMR, one can also obtain quantitative information. Depending on the sample, NMR can measure relative quantities of components in a mixture as low as 0.1–1% in the solid state. NMR limits of detection are much lower in the liquid state, often as low as 1000:1 down to 10,000:1. Internal standards can be used to translate these values into absolute quantities. Of course, the limit of quantitation is not only dependent on

Comprehensive Analytical Chemistry 47
S. Ahuja and N. Jespersen (Eds)
Volume 47 ISSN: 0166-526X DOI: 10.1016/S0166-526X(06)47010-0

the type of sample but also on the amount of sample. While, not as mass sensitive as other analytical techniques, NMR has dramatically improved in sensitivity in recent years [1].

A standard liquid NMR sample contains approximately 500 μl of solvent in a 5 mm diameter glass tube. Typical amounts of sample for this configuration range from 1 to 20 mg depending on the amount and solubility of the sample available. NMR hardware accommodating smaller diameter sample tubes enables the detection of much smaller samples. Common commercially available liquid NMR tubes range from 1 to 8 mm in diameter. Amounts of detectable sample for these configurations can be as low as hundreds of nanograms for proton NMR detection [2]. Carbon sensitivity is on the order of 100 times worse, so detection limits are usually limited to tens of micrograms.

The amount of sample required for NMR of solids is much greater in part because the apparent signal-to-noise ratio is significantly reduced. This is due to much broader line shapes, easily an order of magnitude wider than those observed in equivalent spectra of liquids. A standard solid NMR sample is a powder packed tightly into a small zirconia rotor and sealed with end caps. As for liquids, solid sample configurations are described in terms of their diameter, in this case the diameter of the sample rotor. Common and commercially available rotors range from 2.5 to 7 mm in diameter. Amounts of sample for these configurations depend on the sample and its density and typically range from 30 mg up to 500 mg.

This range of sample options is the result of rapid technological development throughout the history of NMR. Traditionally, NMR probes were standardized at 5 mm for many years. The advent of pulsed Fourier transform (FT) NMR spectrometers began to change that paradigm. The desire to acquire ^{13}C NMR spectra began with the commercial availability of instruments such as the Varian CFT-20 (later designated as an FT-80) spectrometer in the early 1970s. These instruments were equipped with 8 mm diameter probes permitting larger samples of material to be studied and the data to be acquired more rapidly. By today's standards, the CFT-20 and FT-80 instruments are primitive, but at that time, they opened the door for the acquisition of ^{13}C spectra. The other advantage of FT-NMR instrumentation arose in the data sampling. The entire ^{13}C spectrum is sampled simultaneously with the application of a radiofrequency (RF) pulse capable of exciting from approximately 0 to 200 ppm in the ^{13}C frequency window. Following pulsed excitation, the resultant ^{13}C spectrum can be recorded as a free induction decay (FID) (Fig. 10.1). Thus, using an FT-NMR instrument, the entire ^{13}C

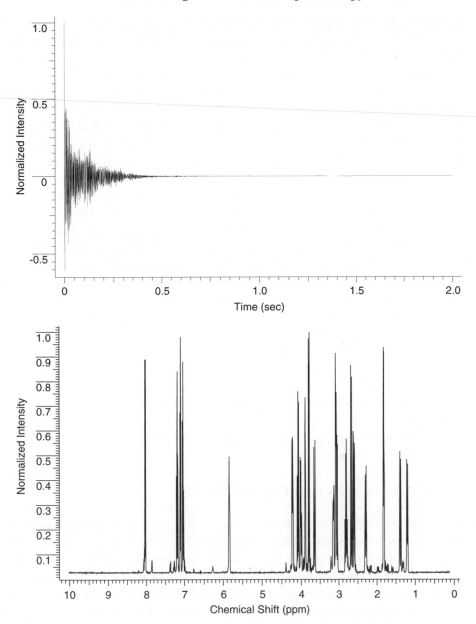

Fig. 10.1. The top panel shows the free induction decay (FID) acquired for a sample of strychnine (**1**) at an observation frequency of 500 MHz. The spectrum was digitized with 16 K points and an acquisition time of ~2 s. Fourier transforming the data from the time domain to the frequency domain yields the spectrum of strychnine presented as intensity versus frequency shown in the bottom panel.

spectrum can be sampled in a few seconds as opposed to tens or even hundreds of seconds required for a single transient by frequency swept NMR methods. After a delay to allow for sample relaxation processes, the sample can be pulsed again and the acquisition repeated.

NMR continues to be a very powerful analytical technique, and we touch on a variety of aspects of this throughout this chapter. We begin with an overview of the theory and basic principles of NMR. This includes a discussion of nuclear spin and the gyromagnetic ratio, as well as the Boltzmann distribution and energy levels and how all these factors collectively relate to detecting NMR signals. We review what actually happens when a radio frequency pulse is applied to a nuclear spin in a magnetic field. We then see how electronic shielding and different types of nuclear coupling influence this response. The theory and basic principles section is followed by an analysis of modern NMR instrumentation including magnet technology, quadrature detection, NMR probes and a variety of hardware accessories. Next, we look at the common NMR methods being used today for liquids. This spans simple proton and carbon experiments through much more sophisticated multidimensional, homonuclear and heteronuclear experiments. Next, we cover standard NMR experiments used for solid samples including both one-and-two (1D-, 2D-) dimensional experiments and practical considerations for selecting solids experiments versus liquids experiments. Finally, we work through specific case studies to better understand how NMR spectroscopy is applied. Review questions are provided at the end of the chapter to test one's knowledge of the material presented.

10.2   THEORY AND BASIC PRINCIPLES

NMR spectroscopy is a field of investigation based on the behavior of the nuclei in a molecule when subjected to an externally applied magnetic field. Nuclei spin about the axis of the externally applied magnetic field and consequently possess an angular momentum. The angular momentum can be expressed in units of Planck's constant and quantized as a function of a proportionality constant, I, that can be either an integer or a half-integer. I is referred to as the spin quantum number, or more simply as the nuclear spin. Some nuclei have a spin quantum number $I = 0$. These nuclei have an even atomic number and even mass. Examples commonly encountered for small molecules include ^{12}C, ^{16}O and ^{32}S. These nuclei cannot exhibit a magnetic resonance response under any circumstances. The group of nuclei most commonly exploited in the structural characterization of small molecules by NMR methods are nuclei with a spin of one half, which include: ^{1}H, ^{13}C, ^{19}F, ^{31}P and ^{15}N. Frequently encountered spin $I = 1$ nuclei include ^{2}H and ^{14}N. Examples of spin $I > 1$ nuclei include ^{10}B, ^{11}B and ^{23}Na.

Spin $I > 0$ nuclei possess a magnetic dipole or dipole moment, μ, which arises from a spinning, charged particle. Nuclei that have a non-zero spin will also have a magnetic moment, and the direction of that magnetic moment is collinear with the angular momentum vector associated with the nucleus. This can be expressed as

$$\mu = \gamma \boldsymbol{p} \qquad (10.1)$$

where γ is a proportionality constant known as the gyromagnetic ratio and $\boldsymbol{p}$ is a multiple of Planck's constant.

Next, consider what happens when a magnetic moment experiences an externally applied magnetic field, B_o. The interaction of μ with the magnetic field, B_o, produces torque, causing the magnetic moment to precess about the axis of the externally applied field. The precession frequency is a function of the externally applied field and the nucleus. As an analogy, consider the motion of a spinning gyroscope in the Earth's gravitational field. For a proton, ^{1}H, in an applied 2.35 T magnetic field, the precession frequency is ~100 MHz. In the same externally applied field, other nuclei that have different gyromagnetic ratios, γ, e.g., ^{13}C or ^{19}F, will precess at characteristically different frequencies. In a 2.35 T magnetic field, ^{13}C and ^{19}F nuclei will precess at ~25 and ~94 MHz, respectively. This characteristic precession frequency is known as the Larmor frequency of the nucleus.

Nuclei aligned with the axis of the externally applied magnetic field will be in the lowest possible energy state. Thermal processes oppose this tendency, such that there are two populations of nuclei in an externally applied magnetic field. One is aligned with the axis of the field and another, which is only slightly smaller, is aligned opposite to the direction of the applied field. The distribution of spins between these two energy levels is referred to as the Boltzman distribution, and it is this population difference between the two levels that provides the observable collection of spins in an NMR experiment. This difference is very small compared to the total number of spins present, which leads to the inherent insensitivity of NMR.

When an NMR experiment is performed, the application of a RFpulse orthogonal to the axis of the applied magnetic field perturbs the Boltzmann distribution, thereby producing an observable event that is governed by the Bloch equations [3]. Using a vector representation, the

Boltzmann excess population will correspond to a vector aligned with the axis of the externally applied magnetic field, B_o. Next, assume that sufficient energy is applied to tip or rotate that vector ensemble of spins from the z-axis into the xy plane. This would correspond in NMR terms to the application of a 90° RF pulse. When the application of that pulse is completed, the vector that has been rotated into the xy plane will continue to precess about the axis of the externally applied field (z-axis), generating an oscillating signal in a receiver coil as the vector rotates about the z-axis at its characteristic Larmor frequency. As a function of two-time constants, the spin–spin or transverse relaxation time, T_2, and the spin-lattice relaxation time, T_1, the signal from the magnetization vector will decay back to an equilibrium condition along the B_o axis, and the process can then be repeated. The decaying signal recorded by the receiver coil is processed through an analog-to-digital converter (ADC) and stored in memory. When the NMR experiment is finished, the stored time domain signal from the decaying magnetization in the xy plane can be converted to the frequency domain using a Fourier transformation, affording a representation of the NMR data in a histogram format (intensity versus f'requency) that is the familiar form of an NMR spectrum (Fig. 10.2).

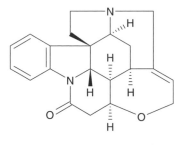

1

The location of NMR signals in a spectrum, which is known as a signal's "chemical shift," is a function of the chemical environment of the sampled nuclei. For solution state ^{1}H and ^{13}C NMR experiments, the reference standard is tetramethylsilane (TMS), which has an accepted chemical shift of 0.00 ppm. Most proton signals appear to the left or "downfield" of TMS. Aliphatic hydrocarbon signals will generally be grouped nearer the position of TMS and are said to be "shielded" relative to vinyl or aromatic signals that are as a group referred to as "deshielded." Chemical moieties involving heteroatoms, e.g., $-OCH_3$, $-NCH_2-$, typically will be located in a region of the NMR spectrum between the aliphatic and vinyl/aromatic signals.

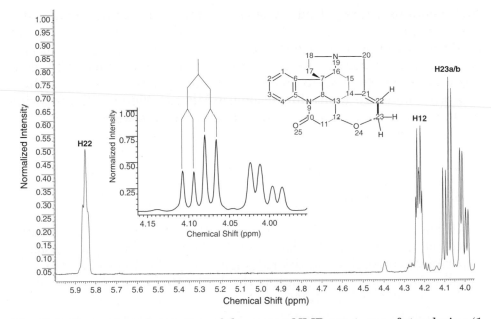

Fig. 10.2. Expansion of a portion of the proton NMR spectrum of strychnine (1 inset structure). The full proton spectrum is shown in Fig. 10.1. The resonances for the H22 vinyl proton and the H12 and H23 oxygen-bearing methine and methylene resonances, respectively, are shown. The inset expansion of the H23 methylene protons shows a splitting diagram for this resonance. The larger of the two couplings is the geminal coupling to the other H23 resonance and the smaller coupling is the vicinal coupling to the H22 vinyl proton.

In addition to the chemical shift information, an NMR spectrum may also contain coupling information. The types of couplings frequently present in NMR experiments include scalar (J) couplings between high-abundance nuclei such as protons, dipolar couplings that are important for cross-relaxation processes and the determination of nuclear Overhauser effect (NOE) (described later in this chapter), and quadrupolar coupling associated with quadrupolar nuclei ($I > 1/2$).

For NMR of liquids, scalar (J) couplings in proton spectra provide information about the local chemical environment of a given proton resonance. Proton resonances are split into multiplets related to the number of neighboring protons. For example, an ethyl fragment will be represented by a triplet with relative peak intensities of 1:2:1 for the methyl group, the splitting due to the two neighboring methylene protons, and a 1:3:3:1 quartet for the methylene group, with the splitting due to the three equivalent methyl protons. More complex molecules, of course, lead to considerably more complicated spin-coupling patterns.

Prior to the advent of homonuclear 2D-NMR experiments, it was necessary to rigorously interpret a proton NMR experiment and identify all of the homonuclear couplings to assemble the structure. Alternatively, there are multi-dimensional NMR experiments that provide similar information in a more readily interpretable way.

10.3   INSTRUMENTATION

10.3.1   Spectrometer overview

NMR spectrometers in today's research laboratories are sophisticated pieces of instrumentation that are capable of performing a myriad of experiments to analyze questions ranging from the molecular structure of unknown organic compounds to the different crystal forms contained within a solid. While an NMR spectrometer is quite complex, it is comprised of a few key components. An NMR spectrometer in its most basic form consists of the following: (i) a magnet, (ii) shims (iii) a RF generator and a receiver—probe, and (iv) a receiver.

10.3.1.1   Magnets

Superconducting magnets are used in modern NMR instruments to achieve the high-magnetic fields required. The basic design of these magnets consist of a large coil of very sophisticated wire in which an electric current flows thereby inducing the magnetic field. The wire that is used has been developed to remove the resistance at very low temperatures ($< -267°C$). This is the property that makes the wire superconducting. Therefore, once the current has been supplied to the wire, the source can be removed, and the magnet will remain energized, in principle, indefinitely. The wire is formed into coils to induce a linear magnetic field. The design of the "can" that is associated with an NMR instrument is almost solely to accommodate the cryogenic liquids used to keep the magnet coils at a reduced temperature. The "can" consists of two main dewars that hold the cryogens as shown in Fig. 10.3. The coils are submersed in a bath of liquid helium (B). To reduce the boiling rate of the liquid helium, this dewar is surrounded by a bath of liquid nitrogen (C). The magnetic field strength is often spoken of in terms of the Larmor frequency of protons. Magnetic fields operating at 18.8 T correspond to a ^{1}H Larmor frequency of 800 MHz.

Fig. 10.3. Pictures of a dissected NMR dewar for a superconducting magnet. Pictures taken from http://www.joel.com./nmr/mag_view/magnet_destruction. html, permission not obtained.

10.3.1.2 Shims

Fast isotropic molecular tumbling in liquids gives rise to very narrow lines in the NMR spectrum. One can typically expect to see lines as narrow as 1–10 Hz for NMR spectra of liquids. To achieve these line widths, the external magnetic field experienced by the sample must be extremely homogenous. Today's superconducting magnets cannot themselves produce the required field homogeneity that is needed. To overcome a magnet's inhomogeneity issues, electric coils, or "shim coils," are placed in various geometries around the portion of the magnetic field encompassing the sample area. Current is applied through these shim coils to induce small, spatially distinct magnetic fields to compensate for the overall magnetic field inhomogeneity.

10.3.1.3 RF generation

A key feature that enables the transition of energy states in spins is the generation of an applied magnetic field, called B_1. This magnetic field is induced through RF pulses that are applied orthogonal to the static field, B_o. This RF field is most often referred to as an RF pulse. This pulse is simply a gated current (irradiation) that has both amplitude and duration. The applied B_1 field excites the nuclei with enough energy to tip the nuclear magnetization away from alignment with the static B_o field. The amplitude and duration of the RF pulse are those that allow a specific "tip angle" of the nuclear magnetization vector to be achieved. The RF pulses are typically generated in a frequency generator and are, usually, only a few milliwatts for NMR of liquids. To provide a useful B_1 field, amplification is therefore required. Typical ^{1}H amplifiers on modern NMR spectrometers are capable of amplifying these RF pulses up to nominally 100 W.

10.3.1.4   The receiver

Nuclear spins are like small magnets that collectively induce a detectable electric current when they align along the y-axis of the magnetic field. This electric current is detected by the coil of the NMR probe and is exquisitely small, on the order of microvolts. This signal is therefore amplified in order to be more easily and precisely detected. A preamplifier is used to enhance the weak sample signal that comes from the RF probe coil. Once the signal is amplified, it is sent to a receiver, where it is then passed on to an ADC to digitize the incoming analog signal.

10.3.2   Quadrature detection

One of the limitations to the FT is that it cannot discriminate between positive and negative frequencies. Therefore, if the digitized signal is detected by a single channel, the observed spectrum is a mirror image set around the irradiation frequency. There are several ways around this phenomenon, one of which is quadrature detection [4]. If observing signal from a single channel, one would only observe a cosine contribution of the signal and therefore be unable to differentiate sign. However, if two detectors 90° out of phase with respect to one another are utilized, then the sinusoidal contribution can also be collected. These signals are received and digitized separately. They make up the real (cosine) contribution and the imaginary (sine) contribution of the complex data that make up the FID. With both the real and imaginary signal contributions, the sense of precession can be determined, and therefore a single resonance is observed. This allows one to set the irradiation frequency in the middle of the spectrum and properly sample the entire spectral width.

10.3.3   Probes

Larger format probes, e.g., 12, 18 and 22 mm facilitate the study of rare nuclei such as ^{15}N and ^{17}O. "X-nucleus" or "broadband" multinuclear

NMR probes are designed with the X-coil closest to the sample for improved sensitivity of rare nuclei. Inverse detection NMR probes have the proton coil inside the X-coil to afford better proton sensitivity, with the X-coil largely relegated to the task of broadband X-nucleus decoupling. These proton optimized probes are often used for heteronuclear shift correlation experiments.

Smaller diameter probes reduce sample volumes from 500 to 600 µl typical with a 5 mm probe down to 120–160 µl with a 3 mm tube. By reducing the sample volume, the relative concentration of the sample can be correspondingly increased for non-solubility limited samples. This dramatically reduces data acquisition times when more abundant samples are available or sample quantity requirements when dealing with scarce samples. At present, the smallest commercially available NMR tubes have a diameter of 1.0 mm and allow the acquisition of heteronuclear shift correlation experiments on samples as small as 1 µg of material, for example in the case of the small drug molecule, ibuprofen [5]. In addition to conventional tube-based NMR probes, there are also a number of other types of small volume NMR probes and flow probes commercially available [6]. Here again, the primary application of these probes is the reduction of sample requirements to facilitate the structural characterization of mass limited samples. Overall, many probe options are available to optimize the NMR hardware configuration for the type and amount of sample, its solubility, the nucleus to be detected as well as the type and number of experiments to be run.

Cryogenically cooled NMR probes cool the probe electronics to reduce the thermal noise thereby effectively increasing the signal-to-noise ratio. These probes afford a three- to four-fold sensitivity enhancement over a conventional NMR probe in the same configuration. This makes them ideally suited for the structural characterization of small samples, proteins, and other samples requiring high sensitivity to minimize data acquisition times. A typical hardware configuration uses a closed-loop helium refrigeration system capable of maintaining the probe electronics at an operating temperature of 20–25 K. The sample itself is isolated so that it can remain at ambient temperature or any other standard temperature setting desired.

10.3.4   Accessories

Aside from the basic components of the spectrometer as mentioned above, there are many optional accessories that should be considered for establishing an appropriate NMR hardware configuration. Variable temperature (VT) controllers and gradient amplifiers are standard features on today's NMR spectrometers. Additional accessories include sample changers, automatic probe tuning capabilities, liquid chromatography (LC) and other related capabilities.

VT controllers should be considered an essential part of the modern NMR spectrometer. The effects of chemical shifts, equilibrium kinetics and the homogeneity of the sample are all affected by temperature fluctuations and can have significant impact of the resulting data acquired. Most commercial VT units have very strict tolerances that allow only very slight temperature fluctuations within the NMR probe. These units typically utilize a source of nitrogen gas that flows through the probe and over the sample tube. The probe contains a software controlled heater coil that accurately monitors the amount of current that flows through the heater coil and thereby mediates the temperature. While the effects of temperature fluctuation can be deleterious to the quality of the NMR data, it is often the case that the temperature needs to be raised or lowered to take advantage of a particular temperature-dependent feature. An example would be a spectrum with quite broad lines due to intermediate slow exchange on the NMR timescale. In favorable cases, broad peaks at room temperature become narrow due to coalescence of interconverting species when the temperature is raised to an appropriate level. An additional example would be the study of a mechanism in an organic chemistry reaction, which is often run under varied temperatures to determine the kinetics of the reaction. This type of study is a very powerful demonstration of the power and versatility of NMR spectroscopy.

Most modern NMR experiments contain pulsed field gradient elements. The basic premise behind using pulsed field gradients is that when applied, the magnetic field becomes spatially inhomogeneous. This field inhomogeneity dephases the nuclear magnetization, and it is therefore effectively rendered unobservable. Simply applying an additional gradient pulse, thereby refocusing the magnetization, can reverse this signal dephasing. Pulsed field gradients are typically used to selectively filter out undesired signals in an NMR experiment through a process known as coherence pathway selection. Arguably, the implementation of the pulsed field gradients is one of the key milestones in NMR spectroscopy in the last ~20 years. These experimental building blocks provide superior results when compared to the alternative approach of cycling the direction of the RF pulses. In addition, gradient pulses can be used to acquire a spatial image of a sample.

This image is typically referred to as a gradient field map and is used, for example, when executing gradient shimming routines to ensure optimal B_o homogeneity. This type of shimming routine is advantageous for running samples under automation routines.

An automatic probe tuning and matching (ATM) accessory allows one to automatically tune the NMR probe to the desired nuclei's resonant frequency and match the resistance of the probe circuit to $50\,\Omega$ [7]. Traditional NMR instruments are designed so that one must perform these adjustments manually prior to data acquisition on a new sample. The advent of the ATM accessory allows the sampling of many different NMR samples without the need for human intervention. The ATM in conjunction with a sample changer enables NMR experiments to be conducted under complete automation. The sample changers are designed so that once the samples are prepared, they are placed into the instrument's sample holders. Data are then acquired under software control of both the mechanical sample delivery system as well as the electronics of the spectrometer.

The hyphenation of LC and NMR spectroscopy (LC–NMR) has become very popular in recent years for applications such as drug metabolite screening [8]. The basic principle of an LC-NMR instrument is to perform in-line separations of analytes and flow the subsequent eluent into a specially designed probe. These probes, often referred to as flow probes, are designed to have a static NMR sample cell fixed within the RF coil. The flow cell is coupled with chemically inert tubing that allows the sample to flow through the coil. The NMR signal of the flowing liquid is then detected. The overriding practical issue with LC–NMR is to be able to transfer enough sample mass from the LC run to the flow cell. The introduction of solid-phase extractions and other peak trapping techniques have helped broaden the utility of LC–NMR. Peak trapping onto cartridges, filled with a stationary phase with appropriate retention properties, allows multiple injections and subsequent isolations of a single chromatographic peak to be trapped within this cartridge [9]. This "stacking" of peaks allows a significant increase in total mass to be transferred to the flow cell and overcome some of the inherent sensitivity limitations associated with LC–NMR. Additional advances in hyphenated NMR techniques include LC–NMR/MS, LC–SPE–NMR (where SPE stands for solid phase extraction) and its mass spectrometer (MS) derivative [10,11]. The coupling of a MS to the LC–NMR system enables the collection of both NMR data and MS data on the analyte of interest.

10.4   NMR METHODS FOR LIQUIDS

10.4.1   1D Proton NMR methods

10.4.1.1   *Magnetically equivalent nuclei*

Magnetically equivalent nuclei are defined as nuclei having both the same resonance frequency and spin–spin interaction with neighboring atoms. The spin–spin interaction does not appear in the resonance signal observed in the spectrum. It should be noted that magnetically equivalent nuclei are inherently chemically equivalent, but the reverse is not necessarily true. An example of magnetic equivalence is the set of three protons within a methyl group. All three protons attached to the carbon of a methyl group have the same resonance frequency and encounter the same spin–spin interaction with its vicinal neighbors, but not with its geminal counterparts. The resultant is a resonance that integrates for three protons showing coupling to its adjacent nuclei.

10.4.1.2   *Pascal's triangle*

J-Coupling (a.k.a. "spin–spin coupling" or "scalar coupling") is an interaction that arises from a magnetic interaction of nuclei through the physical connection of a chemical bond. The syntax commonly used for this interaction is $^{n}J_{AB}$, where A and B are the two nuclei impacted by the interaction, J, and n is the number of chemical bonds separating A and B. For example, $^{2}J_{CH}$ represents a J-coupling between a carbon and proton separated by two bonds. Scalar coupling is a fundamental physical phenomenon observed in NMR, one of the innate primary forces that act on an irradiated collection of spins is J-coupling. This interaction is utilized extensively in 2D methods for correlation spectroscopy. The frequency of this interaction is usually described in Hertz and, in 1D ^{1}H data, measured directly from the multiplet. The intensities of these lines are given by a binomial expansion, or more conveniently by Pascal's triangle (see Fig. 10.4). The extent of the observed splitting is governed by $M = (n+1)$, where M is the multiplicity and n is the number of adjacent nuclei. This holds true for all spin $= 1/2$ nuclei. J-coupling of adjacent spins is typically referred to as a spin system. A spin system can be two adjacent atoms or a series of contiguous coupled spins.

10.4.1.3   *Table of proton chemical shifts*

Proton chemical shifts are dictated directly by the chemical environment in which they reside. This can be through neighboring atoms that

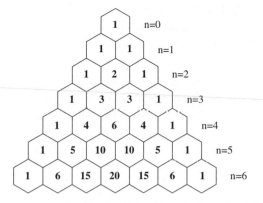

Fig. 10.4. Illustration of Pascal's triangle only showing ratios to $n = 6$ according to $M = (n+1)$ where M is the multiplicity and n is the number of scalar coupled nuclei. For example, a proton adjacent to three protons ($n = 3$) would appear as a quartet ($M = 4$) with relative peak intensities of 1:3:3:1.

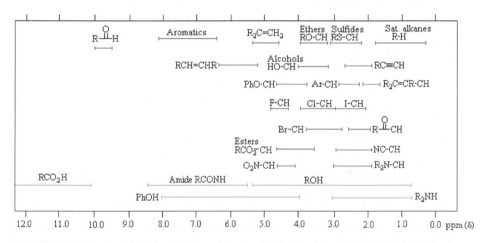

Fig. 10.5. Diagram demonstrating the general resonance frequencies of typical ^{1}H's in small organic molecules.

induce an electronic shielding effect, such as electron donating groups, which will cause a downfield shift (larger ppm values), or the effects of the surrounding solvent. Proton chemical shifts are solvent dependent and may have a slightly different value in different solvents. Of interest to NMR spectroscopists is that while the chemical shift dispersion is quite small (~15 ppm) relative to other magnetically active nuclei (^{13}C has a shift dispersion of ~200 ppm), the observed nuclei resonate in particular regions within this narrow-spectral-range based on their electronic environment. These ranges can be seen in Fig. 10.5.

10.4.1.4   NOE experiments

The acronym NOE is derived from the nuclear Overhauser effect. This phenomenon was first predicted by Overhauser in 1953 [12]. It was later experimentally observed by Solomon in 1955 [13]. The importance of the NOE in the world of small molecule structure elucidation cannot be overstated. As opposed to scalar coupling described above, NOE allows the analysis of dipolar coupling. Dipolar coupling is often referred to as through-space coupling and is most often used to explore the spatial relationship between the two atoms experiencing zero scalar coupling. The spatial relationship of atoms within a molecule can provide an immense amount of information about a molecule, ranging from the regiochemistry of an olefin to the 3D-solution structure. An example of a regiochemical application is illustrated below for the marine natural product jamaicamide A (2) in Fig. 10.6 [14].

A simplified explanation of the NOE is the magnetic perturbations induced on a neighboring atom resulting in a change in intensity. This is usually an increase in intensity, but may alternatively be zero or even negative. One can envision saturating a particular resonance of interest for a time t. During the saturation process a population transfer occurs to all spins that feel an induced magnetization.

There are two types of NOE experiments that can be performed. These are referred to as the steady-state NOE and the transient NOE. The steady-state NOE experiment is exemplified by the classic NOE difference experiment [15]. Steady-state NOE experiments allow one to quantitate relative atomic distances. However, there are many issues that can complicate their measurement, and a qualitative interpretation is more reliable [16]. Spectral artifacts can be observed from imperfect subtraction of spectra. In addition, this experiment is extremely susceptible to inhomogeneity issues and temperature fluctuations.

1D-transient NOE experiments employing gradient selection are more robust and therefore are more reliable for measuring dipolar

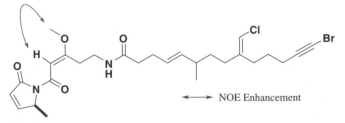

Fig. 10.6. Structure of jamaicamide A (2) showing a key NOE defining the geometry of a tri-substituted olefin.

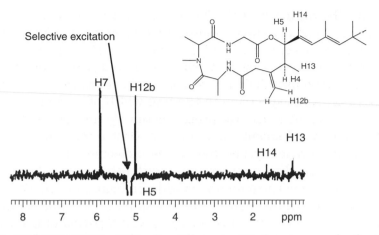

Fig. 10.7. DPFGSE NOE of H5 of antillatoxin (**3**) showing good selection and clean NOE enhancements.

coupling interactions [17]. Shaka *et al.* published one such 1D-transient NOE experiment that has, in most cases, replaced the traditional NOE difference experiment [18]. The sequence dubbed the double-pulsed field gradient spin echo (DPFGSE) NOE employs selective excitation through the DPFGSE portion of the sequence [19]. Magnetization is initially created with a 90° ^{1}H pulse. Following this pulse are two gradient echoes employing selective 180° pulses. The flanking gradient pulses are used to dephase and recover the desired magnetization as described above. This selection mechanism provides very efficient selection of the resonance of interest prior to the mixing time where dipolar coupling is allowed to buildup. An example of this experiment is shown in Fig. 10.7 for antillatoxin (**3**) [20].

10.4.1.5 Relaxation measurements
Relaxation is an inherent property of all nuclear spins. There are two predominant types of relaxation processes in NMR of liquids. These relaxation processes are denoted by the longitudinal (T_1) and transverse (T_2) relaxation time constants. When a sample is excited from its thermal equilibrium with an RF pulse, its tendency is to relax back to its Boltzmann distribution. The amount of time to re-equilibrate is typically on the order of seconds to minutes. T_1 and T_2 relaxation processes operate simultaneously. The recovery of magnetization to the equilibrium state along the z-axis is longitudinal or the T_1 relaxation time. The loss of coherence of the ensemble of excited spins (uniform distribution) in the x-, y-plane following the completion of a pulse is transverse or T_2

275

relaxation. The duration of the T_1 relaxation time is a very important feature as it allows us to manipulate spins through a series of RF pulses and delays. Transverse relaxation is governed by the loss of phase coherence of the precessing spins when removed from thermal equilibrium (e.g., a RF pulse). The transverse or T_2 relaxation time is visibly manifest in an NMR spectrum in the line width of resonances; the line width at half height is the reciprocal of the T_2 relaxation time. These two relaxation mechanisms can provide very important information concerning the physical properties of the molecule under study, tumbling in solution, binding or interaction with other molecules, etc.

T_1 relaxation measurements provide information concerning the time constant for the return of excited spins to thermal equilibrium. For spins to fully relax, it is necessary to wait a period of five times T_1. To accelerate data collection, in most cases one can perform smaller flip-angles than 90° and wait a shorter time before repeating the pulse sequence. Knowing the value of T_1 proves to be very useful in some instances, and it is quite simple to measure. The pulse sequence used to perform this measurement is an inversion recovery sequence [21]. The basic linear sequence of RF pulses (an NMR pulse sequence) consists of a 180-τ-90-acquire. The delay τ is incrementally increased, and the frequency domain data are plotted versus τ to look for the maximum signal recovery (greatest peak intensity). The 180° pulse simply inverts the magnetization, and after waiting for the delay, τ a 90° pulse is used to place the magnetization into the x–y plane for detection. Therefore, by varying τ one will identify a value that is sufficiently long to allow maximum signal intensity.

T_2 relaxation can be a difficult parameter to measure accurately. There are several ways to accomplish this task. The experiments that are employed for this measurement are all variants of the classic spin-echo experiment [22,23]. The basic sequence consists of a 90-τ-180-τ-90-acquire. The magnetization is first placed in the x–y plane with the application of a 90° pulse. Following this pulse is a delay τ that allows the magnetization vector to dephase in the x–y plane. A 180° pulse is then applied to invert the magnetization, and an additional period of τ, identical in duration to the first, refocuses the magnetization vector prior to acquisition. By varying τ one can measure the T_2 relaxation time. Variants of this approach are the Carr–Purcell and CPMG pulse sequences [24]. As opposed to the relaxation process described by T_1 (a return to thermal equilibrium along the z-axis), T_2 measures the relaxation of magnetization caused by dephasing in the x–y plane. Therefore, T_2 is an extremely important parameter that dictates the amount of time one can manipulate the spins before the magnetization is completely dephased, and thus not observable.

10.4.2   1D Carbon-NMR methods

10.4.2.1   Proton decoupling

Carbon-13, or ^{13}C, is a rare isotope of carbon with a natural abundance of 1.13% and a gyromagnetic ratio, γ_C, that is approximately one quarter that of ^{1}H. Early efforts to observe ^{13}C NMR signals were hampered by several factors. First, the 100% abundance of ^{1}H and the heteronuclear spin coupling, $^nJ_{CH}$ where $n = 1 - 4$, split the ^{13}C signals into multiplets, thereby making them more difficult to observe. The original efforts to observe ^{13}C spectra were further hampered by attempts to record them in the swept mode, necessitating long acquisition times and computer averaging of scans. These limitations were circumvented, however, with the advent of pulsed FT–NMR spectrometers with broadband proton decoupling capabilities [25].

Broadband ^{1}H decoupling, in which the entire proton spectral window is irradiated, collapses all of the ^{13}C multiplets to singlets, vastly simplifying the ^{13}C spectrum. An added benefit of broadband proton decoupling is NOE enhancement of protonated ^{13}C signals by as much as a factor of three.

Early broadband proton decoupling was accomplished by noise modulation that required considerable power, typically 10 W or more, and thus caused significant sample heating. Over the years since the advent of broadband proton decoupling methods, more efficient decoupling methods have been developed including globally optimised, alternating phase rectangular pulse (GARP), wideband uniform rate and smooth truncation (WURST), and others [26]. The net result is that ^{13}C spectra can now be acquired when needed with low-power-pulsed decoupling methods and almost no sample heating.

10.4.2.2   Magnetization transfer

The earliest of the magnetization transfer experiments is the spin population inversion (SPI) experiment [27]. By selectively irradiating and inverting one of the ^{13}C satellites of a proton resonance, the recorded proton spectrum is correspondingly perturbed and enhanced. Experiments of this type have been successfully utilized to solve complex structural assignments. They also form the basis for 2D-heteronuclear chemical shift correlation experiments that are discussed in more detail later in this chapter.

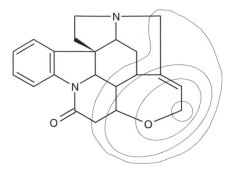

Fig. 10.8. Spherical environments surrounding the 23-position of strychnine whose effects would be incorporated into the calculation of the C-23 chemical shift using a HOSE code approach.

10.4.2.3 Empirical chemical shift calculations

Vast tabulations of ^{13}C chemical shift data have been assembled in computer searchable form. These databases form the basis for ^{13}C chemical shift prediction algorithms. For the most part, carbon chemical shifts can be calculated using what is referred to as a Hierarchically Ordered Spherical Environment (HOSE) code approach [28]. To calculate a given carbon's chemical shift, the influence of each successive spherical shell is applied to the starting chemical shift for that carbon to calculate its overall chemical shift. Typically, programs will calculate shifts for 3 or 4 layers, beyond which the effects of most substituents are negligible. The spherical layers surrounding the 23-position of strychnine are shown in Fig. 10.8.

10.4.2.4 Standard 1D experiments

Of the multitude of 1D ^{13}C NMR experiments that can be performed, the two most common experiments are a simple broadband proton-decoupled ^{13}C reference spectrum, and a distortionless enhancement polarization transfer (DEPT) sequence of experiments [29]. The latter, through addition and subtraction of data subsets, allows the presentation of the data as a series of "edited" experiments containing only methine, methylene and methyl resonances as separate subspectra. Quaternary carbons are excluded in the DEPT experiment and can only be observed in the ^{13}C reference spectrum or by using another editing sequence such as APT [30]. The individual DEPT subspectra for CH, CH_2 and CH_3 resonances of santonin (4) are presented in Fig. 10.9.

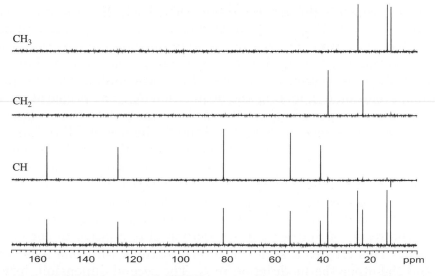

Fig. 10.9. Multiplicity edited DEPT traces for the methine, methylene and methyl resonances of santonin (**4**). Quaternary carbons are excluded in the DEPT experiment and must be observed in the ^{13}C reference spectrum or through the use of another multiplicity editing experiment such as APT.

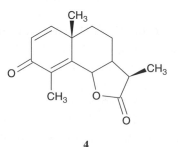

4

10.4.3   Homonuclear 2D methods

10.4.3.1   Basic principles of 2D

2D-NMR methods are highly useful for structure elucidation. Jeener described the principles of the first 2D-NMR experiment in 1971 [31]. In standard NMR nomenclature, a data set is referred to by one, i.e., less than the total number of actual dimensions, since the intensity dimension is implied. The 2D-data matrix therefore can be described as a plot containing two frequency dimensions. The inherent third dimension is the intensity of the correlations within the data matrix. This is the case in "1D" NMR data as well. The implied second dimension actually reflects the intensity of the peaks of a certain resonance frequency (which is the 1st dimension). The basic 2D-experiment consists of a series of 1D-spectra. These data are run in sequence and have a variable delay built into the pulse program that supplies the means for the second dimension. The variable delay period is referred to as t_1.

Most 2D-pulse sequences consist of four basic building blocks: preparation, evolution (t_1), mixing and acquisition (t_2). The preparation and mixing times are periods typically used to manipulate the magnetization (a.k.a. "coherence pathways") through the use of RF pulses. The evolution period is a variable time component of the pulse sequence. Successive incrementing of the evolution time introduces a new time domain. This time increment is typically referred to as a t_1 increment and is used to create the second dimension. The acquisition period is commonly referred to as t_2. The first dimension, generally referred to as F_2, is the result of Fourier transformation of t_2 relative to each t_1 increment. This creates a series of interferograms with one axis being F_2 and the other the modulation in t_1. The second dimension, termed F_1, is then transformed with respect to the t_1 modulation. The resultant is the two frequency dimensions correlating the desired magnetization interaction, most typically scalar or dipolar coupling. Also keep in mind that there is the "third dimension" that shows the intensity of the correlations.

When performing 2D-NMR experiments one must keep in mind that the second frequency dimension (F_1) is digitized by the number of t_1 increments. Therefore, it is important to consider the amount of spectral resolution that is needed to resolve the correlations of interest. In the first dimension (F_2), the resolution is independent of time relative to F_1. The only requirement for F_2 is that the necessary number of scans is obtained to allow appropriate signal averaging to obtain the desired S/N. These two parameters, the number of scans acquired per t_1 increment and the total number of t_1 increments, are what dictate the amount of time required to acquire the full 2D-data matrix. 2D-homonuclear spectroscopy can be summarized by three different interactions, namely scalar coupling, dipolar coupling and exchange processes.

10.4.3.2 Scalar coupled experiments: COSY and TOCSY

The correlated spectroscopy (COSY) experiment is one of the most simple 2D-NMR pulse sequences in terms of the number of RF pulses it requires [32]. The basic sequence consists of a 90-t_1-90-acquire. The sequence starts with an excitation pulse followed by an evolution period and then an additional 90° pulse prior to acquisition. Once the time domain data are Fourier transformed, the data appear as a diagonal in the spectrum that consists of the ^{1}H chemical shift centered at each proton's resonance frequency. The off-diagonal peaks are a result of scalar coupling evolution during t_1 between neighboring protons. The data allow one to visualize contiguous spin systems within the molecule under study.

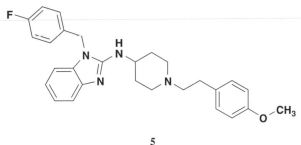

5

In addition to the basic COSY experiment, there are phase-sensitive variants that allow one to discriminate the active from the passive couplings allowing clearer measurement of the former. Active couplings give rise to the off-diagonal crosspeak. However, the multiplicity of the correlation has couplings inherent to additional coupled spins. These additional couplings are referred to as passive couplings. One such experiment is the double quantum-filtered (DQF) COSY experiment [33]. Homonuclear couplings can be measured in this experiment between two protons isolated in a single spin system. Additional experiments have been developed that allow the measurement of more complicated spin systems involving multiple protons in the same spin system [34]. The 2D representation of the scalar coupled experiment is useful when identifying coupled spins that are overlapped or are in a crowded region of the spectrum. An example of a DQF–COSY spectrum is shown in Fig. 10.10. This data set was collected on astemizole (**5**).

Total correlation spectroscopy (TOCSY) is similar to the COSY sequence in that it allows observation of contiguous spin systems [35]. However, the TOCSY experiment additionally will allow observation of up to about six coupled spins simultaneously (contiguous spin system). The basic sequence is similar to the COSY sequence with the exception of the last pulse, which is a "spin-lock" pulse train. The spin lock can be thought of as a number of homonuclear spin echoes placed very close to one another. The number of spin echoes is dependent on the amount of time one wants to apply the spin lock (typically ~60 msec for small molecules). This sequence is extremely useful in the identification of spin systems. The TOCSY sequence can also be coupled to a heteronuclear correlation experiment as described later in this chapter.

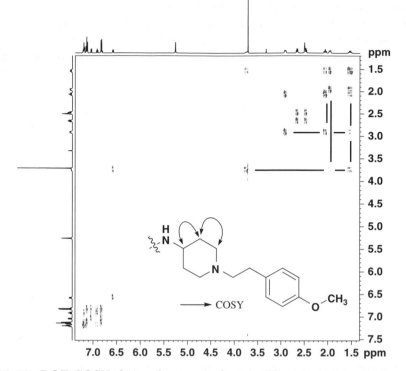

Fig. 10.10. DQF-COSY data of astemizole (**5**). The black bars indicate the contiguous spin system drawn with arrows on the relevant portion of astemizole.

10.4.3.3   Scalar coupled experiments: INADEQUATE

2D-homonuclear correlation experiments are typically run using ^{1}H as the nucleus in both dimensions. This is advantageous, as the sensitivity for protons is quite high. However, there are 2D-homonuclear techniques that detect other nuclei. One such experiment is the INADEQUATE experiment [36]. It is quite often used in solid-state NMR (SSNMR) as will be discussed later in this chapter. Its insensitive nature arises from the low-natural abundance of ^{13}C and the fact that one is trying to detect an interaction between two adjacent ^{13}C atoms. The chance of observing this correlation is 1 in every 10,000 molecules. Given sufficient time or isotope-enriched molecules, the INADEQUATE provides highly valuable information. The data generated from this experiment allow one to map out the entire carbon skeleton of the molecular structure. The only missing structural features occur where there are intervening heteroatoms (e.g., O, N).

10.4.3.4 Dipolar coupled experiments: NOESY

The 2D experiments described thus far rely solely on the presence of scalar coupling. There are other sequences that allow one to capitalize on the chemical shift dispersion gained with the 2nd dimension for dipolar coupling experiments as well. One example in this category is the 2D NOESY pulse sequence [37]. The pulse sequence for the 2D NOESY experiment is essentially identical to that of the DQF COSY experiment. The basic pulse sequence consists of a 90-t_1-90-τ_m-90-acquire. The magnetization is first excited by means of a $90°$ pulse, followed by the evolution period, t_1. After the evolution period, a second $90°$ pulse is applied to prepare for the NOE mixing time, τ_m. During this mixing time, an NOE is allowed to build up and then subsequently sampled after bringing the magnetization back to the x–y plane. The resulting data matrix is reminiscent of a COSY spectrum in that there is a diagonal represented by the 1D ^{1}H spectrum in both frequency dimensions. In sharp contrast to the scalar coupled cross-peaks in the COSY experiment, NOESY provides off-diagonal responses that correlate spins through-space. This sequence is used extensively in the structure characterization of small molecules for the same reason as its 1D counterparts. The spatial relationship of ^{1}H atoms is an invaluable tool.

The data represented in this correlation spectrum are such that the diagonal peaks are phased a particular way (typically down) and the corresponding off-diagonal peaks representing dipolar coupling responses are $180°$ out of phase (assuming positive NOE enhancements). In some cases, peaks of the same phase of the diagonal are observed. These peaks are most commonly associated with chemical exchange and if not properly identified can lead to incorrect interpretation of the data. The similarity of the NOESY to the COSY also causes some artifacts to arise in the 2D-data matrix of a NOESY spectrum. The artifacts arise from residual scalar coupling contributions that survive throughout the NOESY pulse sequence. These artifacts are usually quite straightforward to identify, as they have a similar anti-phase behavior as can be seen for the DQF COSY data, Fig. 10.10. Keeler *et al.* reported a method for removing these scalar artifacts through the use of gradient pulses [38]. An example of a NOESY data set is shown in Fig. 10.11, recorded using astemizole (**5**) and a 500 ms mixing time. The mixing time used during a NOE type of experiment is to allow the build-up of the NOE. Selection of appropriate mixing times is a crucial aspect to consider when setting up a dipolar coupling experiment. Incorrect choice could result in no NOE and hence no correlations.

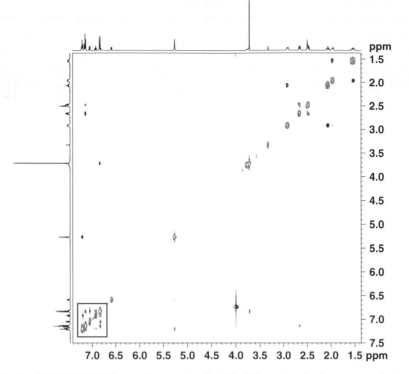

Fig. 10.11. 2D NOESY data of astemizole (**5**). The mixed phase correlations in the aromatic region (boxed) are examples of the artifacts described in the text.

10.4.4 Heteronuclear 2D methods

10.4.4.1 2D J-Resolved experiments—A historical perspective

The earliest 2D-NMR methods are the 2D J-resolved experiments [39]. These experiments display ^{13}C chemical shift information along the F_2-axis with heteronuclear coupling constant information displayed orthogonally in the F_1 dimension. The pulse sequence for a 2D J-resolved experiment is shown in Fig. 10.12, and the experimental results for santonin (**4**) are shown in Fig. 10.13. While the 2D J-resolved experiment can provide heteronuclear coupling constant information, the net chemical structure information obtained for the time invested in acquiring the data is relatively low in comparison, for example, to heteronuclear chemical shift correlation experiments. Consequently, the heteronuclear 2D J-resolved NMR experiment is seldom used at present unless there is a specific need to access heteronuclear coupling constant information for numerous resonances in a given molecule.

284

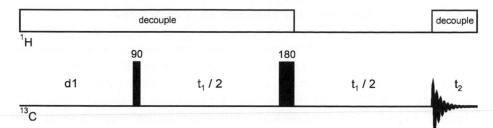

Fig. 10.12. Pulse sequence for amplitude modulated 2D J-resolved spectros-copy. The experiment is effectively a spin echo, with the ^{13}C signal amplitude modulated by the heteronuclear coupling constant(s) during the second half of the evolution period when the decoupler is gated off. Fourier transformation of the 2D-data matrix displays ^{13}C chemical shift information along the F_2 axis of the processed data and heteronuclear coupling constant information, scaled by J/2, in the F_1 dimension.

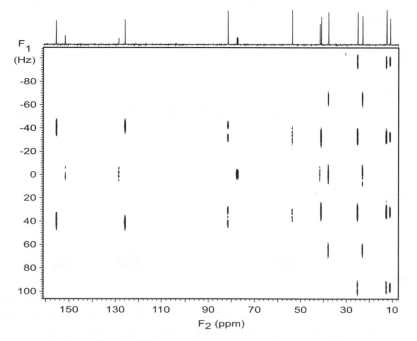

Fig. 10.13. 2D J-resolved NMR spectrum of santonin (4). The data were ac-quired using the pulse sequence shown in Fig. 10.12. Chemical shifts are sorted along the F_2 axis with heteronuclear coupling constant information displayed orthogonally in F_1. Coupling constants are scaled as J/2, since they evolve only during the second half of the evolution period, $t_1/2$. ^{13}C signals are amplitude modulated during the evolution period as opposed to being phase modulated as in other ^{13}C-detected heteronuclear shift correlation experi-ments.

10.4.4.2 Direct heteronuclear chemical shift correlation

Conceptually, the 2D J-resolved experiments lay the groundwork for heteronuclear chemical shift correlation experiments. For molecules with highly congested ^{13}C spectra, ^{13}C rather than ^{1}H detection is desirable due to high resolution in the F_2 dimension [40]. Otherwise, much more sensitive and time-efficient proton or so-called "inverse"-detected heteronuclear chemical shift correlation experiments are preferable [41].

The first of the proton-detected experiments is the Heteronuclear Multiple Quantum Correlation HMQC experiment of Bax, Griffey and Hawkins reported in 1983, which was first demonstrated using ^{1}H–^{15}N heteronuclear shift correlation [42]. The version that has come into wide-spread usage, particularly among the natural products community, is that of Bax and Subramanian reported in 1986 [43]. A more contemporary gradient-enhanced version of the experiment is shown in Fig. 10.14 [44].

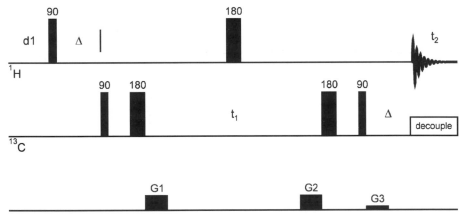

Fig. 10.14. Gradient-enhanced HMQC pulse sequence described in 1991 by Hurd and John derived from the earlier non-gradient experiment of Bax and Subramanian. For ^{1}H-^{13}C heteronuclear shift correlation, the gradient ratio, G1:G2:G3 should be 2:2:1 or a comparable ratio. The pulses sequence creates heteronuclear multiple quantum of orders zero and two with the application of the 90° ^{13}C pulse. The multiple quantum coherence evolves during the first half of t_1. The 180° proton pulse midway through the evolution period decouples proton chemical shift evolution and interchanges the zero and double quantum coherence terms. Antiphase proton magnetization is created by the second 90° ^{13}C pulse that is refocused during the interval Δ prior to detection and the application of broadband X-decoupling.

The gradient HMQC (GHMQC), sequence shown in Fig. 10.14 can be used to detect ^{1}H–^{13}C heteronuclear shift correlations and, by changing the gradient ratio to 5:5:1, for example, can also be used to detect direct ^{1}H–^{15}N heteronuclear correlations. While HMQC and GHMQC are still in wide usage, particularly among natural products chemists, it is at a disadvantage relative to the Heteronuclear Single Quantum Coherence HSQC experiment originally described by Bodenhausen and Ruben [45]. The effective F_1 resolution in the HMQC experiment, regardless of which variant is used, suffers due to homonuclear coupling modulation during the evolution period, which leads to broadening of responses in the F_1 frequency domain as noted by various workers [46].

The HSQC experiment is based on single rather than multiple quantum coherence during the evolution time, t_1. The contemporary multiplicity-edited gradient HSQC pulse sequence is shown in Fig. 10.15. Relative to the much simpler HMQC pulse sequence, the HSQC

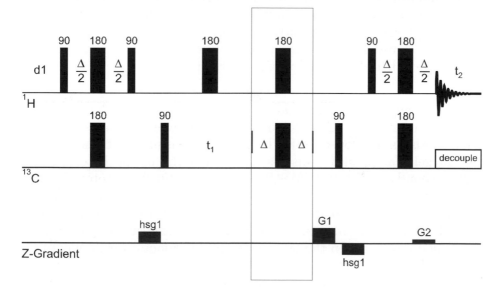

Fig. 10.15. Pulse sequence for the multiplicity-edited gradient HSQC experiment. Heteronuclear single quantum coherence is created by the first INEPT step within the pulse sequence, followed by the evolution period, t_1. Following evolution, the heteronuclear single quantum coherence is reconverted to observable proton magnetization by the reverse INEPT step. The simultaneous 180° ^{1}H and ^{13}C pulses flanked by the delays, $\Delta = 1/2(^1J_{CH})$, edits magnetization inverting signals for methylene resonances, while leaving methine and methyl signals with positive phase (Fig. 16A). Eliminating this pulse sequence element affords a heteronuclear shift correlation experiment in which all resonances have the same phase (Fig. 16B).

experiment has multiple 180° pulses applied to both ^{1}H and the hetero-nucleus. This contrasts with the single 180° proton refocusing pulse used in the HMQC experiment. If there is a downside to the HSQC experiment and the improved F_1 resolution that it offers relative to HMQC experiments, it would be that the experiment is more sensitive to proper calibration of the 180° refocusing pulses used for both protons and the heteronucleus.

Using strychnine (**1**) as a model compound, a pair of HSQC spectra are shown in Fig. 10.16. The top panel shows the HSQC spectrum of strychnine without multiplicity editing. All resonances have positive phase. The pulse sequence used is that shown in Fig. 10.15 with the pulse sequence operator enclosed in the box eliminated. In contrast, the multiplicity-edited variant of the experiment is shown in the bottom panel. The pulse sequence operator is comprised of a pair of 180° pulses simultaneously applied to both ^{1}H and ^{13}C. These pulses are flanked by the delays, $\Delta = 1/2(^1J_{CH})$, which invert the magnetization for the methylene signals (red contours in Fig. 10.16B), while leaving methine and methyl resonances (positive phase, black contours) unaffected. Other less commonly used direct heteronuclear shift correlation experiments have been described in the literature [47].

10.4.4.3 Long-Range heteronuclear shift correlation methods

There are numerous ^{13}C detected long-range heteronuclear shift correlation methods developed [48]. The primary reason that these methods have largely fallen into disuse is because of the of the heteronuclear multiple bond correlation (HMBC) experiment [49]. The proton-detected HMBC experiment and its gradient enhanced variant (GHMBC) offer considerably greater sensitivity than the original heteronucleus-detected methods. The signal intensity of the GHMBC experiment depends on the congruence between the optimization of the long-range delay in the

Fig. 10.16. (A) GHSQC spectrum of strychnine (**1**) using the pulse sequence shown in Fig. 10.15 without multiplicity editing. (B) Multiplicity-edited GHSQC spectrum of strychinine showing methylene resonances (red contours) inverted with methine resonances (black contours) with positive phase. (Strychnine has no methyl resonances.) Multiplicity-editing does have some cost in sensitivity, estimated to be ~20% by the authors. For this reason, when severely sample limited, it is preferable to record an HSQC spectrum without multiplicity editing. Likewise, there is a sensitivity cost associated with the use of gradient based pulse sequences. For extremely small quantities of sample, non-gradient experiments are preferable.

Nuclear magnetic resonance spectroscopy

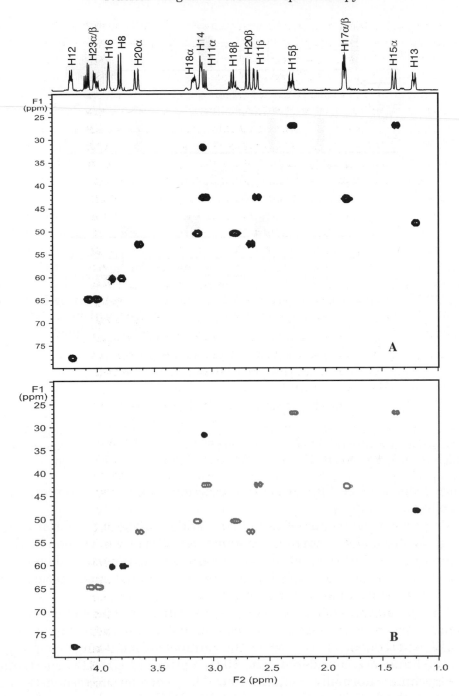

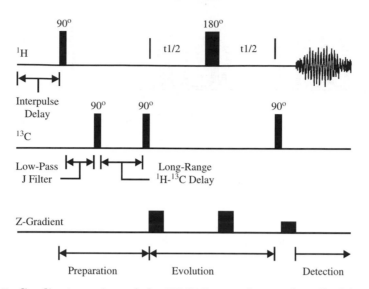

Fig. 10.17. Gradient version of the HMBC experiment described by Bax and Summers [49]. The gradient version shown is that of Hurd and John [44]. The original version of the experiment did not employ either gradients or the low-pass J-filter as shown. Typically the low-pass J-filter is optimized as a function of the average $^1J_{CH}$ coupling constant, typically $\sim$145 Hz The long-range delay is optimized for the average $^nJ_{CH}$ coupling where $n = 2 - 4$, which is usually in the range of 6–10 Hz Optimization of this delay for 6 and 10 Hz is frequently seen in literature reports. The gradient ratios for ^1H-^{13}C experiments are most typically set to 2:2:1 or 5:3:4 for ^1H-^{13}C or 5:5:1 for ^1H-^{15}N long-range experiments. For sample-limited problems, it is advantageous to rely on phase cycling rather than gradient coherence selection to improve the sensitivity of the experiment, which is typically regarded as about 1/4 that of the HSQC experiment.

pulse sequence (see Fig. 10.17) to the actual long-range heteronuclear coupling constant.

To provide the means of sampling a wider range of potential long-range heteronuclear coupling constants, an alternative version of this heteronuclear shift correlation experiment is called the accordion-optimized HMBC, or ACCORD–HMBC [50]. Additional modifications of this experiment are also available [51].

The advantage inherent to using an accordion–optimized long-range experiment over the single or "statically"-optimized delay in the HMBC experiment is seen when the number of long-range couplings of strychnine are considered. The 10 Hz optimized GHMBC spectrum of strychnine normally contains four $^4J_{CH}$ correlation responses. In contrast, the 2–25 Hz optimized ACCORD–HMBC experiment contains 17 $^4J_{CH}$ correlations, as shown by (**6**).

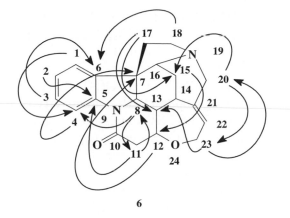

6

Additional heteronuclear long-range experiments include the BIRD–HMBC experiment and the broadband HMBC experiment [52]. Experimental variants capable of simultaneously recording direct and long-range heteronuclear correlations include the HMSQ and MBOB experiments [53].

Another class of heteronuclear long-range experiments focus on sorting long-range coupling constants. One such example is the XH correlation with fixed evolution pulse sequence [54]. This prototypical heteronucleus-detected long-range experiment differentiates two-bond from n-bond (n > 2) long-range correlations to protonated carbons. Inverse-detected experiments of this type include $^2J^3J$-HMBC, ^nJ-HMBC, HMBC–RELAY and H2BC [55]. While all of these various $^2J/^nJ$ HMBC variants work for protonated carbons, unfortunately, none of them are as yet capable of differentiating ^2J from ^nJ long-range correlations to quaternary carbon resonances. Likewise, none of the experiments presently available is capable of further differentiation of, for example, ^3J from ^nJ correlations where $n > 3$. Proton-detected long-range heteronuclear shift correlation experiments, irrespective of which one is used, provide a high-sensitivity method for accessing long-range heteronuclear connectivity information vital to structure elucidation efforts.

10.4.4.4 Hyphenated heteronuclear shift correlation methods
Many of the experiments described thus far can be used as building blocks to create more sophisticated pulse sequences. The earliest variants of hyphenated NMR experiments, such as HC–RELAY, were heteronuclear detected [56]. More recent hyphenated heteronuclear shift correlation methods are based on proton detection and include

experiments such as HXQC–TOCSY (X = S – single or M – multiple) or
HXQC–NOESY/ROESY [57]. The HSQC portion of the experiment
is used to sort the information afforded by the hyphenated segment of
the experiment as a function of carbon chemical shift in the F_1
frequency domain. The HMQC- or HSQC-TOCSY experiment is relatively insensitive, offering about 1/4 the sensitivity of a ^{1}H-^{13}C HMBC
experiment. The -NOESY/ROESY experiments have correspondingly
lower sensitivity than their TOCSY counterparts, since the NOESY or
ROESY responses being sorted as a function of the carbon chemical
shift of the directly bound carbon are typically only a few percent of the
intensity of the direct response. Of course, in principle any NMR active
heteronucleus can be used instead of carbon. One application of the
HMQC–TOCSY is for the natural products community. These experiments provide a powerful means of sorting proton–proton connectivity
information for complex molecules with highly overlapped proton spectra that would be otherwise difficult to disentangle with conventional
homonuclear 2D experiments such as COSY and TOCSY.

The Inverted Direct Response (IDR)-HSQC–TOCSY pulse sequence
is shown in Fig. 18 [58]. The experiment begins with an HSQC segment

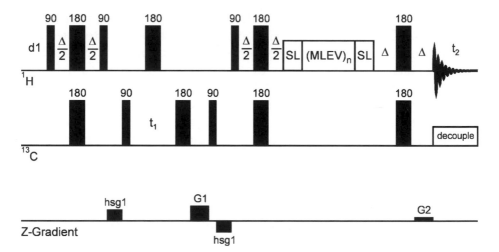

Fig. 10.18. IDR (Inverted Direct Response)—HSQC-TOCSY pulse sequence.
The experiment first uses an HSQC sequence to label protons with the chemical shift of their directly bound carbons, followed by an isotropic mixing period that propagates magnetization to vicinal neighbor and more distant
protons. The extent to which magnetization is propagated in the experiment is
a function of both the size of the intervening vicinal coupling constants and
the duration of the mixing period. After isotropic mixing, direct responses are
inverted by the experiment and proton detection begins.

that establishes coherence between protonated carbons and their directly bound protons. After carbon chemical shift "labeling" in the HSQC portion of the experiment, proton magnetization is propagated from the directly bound protons to their vicinal neighbor protons as a function of an isotropic mixing period. The number of bonds through which magnetization is transferred is a function of the size of the intervening coupling constant and the duration of the mixing period. Following propagation, magnetization is edited to invert direct responses and then detected.

As an example, consider the complex polyether marine toxin brevetoxin-2 (**7**) [59]. The proton NMR spectrum of this molecule, even at 600 MHz, has considerable overlap making the establishment of proton–proton connectivity information from a COSY or TOCSY spectrum difficult at best. In contrast, the IDR–HSQC–TOCSY spectrum presented in Fig. 10.19, in conjunction with an HMBC spectrum, allows the total assignment of the proton and carbon resonances of the molecule.

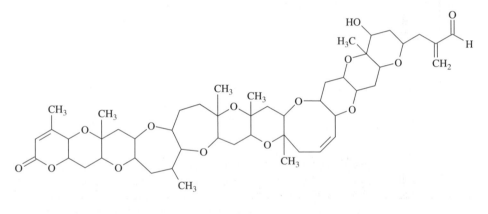

7

10.5   NMR OF SOLIDS METHODS

10.5.1   Introduction to solid-state NMR

In general, liquid-state NMR is preferable as a starting point over SSNMR for several compelling reasons. The primary reason is due to the ability of liquid-state NMR to yield much narrower lines. In practice, liquid-state NMR line shapes are typically an order of the magnitude or more narrower than the solid state. This provides much greater spectral resolution and often an apparently greater signal-to-noise ratio for liquids spectra as compared with solids spectra. NMR

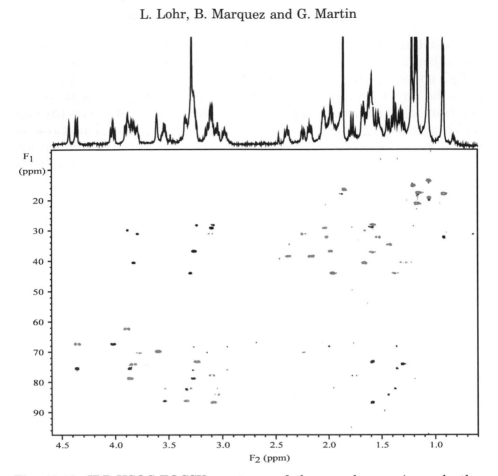

Fig. 10.19. IDR-HSQC-TOCSY spectrum of the complex marine polyether toxin brevetoxin-2 (**7**). The data were recorded overnight using a 500 μg sample of the toxin (MW = 895) dissolved in 30 μl of d_6-benzene. The data were recorded at 600 MHz using an instrument equipped with a Nalorac 1.7 mm SMIDG probe. Direct responses are inverted and identified by red contours; relayed responses are plotted in black. The IDR-HSQC-TOCSY data shown allows large contiguous protonated segments of the brevetoxin-2 structure to be assembled, with ether linkages established from either long-range connectivities in the HMBC spectrum and/or a homonuclear ROESY spectrum.

of liquids spectra can also be less complex to interpret than solids spectra, since some interactions, such as dipolar couplings, are averaged to zero in the liquid state. In addition, the hardware requirements for a SSNMR spectrometer are arguably more demanding in many ways than for a NMR of liquids spectrometer. For instance, much higher RF power levels are necessary to generate very short pulse widths, and highly precise pulse phase transitions and very linear power amplifiers are required for many of today's sophisticated SSNMR pulse sequences.

So, given that SSNMR may yield poorer spectral quality, may provide data that are much more difficult to interpret, and may pose higher requirements on instrument hardware, why would one choose to run NMR experiments on solid samples? There are a variety of cases in which the liquid state cannot provide the desired information. In general, when one wishes to probe properties specific to the solid state that cannot be observed in the liquid state, then SSNMR is used. For example, distance measurements can be investigated by measuring dipolar couplings, which are only observable in the solid state. Spatial tensor information can be derived from single crystal samples. Solid-state NMR can also be used to distinguish amorphous versus crystalline material as well as distinguishing different polymorphs. Solvation states can also be measured by SSNMR and not by NMR of liquids. One may also want to study solid-phase interfaces such as in a multi-component solid like a controlled release drug tablet. Another potential reason for working in the solid state is if the sample to be studied is highly insoluble in any usable solvent, or the sample is not stable in solution. These types of investigations are uniquely suitable to the solid state and have a wide array of practical applications. These include, for instance, pharmaceuticals, petrochemicals and other fuels, pigments, polymers, catalysts, cellulose and more [60].

These differences between liquids and NMR of solids are primarily due to molecular mobility. In the solid state, molecules are held rigidly in place by the crystal lattice and cannot tumble freely in space. All nuclear interactions are therefore theoretically observable, and this is why SSNMR spectra are very rich in molecular information. Such interactions include dipolar coupling, quadrupolar coupling and anisotropic chemical shift in addition to the isotropic chemical shift and J-coupling observed in NMR of liquids spectra. The challenge for the SSNMR spectroscopist is to disentangle the many different interactions that collectively contribute to the observed NMR spectrum, and this spectral selectivity forms the basis of modern SSNMR techniques. Typically this interplay of molecular interactions gives rise to very broad spectral resonances. In solution, fast isotropic molecular tumbling averages out most of the nuclear interactions to zero, thus yielding relatively narrow spectral peaks. These averaged interactions include dipolar coupling, for instance, which relates to distance measurements and other spatial relationships on the molecular level, and anisotropic chemical shift, which shows a different chemical shift for each molecular orientation in the magnetic field. What remain in NMR of liquids spectra are only isotropic chemical shift interactions and J-coupling. Solution-state NMR spectra are therefore typically much simpler and easier to interpret than the corresponding SSNMR spectra. Also, the basic-solution NMR experiments are for the most part simpler to execute and less demanding on the instrument hardware, since complex selectivity schemes are not required to simplify the spectrum.

Given that one wishes to probe a solid-state effect, when would one choose SSNMR over other possible characterization techniques? One major advantage that affords over other solids characterization techniques including powder X-ray diffraction, Fourier Transform Raman spectroscopy and infrared spectroscopy, is its exquisite sensitivity to small nuances in a chemical environment arising from differences in the crystal packing. Toward this end, SSNMR can often detect the presence of multiple crystal forms in a sample, as well as the number of molecules in a crystalline unit cell, when results from alternative techniques are ambiguous. Similarly, it is a valuable tool for investigating solid-phase transitions. SSNMR is also well suited for studying complex mixtures, since the spectral dispersion and other experimental techniques often provide enough selectivity to observe the compound of interest. This differs from infrared and Raman spectroscopy, for instance, which are usually challenging to interpret for heterogeneous systems. Furthermore, as is the case for solution NMR, SSNMR selectively detects various nuclei, such as hydrogen, carbon and fluorine, and this provides an additional selection parameter when studying complex samples. Another advantage of SSNMR over other techniques, depending on the question to be investigated, is its lack of sensitivity to particle size. That is to say that NMR is most influenced by the local environment rather than by long-range order. By measuring line widths and relaxation parameters, SSNMR can also evaluate the degree of sample crystallinity as well as chemical interactions between components in a sample. VT SSNMR can be used to probe molecular dynamics. Of course, the nondestructive nature of NMR in general makes it a preferred choice for precious samples.

10.5.2   Basic principles of SSNMR

In general, NMR properties of a molecule are tensor properties. That is, their value depends upon the spatial relationship of the molecule to the applied magnetic field. As such, each property can be described using three principal components, plus three angles to specify the orientation of the molecule within the applied magnetic field. For all such inter-actions, the magnitude of the interaction depends upon $(3\cos^2\theta-1)$, where θ is the angle between the magnetic field and a significant molecular direction. This could be, for instance, one of the three prin-cipal component tensor directions. In the liquid state, fast isotropic molecular tumbling causes $(3\cos^2\theta-1)$ to go to zero, so that no aniso-tropic interactions are present. In the solid state, these interactions can dominate the observed spectrum. Anisotropic interactions include ani-sotropic chemical shift (a.k.a. "chemical shielding anisotropy") which effectively yields a different chemical shift value for each molecular orientation in the sample, dipolar coupling that is related to inter- and intra-molecular spatial relationships, and quadrupolar coupling which applies only to molecules with spins greater than 1/2. The angle θ can take on all possible values, representing all possible orientations in space, and each value yields a different interaction value, e.g., dipolar-coupling constant or anisotropic chemical shift. Therefore, a range of values is observed for powdered or amorphous samples, thus yielding broad spectral resonances. (For a single crystal sample, the molecules are all at a single orientation in the magnetic field, so only one value arises for the chemical shift, and one value for the dipolar coupling constant.)

The dipolar interaction is inversely proportional to the cube of the distance between the two interacting nuclei. Since this is a through-space interaction, the two nuclei do not need to be directly bonded to-gether to observe the interaction. Dipolar-coupling constants therefore provide a measure of internuclear distances. It is important to note that these measurements provide the average distance between the nuclei over any motion of those nuclei. If temperature significantly affects the molecular motion, then the average distance measured will also be af-fected. This therefore makes the NMR measurement of the average nu-clear distance an appropriate compliment to static distances measured by diffraction techniques. Owing to molecular mobility, these two dis-tance values may not be the same for a given sample. For dipolar in-teractions between protons, the range of dipolar-coupling values may be in excess of 100 kHz. For heteronuclear dipolar interactions such as be-tween protons and carbons, the dipolar interactions are typically on the order of 30 kHz. Chemical shift anisotropy for carbons is commonly on the order of 20 kHz and is proportional to the size of the magnetic field.

Representative spectral patterns are shown in Fig. 10.20 for a static powder sample. The three spectra show the impact that the degree of chemical shift symmetry in space has on the observed spectrum.

297

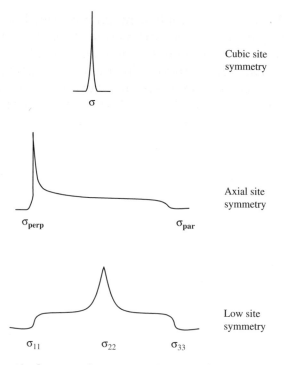

Fig. 10.20. Theoretical spectral patterns for NMR of solid powders. The top trace shows the example of high symmetry, or cubic site symmetry. In this case, all three chemical shift tensor components are equal in value, σ, and the tensor is best represented by a sphere. This gives rise to a single, narrow peak. In the middle trace, two of the three components are equal, so the tensor is said to have axial site symmetry. This tensor is best represented by an ellipsoid and gives rise to the assymetric lineshape shown. If all three chemical shift components are of different values, then the tensor is said to have low-site symmetry. This gives rise to the broad pattern shown in the bottom trace.

As stated above, the chemical shift can be represented by three principal tensor components. If these three components are all equal in magnitude, then the chemical shift tensor is said to have a high degree of symmetry, or "cubic site symmetry." The tensor could be envisioned as a perfect sphere in this case. This gives rise to a single, narrow resonance in the spectrum. If, on the other hand, two of the three components are equal in value, then the tensor is said to have "axial site symmetry." In this case, the tensor is best represented by an ellipsoid. The contribution to the observed spectrum would be a peak with a significantly elongated edge. If all three of the principal tensor components are different, then the tensor has "low site symmetry."

In this case, the contribution to the observed spectrum would be quite broad, and its general shape is represented in the bottom of Fig. 10.20.

In many cases, these broad contributions to the observed SSNMR spectrum complicate one's ability to extract the information of interest, so experimental approaches have been developed to eliminate or reduce these spectral contributions. One such approach involves spinning the sample about a particular axis at very fast speeds. The explanation for this is actually quite straightforward. Recall that these anisotropic interactions depend upon the term $(3 \cos^2 \theta - 1)$, where θ is the angle between the magnetic field and a significant molecular direction. If one wants these interactions to go away, simply set this mathematical expression equal to zero, and solve for the angle θ. The result is 54.74°, which is called the "magic angle," and the technique is known as "magic angle spinning" or "MAS." In fact, there is no magic about it. This is the angle between any side of a cube and its diagonal. By rotating the sample rapidly about this angle, the three principal component directions average to zero, so the anisotropic contributions go to zero (see Fig. 10.21). Effectively the fast, random tumbling of molecules present in solution is artificially reintroduced for solids using this technique.

In practice, the spinning rate of the sample about the magic angle needs to be fast relative to the observed static line width of the sample. So, for example, if the observed anisotropic line width is 20 kHz, then the sample must spin at faster than 20 kHz, or 20,000 revolutions per

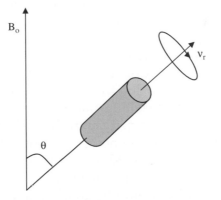

Fig. 10.21. Magic angle spinning (MAS) diagram. B_o represents the direction of the magnetic field. The sample rotor is rotated rapidly at a velocity, v_r, and an angle, θ, relative to the magnetic field. The angle is set to 54.74°, equivalent to the diagonal of a cube, in order to average dipolar interactions to zero, thus narrowing the observed spectral lines. In practice, typical spinning speeds are on the order of tens of kiloHertz.

second, in order to average out this contribution to yield a narrow spectral resonance. If the spinning speed is not fast enough, then the spinning rate of the sample rotor will be observed in the spectrum as "spinning side bands." These spectral artifacts occur at multiples of the spinning speed and are symmetric about each isotropic peak in the spectrum. The isotropic peak is therefore often commonly referred to as the "center band," since it lies in the center of the spinning side bands. Spinning side bands can be readily discerned either by calculating their positions based on the spinning speed, or by varying the spinning speed and observing which peaks move. For example, suppose the spinning rate is set to 7 kHz. On an 11.7 T magnet, this corresponds to 55.7 ppm, since 1 ppm = 125.6 Hz at this field strength. The spinning side bands will therefore be offset from each center band by multiples of 55.7 ppm. The intensity of the side bands depends upon the spinning speed as well, where faster speeds diminish the size of the side bands. The intensity also depends upon the magnitude of the anisotropic interaction to be overcome. Sample rotors are most commonly made out of zirconia, and a polymeric material called Kel-F is used to make the drive tips and end caps to seal the rotor for each sample.

The effect of MAS is significant on the observed spectrum. If the interaction to be averaged by MAS is a homogeneous dipolar interaction between protons, then as the spinning speed is increased, the spectrum changes from a single, broad resonance to a narrower isotropic center band flanked by low intensity spinning side bands. If the interaction to be averaged out is an inhomogeneous chemical shielding tensor interaction, then axial site symmetry is observed in the static spectrum, and increasing the spinning speed narrows this contribution to the isotropic chemical shift plus low intensity spinning side bands.

In practice, carbonyl and aromatic carbons tend to give rise to very intense side band resonances due to their large chemical shielding anisotropies. For the same reasoning, methine and methylene carbons usually have relatively low intensity spinning side bands. Methyl resonances typically do not generate observable side bands, because they have a very small degree of chemical shielding anisotropy due to the rapid rotation of the methyl group about its axis.

As is the case for NMR of liquids, another important consideration for SSNMR spectroscopy is relaxation. There are different types of relaxation present during a SSNMR experiment. The spin-lattice relaxation, T_1, dictates how fast one can repeat scans. This time between experiments must be set greater than five times T_1 in order for complete relaxation to occur. Otherwise, the full signal intensity will not be collected. In the solid state, this relaxation time can vary widely from seconds to hours or even longer. Similar to T_1 is the spin-lattice relaxation time in the rotating frame of reference, or "$T_{1\rho}$." This relaxation time determines the decay of carbon signal for some heteronuclear experiments. For solids, $T_{1\rho}$ is commonly on the order of tens of milliseconds. Finally, T_2 relaxation between nuclear spins is what determines the spectral line width. It is typically on the order of tens of microseconds for non-spinning solid samples.

10.5.3   Common SSNMR experiments

The vast majority of liquid-state NMR experiments are based upon the detection of hydrogen nuclei, or protons. This is for several reasons including its high natural abundance and gyromagnetic ratio that make it a very sensitive nucleus to detect by NMR, its nuclear spin of 1/2, which yields the simplest possible spectral line shapes, and its common presence in a great range of molecules of interest. This may seem like a logical choice then for SSNMR as well. Unfortunately, the very strong dipole–dipole interactions that exist between the numerous protons in a typical sample cause extensive line broadening in the observed NMR spectrum. While this dipolar contribution is averaged out in solution due to isotropic molecular tumbling, in solids it can broaden resonances far beyond the chemical shift range of protons. Recall that homogeneous dipolar couplings are on the order of 100 kHz or more, while common proton chemical shifts are in the range of approximately 0–10 ppm, or upto 6 kHz on a 600 MHz spectrometer. In order to average out this anisotropic contribution, the sample rotor needs to spin at the magic angle faster than this (>100 kHz), which is not achievable with today's technology. Typical sample rotor spinning speeds are roughly an order of magnitude slower than that.

Alternative approaches to average out dipolar couplings therefore been developed to be used instead of, or more commonly, in combination with fast sample spinning. These approaches rely on manipulation of the nuclear spins directly rather than rotating the entire sample at once. One of the original techniques developed is called "Combined Rotation and Multiple Pulse Spectroscopy" (CRAMPS) [61]. A more modern method, called "Frequency Switched Lee–Goldberg" (FSLG) spectroscopy, improves upon the results obtainable using the CRAMPS approach [62]. Nonetheless, even these line-narrowing techniques still yield proton line widths of about 0.5–1.0 ppm, compared to 0.01–0.001 ppm readily achievable by solution NMR.

By far the most popular approach to SSNMR has therefore been to look at less naturally abundant nuclei in order to greatly reduce the line broadening contribution of homogeneous dipolar coupling to the spectrum. Carbon-13 SSNMR and to a lesser extent 15-nitrogen are particularly common nuclei to study due to their presence in so many materials of interest. Compared with protons, these nuclei have very low natural abundance, so the number of nuclei to detect, and hence the sensitivity, is greatly reduced. Most of these experiments therefore focus on enhancing the sensitivity of the low-abundance nucleus to be detected. A key experimental building block that addresses this issue is called "cross-polarization" CP [63]. The fundamental strategy behind CP is to create a very large amount of magnetization by irradiating a group of high abundance nuclei, typically protons, and transferring this magnetization to a group of low abundance nuclei, typically carbon or nitrogen, in order to increase the magnetization and thus the observed signal intensity of the low abundance nuclei. As shown in Fig. 10.22, heteronuclear dipolar coupling enables this transfer of nuclear magnetization from protons to carbons, while homonuclear dipolar coupling between the protons enables the redistribution of spin energy between protons through spin diffusion.

In practice, CP is achieved using the pulse sequence shown in Fig. 10.23. This shows the case for protons as the high abundance nuclei and carbons as the low abundance nuclei. First a standard 90° pulse is applied to the protons to create the initial magnetization. Then a pair

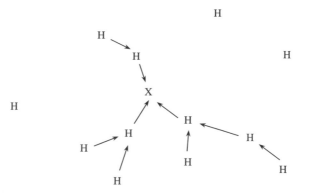

Fig. 10.22. Diagram showing the cross-polarization from protons, "H," to a heteronucleus, "X," such as carbons. Heteronuclear dipolar coupling enables the transfer of magnetization from H to X, such as protons to carbons. Homonuclear dipolar coupling between the abundant protons enables the redistribution of proton spin energy through spin diffusion.

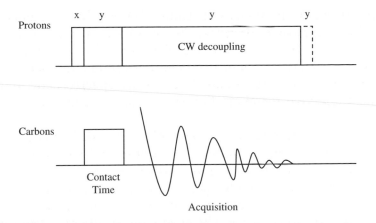

Fig. 10.23. Cross-polarization pulse sequence. The high abundance nuclei, such as protons, are first irradiated with a standard 90° pulse to create the initial magnetization. A special pair of spin-locking pulses is applied during a period called the contact time in order to transfer the magnetization from the protons to the low abundance nuclei, such as carbons. Protons are then decoupled from carbons during the acquisition of the carbon signal. In the case of protons and carbons, cross-polarization can enhance the observed carbon signal by as much as four-fold.

of special spin-locking pulses is applied. This pair of pulses must meet the requirements of the "Hartmann–Hahn" match condition [64].

$$\gamma_H B_H = \gamma_C B_C \qquad (10.2)$$

In this equation, γ_H is the gyromagnetic ratio of the high abundance nucleus, B_H is the applied field for the high abundance nucleus, γ_C the gyromagnetic ratio of the low abundance nucleus, and B_C the applied field for the low abundance nucleus. This spin-locking pulse pair is followed by a decoupling scheme to decouple the protons from carbon during the acquisition of the carbon signal.

The duration of the spin-locking pair of pulses is known as the "contact time" or CP time ("T_{CP}") and is typically set for between 1 and 10 ms. The overall sensitivity gain achievable is equal to the ratio of the gyromagnetic ratios. For protons and carbons, $\gamma_H/\gamma_C = 4$, so there is a four-fold potential gain in the signal-to-noise ratio. The potential enhancement is approximately 10-fold for protons and nitrogen. There is yet another sensitivity benefit to be gained by using CP. The low abundance nucleus relaxes according to the T_1 rate of the high abundance nucleus, rather than by its own relaxation rate. Since protons typically relax much more quickly than other nuclei do, the recycle time between experimental scans ($\sim$5 times T_1) can be shortened significantly. More scans can therefore be acquired in a given amount of time, so the signal increases. MAS is often combined with CP to achieve both line narrowing and sensitivity enhancement, respectively. This combination of techniques is called "CPMAS" spectroscopy. In general, CP is most effective for short heteronuclear distances and rigid local environments.

There is one drawback of CP, and that is the difficulty it creates in quantifying spectral contributions. The observed carbon signal originates from the transfer of magnetization from the protons. The efficiency of this transfer depends upon the local molecular mobility as well as the distances between proton and carbon nuclei. As a result, the intensity of each peak in the CP spectrum is affected differently depending upon these parameters, so using the peak intensities is challenging to identify the number of carbons giving rise to each peak. Alternatively, one could use both peak heights and line widths to try to quantitate a spectrum, but accurate line width measurements are usually quite challenging in a SSNMR spectrum. Peak area integration may also be attempted. However, this is commonly thwarted by significant overlap of resonances. Spectral deconvolution is another approach to quantifying solids spectra, but in this case, one needs to choose the relevant line shape, i.e., percent Gaussian versus Lorentzian, and this is not intuitive. Other factors affecting the relative peak intensities of heterogeneous samples include T_1 and $T_{1\rho}$ for protons in the sample, line widths (influenced by T_2), and spinning side bands, which steal intensity from their corresponding center band. The limit of quantitation for natural abundance ^{13}C CPMAS spectra using an internal standard is approximately 1% by current methods. This compares quite favorably with other solid-state techniques such as powder X-ray diffraction and infrared or Raman spectroscopy.

In reference to the CP pulse sequence, it was mentioned that decoupling of the abundant nuclei from the rare nuclei is applied during the acquisition period. Without this "dipolar dephasing," protonated carbons such as methines and methylenes would yield broad resonances due to the dipolar coupling to protons, and these resonances would decay very rapidly. These signals are therefore lost before the signal is acquired. Non-protonated carbons, i.e., quaternaries, would not be broadened by proton–dipolar coupling, so they would yield relatively sharp, long-lived resonances readily detected without dipolar dephasing. Methyl carbons undergo fast molecular motion about their primary axis, and this molecular motion narrows the resulting resonance, such that they would also be observed to some extent without dipolar dephasing.

Dipolar dephasing experiments thus make an excellent compliment to CP experiments with variable contact times when one wishes to perform spectral editing. A logical approach for discerning different carbon types is as follows. The standard ^{13}C CPMAS spectrum shows all the carbon resonances present for the sample. If necessary, one can acquire the same spectrum at two different spinning speeds to properly identify the isotropic center band peaks. In an interrupted decoupling experiment, the proton decoupling power is turned off for a short period of time immediately following the CP contact time. This effectively cancels the signals from all protonated carbons except methyl groups due to their fast rotation. In other words, methine and methylene spectral contributions are eliminated, and only quaternary and methyl resonances are observed. In a separate experiment, the standard CP sequence is applied using only a very short contact time. This enhances predominantly the protonated carbon signals, so quaternary signals are relatively attenuated. By using this approach in combination with any structural information that may be available from the corresponding solution state sample, one can begin to assign a chemical structure for a solid sample.

For heteronuclear SSNMR, MAS, CP, dipolar dephasing and spectral editing form the building blocks for a wide array of diverse applications. Some SSNMR pulse sequences that build upon these experiments as a foundation include the following. "Recoupled Dipolar" spectroscopy, or "REDOR," is a triple resonance experiment that probes heteronuclear dipolar coupling constants [65]. This is particularly important in distance measurements and is frequently employed for protein NMR applications. Domain selection techniques often rely on filtering the proton signal to the targeted domain and then cross-polarizing to the rare nucleus such as carbon-13 for detection. Such filtering approaches generally eliminate spectral contributions from specific nuclei based either on dipolar coupling strength or differences in relaxation rates.

10.5.4 Practical considerations for SSNMR

There are several practical aspects to consider before launching a SSNMR experiment. As was mentioned at the beginning of this section, SSNMR experiments are usually much more demanding on the instrument hardware than NMR of liquids experiments. Because of the type of high power capacitors and other electronics necessary to perform

SSNMR experiments, the standard NMR of solids spectrometer uses a wide bore vertical superconducting magnet. While this may sacrifice some magnetic field homogeneity compared with an equivalent narrow-bore magnet, this is not an issue for SSNMR, which is dominated by much broader lines than for solution NMR. Perhaps the first question to address then is what magnetic field strength is appropriate to use. In many cases, a higher field strength is advantageous. This is true for detecting protons, quadrupolar nuclei (spin $> 1/2$) and nuclei with a low gyromagnetic ratio. However, higher magnetic fields may be more challenging for dipolar dephasing of spin 1/2 nuclei.

The next most important question to address is what type of probe to use. There are many available choices. A double resonance probe with CPMAS capabilities can handle most standard SSNMR applications. Of course, this would not include triple resonance experiments such as REDOR. Next one should identify which nuclei will be detected and ensure that the probe can cover these frequency ranges. Beyond this, a sample rotor size must be specified. These sizes are typically quoted in outer diameter in millimeters and commonly range from 2 to 15 mm. If sensitivity is a primary concern for the sample to be studied, then using a larger rotor size provides greater sample volume and therefore a larger signal to detect. Unfortunately there is a trade-off between the rotor size and how fast it can practically spin. Therefore, if line-narrowing techniques and spectral resolution are the primary concern, then a smaller rotor that can spin at the magic angle faster is the better choice. Finally, many experiments require high power pulses, particularly to decouple large dipolar coupling. In this case, very fast pulse widths are desirable. This is an important consideration depending upon what nuclei are under investigation. For example, MAS line widths for carbon-13 and phosphorus-31 are typically on the order of 30 Hz, but closer to 400 Hz for protons and 200 Hz for fluorine-19. Table 10.1 summarizes some approximate values of these parameters for various probes on the market today.

TABLE 10.1

Representative commercial CPMAS SSNMR probe specifications

Rotor O.D. (mm)	14	7.5	5.0	4.0	3.2	2.5
Sample volume (µl)	2100	450	160	52	20	11
Fastest spin rate (kHz)	4.5	7	12	18	25	30
Shortest 90° pulse width (µs)	7	4	3	2	1.7	1.5

10.6 APPLICATIONS

10.6.1 Overview

The diversity and depth of these technological advances in the field of NMR are invaluable for a wide variety of applications. Prudent decision-making is necessary to select the most advantageous NMR technologies and approaches specific for each individual application. Such selections are often made in order to overcome limitations in sensitivity or resolution, or both. The following case studies demonstrate representative examples of how modern NMR technologies and techniques have been employed to solve real-life scientific challenges.

10.6.2 Case studies for liquids

In the characterization of low-level pharmaceutical impurities. Prudent decision-making is necessary to select the most advantageous technologies and capabilities specific for each project. Such selections are made to overcome extreme limitations in sensitivity or resolution, or both, that standard techniques would afford. The following two case studies covers show how new NMR technologies and techniques were employed to solve real pharmaceutical challenges.

10.6.2.1 Case study 1

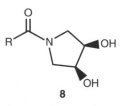

8

The first example involves the drug substance represented as Structure 8. Multiple related degradation products were detected using HPLC, and LC/MS demonstrated the mass of each to be significantly larger than the parent species. Based on the UV response, these degradants were determined to be above 0.1%, but significantly below 1% of the sample mass. Unfortunately, attempts to isolate the individual degradants for standard NMR analyses were unsuccessful. The isolated species proved unstable over prolonged periods. LC–NMR was therefore selected to resolve this complex mixture. Additionally, deuterated solvents were used to minimize solvent contributions and improve detection of the targeted low-level signals. Selected 1D-versions of standard

2D-correlation spectra were collected. This information was used with empirical spectral simulations of proton and carbon spectra to elucidate the structures. The necessary information was thus provided to demonstrate that multiple dimer-like structures were formed through bonding of residual synthetic precursor to each of the hydroxyl sites of the drug substance itself. Three distinct dimers were identified. These species were tracked by LC–NMR, and two were shown to interconvert over time. Analogous trimer structures were also evident at lower levels.

10.6.2.2 *Case study 2*

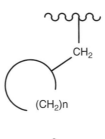

9

The next liquids example demonstrates a different pharmaceutical mixture problem that could not be solved using LC–NMR. The unexpectedly reactive portion of the parent structure, shown as Structure 9, was an aliphatic ring attached to a methylene group. A process related impurity in the drug substance was detected at the 1% level. Identification of the impurity structure was required in this case for proper drug safety assessment. Unfortunately, isolation via preparative HPLC was complicated by the structural similarity of the impurity with the parent drug species. The mass of the impurity was identical to that of the drug substance itself, and the retention time was extremely similar. Unlike in the previous example, LC could not resolve the impurity from the parent, so HPLC–NMR was not a viable approach. Instead, the high resolution afforded by a high field magnet, along with the corresponding sensitivity enhancement and field stability, enabled the acquisition of useful data on approximately 1 mg of sample containing both the parent and target impurity. Comparing the data for the impurity to that for the parent, the original methylene neighboring the aliphatic ring was no longer evident in the DEPT-HSQC spectrum. Instead, a new methylene resonance was observed, and long-range couplings confirmed its placement within the aliphatic ring. This unexpected ring enlargement was contrary to the structure proposed based upon plausible synthetic routes, which showed the reactive site to be at a different location within the molecule. These results thus revealed a case of the unusual chemistry often exhibited for low-level impurities.

10.6.3 Case studies for solids

10.6.3.1 *Case study 1*

As is also the case for NMR of liquids, unexpected results can be observed for the solid state as well. In the following case study, the effects of a pharmaceutical manufacturing process on the final formulation of a drug candidate were studied by solid-state NMR. The drug formulation was comprised of a very low level of the drug substance or "active pharmaceutical ingredient" (API), amidst a complex mixture of multiple excipients. In order to simplify the problem of analyzing a complex mixture, standard samples of each of two polymorphs of the API alone were separately analyzed by solid-state NMR. Carbon-13 CP magic angle spinning (CPMAS) solid-state NMR spectra of the solvated crystalline form were acquired on a 500 MHz wide-bore NMR spectrometer (see Fig. 10.24). The polymorph used in the final formulation is "Form

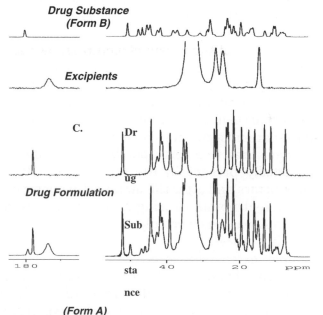

Fig. 10.24. ^{13}C CPMAS SSNMR spectrum of drug formulation (bottom) together with the spectrum of the solvated form A of the API (second from the bottom), formulation excipients (second from the top) and the solvated form B of API (top).

A," which is shown in the second spectrum from the bottom in Fig. 10.24. Comparing the carbon spectrum of Form A to that of a different crystalline form, "Form B," shown in the topmost spectrum of Fig. 10.24, reveals several notable differences in both chemical shifts and line shapes. This, in fact, is often the case for different polymorphs of the same chemical entity, and it enables a clear distinction between the various crystalline forms when present in a mixture. A ^{13}C CPMAS spectrum of a standard sample of the excipients, in the appropriate proportions for the formulation, was also acquired for reference. This is shown as the second spectrum from the top in Fig. 10.24.

Having collected the appropriate reference spectra, the full formulation spectrum can now be interpreted. It is shown as the bottom spectrum in Fig. 10.24. Comparing the excipients reference spectrum to the API reference spectrum reveals that signals arising from the excipients overlap with some of the signals arising from the drug substance. Nonetheless, there are still several API peaks that each exhibits a unique chemical shift. This is quite common for pharmaceutical formulations, since many of today's drug substances contain aromatic groups, while typical excipient blends are predominantly aliphatic in nature. By comparing the full formulation spectrum to each of the reference spectra, API peaks can clearly be distinguished from the excipient resonances. These match the chemical shifts observed for the reference sample of Form A. The expected Form A is therefore the predominant crystalline form of API present in the formulation. However, additional lower level peaks attributable to drug substance are also observed. This can be seen, for example, in the aromatic region of the full formulation spectrum. Referring again to the reference spectra, it's evident that this corresponds to a significant amount of Form B unexpectedly present in the formulation. By subsequently studying samples before and after the manufacturing process, as well as samples from different manufacturing processes, it was demonstrated that some of Form A was unexpectedly converted to Form B during the drug manufacturing process.

In this example, extensive use of the available reference samples greatly simplified an otherwise highly complex sample mixture. In addition, utilization of a high-field, wide-bore magnet afforded the necessary spectral resolution to easily resolve the two crystalline forms from each other, as well as from the excipients. A lower field magnet, or a less optimized system, may not have been able to differentiate the two forms, and this significant potential impact on the drug's overall bioavailability may have been overlooked.

10.6.3.2   Case study 2

The next SSNMR case study also involves a pharmaceutical formulation. However, this case investigates a different type of formulation issue and uses alternative NMR technologies. In this example, the project team suspected a possible strong interaction between the drug substance and the excipients in the formulation. As in the previous example, ^{13}C CPMAS spectra were again collected for a reference sample. This sample was a physical mixture of the drug substance and the excipients in the appropriate formulation proportions. However, this standard sample did not undergo the formulation manufacturing process. This reference spectrum of the physical mixture is shown as the bottom left spectrum in Fig. 10.25. The ^{13}C CPMAS spectrum of the actual drug formulation was also acquired and is shown as the top left spectrum in Fig. 10.25.

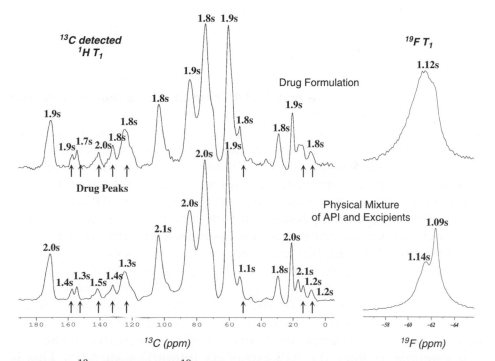

Fig. 10.25. ^{13}C CPMAS and ^{19}F MAS SSNMR spectra of the drug formulation (top) and the physical mixture of API with excipients (bottom). Shown above each peak are the ^{1}H T_1 relaxation times calculated from the ^{13}C detected ^{1}H T_1 relaxation experiment (left side of the figure) as well as ^{19}F T_1 (right side of the figure).

A comparison of the carbon SSNMR spectra of the manufactured formulation to that of the equivalent physical mixture, both shown on the left of Fig. 10.25, shows no significant differences. The chemical shift and line shape differences between the top and bottom carbon spectra in the figure are minor and thus do not themselves prove an interaction between the API and excipients. Small spectral differences such as these may arise from minor fluctuations in sample temperature, for instance. One may, albeit incorrectly, conclude at this point that no drug-excipient interactions exist. However, as we shall soon see, it is risky to make such conclusions based on the lack of an observed change.

To continue the investigation, carbon detected proton T_1 relaxation data were also collected and were used to calculate proton T_1 relaxation times. Similarly, ^{19}F T_1 measurements were also made. The calculated relaxation values are shown above each peak of interest in Fig. 10.25. A substantial difference is evident in the proton T_1 relaxation times across the API peaks in both carbon spectra. Due to spin diffusion, the protons can exchange their signals with each other even when separated by as much as tens of nanometers. Since a potential API-excipient interaction would act on the molecular scale, spin diffusion occurs between the API and excipient molecules, and the protons therefore show a single, uniform relaxation time regardless of whether they are on the API or the excipients. On the other hand, in the case of a physical mixture, the molecules of API and excipients are well separated spatially, and so no bulk spin diffusion can occur. Two unique proton relaxation rates are then expected, one for the API and another for the excipients. This is evident in the carbon spectrum of the physical mixture shown on the bottom of Fig. 10.25. Comparing this reference to the relaxation data for the formulation, it is readily apparent that the formulation exhibits essentially one proton T_1 relaxation time across the carbon spectrum. This therefore demonstrates that there is indeed an interaction between the drug substance and the excipients in the formulation.

Fortuitously for this project, the drug substance, and not the excipients, contained a fluorine moiety. Fluorine-19 MAS spectra were therefore also acquired at 500 MHz on the two samples, and they are shown to the right of the corresponding carbon spectra for each sample in Fig. 10.25. The fluorine-19 chemical shifts are sensitive enough in this example to show the API-excipients interaction directly. This is evident from the dramatic change in spectral line shape.

In this case study, it was demonstrated that conclusions based on the lack of observations, such as no change observed for the carbon spectra, could be misleading. It is much more compelling to identify an experimental approach that leads to a decisive conclusion. This was demonstrated here by using a series of proton relaxation measurements. This example also showed that the availability of alternative technologies, in this case ^{19}F SSNMR capabilities, could greatly simplify the problem. Also, since fluorine-19 is a much more sensitive nucleus that carbon-13, plus a single directly detected spectrum sufficed rather than a series of measurements and calculations, this afforded a significant overall time savings in the data acquisition and problem-solving process.

REFERENCES

1 G.E. Martin, Small volume and high sensitivity NMR probes. In: G.A. Webb (Ed.), *Annual Reports NMR Spectroscopy*, Vol. 56, Elsevier, Amsterdam, 2005, pp. 1–99; G.E. Martin, Cryogenic NMR probes: Applications. In: D.M. Grant and R.K. Harris (Eds.), *Encyclopedia of Nuclear Magnetic Resonance*, Vol. 9, Wiley, Chichester, 2002, pp. 33–35; G.E. Martin, Microprobes and methodologies for spectral assignments: Applications. In: D.M. Grant and R.K. Harris (Eds.), *Encyclopedia of Nuclear Magnetic Resonance*, Vol. 9, Wiley, Chichester, 2002, pp. 98–112.

2 G.E. Martin, R.C. Crouch and A.P. Zens, *Magn. Reson. Chem.*, 36 (1998) 551–557 for 1.7 mm sensitivity; G. Schlotterbeck, A. Ross, R. Hochstrasser, H. Senn, T. Kuhn, D. Marek and O. Schett, *Anal. Chem.*, 74 (2002) 4464 for 1 mm Bruker probe performance; M.E. Lacey, R. Subramanian, D.L. Olsen, A.G. Webb and J.V. Sweedler, *Chem. Rev.*, 99 (1999) 3133–3152 General review of small volume probe performance.

3 F. Bloch, W.W. Hensen and M. Packard, *Phys. Rev.*, 69 (1946) 127.

4 A.G. Redfield and R.K. Gupta, *Adv. Magn. Reson.*, 5 (1971) 81.

5 G. Schlotterbeck, A. Ross, R. Hochstrasser, H. Senn, T. Kuhn, D. Marek and O. Schett, *Anal. Chem.*, 74 (2002) 4464.

6 D.L. Olson, M.E. Lacey and J.V. Sweedler, *Science*, 270 (1995) 1967; M.E. Lacey, R. Subramanian, D.L. Olsen, A.G. Webb and J.V. Sweedler, *Chem. Rev.*, 99 (1999) 3133–3152; G.E. Martin, Small volume and high sensitivity NMR probes. In: G.A. Webb (Ed.), *Annual Reports NMR Spectroscopy*, Amsterdam, 2005, pp. 1–99.

7 Bruker-Biospin Inc, 15 Fortune Drive, Billerica, MA, 01821.

8 K. Albert (Ed.), *On-line LC-NMR and related techniques*, Wiley, West Sussex, England, 2002.

9 V. Exarchou, M. Krucker, T.A. van Beek, J. Vervoort, I.P. Gerothanassis and K. Albert, *Magn. Reson. Chem.*, 43 (2005) 681.

10 O. Corcoran and M. Spraul, *Drug Disc. Today*, 8 (2003) 624.

11 M. Godejohann, Li-H. Tseng, U. Braumann, J. Fuchser and M. Spraul, *J. Chrom. A*, 1058 (2004) 191.

12 A.W. Overhauser, *Phys. Rev.*, 89 (1953) 689; A.W. Overhauser, *Phys. Rev.*, 92 (1953) 411.

13 I. Solomon, *Phys. Rev.*, 99 (1955) 559.

14 D.J. Edwards, B.L. Marquez, L.M. Nogle, K. McPhail, D.E. Goeger, M-A. Roberts and W.H. Gerwick, *Chem. Biol.*, 11 (2004) 817.

15 R. Richarz and K. Wûthrich, *J. Magn. Reson.*, 30 (1978) 147.

16 D. Neuhaus and M. Williamson, *The nuclear Overhauser effect in structural and conformational analysis*, 2nd ed., Wiley, New York, 2000.

17 T.D.W. Claridge, High-resolution NMR techniques in organic chemistry. In: J.E. Baldwin, F.R.S. Williams and R.M. Williams (Eds.), *Tetrahedron Organic Chemistry Series*, Vol. 19, Elsevier, Oxford, UK, 1999.

18 K. Stott, J. Keeler, Q.N. Van and A.J. Shaka, *J. Magn. Reson.*, 1125 (1997) 302.

19 K. Stott, J. Stonehouse, J. Keeler, T.-L. Hwang and A.J. Shaka, *J. Am. Chem. Soc.*, 117 (1995) 4199.

20 W.I. Li, B.L. Marquez, T. Okino, F. Yokokawa, T. Shioiri, W.H. Gerwick and T.F. Murray, *J. Nat. Prod.*, 67 (2004) 559.

21 E.L. Hahn, *Phys. Rev.*, 76 (1949) 145.

22 E.L. Hahn, *Phys. Rev.*, 80 (1950) 580.

23 H.Y. Carr and E.M. Purcell, *Phys. Rev.*, 94 (1954) 630.

24 S. Meiboom and D. Gill, *Rev. Sci. Instrum.*, 29 (1958) 688.

25 G.C. Levy and G.L. Nelson, *Carbon-13 Nuclear Magnetic Resonance for Organic Chemists*, Wiley, New York, 1972; J.B. Stothers, *Carbon-13 NMR Spectroscopy*, Academic Press, New York, 1972.

26 A.J. Shaka and J. Keeler, *Prog. Nucl. Magn. Reson. Spectrosc.*, 19 (1987) 47–129.

27 K.G.R. Pachler, P.S. Steyn, R. Vleggaar and P.L. Wessels, *J. Magn. Reson.*, 12 (1973) 337.

28 C. Lingran and W. Robein, *Fres. J. Anal. Chem.*, 344 (1992) 214–216.

29 D.M. Doddrell, D.T. Pegg and M.R. Bendall, *J. Magn. Reson.*, 48 (1982) 323–327; D.M. Doddrell, D.T. Pegg and M.R. Bendall, *J. Chem. Phys.*, 77 (1982) 2745–2752.

30 S.L. Patt and J.N. Shoolery, *J. Magn. Reson.*, 46 (1982) 535–539.

31 J. Jeener, Proceedings of the *Ampere International Summer School*, Basko Polje, Yogoslavia, 1971 (Unpublished Proposal).

32 W.P. Aue, E. Bartholdi and R.R. Ernst, *J. Chem. Phys.*, 64 (1976) 2229–2246; A. Bax, R. Freeman and G.A. Morris, *J. Magn. Reson.*, 42 (1981) 164–168.

33 U. Piantini, O.W. Sørensen and R.R. Ernst, *J. Am. Chem. Soc.*, 104 (1982) 6800.

34 C. Griesinger, O.W. Sørensen and R.R. Ernst, *J. Am. Chem. Soc.*, 107 (1985) 6394.

35 L. Braunschweiler and R.R. Ernst, *J. Magn. Reson.*, 53 (1983) 521.

36   A. Bax, R. Freeman and T.A. Frenkiel, *J. Am. Chem. Soc.*, 103 (1981) 2102.

37   J. Jeener, B.H. Meier, P. Bachmann and R.R. Ernst, *J. Chem. Phys.*, 71 (1979) 4546. Please note this is the first appearance of the NOSY sequence; however, this publication deals with chemical exchange. For a quite comprehensive text on NOE please see the above reference by Neuhaus and Williamson..

38   M.J. Thrippleton and J. Keeler, *Angew. Chem. Int. Ed.*, 42 (2004) 3938.

39   G. Bodenhausen, R. Freeman and D.L. Turner, *J. Chem. Phys.*, 65 (1976) 839.

40   W.F. Reynolds, S. MacLean, H. Jacobs and W.W. Harding, *Can. J. Chem.*, 77 (1999) 1922–1930.

41   L. Müller, *J. Am. Chem. Soc.*, 101 (1979) 4481; G. Bodenhausen and D.J. Ruben, *Chem. Phys. Lett.*, 69 (1980) 185–189.

42   A. Bax, R.H. Griffey and B.L. Hawkins, *J. Am. Chem. Soc.*, 105 (1983) 7188.

43   A. Bax and S. Subramanian, *J. Magn. Reson.*, 67 (1986) 565–569.

44   R.E. Hurd and B.K. John, *J. Magn. Reson.*, 91 (1991) 648–653; J. Ruiz-Cabello, G.W. Vuister, C.T.W. Moonen and P.C. van Zilj, *J. Magn. Reson.*, 100 (1992) 282–302.

45   G. Bodenhausen and D.J. Ruben, *Chem. Phys. Lett.*, 69 (1980) 185–189.

46   W.F. Reynolds, S. McLean, L.-L. Tay, M. Yu, R.G. Enriquez, D.M. Estwick and K.O. Pascoe, *Magn. Reson. Chem.*, 35 (1997) 455–462.

47   G.E. Martin, Qualitative and quantitative exploitation of heteronuclear coupling constants. In: G.A. Webb (Ed.), *Annual Report NMR Spectroscopy*, Vol. 46, Academic Press, New York, 2002, pp. 37–100.

48   G.E. Martin and A.S. Zektzer, *Magn. Reson. Chem.*, 26 (1988) 631–652.

49   A. Bax and M.F. Summers, *J. Am. Chem. Soc.*, 108 (1986) 2093–2094.

50   R. Wagner and S. Berger, *Magn. Reson. Chem.*, 36 (1998) S44.

51   G.E. Martin, Qualitative and quantitative exploitation of heteronuclear coupling constants. In: G.A. Webb (Ed.), *Annual Report NMR Spectroscopy*, Vol. 46, Academic Press, New York, 2002, pp. 37–100.

52   A. Meissner and O.W. Sørensen, *Magn. Reson. Chem.*, 38 (2000) 981.

53   R. Berger, C. Schorn and P. Bigler, *J. Magn. Reson.*, 148 (2001) 88.

54   W.F. Reynolds, D.W. Hughes, M. Perpick-Dumont and R.G. Enriquez, *J. Magn. Reson.*, 63 (1985) 413.

55   V.V. Krishnamurthy, D.J. Russell, C.E. Hadden and G.E. Martin, *J. Magn. Reson.*, 146 (2000) 232; T. Spang and P. Bigler, *Magn. Reson. Chem.*, 41 (2003) 177–182; T. Spang and P. Bigler, *Magn. Reson. Chem.*, 42 (2004) 55–60; N.T. Nyberg, J.Ø. Duus and O.W. Sørensen, *J. Am. Chem. Soc.*, 127 (2005) 6154–6155.

56   P.H. Bolton, *J. Magn. Reson.*, 48 (1982) 336–340; P.H. Bolton and G. Bodenhausen, *Chem. Phys. Lett.*, 89 (1982) 139–144; A. Bax, *J. Magn. Reson.*, 53 (1983) 149–153; H. Kessler, M. Bernd, H. Kogler, J. Zarbock,

O.W. Sørensen, G. Bodenhausen and R.R. Ernst, *J. Am. Chem. Soc.*, 105 (1983) 6944–6952; P. Bigler, W. Ammann and R. Richarz, *Org. Magn. Reson.*, 22 (1983) 109–111.

57 P.H. Bolton, *J. Magn. Reson.*, 62 (1985) 143–146; L. Lerner and A. Bax, *J. Magn. Reson.*, 69 (1986) 375–380; D. Brühwiler and G. Wagner, *J. Magn.Reson.*, 82 (1989) 193–197; K. Sohn and S.J. Opella, *J. Magn. Reson.*, 82 (1989) 193–197; J. Kawabata, E. Fukushi and J. Mizutami, *J. Am. Chem. Soc.*, 114 (1992) 1115–1117.

58 T. Domke, *J. Magn. Reson.*, 95 (1991) 174–177; R.C. Crouch, T.D. Spizter and G.E. Martin, *Magn. Reson. Chem.*, 30 (1992) S71–S73; R.C. Crouch, A.O. Davis and G.E. Martin, *Magn. Reson. Chem.*, 33 (1995) 889–892.

59 R.C. Crouch, G.E. Martin, R.W. Dickey, D.G. Baden, R.E. Gawley and K.S. Rein, *Tetrahedron*, 51 (1995) 8409–8422.

60 R.K. Harris, *Nuclear Magnetic Resonance Spectroscopy* Chapter 6, Longmans, 1997; C.A. Fyfe, *Solid State NMR for Chemists*, C.F.C. Press, 1983; M.J. Duer, *Introduction to Solid-State NMR Spectroscopy*, Blackwell Science, London, 2004; M.J. Duer, *Solid-State NMR Spectroscopy: Principles and Applications*, Blackwell Science, London, 2002.

61 R.E. Taylor, R.G. Pembleton, L.M. Ryan and B.C. Gerstein, *J. Chem. Phys.*, 71 (1979) 4541.

62 A. Bielecki, A.C. Kolbert and M.H. Levitt, *Chem. Phys. Lett.*, 155 (1989) 341–346.

63 A. Pines, M.G. Gibby and J.S. Waugh, *J. Chem. Phys.*, 59 (1973) 569.

64 S.R. Hartmann and E.L. Hahn, *Phys. Rev.*, 128 (1962) 2042.

65 T. Guillon and J. Schaefer, *J. Adv. Mag. Res.*, 13 (1989) 57–83; K.L. Wooley, C.A. Klug, K. Tasaki and J. Schaefer, *J. Am. Chem. Soc.*, 119 (1997) 53–58; G. Tong and J. Schaefer, *J. Am. Chem. Soc.*, 121 (1999) 5238–5248; J. Huang, K.L. Wooley and J. Schaefer, *Macromolecules*, 34 (2001) 544–546.

REVIEW QUESTIONS

A. What are the relative peak intensities of a proton multiplet arising from a proton adjacent to five neighboring protons? (see Fig. 10.4).

B. Assign the proton resonances of dibromopropionic acid. (See Fig. 10.26.)

C. Calculate the relative sensitivity of 19-fluorine and 15-nitrogen versus protons.

D. Internet search question: Who won the Nobel Prize for their work in NMR? What field was it in?

Nuclear magnetic resonance spectroscopy

PROTON OF dibromopropionic acid in CDCl3
TEMP=25C, Liquids 600 MHz NMR Instrument

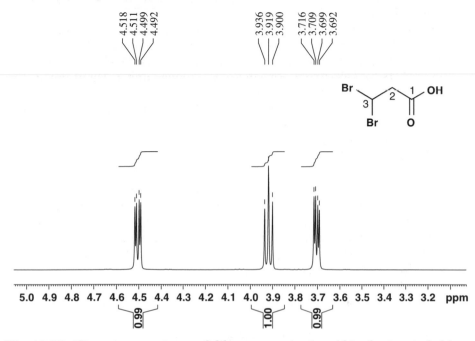

Fig. 10.26. 1D proton spectrum of dibromopropionic acid in deuterated chloroform.

Chapter 11

Mass spectrometry

David J. Burinsky

11.1   INTRODUCTION

Mass spectrometry is arguably one of the most versatile analytical measurement tools available to scientists today, finding application in virtually every discipline of chemistry (i.e., organic, inorganic, physical and analytical) as well as biology, medicine and materials science. The technique provides both qualitative and quantitative information about organic and inorganic materials, including elemental composition, molecular structure and the composition of mixtures. The origin of the technique can be traced to the work of physicists such as J.J. Thomson (1912) [1], F.W. Aston (1919) [2] and A.J. Dempster (1918) [3] in the early part of the 20th century. By the 1940s, the technique was finding application in the analysis of hydrocarbon fractions in the petroleum industry, which led to improvements in the instrumentation and methodology. The commercialization of more reliable instrumentation in the 1950s spurred utilization of the technique by chemists for the characterization of an increasing variety of organic compounds. Subsequent decades brought about the combination or "hyphenation" of the technique with the powerful separation capabilities of gas chromatography (GC-MS) and liquid chromatography (LC-MS), the development of new kinds of mass analyzers and the introduction of revolutionary new ionization techniques. These new ionization techniques (introduced largely in the past 15 years) are primarily responsible for the explosive growth and proliferation of mass spectrometry as an essential tool for biologists and biochemists.

The mass spectrometer converts neutral molecules into charged particles (either positive or negative ions) and sorts them according to their respective mass-to-charge (m/z) ratios. A graphical presentation of the relative abundances of the various ionic species, as a function of their m/z value, is a mass spectrum. The appearance of a mass

Comprehensive Analytical Chemistry 47
S. Ahuja and N. Jespersen (Eds)
Volume 47 ISSN: 0166-526X DOI: 10.1016/S0166-526X(06)47011-2

spectrum, both in terms of which ionic species are observed, as well as their abundances, serves as the basis for compound identification. The information derived from a mass spectrum is often combined with that from other analytical techniques, such as infrared spectroscopy and nuclear magnetic resonance spectroscopy, to generate structural assignments for organic molecules. The attributes of mass spectrometry that make it a versatile and valuable analytical technique are its sensitivity (e.g., recently, a detection limit of approximately 500 molecules, or 800 yoctomoles $= 800 \times 10^{-24}$ mol has been reported) and specificity in detecting or identifying unknown compounds. The sensitivity of the technique is attributable to the filtering action of the analyzers with the associated elimination of background interferences, and the significant signal amplification (on the order of six orders of magnitude) achievable with modern electron multiplier detectors. Specificity is a product of both the molecular mass information carried by the parent ion (i.e., the ionized neutral analyte molecule) and the aspects of molecular structure that can be inferred from the reproducible and characteristic fragmentation processes. In addition to these analytical attributes, mass spectrometry continues to contribute to the fundamental understanding of gas-phase unimolecular chemistry, which has implications in the study of areas such as atmospheric chemistry and interstellar phenomena, among others.

Despite the apparent simplicity of the technique—create ions, separate (measure) ions according to m/z ratio and detect ions—there is no one best way to accomplish these operations. A variety of components and configurations have been (and continue to be) developed for this purpose. Each lends itself to the investigation of certain problems or phenomena, and brings with it certain limitations. Several elements are common to most mass spectrometers (Fig. 11.1) and include the vacuum system that encloses most (if not all) of the other components, an interface or sample introduction/handling system, the ion source, one or more analyzer(s), an ion collection/detection system and a computer system (for instrument control and operation in addition to data collection, manipulation and display). In order for ions to be effectively manipulated by electric or magnetic fields (analyzed) and collected with high efficiency and minimal loss either due to neutralization or scattering, they must traverse a collision-free path. Maintaining the internal pressure of a mass spectrometer at less than 10^{-4} Pa (10^{-6} torr) provides a mean-free path for the ions that make detrimental collisions unlikely. Ions can be easily moved and manipulated by means of electrical fields within the mass spectrometer (acceleration, focusing, etc.),

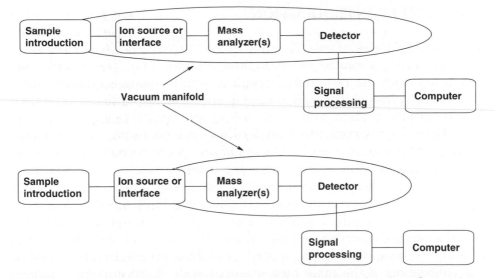

Fig. 11.1. Conceptual diagrams of a mass spectrometer showing the various functional components. The top diagram represents instruments that employ conventional modes of ionization such as EI or CI. In such instruments, the sample introduction process (for example, direct insertion probe) bridges the atmospheric pressure/high-vacuum interface. The bottom diagram represents instruments that employ the recently developed API techniques such as ESI. Ions are formed outside the vacuum envelope of the instrument and transported into the instrument through the API interface.

and their m/z ratio can be determined on the basis of kinetic energy, momentum or velocity, depending on the type of analyzer(s) being employed. The final step in the analysis process is the detection of the ions through some measurement of charge, which in most instrument types takes the form of a transformation of the ion into a cascade of electrons that create a measurable signal (current) in the detection circuitry.

In the sections that follow, the details of some of the many components and configurations of modern mass spectrometers will be presented. Admittedly, the extensive breadth and depth of the available information makes comprehensive coverage impractical, if not impossible. Consequently, the focus of this account will be the key elements of organic or molecular mass spectrometry (rather than inorganic or atomic mass spectrometry [4–8]) including important instrumental configurations and the types of spectra (and information) produced by these instruments. Emphasis will be given to the most prevalent instrument configurations and experimental methods, which represent the current practice of molecular mass spectrometry.

11.2 SAMPLE INTRODUCTION

Traditionally, gases or volatile liquids were introduced into a mass spectrometer using a heated reservoir and transfer line. The sample was introduced into the reservoir through a septum using a syringe, being subsequently leaked (through a small diameter orifice) into the ion source in a controlled fashion so as to maintain an acceptable operating pressure. Both the reservoir and transfer line were maintained at a temperature high enough to vaporize the liquid and prevent condensation prior to it reaching the ion source. Since the complexity of a mass spectrum increases with the number of components being ionized at any given time, the utility of such introduction devices was limited when gaseous or liquid mixtures required analysis. One of the few remaining uses of a liquid sampling reservoir on modern instruments is for the introduction of a known compound (such as perfluorotributylamine, PFTBA) used in the calibration of the mass measurement scale. Most current commercially available mass spectrometers have limited sample introduction capability for gaseous or liquid samples other than through the gas or liquid chromatograph that often comes as an integral part of the system.

Solid samples, with low vapor pressures or melting/boiling points of less than 350–400°C (at reduced pressure), can be introduced directly into the source of a mass spectrometer using a probe. Depending on the volatility and thermal stability of the compound(s), either slow or rapid ballistic heating of the probe can affect sufficient volatilization of the compound. As the relative molecular mass (M_r) [9] and polarity of compounds increases, so does the likelihood of thermal decomposition prior to volatilization. This leads to a mass spectrum not of the molecule of interest, but rather of its thermolysis products. Two possible solutions exist for this problem. The analyte can be made more volatile through a simple chemical modification (derivatization) or a special, rapid-heating probe [10] can be used. Unlike the standard direct insertion probe that heats liquid or solid samples to temperatures of 350–400°C over the course of a minute or more, the direct exposure probe reaches temperatures in excess of 800°C in 1 s or less. The two probes also differ in how the sample is held during heating. The standard direct insertion probe holds the sample material in a cup or crucible (either glass or stainless steel), while the direct exposure probe [11] has the sample applied as a thin layer (from solution or suspension) to an exposed, low thermal mass tungsten or rhenium filament. The rapid heating results in a rate of vaporization that exceeds the rate of thermal decomposition, even for very polar compounds.

Compounds that are solids at room temperature, but prone to decomposition upon heating, cannot be introduced successfully into a mass spectrometer (for subsequent ionization) employing standard methods. There are methodologies that rely on the intentional pyrolysis [12] of complex organic polymers (e.g., humic substances such as coal, soil, industrial elastomeric polymers, etc.), but this is not the generally desired outcome when introducing samples into a mass spectrometer. Controlled pyrolysis results in characteristic mass spectra of the thermal fragments. These spectra can be analyzed using principal component analysis [13] allowing for the grouping or categorization of sample types according to similar trends or characteristics. However, this approach does not provide desired identity and structure information for intact biological molecules such as peptides, proteins or oligonucleotides. These types of compounds must be transferred from the solid state to the gaseous state by other means.

11.3   IONIZATION METHODS

The formation of ions is the starting point for every mass spectrometric analysis. The ionization event also dictates the appearance of the resulting mass spectrum, primarily as a consequence of the energetics of the event. In general terms, ionization methods fall into two general types: gas phase and desorption. Gas-phase ionization techniques depend on the analyte molecules either starting in the gas phase or being able to undergo a phase transition from the solid or liquid phase to the gas phase at reasonable temperatures and pressures (i.e., avoiding pyrolysis). As discussed previously (see Section 11.2), this phase transition is often achieved through the introduction of thermal energy, which can substantially decompose the analyte if the polarity (degree of hydrogen bonding) of the compound is high. Molecular mass also factors into the suitability of a gas-phase ionization method, since the volatility of even relatively non-polar compounds decreases substantially with increasing molecular mass. This factor generally limits the utility of these techniques to compounds having masses less than 1000 Da. Once vaporized, ions are generated as a result of some interaction of the gaseous analyte molecules with a source of energy (e.g., energetic electrons, photons, etc.) or some reactive species such as a strong gas-phase ionic acid or base that can donate or abstract a proton. Traditional gas-phase ionization methods, such as electron ionization (EI) and chemical ionization (CI), are conducted within the vacuum chamber of the mass spectrometer, since many of the processes require the low-pressure environment. However, there have been several ionization techniques (atmospheric pressure chemical or photoionization, APCI or APPI) introduced in recent years capable of producing ions from gaseous analytes at atmospheric pressure, outside the vacuum chamber of the mass spectrometer. In contrast, there are a variety of ionization methods that do not depend on volatilization of the analyte molecules prior to ionization. In these *desorption* ionization techniques, the ionization and phase-transition event (transfer of the analyte molecules from either the solid phase or liquid phase to the gas phase) are integrated, occurring so closely together in either space or time as to be indistinguishable. Because desorption ionization techniques do not depend on thermal energy to volatilize the analyte, they are particularly well suited for the analysis of large thermally fragile biomolecules.

Methods of ionization can also be categorized according to the amount of energy they impart to ions in the formation process, with those techniques that transfer lesser amounts of energy into the newly formed ions being referred to as *soft* ionization methods. An indication of whether an ionization event is either hard or soft is the presence or absence of fragment ions in the mass spectrum. Soft ionization methods impart less energy into the ionized molecule producing mass spectra that display a strong response for the intact parent ion (from which the molecular mass of the analyte is easily deduced), but few if any fragment ion signals. On the other hand, higher energy or hard ionization methods produce mass spectra that generally contain a variety of fragment ions, which are indicative of the parent ion structure. In some cases, the parent ion signal is absent from the mass spectrum, making it difficult to determine the molecular mass of the analyte.

11.3.1 Electron ionization (EI)

The EI source has been the most widely used ion source over the past 60 years and continues to be the method of choice for the analysis (either qualitative or quantitative) of small- to medium-sized volatile organic compounds. The inherent reproducibility of the mass spectra has enabled the assembly of large spectral libraries. Computers associated with current generation instruments can efficiently (in a few seconds) search an unknown mass spectrum against tens of thousands of reference spectra in order to aid in the identification of an analyte. The general scheme of an EI source includes the introduction of the vaporized analyte molecules into the ionization chamber, exposure of those molecules

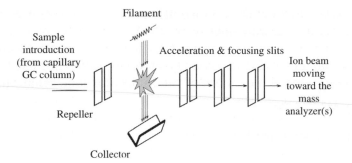

Fig. 11.2. Diagram of the components of an EI source. Gaseous samples (gases or vaporized liquids or solids) are introduced into the ionization chamber using a reservoir with molecular leak, direct insertion probe or as an eluent from a GC column. The collimated stream of 70 eV electrons interacts with the neutral analyte molecules to generate stable radical cations ($M^{+\bullet}$) and unstable radical cations ($M^{+\bullet}*$) that undergo dissociation reactions to form the characteristic fragment ions observed in many EI mass spectra.

to a collimated beam of electrons creating ions and extraction of the ions from the chamber into the analyzer region of the instrument (Fig. 11.2). As described previously (see Section 11.2), applications involving the analysis of gaseous samples employed a heated reservoir and transfer line that terminated in a very small diameter orifice through which the gas molecules would leak into the ionization chamber. In most cases today, gaseous samples (either gases or vaporized liquids) are delivered to the ionization chamber through a capillary GC column as part of an on-line separation/analysis scenario. Analogously, solid compounds that are amenable to vaporization by ballistic heating (and not prone to thermal decomposition) are delivered directly to the ionization chamber on the tip of a probe. Once in the ionization chamber, the neutral analyte vapor molecules interact with a focused stream of electrons. Passing a current through a coiled or ribbon filament (typically W or Re) generates this stream of electrons. These electrons are accelerated toward a "collector" using a potential offset between the filament and the collector. In many sources there is also a set of small permanent magnets that aid in the focusing of the electron beam. While the potential offset between filament and collector is often variable (typically in the range of 0–500 V), it is the agreed upon standard of 70 electron volts (eV) that enables comparison of spectra obtained in any laboratory with those contained in spectral libraries. Any close encounter between an analyte molecule and a 70 eV electron results in energy transfer to the molecule (referred to as a vertical or Franck-Condon transition), excitation (both electronic and vibrational) and formation of a radical cation or molecular ion, $M^{+\bullet}$. At an electron energy of 70 eV, formation of the molecular ion is also generally accompanied by the formation of fragment ions (Figs. 11.3 and 11.4) [14]. If lower electron energies are used, fragmentation can be reduced (i.e., more of the molecular ions remaining intact).

$$M + e^- \rightarrow M^{+\bullet*} + 2e^-$$

$$M^{+\bullet*} \rightarrow F_1^{+\bullet} + F_2^+ + F_3^+ + \dots$$

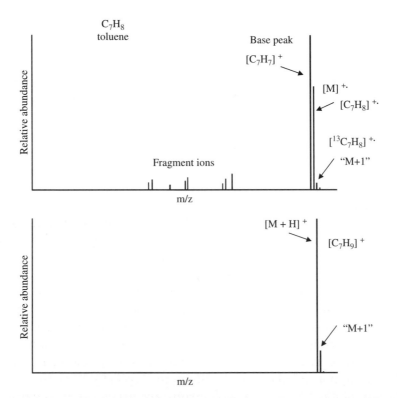

Fig. 11.3. Electron ionization and methane CI mass spectra of toluene. The key features of the respective mass spectra are labeled. Spectral interpretation is based on recognition and understanding of these key features and how they correlate with structural elements of the analyte molecule of interest. The signal representing the most abundant ion in a mass spectrum is referred to as the base peak, and may or may not be the molecular ion peak (which carries the molecular mass information). CI spectra provide confirmation of molecular mass in situations where the EI signal for the molecular ion ($M^{+\bullet}$) is weak or absent. The CI mass spectrum provides reliable molecular mass information, but relatively little structural information (low abundance of the fragment ions). Compare with Fig. 11.4.

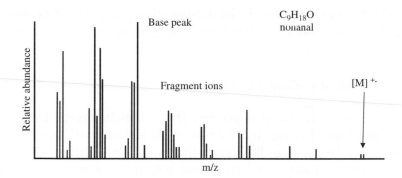

Fig. 11.4. Electron ionization mass spectrum of nonanal. Unlike the previous example (toluene, Fig. 11.3), this 9-carbon alkyl aldehyde displays extensive fragmentation and a very low abundance molecular ion at m/z 142. The extensive degree of fragmentation exhibited by many compounds under EI conditions makes manual interpretation complex and tedious. Consequently, computerized searches of spectral libraries find extensive use in compound identification.

Once formed, all ions are extracted from the ionization chamber and accelerated into the analyzer portion of the instrument using a series of electrostatic elements (repeller, lenses, etc.). The type of mass analyzer dictates what shaping of the ion beam is required, which is achieved through the use of slits or other focusing elements. For example, in a double-focusing sector mass spectrometer the ion beam takes the shape of a ribbon while passing through a series of rectangular shaped slits and lenses. In contrast, the shape of the ion beam in an instrument having one or more quadrupole mass filters is cylindrical due to the use of apertures and lenses with circular openings. The potentials at which these electrostatic elements are held are also a function of the analyzer type. Sector and time-of-flight (TOF) mass spectrometers operate with potential differences of 3,000–40,000 V between source and analyzer, whereas quadrupole instruments operate in a regime of potential differences on the order of 10 V. Regardless of the magnitude, the kinetic (translational) energy of the ions that enter the mass analyzer can be described by Eqs. (11.1) and (11.2).

$$KE = qV = zeV \qquad (11.1)$$

$$KE = \frac{1}{2}mv^2 \qquad (11.2)$$

where KE is kinetic energy, V the accelerating potential, e the electronic charge (1.6×10^{-19} C), z the number of charges, m the ion mass and v the ion velocity.

11.3.2 Chemical ionization (CI)

Chemical ionization [15] is a process that imparts charge to a gaseous analyte molecule through interaction with a "reagent" ion (ion/molecule reaction). Being classified as a soft ionization technique, CI is often used as a compliment to EI in either establishing or confirming the molecular mass of an analyte (Fig. 11.3). Reagent ions typically are formed by electron bombardment of a reagent gas, in the same way that analyte ions are produced in EI. However, there are several important differences. In order to produce the desired reagent ions and to then have the reagent ions encounter and react with the analyte molecules, significantly higher pressures are required within the ionization chamber. The required pressures are attained through modifications of the ionization chamber that result in a more enclosed "gas-tight" configuration (reduction of exit slit width, reduction in the diameter of the electron entrance and exit apertures, etc.). Taken together these changes allow localized pressures of up to 133 Pa (1 torr) to be maintained within the ionization chamber, while the pressure within the surrounding ion source region of the mass spectrometer is maintained in the 10^{-2} or 10^{-3} Pa (10^{-4} or 10^{-5} torr) range. This relatively high pressure requires significantly higher electron energies (100–500 V) in order to penetrate the reagent gas and initiate the primary ionization event. The increased pressure results in very short mean-free paths (and thus very short times between collisions) for the ions initially formed by EI. Being very reactive, the initially formed ions rapidly produce secondary products (ions) when they come into contact with other neutral reagent gas molecules through processes such as hydride abstraction or addition with elimination. It is these secondary (and in some cases tertiary or higher order) ion/molecule reaction products that serve as the important reagent ion species in CI. The high number density of reagent gas molecules in the ionization chamber not only is necessary to achieve the higher order ion/molecule reactions, but also provides for the stabilization of the resulting ions through a process known as collisional cooling. Some commonly used CI reagent gases are hydrocarbons such as methane and isobutane, ammonia, water, methanol, hydrogen, dichloromethane, oxygen, nitrogen and argon, among a myriad of other possibilities. The wide variety of potential reagent gases (and resulting reagent ions) [16] provides tremendous flexibility in both the type of analyte ions that can be formed as well as the degree of fragmentation (or lack thereof) that is observed in a CI mass spectrum. Formation of even-electron cations is accomplished through the use of gas-phase acids (reagent ions) that can readily transfer a proton to a basic analyte molecule or gas-phase bases that can abstract a proton from an acidic molecule. For any given analyte, the difference between its gas-phase basicity and that of the corresponding neutral of the reagent ion dictates the maximum amount of internal energy that can be transferred during the protonation event. Thus, by changing the reagent gas from hydrogen (H_3^+) to methane ($CH_5^+/C_2H_5^+$) to isobutane ($C_4H_9^+$) to ammonia (NH_4^+) one can substantially alter the appearance of a CI mass spectrum (from many fragment ions to very few, if any). Summarized in the expressions below are the reactions leading to the formation of the primary reagent ions for methane as well as some of the other reactions (negative CI, charge exchange, hydride transfer, adduct formation, etc.) that can be carried out under CI conditions.

$$CH_4 + e^- \rightarrow CH_4^{+\bullet}$$

$$CH_4^{+\bullet} + CH_4 \rightarrow CH_5^+ + CH_3^{\bullet}$$

$$CH_4^{+\bullet} \rightarrow CH_3^+ + H^{\bullet}$$

$$CH_3^+ + CH_4 \rightarrow C_2H_5^+ + H_2$$

Protonation	$C_2H_5^+ + M \rightarrow C_2H_4 + [M+H]^+$
Deprotonation	$OH^- + M \rightarrow H_2O + [M-H]^-$
Anion attachment	$[CH_2Cl_2]Cl^- + M \rightarrow$
	$CH_2Cl_2 + [M+Cl]^-$
Charge exchange (radical cation formation)	$N_2^{+\bullet} + M \rightarrow N_2 + M^{+\bullet}$
Electron capture (radical anion formation)	$CO_2 + e^- \rightarrow$ thermal energy
	$e^- + M \rightarrow M^{-\bullet}$

11.3.3 Atmospheric pressure chemical ionization (APCI)

Atmospheric pressure chemical ionization [17] shares many of the attributes of classical CI. The technique has gained great popularity in recent years as one of the simple and rugged techniques for coupling liquid chromatography (HPLC) with mass spectrometry. The most significant difference between APCI and the traditional CI technique is the environment in which the two ionization events occur—with the APCI reaction, taking place at (or near) atmospheric pressure outside the vacuum manifold of the mass spectrometer. Much of the vaporized HPLC eluent can be dissipated and pumped away significantly more easily, if done prior to entering the mass spectrometer. A corona discharge (a needle held at a potential of several thousand volts) rather than a filament initiates the production of primary ($N_2^{+\bullet}$) and secondary reagent ions (H_3O^+, CH_3CNH^+, $CH_3OH_2^+$ or $C_4H_8OH^+$). One of the drawbacks of the APCI technique is the limited variety of reagent ions that can be generated since the technique depends on the composition of the HPLC eluent as the reagent gas source. Typical choices of organic modifier in reverse-phase HPLC separations are water-miscible solvents such as methanol, acetonitrile or tetrahydrofuran. Because the prevalent APCI reactions are protonation or deprotonation, the relatively low gas-phase acidity of the typical APCI reagent ions (when compared to the hydrocarbon species CH_5^+ and $C_2H_5^+$ produced in CI) reduces both the range of compounds that can be ionized using APCI as well as the response (sensitivity) of certain classes of compounds. Analyte compounds possessing gas-phase basicities lower than those of the reagent gas neutrals (conjugate bases) will not undergo the proton-transfer reaction necessary for the formation of the $[M+H]^+$ ion.

11.3.4 Atmospheric pressure photoionization (APPI)

Atmospheric pressure photoionization [18] is a new technique that employs an ultraviolet (UV) lamp in place of the corona discharge needle used in APCI. The radiation from the UV lamp (typically in the range of 10 eV) possesses sufficient energy to ionize many classes of organic analyte molecules, but not the typical reverse-phase HPLC eluent components. This selectivity results in a mode of ionization that may be amenable to compounds lacking functional groups of sufficient basicity to accept an ionizing proton from a reagent ion. When used in combination with aprotic solvents, the APPI technique generates radical cations ($M^{+\bullet}$) much like charge-exchange CI and EI. The presence of aprotic solvents often results in the formation of protonated analyte molecules ($[M+H]^+$) through what is likely a hydrogen atom abstraction mechanism (having been referred to as photo-initiated APCI). The ionization response for molecules having a low photoionization cross section (probability) has been shown to be enhanced by the use of a dopant that is introduced into the vaporized plume of analyte molecules. The dopant is selected on the basis of a high-UV absorptivity and serves as a charge-transfer reagent. Common dopants include acetone, toluene and anisole.

$$M + hv \rightarrow M^{+\bullet}$$

$$M^{+\bullet} + S \rightarrow [M + H]^+ + [S - H^\bullet]$$

$$D + hv \rightarrow D^{+\bullet}$$

$$M + D^{+\bullet} \rightarrow M^{+\bullet} + D$$

$$D^{+\bullet} + S \rightarrow [D + H]^+ + [S - H^\bullet]$$

$$M + [D + H]^+ \rightarrow [M + H]^+ + D$$

where M is the analyte, S the solvent and D the dopant (a photo-active compound intentionally added to the solvent to enhance photo-ionization).

The number of published reports detailing the APPI technique is still relatively small. Some preliminary accounts indicate lower absolute signal intensities when compared to APCI, but this is accompanied by a significant reduction in the intensity of background signals for overall improvement in the signal-to-noise (S/N) ratio. The net outcome is comparable or better detection limits for a limited set of test compounds significantly promise for future applications.

11.3.5 Field ionization (FI)

Field ionization [19,20] is a specialized technique that generates radical cations of low internal energy. Thus, it produces mass spectra that display abundant $M^{+\bullet}$ ions indicative of the molecular mass of the analyte with little or no fragmentation. Much like EI, the FI technique depends on the introduction of gaseous analyte molecules into the ionization chamber from a heated reservoir or GC column. Ionization is affected not by a stream of electrons but rather by a high electric field. When the analyte molecules pass very close to the FI emitter, the magnitude of the force created by the high electric field (approximately 10^7–10^8 V/cm) is sufficient to directly displace an electron from the molecule, producing a radical cation. The FI emitter is typically one or more sharp tips, small diameter wires or a sharp blade. The most common emitters are small (microscopic) diameter tungsten wires (diameter of approximately $10 \, \mu m$) upon which have been "grown" many thousands of microscopic carbon needles (dendrites) by the pyrolysis of an organic vapor.

11.3.6   Electrospray ionization (ESI)

Electrospray ionization [21] is one of the most widely utilized ionization techniques employed today for the analysis of thermally fragile molecules. As such, it has assumed an important role in the analysis of biologically important molecules. ESI is a *desorption* ionization technique. This means that ions are formed before or during the transition from the liquid phase and need not be volatilized in advance of the ionization event (as is the case for EI, CI, etc.). Like APCI and APPI, ESI occurs at atmospheric pressure outside the vacuum chamber of the mass spectrometer (Fig. 11.5). A solution of the analyte passes through

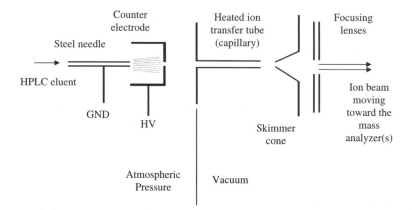

Fig. 11.5. Diagram illustrating the components of an ESI source. A solution from a pump or the eluent from an HPLC is introduced through a narrow gage needle (approximately 150 µm i.d.). The voltage differential (4–5 kV) between the needle and the counter electrode causes the solution to form a fine spray of small charged droplets. At elevated flow rates (greater than a few µl/min up to 1 ml/min), the formation of droplets is assisted by a high velocity flow of N_2 (pneumatically assisted ESI). Once formed, the droplets diminish in size due to evaporative processes and droplet fission resulting from coulombic repulsion (the so-called coulombic "explosions"). The preformed ions in the droplets remain after complete evaporation of the solvent or are ejected from the droplet surface (ion evaporation) by the same forces of coulombic repulsion that cause droplet fission. The ions are transformed into the vacuum envelope of the instrument and to the mass analyzer(s) through the heated transfer tube, one or more skimmers and a series of lenses.

a capillary tube (typically fused silica or stainless steel) at flow rates ranging from 1 to 1000 µl/min. The tip of the capillary tube is held at a potential of several kilovolts with respect to a counter electrode causing a fine spray of tiny charged droplets to form when the solution flow rate is low (a few microliters per minute). As flow rates rise above a few microliters per minute, it is necessary to pneumatically assist droplet formation using a high velocity flow of nitrogen. This auxiliary gas flow produces shearing forces in the liquid stream that results in more efficient droplet formation. Once formed, the droplets begin to decrease in size due to evaporation as they move toward the entrance aperture of the mass spectrometer. The droplets continue to decrease in size through desolvation and spontaneous fission (coulombic explosion) caused by increasing charge density within the droplet. As the droplets continue to shrink, ions are either desorbed into the gas phase from the surface (ion evaporation) or simply loose the remainder of the solvent molecules through evaporative processes. The result is a gas-phase ion, having been produced directly from the analyte solution without the addition of thermal energy. A typical electrospray mass spectrum can exhibit a number of signals, particularly for analytes such as large biomolecules like polypeptides and proteins that possess many basic functional groups (i.e., various nitrogen-containing functional groups). However, unlike a hard ionization technique such as EI, these signals are not structurally diagnostic fragment ions but rather are various populations of the intact ionized molecule, each bearing a different number of protons and representing a distinct charge state (e.g., $[M+H]^{+}$, $[M+2H]^{2+}$, $[M+3H]^{3+}$, $[M+4H]^{4+}$, $[M+5H]^{5+}$, $[M+6H]^{6+}$, ...). Thus, proteins of considerable size (and mass) can be analyzed using instruments not specifically designed for (or capable of) the analysis of high mass species. The multiple charging of these molecules reduces their "effective mass" such that their spectra fall within the attainable m/z range of standard instrumentation. For example, standard quadrupole mass spectrometers are able to detect ions having maximum m/z ratios of 2000–4000. Since most ionization techniques produce singly charged ions ($z = 1$), this translates to a mass range of not greater than a few thousand Daltons. However, if a 50,000 Da protein contains 50 basic amino acid residues that can be protonated in solution, the ESI mass spectrum of that molecule will display an envelope of multiply charged ions in the m/z range of 1000 ($m = 50,000$ and $z = 50$, therefore, $m/z = 1000$). The molecular mass of the protein can easily be calculated [22] using a simple algebraic equation that contains

D.J. Burinsky

the m/z assignments of two adjacent charge state peaks in a mass spectrum as illustrated in the following example:

Two adjacent signals in the ESI mass spectrum of a pure protein have m/z values of 1428.6 and 1666.7, respectively. Since adjacent signals differ in the value of their respective charge states by ± 1, the following expressions can be solved for the value of one of the charge states (z_2 in this case) and then for the molecular mass of the protein analyte, m.

$$m/z_1 = 1666.7$$
$$m/z_2 = 1428.6$$
$$m = 1428.6z_2$$
$$m = 1666.7z_1 \text{ or } 1666.7(z_2 - 1), \text{ since } z_1 = z_2 - 1$$
$$1428.6z_2 = 1666.7(z_2 - 1)$$
$$1428.6z_2 = 1666.7z_2 - 1666.7$$
$$238.1z_2 = 1666.7$$
$$z_2 = 7$$
$$\text{therefore}, m = 10,000 \text{ Da}$$

In practice, modern data-processing software applications perform these calculations very rapidly [23,24]. These algorithms are capable of "deconvoluting" the relatively simple envelope of peaks generated by ESI of a single protein, or complex spectra produced by protein mixtures.

11.3.7 Matrix-assisted laser desorption ionization (MALDI)

Another ionization technique that has made significant contributions to the analysis of large biologically important molecules is MALDI (Fig. 11.6) [25]. The technique is so named because the creation/desorption of protonated molecules (primarily singly and doubly charged ions) is facilitated by the presence of an excess auxiliary or matrix compound. The matrix appears to play several roles. These include the following: (i) absorption of the laser radiation, (ii) rapid heating and sublimation resulting in desorption of the analyte molecules from the solid phase into the gas phase, (iii) providing the ionizing "environment" in which the analyte molecules obtain their charge, and (iv) dissipation of excess energy that might otherwise cause the fragile analyte molecules to fragment or decompose. The list of compounds that have proven useful in the role of matrix is surprisingly short. They are

Mass spectrometry

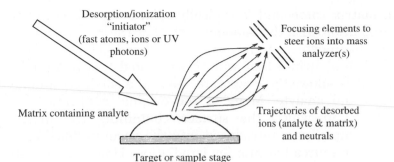

Fig. 11.6. Diagram depicting desorption ionization (MALDI, FAB or SIMS). The operating principles of the three techniques are similar. The initiating event is exposure of the analyte to a beam of photons, atoms or ions. In order to prevent damage to the fragile analyte molecules and enhance the conversion of the involatile molecules into gas-phase ions, a "matrix" is employed. For MALDI, the matrix compounds are UV absorbing compounds such as hydroxycinnamic acid. The most commonly used FAB matrix was glycerol and ammonium chloride was employed by some investigators in SIMS experiments (although at low ion beam fluxes molecular species could be effectively ionized for many analytes with minimal evidence of damage by the primary ion beam).

all aromatic, relatively small molecules that absorb strongly at one or more wavelengths (most commonly 266, 337 and 355 nm), characteristic of the radiation emitted by commercially available pulsed UV lasers (e.g., N_2 lasers). Some of the more commonly used matrix compounds are listed below.

Compound name (absorbing wavelengths)	CAS number
2,5-dihydroxybenzoic acid (266, 337 and 355 nm)	[490-79-9]
2-aminobenzoic acid (266, 337 and 355 nm)	[118-92-3]
3,5-dimethoxy-4-hydroxy-3-phenylacrylic acid (266, 337 and 355 nm)	[530-59-6]
4-hydroxy-3-methoxy-3-phenylacrylic acid (266, 337 and 355 nm)	[537-98-4]
3,4-dihydroxy-3-phenylacrylic acid (266, 337 and 355 nm)	[331-39-5]
Nicotinic acid (266, 337 and 355 nm)	[59-67-6]
Pyrazinecarboxylic acid (266 nm)	[98-97-5]
3-nitrobenzylalcohol (266 nm)	[619-25-0]
4-hydroxy-3-methoxybenzoic acid (266 nm)	[121-34-6]
3-aminopyrazine-2-carboxylic acid (337 nm)	[5424-01-1]

The matrix compound is typically mixed together in an aqueous/organic solution with the analyte, such that the relative concentration of matrix to analyte is on the order of 5000 or 10,000 to 1. The solution is applied to a surface that will be irradiated by the laser beam and the solvent is allowed to evaporate, leaving a solid, crystalline deposit of matrix and analyte. Many of the original applications described instruments in which the surface containing the dried matrix/analyte "sample" was introduced into the vacuum housing of a mass spectrometer source housing for irradiation. However, recently it has been demonstrated that MALDI can be successfully carried out at atmospheric pressure (outside the vacuum chamber of the mass spectrometer), much in the same way as the ESI, APCI and APPI techniques.

The MALDI technique enables the analysis of molecules of significant molecular mass (150 kDa and beyond) [26] when coupled with an appropriate mass analyzer, produces mass spectra that are relatively simple. These spectra typically exhibit signals indicative of the molecular mass (predominantly $[M+H]^+$ with lower abundance $[M+2H]^{2+}$) and cluster species comprised of two or more analyte molecules bound together by weak ionic forces (e.g., $[2M+H]^+$, $[3M+H]^+$, etc.). Details about the mechanism(s) responsible for desorption and ionization in MALDI are only partially understood. Some researchers have described the desorption event as being driven by extremely rapid heating of the matrix, which causes sublimation. Since the analyte molecules are present as only a minor component of the material being vaporized by the laser heating, they are lifted into the gas phase along with the "eruption" of matrix molecules. Whether the analyte molecules are pre-charged in solution prior to being desorbed (notice that most of the matrix compounds are organic acids that could protonate basic biomolecules in solution before evaporation of the solvent) or become charged as a function of gas-phase processes occurring in the region immediately adjacent to the surface during the matrix desorption is a matter of speculation. Rationalizations for the formation of doubly charged species and cluster species are consistent with either line of reasoning.

11.3.8 Fast-atom bombardment (FAB) ionization

Fast-atom bombardment ionization [27,28] was introduced in the early 1980s and was the first widely used desorption ionization technique. Its universal appeal stemmed from its ability to ionize polar, high-molecular-mass compounds that had previously been beyond the reach of mass spectrometric analysis, as well as its ease of implementation and use. FAB shares certain similarities with MALDI in that the sample is dispersed as a dilute component of a matrix with desorption of analyte ions resulting from a violent disruption of that matrix. For FAB, the matrix is a viscous polar liquid in which the analyte will dissolve (i.e., little or no signal is obtained if the sample forms a slurry or suspension with the liquid matrix) and that will have some reasonable lifetime under high vacuum conditions. The most common FAB solvent is glycerol, although thioglycerol and nitrobenzylalcohol also found utility depending on the solubility of particular analytes. A typical experiment involves mixing of a small amount of analyte (perhaps tens to hundreds of micrograms) with a few microliters of glycerol. The mixture is most often introduced into the mass spectrometer source on a small stainless steel target (having an area of a few square millimeters) using a probe. Once positioned in the source, the target is exposed to a kilovolt energy beam of atoms, typically xenon or argon. The atom beam that bombards the sample-containing matrix is produced by neutralization (via resonance charge exchange) of an ion beam. The ions are formed by electrical discharge and accelerated (to kinetic energies on the order of a few kilovolts) toward the target, being neutralized by interactions with other gaseous xenon atoms in the locally high-pressure region of the atom "gun." Because the neutralized ions retain their original velocity, they are referred to as "fast atoms." Atoms formed by the neutralization of 4 keV xenon ions have a velocity of approximately 2400 m/s and are used as the bombarding particles rather than ions in order to avoid charge accumulation at the surface of the matrix. Charge accumulation has a detrimental impact on the formation and desorption of analyte ions. Like MALDI, FAB produces predominantly molecular ions that are either desorbed as ions preformed in the matrix under solution-phase conditions or result from gas-phase processes taking place just above the matrix surface. The preponderance of matrix (e.g., glycerol) molecules spares most analyte molecules direct exposure to the high-energy bombarding particles, thereby minimizing direct energy transfer that would result in fragment ions. The bombarding particles cause significant disruption to the matrix surface through momentum transfer and localized heating. These processes are responsible for the sputtering or desorption of both analyte molecules and/or ions into the gas phase along with the many matrix molecules (and ions). Unlike the MALDI solid-target surface, the liquid FAB matrix continually regenerates its surface during the analysis. The types and size (or mass) of molecules that have been analyzed successfully using FAB is more limited than some of the other techniques described previously (i.e., ESI and MALDI). The practical molecular mass cut-off for FAB has been observed to be in the range of 10,000 Da with many applications describing the analysis of molecules with molecular masses of 3000 Da or less.

11.3.9 Ion bombardment or secondary ion mass spectrometry (SIMS)

Closely related to the technique of FAB, ion bombardment or SIMS [29] employs many of the same principles. The technique finds most of its applications in the analysis of inorganic substances where energy deposition and disposition are less of a concern. The bombarding particles need not be neutralized (converted to atoms) since most of the targets/samples that are interrogated using this technique are conductive, thereby eliminating the problem of charging. Inorganic SIMS is often used to gather compositional and spatial (imaging) information about species present on the surface of the sample of interest. But the ion beam can also be used to remove layers of atoms or molecules from the surface (sputtering) in order to reveal and probe compounds or elements that may reside below the sample surface. In addition to the inorganic applications (generally referred to as "dynamic" SIMS, particularly when high-flux ion beams are used to ablate the sample surface), SIMS has also found limited application in the analysis of organic compounds. Unlike its inorganic analysis counterpart, organic or "static" SIMS employs ion beams of lower flux, thereby causing significantly less surface damage. Static SIMS can also use a solid or liquid matrix for all the same reasons as FAB and MALDI. The SIMS matrix absorbs most of the energy imparted by the bombarding ions (preventing extensive fragmentation of the organic analyte) and assists in the process of liberating (desorbing) the ions of interest from the surface into the gas phase. One notable advantage of the use of ions as bombarding particles (rather than atoms) is that ions can be electrostatically manipulated allowing for control of the ion beam both in terms of its position and direction as well as the shape/size of the beam itself. Thus, an ion beam can be focused down to a diameter of approximately $100\,\mu m^2$ to provide excellent spatial resolution (i.e., secondary/analyte ions are being liberated from a very specific location within the sampling area). Combining good spatial resolution (small spot size) with the ability to scan or raster the ion beam across a sampling area with a high degree of accuracy and precision, SIMS has demonstrated capabilities for imaging applications [30].

11.3.10 Plasma desorption (PD) ionization

Plasma desorption ionization [31] occupies an important place in the history of mass spectrometry as the vanguard for important techniques such as FAB ionization and MALDI. PD ionization was extremely important in that it demonstrated the feasibility of ionizing biological molecules that previously had been inaccessible to mass spectrometric analysis. The technique produces intact molecular ions in a plasma that forms near a surface that has been coated with a mixture of sample and matrix. The production of the energetic plasma is accomplished using a radioactive decay source. The radioactive source (Californium, ^{252}Cf) emits high-energy fission fragments that penetrate the sample-containing foil from the opposite side. The high-energy particle desorbs a significant amount of matrix and sample material as it passes through the foil and causes electronic excitation among the desorbed species that it encounters. This excitation initiates an ionization cascade (like chemical ionization) that results in the formation of the analyte ions. Like all desorption techniques, the large concentration of matrix molecules dissipates much of the available energy, resulting in mass spectra that are dominated by intact molecular ions.

11.3.11 Field desorption (FD) ionization

Field desorption ionization [32] is a variation of the FI technique described previously for the analysis of gaseous samples. Being conceptually similar, the major differences between the two techniques are the sample type and how that sample is introduced into the ion source of the mass spectrometer. In field desorption, the wire emitter (containing the many tiny sharp points or edges) has the sample applied directly to it from a solution. Once the solvent evaporates, the loaded emitter is admitted to the ion source on a probe tip. Ionization of the sample is affected by application of the same high voltage used in the FI technique. In some cases, heating of the sample enhances the ionization response. Sample heating is accomplished by application of a current through the wire emitter (resistive heating). Of course, if heating of the sample is too aggressive thermal decomposition can occur, since the samples most often analyzed by this technique (like all of the desorption techniques) are thermally labile and prone to decomposition. Ion formation in field desorption may result from the expulsion of preformed ions directly from the surface, as a consequence of the very high electric field strength, or may be the result of gas-phase processes occurring in an ionization cascade very near to the emitter surface. The mass spectra produced by this technique resemble those from other desorption techniques in that they display protonated molecules ($[M+H]^+$) rather than the radical cations ($M^{+\bullet}$) observed in FI spectra.

11.4 MASS ANALYZERS

The heart of the mass spectrometer is the mass analyzer [33], which is the device that separates the ions created in the ion source according to their m/z ratios. The ability of an analyzer to distinguish an ion of mass m from an adjacent ion differing in mass by Δm is the measure of either *resolution*, $R = \Delta m/m$ or *resolving power*, RP $= m/\Delta m$ [34,35]. The degree of separation of two adjacent peaks in a mass spectrum is defined on the basis of the fractional height of the remaining valley between them (e.g., 10% valley or 50% valley), assuming the peak shape is Gaussian. Some mass analyzers produce peak shapes that are not Gaussian (e.g., quadrupole mass filters have imperfect rectangular peak shapes and ion cyclotron resonance (ICR) spectrometers produce Lorentzian shaped peaks), so comparison of stated values for either R or RP can be challenging. Another complication is that the peak width for some analyzers is constant over the mass range (quadrupole mass filters), while for others the peak width varies with m/z ratio. This yields what is in effect a variable resolution for quadrupoles since the Δm term in the above expressions remains constant. Consequently, the figure of merit for R (or RP) in these instruments is typically specified simply as the peak width at half height (e.g., quadrupoles are often operated such that each peak for every singly charged ion is nominally separated from the next, resulting in what is referred to as "unit" resolving power). Separation of each adjacent signal in the mass spectrum from the next equates to a peak width of approximately 0.7 m/z units at half height. Representative values of resolving power for commercially available mass spectrometers (based on various types of mass analyzers) are shown below. While many of these devices are capable of operating at resolving powers in excess of those listed in the table, these values are typical of normal or routine operation.

Mass analyzer	Representative resolving power values
Quadrupole mass filter	0.7 m/z full width half height (e.g., RP = 100 @ m/z 100 and 2000 @ m/z 2000)[a]
Three-dimensional quadrupole ion trap	Similar to quadrupole mass filters under routine conditions (at reduced scan rates, RP = 5000 or greater is achievable over limited m/z ranges)[b]
Two-dimensional (linear) quadrupole ion trap	Similar to quadrupole mass filters under routine conditions (at reduced scan rates, RP = 15,000 or greater is achievable over limited m/z ranges)[c]
Time-of-flight (linear)	1000–5000
Time-of-flight (reflectron)	1000–15,000
Fourier transform ion cyclotron resonance	100,000–500,000 or greater[d]
Double focusing magnetic sector	1000–100,000+

[a]The latest generation quadrupole mass filters are capable of peak widths on the order of 0.1 m/z, full width half maximum, for a RP = 10,000 @ m/z 1000.
[b]The 3D quadrupole ion trap is capable of extremely high resolving powers at reduced scan rates [36].
[c]One commercially available linear (2D) ion trap is capable of up to 30,000 resolving power (for m/z 1500), which is analogous to a constant peak width of 0.05 full width at half maximum (FWHM).
[d]Depending on instrument pressure and the duration of the free induction decay.

Resolution (or resolving power) plays an important role in mass spectrometry for applications requiring the characterization of very similar chemical species. The ability to detect and accurately measure the m/z ratio of a particular ion depends directly on the resolving power of the mass analyzer. For example, if a sample contains two isobaric compounds (i.e., having the same nominal molecular mass but different elemental formulae) the difference in the exact masses of the molecular ions will be much less than 1 m/z unit. Any mass analyzer possessing a nominal resolving power (e.g., RP < 1000) will register only one peak in the mass spectrum of such a binary mixture. Attempts to measure the exact mass of either of the molecular ions (i.e., to deduce the elemental composition of the molecular ions) will be unsuccessful, since the value derived by measuring the composite peak will be some weighted average value of the two contributing compounds (rather than the true value for either).

Example:
Two isobaric peptides (Ala-Arg-Asn-Gly Leu-Phe-Pro-Tyr, **1** and Ala-Arg-Orn-Gly Leu-Phe-Pro-Tyr, **2**) are ionized concomitantly in the ESI source of a mass spectrometer producing a single signal at the nominal mass (m/z) of 937.5 for the singly charged ion.

$$m/z \text{ of } [M_1H]^+ = 937.4891 \ (C_{44}H_{65}N_{12}O_{11}^+)$$

$$m/z \text{ of } [M_2H]^+ = 937.5254 \ (C_{45}H_{69}N_{12}O_{10}^+)$$

$$m/\Delta m = 25,827$$

Therefore, a mass analyzer capable of resolving powers of 26,000 or better is required to reveal the presence of two ions (representing two distinct compounds). If an analyzer with RP $<$ 26,000 is used, the presence of a second compound would be overlooked. Additionally, attempts to measure the exact m/z value of the peak will not provide a true value, but rather some average value (e.g., $m/z \cong 937.5072$, assuming the compounds produce signals of comparable intensity) that will be of little use in identifying either unknown compound. Ambiguous results such as these only complicate qualitative analysis efforts.

Resolving power is not only important for distinguishing between isobaric ions (i.e., having the same nominal mass but distinct elemental formulae), but also for distinguishing the charge state of multiply charged ions generated by ionization techniques such as ESI. The spacing of signals or peaks in a mass spectrum is inversely proportional to the number of charges, or the value of z. In the case of singly charged ions where $z = 1$ (the predominant species generated by many ionization techniques), 1 m/z unit separates the signals for the adjacent signals in the isotopic distribution observed in the mass spectrum. Thus, the signals in the isotopic distribution for a singly charge molecular ion ($[M+H]^+$) are cleanly resolved by a mass analyzer capable of "unit" resolving power at the m/z value of the ion. When the value of z increases, as is the case for many of the multiply charged signals observed in the ESI mass spectrum of a polypeptide, the signals for adjacent isotopes are no longer spaced 1 m/z unit apart. Instead, the isotopes of doubly charged ions are separated by 0.5 m/z units (said another way, there are two isotopic peaks per m/z unit), the isotopes for triply charged ions are separated by 0.33 m/z units, and so on. For reasonably large polypeptides or proteins that bear many charges (upward of 60 charges for the protein bovine serum albumin, M_r 66 kDa), a mass analyzer capable of a RP $> 66,000$ is required to separate all the isotopic peaks from one another, and thereby recognize/identify the charge state.

Achieving high resolving power and high m/z measurement accuracy is one way of decreasing uncertainty when the determination of unknown analyte identity is the object of an experiment. But like many techniques, an increase in experimental or interpretive confidence does not come without some cost (e.g., instrument size, complexity, price, etc.). However, exact m/z measurements (and their associated elemental formula information) are but one type of information that can be derived from mass spectrometers. In the sections that follow, a variety of mass analyzers will be described in terms of their basic principles, functionality and applications.

11.4.1 Magnetic sector analyzer

The first mass spectrometers of the early physicists were based on mass analyzers that separated ions by passing them through a magnetic field. The ribbon-like beam of ions (the shape of the ion beam is defined by the shape of the electrostatic elements and slits through which it passes) leaves the ion source with all ions having the same kinetic energy. As described in Eq. (11.3), this kinetic (translational) energy is imparted to the ions by the electric field through which they pass as they leave the source (accelerating potential), with the source typically held at some high potential (several thousand volts) and the entrance slit of the mass analyzer at ground potential.

$$KE = qV = zeV = \frac{1}{2}mv^2 \tag{11.3}$$

Given that all of the ions have the same charge (assuming singly charged ions) and the same kinetic energy, the heavier ions will have lower velocities than the lighter ions (e.g., at KE $= 10$ keV, $v_{m/z\ 100} = 4393$ m/s and $v_{m/z\ 500} = 1965$ m/s). When they pass through the magnetic field, the ion trajectories will be altered to varying degrees—with the paths of the heavier ions being deflected more than the paths of the lighter ions due to the varying amounts of time they are subjected to the force of the magnetic field. The relevant equations of motion for ions in a magnetic field are shown below, where the magnetic force exerted on the ions is being balanced by the centripetal force required for the ions to traverse the sector (without colliding with the analyzer walls) and focus on an exit slit, as seen in Eqs. (11.4) and (11.5). Rearranging equations and substituting into Eq. (11.3) above yields Eq. (11.6) that describes how the m/z ratio depends on the instrumental variables of the mass spectrometer.

$$F_{\text{magnetic}} = Bzev \tag{11.4}$$

$$F_{\text{centripetal}} = \frac{mv^2}{r} \tag{11.5}$$

$$\frac{m}{z} = \frac{B^2 r^2 e}{2V} \tag{11.6}$$

where F_{magnetic} is the force exerted by the magnetic field, e the electronic charge, v the velocity, m the mass, z the number of charges, B the magnetic field strength, $F_{\text{centripetal}}$ the centripetal force of the ion, r the radius of curvature and V the accelerating potential.

As can be seen from the expression, the m/z ratio of the ions focused onto the exit slit is a function of three important variables: B, r and V. A mass spectrum can be recorded (i.e., successive ions of increasing or decreasing m/z that have been dispersed by the magnetic field according to their momentum, mv, can be swept across the exit slit and onto the detector) by holding two of the three variables constant and varying the third. In practice, varying the strength of the magnetic field is the most common method for obtaining mass spectra in magnetic sector instruments, but scanning of the accelerating potential can also be done. Because the physical shape and size of the analyzer are both fixed, the third variable, r, is rarely changed (i.e., only in the case of planar detectors such as photographic plates). Such a single magnetic sector mass analyzer [37] will have a maximum achievable resolving power of approximately 2000.

The assumption about the kinetic energies of all ions leaving the source region being equal is not strictly true. The neutral analyte molecules possess some inherent distribution of kinetic energy (Boltzmann) prior to ionization, which they retain. Imperfections in the electrical fields that accelerate the ions contribute an additional kinetic energy spread to the ion population that enters the magnetic sector analyzer. The upshot of this energy spread is some minor variation in the momentum of the ions and this translates into broader peak shapes at the detector, or reduced spectral resolution (or resolving power). Passing the ions through an energy filter prior to momentum analysis (Fig. 11.7) can reduce the inherent uncertainty or spread in the kinetic energies of the ions entering the magnetic analyzer. The most commonly used energy filter is an electrostatic analyzer (ESA) comprised of two curved parallel plates (electrodes) between which a potential is applied. The trajectories of ions passing through the ESA are altered (deflected) according to equations of motion similar to those described previously for a magnetic field. The ESA electrical field is matched to the accelerating potential of the instrument, so only ions with a very precise kinetic energy pass into the magnetic sector. This double focusing of the ions, energy focusing in the ESA and directional focusing in the magnetic sector, produces the spectral resolution ($5000 \leqslant$ RP $\leqslant 100,000$) necessary for the separation of isobaric ions. Separation

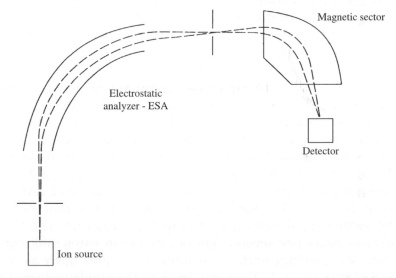

Fig. 11.7. Diagram showing the fundamental components of a double-focusing forward geometry sector mass spectrometer. Ions are accelerated to kilovolt translational energies as they exit the ion source. The ESA provides energy focusing by allowing only ions of a very narrowly defined kinetic energy distribution to pass into the magnetic sector. The ion beam is dispersed according to m/z value as it enters the magnetic sector. The trajectories of lighter ions are altered (deflected) more than those of the heavier ions. Only ions of a particular m/z ratio transit the exit slit successfully and strike the detector. Ions of successively higher m/z are brought to focus at the detector by increasing the strength of the magnetic field.

of isobaric species, in combination with high-stability electronic components, leads to the very accurate and precise (on the order of a few parts per million) m/z ratio measurements required for deducing elemental formulae. The configuration of both single- and double-focusing mass spectrometers has changed over time, with various geometries being adopted as improvements became available, or for specific experimental needs. Commercial instruments from various manufacturers have had components of varying size (radius—10 cm to more than 30 cm or more) and shape (defined by the angle—ranging from approximately 30–90°) with some of the configurations being described by the names of their developers (e.g., Mattauch-Herzog [38], Nier-Johnson [39], etc.).

11.4.2 Quadrupole mass filter

The quadrupole mass filter [40] is perhaps the most widely used type of mass analyzer, primarily due to its compact size, modest cost, linear scan relationship, versatility and ease of use. Whereas the analyzers of a double focusing instrument are large (thousands of cubic centimeters in volume weighing hundreds of kilograms), a quadrupole mass filter is a mere 15–20 cm in length (diameter < 3 cm) and weighs considerably less than a kilogram. The device was invented in the 1950s and quickly found utility in chemical analysis applications that required mass spectrometers to acquire spectra quickly. The magnet sector instruments of the day had limited scan rates due to hysteresis. While modern magnet technology produces analyzers that are capable of very rapid scan rates compared to their predecessors (e.g., a range of several hundred m/z in less than 1 s), modern quadrupole mass filters outperform magnetic sectors in this regard, being capable of scan rates on the order of 10,000 m/z units per second. These rapid scan rates make quadrupoles ideal for coupling with separation techniques such as gas and liquid chromatography. As the name implies, the quadrupole mass filter (Fig. 11.8) is comprised of four electrodes (referred to as rods, although electrodes that have a hyperbolic shape produce the highest quality electrical fields) that are arranged parallel to each other over their entire length. The opposing sets of rods are electrically connected and both DC and radio frequency (RF) voltages are applied, with the DC voltages having opposite polarity and the RF signals being 180° out of phase ($-U_{dc}-V_{rf}\cos\Omega t$ and $U_{dc}+V_{rf}\cos\Omega t$), with the two-dimensional

Mass spectrometry

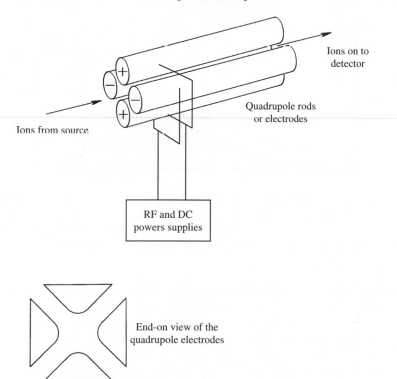

Fig. 11.8. Diagram showing the important elements of a quadrupole mass filter. The four parallel electrodes receive both DC and RF voltages of increasing amplitude during a scan function, causing ions of sequential m/z ratios to attain stable trajectories and be transmitted through the device to the detector. Ions having all other m/z ratios display unstable trajectories and are lost by striking the electrode surfaces or by leaving the device radially. The inset in the lower left portion of the diagram shows an end-on view of the electrodes and the ideal hyperbolic shape of their surfaces. The hyperbolic surfaces create near ideal quadrupolar fields within the device. Round rods approximate the ideal situation but generate imperfect fields that lower the performance of the device.

(2D) potential defined by Eq. (11.7):

$$\phi_{xyzt} = (V \cos \Omega t + U)\frac{(x^2 - y^2)}{r_0^2} \tag{11.7}$$

where r is the inside radius of the four electrode array.

Ramping of both DC and RF voltages in a simultaneous fashion produces resonant or stable trajectories (Fig. 11.9) for ions of sequential

347

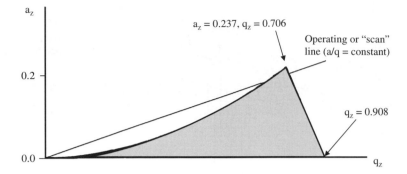

Fig. 11.9. The stability diagram for a quadrupole mass filter. The shaded area represents solutions for the Mathieu equations that result in stable ion trajectories through the device. The area outside the boundaries represents solutions for the equations that produce unstable trajectories.

m/z ratios. These ions pass through the device to a detector where they produce a signal that is transformed into a mass spectrum. Non-resonant ions do not achieve a stable trajectory and are lost to collisions with the electrodes or other surfaces inside the instrument. It is because of these time-dependent forces acting on the ions to produce stable or unstable trajectories that quadrupole mass filters are referred to as dynamic analyzers. The performance of the device with respect to m/z range, resolving power and transmission efficiency (sensitivity) hinges on the length of the electrodes, the RF and the radius of the electrodes. Mass (m/z) range increases with decreasing electrode radius and resolving power increases with increasing electrode length or frequency. In contrast to the magnetic sector instruments described previously, quadrupole mass filters are relatively insensitive to minor differences either in the kinetic energies or initial trajectories of incoming ions. The typical kinetic (translational) energies for ions traversing quadrupole mass filters is on the order of 10 eV, considerably less than the 3–10 keV kinetic energies characteristic of ions traveling through sector analyzers.

11.4.3   Ion traps

11.4.3.1   Three-dimensional quadrupole ion trap
Quadrupole (RF) ion traps are the newest of the commercially available mass analyzers, despite having been invented at about the same time as the quadrupole mass filter, nearly 50 years ago. The Paul ion trap

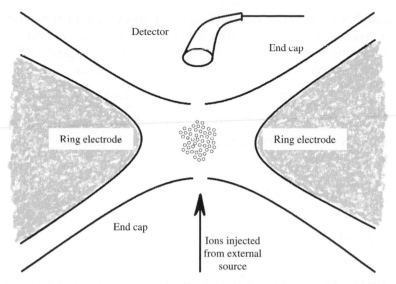

Fig. 11.10. Diagram illustrating the inner surfaces of the primary components of a Paul (3D) quadrupole ion trap. Ions generated by an external source are injected into the trap through an aperture in one of the end caps. Scan functions for isolating ions in the trap, exciting the mass selected ions to induce unimolecular dissociation, and ejecting ions from the trap (for detection) are implemented through the application of DC and RF voltages to the ring electrode.

(Fig. 11.10) [41] is a three-dimensional (3D) device that employs many of the same operating principles as the 2D quadrupole mass filter. Instead of four parallel electrodes, the ion trap has three electrodes—two hyperbolic end caps (separated by distance $2z_0$) and a centrally located toroidal "ring" electrode (of radius r_0). Most commercially available 3D ion traps have r_0 values of either 0.707 or 1.00 cm. Like ideal quadrupole rods, the surfaces of the three ion trap electrodes are hyperbolic (a requirement to produce quadrupolar electrical fields within the devices). The trap is generally combined with both an external ion source (although ions can be generated within the ion trap) and an external detector (electron multiplier). Ions are admitted to the device for analysis and leave the device for detection through apertures in the end-cap electrodes. As the name implies, the ion trap is unique among analyzers discussed thus far in that it has the capacity not only to separate ions according to their m/z ratio, but also hold or store ions for periods of milliseconds to seconds. This unique ability of trapping devices to store ions provides access to a wide array of possible experiments. The equations of motion for an ion inside a quadrupole trap are considerably more complex than those describing the trajectories of ions through a magnetic sector. The forces acting upon an ion within the trap result in a 3D oscillating trajectory (described as a Lissajous figure) due to the shape of the potential field, as described by the following Eq. (11.8):

$$\phi_{xyz} = (U - V \cos \Omega t) \frac{(x^2 + y^2 - 2z^2)}{(r_0^2 + 2z_0^2)} + \frac{U - V \cos \Omega t}{2} \tag{11.8}$$

The perpendicular forces acting on the ions can be described by Eqs. (11.9–11.11):

$$\left(\frac{m}{e}\right)\ddot{x} + (U - V \cos \Omega t)\frac{x}{(r_0^2 + 2z_0^2)} = 0 \tag{11.9}$$

$$\left(\frac{m}{e}\right)\ddot{y} + (U - V \cos \Omega t)\frac{y}{(r_0^2 + 2z_0^2)} = 0 \tag{11.10}$$

$$\left(\frac{m}{e}\right)\ddot{z} - 2(U - V \cos \Omega t)\frac{z}{(r_0^2 + 2z_0^2)} = 0 \tag{11.11}$$

which are all examples of the Mathieu equation, Eq. (11.12), with the general form

$$\frac{d^2u}{d\xi^2} + (a_u - 2q_u \cos 2\xi)u = 0 \tag{11.12}$$

where $u = x$, y, z and $\xi = \Omega t/2$, leading to the Mathieu parameters, a_z and q_z, which relate the experimental variables to the time-dependent variables t and Ω, as shown in Eqs. (11.13) and (11.14):

$$a_z = -\frac{16eU}{m(r_0^2 + 2z_0^2)\Omega^2} \tag{11.13}$$

$$q_z = \frac{8eV}{m(r_0^2 + 2z_0^2)\Omega^2} \tag{11.14}$$

It is these Mathieu parameters that determine whether the motion (trajectory) of a given ion will be stable or unstable, making them fundamental to ion-trap operation. The solutions to the Mathieu equations can be presented graphically as a series of curves denoting regions of stability and instability of ions in terms of the Mathieu parameters. A particularly important region of the graphical representation of (a_z, q_z) "space" (Fig. 11.11) is where ions are simultaneously stable in both the radial (r) and axial (z) dimensions within the trap. Selection of appropriate DC (U) and RF (V) voltages, angular frequency of the RF voltage (Ω),

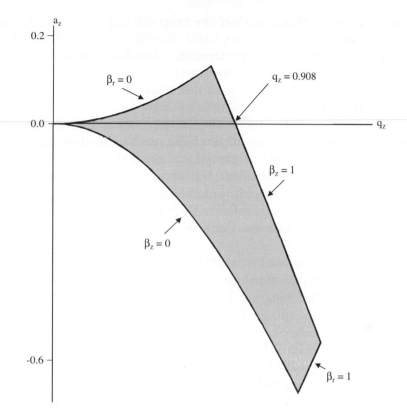

Fig. 11.11. The stability diagram for a 3D quadrupole ion trap. The area enclosed within the boundaries (shaded gray) represents solutions to the Mathieu equations that result in stable trajectories for ions within the trap. The area outside the boundaries represents solutions for the equations that produce unstable trajectories.

and trap dimensions (radius, r, and separation distance between end caps, $2z$) leads to the conditions under which ions can be trapped, isolated (mass selected), excited and mass analyzed (ejected) in a quadrupole ion trap. Increasing the amplitude of the RF voltage while maintaining zero volts DC on the end caps ($a_z = 0$) produces a "mass selective" scan of the trap. Ions are ejected from the device when their $q_z = 0.908$. The m/z ratio of an ion being ejected under these conditions can be calculated using Eq. (11.15)

$$\frac{m}{z} = \frac{8eV}{(r_0^2 + 2z_0^2)q_z\Omega^2} \tag{11.15}$$

351

Because ions are confined within the trap during the experiment, the device overcomes the major intrinsic drawback of quadrupole mass filters (or any "beam"-type instrument)—the loss of a significant portion of the ions created by the ion source. Interestingly, despite being first described in the 1950s, it was not until the early 1980s that important modifications to the original concept (i.e., use of helium buffer gas to collisionally cool the ions, "stretching" of the trap dimensions to eliminate compound-specific mass shifts, and operation in the mass-selective instability mode) produced a device that was commercially viable. Since the introduction of the commercial 3D quadrupole ion trap, the versatility of the instrument has been demonstrated in numerous examples, with particularly important applications coming in the area of peptide and protein analysis [42].

11.4.3.2 *Two-dimensional (linear) quadrupole ion trap*

The newest commercial mass analyzer is the two-dimensional (2D) or linear ion trap [43]. Designed to improve performance and eliminate some shortcomings of the 3D ion trap, this analyzer possesses many of the physical attributes and operational principles of both the quadrupole mass filter and the 3D ion trap. The device bears a strong physical resemblance to a quadrupole mass filter (four parallel electrodes with hyperbolic surfaces) and like the 3D trap, it confines and preserves the majority of ions that enter the device. The 2D device is superior to the 3D trap by virtue of its higher trapping efficiency and increased dynamic range, due principally to a larger internal volume. There are two distinct modes of operation for a 2D ion trap. One mode features mass selective axial ejection of ions, where the trapped ions enter and exit the trap on a path parallel to the electrodes. Ions are injected into one end of the trap from an external ion source and exit the opposite end of the trap when the frequency of an auxiliary RF potential applied to the exit lens matches the radial secular frequency of ions having a particular m/z ratio. The Mathieu equation, shown in Eq. (11.16), describes radial ion motion within the linear trap, but differs from those shown previously for the 3D ion trap.

$$\phi_{xyzt} = (U - V \cos \Omega t)\frac{(x^2 - y^2)}{r_0^2} \qquad (11.16)$$

The forces acting on the ions can be described by Eqs. (11.17–11.19):

$$\frac{d^2x}{dt^2} = -\frac{2e}{mr_0^2}(U - V \cos \Omega t)x \qquad (11.17)$$

$$\frac{d^2y}{dt^2} = \frac{2e}{mr_0^2}(U - V\cos\Omega t)y \qquad\qquad (11.18)$$

$$\frac{d^2z}{dt^2} = 0 \qquad\qquad (11.19)$$

The Mathieu q parameter is given in Eq. (11.20), and the Mathieu a parameter has a value of zero in the absence of a DC potential being applied to the rods.

$$q = \frac{4eV}{mr_0^2\Omega^2} \qquad\qquad (11.20)$$

Studies conducted to characterize the attributes of this axial ejection device have produced results indicating that the theoretical storage capacity of a linear trap is approximately 40 times greater than the typical ($r_0 = 0.707\,\mathrm{cm}$) 3D ion trap. Increasing the storage capacity of the ion trap was an important driver in the evolution of 2D from 3D ion traps as commercially available mass analyzers because of the impact of reduced space charge on performance and capability. The effects of space charge can significantly diminish many aspects of instrument performance including resolving power, mass accuracy, sensitivity and dynamic range. Because of the estimated 20% extraction efficiency of the device, not all of the increase in ion capacity is translated into improved sensitivity for this analyzer, but even an eight-fold sensitivity increase is not insignificant.

An alternative operational mode of a 2D ion-trap mass analyzer features the radial ejection of ions that have entered the trap along the axis of the electrodes. To accommodate radial ion ejection, an exit slot must be cut in each of the two opposing sets of electrodes to allow the ions to leave the trap, and two detectors must be appropriately located. The estimated efficiency with which this type of trap can be cleared of ions during a mass scan is greater than 80%. Together, the scan-out efficiency and the significantly larger ion storage capacity make for an analyzer that is more than an order of magnitude more sensitive that the 3D ion trap. One potential challenge to the radial ejection mode of operation is the need for exceptionally high mechanical tolerances in the fabrication and construction of the device. Relatively minor deviations from parallel (greater than $20\,\mu\mathrm{m}$) in the alignment of the electrodes can result in reduced resolving power for the device. While neither of the 2D ion traps (axial ejection or radial ejection) is capable of the accuracy or precision required for exact mass measurements, at reduced scan rates the resolving power can be increased sufficiently to provide the separating power necessary to distinguish isotopic patterns of multiply charged ions, thereby permitting determination of charge state.

11.4.3.3 Ion cyclotron resonance analyzer (Penning ion trap)

The two ion trapping devices described previously rely solely on voltages (RF and DC) to trap, confine, excite and eject ions. A Penning ion trap employs a static magnetic field in addition to DC and RF voltages for the trapping, manipulation and detection of ions. Ions within the magnetic field experience the Lorentz force, causing those having a particular m/z ratio to move with a cyclotron motion at a specific frequency (cyclotron frequency). Together, all ions (of various m/z ratios) within the magnetic field adopt the circular motion in a plane that is perpendicular to the magnetic field, with each m/z ion population possessing a characteristic cyclotron frequency as given by Eq. (11.21):

$$\omega_{\text{cyclotron}} = \frac{v}{r} = \frac{zeB}{m} \tag{11.21}$$

where v is the radial velocity, r the radius of the ion orbit, B the magnetic field strength, e the electronic charge, m the mass of the ion and z the number of charges. Since the frequency for a given ion population in a fixed magnetic field is constant, any increase in the velocity of ions having a given m/z ratio must be accompanied by an increase in the radius of their orbit. Ions that are trapped in such a magnetic field will absorb energy from an RF electric field provided the frequency of the RF voltage matches the cyclotron frequency. The Penning trap employed in many ICR mass analyzers is a cubic arrangement (Fig. 11.12) of electrically isolated electrodes or plates, being commonly referred to as a "cell" (generally measuring a few centimeters on a side). The cell is located within the field of a superconducting magnet with typical field strengths on the order of 7 or 9.4 T (tesla). In practice, the excitation RF voltage is applied across opposing electrodes (plates) of the ICR cell, causing the orbital radius of a particular population of ions to increase. Because there are generally many ions of a given m/z ratio present, the entire population absorbs the RF energy and attains the new orbit, leaving all ions of all other masses unaffected. The population of ions that absorbs the energy achieves a coherent motion, meaning all of the ions are moving together and all are in phase with the RF electric field. As the orbital radius of any ion population increases, they eventually come into proximity with the walls or plates of the ICR cell.

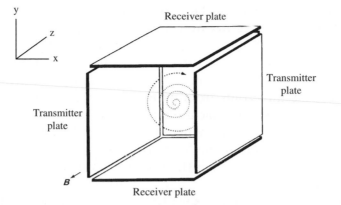

Fig. 11.12. Diagram showing the classic cubic ICR cell. Ions are injected into the cell from an external ion source through an aperture in the front trapping plate (parallel to the applied magnetic field along the z-axis; the front trapping plate has been omitted from the diagram for clarity). Excitation of the ions (either for detection or as a prelude to CID) is accomplished through the application of a wide-band frequency pulse (sweep) to the transmitter plates. As the ions of each m/z value attain coherent motion and an orbital path of increasing diameter (as shown in the diagram), they come into proximity with the receiver plates and induce a capacitive current (image current). This signal is recorded and converted to the familiar m/z vs. intensity representation of a mass spectrum (Fourier transform).

The circular motion of the ions induces a capacitive "image" current that possesses the same frequency as the orbiting ions (the cyclotron frequency), which can be measured with high accuracy and precision. As shown in Eq. (11.21), the frequency of the image current is inversely proportional to the m/z ratio of the ions in the orbiting population, and the magnitude of the current is proportional to the number of ions (charges) in the orbiting ion packet. Since the coherent nature of these ion packets begins to dissipate exponentially as soon as the RF field is removed, the image current also decays away in an exponential fashion over the course of time (fractions of a second to a few seconds). The rate of decay depends on the pressure within the ICR cell, since collisions with residual gas molecules are a primary mechanism by which the excitation energy is lost. The duration of the decay period is inversely proportional to the pressure within the instrument. The time-domain free-induction decay (FID) signal (signal intensity vs. time) is converted to a frequency-domain spectrum (signal intensity vs. frequency) through the application of a Fourier transform. Since the frequencies can be assigned to specific m/z ratios by using the cyclotron equation (i.e., the frequency axis can be converted to a m/z axis), the Fourier transform of the FID results in a mass spectrum.

The ICR mass analyzer, like other ion traps, originally analyzed ions that were formed within the trap using either EI or CI processes. As other ionization methods became important and prominent, means of introducing ions into traps from external sources were developed and today that represents the typical scenario. Ions are pulsed into the trap and retained within the cell by potentials on the order of 1–5 V applied to the plates. Detection of the ions (i.e., generation of the mass spectrum) is accomplished through the application of a brief RF pulse that increases in frequency (linearly). This frequency "sweep" causes all ions of all m/z ratios in the cell to absorb RF energy, increase in orbital radius and produce image currents. The recorded current (the FID) is comprised of all frequencies and all amplitudes, and cannot be interpreted manually. Only after the Fourier transformation is the mass spectrum revealed. The resolving power of the ICR analyzer is a function of the precision and accuracy with which the cyclotron frequencies of the ions can be measured. Typical values range from 50,000 to 500,000 ($m/\Delta m$) with values of greater than 1,000,000 having been reported. This makes the ICR analyzer (also commonly referred to as a Fourier transform mass spectrometer or FTMS [44]) the highest resolution mass analyzer currently available. The implications of ultra-high resolving power, measurement accuracy and measurement precision (i.e., low to subppm) manifest themselves in the direct analysis of complex mixtures (i.e., the ability to separate chemical entities solely by m/z without chromatography), the characterization of large biomolecules such as proteins (i.e., the recognition and identification of fragment ions with very high charge states) [45] and the unequivocal determination of elemental formulae. However, this exceptional capability comes at a price. The cost of FTMS instruments can be substantially higher than that of those based on other types of mass analyzers.

11.4.4 Time-of-flight mass analyzer

Perhaps the simplest mass analyzer of all, the TOF mass spectrometer [46] has experienced a reemergence in the past several years. Like the 3D quadrupole ion trap, the TOF analyzer has come to commercial prominence several decades after its initial introduction. The limitations of electronic components in the 1960s constrained the capabilities of the instrument, limiting its mass range and resolving power. The TOF analyzer operates in a pulsed mode, requiring either a pulsed ion source like a laser (e.g., MALDI) or a "continuous" ion source (e.g., ESI) that has the stream of ions modulated by gating circuitry (repelling grid) or a repeller (pusher). Modulation or "gating" of the ion beam is not required for laser sources that emit their radiation at a specific repetition rate (discounting certain infrared lasers that operate in a continuous mode). The resulting beam of ions is not continuous as in some other instruments discussed thus far, but rather is comprised of discrete "packets" of ions. All of the ions present in the source at the moment the potential on the repeller (pusher) is applied are instantly accelerated toward the detector. The duration of the accelerating potential "pulse" is typically very short (approximately 33 μs for an instrument that operates with a pusher repetition rate of 30 kHz). Since ions of various masses (m/z ratios) are present in the ion source, the composition of each ion packet will be representative of the ion source composition at the instant of sampling. All of the ions experience the same accelerating potential and consequently have the same kinetic energy, but different velocities since KE $= \frac{1}{2}mv^2$. If the ions are permitted to travel through an evacuated region of fixed length (to minimize the probability of neutralizing or scattering collisions), in the absence of any electric or magnetic fields they will begin to separate according to their masses, as lighter ions will have higher velocities than heavier ions. A flight path of fixed length and differing velocities results in flight times (velocity = distance/time) for the ions that are proportional to the square root of their masses (m/z ratios), as shown in Eq. (11.22):

$$t = \left(\frac{m}{2eV}\right)^{1/2} D \qquad (11.22)$$

where t is the flight time, m the mass of the ion, e the electronic charge, V the accelerating potential and D the length of the drift tube. For example, singly charged ions with the m/z ratios of 1000 and 1001, respectively, will have flight times of 16.0969 and 16.1049 μs (differing by only 8 ns) assuming a 1 m length for the flight path and an accelerating potential of 20 kV. More impressive yet is the ability of these instruments to distinguish flight times of ions that are separated by fractions of an m/z unit. When operating at "high" resolving power (RP = 10,000), a TOF instrument is distinguishing and recording flight times for sequentially arriving ions that differ by less than 0.6 ns (assuming two isobaric ions with m/z values of 500.0000 and 500.0500, respectively, a 1 m flight path and 20 kV accelerating potential). The high-speed electronics required for this task only became available within the past 10–15 years, and have been the driving force behind the reemergence and widespread use of a mass analyzer viewed as having only limited capabilities when it was first introduced four decades ago. In reality, simple linear TOF analyzers are not capable of resolving powers of the magnitude described in the previous example. In order to achieve high resolving powers and to make *m/z* measurements with high accuracy and precision, the TOF analyzer requires ions that have a very narrow kinetic energy spread, in order to minimize the uncertainty in the recorded flight times. This reduction in the kinetic energy spread (energy focusing) of the ion packets is achieved through retarding and then reversing the velocities of the ions (decelerating and re-accelerating) within the electric field of a reflectron (Fig. 11.13) or ion mirror [47]. This process preserves most of the ions and therefore is considerably different from that employed in double-focusing magnetic sector instruments, where the ESA deflects and eliminates ions of non-optimal kinetic energy from the ion beam, diminishing the signal intensity. It is the focusing effect of the reflectron and the absence of slits or apertures in TOF analyzers that produces the exquisite sensitivity for which the instruments are known. The other notable feature of

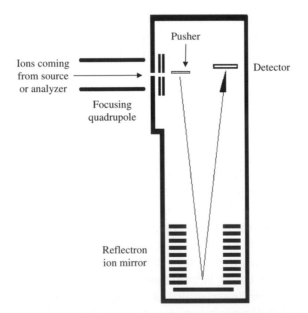

Fig. 11.13. Diagram of a TOF mass analyzer (with reflectron). Ions enter from an external source and are accelerated (orthogonally) by the pusher electrode toward the reflectron. The reflectron (ion mirror) retards, reverses and reaccelerates the ions back toward the micro-channel plate detector.

TOF analyzers is the theoretically infinite m/z range. Whereas most of the other analyzers discussed in this section operate over very finite ranges of m/z values, due to the limited range of electrical potentials, magnetic field strengths and radio frequencies that can be practically employed in these instruments; the TOF analyzer has few such limitations. That being said, ever increasing flight times of larger and larger (more massive) ions require lower sampling rates. Some commercially available instruments based on TOF analyzers have a maximum m/z range on the order of 100,000–150,000.

11.5   DETECTORS

The first and simplest detector used in mass spectrometry was the Faraday cup. This device is comprised of an electrode that collects incoming ions, and in so doing generates a current as electrons flow from ground to the electrode surface to neutralize the positive ions. The voltage drop across a resistor through which the current flows is amplified and recorded as the ion signal. Current generated by the device is proportional to the number of charges (ions) striking the electrode. Despite being relatively unsophisticated, the Faraday cup produces a response that is not dependent on any characteristic of the ions striking it, such as their kinetic energy or mass. They are rugged, reliable, inexpensive, exhibit very little electronic noise and demonstrate constant sensitivity. Their major liability is slow response, generally due to delays in the amplification circuitry. This characteristic limits their utility in situations where signal intensity fluctuates rapidly (such as chromatographic separations).

The most widely used mass spectrometric detector is the electron multiplier. This device affects a first stage of signal amplification by exploiting secondary-electron emission. Ions impact upon a conversion dynode, a surface comprised of a material having a high work function (i.e., easily liberates electrons when struck by high-energy particles) that is held at a high electrical potential. As the ions approach the dynode, they are accelerated into the surface impacting with energies of several thousand electron volts. These impact events cause the dynode to emit electrons (referred to as secondary electrons) in numbers that are proportional to the number of impacting ions. If the electron multiplier is of the conventional or discrete dynode type, the secondary electrons liberated from the first dynode are focused and accelerated toward a second dynode element. Each of the first-generation secondary electrons strikes the surface of the second dynode and

D.J. Burinsky

liberates a number of second-generation electrons. The stream of second-generation electrons is subsequently accelerated toward and impacts the third dynode element of the device with each of these electrons causing the emission of several third-generation secondary electrons. This process is repeated for every stage of the device (which is typically between 10 and 20) until the "amplified" current produced by the cascade of electrons is collected at the anode of the electron multiplier and passed on for additional amplification within the detection circuitry of the instrument. The dynodes are electrically isolated from one another by a chain of resistors that produce the potential difference between each successive element, resulting in the acceleration and focusing of the electrons.

In many applications, discrete dynode electron multipliers have been replaced by a less costly continuous dynode design. These conical-shaped devices (Fig. 11.14) are fabricated from resistive glass (doped

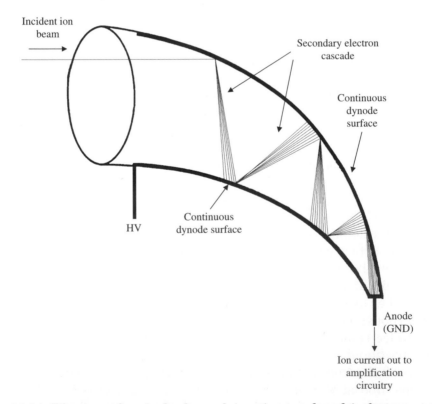

Fig. 11.14. Diagram of a single-channel (continuous dynode) electron multiplier. Gains of 10^5 or 10^6 are achievable with modern electron multipliers.

with lead), which enables them to maintain a potential gradient in a similar fashion to their discrete dynode counterparts. Ions enter the open end of the multiplier striking the surface and liberating secondary electrons. The secondary electrons are attracted back toward the multiplier surface as a consequence of the potential gradient, resulting in a second burst of electrons from the surface. This sequence is repeated several more times until the electrons emitted from the final electron-surface collision reach the anode. The curved shape of the continuous dynode multiplier (referred to as a channel electron multipliers, CEM, or the trade name Channeltron®) and its positioning with respect to the angle of the incident ion beam prevents undesirable deep penetration of the positive ions (resulting in a reduction of gain) and spurious signals from positive ions or secondary ionization of residual gas molecules. Both types of electron multipliers exhibit a gain in the range of 10^5–10^6 (i.e., approximately 1 million electrons are collected at the anode for each ion that strikes the front of the multiplier). In addition to being less costly, the channel electron multipliers are somewhat more forgiving if exposed to atmosphere while at high voltage. Despite being an indispensable component of modern mass spectrometers, electron multipliers are known to exhibit varying degrees of response discrimination depending on the attributes (velocity, mass or charge) of the incident ions, and they are subject to overloading or saturation if they experience excessive (greater than 10 nA) currents.

A variation of the single-channel electron multiplier (continuous dynode) configuration is used in conjunction with TOF mass analyzers. The multiple- or micro-channel electron multiplier array (also referred to as micro-channel plate or MCP detector) consists of a large collection of glass capillaries (10–25 μm inner diameter) that are coated on the inside with an electron-emissive material. Like the single-channel version, the capillaries are biased at a high voltage so that when an ion strikes the inside wall it creates an avalanche of secondary electrons. The cascading effect of consecutive electron liberation events creates a gain of approximately 10^3–10^4 and produces a current pulse output. The densely packed honeycomb arrangement of channels (typically hundreds per square inch) was designed specifically for 2D imaging applications. The use of the device in TOF instruments takes advantage of its flat surface, which can be positioned perpendicular to the incident ion beam—this being necessary to preserve the velocity resolution of the ions (and thus mass resolving power and the ability to obtain exact mass measurements).

11.6   COMPUTERS

Today, it is a foregone conclusion that mass spectrometers, whether commercially produced or constructed in research laboratories, are computerized. In fact, the computerization of mass spectrometry over the past two decades has increased proportionally with the capability of computers and inversely with their size. It is now common for a mass spectrometer to have not one but several computers associated with it. In addition to the main computer (typically a desktop personal computer, PC) that executes the overall instrument control, data acquisition and data-processing/display activities, there will also be several the so-called "on-board" microcomputers. These microprocessors are embedded within the instrument electronics chassis and are specifically dedicated to important sub-processes of either instrument control or data acquisition. The high processing speeds with which computers execute instructions provide today's mass spectrometers with the capability to acquire and store exceptionally large, complex data sets. Perhaps most importantly, the computing power of these processors and the sophistication of the software are both now great enough that changes can be made to various experimental parameters during the course of an experiment without operator intervention. Such semi- or fully automated data acquisition scenarios that are driven in real time by actual experimental results can lead to substantial gains in sample analysis throughput, productivity and the quality of information obtained. The main PC provides the means by which the operator interacts with the instrument, through the keyboard and some graphical user interface. The various experimental parameters are entered and monitored through the user interface. Once the parameters are entered and stored, the experiment is initiated and the PC receives and stores the digital data stream from the instrument. There are a number of transformations that occur in the interval between the digitization of the current generated by the instrument detector and the final display of the data in a human interpretable form (i.e., as a chromatogram or mass spectrum). The raw signal coming from many mass spectrometers (the ICR spectrometer being a notable exception) is recorded as a series of x, y pairs containing time-intensity information. The time-intensity data is converted to mass-intensity data (the most common format in which mass spectral data are displayed) through correlation with a calibration spectrum that contains signals representing ions of known mass (m/z). The intensities of all signals in the mass spectrum are normalized, with the most intense signal being assigned a value of 100% relative abundance. In addition to the most common display format of the intensity (relative abundance) vs. m/z histogram, the computer also readily displays the mass spectral data in tabular form. The use of mass spectrometers in combination with various separation methods (e.g., GC, liquid chromatography, capillary electrophoresis, etc.) requires data display formats similar to those used by other types of detectors. One of the most common formats displays the summed total intensity (or some portion of the intensity according to one or more selected m/z values) of the mass spectral signal as a function of time producing a recognizable "chromatogram." However, it is important to realize that such a data set is 3D, with the x and y axes containing the intensity and time information (as in every chromatogram), and the z-axis containing either a partial or complete mass spectrum.

The other major role of computerization in mass spectrometry is library [48] or database searching [49]. Once spectral information is acquired and stored, it can be compared with large arrays of previously acquired spectra for the purpose of compound identification. The original application of spectral database searching was in association with EI mass spectra. Because of standardized experimental conditions (i.e., 70 eV ionizing energy), spectral reproducibility was sufficiently high that the mass spectrum of a compound acquired in virtually any laboratory could be matched with the mass spectrum of the same compound acquired in any other laboratory. Therefore, once the identity of a compound had been established unequivocally (employing multiple analytical techniques) and its mass spectrum recorded in a spectral library (database), the need for subsequent (and repetitive) reinterpretation of the same mass spectrum was eliminated. When that (or any other) mass spectrum is next encountered in any laboratory, a computer-aided search of the mass spectral database quickly establishes a match between the two mass spectra, leading to extremely rapid compound recognition. The second and more recent application of computerized database searching has been in the area of protein identification. Because of the very regular and predictable fragmentation patterns displayed by peptides upon collision-activation, the primary amino acid sequence of moderately sized peptides can be determined routinely. For mixtures of peptides generated through enzymatic digestion of proteins, this sequence information can be searched against large protein databases. The resulting ranked list of possible proteins that contain the peptide of interest can be used to quickly confirm the identity of a particular protein or narrow the list of possibilities.

11.7   SEPARATION TECHNIQUES

As alluded to in Section 11.1, some of the most important developments in the field of mass spectrometry have been the result not of the invention or design of a new mass analyzer or ion source, but rather the integration of mass spectrometry with other important analytical techniques. The most notable of these "marriages" has been between mass spectrometry and important separation techniques such as gas chromatography (GC/MS) and liquid chromatography (LC/MS) (see Chapter 19). These combinations are intuitively logical since together the techniques can address some of the shortcomings experienced by each when used individually. Mass spectrometry is a technique capable of producing ions from a wide variety of analytes, including mixtures of compounds (or specific compounds present in the midst of complicated sample matrices). However, such spectra exhibit a complex appearance that reflects the nature of the sample and makes spectral interpretation (for either qualitative or quantitative purposes) difficult at best. Chromatographic methods have historically used simple, inexpensive and robust non-specific detectors (e.g., thermal conductivity, flame ionization, refractive index, evaporative light scattering, UV-visible spectrophotometer, etc.). So, while the elution of analytes from the separation medium could be visualized and the quantity of analyte determined through an integration of the recorded peak area, there was no immediate or definitive means of recognizing or identifying the separated compounds. Joining chromatographic separations together with mass spectrometric detection was a logical solution to some of these challenges, but required some fundamental barriers to be overcome before becoming a reality. In the sections that follow, specific details about the integration of these two widely practiced separations methodologies with mass spectrometry will be discussed. A comprehensive discussion covering the coupling of mass spectrometry with other separation techniques (e.g., supercritical fluid chromatography [50], capillary electrophoresis [51], etc.) is beyond the scope of this account.

11.7.1   Gas chromatography

The use of a mass spectrometer as a detector for a gas chromatograph overcomes some of the shortcomings encountered when employing either of the techniques individually. The superimposition of information caused by ionization of multiple compounds can produce complicated and uninterpretable mass spectra, particularly when using EI.

Conversely, gas chromatographic analysis of a complex mixture using one of the many available non-specific detectors (i.e., flame ionization or thermal conductivity) produces a trace that lacks information about the identity of the eluting compounds. Even under the best of circumstances, correlation of retention times between known standards and unknown analytes can be less than reliable. However, when GC and mass spectrometry are used together they can provide both qualitative and quantitative information about numerous components of a complex mixture that may be present at levels of a picomole or less per analyte. The union of GC and mass spectrometry [52] was facilitated by the commonality shared by the techniques. Vaporization of analytes in the injector port, as a prelude to partitioning interactions between the helium carrier gas and the stationary phase material of the GC column, also provides the appropriate and requisite phase transition (to the gas phase) required for either EI or classical CI. However, the relatively large volumes of helium carrier gas within which the analytes are entrained posed challenges for mass spectrometer vacuum systems, particularly in the early days of the techniques when large diameter, packed GC columns having flow rates of 10+ ml/min were the norm. The difference between the optimal operating pressure for most mass spectrometers (10^{-3}–10^{-4} Pa, or 10^{-5}–10^{-6} torr) and the terminus of a GC column (10^5 Pa or 760 torr) leads to an inevitable increase in pressure inside the mass spectrometer—with an associated loss of performance due to scattering and neutralization. Solutions to the pressure differential problem included the jet separator and the membrane separator. These devices affected a reduction in the amount of helium entering the mass spectrometer ion source by taking advantage of (i) the high diffusivity of carrier gas with respect to the gaseous analytes or (ii) the preferential permeability of silicone elastomers to organic compounds. As illustrated in Fig. 11.15, in the jet separator the GC effluent expands through a nozzle and is directed at an adjacent orifice. The "free jet" expansion of the effluent stream results in much of the helium being removed by a vacuum pump and most of the sample proceeding to the ion source. An alternative device, the membrane separator (also shown in Fig. 11.15) is comprised of a thin silicone membrane that serves as the interface between the atmospheric pressure of the GC and the vacuum of the mass spectrometer ion source. As the GC eluent passes across the membrane the organic analytes permeate through it and into the vacuum system of the mass spectrometer, while the inert helium carrier gas does not. With the advent of improved differential vacuum pumping and capillary GC columns, the

D.J. Burinsky

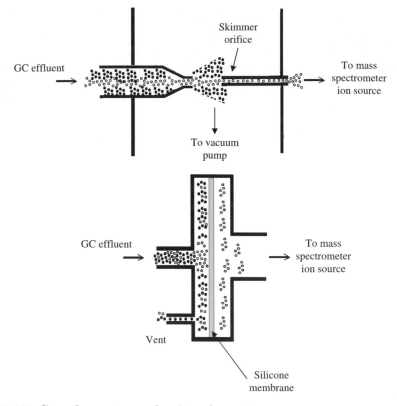

Fig. 11.15. Gas chromatography interfaces (jet separator, top; membrane separator, bottom). In the jet separator, momentum of the heavier analyte molecules causes them to be sampled preferentially by the sampling orifice with respect to the helium carrier gas molecules (which diffuse away at a much higher rate). In the membrane separator, the analyte molecules are more soluble in the silicone membrane material leading to preferential permeability. Helium does not permeate the membrane with the same efficiency and is vented away.

need for such "interfaces" between the GC column and the mass spectrometer has been eliminated. Today's capillary GC columns (0.10 mm, 0.25 mm or 0.32 mm i.d.) operate at helium flow rates that enable them to be inserted directly into the mass spectrometer ion source without compromising performance. Placement of the terminus of the capillary GC column directly in the ion source does however have a minor impact on chromatographic performance. The presence of vacuum (rather than atmospheric pressure) can degrade chromatographic resolution and peak shape. This deleterious effect is mitigated to some degree by modern GC instrumentation equipped having electronic pressure control capability that maintains constant carrier gas flow despite changing conditions such as column temperature (or pressure). In addition to the impact on chromatography performance, the rapidly changing sample concentration within the mass spectrometer ion source (as a consequence of analyte elution from the column) produces mass spectra that have distorted relative peak intensities, referred to as spectral "skewing." For a capillary GC separation, typical peak widths are on the order of 10 s or less. Depending on the rate at which the mass spectrometer scans across the mass range of interest, the sample concentration will not only be different from one mass spectrum to the next, but it will also be different between the low mass end and the high mass end of a spectrum. Such spectral distortion can have extremely negative consequences for qualitative analyses where spectral library searching is important for compound recognition or identification. Obviously, library matching is only effective if the spectra of unknown compounds correlate closely with those that populate the reference library. The mass spectrometer scan rate also has implications for quantitative analyses since an insufficient number of data points (mass spectra) obtained during the elution profile of an analyte reduces both the accuracy and precision of the quantitative measurement, since it negatively impacts integration of peak area. Fortunately for modern-day practitioners, issues associated with mass spectrometer data acquisition or scan rates are becoming less of an issue as instruments become increasingly capable. Many current instruments scan or acquire data across the desired mass range at rates substantially less than 1 ms per m/z unit.

Two particular issues that require attention when conducting GC/MS experiments (that are of little or no consequence when using the techniques individually) are the impact of reactive surfaces and the contribution of the GC column stationary phase material to the mass spectral background. The physical transfer of the vapor-phase analytes from the GC column to the mass spectrometer ion source requires that all surfaces with which the organic compounds come into contact are both chemically inert and maintained at a temperature that prevents condensation. Obviously, if a vapor-phase analyte comes into contact with a "cold spot" anywhere along its path between the GC oven and the ion source condensation can occur, thereby reducing the amount of that compound that reaches the ion source. To prevent such losses of sample, the transfer region between the GC and the mass spectrometer is typically maintained at high temperatures (250–300°C). An unintended outcome of exposure to such high temperatures for some classes of organic compounds (e.g., steroids and saccharides) is thermal decomposition. Ensuring that the surfaces with which compounds come into contact are chemically inert can minimize this decomposition. Stainless steel and silanized glass both perform well with respect to their general lack of reactivity. The issue of surface reactivity is diminished substantially when the GC column is inserted directly into the ion source, since the silane-based stationary phase coating of capillary columns provides reasonable protection against thermal decomposition. Direct threading of the capillary GC column into the ion source of the mass spectrometer provides for the most efficient transfer of sample from the GC to the MS; however, this means of direct interfacing can be somewhat tedious when it comes to changing or replacing the GC column. A common and convenient practice employed by many practitioners is the use of a zero-dead volume connector within the GC oven between the analytical GC column and an appropriate length of silanized fused silica capillary. The fused silica "transfer line" is then threaded through a heated transfer tube and into the ion source *in lieu* of the column itself.

GC column stationary phase coatings have undergone substantial developments and improvements over the past decade. Many capillary columns are now available with bonded or cross-linked stationary phase coatings that substantially reduce their volatility and with it the amount of stationary phase that is volatilized into the ion source. Even so, prolonged periods of time at the temperatures typically encountered in GC analyses (often greater than 250°C) lead to a near constant volatilization of some of the stationary phase coating (often referred to as column "bleed"). Consequently, every mass spectrum will contain ions attributable to the volatile components (or their thermolysis products) of the column phase (provided that the compounds ionize under the conditions of the experiment). The presence of these "background" ions can complicate the interpretation of mass spectra obtained for low-level analytes since the signal intensities of the interfering ions may equal or surpass those of the analyte. Fortunately, the software applications that acquire and process data for the current generation of instruments all possess the ability to "background subtract" and produce a processed spectrum that is devoid of many (if not all) of the background ions. In addition to the impact of the stationary phase bleed on the appearance and quality of mass spectra, the continual exposure (contamination) of the ion source and associated electrostatic elements diminishes the performance of the instrument causing poor peak shape and reduced sensitivity.

11.7.2 Liquid chromatography

The proliferation of mass spectrometric detectors in the past several years has sparked considerable commentary regarding the concept of a "universal" detector for HPLC separations. The fact that modern mass spectrometers equipped with atmospheric ionization sources (i.e., ESI or APCI) exhibit exquisite sensitivity for many classes of compounds that are also amenable to separation by HPLC is indisputable. However, to state that a mass spectrometer (or any other detector) is capable of detecting all molecules that can be eluted from an HPLC column seems to be an oversimplification. Certainly, when using mass spectrometric detection in tandem with the ubiquitous UV-visible spectrophotometric detector there often can be an empirical correlation between the respective outputs of the two detectors (i.e., each produces a response upon the elution of an analyte from the HPLC column). What must not be overlooked, however, is the highly selective nature of many analytical detectors. The response generated by a molecule passing through the flow cell of a typical HPLC/UV-visible detector is related directly to the ease with which electronic excitation can be induced by photons of a particular energy regime (190–600 nm). In contrast, detectability in the mass spectrometer relies on the intrinsic molecular properties of solution- or gas-phase basicity (or acidity)—the primary attribute(s) responsible for ionization (protonation or deprotonation) in LC/MS applications. Practitioners of HPLC are probably familiar with many examples of compounds that are undetectable when using the popular UV-visible detector because they lack an appropriate chromophore. This phenomenon of detector "transparency" to a particular class of organic compounds is no less problematic for the mass spectrometric detector when compounds lacking an "ionophore" (i.e., a suitably basic or acid functional group) are being analyzed. For example, an equimolar binary mixture of tributylamine and stilbene (1,2-diphenylethylene) would produce vastly different chromatograms depending on the detector being employed. The UV-visible detector (monitoring at 295 nm) would produce a strong response for the stilbene (due to its high molar absorptivity, $\varepsilon = 25,000$), but little or no signal for the alkylamine, since it lacks the requisite electronic "structure" (e.g., no $\pi \rightarrow \pi^*$ transitions). Conversely, the very high basicity of the alkylamine would favor protonation (formation of an $[M+H]^+$ ion), thereby producing an intense mass spectrometric response. Stilbene, on the other hand, would be invisible to the mass spectrometer since it lacks any sufficiently basic (or acid) functional group to undergo

369

ionization. While this trivial example is contrived, it should not be difficult to imagine that a reasonably complex mixture (real-world analytical sample) could easily contain numerous compounds having a wide diversity of chemical structures and properties. Some of the compounds may be very strong organic acids (carboxylic or sulfonic acids) or bases (primary, secondary or tertiary amines), some may be highly conjugated hydrocarbons (carotene or polycyclic aromatics) and others may be electroactive in nature (hydroquinone and catecholamine). It seems improbable that any currently available HPLC detector would be capable of "universal" detectability for all of the compounds in such a sample. Nevertheless, the importance of the mass spectrometric detector should not be minimized since it has brought many of the same advantages to HPLC as it did to GC several decades earlier, particularly for compounds of biological interest and relevance.

The joining of mass spectrometry and liquid chromatography proved to be intrinsically more difficult than the union with GC. This was due to the challenges associated with extremely rapid vaporization of significant quantities of solvent (HPLC mobile phase) and the associated gas load that would be imposed on the high vacuum pumping system of the mass spectrometer. The dynamic linkage of HPLC and mass spectrometry needed to achieve continuous operation has taken many forms over the past two decades, including direct liquid introduction (DLI), moving belt, thermospray (TSP), particle beam (PB), continuous flow fast-atom bombardment (CF-FAB), atmospheric pressure chemical ionization (APCI), and electrospray ionization (ESI). The last two of these methods have been described previously (see Section 11.3) and constitute the current "state-of-the-art" of interfacing HPLC with mass spectrometry. By virtue of the explosive proliferation of mass spectrometers in the past few years as detectors for HPLC, these two techniques are also perhaps the most prevalent interfaces, having largely supplanted all of their predecessors. One of the key factors responsible for the widespread use of the atmospheric pressure ionization (API) techniques is the comparative ease with which the bulk of the volatilized HPLC solvent can be removed (separated from the analyte) if this operation is conducted outside of the mass spectrometer vacuum chamber. Many previous attempts at interfacing HPLC with mass spectrometry had attempted to conduct the evaporation and removal of the solvent in the high vacuum environment of the mass spectrometer ion source, with varying degrees of success. The breakthrough realization that propelled the API techniques to the forefront was that ions could be efficiently formed at atmospheric pressure and then introduced into the mass spectrometer ion source using electrostatic manipulation, thereby leaving the vast majority of the vaporized HPLC eluent (neutral molecules) outside of the mass spectrometer. In the sections that follow are brief descriptions of some of the precursors of today's most popular and prevalent HPLC interfaces—ESI and APCI. While most of these techniques and approaches are no longer in use, they provide an interesting historical perspective on the evolution of scientific instrumentation in response to an important area of research.

11.7.2.1 Direct liquid introduction

Direct liquid introduction (DLI) [53] was one of the earliest attempts at bridging the gap between mass spectrometry and HPLC. Simple in principle, the technique employed a small diameter orifice (or capillary) through which a portion of the HPLC effluent passed as it entered the mass spectrometer source. Since the vaporization of the mobile phase produced a large gas load for the mass spectrometer pumping system, only a small amount of condensed liquid could be introduced into the mass spectrometer—by either splitting the HPLC eluent flow (perhaps 100:1) or utilizing low-flow HPLC techniques such as micro-HPLC (flow rates of approximately $10 \, \mu l/min$). The passage of the eluent stream through the orifice formed a stream of tiny droplets that were to be desolvated by the heat and vacuum of the mass spectrometer ion source. Once the solvent had been eliminated (removed by the vacuum pumps), the analyte molecules were then ionized by classical chemical ionization processes with the background pressure of the vaporized HPLC mobile phase serving a reagent gas. Subsequent developments included the introduction of a jet interface that further reduced the gas load within the ion source making possible the acquisition of EI mass spectra. Limitations of the various iterations of the DLI interface were the necessity for low flow rates (limited the amount of analyte that could be sampled, which translated into limitations of detection limits) and the tendency for the small diameter orifice (or capillary) to become clogged.

11.7.2.2 Moving belt

Perhaps the most mechanically complex solution ever developed for uniting HPLC with mass spectrometry was the moving belt interface [54]. The heart of this system was a mechanically driven continuous belt (analogous to an escalator or moving walkway) to which the HPLC eluent was applied. The majority of the mobile phase was evaporated by a heat source (ideally hot enough to vaporize the solvents but not to induce vaporization or pyrolysis of analytes). After passing the heat source, the belt (carrying the analytes) then moved through one or more vacuum locks (comprised of seals that allowed the moving belt to pass, but retained a pressure differential by means of one or more stages of pumping) as it traveled from atmospheric pressure to the high vacuum environment of the mass spectrometer source. Once inside the vacuum envelope of the instrument, the belt (and analytes) next encountered a high-capacity heating device capable of very rapidly volatilizing (flash vaporization) the analytes in the ion source. The vapor-phase analyte molecules could then be ionized using either EI or conventional CI just as if they had been delivered to the ion source by a direct insertion probe or a GC column. After liberation of the analytes, the belt continued along its path and was next heated to a high temperature in order to remove any residual organic compounds making the surface ready to receive the next deposition of solvent and analyte. Cleaning of the wire (making it continually reusable) helped to reduce or eliminate cross-contamination between various analytes applied to the belt over time. The belt passed through another set of vacuum lock(s), moving out of the mass spectrometer, through some physical wipers to remove the remainder of any residue, and eventually back to the position where the HPLC eluent was being continually applied to the surface. These interfaces were reasonably efficient in transferring analyte from the HPLC column to the mass spectrometer (estimated to be approximately 50% in some instances) and maintained chromatographic integrity. The belt transported material into the mass spectrometer at a linear velocity of approximately 2–3 cm/s. The range of molecules that were amenable to analysis using this device was similar to those of other devices that required thermal energy to convert solid or liquid analytes to vapors that could undergo either EI or CI. An additional (and considerable) limitation of these devices was their mechanical complexity, which made them challenging to operate and maintain.

11.7.2.3 Particle beam

The particle beam interface [55] borrowed and built upon some of the key elements and concepts of its predecessors. Eluent from the HPLC was nebulized into a spray of small droplets by a flow of helium. The spray of droplets entered a heated chamber where evaporative processes further reduced the droplet size creating an aerosol. The next step of the process involved the spraying of the aerosol (i.e., the mixture of analyte particles, residual solvent molecules and helium) through an orifice into an evacuated chamber. The vacuum chamber contained a device that very much resembled the jet separator discussed previously in Section 11.7.1. Although differing in details and materials of construction, the momentum separator of the particle beam interface operated on the very same principles as the GC jet separator—the heavier analyte particles would possess sufficient momentum as a result of the free-jet expansion to successfully traverse the distance to the sampling cone leading to the source of the mass spectrometer. Conversely, lighter solvent molecules and helium atoms would have their flight paths altered (and thus not be collected by the sampling cone) by the turbulent flow within the momentum separator, and be pumped away. Once inside the source of the mass spectrometer, the analyte particles impacted a hot surface that served to volatilize the compounds so that they could be ionized using either EI or CI. The particle beam interface demonstrated varying levels of performance with different classes of analyte molecules. Compound volatility was cited as an important factor in examples where thermal decomposition was observed in particle beam mass spectra or when quantitative responses were not linear with increasing or decreasing analyte concentration.

11.7.2.4 Thermospray

Unlike the HPLC/MS interfaces mentioned thus far, the thermospray (TSP) interface [56] was the first not to require vaporization of the analyte molecules as a prelude to ionization. Thermospray ionization was a true desorption ionization technique in that ions formed in solution were desolvated in a heated region that was evacuated by a rough pump. Ions were formed in solution and released into the gas phase through evaporation (assisted by heating of the mobile phase as it was sprayed) of the solvent within which they were entrained. Once liberated from solution, the ions were sampled orthogonally by a sampling cone and passed into the mass spectrometer ion source. The residual solvent molecules passed by the sampling orifice and were removed from the vacuum system by a rotary pump working together with a cold trap (using dry ice or liquid nitrogen). The latent heat of vaporization dissipated most of the thermal energy that was applied to the HPLC eluent to desolvate the analyte ions. Consequently, thermally fragile compounds could be analyzed successfully with little or no pyrolysis products being observed in their mass spectra.

The orthogonal design of the interface (sampling ions into the mass spectrometer at 90° to the primary path of the spray) allowed HPLC flow rates approaching 2 ml/min to be sampled effectively, even when the aqueous composition of the mobile phase was high. The lack of any external means of ionization (such as the filament used in EI or classical CI) required the droplets formed in the initial stages of spray formation to carry charge, which could be retained by the analyte molecules once the solvent molecules had been removed. Often, this residual charge was provided by buffers or other electrolytes that were present in the HPLC mobile phase for the purposes of the separation (e.g., ammonium acetate, acetic acid, etc.) or were added post-column if their presence would negatively perturb the separation. It was not uncommon for the desolvation process to be less than complete for various classes of compounds, leading to solvated ions in thermospray ionization mass spectra such as $[M+H+H_2O]^+$, $[M+H+NH_3]^+$ or $[M-H+CH_3CO_2H]^-$. When compared to ESI, thermospray exhibited less sensitivity and a more limited operational domain in terms of the size (molecular mass) of molecules that could be ionized and detected using the technique. Interestingly, at least one commercial manufacturer of the thermospray interface equipped their instrument with a corona discharge needle that was located in the vicinity of the ion source sampling cone. Application of a high voltage (several thousand volts) to the needle resulted in a corona discharge that cascaded into a plasma containing ions and electrons capable of causing ionization of analyte molecules as they passed through this region of the interface. Referred to at the time as "discharge assisted" thermospray, in retrospect this mode of operation bore many striking similarities to the current APCI ion source.

11.7.2.5 Continuous-flow FAB

Discussed earlier in the *Ionization* section, FAB was most commonly used in what was referred to as a "static" mode, in which the sample was dissolved in a suitable matrix and applied to a target that was inserted into the ion source using a probe. Once the analysis was complete (or the matrix evaporated or was sputtered away), the process was repeated as necessary. The initial FAB targets that supported the glycerol/sample droplet were solid, stainless surfaces. The earliest attempts at using FAB as an ionization mode for use in conjunction with HPLC introduced the eluent into the mass spectrometer by means of the probe that supported the target. The target was modified in order to allow the HPLC eluent to come into contact with the bombarding atom beam. In order to reduce the rate of solvent evaporation and provide a suitable surface across which the sample could be dispersed (creating a scenario that resembled the conditions created by the glycerol droplet in the static experiment), the solid stainless steel target used for static FAB experiments was replaced with a porous glass frit. The HPLC eluent was delivered through the probe to the backside of the frit where it diffused through to the side being bombarded by the fast atom beam. A second version of the CF-FAB interface [57] used a reasonably standard target surface through which a small diameter hole was drilled. The HPLC eluent was pumped to the target surface through a micro-bore capillary tube. In order to maintain a suitable sampling environment (i.e., a continually refreshed sampling area and adequate energy dissipation to prevent decomposition of thermally labile analyte compounds), the HPLC mobile phase was mixed with a small amount of glycerol. At levels of 5–10%, there was a sufficient amount of glycerol to provide optimum FAB ionization conditions for analytes as they eluted from the HPLC column and reached the target surface. Best results were obtained when the rate of solution delivery to the target was equal to the rate of evaporation of the eluent (water plus organic modifier) and glycerol. The need to balance the flow of HPLC eluent with the rate of the evaporative processes made HPLC separations that employed gradient elution conditions difficult (due to the continually changing composition of the mobile phase). Additionally, the low flow rates required by the technique favored either micro-bore chromatography or splitting of the eluent stream. Both approaches diminished the achievable sensitivity exhibited by the technique.

11.8 TANDEM MASS SPECTROMETRY (MS/MS AND MS^n)

Tandem mass spectrometry (also referred to as MS/MS or MS^n) [58,59] covers a broad compilation of techniques and instruments. In principal, it can be thought of as being analogous to either GC/MS or LC/MS in that it involves the joining together of two (or more) mass spectrometers or mass spectrometric experiments. The specific reasons for combining mass spectrometers together are many and varied, but the fundamental driver for the evolution of the techniques and the instrumentation has been the quest for new knowledge and information about ions and their structures. The rise of tandem mass spectrometry correlates with the introduction and proliferation of the so-called "soft" ionization techniques. When EI was the prevalent ionization technique, all of the information reflecting the competitive and consecutive unimolecular dissociation processes of the parent or molecular ion was displayed in the mass spectrum. The appearance (or absence) of the particular ion signals, as well as their relative abundances, was dictated by thermodynamic and kinetic factors associated with the structure of the ion and the amount of energy imparted to it by the ionization event. The advent of CI, FAB, thermospray and their successors (ESI and APCI) enabled intact molecules of increasing size (mass) and complexity (and biological importance) to be analyzed. But, the appearance of the resulting mass spectra was significantly different—containing only molecular mass information and little if any of the structural information that had been contained within the fragment ion-rich EI mass spectra. In the late 1960s and mid-1970s, it was recognized that the internal energy responsible for structurally diagnostic ion fragmentation processes, which had been eliminated through the use of soft ionization techniques, could be reintroduced by converting some portion of ion kinetic energy into internal energy. This energy conversion was accomplished initially through collisions of the ions with neutral gas molecules (collisional activation) in the region between two analyzers. By physically separating (in both space and time) the ionization event occurring in the source from the energizing or collisional "activation" event occurring after a first stage of mass analysis, labile compounds could be ionized successfully and subsequently dissociated to yield structurally diagnostic fragment ions. Thus, a mass spectrum could be obtained for any/every ion population that could be generated from a given sample. Over the years there have been a multitude of double, triple and multiple analyzer mass spectrometers (MS/MS, MS/MS/MS, MS^4, etc.) constructed for studies of ion structure, energetics and kinetics using probing interactions with gases, surfaces, photons, ions and electrons. While a comprehensive description of all of the various combinations of mass analyzers is clearly beyond the scope of this account, listed in the table below are some of the more noteworthy examples (in approximate chronological order of their development). Each configuration has advantages or disadvantages when compared to the others, including m/z range, collision energy, mass resolving power, mass accuracy, ease of automation, software sophistication (instrument control and data acquisition), size, cost, operational complexity and data acquisition speed.

Examples of commercialized hybrid or multiple analyzer mass spectrometers	
BE	
EBE	
BEEB	B = magnetic sector
EBQ	E = electrostatic analyzer
QQ*Q	Q = quadrupole mass filter
QQ*(TOF)	3DQIT = three-dimensional (Paul) quadrupole ion trap
QQ*(ICR)	2DQIT = two-dimensional (linear) quadrupole ion trap
EB(3DQIT)	ICR = ion cyclotron resonance spectrometer
(TOF)(TOF)	TOF = time-of-flight analyzer
QQ*(2DQIT)	
(2DQIT)(ICR)	

*Note: These quadrupole elements do not function as mass analyzers; depending on the manufacturer, this rf-only focusing element (which also serves as the region where the collisional activation event takes place) has taken the form of any one of a variety of multiplole elements (i.e., quadrupole, hexapole, octapole, etc.), despite being generically referred to as a "quadrupole."

With the introduction and popularization of ion trap instruments, the very nature of the sequential, multi-stage "MS/MS" experiment changed from being tandem "in space" to being tandem "in time." Unlike a multiple analyzer instrument comprised of sectors, quadrupoles or combinations of these or other analyzers, all experiments conducted with an ion trap instrument take place within a single analyzer. Instead of a continuous stream of ions physically passing from one analyzer to the next, ions are injected (gated) into and retained within the ion trap for the entire duration of the multi-stage experiment. The various interrogation and measurement activities take place in the same physical space being separated only in time. An obvious advantage of conducting experiments in this way is that the processes are highly efficient since fewer ions are lost during the course of the experiment. In fact, one literature report described a sequential experiment comprised of as many as 10 stages of mass analysis (i.e., MS^{10}). However, in practice, it is far more probable that 3–4 stages of mass analysis can be achieved in a sequential manner before the ion signal dissipates. The other obvious benefit of experiments conducted sequentially "in time" is the tremendous simplification of the instrument, with the associated cost savings (since analyzers, vacuum pumps and other associated hardware is very expensive). However, there are also certain disadvantages associated with the trapping instruments—particularly with respect to the types of experiments that can be conducted.

By far, the most prevalent scan mode conducted in tandem mass spectrometers (of any kind) is the product ion scan. This experiment records the m/z values and relative abundances of ions that are the result of energy-induced dissociation reactions of a population of ions (precursor ions) having a specific m/z value. The selection of the precursor ions (i.e., exclusion of all other m/z values) can be accomplished using any of a variety of mass analyzers, and depending upon the device used it can be accomplished at either nominal mass resolving power (where no distinction between isobaric ions is possible) or high resolving power. After selection, the precursor ion population can be energized (assuming this is required) by a variety of means including collisions of the ions with neutral gas molecules, collisions of the ions with a solid surface or irradiation of the ions with laser light. The energized ions quickly dissociate into the same kinds of fragment ions observed for processes that occur in certain ion sources (caused by energy in excess of that required to achieve ionization). A second mass analyzer (or a second timed analyzing scan function) is then employed to record the m/z values and abundances of the fragment ions. This product ion spectrum can then be matched against previously obtained or related spectra, or interpreted directly to deduce the structure of the mass-selected precursor ions. However, there are a number of other "MS/MS" scan modes [60] that can be implemented in instruments comprised of various mass analyzers. Two of these experiments are of sufficient utility to warrant mention here (Fig. 11.16). The constant neutral loss scan and precursor ion scan are most often implemented using triple quadrupole instruments and are used as screening or survey scans. In contrast to a product ion scan, where a known m/z value is specifically and intentionally chosen for the first stage of analysis on the basis of some previous knowledge or information, the neutral loss and precursor ion scans are applied to complex samples in order to uncover the presence of unknown related compounds. The premise upon which these two scans are based is that homologous series of compounds (e.g., various metabolites of a medicine or a family of therapeutic natural products) which will produce

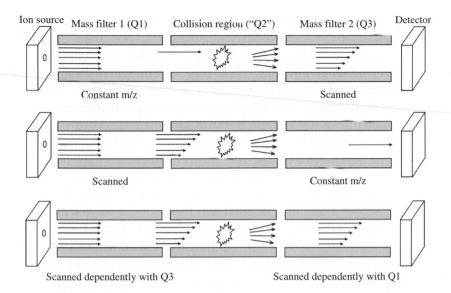

Fig. 11.16. Representation of three tandem mass spectrometry (MS/MS) scan modes illustrated for a triple quadrupole instrument configuration. The top panel shows the attributes of the popular and prevalent product ion CID experiment. The first mass filter is held at a constant m/z value transmitting only ions of a single m/z value into the collision region. Conversion of a portion of translational energy into internal energy in the collision event results in excitation of the mass-selected ions, followed by unimolecular dissociation. The spectrum of product ions is recorded by scanning the second mass filter (commonly referred to as "Q3"). The center panel illustrates the precursor ion CID experiment. Ions of all m/z values are transmitted sequentially into the collision region as the first analyzer (Q1) is scanned. Only dissociation processes that generate product ions of a specific m/z ratio are transmitted by Q3 to the detector. The lower panel shows the constant neutral loss CID experiment. Both mass analyzers are scanned simultaneously, at the same rate, and at a constant m/z offset. The m/z offset is selected on the basis of known neutral elimination products (e.g., H_2O, NH_3, CH_3COOH, etc.) that may be particularly diagnostic of one or more compound classes that may be present in a sample mixture. The utility of the two compound class-specific scans (precursor ion and neutral loss) is illustrated in Fig. 11.17.

either product ions of common mass or eliminate the same neutral molecules as a result of the collision-induced dissociation (CID) process. By scanning the two analyzers of a tandem mass spectrometer such as a triple quadrupole instrument so that a signal is recorded only when the specific product ion appears or when the specific neutral is eliminated, the chromatographic trace of a complicated biological sample can be simplified. Families or classes of compounds of potential interest or importance can then be targeted for further analysis (i.e., product ion analysis) or isolation. Operationally, the precursor ion scan is the "inverse" of the product ion scan. The second mass analyzer is set to a specific m/z value, while the first mass analyzer is scanned. This means that all ions of all masses produced in the mass spectrometer source are transmitted sequentially (one m/z value after the next) to the collision region of the instrument where they are energized and undergo unimolecular dissociation. Because the second analyzer is set to pass only ions having a specific, pre-determined m/z value, the vast majority of the fragment ions do not reach the detector. Only when one of the precursor ions undergoes a dissociation reaction that produces a fragment ion having the specified m/z value does the detector record a signal. For example, an extract of tree bark suspected of containing one or more members of a family of compounds thought to have medicinal properties is first analyzed using HPLC with a mass spectrometric detection. The resulting chromatogram (total ion chromatogram) would likely reveal numerous compounds. The mass spectrum of each of the many compounds would display $[M+H]^+$ and possibly adduct ions, but there would be no way to easily or quickly establish which compounds might be related to one another or merit further investigation. A time-consuming and laborious approach might be to reanalyze the sample acquiring a product ion CID spectrum for every ion observed in every peak in the chromatogram, followed by painstaking analysis of all of the spectra in search of common characteristics. A much more efficient means of accomplishing the same outcome is to employ precursor ion and neutral loss scans. Hypothetically, if two classes of compounds thought to be present in the extract were known to produce either benzyl product ions ($C_7H_7^+$, m/z 91) in one case, or to eliminate neutral dimethylamine (C_2H_7N) in another, the following scan descriptors would immediately reveal possible compounds possessing those characteristics.

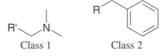

Class 1   Class 2

constant neutral loss scan:
analyzer 1—scans selected m/z range at typical scan rate (e.g., 1 ms per m/z unit)

analyzer 2—synchronized with analyzer 1 (identical scan rate) but offset by 45 m/z units

(e.g., when analyzer 1 is passing m/z 155, analyzer 2 is passing m/z 110, etc.)

precursor ion scan:

analyzer 1—scans selected m/z range at typical scan rate (e.g., 1 ms per m/z unit)

analyzer 2—does not scan, set to pass m/z 91 at all times

(e.g., when analyzer 1 is passing any m/z value, analyzer 2 is passing m/z 91)

The three simulated chromatograms in Fig. 11.17 illustrate this point. The top chromatogram displays a significant number of compounds present in the hypothetical bark extract. However, the complexity of the data set is reduced substantially by the selective data acquisition modes described above. The middle trace represents the constant neutral loss mass chromatogram (neutral loss of 45 Da for dimethylamine). Analogously, the bottom trace represents the precursor ion spectrum for m/z 91 (benzyl cation). Further investigation of each of the peaks in these respective "functional group specific" chromatograms is now possible.

11.8.1 Applications

Tandem mass spectrometry originated as a tool for interrogating interesting and unique ion structures of fundamental interest. Unlike traditional EI mass spectrometry, which produced spectra reflective of the behavior and unimolecular chemistry of the radical cations formed from a volatile organic analyte, tandem mass spectrometry enabled researchers to probe the structure and energetics of even-electron ions, such as protonated and deprotonated molecules, as well as radical cations that had no stable neutral precursors (i.e., resulting from a previous ionic decomposition reaction). However, the ability of tandem mass spectrometry to directly analyze components of complex mixtures of organic and biological compounds is most responsible for its widespread use today as a practical analytical tool. Early experiments demonstrated the ability of the technique to cope with very complicated analytical matrices without the need for previous chromatographic separation of the analytes (by either GC or HPLC). The entire sample would be volatized into a CI source and each component (analyte) of the mixture would yield a signal at a distinct m/z value (e.g., an $[M+H]^+$ ion).

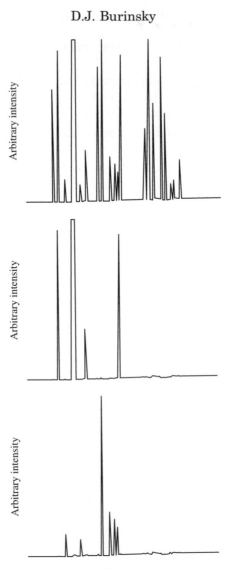

Fig. 11.17. Simulated mass chromatograms resulting from precursor ion and constant neutral loss tandem mass spectra (middle and bottom traces), illustrating the selectivity that those MS/MS scan modes can bring to chromatographic analyses. The top trace in the figure represents a total ion chromatogram obtained using a conventional single stage of mass analysis.

The first mass analyzer was used to select a given *m/z* value and those ions would be transmitted to the collision region where they would undergo collisional activation. Subsequent decomposition of the excited ions resulted in characteristic fragment ions (with equally characteristic relative abundances) that were recorded as a CID product ion spectrum. Comparison with a similar spectrum obtained from an authentic standard or a structurally similar molecule leads to identification of the unknown compound in the original sample. Because ions traverse the distances within mass spectrometers on the microsecond to millisecond timescale, and voltages can be changed very quickly, these experiments take very little time to conduct. This drastic difference in experimental time scale, when compared to a typical GC/MS or LC/MS analyses that requires tens of minutes, led some early observers to believe that tandem mass spectrometry would replace the other techniques. Clearly, this has not happened. Reasons for this overly optimistic prediction include factors such as ion suppression (i.e., the simultaneous presence of multiple compounds in the ion source significantly reduces the ionization response of the analyte of interest) and the difficulty of detecting low-level unknown compounds in a mass spectrum containing hundreds of ion signals of varying intensity. In fact, today tandem mass spectrometry is used almost exclusively with HPLC (and to a lesser degree GC) in order to take advantage of the powerful analytical attributes of both techniques and minimize the weaknesses of each. The types and varieties of samples that are analyzed using MS/MS techniques vary widely. Soft ionization techniques such as the API methods have led to explosive growth in the utilization of mass spectrometry in medical and biomedical research. Tandem mass spectrometry has figured prominently in many of these advances through the qualitative analysis of proteins [61], carbohydrates [62], oligonucleotides [63] and lipids [64]. Quantitative methodologies have been applied to studies of the metabolism of a wide variety of small- to medium-sized analytes [65] as well as to proteins using ICAT technology [66].

11.9 QUALITATIVE ANALYSIS

All analytical techniques are designed to provide the answer to one or both of the two important questions: "what is it?" and "how much is there?" Mass spectrometry possesses attributes that allow it to contribute answers to both of these questions. The nominal (integral) m/z ratio of the molecular ion can sometimes be sufficient to identify a chemical compound, particularly if there is additional information available (either from the mass spectrum itself or another analytical technique). The presence of other signals in a mass spectrum attributable to isotopomers and fragment ions (each with a distinct m/z value and relative abundance) combined with the molecular mass information adds certainty to the proposed identification, since fragment ions are often characteristic of particular functional groups being present in the molecule [67,68]. If the available instrumentation is capable of separating closely related molecular species (high resolving power) and possesses the capability to measure m/z values with high accuracy and precision (exact mass measurements), then possible elemental formulae can be proposed for the ions of interest. Together with previous knowledge (stored in the form of a spectral library or database entry), information about a structurally similar analogue or data from one or more complimentary analytical techniques (e.g., NMR spectroscopy, IR spectroscopy, etc.), the mass spectral information can lead to a confident conclusion about the identity of the compound in question. With the current arsenal of ionization sources and analyzers, the range and variety of compounds amenable to identification by mass spectrometry is virtually limitless. Mass spectrometers can identify compounds present at low levels ranging from the simplest known (such as methane or helium) up to non-covalent protein complexes weighing several hundred thousand Daltons.

Determination of molecular mass from the information displayed in a mass spectrum is usually straightforward, provided the molecular or parent ion can be identified with confidence. Often, the signal with the largest m/z value and the highest abundance represents the intact molecule and thereby carries the molecular mass information. However, depending on the mode of ionization, this may not always be true. For example, EI may not produce a molecular ion because of unfavorable thermodynamics or kinetics (i.e., the molecular ion is considerably less stable than one or more of the unimolecular dissociation products). Other ionization modes may produce cluster ions (non-covalent association complexes of the analyte molecule with itself or other species such as solvents or metal cations) that will have m/z values greater than that of the analyte molecule. While these adduct ions have the potential for causing interpretive confusion, an understanding and recognition of the underlying processes can actually enhance the certainty associated with a given molecular mass assignment. In addition to the nominal mass information displayed in many mass spectra, the relative abundances of the isotopomers (the signals immediately adjacent to the base peak in the isotopic cluster of a molecular ion signal) can also provide useful information, thus enabling certain putative empirical formulae to be eliminated from consideration. Compounds containing particular elements (such as chlorine, bromine, sulfur, boron or various transition metals) with unique and recognizable isotopic distributions can provide considerable insight into their elemental composition. A few examples of elements possessing recognizable isotopic distributions are shown below along with the relative abundances of the important isotopomers.

^{35}Cl, ^{37}Cl – 100:32 ^{79}Br, ^{81}Br – 100:97
^{32}S, ^{34}S – 100:4.5 ^{10}B, ^{11}B – 25:100

^{63}Cu, ^{65}Cu–100:45

Even organic compounds of reasonably ordinary composition (containing only the elements of carbon, hydrogen, nitrogen and oxygen) can be distinguished from one another according to the relative abundance of the naturally occurring ^{13}C isotope in their mass spectra. For example, two compounds having the same nominal molecular mass display relative abundances for the molecular ion isotopic distributions as shown below. Recognizing the significant difference in the peak heights (relative abundances) of the "M+1" signals (6.9% vs. 12.2%) makes it possible to distinguish between two potential elemental compositions for the unknown compound.

$[C_6H_{12}O_6]H^+$		$[C_{11}H_{16}O_2]H^+$	
m/z	Relative abundance (%)	m/z	Relative abundance (%)
181	100	181	100
180	6.9	180	12.2
179	1.4	179	1.1

In addition to the molecular mass of the ionized molecule, the masses of fragment ions (whether produced in an EI ion source or by some other mode of excitation such as collisional activation) constitute another very important source of information in the qualitative analysis of organic compounds. Nearly, all organic ions dissociate (provided they possess sufficient internal energy) systematically and reproducibly to yield characteristic fragment ion signals in their mass spectra. The m/z ratios of the ions observed in the mass spectrum, as well as the masses of neutrals and radicals represented by the spacing between the peaks, provide information about the various kinds of structural subunits that comprised the precursor ion. The information contained in a mass spectrum also can sometimes provide insight into the spatial relationships between various structural subunits (i.e., connectivity), but not always. In this sense, mass spectrometry is more similar to infrared spectroscopy (excellent functional group characterization) than nuclear magnetic resonance (NMR) spectroscopy, which establishes connective relationships between atoms and structural subunits. For example, the EI mass spectrum of a molecule such as *p*-aminomethylbenzylalcohol ($H_2NCH_2C_6H_4CH_2OH$) would be expected to display signals associated with the elimination of water (or formaldehyde) and ammonia from the molecular ion, along with other fragment ions characteristic of a sub-stituted aromatic ring system. However, none of the information contained within the mass spectrum would permit an *a priori* deter-mination of the substitution pattern of the subunits around the ben-zene ring (i.e., *ortho-*, *meta-* or *para-*). Certain ions are easily recognized as indicators of particular structural subunits—e.g., *m/z* 43, CH_3CO^+, *m/z* 77, $C_6H_5^+$, *m/z* 57, $C_4H_9^+$, *m/z* 91, $C_7H_7^+$, *m/z* 69, CF_3^+, although the utility of such "milepost" ions diminishes as molecules become larger and more complex. As discussed in the context of isotopomers, relative abundances of fragment ions contribute additional information to the structure elucidation process, but are generally less important than the *m/z* values. Because relative abundances are virtually impossible to predict *a priori*, they are of greatest use when the mass spectrum of an analyte can be compared directly to that of an authentic standard (ei-ther experimentally or through computerized library matching). Differ-ences in relative abundances can be ascribed to isomeric ion structures, but differentiation of the isomeric species by mass spectrometry alone is neither simple nor common.

One particular class of molecules that lends itself particularly well to characterization by tandem mass spectrometry methods is peptides. *De novo* amino acid sequencing of small peptides (M_r of approxi-mately 1500–2000) has become routine and is used successfully in conjunction with protein database searching to identify a wide variety of important proteins associated with cellular functions and various disease states. Isolation of either singly ($[M+H]^+$) or doubly charged ($[M+2H]^{2+}$) peptide ions is followed by collisional activation and recording of the product ion spectrum. Because of the well-known molecular masses of naturally occurring amino acids and the tendency of peptide ions to fragment predictably at the peptide amide linkages, the CID product ion spectra of peptide ions often yield signals that directly reveal the primary amino acid sequence. While used widely with great success, the technique is not without some shortcomings. Certain amino acids (particularly proline) can terminate the fragmen-tation cascade, leaving gaps in the information content derived from spectrum. Other amino acids such as the isobars glutamine and lysine or the isomers leucine and isoleucine produce ambiguous interpretations due to mass overlap.

Electron ionization mass spectra are well suited to library (database) searching as a rapid means of compound identification by virtue of the reproducible nature of 70 eV EI mass spectra. Despite the fact that the largest commercially available library contains upward of 150,000 entries, an obvious caveat is that a mass spectrum for the compound being analyzed must be contained within the library in order for a match to be possible. In situations where the mass spectrum of the analyte is not included in the library, compounds having similar structures will be suggested as "partial" matches allowing the analyst to incorporate the information about structural analogues into their final interpretation. Unfortunately, the nearly infinite diversity in tuning parameters and differences in instrument characteristics (i.e., a wide variety of analyzer types, energy regimes and time scales) has made the compilation of a "universal" CID product ion spectral library impractical to date. Individual researchers are able to compile customized libraries of spectra for their own compounds, but widespread dissemination of such information is rare.

11.10   QUANTITATIVE ANALYSIS

Mass spectrometry has a long history as a technique that is well suited to quantitative determinations. While used infrequently today, the original quantitative approach obtained analyte concentrations directly from peak heights of signals in EI mass spectra. Calibration curves were constructed using the measured peak heights of signals at unique m/z values vs. analyte concentration (partial pressure of the compound admitted to the ion source). The accuracy of the results was improved if an internal standard was incorporated into both calibration standards and samples containing unknown amounts of analyte. The amount of unknown compound present was deduced from a "calibration curve" on which the peak height ratio (analyte response/internal standard response) was plotted as a function of analyte concentration. Because the internal standard (often an analogue of the analyte molecule into which one or more stable isotopes such as ^2H, ^{13}C or ^{15}N had been incorporated) undergoes the exact sample preparation and experimental conditions as the analyte, the often significant sources of indeterminate error are cancelled out. If an isotopically labeled standard is not available, it is possible to use a homologue of the analyte that produces one or more fragment ions that are reasonably close in m/z value to ions coming from the analyte. However, this approach is less desirable than using isotopically labeled standards.

By far, the most prevalent type of quantitative mass spectrometric analysis used today employs the mass spectrometer as a very sensitive and specific detector for a chromatographic or electrophoretic separation. In this context, the mass spectrometer serves the same purpose as any other detector, with the advantages of excellent sensitivity and specificity. The simplest version of this experiment employs the mass spectrometer as a "single-channel" detector. The instrument is tuned to transmit ions to the detector that possess a specific nominal m/z value. A recording of ion current intensity (y-axis) as a function of time (x-axis) yields the elution profile of analytes from the chromatographic medium, which we recognize as the characteristic GC or HPLC chromatogram. Like similar chromatograms produced using other detectors, the peak area is proportional to the quantity of analyte present in the sample. While this "single-ion monitoring" (SIM) experiment provides considerably more specificity than other popular non-specific detectors such as refractive index and UV/Vis for HPLC or thermal conductivity and flame ionization detectors for GC, it is still subject to false-positive responses. These spurious signals arise from the inevitable presence of interfering compounds that generate isobars (same nominal mass but different elemental composition) or isomers (same elemental composition) having the same nominal m/z value as the analyte of interest (either molecular ions or fragment ions). If the mass spectrometer being employed as the chromatographic detector is capable of either high resolving power or tandem mass spectrometry scanning modes, then some of these false-positive responses can be eliminated. Increasing the resolving power of the mass spectrometer increases the specificity of the experiment by excluding some of the interfering isobaric species. Because these isobars have different elemental compositions (and therefore different exact masses), they can be differentiated from the analyte ion as the resolving power of the experiment increases. Alternatively, additional specificity can be introduced into the mass spectrometric detection scheme if a tandem or MS/MS instrument is employed as the detector rather than a single mass analyzer, even if both of mass analysis are low resolving power stages. Another entire "dimension" of selectivity can be achieved by selecting not only the m/z value of the molecular ion (in the first stage of mass analysis), but also the m/z value of a particularly diagnostic (or characteristic) fragment ion (in the second stage of mass analysis). In this way, the "selected reaction monitoring" (SRM) acquisition mode exploits the unique, intrinsic unimolecular dissociation chemistry of the molecular ion to eliminate potential interferences and their associated false-positive responses. For example, if a sample contains the antidepressant amitriptyline, it would be expected to yield an $[M+H]^+$ ion at m/z 278 under ESI conditions.

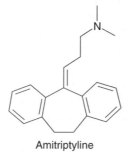

Amitriptyline

A single-stage mass spectrometer could be set to monitor only that m/z value during the course of an HPLC analysis and should produce a chromatogram that contains a single response at the appropriate retention time for amitriptyline. In reality, if the sample were a complex biological matrix (e.g., urine, plasma, etc.) it is possible that the chromatogram could contain more than a single response, with the false-positive peaks in the chromatogram being attributed to unknown compounds that also generate m/z 278 ions (either $[M+H]^+$ ions, adduct ions or fragment ions). However, the selectivity of the analysis can be improved substantially by taking advantage of the fact that protonated amitriptyline will dissociate upon collisional activation to yield a product ion at m/z 233 ($278^+ \rightarrow 233^+$, elimination of dimethylamine). Specifying the precursor ion mass, the fragmentation transition (product ion mass) and a known chromatographic retention time requires that three "orthogonal" selection criteria be satisfied in order to generate a positive response. The multiplicative discriminating power of these three orthogonal "gates" is tremendous, with the net result being a drastic reduction in the number (and probability) of false-positive responses for any chromatographic separation. For additional selectivity, other dissociation reactions of the same $[M+H]^+$ ion may be monitored simultaneously in what is described as a "multiple reaction monitoring" or MRM data acquisition mode. However, this nomenclature can be somewhat confusing since the monitoring of a single dissociation transition from each of several different $[M+H]^+$ ions within the same chromatographic analysis can also be described as multiple reaction monitoring. The specificity conveyed by a particular unimolecular transition between precursor and product ions can be leveraged using any one of the three most common MS/MS scan modes described earlier (i.e., product ion scans, precursor ions scans or neutral loss scans), provided the configuration and type of mass analyzer(s) being employed is capable of executing the scan type of interest. For instance, a triple quadrupole mass spectrometer (or any other tandem "in space" or beam instrument) is capable of executing any of these three scan modes, while tandem "in time" instruments (ion traps) are not. Once the chromatographic data are acquired, the processing and presentation of the results or conclusions of the quantitative determination are conducted as for any other method or detector that might be used for the purpose (i.e., plotting of peak areas or peak area ratios vs. analyte concentration).

REFERENCES

1 J.J. Thomson, A new method of chemical analysis, *Chem. News J. Industr. Sci.*, 103 (1912) 265–268.
2 F.W. Aston, A magnetic mass spectrometer, *Phil. Mag.*, 38 (1919) 707–712.
3 A.J. Dempster, Positive-ray abundance spectrometry, *Phys. Rev.*, 1 (1918) 316–321.
4 G.I. Ramendik, I.V. Kinayeva and D.A. Tyurin, Quantitative mass spectrometric element determination without reference materials using several internal standards, *Fresenius J. Anal. Chem.*, 354 (1996) 150–158.
5 R.S. Houk, Elemental and isotopic analysis by inductively coupled plasma mass spectrometry, *Acc. Chem. Res.*, 27 (1994) 333–339.
6 J.W. Olesik, Fundamental research in ICP/OES and ICP/MS, *Anal. Chem.*, 68 (1996) 469A–474A.
7 N. Jakubowski, D. Stuewer and W. Vieth, Performance of a glow discharge mass spectrometer for simultaneous multielement analysis of steel, *Anal. Chem.*, 59 (1987) 1825–1830.
8 S. DeGendt, R.E. Van Grieken, S.K. Ohorodnik and W.W. Harrison, Parameter evaluation for the analysis of oxide-based samples with radio frequency glow discharge mass spectrometry, *Anal. Chem.*, 67 (1995) 1026–1033.
9 B.N. Taylor, Guide for the use of the International System of Units (SI), pp. 24–25, *NIST Special Publication 811* (1995 Edition), U.S. Government Printing Office, Washington, 1995 (http://physics.nist.gov/Pubs/SP811/sec08.html).

10 R.J. Cotter, Mass spectrometry of nonvolatile compounds by desorption from extended probes, *Anal. Chem.*, 52 (1980) 1590A–1606A.

11 V.N. Reinhold and S.A. Carr, Desorption chemical ionization with polyimide coated wires, *Anal. Chem.*, 54 (1982) 499–503.

12 H.L.C. Meuzelaar, W. Windig, A.M. Harper, S.M. Huff, W.H. McClennen and J.M. Richards, Pyrolysis mass spectrometry of complex organic materials, *Science*, 226 (1984) 268–274.

13 A.C. Tas and J. van der Greef, Pyrolysis-mass spectrometry under soft ionization conditions, *Trends Anal. Chem.*, 12 (1993) 60–66.

14 J. Budzikiewicz, C. Djerassi and D.H. Williams, *Mass Spectrometry of Organic Compounds*, Holden-Day, San Francisco, 1967.

15 A.G. Harrison, *Chemical Ionization Mass Spectrometry*, CRC Press, Boca Raton, 1992.

16 D.F. Hunt, Reagent gases for chemical ionization mass spectrometry, *Adv. Mass Spectrom.*, 6 (1974) 517–522.

17 A.P. Bruins, Atmospheric pressure ionization-mass spectrometry I. Instrumentation and ionization techniques, *Trends Anal. Chem.*, 13 (1994) 37–43.

18 A. Raffaelli and A. Saba, Atmospheric pressure photoionization mass spectrometry, *Mass Spectrom. Rev.*, 22 (2003) 318–331.

19 R.P. Lattimer and H.R. Schulten, Field ionization and field desorption mass spectrometry: past, present and future, *Anal. Chem.*, 61 (1989) 1201A–1215A.

20 M. Anbar and W.H. Aberth, Field ionization mass spectrometry: new tool for the analytical chemist, *Anal. Chem.*, 46 (1974) 59A–64A.

21 R.D. Smith, K.J. Light-Wahl, B.E. Winger and D.R. Goodlett, Electrospray ionization. In: T. Matsuo, R.M. Caprioli, M.L. Gross and Y. Seyama (Eds.), *Biological Mass Spectrometry: Present and Future*, Wiley, New York, 1994, pp. 41–74.

22 M. Labowsky, C. Whitehouse and J.B. Fenn, Three-dimensional deconvolution of multiply charged spectra, *Rapid Commun. Mass Spectrom.*, 7 (1993) 71–84.

23 M.W. Senko, S.C. Beu and F.W. McLafferty, Automated assignment of charge states from resolved isotopic peaks for multiply charged ions, *J. Am. Soc. Mass Spectrom.*, 6 (1995) 52–56.

24 G. Neubauer and R.J. Anderegg, Identifying charge states of peptides in liquid chromatography/electrospray ionization mass spectrometry, *Anal. Chem.*, 66 (1994) 1056–1061.

25 M. Karas, U. Bahr and U. Giessman, Matrix-assisted laser desorption ionization mass spectrometry, *Mass Spectrom. Rev.*, 10 (1991) 335–357.

26 F. Hillenkamp and M. Karas, Mass spectrometry of peptides and proteins by matrix-assisted ultraviolet laser desorption/ionization. In: J.A. McCloskey (Ed.), *Methods in Enzymology, Vol. 193: Mass Spectrometry*, Academic Press, New York, 1990, pp. 280–295.

27 M. Barber, R.S. Bordoli, G.J. Elliott, R.D. Sedgwick and A.N. Tyler, Fast atom bombardment mass spectrometry, *Anal. Chem.*, 54 (1982) 645A–657A.

28 D. Kidwell, M. Ross and R. Colton, A mechanism of ion production in secondary ion mass spectrometry, *Int. J. Mass Spectrom. Ion Proc.*, 78 (1987) 315–328.

29 S. Pachuta and R.G. Cooks, Mechanisms in molecular secondary ion mass spectrometry, *Chem. Rev.*, 87 (1987) 647–669.

30 N. Winograd, Prospects for imaging TOF-SIMS: from fundamentals to biotechnology, *Appl. Surf. Sci.*, 203/204 (2003) 13–19.

31 B. Sundqvist and R.D. Macfarlane, ^{252}Cf plasma desorption mass spectrometry, *Mass Spectrom. Rev.*, 4 (1985) 421–460.

32 G.W. Wood, Field desorption mass spectrometry: applications, *Mass Spectrom. Rev.*, 1 (1982) 63–102.

33 R.G. Cooks, S.H. Hoke II, K.L. Morand and S.A. Lammert, Mass spectrometers: instrumentation, *Int. J. Mass Spectrom. Ion Proc.*, 118/119 (1992) 1–36.

34 K. Biemann, High resolution mass spectrometry, *Adv. Anal. Chem. Instr.*, 8 (1970) 185–221.

35 J.A. Yergey, R.J. Cotter, D. Heller and C. Fenselau, Resolution requirements for middle molecule mass spectrometry, *Anal. Chem.*, 56 (1984) 2262–2263.

36 J.C. Schwartz, J.E.P. Syka and I. Jardine, High resolution on a quadrupole ion trap mass spectrometer, *J. Am. Soc. Mass Spectrom.*, 2 (1991) 198–204.

37 J.R. Trainor and P.J. Derrick, Sectors and tandem sectors. In: M.L. Gross (Ed.), *Mass Spectrometry in the Biological Sciences: A Tutorial*, Kluwer Academic Publishers, Dordrecht, The Netherlands, 1992, pp. 1–27.

38 J. Mattauch, A double-focusing mass spectrograph and the masses of ^{15}N and ^{18}O, *Phys. Rev.*, 50 (1936) 617–623.

39 E.G. Johnson and A.O. Nier, Angular aberrations in sector-shaped electromagnetic lenses for focusing beams of charged particles, *Phys. Rev.*, 91 (1953) 10–17.

40 P.H. Dawson, Quadrupole mass analyzers: performance, design and some recent applications, *Mass Spectrom. Rev.*, 5 (1986) 1–37.

41 R.G. Cooks and K.A. Cox, Ion-trap mass spectrometry. In: T. Matsuo, R.M. Caprioli, M.L. Gross and Y. Seyama (Eds.), *Biological Mass Spectrometry: Present and Future*, Wiley, New York, 1994, pp. 179–197.

42 S.A. McLuckey, G.J. Van Berkel, G.L. Glish and J.C. Schwartz, Electrospray and the quadrupole ion trap. In: R.E. March and J.F.J. Todd (Eds.), *Aspects of Ion Trap Mass Spectrometry*, CRC Press, Boca Raton, 1995, pp. 89–141.

43 J.C. Schwartz, M.W. Senko and J.E.P. Syka, A two-dimensional quadrupole ion trap mass spectrometer, *J. Am. Soc. Mass Spectrom.*, 13 (2002) 659–669.

44 M.B. Comisarow and A.G. Marshall, The early development of Fourier transform ion cyclotron resonance (FT-ICR) spectroscopy, *J. Mass Spectrom.*, 31 (1996) 581–585.

45 H. Wollnik, Time-of-flight mass analyzers, *Mass Spectrom. Rev.*, 12 (1993) 89–114.

46 B.A. Mamyrin, Laser assisted reflectron time-of-flight mass spectrometry, *Int. J. Mass Spectrom. Ion Proc.*, 131 (1994) 1–19.

47 F.W. McLafferty, High-resolution tandem FT mass spectrometry above 10 kDa, *Acc. Chem. Res.*, 27 (1994) 379–386.

48 N.A.B. Gray, A. Buchs, D.H. Smith and C. Djerassi, Application of artificial intelligence for chemical inference. Part XXXVI. Computer assisted structural interpretation of mass spectral data, *Helv. Chim. Acta*, 64 (1981) 458–470.

49 J.R. Yates, Mass spectrometry and the age of the proteome, *J. Mass Spectrom.*, 33 (1998) 1–19.

50 J.D. Pinkston and T.L. Chester, Guidelines for successful SFC-MS, *Anal. Chem.*, 67 (1995) 650A–656A.

51 Z. Zhao, J.H. Wahl, H.R. Udseth, S.A. Hofstadler, A.F. Fuciarelli and R.D. Smith, Online capillary electrophoresis-electrospray ionization mass spectrometry of nucleotides, *Electrophoresis*, 16 (1995) 389–395.

52 W.H. McFadden, Mass-spectrometric analysis of gas-chromatographic eluents, *Adv. Chromatogr.*, 4 (1967) 265–332.

53 J.D. Henion, A comparison of direct liquid introduction LC/MS techniques employing microbore and conventional packed columns, *J. Chromatogr. Sci.*, 18 (1980) 101–115.

54 D.E. Games, P. Hirter, W. Kuhnz, E. Lewis, N.C.A. Weerasinghe and S.A. Westwood, Studies of combined liquid chromatography-mass spectrometry with a moving-belt interface, *J. Chromatogr.*, 203 (1981) 131–138.

55 R.D. Voyksner, C.S. Smith and P.C. Knox, Optimization and application of particle beam high-performance liquid chromatography/mass spectrometry to compounds of pharmaceutical interest, *Biomed. Environ. Mass Spectrom.*, 19 (1990) 523–534.

56 C.R. Blakley and M.L. Vestal, Thermospray interface for liquid chromatography/mass spectrometry, *Anal. Chem.*, 55 (1983) 750–754.

57 L.J. Deterding, M.A. Moseley, K.B. Tomer and J.W. Jorgenson, Coaxial continuous flow fast atom bombardment in conjunction with tandem mass spectrometry for the analysis of biomolecules, *Anal. Chem.*, 61 (1989) 2504–2511.

58 F.W. McLafferty (Ed.), *Tandem Mass Spectrometry*, Wiley, New York, 1983.

59 K.L. Busch, G.L. Glish and S.A. McLuckey (Eds.), *Mass Spectrometry/Mass Spectrometry*, VCH Publishers, New York, 1988.

60 J.C. Schwartz, A.P. Wade, C.G. Enke and R.G. Cooks, Systematic delineation of scan modes in multidimensional mass spectrometry, *Anal. Chem.*, 62 (1990) 1809–1818.

61 M.E. Bier, Analysis of proteins by mass spectrometry. In: G.C. Howard and W.E. Brown (Eds.), *Modern Protein Chemistry*, CRC Press, Boca Raton, 2002, pp. 71–102.

62 Z. Zhou, S. Ogden and J.A. Leary, Linkage position determination in oligosaccharides: mass spectrometry (MS/MS) study of lithium-cationized carbohydrates, *J. Org. Chem.*, 55 (1990) 5444–5446.

63 C.L. Andrews, A. Harsch and P. Vouros, Analysis of the *in vitro* digestion of modified DNA to oligonucleotides by LC/MS and LC/MS/MS, *Int. J. Mass Spectrom.*, 231 (2004) 169–177.

64 W.J. Griffiths, Tandem mass spectrometry in the study of fatty acids, bile acids, and steroids, *Mass Spectrom. Rev.*, 22 (2003) 81–152.

65 G. Hopfgartner and E. Bourgogne, Quantitative high-throughput analysis of drugs in biological matrices by mass spectrometry, *Mass Spectrom. Rev.*, 22 (2003) 195–214.

66 D.K. Han, J. Eng, H. Zhou and R. Aebersold, Quantitative profiling of differentiation-induced microsomal proteins using isotope-coded affinity tags (ICAT) and mass spectrometry, *Nat. Biotech.*, 19 (2001) 946–951.

67 J.L. Holmes, Assigning structures to ions in the gas phase, *Org. Mass Spectrom.*, 20 (1985) 169–183.

68 J.H. Bowie, The fragmentations of even-electron organic negative ions, *Mass Spectrom. Rev.*, 9 (1990) 349–379.

REVIEW QUESTIONS

1. What two attributes of mass spectrometry make it a versatile and valuable analytical technique? (sensitivity and specificity).

2. In additional to analytical chemistry, to what other area of chemistry has mass spectrometry made significant contributions? (gas-phase unimolecular chemistry).

3. What are the six basic elements of a mass spectrometer? (sample introduction, ion source, mass analyzer, detector, signal processing and computer).

4. How are low vapor pressure solid samples introduced into the source of a mass spectrometer? (direct insertion probe).

5. What dictates the appearance of a mass spectrum? (structure of the neutral/ion and the energetics of the ion formation event).

6. Increasing molecular mass eventually limits the effectiveness of gas-phase ionization methods. What is the practical upper limit? (approximately 1000 Da).

7. Why are desorption ionization techniques well suited to the analysis of biomolecules? (they do not depend on thermal energy to volatilize analytes).

8. What features of a mass spectrum indicate that a "soft" ionization technique has been employed? (few if any fragment ions).

9. What are the respective kinetic energy regimes of sector and quadrupole mass spectrometers? (thousands of electron volts and tens of electron volts).

10. What are the three commonly used chemical ionization reagent gases? (methane, isobutane, water, ammonia, methanol, hydrogen, oxygen, nitrogen, etc.).

11. What is one of the disadvantages of APCI? How does this disadvantage manifest itself? (only a limited variety of reagent ions can be produced from common HPLC eluents; many of those eluents produce reagent ions that are relatively weak gas-phase acids, which limits the range of compounds that can be ionized effectively).

12. What role does the dopant play in APPI? (Ionization of molecules having a low photoionization cross section (probability) has been shown to be enhanced by the use of a dopant that is introduced into the vaporized plume of analyte molecules; the dopant is selected on the basis of its high UV absorptivity and serves as a charge transfer reagent).

13. What characteristic of ESI permits large biomolecules ($M_r >$ 10,000) to be analyzed using mass spectrometers having modest mass range capabilities? (multiple charging reduces the "effective mass" to within the accessible mass range of conventional mass analyzers).

14. Two adjacent signals in an electrospray mass spectrum of a pure protein sample have m/z values of 893.9 and 834.3. Calculate the charge state of each signal and the mass of the neutral protein. (+14 and +15; 12,500 Da).

15. Name two commonly used MALDI matrix compounds. (3,5-dimethoxy-4-hydroxy-3-phenylacrylic acid, 2,5-dihydroxybenzoic acid ... among others).

16. What is a major advantage of using fast atoms rather than ions as bombarding particles in desorption ionization using a liquid matrix (i.e., FAB)? (eliminates charging, which can lead to signal instability).

17. What are the two advantages of using ions as bombarding particles in desorption ionization (SIMS)? (an ion beam can be focused for improved spatial resolution and can be rastered for imaging applications).

18. What mass analyzer exhibits the highest achievable mass resolving power? (ICR spectrometer, a.k.a. FTMS).

19. What is one of the significant advantages of a linear (2D) ion trap mass analyzer when compared to a 3D (Paul) ion trap? (increased ion capacity, which leads to improved sensitivity).

20. What are the three most common tandem mass spectrometry (MS/MS) scan modes? (product ion scan, precursor ion scan, constant neutral loss scan).

Chapter 12

Theory of separation methods

C.H. Lochmüller

12.1 INTRODUCTION

In the strictest meaning of the word, there is no theoretical basis for separation methods but rather models. Models are, in most cases, a mixture of empirical and chemical kinetic/chemical thermodynamics. There are many examples of attempts to relate known physical and chemical properties to observe and develop a "predictive theory". The goal is both an improvement in understanding of the underlying phenomena and an efficient approach to method development. This latter goal is of greater interest to the vast majority of those who make use of separation methods, as their interest is in either obtaining relatively pure material or in isolating components of a chemical mixture in time and space for quantitative measurement. If a separation method can also provide complimentary information leading to qualitative identification, so much the better.

This chapter seeks to give the "user" of chemical separation methods the beginnings of a basis for understanding the methods described in this book and the ability to recognize "normal behavior" and to distinguish "anomalous behavior". Justice to all the important theory would require several volumes of substantial size and especially if historical justice were to be given to the development of current models. At times the chapter's content will seem more conversational than hard scientific and the choice of style in any given instance, is deliberate. Stories are part of the history of separation methods, after all.

12.2 THE GREAT DIVIDE

Julius Caesar tells us "All Gaul is divided into three parts". Separation methods can be divided into two categories: equilibrium and

Comprehensive Analytical Chemistry 47
S. Ahuja and N. Jespersen (Eds)
Volume 47 ISSN: 0166-526X DOI: 10.1016/S0166-526X(06)47012-4

non-equilibrium. The equilibria spoken of in this case are "chemical" rather than physical in the vast majority of cases.

They are the kind of equilibria spoken of as "ionization in solution" or between two phases, such as gas and liquid in introductory chemistry courses. Finding the point between a knife tip and the end of its' handle that when placed on a finger leaves the knife neither tipping toward one end or the other, is the location of the "center of gravity". Supported at that point the knife is said to be "in balance" and that balance is a physical equilibrium. It is based on simple mechanics and the concept of mass and length. Chemical equilibria arise from not just mass but the underlying properties of molecules and their interaction with each other. A sealed half-full bottle of gin contains a liquid which is ~40% by volume ethyl alcohol and 60% water. If we call the liquid the condensed phase, the vapor phase is the gaseous region above the condensed phase. The composition of the vapor phase is found to be different than the condensed phase and richer in alcohol than water. At constant temperature, the mole fraction of alcohol in the vapor phase is higher than water. Why? Ethanol is more volatile than water. True and it is also true that the composition of ethanol in the vapor phase will reach a constant value in time. Separation theory attempts to understand why such a chemical equilibrium occurs between the two phases and why, if methanol is substituted for ethanol, the alcohol content in the vapor phase is higher. The fact that the equilibrium is established over time is a kinetic consideration, while the value of the equilibrium constant and its' temperature dependence is a question of chemical thermodynamics.

12.2.1 Physical separation methods

Physical separation methods can be based on equilibrium considerations, but the majority are not. Ordinary filtration is an example of a non-equilibrium, physical method and so is ordinary centrifugation—e.g.—the separation of a precipitate from the suspending liquid using an artificial gravity field. There are separation methods, which are called "filtration" which are not such as gel filtration. Ultracentrifugation in a salt gradient is a physical equilibrium method.

Separation of desirable components of a mixture is not a new desire on the part of humanity. Removal of water-borne particles including sand, silt, and bacteria is desirable for both sensory appeal and health. The use of sand as a filtration medium for such filtration is very old. The use of a pan to separate gold dust from silica sand by gold miners is a use of the large difference in density to wash the lighter silica away from the much denser gold, which resembles black sand. Not filtration but certainly a non-equilibrium, physical method.

Molecules which exhibit optical activity are molecules which have a "handedness" in their structure. They are "chiral". Chemists often have reasons to obtain chemical pure aliquots of particular molecules. Since the chirality of molecules can influence biological effect in pharmaceuticals, the chiral purity of a drug substance can pose a challenge both in terms of obtaining the molecules and in assaying the chiral purity by instrumental methods. While diastereomers can have different physical properties including solubility, enantiomers have the same physical properties and the same chemical composition. How then to separate optically active molecules?

The first example of the deliberate separation of optically active molecules is appropriate as an example of physical separation in the clearest sense of the term. The molecules are referred to as optically active because polarized light interacts differently with right- and left-handed molecules. In the case of simple diastereomers the RR and SS forms are enantiomers while the RS and SR forms are not. The separation of the latter and former was first done under a microscope using crossed polarizers and the crystals which were seen were separated from those that caused little or no rotation of plane-polarized light by hand using tweezers. A truly physical separation of chemical species using a physical property of chemical origin!

One could go on with examples such as the use of a shirt rather than sand reduce the silt content of drinking water or the use of a net to separate fish from their native waters. Rather than that perhaps we should rely on the definition of a chemical equilibrium and its' presence or absence.... Chemical equilibria are dynamic with only the illusion of static state. Acetic acid dissociates in water to acetate-ion and hydrated hydrogen ion. At any instant, however, there is an acid molecule formed by recombination of acid anion and a proton cation while another acid molecule dissociates. The equilibrium constant is based on a dynamic process. Ordinary filtration is not an equilibrium process nor is it the case of crystals plucked from under a microscope into a waiting vial.

12.2.2   Chemical separation methods

Most, but not all, chemical separation methods are based on equilibrium in the sense of chemical equilibrium. Clearly, solubility is a chemical question but formation of a precipitate and filtration is a physical separation, which happens to use a favorable K_{sp} equilibrium

C.H. Lochmüller

TABLE 1

Possible combinations and typical separation methods using them in a not-all-inclusive list

Interface type	Separation method(s)
Gas–gas	Possible but no common examples
Gas–liquid	Gas chromatography; distillation
Gas–solid	Gas chromatography; gas masks
Liquid–liquid	Extraction; liquid chromatography
Liquid–solid	Liquid chromatography; solvent drying
Solid–solid	Possible but no common examples

value to begin the process. Distillation and extraction are indeed chemical separation methods based on equilibrium and differences in equilibrium constants.

Chemical separations are often either a question of equilibrium established in two immiscible phases across the contact between the two phases. In the case of true distillation, the equilibrium is established in the reflux process where the condensed material returning to the pot is in contact with the vapor rising from the pot. It is a *gas–liquid interface*. In an extraction, the equilibrium is established by motion of the solute molecules across the interface between the immiscible layers. It is a *liquid–liquid interface*. If one adds a finely divided solid to a liquid phase and molecules are then distributed in equilibrium between the solid surface and the liquid, it is a *liquid–solid interface* (Table 1).

12.3 CHEMICAL EQUILIBRIA AND THE PROPERTIES OF THE EQUILIBRIUM CONSTANT

Chemistry deals with molecules not atoms. True thermodynamics knows no molecules with much less properties of molecules derived from chemical effects. Its' origin is in such concerns as heat flow and the heat equivalent of mechanical work. Most of us have heard in physical chemistry about how it was the drilling of cannon barrels that created the connection between work and heat energy. One can take entire semester course on thermodynamics in physics and in engineering and never deal with the solution thermodynamics, which often dominates chemistry courses. To the extent that thermodynamics has been used in developing a theory for separation methods, it is almost entirely chemical thermodynamics.

400

In talking about thermodynamics and the properties of chemical equilibrium constants it is very difficult for chemists to avoid attempts to include the influence of forces between molecules on the magnitude of the equilibrium constant and differences between observed equilibrium constants. For the purpose of this chapter, it is convenient to talk first about the equilibrium constant and the macroscopic properties of matter which affect it first. Next, the reader will be introduced to the concept of forces between molecules, their relative magnitude and influence is separations.

12.3.1   Thermodynamics and kinetics in separations

If all that is achieved, in an attempted separation, is the transfer of the components of the mixture (atoms and/or molecules) from one place to another, no separation is achieved. The result is just mass transport. An analogy would be carrying a bag of mixed groceries from the store to a car. The tomatoes, onions, garlic and meat are still in the bag. What is needed is a means to have the components move to the end of an experiment but at different apparent rates of speed. For most equilibrium separation methods, the answer lies in controlling the thermodynamics of the system. Molecules or atoms are made to move in one of the two immiscible phases while undergoing transfer to the other, *stationary phase* and quickly back to the *mobile phase*. If the net effects in the molecules or atoms move at different rates depending on their chemical character, some will move faster and others more slowly. The goal would be to have molecules of type A moving faster than type B and both faster than type C and so on. Ideally, it would be best if they remained in tight "packets" as the forward motion of the mobile phase caused them to migrate in space and the difference in net rate moved the packets further and further apart as time passed.

The simplest rules of thermodynamics suggest that energy must be expended to do work—"You cannot get something for nothing," and that even if work is done some energy is forever lost to useful work— "You cannot even get what you paid for". And that this entropy effect is such that the entropy of the universe is forever driving toward a maximum—"Nature spontaneously falls into a mess!" Humor aside, the consequence is that any narrow packet as described above will spread over space in an attempt to make the local and universal mole fraction of A, B or C ... the same everywhere.

Put in ordinary terms, the more successful we are in causing a separation, the more propensities there are for a re-mixing of the components. There are many ways this can occur but there are a fewer number of important routes to mixing. It seems reasonable that we examine these before we consider all the possible ways in which thermodynamics can be controlled in general terms. In almost all equilibrium separation systems, the separation can occur either in a packed bed of particles or fibers or in an open channel or tube. The stationary phase is either coated on the walls of the channel or on the particles/fibers of the packed bed. If there were no mixing mechanisms an infinitely narrow packet containing the components would become a series of infinitely narrow packets of pure components moving at different velocities toward the end of the packed bed or tube.

12.3.2 Mechanisms for band broadening

How to best describe this broadening we expect to occur? One way is by analogy to random error in measurements. We know or assume there is a truly correct answer to any measurement of quantity present and attempt to determine that number. In real measurements there are real, if random, sources of error. It is convenient to talk about standard deviation of the measurement but actually the error in measurement accumulates as the sum of the square of each error process or variance producing mechanism or total variance = sum of the individual variances. If we ignore effects outside the actual separation process (e.g. injection/spot size, connecting tubing, detector volume), this sum can be factored into three main influences:

1. The random path of molecules taken in a packed bed (called Random Path Term).
2. Ordinary diffusion along the length of the path taken by a packet of molecules (called Longitudinal Molecular Diffusion).
3. Departure from true equilibrium of the molecules moving over the stationary phase and moving in and out of the stationary phase (called Resistance to Mass Transfer).

If we give variance then

$$\sigma^2[\text{total}] = \sigma^2[\text{RPT}] + \sigma^2[\text{LMD}] + \sigma^2[\text{RTMT}]$$

12.3.3 Multipath processes

If flow occurred in an open channel with no particles or fibers and if there were no other mixing mechanisms, all particles would transit the same distance from beginning to end. In a packed bed, each time a molecule or atom encounters a particle or fiber it must go around it to continue on. It is analogous to encounter a tree in a field—one either walks around it to the right or the left and that is equivalent to flipping a coin. Some molecules will encounter more particles than others as illustrated in the following scheme where each encounter causes a chance in direction and the path of a hypothetical molecule is traced by a line (Scheme 1).

Since the paths are random and the number of collisions large, the effect is to make the paths random and distributed as a Gaussian distribution. If you are familiar with the Random Walk model, this can be considered as analogous Random Path Model. The actual situation is a bit more complicated but for practical purposes the intuitive insights that the degree of broadening will be influenced by particle size d_p (and, thus, distance between particles or fibers) and the geometry of the packing. The variance produced per unit length l of the bed traversed (σ^2/l) can be given the symbol H:

$$H = 2\lambda d_p$$

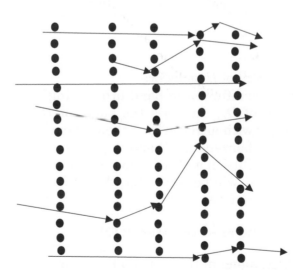

Scheme 1. Traces of the path of molecules in a packed bed.

The symbol λ in this case was first introduced by Van Deemter to account for packing inhomogeneity.

12.3.4 Broadening by diffusion

If a band of molecules or atoms are placed in a container such as a tube at the center of its length, the molecules will diffuse in the direction of lower concentration. Clearly this occurs in all directions but in a tube the walls are a limit and so our concern is along the axis of movement caused by the mobile phase and this is referred to as *longitudinal molecular diffusion*.

The Einstein equation for diffusion along a single axis states that the process is Gaussian and the variance is time dependent:

$$\sigma^2 = 2Dt$$

where D is a characteristic diffusion coefficient. The time a sample spends in the moving phase on its transit from the beginning to the end of the bed is proportional not to the flow rate but the flow velocity of the mobile phase, which we can call u. If the time to transit a distance l is l/u then for a unit distance $t = 1/u$. It can be seen then that the variance contribution per unit length in an open tube is $\sigma^2/l = 2D_m/u$ where D_m is the diffusion coefficient for the molecules of interest in the moving or mobile phase. Van Deemter added a factor γ which has the value of 1 in open space and less than 1 in a packed bed of particles or fibers. The final equation becomes

$$H = 2\gamma D_m/u$$

when we use the symbol H as before.

How important is diffusion in band broadening? Clearly it depends on the magnitude of D_m. In gases the value of D is relatively large compared to liquids with nitrogen in air being $1\,cm^2\,s^{-1}$. The value for a small molecule like benzene in ethanol liquid is $\sim 10^{-4}\,cm^2\,s^{-1}$. Longitudinal molecular diffusion is clearly more of concern in gaseous mobile phase systems but as will be seen later, diffusion is important in condensed or liquid systems as well.

12.3.5 Resistance to mass transfer (RTMT)

A molecule carried along in the mobile phase must contact the stationary phase if the system is an equilibrium separation method. That means that a molecule moving between particles must sense the chemical potential in the direction of the stationary phase and move toward it.

The diving force is the Gibbs free energy change and the mechanism is diffusion. This is, by analogy, akin to dropping a bowling ball from a moving airplane in that the approach of the molecule will take a parabolic path to the stationary phase moving forward because of the flow of the mobile phase and "down" by diffusion rather than gravity. The Van Deemter expression for this broadening in terms of variance due to mobile phase resistance to mass transfer to the stationary phase is

$$\sigma^2_{RM} = \left(f(k)d_p^2/D_m \right) u$$

where $f(k)$ is a function related to that fraction of molecules in the moving phase because of the equilibrium.

A solute entering onto or into a stationary phase must re-enter the moving phase when the molecules in the moving phase move on. For molecules that are adsorbed on a surface, this leaving can be a rapid process with little variation in how fast any given molecule returns. When the stationary phase is a condensed liquid film, there can be significant differences depending on how near to the surface interface the molecule is at the time when reentry into the mobile phase is required to maintain equilibrium. Here the orders of magnitude show a smaller value of Ds—the diffusion coefficient of the molecule in the liquid stationary phase—compared to the gas phase is again important and an expression analogous to the moving phase RTMT can be written.

An expression can be written which lumps together RTMT in both mobile and stationary phase and sums the variance contributions which can be written as

$$H = A + B/u + Cu$$

Here, A is the random path, B the longitudinal molecular diffusion and C the RTMT contributions with the velocity of the mobile phase u shown separately. H is referred to as the "Height Equivalent to a Theoretical Plate" and is terminology borrowed from distillation. While the distillation HETP is not truly applicable, the terminology has persisted. It can be shown that the H in this expression is the equivalent to variance/unit length. This is the expression introduced by Van Deemter and co-workers in 1956 in a discussion of band broadening.

12.3.6 Practical consequences

The Van Deemter equation contains a linear term A influenced only by particle size and geometry of packing of the particles. It is a constant and sets a lower limit for variance production. The diffusion term B is inversely proportional to velocity and will become asymptotic to the value of A at large u suggesting that high velocities are best. It is unfortunate for practical analysts and analysis time that the C term is directly dependent on velocity. This means attempts to speed up analysis by increasing flow velocity will become a trade off in band broadening vs. analysis time. Smaller particle size will reduce the A term and improve the mobile phase contribution to C but forcing a gas or liquid through a bed of tinier particles will lead to a need for higher pressures at the inlet, compression of a gas and wear on pumps for liquid delivery. Better, more uniform packing will help as would particles that are spherical and of quite narrow particle size distribution. Modern packings are spherical and of narrow particle size distribution leading to more permeable and lower variance contribution. Fibrous beds such as paper or the fiber-webs of polymer gels are difficult to produce in high uniformity of fiber-gap and fiber-diameter at present.

12.4 A CHEMIST'S VIEW OF EQUILIBRIUM CONSTANTS

A plug of molecules of the same kind moving under the flow of the mobile phase and in equilibrium with a stationary phase will require a volume of mobile phase V_r to be washed from one end of a separator to the other. A packed tube (or column) full of particles has an open total volume occupied by mobile phase. If there is no equilibrium—if the equilibrium constant is $K = 0$—it is void volume that is required for an infinitely narrow plug through the system. It is more common to define that volume V_0 as the volume here half the plug has eluted. If K is finite and non-zero, the net effect is for the plug to move more slowly and for it to take a volume larger than V_0 to achieve elution.

The retention of the band or "peak" beyond what V_0 predicts depends on the magnitude of the equilibrium constant and logically on the volume V_s or area A_s of the stationary phase. The equation of importance is $V_r = V_0 + K V_s$ and the net retention $V_r' = K V_s$. Two main factors influence the value of the equilibrium constant and these are the chemical nature of the mobile and stationary phases. Chemistry is molecules and while true thermodynamics knows no molecules or forces between molecules, chemists think in terms of molecular properties. Among those properties, there is a consideration of the kinds of forces that exist between molecules. Granted that thermodynamics are energy not force considerations but it is useful to understand the main forces involved in the interaction between molecules. Put another way, it is the chemist who asks why the boiling point of water (a small molecule) is 100°C but if we methylate water—

$$H-O-H \Rightarrow CH_3-O-H \Rightarrow CH_3-O-CH_3$$

– the boiling point changes to ~30°C on the first methylation and even more so in the dimethyl case. One way to think about that observation is in terms of forces between molecules.

The concept of forces between molecules was an area of intense interest in the early third of the 20th century. The focus at that time was not on explaining chemical separation concepts. The focus, rather, was on explaining why liquids exist, why solids exist above absolute zero, why members of homologous series have incremental changes in their boiling points depending on—e.g.—carbon number. Even more challenging was the question of why helium can be liquefied! Among the scientists who pondered this question are Nobel laureates—e.g. Debye—and near Nobel laureates—e.g.—London.

We are all familiar with forces between bodies on a macroscopic scale. If one sits at their desk on earth pondering the view and inadvertently lets go of their lunch sandwich it falls until it lands on your manuscript, your lap or the floor—right? Better yet, a cannonball and a softball fall at the same rate. It is the definition of "gravity". Long after the Tower of Pisa experiments, we discovered why the moon is in seemingly stable orbit around the earth and the earth around the sun is related to forces between bodies. We are also familiar with electrostatic forces and their effects. Dust from the air in the room is attracted to the screen of everyone's TV because of the electrostatic charge it develops. It would seem reasonable that attraction occurs on a smaller scale and even on a molecular scale that does not involve energies on the order of true chemical bonds.

12.4.1 Categories of forces

It is possible to make a list of forces between molecules in the order of weakest to strongest:

- *Dispersion forces*. These are weak attractions caused by instantaneous fluctuation of the electron distribution in molecules and even atoms. They were first posed by Fritz London whose focus was on helium liquefaction. Such London forces fall off with the sixth power of the distance of separation. Any individual fluctuation creates a +/− local charge and that instantaneous dipole can interact with other such instantaneous dipoles nearby. The important

insight is that while weak individually the sum of all these interactions occurring at any moment is a significant component of the
internal energy of the system. In a homologous series like pentane,
hexane, heptane, octane each additional—CH_2- group adds one
more possibility of a local fluctuation, which on average increases
the overall chance of attraction between molecules. The result
should be an increment in the boiling point of these liquids that is
monotonic and this is what is observed. (The reader is invited to
make a plot of boiling point vs. carbon number for the alkanes
beginning with propane and ending with dodecane.)

- *Dipole–dipole forces.* If there is some substituent on a molecule,
which is sufficiently electronegative to cause a permanent polarization
of a chemical bond, this creates a permanent dipole. The results are
referred to as Debye forces. There are untold examples but a simple
one to consider is a comparison of the properties of benzene and
chlorobenzene. Benzene is non-polar in the sense we use the word
here. Chlorobenzene is polar. The reason is that chlorine is more
electronegative than carbon and the electron distribution in chlorobenzene is determined by the electron-withdrawing effect of the chlorine substituent. The result is a permanent molecular dipole, which
can react with other permanent dipoles and cause larger changes in
the observed melting and boiling point that the difference in molecular
weight between benzene and chlorobenzene would predict.

What about dichlorobenzenes? Substitution of another hydrogen by
chlorine creates another local dipole and a more polar molecule. Certainly 1,2 and 1,3-dichlorobenzene are more polar but 1,4-dichloro-
benzene poses a quandary. It is what chemist call a polar molecule but
the opposing chlorines result in a net molecular dipole of zero. Right?
Perhaps it is best for chemists to think in terms of "bond-dipoles"
rather than molecular dipoles and in many cases this is the case in
chemical separation discussions. (Reader should look up the properties
of the dichlorobenzenes, such as melting and boiling point and liquid
density.)

- *Dipole-induced dipole forces.* A molecule with a strong molecular or
bond dipole can induce a dipole in a molecule nearby that is polarizable. These Keesom forces have the same inverse 6th power
dependence with distance. An example could be the interaction of
chlorobenzene with naphthalene.

- *Hydrogen bond forces.* The hydrogen bond is a special case of dipole–dipole interaction but is often spoken of separately because most
hydrogen bonds are more energetic than other dipole interactions.

Hydrogen bonds involve a hydrogen donor and a hydrogen bond receptor—e.g.—ethanol and acetone.

12.4.2 Use of forces in manipulation of separations

There are many examples of the deliberate use of chemical intuition and understanding of forces between molecules in getting selective interactions that enhance separation. An entire book could be written just cataloging examples. It is more appropriate here to give a few stationary phase examples.

Aromatic molecules with no polar substituent include benzene derivatives or other, more polyaromatic molecules, such as naphthalene, phenanthrene, and anthracene. These are polarizable. Paraffins are not polarizable by comparison. In gas–liquid systems, aromatic molecules will show stronger interactions with polar stationary phases that paraffins of comparable boiling point and, thus, polar stationary phases can aid in improving separation of substituted aromatics.

A more familiar example of deliberate design of a stationary phase is familiar to organic chemists. It is called "argentation" and is done by treating a packing material with silver nitrate and using a mobile phase in which the silver salt is not soluble. Silver forms adducts with olefins and the adducts formed with *cis* geometric isomers is more stable than the corresponding *trans* isomer. The result is that *cis*-olefins show larger equilibrium constants for stationary phases that contain silver ions and are thus retained longer.

12.4.3 Properties of equilibrium constants

It is a great deal of work to actually determine a true equilibrium constant and most chemical separation methods speak in terms of values which are proportional to the actual equilibrium constant. At constant flow, the time that a given type of molecule is retained is related to the time for the void volume to pass after the sample is placed in a column or on a plate with the addition of the time for the net retention volume. If the flow remains constant, the temperature of the separation remains constant and no stationary phase is gained or lost, one can attempt qualitative identification using retention times. It is more reasonable to calculate the ratio of net retention volume to the void volume and call the result partition factor or capacity factor, k'.

$$k' = (t_{\mathrm{r}} - t_0)/0$$

Since k' is proportional to K, mathematical treatments appropriate for equilibrium constants apply to the capacity factor k'. Plots of log k' vs. $1/T$ where T is in degrees Kelvin for any given molecule should yield a straight line over reasonable ranges of temperature.

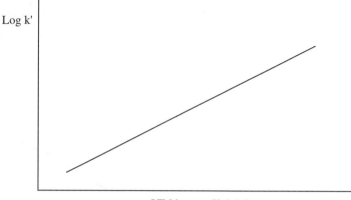

Since members of a homologous series have incremental boiling point differences and if the amount of any homolog in the moving gas phase is related to vapor pressure at the temperature of the experiment, plots of log k' vs. carbon number should also be a straight line. (The enthalpy of vaporization increases monotonically with carbon number.) This in fact is observed in gas–liquid equilibrium separation systems. It is the basis of "retention index" systems pioneered by Kovats for qualitative identification.

12.4.4 Choosing mobile and stationary phases

Mobile phases can be gases or liquids and stationary phases can be solids or liquids. If as suggested K and thus k' is proportional to vapor pressure in gas–liquid and gas–solid systems then manipulation of the apparent volatility is dependent on non-ideal behavior in the gas mobile phase and in the liquid phase in gas–liquid systems (the most common). Put in other terms k' is proportional to $1/\gamma p$ where γ is the activity coefficient of the sample molecules in the liquid and p the vapor pressure at the temperature of the experiment. Except at very high density, the contribution of the gas phase to non-ideality is small and fugacity coefficients are near 1. Changing mobile phase from helium, to hydrogen, to nitrogen to carbon dioxide changes k' by a few percent. The power is in the choice of stationary phase and, in part, is why there are so many stationary phases for gas–liquid separations. (Truthfully, many of the thousands of reported stationary phases are of equal value.)

In liquid–solid and liquid–liquid systems, manipulation of mobile phase properties is the dominant approach. Mixtures of poor and good solvents for the types of molecules being separated are prepared and retention observed. There are fewer stationary phases for liquid mobile phase systems for this reason and at least one other.

It is desirable that the equilibrium constant for a "solute" be not zero or very large lest there be no net retention or near infinite retention. The "catch" comes in the fact that liquids, which are relatively good solvents for a given type of molecule are also solvents for each other. This means the risk involved is by washing off the stationary phase with the mobile phase. Yet liquid–liquid methods offer much promise for relatively non-volatile but soluble molecules and their separation of one from the other. The discovery of liquid–liquid chromatography earned Martin and Synge the Nobel Prize when they applied it to amino acids with water mobile phases and organic liquid stationary phases.

The solution to stationary phase solubility is solved for most cases by covalently attaching the molecules to the solid chromatography packing. The chemistry used for the vast majority of such phases in current use is an adaptation of the sylilation chemistry used to modify window glass surfaces. The surface of silica gel contains Si-OH functions and these can be either esterified by cooking the silica with alkanols or by reaction with chloro- or alkoxysilanes. The most common attached molecules are propyl, octyl, and octadecyl hydrocarbons. This can either be thought of as converting the hydrophilic silica surface to a carbon-like one or as a very thin film of hydrocarbon. These materials are stable to moderate pH water-based mobile phases. They are the basis of "reversed-phase" method where the moving phase is water or water-methanol, water acetonitrile, and water tetrahydrofuran mixtures (polar) and a hydrocarbon-like stationary phase.

Further Reading

R.P.W. Scott, *Liquid Chromatography Column Theory*, Wiley, UK, ISBN 0 471 93305 8, 1992.
A.J.P. Martin and R.L.M. Synge, *Biochem. J.*, 35 (1941) 358.
Roger M. Smith, *Retention and Selectivity in Chromatography*, Elsevier, Amsterdam, ISBN 0-444-81539-2, 1995.

C.H. Lochmüller

REVIEW QUESTIONS

1. If gas–liquid and gas–solid separations are dependent on the "saturation vapor pressure" of the chemical component undergoing equilibration: (a) What is the expected effect when the temperature of the system is raised? (b) If the system is a gas-liquid system sketch what a plot of log V_r' vs. $1/T$ would look like including when the T is below the freezing point of the stationary phase. (c) Why might it be better to sample the vapor phase above a solution as a sample to determine trace materials in the solution?

2. Does solubility in the mobile phase influence net retention? Polyethylene glycols are more soluble in cold solvents than warm. What is the noted effect if the temperature of the liquid chromatography separation of poly ethers is raised?

3. The addition of a detergent such as sodium sulfate can cause a protein to completely unfold. What would be the effect of adding sodium dodecylsulfate to the mobile phase of a protein size separation? Explain why?

4. Separation scientists speak of a "general elution problem" when asked to develop a universal separation method using chromatography. What is the problem? (Hint: you may need to consult works by Snyder or Heftmann in the library.)

Chapter 13

Thin-layer chromatography

Pamela M. Grillini

13.1   INTRODUCTION

Thin-layer chromatography (TLC) is one of the most popular and widely used separation techniques because of its ease of use, cost-effectiveness, high sensitivity, speed of separation, as well as its capacity to analyze multiple samples simultaneously. It has been applied to many disciplines including biochemistry [1,2], toxicology [3,4], pharmacology [5,6], environmental science [7], food science [8,9], and chemistry [10,11]. TLC can be used for separation, isolation, identification, and quantification of components in a mixture. It can also be utilized on the preparative scale to isolate an individual component. A large variety of TLC equipment is available and discussed later in this chapter.

High-performance thin-layer chromatography (HPTLC) has become widely used and while it follows the same principles as TLC, it makes use of modern technology including automatic application devices and smaller plates, which allow for better sensitivity.

TLC is related to paper chromatography (PC) as both use a stationary phase and a liquid phase to move the sample [12]. A common example of PC is the separation of black ink into its individual colors. Because the individual molecules behave differently when exposed to a solvent such as water or isopropyl alcohol, they are retained on the paper at different intervals, creating a visible separation of the individual components. This helps to identify each component in a mixture.

13.2   THEORY AND BASIC PRINCIPLES

In TLC, the sample is applied as a small spot or streak to the marked origin of stationary phase supported on a glass, plastic, or metal plate. The sample solvent is allowed to evaporate from the plate that is then placed in a closed chamber containing a shallow pool of mobile phase at

Comprehensive Analytical Chemistry 47
S. Ahuja and N. Jespersen (Eds)
Volume 47 ISSN: 0166-526X DOI: 10.1016/S0166-526X(06)47013-6

the bottom. The mobile phase moves through the stationary phase by capillary forces. The components of the sample mixture migrate at different rates during movement of the mobile phased through the stationary phase. When the mobile phase has moved an appropriate distance, the plate is removed from the chamber and the solvent front is marked. Mobile phase is evaporated from the plate by drying the plate at room temperature, by forced airflow, or in a heated oven. If the components of the mixture are not naturally colored or fluorescent, a derivatizing reagent is applied to visualize the bands. Sometimes, more than one detection technique is used to ensure the detection of all components in the mixture [13,14].

The R_f value is a convenient way to express the position of the substance on a developed plate. It is calculated as follows: $R_f =$ distance of component from origin/distance of solvent front from origin.

A variety of sorbents have been used as the stationary phase in TLC including silica gel, cellulose, alumina, polyamides, ion exchangers, chemically modified silica gel, and mixed layers of two or more materials, coated on a suitable support. Silica gel is by far the most commonly used sorbent supported on a glass plate. Currently, in the pharmaceutical industry, commercially available precoated HPTLC plates with fine particle layers are commonly used for fast, efficient, and reproducible separations. The choices of mobile phase range from single component solvent systems to multiple-component solvent systems with the latter being the most common. The majority of TLC applications are normal phase, which is also a complementary feature to high-performance liquid chromatography (HPLC) (see Chapter 15) that uses mostly reverse-phase columns.

Tables are available to list the level of interaction of the sorbent and mobile phase [15]. The higher the mobile phase strength, the greater the interaction with the sorbent, thus the greater the R_f of the solute. Usually, one tries to select a mobile phase that will yield an R_f value of 0.3–0.7, especially if trying to evaluate impurities. In reality, though, single mobile phase systems do not yield the separation desired subtle changes in multiple component mobile phase systems is the rule. Strength is defined as a function of free energy of adsorption on the solid surface. Thus, when a mobile phase has proper strength to give the desired R_f, it may not achieve the desired separation. Numerous references are available to guide in mobile phase selection, while Zweig and Sherma's table is an excellent starting place [15].

While silica plates are a typical place to start, separation of closely related structures can be achieved by switching to a specialty plate, such as amino. In this case, separation is obtained as a result of mobile phase and its interaction with the unusual sorbent.

The migration of each component in a mixture during TLC is a result of two opposing forces: capillary action of the mobile phase and retardation action of the stationary phase. Both forces contribute to achieve differential migration of each component. Developed TLC plates can be detected by various means, based on the nature of the sample. They could be nondestructive (UV densitometer), destructive (derivatizing agents), or the combination of both. The results can be documented by photography and saved electronically for archiving and future reference.

At present, when HPLC is the prevalent analytical method, why would one use TLC? While HPLC is widely used for separation and quantification, TLC remains a valuable and commonly used separation technique because of its features that are complementary to HPLC. The majority of TLC applications use normal-phase methods for separation, whereas reversed-phase methods dominate in HPLC. Some of the most important features of TLC compared to HPLC are briefly discussed here:

1. *Open format of stationary phase and evaluation of the whole sample*: In TLC separation, a mixture is applied to the stationary phase followed by development. It is an open system from separation to detection. In contrast to TLC, HPLC is a closed-column system in which a mixture is introduced into the mobile phase and solutes are eluted and detected in a time-dependent manner. There are times that TLC reveals new and unexpected information about the sample, while that information is lost in HPLC by retention on the column, because of strongly sorbed impurities, early elution, or lack of detection. In addition, TLC has little or less contamination with a disposable stationary phase while in HPLC the column is repeatedly used.

2. *Simple sample preparation*: Samples for TLC separation often involve fewer cleanup steps because every sample is separated on fresh stationary phase, without cross-contamination or carryover. Even strongly absorbed impurities or solid particles in samples are not of much concern. This would be a disaster for HPLC separation, leading to column buildup, decay, and eventually destroying the performance.

3. *High sample throughput*: The simultaneous but independent application and separation of multiple samples in TLC results in

higher sample throughout and less time consumption, as well as lower costs. HPLC cannot compete with TLC in terms of the number of samples processed in a given time period.

4. *Flexible and versatile dissolving solvent and mobile phase*: The choice of the sample solvent is not as critical in TLC as in HPLC because it is removed by evaporation before development of the plate. On the contrary, in HPLC the dissolving solvent chosen must be compatible in terms of composition and strength with the column and mobile phase. The same logic applies to the TLC mobile phase that is completely evaporated before detection. Therefore, the UV-absorbing properties, purity, or acid and base properties of the mobile phase are not as crucial as with HPLC. In addition, there is less solvent waste in TLC than in HPLC.

5. *General and specific detection methods* [13,14]: In TLC, many different detection methods, including inspection under short wave and long wave UV light and sequential application of a series of compatible reagents can be employed on a developed TLC plate. Most well established and routinely used HPLC detection methods still use UV, which cannot always capture all the eluates from column separation.

6. *One-dimensional multiple development and two-dimensional development*: Multiple developments through one or two dimensions can be applied to separate certain components in sequence, with detection at each step. This gives a theoretical increase in the capacity of the spots, so it is ideal for the separation of mixtures with a large number of components. In addition, it is a useful tool to confirm the purity of a given component. Though hyphenated HPLC could serve as a multiple separation technique, TLC takes the lead in this area by its faster separation and choice of different mobile phases and detection methods through each run.

The comparison of TLC and HPLC has been described in a number of publications by a number of authors [16–22].

13.3   INSTRUMENTS AND THEIR USE

In the United States, few vendors exist to obtain HPTLC equipment. CAMAG is a Switzerland based company, and distributes the majority of HPTLC equipment in the United States.

Key instruments required for HPTLC analysis includes a sample application device, similar to the Automatic TLC Sampler 4, a digital camera and photodocumentation system (such as DigiStore) equipped with a light box similar to the Reprostar 3, and for quantitative studies, a densitometer similar to the Scanner 3. In addition, basic lab equipment such as glass twin trough chambers, an oven capable of maintaining up to 100°C temperature, and a hood for spraying plates or TLC spraying cabinet are essential. Additional helpful equipment would include the HPTLC Vario System for method development, and horizontal developing chambers. These are described in more detail later in this section.

In order to utilize TLC as a reliable technique, the sample must be readily soluble in organic solvents such as methanol, isopropanol, or mixtures of methanol and other additives such as ammonium hydroxide and hydrochloric acid. For any type of quantitative testing to be done on a pharmaceutical sample, it must be soluble at a concentration of $\cong 25$ mg/ml. If sonication is necessary to disperse the sample, it may be used. The sample must also be stable in organic solvents for at least 1 h ideally. Validation testing should be conducted throughout the duration of this test to ensure the sample stays in solution, and does not degrade in the solvent of choice.

Samples are usually prepared by weighing a 25 mg sample of drug substance and placing it into an appropriate vial. One millileter of organic solvent, such as methanol, is then added to the vial. The sample is sonicated for 1–2 min, until solution is clear, and now ready to be applied to HPTLC plates.

Limits of Detection (LOD) are not very low when utilizing traditional TLC, however, they improve greatly when using HPTLC. On a typical pharmaceutical sample that is tested at full strength of 25 mg/ml, an LOD of 0.05% or 0.0125 mg/ml can easily be achieved. Often a derivatizing reagent must be utilized, though depending on the compound, an LOD of 0.05% can be achieved using ultraviolet (UV) light at 254 nm.

In order to evaluate a compound for potential impurities or degradants, the shape of the band is often sacrificed in order to achieve the necessary LOD. While distortion of the main band is undesirable, it may become necessary to observe small impurities and the concentration of the main component is increased to achieve the required LOD, as illustrated in Fig. 13.1.

Once the sample is dissolved in a suitable solvent, it should be immediately applied to the HPTLC plate. The Automatic TLC Sampler 4 (ATS 4) is suitable for delivering reproducible spots or bands in a short amount of time. It utilizes a PC and can be programed easily to apply

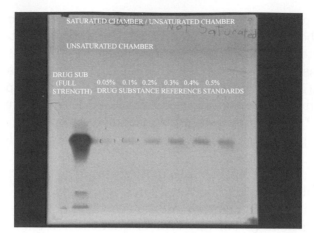

Fig. 13.1. Band shape sacrificed to obtain necessary LOD.

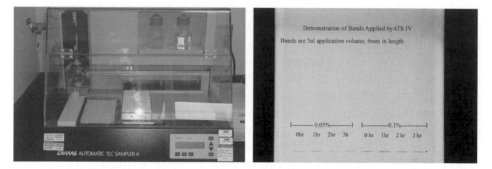

Fig. 13.2. Example of application device and demonstration of the reproducible, tight, bands it can apply.

varying amounts, band sizes, and samples from different vials to one or more HPTLC plates. Typically, a 10×10 cm HPTLC silica plate will hold seven bands that are 7 mm long, and 1–10 µl on average. This applicator can also apply much larger quantities and works on 20×20 cm plates as well. Instrumentation Qualification and Operational Qualification testing can be performed on this equipment to make it suitable for use in a GMP environment. By allowing the ATS 4 to apply the samples to the plate, the analyst is now available to begin the next step, such as mobile phase preparation, so it also saves time. A photograph of an ATS 4 is shown in Fig. 13.2.

After samples have been applied to the plate, the plate then needs to be dried. An oven maintained at 25°C or TLC plate heater is then used

Fig. 13.3. Two views of a typical TLC oven.

to dry the plate for 1–2 min. Figure 13.3 shows one suitable type of oven. After development, the plate needs to dry again to evaporate the solvents from the plate. This is sometimes done at room temperature (25°C) or at 100°C, depending on the compound being evaluated. Plates that have been sprayed with derivatizing agents also need to be activated by a heated oven, thus a large oven is beneficial to TLC testing.

Once the samples on the HPTLC plate have dried, the plate is inserted into a twin trough chamber such as the one shown in Fig. 13.4 for development. Approximately 10 ml of developing solvent is placed on both sides of the tank, and a piece of blotter paper is placed on the trough opposing the HPTLC plate. The blotter paper ensures a saturated environment for the plate to develop.

After the HPTLC plate has been developed, it is dried in the oven (as shown in Fig. 13.3) for approximately 30–60 min. The plate can then be viewed under a light box or Reprostar system as shown below in Fig. 13.5.

This representative light box is equipped with 3 sets of bulbs: 254 nm short wave UV light, 366 nm long-wave UV light, and 302 nm mid wave light. In addition, white light of 400–750 nm is also utilized when viewing derivatized plates. While some compounds respond sufficiently to

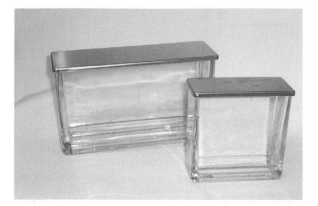

Fig. 13.4. Two sizes of twin trough chambers.

Fig. 13.5. Reprostar light-viewing box.

UV light, others do not. Figure 13.6 demonstrates a plate applied with a UV fluorescent compound and how it appears under 366 nm UV light.

Compounds that do not respond to UV light need to be derivatized. An aerosol device is used to apply derivatizing agents such as Dragendorff's [13] to a HPTLC plate in a very fine mist. Technique is important here as overspraying, underspraying, or uneven spraying will result in poor results. The TLC spraying cabinet or standard lab hood is utilized to capture the vapors that are being sprayed onto the plate. Figure 13.7 demonstrates a standard lab hood suitable for this task.

Once the plate is sprayed with a derivatizing agent, it needs to be dried in the oven as shown in Fig. 13.3. Once dried, the plate is now ready for viewing and photodocumentation. The HPTLC plate is placed in the Reprostar viewing cabinet, the proper light source is selected,

Thin-layer chromatography

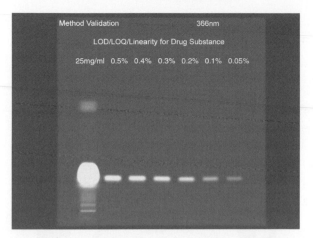

Fig. 13.6. Example of a compound under 366 nm UV light.

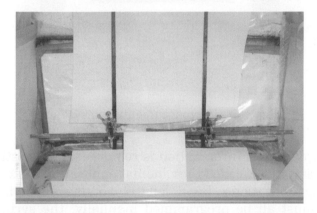

Fig. 13.7. Standard lab hood for spraying plates.

and the digital camera captures the photo. The DigiStore documentation system from CAMAG is a suitable system to document TLC findings electronically, and an appropriate digital printer can be attached to print data to be stored permanently. Figure 13.8 shows a typical photo-documentation system.

Alternatively, TLC plates can also be scanned in the densitometer. This device 'reads' the plate and produces an output similar to a UV scan of the plate. The densitometer is extremely sensitive and it is also nondestructive. A plate can be scanned, the data stored, and then derivatized and documented photographically. An example of a densitometer is shown in Fig. 13.9.

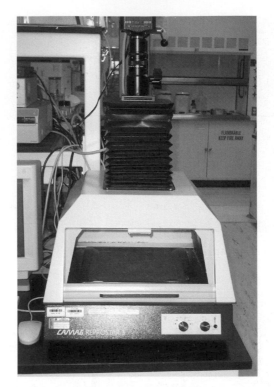

Fig. 13.8. Photodocumentation system.

While more manual intervention is required, a less costly version of the ATS 4 exists and is called the Linomat application system which has the following limitations: Band length, volume applied, number of bands, etc. must all be programmed manually, the syringe must be manually rinsed numerous times before applying a second sample, and it operates at a slower speed than the ATS 4. A photo of the Linomat system is shown in Fig. 13.10.

Derivatizing TLC plates in a standard laboratory hood using an aerosol-spraying device can be cumbersome. An alternative that can be safer and uses less derivatizing agent would be the Immersion device. The TLC plate is attached to the metal bar, an appropriate amount of solution is placed in the tank, and the timer is set for speed and duration of time and then the plate should be immersed. This system uses less solvent and is more reproducible, however, it is not applicable to all derivatizing agents. Some solvents will solubilize the drug substance being evaluated and streak the plate in an adverse manner. An example of this device is shown in Fig. 13.11.

Fig. 13.9. TLC Scanner 3 or densitometer.

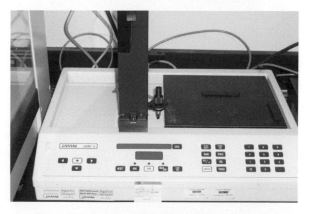

Fig. 13.10. Linomat application device.

In the early stages of method development, when a suitable mobile phase has not yet been identified, the analyst will literally prepare nearly a 100 different mobile phases and plates to obtain the necessary separation. The HPTLC Vario Chamber system allows the user to take one HPTLC plate and cut it into six mini plates and expose them to six mini wells of mobile phase simultaneously. The analyst simply prepares a mixture of the components to be resolved and applies this solution to the six mini plates as a spot. This is a huge time saver since six mobile phases can be evaluated at a time and necessary dilutions can be serially prepared. (see Fig. 13.12).

In traditional TLC experimentation, the chamber is vertical and plates are developed in a vertical fashion. When a large number of

Fig. 13.11. Immersion device.

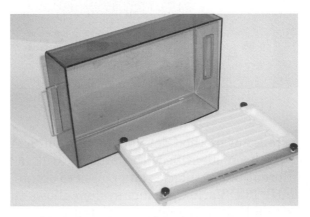

Fig. 13.12. Vario Chamber apparatus.

samples are being evaluated, i.e. a production environment, a horizontal approach to plate development can be very effective and save a large amount of time. In this situation, samples are applied to both ends of an HPTLC plate, and solvent added to the chamber. The plate is then placed horizontally in the device, and glass wicks on each end allow

Fig. 13.13. Horizontal developing chamber.

mobile phase to flow onto the HPTLC plate. When the solvent front travels nearly half way across the plate, the plate is removed, yielding what appear to be two chromatograms on the same plate. An example of this device is shown in Fig. 13.13.

Automated Multiple Development System (AMD2) is a handy device that automatically develops the plates, dries them, and holds the plate in a clean environment for the analyst to document the findings. Several mobile phases can be mixed and preconditioning programs exist to expose the plate to specified solvents prior to development. Upon completion of the development of the plate, the solvent is evacuated and the plate is dried for the predetermined amount of time. The advantage to this system is the user can tend to other tasks without watching the plates develop. The disadvantage is that sample application still needs to occur separate from this unit. An example of this device is shown in Fig. 13.14.

13.4 QUALITATIVE AND QUANTITATIVE APPLICATIONS

TLC and HPTLC can be used for both qualitative and quantitative studies. Five examples where TLC can be utilized qualitatively are discussed in this chapter and include: reaction completion, identity testing, R_f determination, overlapping experiments, and TLC/HPLC correlation experimentation.

In the pharmaceutical industry, it is commonplace to use TLC qualitatively to determine if a reaction has been completed. Fractions are taken at appropriate intervals and applied to the plate and immediately developed and evaluated to determine if most of component A has

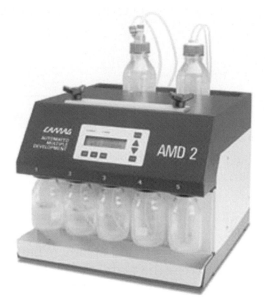

Fig. 13.14. Taken from CAMAG website, an Automated Multiple Development System (AMD2).

converted to component B, for example. Generally, in this situation, it is not necessary to have any standards for comparison purposes as the analyst is simply getting an idea if the reaction has fully completed.

Another common qualitative TLC test in the pharmaceutical industry is identity testing. A standard is applied to the plate side by side with the extracted sample, such as a tablet. The plate is developed in mobile phase and dried, then detected in accordance with the validated procedure. If the R_f value of the sample is consistent with that of the standard, then they are said to be of the same entity, and thus meet the identity test as shown in Fig. 13.15.

R_f determination experiments are useful in the method development phase of TLC. Samples of the drug substance and any known impurities, degradants, or precursors are applied to the HPTLC plate typically at a level of about 0.5% of nominal concentration, or 0.125 mg/ml. During the method development stage, several mobile phases are examined to determine the system that separates all of the components of interest from each other and the drug substance. Since TLC is commonly used for purity testing, it is imperative that all components be resolved from the drug substance. Figure 13.16 shows a successful separation that resolves 5 impurities from the drug substance.

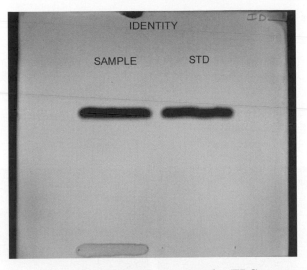

Fig. 13.15. Identity testing by TLC.

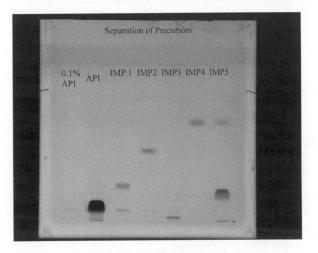

Fig. 13.16. R_f determination.

Figure 13.17 gives an example of some process-related impurities that can be resolved by TLC.

When two impurities have R_f values that are very similar, it can be difficult to determine with certainty that an unknown band is one component over another. In this case, overlap experiments can be very useful. In Fig. 13.18, the left photo shows a standard and a sample that

Drug Substance

Process Related Impurities

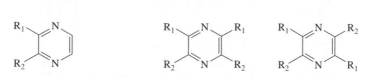

Fig. 13.17. Resolved process-related impurities.

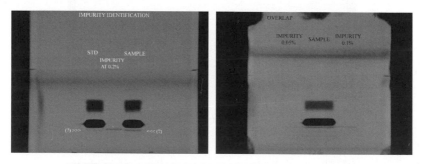

Fig. 13.18. Overlapping experiments can confirm an impurity's identity.

were applied to the same plate, and the question arises whether or not the band seen just below the drug substance is the same impurity in both the standard and sample. On the photo on the left, the impurity was applied in between the two samples at a level of 0.2%. While the R_f values appear to be the same, it is still unclear whether or not this known impurity is definitely the same compound seen in the standard and sample. The photo on the right demonstrates an overlap experiment where the known impurity is applied at two levels, 0.05 and 0.1% and then overlapped with the sample. This photo confirms that the known impurity is in fact the same substance as that seen in the sample. The process is then repeated on another HPTLC plate with the standard to verify the conclusion.

Correlation experiments are routinely done to isolate an impurity seen by TLC and then run it by HPLC to see if the impurity correlates to a peak observed by this technique. This is a lengthy study done by overloading TLC plates with a high concentration of the sample, in order to generate more of the impurity of interest for further testing. The Linomat is usually utilized to produce a 200 µl, 80 mm long band about an inch from the bottom of the 10×20 cm HPTLC plate. This is repeated for about a dozen plates. Figure 13.19 shows how an HPTLC plate appears with one long band applied to it.

Once developed and dried in accordance with the procedure, the plates are then examined under UV light by appropriate detection to find the exact location of the impurity in question as seen in Fig. 13.20.

The band is outlined with pencil and then scraped with a metal spatula (Fig. 13.21) and collected in a beaker. This step is repeated for all 12 plates to increase the concentration of the impurity.

An example extraction procedure is provided as follows; 5–10 ml of dissolving solvent is added to the collected scrapings and the vial is stirred on a stir plate for approximately 10 min. The slurry is then placed in a sonicator for another 10 min, then finally transferred to a

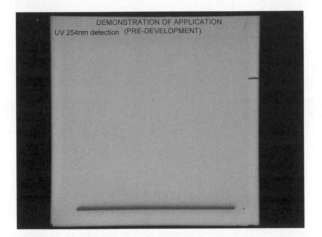

Fig. 13.19. Pre-developed plate for correlation experiment.

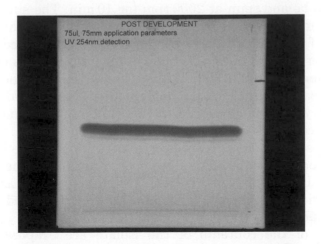

Fig. 13.20. Origin band to be extracted is seen at $R_f = 0.00$.

429

P.M. Grillini

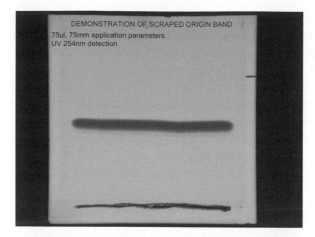

Fig. 13.21. Scraped band from silica plate.

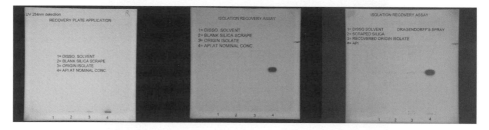

Fig. 13.22. Recovery plates show isolate is recovered.

centrifuge tube and spun at 2000 rpm for 10 min. The silica gravitates to the bottom of the vial while the top layer contains the solubilized compound of interest. The dissolving solvent is evaporated by blowing nitrogen over the vial until only a small amount remains (100–200 µl).

To confirm successful isolation, the original sample is then applied to a fresh plate, and next to it, a small amount of this isolate is applied. This recovery plate is developed in the same manner and upon completion, will confirm the impurity has in fact been isolated. Figure 13.22 demonstrates the samples reapplied to a fresh new plate before development, then this same plate developed, and finally the same plate derivatized. In this final photo, lane 3 is the recovered isolate and it becomes obvious that a portion remains on the origin, and a portion is drug substance.

The rest of the contents of the isolate are evaporated to dryness and the vial is handed off to the analyst running the HPLC study.

The analyst then dissolves this dried isolate in an appropriate HPLC dissolving solvent, injects it onto the HPLC column, and notes its retention time. Further work is done to determine if the retention time of this component matches that of any other known component.

A blank plate is also prepared to rule out any solvent, stationary phase, or mobile phase interference. An HPTLC plate is simply developed in the same manner as the other plates, dried, scraped, and extracted as described above. Figure 13.23 shows a blank plate scraped for HPLC analysis.

This extracted and dried blank sample is also evaluated by HPLC to rule out any extraneous bands due to solvent interaction or stationary phase interference.

In this type of study, one can correlate bands seen on HPLC and TLC. It should be noted that the reverse is easily accomplished as well. An HPLC equipped with a fraction collector can collect an impurity, evaporate off the solvent, and then evaluated on the HPTLC plate to determine if the two impurities are the same.

There are numerous types of TLC testing done in the pharmaceutical industry, which are quantitative in nature. Release testing or purity evaluation, stability testing, and LOD are the three that will be addressed here.

Once a method has been developed and validated for a drug substance, a TLC experiment is conducted to determine the purity of the drug substance. This is done by preparing a drug substance sample at nominal concentration, typically 25 mg/ml, and applying this to the

Fig. 13.23. Scraped blank plate.

plate. In addition, quantitation standards are also prepared by making a stock solution of 0.5% relative to the nominal concentration, or 0.125 mg/ml. This stock solution is diluted appropriately to yield solutions that are 0.4% (0.100 mg/ml), 0.3% (0.075 mg/ml), 0.2% (0.050 mg/ml), and 0.1% (0.025 mg/ml). These standards are also applied to the plate to quantitate any observed impurities.

An example of release testing by TLC is shown in Fig. 13.24. In this example, two more polar bands are observed, one that is quantitated at <0.1% and another at 0.3%. Quantitations are based on visual estimation using the standards on the plate. Alternatively, scanning densitometry can be utilized to obtain a more definitive quantitation of impurities.

Release testing can be used to demonstrate the separation of components such as these in Fig. 13.25.

Stability testing is another quantitative test that is routinely used in TLC. A drug substance or drug product is placed on stability and pulled at an appropriate time point. These samples are examined for visual changes as well as purity and potency changes. TLC testing can be done

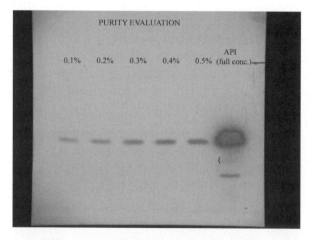

Fig. 13.24. Release testing by TLC.

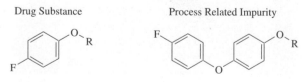

Fig. 13.25. Separation of a drug substance and its impurity.

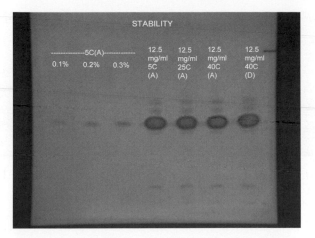

Fig. 13.26. Stability experimentation by TLC.

to determine if any new impurities have appeared while the sample was stressed. For example, four stressed samples are tested and the results are shown in Fig. 13.26. The stressed samples are prepared by making a solution at nominal concentration, usually 25 mg/ml. These stressed samples are then applied to the HPTLC plate. Next to them, quantitation standards are applied of levels 0.1, 0.2, and 0.3% relative to the nominal concentration. These standards are made from the 5°C control sample. Figure 13.26 illustrates five impurities that are seen in all conditions, including the control sample. Had any new impurities been observed, they would be quantitated using the standards on the left side of the plate. If an impurity was more intense in one stressed sample than the control, this too would be quantitated in the same manner.

LOD testing is critical to TLC and validated as part of the method development process. There exists a need to be assured that low-level impurities can always be seen in TLC. When the drug substance is applied at nominal concentration, typically 25 mg/ml, then standards relative to this concentration are prepared and also applied to the plate. An LOD of 0.05% is expected, with a level of quantitation (LOQ) of 0.1%. Appropriate quantitation standards include 0.5, 0.4, 0.3, 0.2, 0.1, and 0.05%. When a detection mode cannot achieve an LOD of 0.05%, other derivatizing agents are examined to find one that will yield the required LOD. Figure 13.27 demonstrates a plate on the left that is detected by UV 254 nm. None of the quantitation standards are visible with this detection. On the right, the same plate is exposed to starch/potassium iodide detection and the sensitivity is greatly increased.

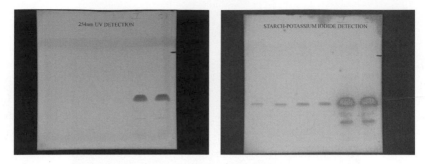

Fig. 13.27. Limit of detection study. Photo on left is 254 nm UV detection, photo on right is starch/KI detection.

Drug Substance   Degradant

Fig. 13.28. Drug substance and degradant respond differently to derivatizing agents.

All quantitation standards are easily observed under these conditions, including the 0.05% LOD standard.

A drug substance and its degradants can respond differently to derivatizing agents as thedid above Fig. 13.28.

System Suitability is another aspect of the LOD testing. With any quantitative TLC experiment, the LOD sample is typically applied in addition to any standards or samples to ensure it can be seen under the conditions of the lab on that given day, and that system suitability has been established.

13.5 SPECIAL TOPICS

Besides all of the previously mentioned examples where TLC can be utilized, there are some unique situations where TLC can be beneficial. Some of these unusual ways to use TLC in the pharmaceutical industry are discussed here:

1. Detection and quantitation of specified impurities
2. Detection of artifacts (2-Dimensional study)
3. Stability of drug substance on plate and in solution

4.   Tablet excipient interactions
5.   Extracted tablet recovery study

The detection of specified impurities is a common test done by TLC when a drug substance or drug product needs monitoring for a given known impurity. Often, this is not a UV responding component, which is the reason TLC is relied upon. The specified impurity is prepared at the appropriate concentrations, typically 0.05, 0.1, 0.2, 0.3, 0.4, and 0.5% relative to the drug substance concentration, which is usually 25 mg/ml. These are applied to the HPTLC plate next to the sample drug substance being evaluated at full strength. The 0.05% level serves as the limit of detection, while the other bands are present to quantitate the known impurity, should it be present. In the case below, the known impurity is observed in the drug substance, which can easily be seen since the R_f values match for the known impurity. It can also be estimated to be present at a level of 0.3% based on the results seen in Fig. 13.29.

In Fig. 13.30, are examples of structures where this may be employed.

A 2-Dimensional experiment is done to determine if there are any TLC artifacts, or what are known as TLC procedure-related effects. In this study, a spot of drug substance at nominal concentration, usually 25 mg/ml, is placed on the HPTLC plate by making one spot by overlaying 3 applications of 5 μl each. This yields one tight spot that contains 15 μl of drug substance solution, rather than a large spot that pools by performing one application. The HPTLC plate is then developed in the usual manner and dried. The plate is then rotated 90° and a second

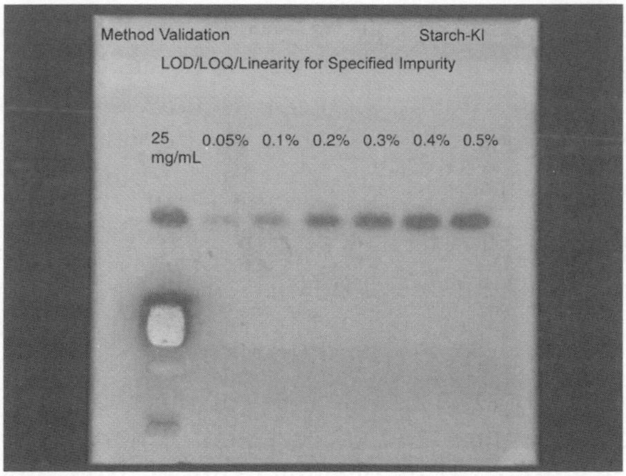

Fig. 13.29. Detection and quantitation of specified impurities.

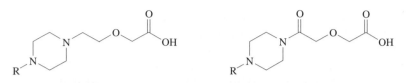

Fig. 13.30. Detection of specified impurity.

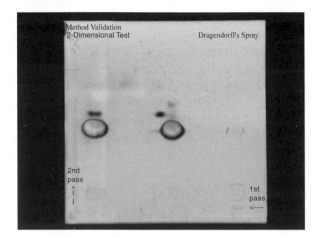

Fig. 13.31. Detection of artifacts.

identical spot of drug substance solution is made by overlaying 3 applications of $5\,\mu l$ each. The plate is then developed in freshly prepared mobile phase and dried again. Examination of the twice-developed plate yields the following possibilities: On the second pass, the drug substance will move on a diagonal line. If there are no bands that have the same R_f value as the drug substance, then the sample is said to be behaving normally. If a band does exist with the same R_f value as drug substance, as is the case in Fig. 13.31, then that origin band is said to be TLC procedure related. This means that all of the drug substance is not being moved from the origin on the first pass through mobile phase; the origin band would not be quantitated since it is ruled as TLC procedure related.

When a TLC method is validated, two of the tests that are done are (1) stability on plate and (2) stability in solution of the drug substance to determine how quickly a sample must be applied to the HPTLC plate and developed before degradation occurs, if it occurs at all. For example, five vials are prepared by placing 25 mg of drug substance in each vial, and labeling them as time 0, 1, 2, 3, and 4 h. The experiment begins by adding 1 ml of dissolving solvent, usually methanol, to the vial labeled 4 h, sonicating, and immediately applying this sample to the plate labeled time on plate. The plate is set aside on the lab bench undeveloped. The 4 h vial is also set aside on the lab bench for the stability in solution plate which is prepared 4 h from this point in the experiment. The time is noted when the first sample was applied. After 1 h, the process is repeated for the 3 h sample, and so on for the 2 and 1 h samples. When the time zero sample time point arrives, the mobile phase is added to the time zero vial as previously mentioned, and applied to the HPTLC plate. This plate is now immediately developed and if at all can yield information of how the drug substance interacts with the silica over time. In Fig. 13.32, the stability of the drug substance on the HPTLC silica plate is demonstrated. It is obvious that the drug begins to degrade sometime between time 0 and 1 h as two bands immediately above the drug substance appear, then increase with intensity with exposure to the plate. As time passes, these two bands increase in intensity, indicating the drug is degrading into these two components. When evaluating this drug substance, it would be noted that degradation is observed and samples should be prepared, immediately applied to the HPTLC plate, and the plate developed quickly to minimize any degradation.

The five vials on the bench top are also then applied to a fresh HPTLC plate to examine any degradation of the drug substance in solution. In this case, the plates looked very similar, indicating similar degradation in solution within 1 hour also.

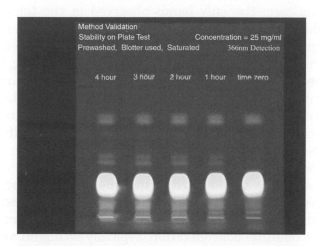

Fig. 13.32. Stability of drug substance on plate.

P.M. Grillini

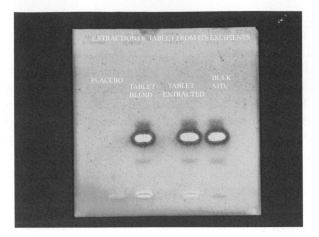

Fig. 13.33. Tablet excipient interactions.

Tablet excipient interactions are occasionally observed when evaluating a drug product for purity. Since there are many excipients in a typical pharmaceutical tablet, known bands need to be identified to make it easier to evaluate for degradation products. Unfortunately, occasionally an inert excipient may react with a derivatizing agent used in TLC making this entity appear as a band that now needs to be identified. In Fig. 13.33, a placebo tablet, an extracted tablet, a handmade tablet blend of all components, and the drug substance standard are all applied to the same HPTLC plate and developed. These results alert the analyst to any excipients that may interfere in the evaluation of the tablet for purity. In this case, the only bands observed in the tablet blend and extracted tablet are the same bands seen in the tablet blend.

If an excipient had been observed, it would need to be identified. In Fig. 13.34, the drug substance standard is applied on lane 1 next to the extracted tablet. The remaining lanes labeled 2–12 are individual excipients in this particular tablet. Only one excipient, number 6, appears and it does in fact have the same R_f value as the band observed in the tablet. This confirmatory test is commonly used to identify interfering excipients. Now this band can be labeled appropriately, rather than mistakenly labeled as a degradant or impurity.

The final step in a tablet purity experiment is to determine if all of the drug substance was extracted from the tablet, and if some adhered to the filter used in the extraction process. A sample of drug substance is prepared at nominal concentration, typically 25 mg/ml and diluted to make a standard that is 0.5% of nominal concentration, or 0.125 mg/ml.

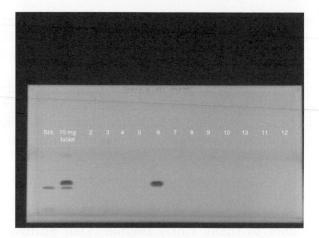

Fig. 13.34. Identification of excipient.

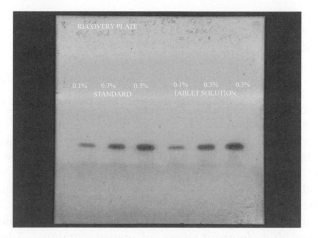

Fig. 13.35. Extracted tablet recovery study.

From this 0.5% standard, two more dilutions are made to yield a 0.3 and 0.1% standard or 0.075 and 0.025 mg/ml, respectively. These three concentrations are applied to the left side of an HPTLC plate, as seen in Fig. 13.35.

In addition, a tablet that is extracted through an appropriate filter such as nylon or PTFE, is collected and diluted to the same active component levels—0.5, 0.3, and 0.1%. These extracted dilutions are applied to the right side of the plate, as seen in Fig. 13.35. After the plate has been developed, it is viewed to determine if the intensity of the bands are consistent between standards and extracted samples. If they appear as in Fig. 13.35, then the tablet is assumed to be fully extracted and no drug substance retained on the filter, or was lost in the extraction process.

In summary, TLC analysis plays a critical role in the drug development process. Many instruments are available, and the technique is used for both quantitative and qualitative testing. R_f determination, identity testing, and stability testing are just a few ways to utilize TLC.

ACKNOWLEDGMENTS

The author wishes to thank Chris Bodnar, Dan Sullivan, and Steve Liguori for their help in contributing to the figures and performing experimental work referenced in this manuscript.

REFERENCES

1   S.L. Abidi, Separation procedures for phosphatidylserines, *J. Chromatogr. B.*, 717(1–2) (1998) 279–293.
2   L.A. Lawton and C. Edwards, Purification of microcystins, *J. Chromatog.*, 912(2) (2001) 191–209.
3   R. Bhushan and J. Martens, Separation of amino acids, their derivatives and enantiomers by impregnated TLC, *Biomed. Chromatogr.*, 15(3) (2001) 155–165.
4   M.V. Reddy, Methods for testing compounds for DNA adduct formation, *Reg. Toxicol. Pharmacol.*, 32(3) (2000) 256–263.
5   G. Alemany, M. Akaarir, A. Gamundi and M.C. Nicolau, Thin-layer chromatographic determinations of catecholamines, 5-hydroxytryptamine, and their metabolites in biological samples—a review, *J. AOAC Int.*, 82(1) (1999) 17–24.
6   A. Roda, F. Piazza and M. Baraldini, Separation techniques for bile salts analysis, *J. Chromatogr. B.*, 717(1–2) (1998) 263–278.
7   J.K. Porter, Analysis of endophyte toxins: fescue and other grasses toxic to livestock, *J. Animal Sci.*, 73(3) (1995) 871–880.
8   H. Oka, Y. Ito and H. Matsumoto, Chromatographic analysis of tetracycline antibiotics in foods, *J. Chromatogr.*, 882(1–2) (2000) 109–133.
9   J. Sherma, Thin-layer chromatography in food and agricultural analysis, *J. Chromatogr.*, 880(1–2) (2000) 129–147.
10 J. Muthing, High resolution thin-layer chromatography of gangliosides, *J. Chromatogr. A.*, 720(1–2) (1996) 3–25.

11 Y. Bereznitski, R. Thompson, E. O'Neill and N. Grinberg, Thin-layer chromatography—a useful technique for the separation of enantiomers, *J. AOAC Int.*, 84(4) (2001) 1242–1251.

12 YES Mag. Canada's Science Magazine for Kids Online, Peter Pipe Publishing, Inc., 2003 [copyright], available: http://www.yesmag.bc.ca, 14 Apr 2003.

13 H. Jork, W. Funk, F. Werner, W. Fischer and H. Wimmer, *Thin-layer Chromatography—Reagents and Detection Methods. Vol 1a: Physical and Chemical Detection Methods: Fundamentals, Reagents I*, VCH, Weinheim, 1990, pp. 464.

14 H. Jork, W. Funk, W. Fischer and H. Wimmer, *Thin-layer Chromatography—Reagents and Detection Methods. Vol 1b: Physical and Chemical Detection Methods: Activation Reactions, Reagent Sequences, Reagents II*, VCH, Weinheim, 1994, pp. 496.

15 J.C. Touchstone and M.F. Dobbins, *Practice of Thin Layer Chromatography*, Wiley, New York, 1978, pp. 103–109.

16 H. Jork and H. Wimmer. Thin-layer chromatography—History and introduction. In: *TLC Report a Collection of Papers*. GIT Verlag, Darmstadt 1986.

17 S.A. Borman, HPTLC: taking off, *Anal Chem.*, 54 (1982) 790A–794A.

18 S.J. Costanzo, High performance thin-layer chromatography, *J.Chem. Educ.*, 61 (1984) 1015–1018.

19 D.C. Fenimore and C.M. Davis, High performance thin-layer chromatography, *Anal. Chem.*, 53 (1981) 252A–266A.

20 F. Geiss, *Fundamentals of Thin Layer Chromatography*, Alfred Huthig Verlag, Heidelberg, 1987, pp. 5–8.

21 T.H. Maugh II., TLC: the overlooked alternative, *Science*, 216 (1982) 161–163.

22 J. Sherma, Comparison of thin layer chromatography with liquid chromatography, *J. Assoc. Off. Anal. Chem.*, 74 (1991) 435–437.

REVIEW QUESTIONS

1. What basic TLC equipment is required to run a simple analysis?
    An application device is critical for reliable sample application, a light-viewing box is also essential to view the bands. Permanent recording of the experiment would also necessitate some type of photodocumentation apparatus.
2. How are impurities described relative to the drug substance?
    R_f value is a mathematical expression to describe the relationship between two components. For example, if a sample is applied at 1 cm from the bottom of the plate, the solvent front

travels to the 8 cm mark, the drug substance travels to the 4 cm mark, and an impurity travels to the 7 cm mark, the R_f values are calculated as follows:

For drug substance: 4-1 cm/8-1 cm = 3 cm/7 cm or 0.43

For impurity: 7-1 cm/8-1 cm = 6 cm/7 cm or 0.86

3. If two components have very similar R_f values, how can their identity be positively determined?

By performing an overlapping experiment, as discussed in Fig. 13.18, two components with similar R_f values can be examined and the proper assignment made. A sample of each component is applied to each side of the plate, while the sample is applied in the middle of the two components, in an overlapping manner. Once developed, the unknown impurity typically can be determined.

4. Level of detection is an important aspect of validation of TLC methods. If an LOD of 0.05% is required, but only 0.1% can be achieved, what can be done to increase the LOD?

Increasing the load on the TLC plate will help increase the LOD. Also, if only ultraviolet detections had been used so far, derivatizing agents should be explored to find one that would increase sensitivity, thus increasing the LOD.

Chapter 14

Gas chromatography

Nicholas H. Snow

14.1   INTRODUCTION

Gas chromatography (GC), now about 50 years old as an instrumental technique, remains one of the most important tools for analysts in a variety of disciplines. As a routine analytical technique, GC provides unparalleled separation power, combined with high sensitivity and ease of use. With only a few hours of training, an analyst can be effectively operating a gas chromatograph, while, as with other instrumental methods, to fully understand GC may require years of work. This chapter provides an introduction and overview of GC, with a focus on helping analysts and laboratory managers to decide when GC will be appropriate for their analytical problem, and then to assist in helping use GC to solve the problem. There is extensive referencing and bibliographic information, to assist in obtaining more detail than can be found in a short chapter.

The term "gas chromatography" refers to a family of separation techniques involving two distinct phases: a stationary phase, which may be a solid or a liquid and a moving phase, which must be a gas that moves in a definite direction. This necessitates that GC, unlike its liquid phase counterparts, is always performed in columns. The chapter begins with a historical overview and discussion of the current status and capabilities of GC, followed by a bibliography and listing of major on-line resources, which are plentiful. The bulk of the chapter provides discussion of the many instrumental considerations and configurations. The analyst must make choices such as packed versus capillary column, type of inlet and injection technique and type of detector. Overviews of these options are presented. There are many options in data handling also available, which must be chosen carefully. Finally, the chapter closes with a summary of major applications in several fields, to give the reader some perspective on the capabilities of GC.

Comprehensive Analytical Chemistry 47
S. Ahuja and N. Jespersen (Eds)
Volume 47 ISSN: 0166-526X DOI: 10.1016/S0166-526X(06)47014-8
443

14.1.1 Historical overview

GC has its roots in distillation and vapor phase extraction, which have their beginnings in the nineteenth century. As we know them today, the fundamental theories of GC have their roots in partition chromatography, described in the Nobel Prize winning work of Martin and Synge in 1940 [1]. The first gas chromatograph was constructed by Martin and James, who showed the first separations using GC in 1952 [2]. Interest in this new technique spread quickly in the 1950s, with several key events, including the invention of many of the instrumental components still in use today: glass capillary columns, flame ionization and electron capture detectors, and the theoretical developments in peak broadening and temperature programming that are still discussed today. As glass capillary columns were difficult to make and use, the period of 1952–1979 saw most of the development related to packed column instrumentation and methods. Many methods developed during that period remain in use, although packed columns now account for only about 10% of the GC market.

The invention of fused-silica capillary columns in 1979 was a pivotal event in the history of GC [3]. Prior to this, capillary columns, which are by far the most widely used today, were relegated primarily to research laboratories and those that could afford the special equipment required to make them. The advent of fused-silica tubing, which provided flexible, easy to make and use columns, changed GC from primarily being based on packed columns to using capillary columns. This was a fundamental change that is still affecting instrument design and method development. The 1980s saw the development of new, highly stable liquid phases for capillary columns and the beginnings of capillary column-specific instrument development. In the 1990s, the manual pneumatics was replaced with solid-state electronically controlled pneumatics that made flows and retention times very highly reproducible. Table 14.1 shows the major historical developments in GC with key references.

14.1.2 Current status

Today's gas chromatograph is a modern, computer-controlled instrument, consisting of an integrated inlet, column oven and detector, with electronically controlled pneumatics and temperature zones. It has an inlet capable of both the split and splitless-injection techniques and it has a highly sensitive (detection limit in the pictogram range) detector

TABLE 14.1

Overview of historical developments in gas chromatography

Date	Event
1941	Partition chromatography described by Martin and Synge
1951	Gas–solid chromatography invented by Cremer
1952	Gas–liquid chromatography invented by Martin and James; Packed columns developed
1956–1958	Theory of capillary column performance described
1958	Kovats retention index proposed
1959	Patent on capillary columns by Perkin-Elmer; inlet splitter
1960	Glass drawing machine developed by Desty
1966	Rohrcschneider introduces phase constants
1979	Fused silica capillary columns invented; programmed temperature vaporization injection
1983	Introduction of megabore columns as alternative to packed columns
1980s	Development of quality testing and deactivation procedures; cross linked polymeric phases; interface to spectroscopic detectors
1990s	Programmed temperature vaporizer, electronic pressure control become commercially available

that is optimized for use with capillary columns. It has the proper fittings and pneumatics for use with fused-silica capillary columns of 10–100 ml and 0.1–0.53 mm inside diameter. It is capable of quantitating analytes in complex mixtures at concentrations from ppb to percent and it is capable of separating hundreds of components in a single analytical run. It has several injections and sampling methods available, including on-line extractions and headspace sampling and it has numerous optional selective detectors, including bench-top mass spectrometry.

14.1.3 Major bibliography

There are numerous major textbooks covering various aspects of GC. For more detail and advanced topics related to GC, the reader is referred to these.

1. H.M. McNair and J.M. Miller, *Basic Gas Chromatography*, 2nd ed., New York, Wiley, 1997.
2. R.L. Grob and E.F. Barry (Eds.), *Modern Practice of Gas Chromatography*, 4th ed., New York, Wiley, 2004.

3. K. Grob, *Split and Splitless Injection in Capillary Gas Chromatography: Concepts, Processes, Practical Guidelines, Sources of Error*, 4th ed., New York, Wiley, 2001.
4. W. Jennings, E. Mittlefehldt and P. Stremple, *Analytical Gas Chromatography*, London, Academic Press/Elsevier, 1997.
5. *Journal of Chromatography* A and B, Elsevier, http://www.elsevier.com.
6. *Journal of Separation Science*, Wiley, http://www.wiley.com, or http://www.separationsnow.com.
7. *Journal of Chromatographic Science*, Preston Publications, http://www.prestonpub.com or http://www.j-chrom-sci.com.
8. *LC-GC Magazine*, Advanstar Publications, http://www.advanstar.com.
9. *Analytical Chemistry* (journal), American Chemical Society, http://www.pubs.acs.org.

14.1.4   Online resources

With the development of the World Wide Web, a large amount of content related to GC and method development is available. Most of the column and instrument manufacturers, as well as publishers and universities have made a wealth of information available. While the list of web sites that follows cannot be exhaustive, the author has found them useful in his own work and they should provide a beginning for any analyst that is looking for web-based information on GC. The URL listings are accurate as of January 2005.

General GC sites:

1. http://chromatographyonline.com. This site is sponsored by Advanstar Communications, publisher of LC-GC Magazine, a useful vendor-supported trade publication that includes troubleshooting and method development help in every issue. The site includes links to all of the vendors that advertise in LC-GC and more.
2. http://separationsnow.com. This site is sponsored by Wiley and includes links to a huge variety of publications and resources, many well beyond Wiley's publications.

Vendor information: Most of the instrument and supply vendors for GC have product lines spanning analytical chemistry and beyond. Many of them have placed extensive libraries of information on-line. To access their information on GC, which is often extensive, you may need to type "gas chromatography" into the search function on the website. Many of the sites require registration, but the information content is available free to registered users. To access vendor sites, the reader is referred to the general sites shown above, which include links to numerous vendor sites. Also, the monthly issues of publications such as American Laboratory and LC-GC and the annual Lab Guide issue of Analytical Chemistry include information from the instrument and column vendors.

14.2 THEORY

The data generated in GC experiments is called a chromatogram, with an example shown in Fig. 14.1. Each peak represents one component of the separated mixture. The retention time (t_R) is indicative of the identity of the analyte; the peak height or peak is related to the amount (mass or concentration, depending on the detector) of the compound that is present. The peak width is also important, as it provides a measure of the efficiency of the separation process and how many peak compounds the method is capable of separating.

14.2.1 Retention times

Retention time is the basic measure used in GC to identify compounds. It is a physical property of the analyte and is dependant on the separation conditions such as temperature, flow rate and chemical composition of the stationary phase. Solubility of the analyte in the stationary phase, which is based on the energy of intermolecular interactions between the analyte and stationary phase, is the most important factor in determining retention time. In Fig. 14.1, the retention

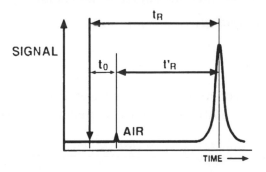

Fig. 14.1. A typical chromatogram showing the data that is generally obtained. Picture courtesy of Prof. Harold McNair.

N.H. Snow

time is defined as the time after the point of injection where the maximum point of the peak occurs. It is the total amount of time that the analyte spends in the column and it is the sum of the amount of time spent in the stationary phase t'_R, termed the adjusted retention time and the amount of time spent in the mobile phase (t_m), termed the gas hold-up time.

$$t_R = t'_R + t_m \tag{14.1}$$

The adjusted retention time provides a measure of the strength of intermolecular interaction between the analyte and the stationary phase, with stronger interactions giving a longer time. The gas hold-up time is derived from the flow rate and the column dimensions and is often measured by injecting a non-retained compound. The retention factor, which represents a ratio of the mass of analyte dissolved in the stationary phase to the mass in the mobile phase, can be calculated from the adjusted retention time and the gas hold-up time.

$$k = \frac{t'_R}{t_m} = \frac{mass_{analyte(sp)}}{mass_{analyte(mp)}} \tag{14.2}$$

As this represents a ratio of masses, if combined with a volume ratio, a ratio of concentrations, which is an equilibrium constant, can be obtained. The volume ratio is called the phase ratio and is the ratio of the volume of the mobile phase to the volume of the stationary phase.

$$\beta = \frac{V_m}{V_s} \tag{14.3}$$

The equilibrium constant is termed the partition coefficient and is a concentration equilibrium constant representing the equilibrium constant for the process of a compound that begins in the vapor (mobile) phase and is dissolved into the solution (stationary) phase.

$$Analyte(mp) \Leftrightarrow Analyte(sp) \tag{14.4}$$

$$K_c = k\beta = \frac{[A(sp)]}{[A(mp)]} \tag{14.5}$$

The equilibrium constant is then connected to the thermodynamics of the mobile phase-stationary phase transfer process using classical expressions.

$$\Delta G^0 = -RT \ln K_c \tag{14.6}$$

$$\Delta G^0 = \Delta H^0 - T\Delta S^0 \tag{14.7}$$

448

If multiple peaks are present, the separation power of the system to separate them can be expressed by calculating the selectivity, which is related to several of the derived parameters, but most critically to the difference in the energetics of the intermolecular interactions.

$$\alpha = \frac{t'_R(2)}{t'_R(1)} = \frac{k_2}{k_1} = \frac{K_{c(2)}}{K_{c(1)}} = f(\Delta\Delta G^0) \tag{14.8}$$

This derivation shows that retention time is dependant on three factors: temperature, energies of intermolecular interactions and flow rate. Temperature and flow rate are controlled by the user. Energies of intermolecular interactions are controlled by stationary phase choice. This theory is also the basis for the popular software programs that are available for computer-assisted method development and optimization [4,5,6,7]. More detailed descriptions of the theory behind retention times can be found in the appropriate chapters in the texts listed in the bibliography.

14.2.2 Gas flows

Because the mobile phase is a gas and is pressure driven, determining the flow rate in GC can be complex, with instrument type and pneumatic configuration playing often confusing roles. From the ideal gas law, $pV = nRT$, it follows that pressure, volume and temperature are inter-related, therefore the volumetric flow rate is dependant on all three. For packed column systems, the flow rate is typically measured using either an electronic or soap-film flow meter connected to the column outlet. In capillary column systems with manual pneumatics, the average volumetric flow rate is measured by injecting a non-retained gas, such as air or methane, determining its retention time and calculating the average linear gas velocity from the column length (cm) and gas hold-up time (s).

$$\bar{v} = \frac{L}{t_m} \tag{14.9}$$

The average volumetric flow rate (ml/min) is calculated from the gas hold-up time, the column length (cm) and the column radius (cm).

$$\overline{F}_c = \frac{\pi r^2 L}{t_m} \tag{14.10}$$

This is the average volumetric flow rate. It is important to remember that, due to gas expansion as the mobile phase moved along the column from high pressure at the inlet to lower pressure at the outlet, the volumetric flow at the inlet is lower and the flow at the outlet is higher.

Systems with electronic pneumatic control use pressure transducers at the inlet and outlet, the column dimensions and physical properties of the carrier gas to determine the gas hold-up time and the flow rate.

$$t_{\mathrm{m}} = \frac{4\eta L^2}{3p_{\mathrm{o}}} \bullet \frac{P^3}{(P^2 - 1)^2} \bullet \frac{32}{d_{\mathrm{c}}^2} \qquad (14.11)$$

The carrier gas viscosity is given as η, L the column length, p_{o} the outlet pressure, P the ratio of inlet pressure to outlet pressure and d_{c} the column diameter. Analysts should take care to be sure that the methods used for determining the flow rate are consistent from instrument-to-instrument and from method-to-method. Otherwise it will be difficult to compare any data that have the flow rate, gas hold-up time or linear carrier gas velocity as a component.

14.2.3 Band broadening

In chromatography, the number of theoretical plates (N) is the most common measure of separation efficiency, which determines the rate at which band broadening will occur during the separation process. In general, a higher number of theoretical plates indicates a more efficient transport process within the instrument and therefore, sharper peaks result. The number of theoretical plates is calculated from isothermal chromatographic data using an expression including the retention time and the peak width at baseline.

$$N = 16\left(\frac{t_{\mathrm{R}}}{W_b}\right)^2 \qquad (14.12)$$

Caution should be used when using theoretical plates to compare columns, as N is highly dependant on the separation conditions, analyte and instrument.

The peak width, therefore, is often as important a measure as the retention time. Broad peaks mean that there can be fewer of them in a chromatogram, therefore less separating power. The sources of peak broadening for packed and capillary columns were described by van Deemter and Golay in the 1950s [8,9]. They describe the height equivalent to a theoretical plate (H), which is a measure of the rate of band broadening, being related to the column length and number of theoretical plates, and to several factors (described below) that depend on the average mobile phase velocity (v); chromatographers generally work to minimize H.

$$H = \frac{L}{N} = A + \frac{B}{v} + Cv \tag{14.13}$$

Band broadening is described as driving from three factors.

- Eddy diffusion (the "A" term) is found in packed columns and derived from the different paths that analyte molecules must take through the column packing. It is minimized by using smaller particles and packing them efficiently. In capillary columns it is minimal enough to be neglected.
- Molecular diffusion (the "B" term) applies to both packed and capillary columns and derives from the fact that all molecules in the gas phase will diffuse into any available space. It is minimized by using an increased flow rate (see the carrier gas velocity in the denominator) and by using a high molecular weight carrier gas.
- Mass transfer (the "C" term), which involves collisions and interactions between molecules, applies differently to both packed and capillary columns. Packed columns are mostly filled with stationary phase so liquid phase diffusion dominates. The mass transfer is minimized by using a small mass of low-viscosity liquid phase. Capillary columns are mostly filled with mobile phase, so mass transfer is important in both the gas and liquid phases. A small mass of low-viscosity liquid phase combined with a low-molecular weight carrier gas will minimize this term.

A sample plot showing the contribution of each term to determining H is shown in Fig. 14.2. An excellent basic discussion of the van Deemter and Golay equations can be found in the text by McNair and Miller [10].

14.2.4 Resolution

Perhaps, the most important goal of all chromatographic method development projects is to obtain the desired resolution (separation) of the analyte peaks in the desired time. On the basis of the above discussions, resolution can be described as resulting from three key factors: set up of the instrument, intermolecular interactions in the column and residence time in the column. First, resolution must come from properly setting up the instrument and operating it correctly. In GC, this is not always trivial, as many operations such as installing a

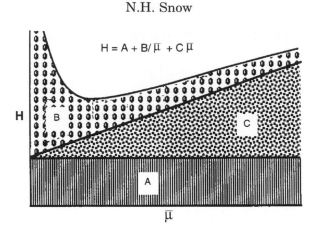

$$H = A + B/\bar{\mu} + C\bar{\mu}$$

Fig. 14.2. van Deemter plot showing contributions of eddy diffusion, molecular diffusion and mass transfer to the rate of band broadening. Picture courtesy of Prof. Harold McNair.

column, ensuring that the instrument has no contamination, and ensuring that correct instrumental parameters are set, require practice and training. Second, the choice of column for a given separation will determine the selectivity, which provides the difference in intermolecular interaction energies that leads to separation. Finally, peak broadening is related to time on the column. This was expressed fundamentally by Purnell [11].

$$R_s = \left(\frac{\sqrt{N}}{4}\right)\left(\frac{\alpha - 1}{\alpha}\right)\left(\frac{k_2}{1 + k_2}\right) \tag{14.14}$$

Resolution may be calculated from a chromatogram as the difference between the retention times divided by the average of the peak widths at the baseline for two adjacent peaks.

$$R_s = \frac{t_{R(2)} - t_{R(1)}}{1/2(W_1 + W_2)} \tag{14.15}$$

A calculated value for resolution greater than 1.5 indicates that the adjacent peaks exhibit baseline resolution; the signal has fully returned to the baseline from the first peak before the second peak begins. Often, a minimum acceptable resolution of 2 is used in method development to ensure that acceptable resolution is maintained, even as the method is transferred among instruments, analysts and laboratories.

14.3   INSTRUMENTATION

A modern gas chromatograph, whether configured for packed or capillary column use, consists of several basic components. All of them must be properly chosen and operated for successful analysis. These are pneumatics and gas-handling systems, an injection device, an inlet, a column oven and column, a detector and a data system. Since the inception of GC in the 1950s, instrumentation has evolved significantly as new techniques and technologies were developed. This section provides an overview of the major components of a modern gas chromatograph, with details about how to choose components based on analytical needs, and applications.

14.3.1   Overview

Figure 14.3 shows a typical capillary gas chromatograph with the major components labeled. This gas chromatograph set up includes compressed gas tanks for the carrier gas (mobile phase) and any necessary detector gases, an auto-injector that employs a micro-syringe for delivering the necessary small sample volumes and an inlet capable of the

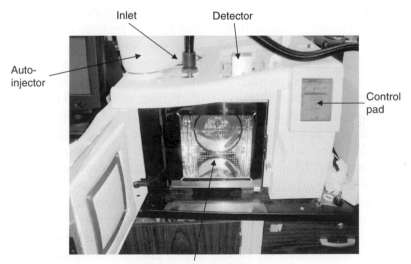

Fig. 14.3. Typical capillary gas chromatograph showing locations of the major components. The black hose that highlights the inlet is the transfer line from a head-space sampler.

split and split-less injection techniques. The column oven typically houses 1–2 capillary columns and this gas chromatograph is equipped with a flame ionization detector, which is the most common for organic analysis. The gas chromatograph is controlled and the data are collected using a computer-based data system. Systems configured for packed columns have a similar appearance, except that the inlet, detector and column oven are designed to accommodate the larger diameter packed columns.

14.3.2 Gas handling and pneumatics

The choice of carrier gas and gas flow control are critical for successful GC. The carrier gas does no more in the separation process than its name implies: it carries the vapor phase analyte molecules along the column. As such, it must be inert, non-toxic, inexpensive, highly pure and must provide efficient transport with minimal band broadening. For packed column GC, nitrogen is the most commonly used carrier gas, followed by helium. For capillary column GC, the most common carrier gas is helium, followed by hydrogen and nitrogen.

Gas handling systems are very important components of any gas chromatograph. For all gas chromatographs, the gas flow should be regulated using a two-stage regulator at the gas cylinder. The regulator should be certified for use with GC and should include a stainless-steel diaphragm, to prevent minor leaks that can allow oxygen to enter the carrier gas flow, causing long-term damage to oxygen-sensitive stationary phases. For packed column systems, the carrier gas flow is typically controlled using a mass-flow controller and metering valve installed between the gas cylinder regulator and the inlet. On capillary column instruments, the gas flows are controlled by a combination of a mass-flow controller and a backpressure regulator; these systems are back-pressure regulated to maintain constant column flow, even when the flow and pressure upstream from the instrument are changed. The need for backpressure regulation will become apparent in the discussions of split and splitless inlets. In modern instruments, the flow controllers are all solid-state and microprocessor-controlled. In instruments manufactured before 1995, the pneumatics is generally manually controlled. Analysts should beware that there are subtle differences in how flows are measured and reported from manual versus electronically controlled pneumatics. This will cause variations in retention times between manual and electronically controlled systems, for nominally the same pressures and flows.

14.3.3 Inlets and sample injection

The inlet is perhaps the least understood of the three zones on a gas chromatograph, yet it is the one that the analyst will most often interact with, and its optimization is critical to obtaining the best separation and quantitation. The type of inlet and the injection technique are related and they are determined by the column type, sample properties and the required analytical sensitivity. For packed columns, the inlet is a relatively simple device; samples are generally injected directly into a short portion of tubing at the column head that is clear of stationary phase. Most often, this is the end of the column itself. The syringe needle easily fits within the bore of the column. For capillary columns, because of their small diameter, the situation is more complex. The syringe needle does not routinely fit within the column. Samples are generally injected into a chamber, separate from the column, vaporized, and finally transferred to the column in the vapor phase. This is a much more complex situation than for packed columns and leads to perhaps the most problematical portion of a gas-chromatographic analysis. This has led to the development of a variety of inlets for packed and capillary GC. The major types are summarized in Table 14.2.

14.3.3.1 *Injection devices*

Whether the injection is conducted using an auto-injector or manually, a micro-syringe is the basic device for handling and injecting liquid samples. For liquid samples, in both packed and capillary GC, the typical sample volume ranges from 0.1–10 µl, with 1 µl being by far the most common. For gas samples, where volumes range from a few microliters to 1 ml or more, a gas-tight syringe or gas sample valve may be used. Finally, solid samples are generally dissolved into an appropriate organic solvent prior to injection as a solution. The micro-syringes that are generally used for injecting liquid samples should be treated as if they are very low-volume pipettes. An auto-injector is capable of injecting very reproducibly, with injection-to-injection precision typically 0.1%, however, from syringe-to-syringe, the volume reproducibility may be ± several percent. With manual injections, precision of ± 1–5% in the resulting peak areas may be expected. There are myriad additional sampling devices that have been used with GC. A detailed description of each is beyond the scope of this chapter, however, they are described briefly at the end.

TABLE 14.2

Common inlet types, advantages and disadvantages

Inlet	Advantages/Applications	Disadvantages
Split	• Simple • Starting point for method development • Use with isothermal and temperature programmed GC • Fast; very sharp peaks	• Choice of glass sleeve not trivial • Limits detection concentration to ppm • Most sample wasted through split vent • Loss of low-volatility, labile analytes
Splitless	• Trace analysis (ppb) possible • Cold trapping and solvent effects provide sharp peaks	• More complex than split • Limited to temperature programming • Several parameters to optimize • Loss of low-volatility, labile analytes
Cool on-column	• Entire sample enters column • Effective trace (ppb) analysis • Little-no sample degradation	• Entire sample enters column • Dirty samples problematic • Automated operation not routine; special syringe
Programmed temperature vaporization (PTV)	• Most versatile inlet • Allows large volume injection • Little-no sample degradation • Effective trace (to sub-ppb) analysis	• Expensive • Requires optimization of many parameters • Not well-known
Direct (packed)	• Simple • No additional pneumatics	• High-thermal mass • Low sensitivity

14.3.3.2 Split inlet

The split inlet was the first developed for capillary GC, arising from two problems: that the syringe needle would not fit directly into the bore of the column and that the very small mass of stationary phase inside a capillary column would be overwhelmed by a full microliter of liquid. The idea behind splitting is simple: that the injected sample would be quickly vaporized, mixed homogeneously with the carrier gas, split, such that a known fraction of the mixture is transferred rapidly to the column, with the remainder going to waste. All of this is accomplished rapidly, often in a few hundred milliseconds, generating very sharp chromatographic peaks. A schematic diagram of a split inlet is shown in Fig. 14.4. The key components are listed below.

- *Septum*. The syringe needle penetrates through the septum to introduce the sample. Septa are typically polymeric materials

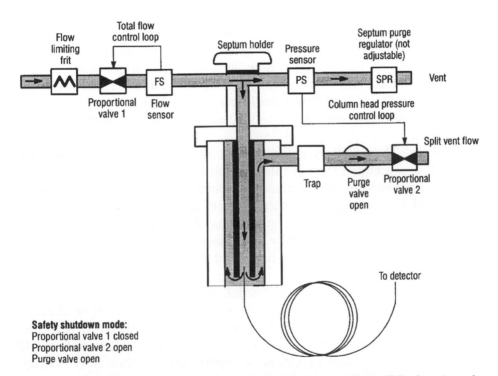

Fig. 14.4. Schematic diagram of a split inlet for capillary GC showing the pneumatics and major components. Reprinted with permission from the 6890 Gas Chromatograph Operating Manual. Copyright (2004), Agilent Technologies.

designed for sustained periods at high inlet temperatures and for multiple needle penetrations. A typical septum will last for about 30–50 injections before it begins to leak. Capillary GC users are cautioned to only use septa that are specially designed for the sustained high temperatures encountered in capillary GC. The septum purge valve provides a low (3–5 ml/min) flow of carrier gas under the septum to carry away any volatile contaminants that may collect on the septum.

- *Glass sleeve.* The injected sample is vaporized inside the glass sleeve, which is used to provide an inert space. The exact design and configuration of the glass sleeve is instrument, injection technique and analyte dependant. There is extensive literature on optimizing choice of glass sleeve. For details, consult the instrument vendor WWW-sites listed in this chapter and the text by K. Grob listed in the bibliography.

- *Split vent.* The sample vapors that do not enter the column are ejected through the split vent. A needle valve on this line regulates the total flow of carrier gas into and from the inlet, generating the split ratio, which determines the portion of sample that enters the column. The split ratio is the ratio of the split vent flow to the column flow and provides a measure of the amount of sample that actually enters the column from the injection. A split ratio of 100:1 indicates that a 1 µl injection from the syringe results in approximately 10 ml of liquid sample reaching the column.

- *Column.* The capillary column must be correctly installed into the inlet, a procedure that is inlet and instrument dependant. In general, the column end is placed a specified distance into the inlet, beyond a fitting at the bottom end of the inlet. Each instrument vendor provides specific instructions on column installation that should be followed carefully. It is especially important to ensure that the seal between the inlet and column is leak-free and not contaminated.

- *Pressure regulator.* The split inlet is back pressure regulated to ensure a constant head pressure, therefore, a steady flow through the column. For capillary columns, the inlet pressure determines the column flow, as per Eq. (14.11). As the inlet operates, the split vent can be opened or closed or upstream gas flow may change; the regulator maintains the desired pressure, therefore the desired column flow.

14.3.3.3 Splitless inlet

A splitless inlet uses the same hardware as a split inlet, so they are typically provided as a single unit. In 1968, the splitless-injection technique was invented accidentally by Grob and Grob [12] who inadvertently left the split vent closed during an injection. At the time, it was believed that injection of a full microliter of liquid solvent would dissolve or destroy most stationary phases. Splitless injection is accomplished by closing the split vent prior to and during the injection and leaving it closed for a time following the injection. This allows nearly the entire injected sample to vaporize and enter the column. The split vent is then opened at high split ratio to remove any sample remaining in the inlet. This results in about 95% of the injected sample reaching the column and is used for dilute samples. A schematic representation of an inlet showing the split vent closed is shown in Fig. 14.5.

14.3.3.4 Additional inlets

For capillary GC, the split/splitless inlet is by far the most common and provides an excellent injection device for most routine applications. For specialized applications, there are several additional inlets available. These include programmed temperature vaporization (PTV) cool on-column and, for packed columns, direct injection. PTV is essentially a split/splitless inlet that has low thermal mass and a heater allowing rapid heating and cooling. Cool injection, which can be performed in both split and splitless mode with the PTV inlet, reduces the possibility of sample degradation in the inlet. Capabilities of the commonly available inlets are summarized in Table 14.3.

14.3.4 Column oven and columns

Many textbooks and experts describe the column as the "heart" of chromatography, as this is where the most separation occurs. Because temperature is the most important user-controllable variable in GC, the column, whether packed or capillary, is operated inside a temperature-controlled oven. The oven and column must be of low thermal mass to allow for rapid heating and cooling. During a chromatographic analysis, the oven may be operated in two modes: isothermal (constant temperature) and temperature programed (linear temperature increase). Typically, depending on manufacturer and power supply capacity, ovens may temperature program up to 30–40°/min. Some specialty ovens for fast GC can change temperature up to 1200°/min.

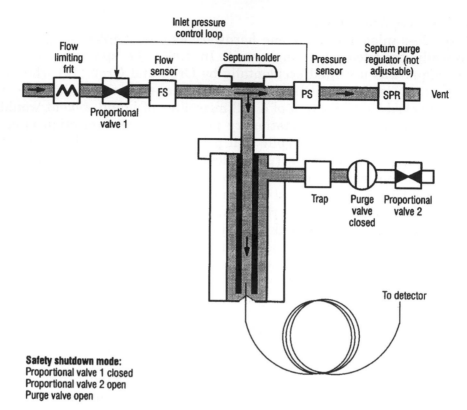

Fig. 14.5. Schematic of splitless inlet with purge valve closed showing the major components. Note the position of the purge valve in comparison to Fig. 14.4. Reprinted with permission from the 6890 Gas Chromatograph Operating Manual. Copyright (2004), Agilent Technologies.

The accuracy of temperature controllers and sensors is typically about 0.1°. Because of their high-thermal mass, packed columns are more often operated isothermally, while due to their low-thermal mass, capillary columns are most often temperature programed. Important parameters related to basic column dimensions are shown in Table 14.4.

14.3.4.1 Packed columns
Packed columns have been in use since the inception of GC and today are used in about 10% of applications, especially in the analysis of very small molecules such as fixed gases and solvents. The dimensions of packed columns are limited by the inlet pressure and fittings of the GC. Typically, packed columns are 6-10 ft long and 1/4 or 1/8-in.

TABLE 14.3

Important column dimension parameters for packed and capillary columns

Variable	Effect on retention time	Effect on peak width
Column length (L)	Retention times longer; peaks are separated further	Peaks broaden with length, but N higher
Column internal diameter (d_c)	None if linear gas velocity the same	Narrower column; sharper peaks
Particle diameter (packed columns) (d_p)	Pressure drop increased with smaller particles; retention time longer at same inlet pressure	Smaller particles; sharper peaks
Liquid phase film thickness (d_f) or phase loading (%)	Thicker film; longer retention	Thinner film; sharper peaks
Column temperature (T)	Increased temperature; decreased retention time	Increased temperature; sharper peaks; effect on N difficult to define; check maximum and minimum temperature before using column

outside diameter tubing are used. This limits separation power. The ability to separate using a packed column is strongly dependant on the analyte-stationary phase intermolecular interactions, so a huge number of stationary phase packing materials have been available. Packed columns are typically made from glass or stainless steel tubing and must be coiled to match the specific instrument into which they will be installed. The stationary phase may be either solid particles or liquid-coated solid particles. If the particles are liquid coated, the phase loading (5–30% by mass) determines the strength of retention, with thicker coatings providing longer retention times. In gas–solid chromatography, high surface area particles will generally provide stronger retention.

14.3.4.2 Capillary columns

Glass capillary columns were invented shortly after GC but did not see wide routine use, as the original glass columns were difficult and expensive to make, requiring an elaborate glass drawing machine, beyond

TABLE 14.4

Summary of important and common capillary column dimensions

Parameter	Values	Characteristics/applications
Length	10 m	• Fast separations • Use with thin films, small inside diameter • Pneumatics and flows most critical
	15–30 m	• Most common • Use with medium film thickness and inside diameter
	100 m	• Slowest • Very high peak capacity; complex samples
Inside diameter	0.1 mm	• High pressure drop required • Lowest sample capacity • Fast separations • Most efficient but usually short column
	0.2, 0.25 and 0.32 mm	• Most common • Combination of good resolution and ease of use
	0.53 mm	• Alternative to packed columns • Used with thick film
Film thickness	0.1 μm	• Thinnest • High boiling analytes • Fastest separations • Most efficient
	0.1–1 μm	• Most common • Compromise between ease of use and efficiency
	>1 μm	• Thickest • Low boiling analytes (gases, solvents)

the capabilities of most laboratories. Prior to the 1979 invention of the fused-silica capillary columns in use today, most GC methods employed packed columns. Thus, routine capillary GC, as it is done today, is a relatively young field, having begun only about 25 years ago. In contrast to packed columns, capillary columns may be of any length, typically 10–100 m, as there are no packing particles. The stationary phase is coated to the inside of the capillary wall in thicknesses ranging from 0.1 to 5 μm. Finally, the inside diameter of a capillary column ranges from 0.1 to 0.53 mm. A summary of the specific column dimensions and their applications is shown in Table 14.4.

14.3.4.3 Liquid phases

There are a variety of liquid phases available for both packed and capillary columns. In packed columns, these are coated onto a solid support; in capillary columns they are coated onto the capillary column wall. At the height of packed column use, there were over 200 liquid phases in common use. With the much higher separating power of capillary columns, this number has dropped to five most common liquid phases with perhaps two dozen specialty materials. A summary of common liquid phases and their application is shown in Table 14.5. The capillary column and instrument vendors also have extensive information on how to properly use packed and capillary columns and on applications available in their literature and on their WWW-sites .

14.3.5 Detectors

In order to perform qualitative and quantitative analysis of the column effluent, a detector is required. Since the column effluent is often very low mass (ng) and is moving at high velocity (50–100 cm/s for capillary columns), the detector must be highly sensitive and have a fast response time. In the development of GC, these requirements meant that detectors were custom-built; they are not generally used in other analytical instruments, except for spectroscopic detectors such as mass and infrared spectrometry. The most common detectors are flame ionization, which is sensitive to carbon-containing compounds and thermal conductivity which is universal. Among spectroscopic detectors, mass spectrometry is by far the most common.

14.3.5.1 Thermal conductivity detector

The thermal conductivity detector (TCD) is a classical detector for both packed and capillary columns. A schematic representation of a modern

TABLE 14.5

Applications of common liquid phases

Stationary phase temperature range	Applications
100% methyl polysiloxane -60–350°C	Alkaloids, amines, drugs, fatty acid methyl esters, hydrocarbons, petroleum products, phenols, solvents, waxes, general purposes
5% phenyl-95%-dimethyl polysiloxane -60–350°C	Alcohols, alkaloids, aromatic hydrocarbons, flavors, fuels, halogenates, herbicides, pesticides, petroleum products, solvents, waxes, general purposes
50% phenyl-50% methyl polysiloxane -60–350°C	Alcohols, drugs, herbicides, pesticides, phenols, steroids, sugars
14% cyanopropylmethyl-86% dimethyl polysiloxane 0–250°C	Alcohols, polychlorinated biphenyls, alcohol acetates, drugs fragrances, pesticides
50% cyanopropylmethyl-50% phenyl polysiloxane 0–250°C	Carbohydrates, fatty acid methyl esters
Trifluoropropyl polysiloxane 0–275°C	Drugs, environmental samples, ketones, nitro-aromatics
Polyethylene glycol 60–225°C	Alcohols, flavors, fragrances, fatty acid methyl esters, amines, acids

TCD is shown in Fig. 14.6. A resistive filament is placed in the stream of the carrier gas effluent. When brought to thermal equilibrium with carrier gas flowing and current applied, the resistance of the filament will cause it to heat. When the carrier gas is mixed with analyte, the insulting power of the gas is reduced, causing the filament to cool, changing its resistance. The filament is placed at one leg of a Wheatstone bridge circuit, which is used to measure an unknown resistance. As the resistance of the filament changes, the circuit generates a measurable current. The TCD is an example of a bulk property detector, as it only depends on the thermal properties of the carrier gas, so it produces a signal as those properties change; it is therefore universal, sensitive, but not equally, to all compounds. Of common gas chromatographic

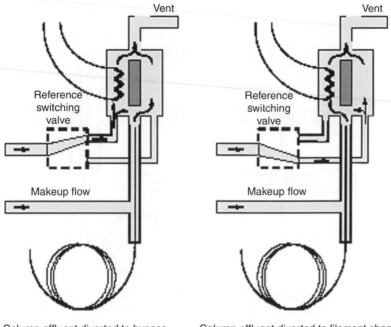

Column effluent diverted to bypass channel. Filament surrounded by reference gas.

Column effluent diverted to filament channel. If sample is present, thermal conductivity rises or falls, depending on gas type

Fig. 14.6. Schematic diagram of a capillary thermal conductivity detector showing both the reference and analytical flows. Reprinted with permission from the 6890 Gas Chromatograph Operating Manual. Copyright (2004), Agilent Technologies.

detectors, the TCD is most universal, but least sensitive, with concentration detection limits of about 10 ppm.

14.3.5.2 Flame ionization detector

For organic compound analysis, which often requires better sensitivity than TCD can provide, flame ionization is the detector of choice. Flame ionization detector (FID) is also one of the classical gas chromatographic detectors and is used with both packed and capillary columns. FID is significantly more complex than TCD, requiring fuel and oxidant gases, along with the carrier gas. In a FID, a hydrogen-air flame, with a temperature approximately 2000°C is used to decompose organic analytes, producing carbon dioxide. The flame provides enough energy to excite a portion of the carbon dioxide, forming the ionized species

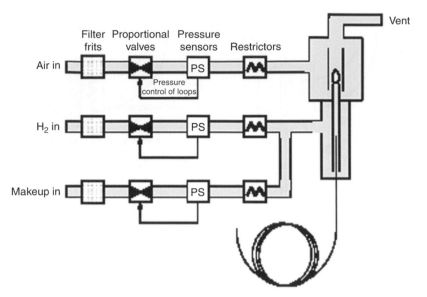

Fig. 14.7. —Schematic diagram of a capillary flame ionization detector showing the gas flows and components. Reprinted with permission from the 6890 Gas Chromatograph Operating Manual. Copyright (2004), Agilent Technologies.

CHO^+, which reacts with water formed in the flame to form H_3O^+, allowing a measurable electrical current to flow across an electrode gap. A schematic representation of an FID is shown in Fig. 14.7, showing the fuel and oxidant flows, flame tip, location of the flame and collector electrode. Like the carrier gas, the fuel (hydrogen) and oxidant (air) gases must be highly pure and carefully flow controlled. For each GC, the manufacturer provides recommendations.

14.3.5.3 Mass spectrometry
In the past decade, as systems have become simpler to operate, mass spectrometry (MS) has become increasingly popular as a detector for GC. Of all detectors for GC, mass spectrometry, often termed "mass selective detector" (MSD) in bench-top systems, offers the most versatile combination of sensitivity and selectivity. The fundamentals of MS are discussed elsewhere in this text. Quadrupole (and ion trap, which is a variant of quadrupole) mass analyzers, with electron impact ionization are by far (over 95%) the most commonly used with GC. They offer the benefits of simplicity, small size, rapid scanning of the entire mass range and sensitivity that make an ideal detector for GC.

One important characteristic of the mass analyzers used in GC is that they do *not* offer high resolution: 0.1–1 Da resolution in the mass range of 10–800 is sufficient.

MS offers the opportunity for both qualitative and quantitative analysis in a single instrument and even in a single experiment. In full-scan mode, mass spectra are continuously scanned as the gas chromatographic analysis proceeds. As shown in Fig. 14.8, GC/MS provides a three-dimensional data set with axes of time, mass-to-charge ratio and

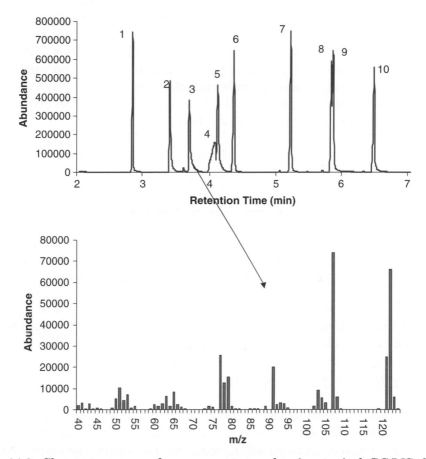

Fig. 14.8. Chromatogram and mass spectrum showing typical GC/MS data. The mass spectrum shown was obtained from the peak indicated with the arrow. Peak identification: 1decane, 2 = 1-octanol, 3 = 2,6-dimethylphenol, 4 = 2-ethylhexanoic acid, 5 = 2,3-dimethylaniline, 6 = dodecane, 7 = decanoic acid, methyl ester, 8 = dicyclohexyl amine, 9 = undecanoic acid, methyl ester, 10-dodecanoic acid, methyl ester.

signal strength. The *total ion chromatogram (TIC)* is a plot of signal strength versus time, including the sum of all of the mass signals. This provides an overall view of all components in the mixture, as all compounds that include mass fragments in a user-selected mass range, are detected. Using the TIC provides a highly sensitive (detection limits in the page range) universal detector. An *extracted ion chromatogram* is a plot of signal strength versus time, including only a single mass-to-charge ratio. This can be used to make the detector instantly selective. The user can choose a single mass, characteristic of a compound of interest, and focus on that. Finally, in a separate experiment, *selected ion monitoring* can be performed. If the analyte is a target compound, with a known mass spectrum, characteristic masses for that compound can be monitored, ignoring the other masses. This provides 1–2 orders of magnitude additional sensitivity, as the reduction in the number of different mass signals also provides a reduction in noise. Further, with the extremely short mass range of a few chosen masses, the acquisition rate of data points is significantly increased, resulting in smoother peaks.

MS as a gas chromatographic detector has become much simpler to operate in the past decade. Nearly all control of the detector is performed through a data system with similar look-and-feel to other chromatographic data systems. Components such as ion sources now have less than 10 parts, making periodic cleaning and maintenance much simpler than in the past. Finally, the pricing of mass selective detectors is now similar to other selective detectors. It is likely that MS will eventually supplant most of the other selective detectors.

14.3.5.4 Additional detectors and summary

Table 14.6, showing a summary of the most important detectors in GC and their major characteristics and applications is shown below.

14.4 DATA SYSTEMS AND HANDLING

The revolution in information technology in the past two decades has transformed analytical chemistry. For GC, data systems and data handling have evolved tremendously, from simple chart recorders, which required manual integration for quantitation, through digital electronic integrators that would provide a print out including retention times, peak heights or areas and calibration data for quantitation, but would not save the results, to computer-based data systems that provide all aspects of automated data analysis, storage and recall. Today's data

TABLE 14.6

Characteristics and figures of merit for detectors

Detector	Basic principle	Detection limit	Linear range $(10^\wedge)$	Applications/ selectivity
Flame ionization (FID)	Generation of ions from carbon in flame	1 pg/s	6	Carbon containing compounds
Thermal conductivity (TCD)	Thermal properties of gas	1 ppb	5	Universal
Electron capture (ECD)	Attraction of electrons to electronegative atoms	10 fg/s	4	Compounds containing electronegative atoms such as Cl, F, Br
Nitrogen phosphorous (NPD)	Attraction of positive ions to N and P	0.5–1 pg/s	4	Compounds containing N and P
Flame photometric (FPD)	Light emission of excited species in flame	2–50 pg/s	3–4	Compounds containing P and S
Photoionization Detector (PID)	UV radiation ionizes column effluent	5 pg/s	7	Aromatics
Sulfur chemi- luminescence (SCD)	Formation and chemiluminescence of SO	0.5 pg/s	5	Sulfur
Electrochemical (ELCD)	Catalytic oxidation or reduction of column effluent	1–5 pg/s	4–5	Sulfur and halogens
Mass selective (full scan)	Ionization, fragmentation and mass separation of column effluent	1 pg/s	4–5	Universal
Mass selective (selected ion monitoring)	Mass spectrometry of analytes with user-selected ions	fg–pg	4–5	Selective depending on chosen ions
Infrared spectrometry	Infrared absorption of eluted analytes	μg	2–3	Compounds with IR chromophores such as aromatics, double bonds

continued

TABLE 14.6 (*continued*)

Detector	Basic principle	Detection limit	Linear range $(10^{\wedge})$	Applications/ selectivity
Atomic emission detector (AED)	Atomic emission of elements	0.1–50 pg/s	4	Element selective as in atomic emission spectrometry

systems are also fully capable of controlling all aspects of instrument operation, data collection and processing.

14.4.1 Chart recorders and integrators

The simplicity of chart recorders and integrators remains useful in many situations, especially in portable systems and instruments used in educational institutions, or where budgets are tight and the laboratory is not subject to regulatory requirements for data management. Chart recorders are analog devices that directly measure the analog output from the gas chromatographic detector. A chromatogram is generated by the chart paper moving at a given speed (for example 1 cm/min) to generate the time (x-axis), underneath a pen that moves in proportion to the input signal, to generate the signal (y-axis) of the plot. Although not in common use today, chart recorders are the simplest and least expensive data systems. A chart recorder remains useful for troubleshooting, as it provides a means for measuring detector output directly, without performing the signal transformations inherent in the other systems.

Electronic integrators were the first digital data systems for GC and provide a simple and dedicated alternative to computer-based data systems. All digital data systems work in the manner pictured in Fig. 14.9. The detector output is an analog electronic signal. First, an analog-to-digital (A/D) converter samples the data and converts the voltage or current output into a digital value. For analog-to-digital converters, the key variable is resolution. Typically, chromatographic integrators and data systems use 22–24 bit A/D converters, meaning that the full signal range can be divided into 2^{22}–2^{24} discrete digital values. This digital value is then stored in computer memory. All of the calculations performed by the computer or integrator are performed on these stored

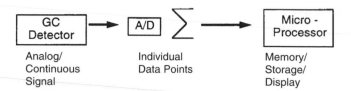

Fig. 14.9. Schematic representation of the analog-to-digital conversion process. The continuous analog signal output by the detector is sampled, converted into digital data points and transferred to the data system.

values. One key advantage is that nearly all chromatographic data systems store the original converted signal as the raw data; the calculations that are performed to do things, such as display the chromatogram, do not change this raw data. Most integrators will perform the necessary calculations for quantitative analysis, but they only have rudimentary ability to store the data for later processing or record keeping.

14.4.2 Computer-based data systems

Chromatographic data systems may either be single computer-single instrument systems, may involve multiple instruments with a single computer, or may include multiple instruments and computers on a client-server network. The main requirements are that the data system must be able to collect and process chromatographic data so that it is useful to both the laboratory originating the data and to the decision makers that use the results. Interfacing a computer-based data system to a gas chromatograph involves basically the same problem as described for integrators above: the analog signal is converted into a digital signal using an analog-to-digital converter, as shown in Fig. 14.10.

Beyond simple data storage and instrument control, modern data systems provide extensive data analysis capabilities, including fitted baselines, peak start and stop tic marks, named components, retention times, timed events and baseline subtraction. Further, they provide advanced capabilities, such as multiple calibration techniques, user-customizable information and reports and collation of multiple reports. If a Laboratory Information Management System (LIMS) is available, the chromatographic data system should be able to directly transfer data files and reports to the LIMS without user intervention. The chapter by McDowall provides a terse but thorough description of the

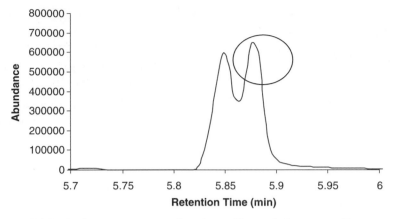

Fig. 14.10. GC/MS chromatogram showing effect of slow sampling rate. Note the peak asymmetry and the non-smooth appearance, especially of the second peak.

requirements for chromatographic data systems in regulated laboratories [13].

14.4.3 Data analysis considerations

Chromatographic data are not overly complex, as a chromatogram is simply a plot of detector signal versus time. However, as data are collected over time and as there are many calculations that are performed for quantitative analysis and for theoretical determinations such as instrument qualification, there are several important considerations in ensuring the best data quality. The major concerns are an understanding of what constitutes raw data when using an electronic data system, data-sampling rate, integration of overlapping peaks and peak asymmetry.

14.4.3.1 Raw data

When using an electronic data system, the analog signal generated by the detector is first converted into a series of digital signals and then transferred to the memory on the data system. The data system then performs operations on the raw data for tasks such as displaying a chromatogram or quantitative calculations. Generally, the original digital signal is permanently stored as a raw data file. When the data analysis portion of the chromatographic software is loaded, the raw data file is accessed and the data analysis parameters set in the method are applied to it. The report is based on a combination of this raw data and the data analysis parameters, which are many, with several of the major parameters described in Table 14.7. Most data systems include auditing functions that do not allow any changes to the original raw data; the transformed results are stored as a separate file. In many regulated laboratories it is important to ensure that the users do not unwittingly or knowingly make changes to raw data files without appropriate audit trails.

14.4.3.2 Sampling rate

Most gas chromatographic detectors have very rapid response to the entry of the eluting analyte, often reaching the equilibrated signal output in a few milliseconds. In capillary GC, the chromatographic peaks are often 1 s wide or less, while approximately 15 data points are needed to define a peak for the best quantitative precision. Therefore, typically, gas chromatographic data systems sample the analog data at 10–20 data points per second, with up to 200 points per second possible for extremely fast applications. As with all detectors, as the sampling rate increases, so does the noise, as the noise components of all signals are additive. The importance of sampling rate is illustrated in the GC/MS chromatogram shown in Fig. 14.10. This mass spectrometer provides approximately 3 data points every 2 s, resulting in the skewed peak shape.

14.4.3.3 Manual integration and peak asymmetry

Peak overlap and distorted peaks are the most common problems affecting quantitative chromatographic analysis. The peak detection and integration algorithms in most data systems are designed assuming fully separated, symmetrical peaks. When peaks are asymmetrical or overlapped, additional data analysis is often performed. The two most common situations are overlapped peaks that require manual integration and asymmetric peaks that require measurement of the degree of asymmetry as a troubleshooting technique.

To determine the area underneath overlapped peaks, the analyst must decide how to determine where the first peak ends and the second peak starts. Overlapped peaks are shown in Fig. 14.11. There are several common manual techniques for determining the peak areas:

1. Dropping a perpendicular from the bottom of the valley to the baseline.
2. Tangent skimming both peaks.
3. Estimating the overlapped portion of each peak.

TABLE 14.7

Common data processing parameters and operations

Parameter	Detector/data system	Function	Effect on data
Chart speed or x-axis scale	Data system	Sets the x-axis scale for the plotted chromatogram	Changes appearance of chromatogram Does not change peak recognition, retention time, area or width Does not affect the raw data
Zero	Data system	Location of the x-axis for the plotted chromatogram	Changes appearance of chromatogram Does not change peak recognition, retention time, area or width Does not affect the raw data
Attenuation	Data system	Vertical scale	Changes appearance of chromatogram (makes peaks shorter or taller) Does not change peak recognition, retention time, area or width Does not affect the raw data
Area reject	Data system	Causes data system to ignore peaks under a given area	Does not change appearance of chromatogramChanges peak recognition (peak with too small are ignored) Does not change retention time, area or width of integrated peaks Does not affect raw data
Threshold	Data system	Parameter for peak start and stop recognition usually based on slope of baseline	Does not change appearance of chromatogram Changes peak recognition (peaks above threshold are integrated, peaks below are not) Changes area and width of integrated peaks (affects location of peak start and stop) Does not change retention time Does not affect raw data

continued

TABLE 14.7 (*continued*)

Parameter	Detector/ data system	Function	Effect on data
Peak width	Data system	Stops integration of peaks that are too sharp	Does not change appearance of chromatogram Changes peak recognition (peaks too sharp are ignored) Does not change peak area, retention time or peak width Does not change raw data
Sampling rate	Detector	Changes rate at which digital data is collected	Changes noise level in chromatogram Does not change peak recognition, retention time, peak width Changes peak area Changes raw data

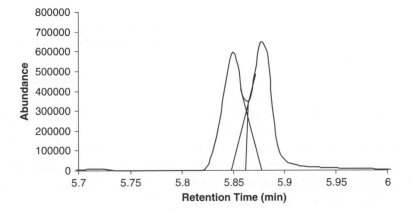

Fig. 14.11. Chromatogram showing overlapping peaks with two separation and integration possibilities: dropping a perpendicular from the valley to the baseline, or using a tangent to each peak. Note that these will result in different areas for both peaks.

It is most important that the same method be applied to all data, so that proper comparisons can be made. It is also important that the method itself be applied consistently, as it often requires the user to manually set parameters such as the point from which a perpendicular

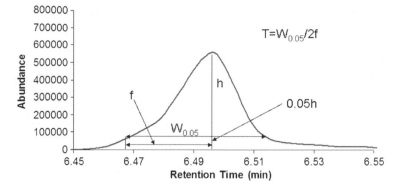

Fig. 14.12. Calculation scheme for determining peak asymmetry.

segment is drawn between the peak valley and the baseline and to manually determine the position of the baseline.

Peak asymmetry is another common problem affecting separating power and quantitative analysis. An example is shown in Fig. 14.12. There are two types of asymmetry:

In tailing, the trailing edge of the peak is affected. Tailing is usually caused by secondary retention either caused by mechanical malfunction (all peaks are tailing) or by undesired additional adsorption of one or more analytes (some peaks are tailing). In fronting, the front edge of the peak is affected. Fronting is usually caused by overload of the analyte, either by too much sample injected or by overload of active sites within the column. Fronting is often seen in later eluting peaks, in long, isothermal separations, where the analyte condenses due to the cool column. In many analytical methods, peak asymmetry is measured and compared with a standard value to determine whether it has become excessive. A typical scheme for measuring peak asymmetry is shown in Fig. 14.12 [14]. A value greater than one indicates tailing and a value less than one indicates fronting.

14.5 APPLICATIONS

GC is one of the most widely applied instrumental techniques. The basis requirement is that the analyte(s) be volatile under conditions within the gas chromatograph and that they be separated from non-volatile matrix components. Misunderstanding of volatility and vapor pressure is the most common misconception about GC; many users assume that GC is only effective for common volatile organic compounds such as solvents. In fact, GC can be effective for compounds that have only a few Torr of vapor pressure at the maximum column temperature. For example, most GC/MS instruments allow analyte molar masses upto 800. To introduce the myriad applications of GC, this section is divided into broad topics: environmental analysis, clinical and forensic analysis, pharmaceutical analysis and food, flavor and fragrance analysis. Several thorough reviews of the applications of GC have recently appeared in the literature and several of the textbooks listed in the bibliography of this chapter include sections describing applications. Finally, most of the instrument and column vendors provide extensive libraries of application notes on their WWW-sites. To access this literature, a user can simply access the site and enter the desired application or topic into the search function. A summary of key applications and references to them appears in Table 14.8.

TABLE 14.8

References to some key applications of gas chromatography

Application type	Analyte(s)/Experiment	Reference(s)
Environmental	Volatile organic compounds	[16,17,18]
	Semi-volatile organic compounds	[19]
	Pesticides	[20]
	Polychlorinated biphenyls	
Petroleum and petrochemical	Crude oil	[21]
	Refinery gases (C1–C4)	[22]
	Naptha-Gasoline (C5–C12)	[23,24,25,26]
	Kerosene-Diesel fuel (C10–C19)	[27,28,29]
	Light gas oil (C12–C21)	[30]
Clinical and pharmaceutical	Amphetamines	[31,32,33]
	Anesthetics	[34,35,36]
	Anti-depressants	[37,38]
	Blood alcohol	[39,40,41]
	Street drugs	[42,43,44]
	Steroids	[45,46,47]
Forensic	Bulk drugs of abuse	[48,49,50]
	Fire debris	[51,52,53]
	Explosives	[54,55]
Physicochemical measurements		[56]

14.6 LOOKING FORWARD

Although is has been in use for over 50 years and has become one of the most widely used routine analysis techniques, GC research remains vibrant and challenging. There are two key areas in which dramatic advancements are being made: sampling and sample introduction methods and multi-dimensional separations. A summary of sampling techniques in use with GC is shown in Table 14.9, including the basic principle of the technique and some key applications. These techniques have become critical in extending the use of GC into the diverse fields described in the Applications section.

Recently, multi-dimensional GC has been used for highly complex separations, especially in analysis of fuels, environmental samples and flavors. Most recently, comprehensive two-dimensional GC, in which samples are continuously taken form the effluent of the (long) first

TABLE 14.9

Summary of major sampling techniques for gas chromatography

Technique	Basic principles	Applications
Syringe injection	Liquid or gas is ejected from syringe directly into inlet	Gases, liquids or solutions; sample clean-up or dilution usually required
Head-space extraction	Vapor above liquid or solid sample is collected and injected	Liquid or solid samples; requires additional instrumentation
Solid phase extraction	Liquid sample passed through cartridge; analytes sorbed then eluted selectively from stationary phase	Liquids that require cleanup such as biological and environmental samples
Solid phase micro-extraction	Small stationary phase coated fiber exposed to sample, then transferred to inlet	Liquid and solid samples
Purge and trap	Inert gas bubbles through sample; vaporized analytes collected on trap and desorbed into GC	Very low-concentration solutions of volatile analytes
Supercritical fluid extraction	Supercritical carbon dioxide extracts analytes from liquid and solid samples	Soils, foods, other complex matrices
Accelerated solvent extraction	Heated, pressurized solvents used to extract analytes form solids and liquids	Soils, foods, complex matrices
Stir-bar sorptive extraction	Phase coated stir bar exposed to sample; extracted analytes desorbed into GC inlet	Trace analysis from liquids
Single drop micro-extraction	Drop of organic solvent exposed to sample on tip of syringe needle	Organic compounds from aqueous solutions

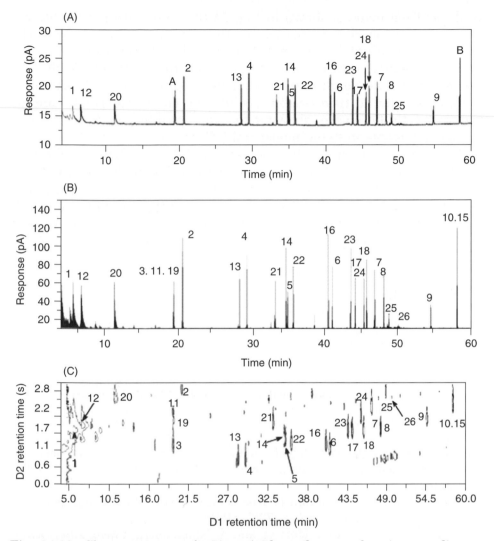

Fig. 14.13. Chromatograms showing single and comprehensive two-dimensional GC of drugs extracted from urine. (A) GC-FID chromatogram, (B) Pulsed GcxGC chromatogram; (C) GCxGC contour plot. Adapted with permission from Ref. [15, Fig. 14.3]. Complete details on separation conditions and analyte identity can be found in the original paper.

column and injected into the (short) second column, has become available. When combined with a rapid mass spectrometer, such as time-of-flight, four dimensions of data can be obtained: first dimension retention time, second dimension retention time, peak height or area and mass spectrum. A comprehensive two-dimensional chromatogram of drugs from urine is shown in Fig. 14.13, along with the traditional single dimension chromatograms of the same sample [15]. Note the large amount of additional information and separation power in the two-dimensional separation, shown as a contour plot. Especially important is that the single dimension chromatogram is so complex that it shows virtually no true baseline, while the two-dimensional contour plot is mostly flat space, showing the huge peak capacity (number of analytes) possible in two-dimensional GC.

GC is now a little more than 50 years old as an analytical technique. While this indicates maturity, and certainly GC is widely used in a huge variety of routine and research applications, there remain many new developments to come. As the technology in instrumentation and columns continually improves, new applications and interesting separation problems are regularly appearing.

REFERENCES

1 A.J.P. Martin and R.L.M. Synge, *Biochem. J*, 35 (1941) 1358.
2 A. James and A.J.P. Martin, *Biochem. J*, 50 (1952) 679.
3 R. Dandeneau, E. Zerenner and J. High Resolut, *Chromatogr. Chromatogr. Commun.*, 2 (1979) 351.
4 DryLab Chromatography Software, http://www.rheodyne.com/products/chromatography/drylab/index.asp.
5 Pro EZ-Gc Method Development Software, http://www.chromtech.net.au/proezgc-2.cfm.
6 N.H. Snow and H.M. McNair, A numerical simulation of temperature programmed gas chromatography, *J. Chromatogr. Sci.*, 30 (1992) 271–275.
7 N.H. Snow, Determination of free energy relationships using gas chromatography, *J. Chem. Educ.*, 73(7) (1996) 592–597.
8 J.J. van Deemter, F. Zuiderweg and A. Klinkenberg, *Chem. Eng. Sci.*, 5 (1956) 271.
9 M.J.E. Golay. In: V. Coates, H. Noebels and I. Fagerson (Eds.), *Gas Chromatography 1957 (East Lansing Symposium)*, Academic Press, New York, 1958, p. 1.
10 E. Barry. In: R. Grob and E. Barry (Eds.), *Modern Practice of Gas Chromatography*, 4th ed, Wiley, New York, 2004, p. 65.
11 J.H. Purnell, *J. Chem. Soc.* (1960) 1268.
12 K. Grob and G. Grob, *J. Chromatogr. Sci.*, 7 (1969) 584.
13 R. McDowall. In: J. Miller and J. Crowther (Eds.), *Analytical Chemistry in a GMP Environment: A Practical Guide*, Wiley, New York, 2000, p. 395.
14 United States Pharmacopeia 24—National Formulary 19 <621> 1914.

15   A. Kueh, P. Marriott, P. Wynne and J. Vine, *J. Chromatogr. A*, 1000 (2003) 109–124.

16   *Test Methods for Evaluating Solid Waste: Physical/Chemical Methods*, USEPA SW846, 3rd ed., Rev. 1, U.S. Government Printing Office, Washington DC, 1996.

17   *Methods for the Determination of Organic Compounds in Drinking Water*, USEPA 600/4-88/039, USEPA, Cincinnati, OH, 1988.

18   B. Kebbekus and S. Mitra, *Environmental Chemical Analysis*, Blackie Academic and Professional, London, 1998.

19   *Determination of Organic Compounds in Drinking Water by Liquid–Solid Extraction and Capillary Column/Mass Spectrometry*, USEPA Method 525.2, Rev. 2, USEPA, Cincinnati, OH, 1995.

20   EPA Method, 608, 8081A, 8082, 508, 508.1, 507.

21   F. Mutelet, G. Ekulu and M. Rogalski, *J. Chromatogr. A*, 969 (2002) 207.

22   M. Kamiński, R. Kartanowicz, D. Jastrzębski and M. Kamiński, *J.Chromatogr. A*, 989 (2003) 277.

23   Á. Stumpf, K. Tolvaj and M. Juhász, *J. Chromatogr. A*, 819 (1998) 67.

24   X. Cheng, L. Huang, X. Fu, P. Li, Z. Hu and L. Cao, *Fuel Process. Technol.*, 85 (2004) 379.

25   P.M.L. Sandercock and E. Du Pasquier, *Forensic Sci. Int.*, 140 (2004) 43.

26   P.M.L. Sandercock and E. Du Pasquier, *Forensic Sci. Int.*, 140 (2004) 71.

27   E. Robert, J.-J. Beboulene, G. Codet and D. Enache, *J. Chromatogr. A*, 683 (1994) 215.

28   M.S. Akhlaq, *J. Chromatogr. A*, 644 (1993) 253.

29   D. Malley, K. Hunter and G. Webster, *J. Soil Contam.*, 8 (1999) 481.

30   R. Alzaga, P. Montuori, L. Ortiz, J. Bayona and J. Albaigés, *J. Chromatogr. A*, 1025 (2004) 133.

31   A. Namera, M. Yashiki and T. Kojma, *J. Chromatogr. Sci.*, 40 (2002) 19.

32   S.-M. Wang, T.-C. Wang and Y.-S. Giang, *J. Chromatogr. B*, 816 (2005) 131.

33   A. Kankaanpää, T. Gunnar, K. Ariniemi, P. Lillsunde, S. Mykkänen and T. Seppälä, *J. Chromatogr. B*, 810 (2004) 57.

34   N. Yang, K. Hwang, C. Shen, H. Wang and W. Ho, *J. Chromatogr. B*, 759 (2001) 307.

35   M. Baniceru, O. Croitoru and S.M. Popescu, *J. Pharm. Biomed. Anal.*, 35 (2004) 598.

36   X-S. Deng and V.J. Simpson, *J. Pharm. Toxicol. Meth.*, 49 (2004) 131.

37   S. Dawling and R.A. Braithwaite, *J. Chromatogr.: Biomed. Appl.*, 146 (1978) 449.

38   H. Fugii and T. Arimoto, *Anal. Chem.*, 57 (1985) 2625.

39   T. Schuberth, *Biol. Mass Spectrom.*, 20 (1991) 699.

40   B. De Martinis, M.A.M. Ruzzene and C.C.S. Martin, *Anal. Chim. Acta*, 522 (2004) 163.

41   C. Winek, W. Wahba, R. Windisch and C. Winek Jr., *Forensic Sci. Int.*, 139 (2004) 1.

42 T. Williams, M. Riddle, S. Morgan and W. Brewer, *J. Chromatogr. Sci.*, 37 (1999) 210.
43 T. Gunnar, S. Mykkänen, K. Ariniemi and P. Lillsunde, *J. Chromatogr. B*, 806 (2004) 205.
44 S. Gentili, M. Cornetta and T. Macchia, *J. Chromatogr. B*, 801 (2004) 289.
45 C. Hamalainen, T. Fotsis and H. Aldercreutz, *Clin. Chim. Acta*, 199 (1991) 205.
46 M.-R. Fuh, S.-Y. Huang and T.-Y. Lin, *Talanta*, 64 (2004) 408.
47 J.B. Quintana, J. Carpinteiro, I. Rodríguez, R.A. Lorenzo, A.M. Carro and R. Cela, *J. Chromatogr.A*, 1024 (2004) 177.
48 I. Barni Comparini and F. Centini, *Forensic Sci. Int.*, 21 (1983) 129.
49 N. Bernardo, M. Pereira, B. Siqueira, M.J.N. de Paiva and P.P. Maia, *Int. J. Drug Policy*, 14 (2003) 331.
50 C. Clark. In: M. Ho (Ed.), *Analytical Methods in Forensic Chemistry*, Ellis Horwood, Chichester, UK, 1990, p. 273.
51 *Standard Practice for Separation and Concentration of Ignitable Liquid Residues in Extracts from Samples of Fire Debris by Gas Chromatography*, ASTM E138701, ASTM, West Conshohocken, PA, 2001.
52 É. Stauffer and J. Lentini, *Forensic Sci. Int.*, 132 (2003) 63.
53 B. Tan, J. Hardy and R. Snavely, *Anal. Chim. Acta*, 422 (2000) 46.
54 R. Waddell, D. Dale, M. Monagle and S. Smith, *J. Chromatogr. A*, 1062 (2005) 125.
55 C. Groom, S. Beaudet, A. Halasz, L. Paquet and J. Hawari, *J. Chromatogr. A*, 909 (2001) 60.
56 M.A. Kaiser and C.R. Dybowski. In: R.L. Grob and E.F. Barry (Eds.), *Modern Practice of Gas Chromatography*, 4th ed., Wiley, New York, 2004, pp. 605–641.

14.7 REVIEW QUESTIONS

Peak	R.T. (min)	Peak width (min)	Area	Area (%)
1	2.856	0.239	98,158,747	9.332
2	3.416	0.217	7,184,362	8.217
3	3.7	0.278	7,479,047	8.555
4	4.125	0.322	15,862,180	18.143
5	4.361	0.172	10,263,279	11.739
6	5.063	0.194	1,49,375	0.171
7	5.228	0.278	10,703,330	12.242
8	5.718	0.128	1,34,294	0.154
9	5.877	0.361	17,991,981	20.579
10	6.497	0.378	9,177,225	10.497

1. Using the retention data and the chromatogram shown in Fig. 14.8, tabulate the following for each peak: retention time (t_R), adjusted retention time (t'_R), retention factor (k), partition coefficient (K_c) and number of theoretical plates (N). The column phase ratio was 250 and the gas hold up time (t_m) was 0.995 min.

2. From the data presented above, use adjacent peak pairs to calculate: selectivity (α) and resolution (R_s).

3. For each of the analytical problems below, choose the following characteristics of a gas chromatograph that will solve the problem: packed or capillary column, inlet, column dimensions and detector. Justify your choice of each. You may need to use additional sources from the bibliography or references.
   a. separation and identification of as many components of 92-octane gasoline as possible,
   b. quantitative determination of 100 ppb anabolic steroids in urine,
   c. trace (ppb) polychlorinated biphenyls in soil,
   d. bulk fats in potato chips,
   e. gaseous impurities greater than 100 ppm in air,
   f. residual solvents (1–100 ppm) in cold medicine tablets, and
   g. trace (ppb or lower) organic contaminants in drinking water.

4. If 10,000 theoretical plates can be generated on a packed column and 250,000 theoretical plates can be generated on a capillary column assuming $k = 5$, what selectivity is required on each column to obtain a resolution greater than 1.5? Use this to explain why 10 times more different liquid phases have been used for packed column GC versus capillary GC.

5. Methane is commonly used as a marker for measuring the gas hold-up time (t_m), which was done on a capillary column 25 m long by 0.25 mm ID by 0.25 µm film thickness. A retention time for methane of 1.76 min was obtained. Determine the average linear gas velocity (v) and the average volumetric flow rate (F_c). Explain how these values differ from the actual velocity and flows at the column inlet and outlet.

6. Explain how you would use a van Deemter plot, as shown in Fig. 14.2 to determine the optimum flow rate for a separation. What are the key variables? When using theoretical plate measures for comparing columns, what experimental conditions must be controlled?

Chapter 15

High-pressure liquid chromatography

Satinder Ahuja

The phenomenal growth in chromatography is largely due to the introduction of the versatile technique called high-pressure liquid chromatography, which is frequently called high-performance liquid chromatography [1]. Both terms can be abbreviated as HPLC; see Section 15.1.1 for discussion as to which term is more appropriate. The technique offers major improvements over the classical column chromatography and some significant advantages over more recent techniques, such as supercritical fluid chromatography (SFC), capillary electrophoresis (CE), and capillary electrochromatography discussed in Chapters 16–17. The number of HPLC applications has increased enormously because a variety of complex samples have to be analyzed to solve numerous problems of scientific interest. Additionally, this demand is being continuously driven by the perpetual need to improve the speed of analysis.

The term liquid chromatography (LC) is applied to any chromatographic procedure in which the moving phase is liquid, as opposed to gas chromatography (GC) where a gas is utilized as a mobile phase (see discussion in Chapter 14). Classical column chromatography (see Section 15.1), paper chromatography—a forerunner of thin-layer chromatography (see Chapter 13), and HPLC are all examples of LC. This should clarify why it is inappropriate to further abbreviate HPLC to LC; unfortunately, it is still commonly done.

The major difference between HPLC and the old column chromatographic procedures is the pressure applied to drive the mobile phase through the column. This requires a number of improvements in the equipment, materials used for separation, and the application of the theory. HPLC offers major advantages in convenience, accuracy, speed, and the ability to carry out difficult separations. Comparisons to and relative advantages of HPLC over classical LC and GC are given below (see Sections 15.1 and 15.2) to provide the reader a better appreciation of this technique. The readers are encouraged to look up an excellent classical text on HPLC by Snyder and Kirkland [2]. A short review of

Comprehensive Analytical Chemistry 47
S. Ahuja and N. Jespersen (Eds)
Volume 47 ISSN: 0166-526X DOI: 10.1016/S0166-526X(06)47015-X

this technique is provided below; for more detailed discussion see references [1–9].

15.1 EVOLUTION OF HPLC

The progress of chromatography remained relatively dormant since its discovery in the early twentieth century until the introduction of partition and paper chromatography in the 1940s, which was followed by development of gas and thin-layer chromatography in the 1950s and various gel or size exclusion methods in the early 1960s. Shortly, thereafter, the need for better resolution and high-speed analyses of nonvolatile samples led to the development of HPLC.

To better understand the evolution of HPLC, let us briefly review classical column chromatography, which is also called packed column or open-bed chromatography. In packed column chromatography, the column is gravity-fed with the sample or mobile phase; the column is generally used only once and then discarded. Therefore packing a column has to be repeated for each separation, and this represents significant expense in time and material. Sample application, if done correctly, requires some skill and time on the part of the operator. Solvent flow is achieved by gravity flow of the solvent through the column, and individual fractions are collected manually. Typical separation requires several hours. Detection and quantification are achieved by the analysis of each fraction. Generally, many fractions are collected, and their processing requires significant time and effort. The results are recorded as a bar graph of sample concentration versus fraction number.

15.1.1 High-pressure (or high-performance) liquid chromatography

Reusable columns, which have a frit at each end to contain the packing, are employed in HPLC so that numerous individual separations can be carried out on a given column. Since the cost of an individual column can be distributed over a large number of samples, it is possible to use more expensive column packing for obtaining high performance and also to spend more time on the careful packing of a column to achieve best results. Precise sample injection is achieved easily and rapidly in HPLC by using either syringe injection or a sample valve. High-pressure pumps are required to obtain the desired solvent flow through these relatively impermeable columns. This has a decided advantage in that controlled, rapid flow of the solvent results in more reproducible operation, which translates to greater accuracy and precision in HPLC analysis. High-pressure operation leads to better and faster separation. Detection and quantification in HPLC are achieved with various types of on-line detectors. These detectors produce a final chromatogram without the operator intervention; thus producing an accurate record of the separation with a minimum effort.

A large number of separations can be performed by HPLC by simply injecting various samples and appropriate final data reduction, although the column and/or solvent may require a change for each new application. Based on these comments, it should be obvious that HPLC is considerably more convenient and less operator-dependent than classical LC. The greater reproducibility and continuous quantitative detection in HPLC allows more reliable qualitative analysis as well as more precise and accurate quantitative analysis than classical LC.

In contrast to classical column chromatography, HPLC requires high pressure for pumping liquids through more efficient columns, and this led to the name high-pressure liquid chromatography. At times, this technique is called high-performance liquid chromatography (which has the same abbreviation, viz., HPLC) because it provides high performance over classical LC. The use of this term should be made judiciously since all applications of HPLC are not necessarily based on high performance. As a result, the term high-pressure liquid chromatography is more appropriate for the majority of applications.

15.2 ADVANTAGES OVER GC, SFC, AND CE

GC provides separations that are faster and better in terms of resolution than the older chromatographic methods (see Chapter 14). It can be used to analyze a variety of samples. However, many samples simply cannot be handled by GC without derivatization, because they are insufficiently volatile and cannot pass through the column or they are thermally unstable and decompose under conditions used for GC separations. It has been estimated that only 20% of known organic compounds can be satisfactorily separated by GC without prior chemical modification of the sample.

15.2.1 Advantages over GC

HPLC is limited by sample volatility or thermal stability. It is also ideally suited for the separation of macromolecules and ionic species of biomedical interest, labile natural products, and a wide variety of other high molecular weight and/or less stable compounds (see Section 15.12).

HPLC offers a number of other advantages over GC and other separation techniques. A greater number of difficult separations are more
readily achieved by HPLC than by GC because of the following reasons:

- More selective interactions are possible with the sample molecule in
  HPLC since both phases participate in the chromatographic process, as opposed to only one (stationary phase) in GC.
- A large variety of unique column packings (stationary phases) used
  in HPLC provide a wide range of selectivity.
- Lower separation temperatures frequently used in HPLC allow
  more effective intermolecular interactions.

Chromatographic separation results from the specific interactions of
the sample molecules with the stationary and mobile phases. These
interactions are essentially absent in the moving gas phase of GC, but
they are present in the mobile liquid phase of HPLC, thus providing an
additional variable for controlling and improving separation. A great
variety of fundamentally different stationary phases have been found
useful in HPLC, which further allows a wider variation of these selective interactions and greater possibilities for separation. The chromatographic separation is generally enhanced as the temperature is
lowered, because intermolecular interactions become more effective.
This works favorably for procedures such as HPLC that are usually
carried out at ambient temperature. HPLC also offers a number of
unique detectors that have found limited application in GC:

- Visible wavelength detectors
- Electrochemical detectors (ECD)
- Refractive index detectors (RI)
- UV detectors
- Fluorescent detectors
- Nuclear magnetic resonance (NMR) as on-line detector.

Even though detectors used for GC are generally more sensitive and
provide unique selectivity for many types of samples, the available
HPLC detectors offer unique advantages in a variety of applications. In
short, it is a good idea to recognize the fact that HPLC detectors are
favored for some samples, whereas GC detectors are better for others. It
should be noted that mass spectrometric detectors have been used
effectively for both GC and HPLC.

HPLC offers another significant advantage over GC in terms of the
relative ease of sample recovery. Separated fractions are easily collected

in HPLC, simply by placing an open vessel at the end of the column. Recovery is quantitative, and separated sample components are readily isolated for identification by supplementary techniques. The recovery of separated sample bands in GC is also possible but is generally less convenient.

15.2.2   Advantages over SFC and CE

Both SFC and CE are emerging separation techniques that offer unique features and advantages (see Chapters 16 and 17). However, HPLC is unsurpassed for the wide variety of samples that can be analyzed by it. For example, it allows addition of a number of modifiers to the mobile phase that cannot be used with SFC. They offer unique selectivity.

Small sample size and sample introduction are significant problems with CE. Furthermore, CE is unable to offer the range of selective separations, including preparative separations, that are possible with HPLC.

15.3   SEPARATION PROCESS

In HPLC, the mobile phase is constantly passing through the column at a finite rate. The sample injections are made rapidly in the dynamic state in the form of a narrow band or plug (see Fig. 15.1), which provides a significant advantage over older stop-flow injection techniques in HPLC. After a sample has been injected as a narrow band, the separation can then be envisioned as a three-step process shown in the figure:

Step 1: As the sample band starts to flow through the column, a partial separation of compounds X, Y, and Z (the components of the sample) occurs.
Step 2: The separation improves as the sample moves further through the column.
Step 3: The three compounds are essentially separated from each other.

At step 3, we see two characteristic features of chromatographic separation.

1. Various compounds (solutes) in the sample migrate at different rates.
2. Each solute's molecules spread along the column.

S. Ahuja

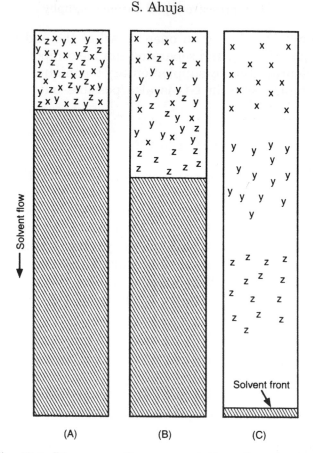

Fig. 15.1. Diagrammatic representation of separation.

Figure 15.1 shows that compound X moves most rapidly and leaves the column first, and compound Z moves the slowest and leaves the column last. As a result, compounds X and Z gradually separate as they move through the column.

The difference in movement rates of various compounds through a column is attributed to differential migration in HPLC. This can be related to the equilibrium distribution of different compounds such as X, Y, and Z between the stationary phase and the flowing solvent(s), or mobile phase. The speed with which each compound moves through the column (u_x) is determined by the number of molecules of that compound in the moving phase, at any moment, since sample molecules do not move through the column while they are in the stationary phase. The molecules of the solvent or mobile phase move at the fastest possible rate except in size exclusion chromatography, where molecular

490

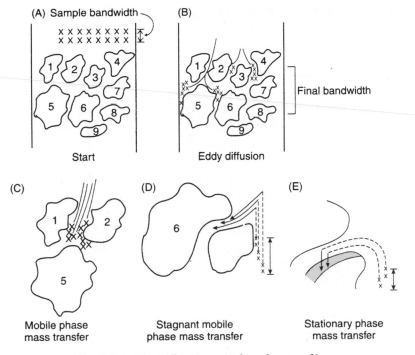

Fig. 15.2. Contributions to band spreading.

size is used to effect the separation. Compounds such as X, whose molecules are found in the mobile phase most of the time, move through the column more rapidly than the molecules of compounds such as Z, whose molecules spend most of the time in the stationary phase, and thus move slowly through the column.

Spreading of molecules such as X along the column is a less desirable characteristic of a chromatographic separation. The molecules start out as a narrow band (see Fig. 15.2A) that gradually broadens as it moves through the column. The differences in molecular migration arise from physical or rate processes. Some of the phenomena leading to molecular spreading on a chromatographic column are discussed below.

15.3.1 Eddy diffusion

One of the processes leading to molecular spreading is caused by multiple flow paths and is called eddy diffusion (see Fig. 15.2B). Within the column, eddy diffusion results from different microscopic flow streams that the solvent follows between different particles. As a result, sample molecules take different paths through the packed bed, depending on which flow streams they follow.

15.3.2 Mass transfer

A second contribution to molecular spreading relating to mobile-phase mass transfer can be seen in Fig. 15.2C. This refers to varying flow rates for different parts of a single flow stream or path between surrounding particles. In Fig. 15.2C, where the flow stream between particles 1 and 2 is shown, it is seen that liquid adjacent to a particle moves slowly or not at all, whereas liquid in the center of the flow stream moves faster. As a result, at any given time, sample molecules near the particle move to a shorter distance, and those in the middle of the flow stream move to a longer distance. Again, this results in a spreading of molecules along the column.

Figure 15.2D shows the contribution of stagnant mobile-phase mass transfer to molecular spreading. With porous column-packing particles, the mobile phase contained within the pores of the particle is stagnant, i.e., it does not move (in Fig. 15.2D one such pore is shown for particle 6). Sample molecules move into and out of these pores by diffusion. Those molecules that happen to diffuse a short distance into the pore and then diffuse out, return to the mobile phase quickly, and move a certain distance down the column. Molecules that diffuse further into the pore, spend more time in the pore and less time in the external mobile phase. As a result, these molecules move to a shorter distance down the column. Again there is an increase in the molecular spreading.

Figure 15.2E shows the effect of stationary phase mass transfer. After molecules of sample diffuse into a pore, they migrate to the stationary phase (shaded region) or become attached to it in some fashion. If a molecule penetrates deep into the stationary phase, it spends a longer time in the particle and travels a shorter distance down the column, just as in Fig. 15.2D. Molecules that spend only a little time moving into and out of the stationary phase return to the mobile phase sooner, and move further down the column.

15.3.3 Longitudinal diffusion

An additional contribution to molecular spreading is provided by longitudinal diffusion. Whether the mobile phase within the column is moving or at rest, sample molecules tend to diffuse randomly in all directions. Besides the other effects shown in Fig. 15.2, this causes a further spreading of sample molecules along the column. Longitudinal diffusion is often not an important effect, but is significant at low mobile-phase flow rates for small-particle columns.

15.4   RETENTION PARAMETERS IN HPLC

The retention parameters in HPLC are similar to those discussed in Chapter 14 on GC. To recap for HPLC, the average velocity of the sample band of X molecules (u_x) depends on M, the fraction of X molecules in the mobile phase, and the average velocity of solvent $(u$, cm/sec):

$$u_x = u\,M \tag{15.1}$$

When the fraction of molecules X in the moving phase is zero $(M = 0)$, no migration occurs and u_x is zero. If the fraction of molecules X in the moving phase is 1 (i.e., all molecules of X are in the mobile phase, $M = 1$), then molecules X move through the column at the same rate as solvent molecules and $u_x = u$. As a result, M is also the relative migration rate of compound X.

15.4.1   Capacity factor

The capacity factor k' is equal to n_s/n_m where n_s is the total moles of X in stationary phase and n_m is the total moles of X in the mobile phase. Based on this relationship, we can formulate the following equations:

$$k' + 1 = \frac{n_s + n_m}{n_m n_m} \tag{15.2}$$

$$= \frac{n_s + n_m}{n_m} \tag{15.3}$$

M can then be related to capacity factor k' as follows:

$$\text{Since } M = \frac{n_m}{n_s + n_m} = \frac{1}{1 + k'} \tag{15.4}$$

Substituting for M in Eq. (15.1), we get

$$u_x = \frac{u}{1 + k'} \tag{15.5}$$

S. Ahuja

15.4.2 Retention time

Based on the following information, we can relate u_x to retention time t_R, the time, a component, is retained in the column, and column length L:

The retention time t_R is the time required for band X to travel the length of column and is generally given as seconds or minutes; the distance is the column length L (in centimeters), and the velocity is that of band X, u_x (cm/sec).

$$t_R = \frac{L}{u_x} \tag{15.6}$$

15.4.3 Zero retention time

The zero retention time t_0 for mobile phase or other unretained molecules to move from one end of the column to the other can be determined as follows:

$$t_0 = \frac{L}{u} \tag{15.7}$$

We can then eliminate L between these last two equations and get

$$t_R = \frac{u t_0}{u_x} \tag{15.8}$$

15.4.4 Relating capacity factor to retention time

The following equation can then be derived by substituting for the value of u_x from Eq. (15.5)

$$t_R = t_0(1 + k') \tag{15.9}$$

This expresses t_R as a function of the fundamental column parameters t_0 and k'; t_R can vary between t_0 (for $k' = 0$) and any larger value (for $k' > 0$). Since t_0 varies inversely with solvent velocity u, so does t_R. For a given column, mobile phase, temperature, and sample component X, k' is normally constant for sufficiently small samples. Thus, t_R is defined for a given compound X by the chromatographic system, and t_R can be used to identify a compound tentatively by comparison with a t_R value of a known compound.

494

TABLE 15.1

Solvent strength of useful solvents for HPLC

Solvent	Silica	Alumina	Selectivity group[*]
n-Hexane	0.01	0.01	—
Chloroform	0.26	0.40	VIII
Methylene chloride	0.32	0.42	V
Ethyl acetate	0.38	0.58	VI
Tetrahydrofuran	0.44	0.57	III
Acetonitrile	0.50	0.65	VI
Methanol	0.7	0.95	II

[*]See Section 15.11.1 for selectivity groups information.

On rearrangement, Eq. (15.9) gives an expression for k'

$$k' = \frac{t_R - t_0}{t_0} \tag{15.10}$$

It is generally necessary to determine k' for one or more bands in the chromatogram to plan a strategy for improving separation. Eq. (15.10) provides a simple, rapid basis for estimating k' values in these cases; k' is equal to the distance between t_0 and the band center, divided by the distance from injection to t_0.

The important column parameter t_0 can be measured in various ways. In most cases, the center of the first band or baseline disturbance, following sample injection, denotes t_0. If there is any doubt on the position of t_0, a weaker solvent (or other unretained compound) can be injected as sample, and its t_R value will equal t_0. A weaker solvent provides larger k' values and stronger sample retention than the solvent used as mobile phase (see Table 15.1 for a list of solvents according to strength).

For example, hexane (see Table 15.1) can be injected as a sample to determine t_0 if chloroform is being used as the mobile phase in liquid–solid chromatography. As long as the flow rate of the mobile phase through the column remains unchanged, t_0 is the same for any mobile phase. If flow rate changes by some factor x, t_0 will change by the factor $1/x$.

15.4.5 Retention volume

Retention in HPLC is sometimes measured in volume units (mL), rather than in time units (s or m). Thus, the retention volume V_R is the total volume of mobile phase required to elute the center of a given band X, i.e., the total solvent flow in time between sample injection and appearance of the band center at the detector. The retention volume V_R is equal to retention time t_R multiplied by the volumetric flow rate F (mL/min.) of the mobile phase through the column.

$$V_R = t_R F \qquad (15.11)$$

The total volume of solvent within the column (V_m):

$$V_m = t_0 F \qquad (15.12)$$

Substituting for F gives

$$V_R = V_m(1 + k') \qquad (15.13)$$

15.4.6 Peak width

The peak width relates to retention time of the peak and the efficiency of a chromatographic column. The efficiency of a column is directly related to the total number of theoretical plates (N) offered by it.

15.4.7 Total number of theoretical plates

Mathematically, the width of a chromatographic peak (t_w) can be related to the number of theoretical plates (N) of the column:

$$N = 16(t_R)^2 \qquad (15.14)$$

The quantity N is approximately constant for different bands or peaks in a chromatogram for a given set of operating conditions (a particular column and mobile phase, with fixed mobile-phase velocity, and temperature). Hence N is a useful measure of column efficiency: the relative ability of a given column to provide narrow bands (small values of t_w) and improved separations.

Since N remains constant for different bands in a chromatogram, Eq. (15.14) predicts that peak width will increase proportionately with t_R, and this is generally found to be the case. Minor exceptions to the constancy of N for different bands exist [1], and in gradient elution chromatography, all bands in the chromatograms tend to be of equal width. Since HPLC peaks broaden as retention time increases, later-eluting bands show a corresponding reduction in peak height and eventually disappear into the baseline. The quantity N is proportional to column length L, so that other factors being equal, an increase in L results in an increase in N and better separation.

15.4.8 Height equivalent of a theoretical plate

The proportionality of N and L can be expressed in terms of the following equation

$$N = \frac{L}{H} \tag{15.15}$$

where H is the height equivalent of a theoretical plate (HETP).

The quantity H (equal to L/N) measures the efficiency of a given column (operated under a specific set of operating conditions) per unit length of the column (see van Deemter's equation in Chapter 14). Small H values mean more efficient columns and large N values. A very important goal in HPLC is to attain small H values that lead to maximum N and highest column efficiencies.

Based on various theoretical and practical observations, we know that the value of H decreases with

- small particles of column packing,
- low mobile-phase flow rates,
- less viscous mobile phases,
- separations at high temperatures, and
- small sample molecules.

15.5   RESOLUTION AND RETENTION TIME

A common but very important goal in HPLC is to obtain adequate separation of a given sample mixture. To achieve this goal, we need to have some quantitative measure of the relative separation or resolution achieved. The resolution, Rs, of two adjacent peaks 1 and 2 is defined as equal to the distance between the center of two peaks, divided by average peak width (see Fig. 15.3):

$$Rs = \frac{(t_2 - t_1)}{\frac{1}{2}(t_{w1} + t_{w2})} \tag{15.16}$$

The retention times t_1 and t_2 refer to the t_R values of peaks 1 and 2, and t_{w1} and t_{w2} are their peak width values. When $Rs = 1$, as in Fig. 15.3, the two peaks are reasonably well separated; that is, only 2% of one peak overlaps the other. Larger values of Rs mean better separation, and smaller values of Rs represent poorer separation. For a given value of Rs, peak overlap becomes more serious when one of the two peaks is much smaller than the other.

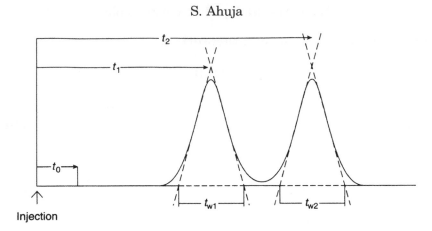

Fig. 15.3. A typical chromatogram.

The resolution value, Rs, of Eq. (15.16) serves to define a given separation. To control separation or resolution, we have to know how Rs varies with experimental parameters such as k' and N. We can derive a simplified relationship for two closely spaced (i.e., overlapping) peaks [1]. Based on Eq. (15.9), $t_1 = t_0 (1+k_1)$ and $t_2 = t_0 (1+k_2)$, where k_1 and k_2 are the k' values of bands 1 and 2.

Since t_1 is approximately equal to t_2, t_{w1} will be approximately equal to t_{w2}, based on Eq. (15.14) if we assume N is constant for both bands, Eq. (15.16) can be written as follows:

$$Rs = \frac{t_0(k_2 - k_1)}{t_{w1}} \tag{15.17}$$

From Eq. (15.14), we know $t_w = 4 t_1/N(1/2)$ or $= 4 t_0 (1+k_1)/N(1/2)$
Substituting this value in Eq. (15.17) gives

$$Rs = \frac{(k_2 - k_1)N^{1/2}}{4(1 + k_1)}$$
$$= \frac{1}{4}\left(\frac{k_2}{k_1} - 1\right)N^{1/2}\left(\frac{k_1}{1 + k_1}\right) \tag{15.18}$$

15.5.1 Separation factor and resolution

The separation factor α is equal to k_2/k_1.

By inserting α and an average value of k' for k_2 and k_1, we can simplify the resolution equation as follows:

$$Rs = \frac{1}{4}(\alpha - 1)N\frac{1}{2}\left(\frac{k'}{1+k'}\right) \tag{15.19}$$

It is important to recognize that a number of assumptions were made in deriving Eq. (15.17) to arrive at the simplified Eq. (15.19). A more fundamental form of resolution expression is given below (see Eq. (15.22) in Section 15.10.1). To get a more accurate equation, the actual values of the peak widths and their respective capacity factors should be used; however, for most practical purposes the above equation or its original form, Eq. (15.16), is satisfactory.

15.6 EQUIPMENT

HPLC equipment has been designed and produced to assure correct volumetric delivery of the mobile phase, including the injected sample, and has low-noise detectors so that low concentrations of samples can be analyzed conveniently. Discussed below, briefly, are some of the important considerations for the HPLC equipment. More detailed discussion can be found in a recent text (see Chapter 3 of reference 3).

15.6.1 Modular versus integrated HPLC systems

A simple system is comprised of an isocratic pump, a manual injector, a UV detector, and a strip-chart recorder. A schematic diagram of an HPLC instrument is shown in Fig. 15.4. This simple configuration is rarely used in most modern laboratories. A typical HPLC system is likely to consist of a multi-solvent pump, an autosampler, an on-line degasser, a column oven, and a UV/Vis or photodiode array detector; all connected to and controlled by a data-handling workstation. Examples of modular and integrated systems are shown in Fig. 15.5. Some of the important instrumental requirements are summarized in Table 15.2.

HPLC systems can be classified as modular or integrated. In a modular system, separate modules are stacked and connected to function as a unit, whereas in an integrated system, modules are built inside a single housing and often share a common controller board. These built-in modules cannot function outside the system; solvent lines and electrical wires are inside the housing. Modular systems are considered easily serviceable since internal components are easily accessible, and the malfunctioning module can be swapped. Integrated systems provide

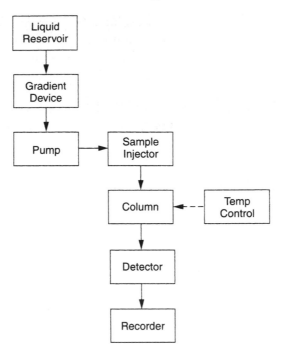

Fig. 15.4. A schematic of HPLC equipment.

better integration of modules; for example, an autosampler can be designed to inject samples right at the beginning of a pump stroke, thus yielding better precision in retention time.

15.6.2 Solvent delivery systems

A modern solvent delivery system consists of one or more pumps, solvent reservoirs, and a degassing system. HPLC pumps can be categorized in several ways: by flow range, driving mechanism, or blending method. A typical analytical pump has a flow range of 0.001–10 mL/min, which handles comfortably the flow rates required for most analytical work (e.g., 0.5–3 mL/min). Preparative pumps can have a flow range from 30 mL/min up to L/m.

Syringe pumps driven by screw mechanisms were popular in the 1960s because of their inherent precision and pulseless flow characteristics. Their disadvantages are higher manufacturing costs and the problems associated with syringe refill cycles. Syringe pumps are currently used in specialized systems for microbore and capillary HPLC.

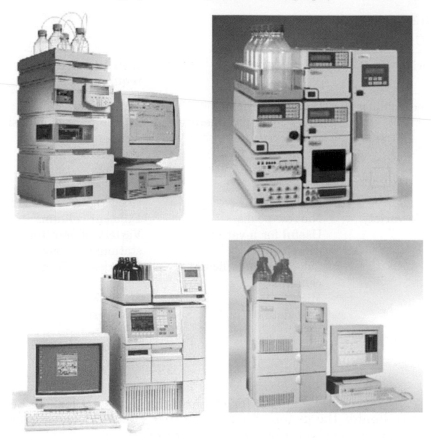

Fig. 15.5. Modular and integrated HPLC systems [3].

Most HPLC pumps today use a reciprocating piston design, as shown in Fig. 15.6. Here, a motorized cam (or a direct screw-drive system) drives a piston back and forth to deliver solvent through a set of inlet and outlet check valves. All wettable components are made from inert materials; examples include stainless steel pump heads, ruby balls, sapphire seats in check valves, sapphire pistons, and fluorocarbon pump seals. Since liquid is delivered only during the inward stroke of the piston, the resulting sinusoidal liquid stream is usually passed through a pulse dampener to reduce pulsation. Another approach is to use a dual-piston in-parallel pump design, where a single motor drives two pistons in separate pump heads. Since the pistons are 180° out of phase, the combined output results in a steadier flow pattern. The digital stepper motor that drives the cam is controlled by a microprocessor

TABLE 15.2

HPLC instrumental requirements

Criteria	Characteristics	Instrumental requirements
Reproducibility	Controls various operational parameters	Controls precisely: mobile-phase composition temperature flow rate detector response
Detectability Sample variety Speed	High-detector response Narrow peaks Useful for a variety of samples Selective and efficient columns High flow rate Fast data output	Good signal-to-noise ratio Efficient columns Variety of detectors and stationary phases Low dead-volume fittings High-pressure pumps Fast-response recorders and automatic data handling

that coordinates the piston speed with other components such as the proportioning valve and the pressure monitor.

Another way to classify pumping systems is based on the achievement of solvent blending; i.e., under low- or high-pressure mixing conditions.

A number of improvements have been made in pumps over the years. The life of the piston seal was improved by better designs, such as more durable spring-loaded seals, self-aligning piston mechanisms, and irrigation systems to eliminate any salt buildup (piston seal wash). Most pumps also have front panel access to the pump heads for easier maintenance. The pump performance often suffers at low flow rates. For example, since a typical piston size is about 100 μL, blending and pulse-free flow are difficult to achieve at a flow rate of <0.2 mL/min or at <2 pump strokes per min. High-end pumps use sophisticated mechanisms such as variable stroke length (20–100 μL), micro-pistons (5–30 μL), and hybrid dual-piston in-series designs to improve performance. Two pistons (often different sizes) are independently driven by separate motors in a dual-piston in-series design. The pre-piston is

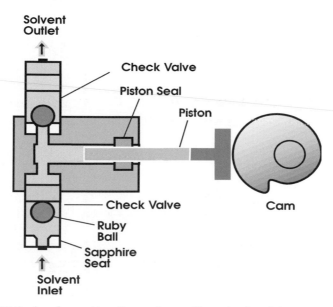

Fig. 15.6. A schematic of a reciprocating single-piston pump [3].

synchronized with the secondary piston to provide smoother flows and higher compositional accuracy. Variations of this design are used in many pumps such as Waters Alliance 2695, Agilent Series 1100, and Shimadzu LC10-AD.

The pumps specifically designed for HPLC are able to provide constant flow of the mobile phase against column pressure up to 10,000 psi. However, it is important to recognize that most HPLC separations are run at pressures lower than 6,000 psi. A comparison of various pumps used for HPLC is given in Table 15.3.

Most chromatographers limit themselves to binary or tertiary gradient systems; however, it should be noted that pumps capable of quaternary gradient are available (see reference 3, Chapter 3) and should be considered in the equipment selection process to allow greater versatility in method development.

15.6.3 Detectors

An HPLC detector is often a modified spectrophotometer equipped with a small flow cell, which monitors the concentration (or mass) of eluting sample components. A number of detectors used in HPLC are discussed below. Most applications utilize absorbance detectors such as UV/Vis or

TABLE 15.3

Comparison of various HPLC pumps

Pump characteristic	Reciprocating						Positive displacement		Pneumatic		
	Simple single-head	Single-head smooth pulse	Simple dual-head	Dual-head, compressibility corrected, smooth pulse	Dual-head, closed loop flow control	Triple head low-volume	Syringe-type	Hydraulic amplifier	Simple	Amplifier	Amplifier with flow control
Resettable	+	+	++	++	++	++	++	++	-	-	+
Drift	+	+	++	++	++	++	++	+	-	+	+
Low Noise	-	+	++	++	++	++	++	++	+	+	++
Accurate	+	+	+	+	++	++	+	+	-	-	+
Versatile	-	+	++	++	+	++	-	+	++	++	+
Low Service	+	+	+	+	+	+	+	+	++	++	-
Durable	+	+	+	+	+	+	+	-	++	++	++
Coast	Low	Moderate	Moderate	High	Very high	Very high	Moderate to high	Moderate	Low	Moderate	High
Consistent flow	Yes	Yes	Yes	Yes	Yes	Yes	Yes	Yes	No	No	No
Consistent pressure	No	No	No	Yes	No	No	No	Yes	Yes	Yes	Yes

Note: Adapted from reference [2].

++ = optimum, + = satisfactory, - = needs improvement.

photodiode array detectors (PDA). Therefore, these detectors are covered in greater detail in this section.

15.6.3.1 UV/Vis absorbance detectors

These detectors monitor the absorption of UV or visible light by analytes in the HPLC eluent. A typical UV/Vis detector consists of a deuterium source and a monochromator (a movable grating controlled by stepper motor to select wavelength through an exit slit) to focus the light through a small flow cell. A dual-beam optical bench is typical for reducing drift. Two photodiodes are used to measure light intensities of sample and reference beams. The observed absorbance is controlled by Beer's Law (see Chapter 5):

$$\text{Absorbance (A)} = \text{molar absorptivity (a)} \times \text{path length (b)}$$
$$\times \text{concentration (c)}$$

Most UV absorption bands correspond to transitions of electrons from $\pi \to \pi*$, $n \to \pi*$, or $n \to \sigma*$ molecular orbitals. Besides aromatic compounds, organic functional groups such as carbonyl, carboxylic, amido, azo, nitro, nitroso, and ketone groups have absorbance in the UV region.

Important performance characteristics of UV/Vis detectors are sensitivity, linearity, and band dispersion. These are controlled by design of the optics and the flow cell—more specifically by spectral bandpass, stray light characteristics, and the volume and path length of the flow cell.

Sensitivity is the most important requirement for any detector, and is influenced by baseline noise. The dual wavelength detection feature is useful for the simultaneous monitoring of active ingredients and preservatives in drug products. The baseline noise can be substantially higher in the dual wavelength mode because this feature is achieved by toggling the monochromator between the two wavelengths. While a wavelength range of 190–700 nm is typical, the sensitivity performance above 400 nm is lower because of lower energy of the deuterium source in the visible region. Some detectors have a secondary tungsten source to augment the deuterium source and extend the wavelength range to 190–1000 nm. A spectral bandwidth of 5–8 nm is typical in HPLC detectors. Spectral bandwidth is defined as the width in nm of the selected wavelength region, and is related to the optical slit width of the spectrometer. Increasing the bandwidth by widening the slit width improves detection sensitivity at the expense of linearity. Wavelength accuracy is an important requirement in instrument calibration. Wider linearity and lower baseline noise are critical for achieving acceptable limits of quantitation (LOQ). Flow-cell design

S. Ahuja

is a critical requirement for increasing sensitivity since signals are proportional to cell path lengths. However, increasing the path length by using a larger flow cell is often detrimental to system band dispersion and leads to extra-column band broadening. Advanced flow-cell designs such as tapered cells, or flow cells equipped with focusing lenses as cell windows, are often used to reduce gradient baseline shifts stemming from refractive index changes in the mobile phase. Most manufacturers offer a number of optional flow cells for applications such as semi-micro, microbore, semi-prep, or HPLC/MS.

Performance sensitivity has improved manyfold in the last two decades. The benchmark noise level of $\pm 1 \times 10^{-5}$ AU/cm, thought at one time to be the physical limit of UV/Vis detection imposed by short-term source fluctuations, thermal flow noise and electronic noise, is now surpassed by many detectors. Extending the linear dynamic range to >2 AU is possible by lowering stray lights in the optical bench. The typical lifetime of the deuterium source is now 1000–2000 h. Many modern detectors have dual- or multiple-wavelength detection and stop-flow scanning features. Most detectors have front panel access to self-aligned sources and flow cells for easy maintenance. Others have self-validation features such as power-up diagnostics, leak sensors, time logs for lamps, built-in holmium oxide filters for wavelength calibration, or filter-wheels for linearity verification.

15.6.3.2 Photodiode array (PDA) detector

A PDA detector provides UV spectra of eluting peaks in addition to monitoring the absorbance of the HPLC eluent. It is the detector of choice for method development and for monitoring impurities. A schematic of a PDA detector (Fig. 15.7) shows where the entire spectrum from the deuterium source or a selected portion passes through the flow cell and is dispersed onto a diode array element that measures the intensity of light at each wavelength. Most PDAs use a charge-coupled diode array with 512–1024 diodes, capable of spectral resolution of ~1 nm. Sophisticated spectral evaluation software allows the convenient display of both chromatographic and spectral data along three axes (absorbance versus wavelength, versus retention time). With multiple windows, a system can display a contour map, a chromatogram at a specified wavelength, and a spectrum of the active ingredient peak in a sample. Most software also allows automated spectral annotations of λ_{max}, peak matching, library searches, and peak purity evaluation.

Further sensitivity enhancements of PDA are likely to stem from advanced flow-cell design using fiber-optic technology to extend the

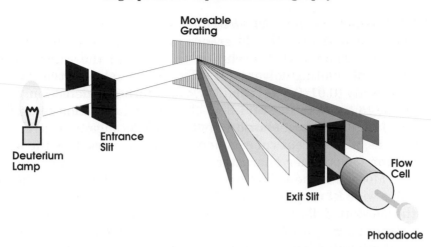

Fig. 15.7. A Schematic of a photodiode array detector [3].

path length without increasing noise or chromatographic band dispersion. This type of flow cell is especially important in micro HPLC for samples of limited availability. Linearity performance at high absorbance (up to 2.0 AU) is important in pharmaceutical purity testing because it allows the injection of a higher concentration of the active ingredient to enhance detection of trace impurities.

15.6.3.3   Fluorescence (Fl) detector

A fluorescence detector monitors the emitted fluorescent light of the HPLC eluent. It is selective and extremely sensitive (pg to fg) to highly fluorescent compounds. A detector consists of a xenon source, an excitation monochromator, an emission monochromator, a square flow cell, and a photomultiplier for amplifying the emitted photons. The xenon source can be a high-power continuous source (150 W) or a pulsed source (<20 W). The pulsed source is becoming popular because it requires less power, has more energy in the far UV region, and allows detection modes such as phosphorescence, chemiluminescence, and bioluminescence. All high-end units have a double monochromator for wavelength programmability. Filters are used in lower-cost units. Sensitivity specification is often quoted by the signal/noise ratio of the Raman band of water which ranges from ~100 in older models to >300 in modern units. Sensitivity can be enhanced by widening the optical slits.

15.6.3.4 Refractive index (RI) detector

An RI detector measures the difference in RI between the sample cell containing the analyte in the eluting solvent and the reference cell (containing only eluting solvent). It offers universal detection but has lower sensitivity (0.01–0.1 µg) than UV/Vis absorbance detection, and is more prone to temperature and flow changes. RI detection is used for components of low chromophoric activities such as sugars, triglycerides, organic acids, pharmaceutical excipients, and polymers. It is the standard detector for polymer characterization in gel permeation chromatography. Modern RI detectors are generally the differential deflection type with a wide RI range of 1.00–1.75 RIU (refractive index unit). They have thermostatted flow-cell assemblies and allow unattended operation via auto-purging the reference flow cell. Sensitivity, baseline stability, and reliability have improved significantly in recent years. The greatest disadvantages are its low sensitivity and its incompatibility with gradient elution.

15.6.3.5 Evaporative light scattering detector (ELSD)

An ELSD converts the HPLC eluent into a particle stream and measures the scattered radiation. It offers universal detection for nonvolatile or semivolatile compounds and has higher sensitivity than the RI detector (in the low ng range) in addition to being compatible with gradient analysis. ELSD is routinely used in combinatorial screening. Response factors are less variable than that of other detectors. An ELSD consists of a nebulizer equipped with a constant temperature drift tube where a counter-current of heated air or nitrogen reduces the HPLC eluent into a fine stream of analyte particles. A laser or a polychromatic beam intersects the particle stream, and the scattered radiation is amplified by a photomultiplier. Manufacturers include Alltech, Polymer Laboratories, Shimadzu, Waters, Sedere, and ESA.

15.6.3.6 Electrochemical detector (ECD)

An ECD measures the current generated by electroactive analytes in the HPLC eluent between electrodes in the flow cell. It offers sensitive detection (pg levels) of catecholamines, neurotransmitters, sugars, glycoproteins, and compounds containing phenolic, hydroxyl, amino, diazo, or nitro functional groups. The detector can be the amperometric, pulsed-amperometric, or coulometric type, with the electrodes made from vitreous or glassy carbon, silver, gold, or platinum, operated in the oxidative or reductive mode. Manufacturers include BSA, ESA, and Shimadzu.

15.6.3.7 Conductivity detector

A conductivity detector measures the electrical conductivity of the HPLC eluent stream and is amenable to low-level determination (ppm and ppb levels) of ionic components such as anions, metals, organic acids, and surfactants. It is the primary detection mode for ion chromatography. Manufacturers include Dionex, Alltech, Shimadzu, and Waters.

15.6.3.8 Radioactivity detector

A radioactivity detector is used to measure radioactivity in the HPLC eluent, using a flow cell. The detection principle is based on liquid scintillation technology to detect phosphors caused by radiation, though a solid-state scintillator is often used around the flow cell [17,31]. This detector is very specific and can be extremely sensitive. It is often used for conducting experiments using tritium or C-14 radiolabeled compounds in toxicological, metabolic, or degradation studies.

15.6.3.9 Nuclear magnetic resonance and mass spectrometric detectors

HPLC/NMR and HPLC/MS are popular techniques that combine the versatility of HPLC with the identification power of NMR or MS (see Chapters 11, 12, and 18).

15.6.4 Data handling

Chromatographic data handling systems range from a strip-chart recorder, an integrator, a PC-based workstation to a server network system designed for the chromatographer's needs. Most PC-based data handling workstations also incorporate full instrumental control of the HPLC system from various manufacturers. Lately, major companies have installed centralized data network systems to ensure data security and integrity.

15.7 MODES OF SEPARATION IN HPLC

Discussed below are various modes of separations in HPLC. Included here is brief coverage of mobile-phase selection for various modes of chromatography and elementary information on mechanism, choice of solvents and columns, and other practical considerations. It should come as no surprise that reversed-phase HPLC is discussed at greater length in this section because it is the most commonly used technique in HPLC (more detailed discussion is provided in Section 15.8). Clearly, a better understanding of separation mechanisms would enable the reader to make better decisions on mobile-phase optimizations (see Section 15.11 later in this chapter).

15.7.1 Adsorption chromatography

Separations in adsorption chromatography result largely from the interaction of polar functional groups with discrete adsorption sites on the stationary phase. The strength of these polar interactions (see Table 15.4) is responsible for the selectivity of the separation in adsorption chromatography. It should be noticed that alcholic, phenolic, and amino groups are highly retained. Adsorption chromatography with relatively nonpolar mobile phases and polar stationary phases is sometimes referred to as "liquid–solid" chromatography.

Adsorption chromatography is generally considered suitable for the separation of nonionic molecules that are soluble in organic solvents. Very polar compounds, those with high solubility in water and low solubility in organic solvents, interact very strongly with the adsorbent surface and result in peaks of poor symmetry and poor efficiency.

Solvent strength and selectivity can be controlled by using binary or higher-order solvent mixtures. The change in solvent strength as a function of the volume percent of the more polar component is not linear. At low concentrations of the polar solvent, small increases in concentration produce large increases in solvent strength; at the other extreme, relatively large changes in the concentration of the polar solvent affect

TABLE 15.4

Adsorption characteristics of various functional groups [39]

Retention	Type of compound
Nonadsorbed	Aliphatics
Weakly adsorbed	Alkenes, mercaptans, sulfides, aromatics, and halogenated aromatics
Moderately adsorbed	Polynuclear aromatics, ethers, nitrites, nitro compounds, and most carbonyls
Strongly adsorbed	Alcohols, phenols, amines, amides, imides, sulfoxides, and acids
Comparative	$F < Cl < Br < I$; cis compounds are retained more strongly than trans; equatorial groups in cyclohexane derivatives (and steroids) are more strongly retained than axial derivatives

the solvent strength of the mobile phase to a lesser extent. Once the optimal solvent strength has been determined for a separation, the resolution of the sample is improved by changing solvent selectivity at a constant solvent strength.

The uptake of water by the column adversely affects separation and leads to irreproducible separations and lengthy column regeneration times. The use of a 100% saturated solvent is undesirable because such liquid–solid chromatographic systems are often unstable. Apparently, under these conditions the pores of the adsorbent gradually fill with water, leading to changes in retention with time and possibly also to a change in the retention mechanism as liquid–liquid partition effects become more important. When silica is the adsorbent, 50% saturation of the mobile phase has been recommended for stable chromatographic conditions [10–12]. Solvents with 50% water saturation can be prepared by mixing dry solvent with a 100% saturated solvent or, preferably, by using a moisture-control system [13]. The latter consists of a water-coated thermostatted adsorbent column through which the mobile phase is recycled during the time required to reach the desired degree of saturation.

It should be noted that a column that has been deactivated with water no longer shows adequate separation properties. Restoring the activity of the column by pumping a large volume of dry mobile phase through the column is a slow and expensive process. Alternatively, reactivation can be accomplished chemically using the acid-catalyzed reaction between water and 2,2-dimethoxypropane, the products of which, acetone and methanol, are easily eluted from the column [14].

In addition to water, virtually any organic polar modifier can be used to control solute retention in liquid–solid chromatography. Alcohols, acetonitrile, tetrahydrofuran, and ethyl acetate in volumes of less than 1% can be incorporated into nonpolar mobile phases to control adsorbent activity. In general, column efficiency declines for alcohol-modulated eluents compared to water-modulated eluent systems.

The retention behavior of a sample solute is dominated largely by its functional groups. As a result, adsorption chromatography has been found useful for different classes of compounds. Isomeric and multifunctional compounds can generally be separated by adsorption chromatography since the relative position of the solute functional groups interacting with the spatial arrangement of the surface hydroxyl groups governs adsorption. This effect leads to a pronounced selectivity of silica gel for positional isomers.

Organic amine compounds can be separated successfully on silica gel columns with good peak symmetry using organic/aqueous mobile phases [15–17]. Solute retention appears to involve both electrostatic and adsorption forces.

15.7.2 Normal bonded phases

Polar bonded phases containing a diol, cyano, diethylamino, amino, or diamino functional group are commercially available; representative structures are shown in Table 15.5. The alkyl nitrile- and alkylamine-substituted stationary phases, when used with a mobile phase of low polarity, behave in a manner similar to the solid adsorbents discussed in the previous section, i.e., the retention of the sample increases with solute polarity, and increasing the polarity of the mobile phase reduces the retention of all solutes. The polar bonded-phase packings are generally less retentive than adsorbent packings, but are relatively free from the problems of chemisorption, tailing, and catalytic activity associated with silica and alumina. The bonded-phase packings respond rapidly to changes in mobile-phase composition and can be used in gradient elution analyses. Adsorbent packings respond slowly to changes in mobile-phase composition because of slow changes in surface hydration, making gradient elution analysis difficult. Because of

TABLE 15.5

Polar bonded phases [39]

Name	Structure	Application
Diol	$-(CH_2)_3\ OCH_2CH(OH)CH_2(OH)$	Surface modifying groups for silica packings used in SEC
Cyano	$-(CH_2)_3CN$	Partition or adsorption chromatography
Amino	$-(CH_2)_n\ NH_2$ $N = 3$ or 5	Adsorption, partition, or ion-exchange chromatography
Dimethyl-amino	$-(CH_2)_3N(CH_3)_2$	Ion-exchange chromatography
Diamino	$-(CH_2)_2NH(CH_2)_2\ NH_2$	Adsorption or ion-exchange chromatography

the above advantages, the polar bonded-phase packings have been proposed as alternatives to microporous adsorbents for separating the same sample type [18].

The alkyl nitrile-substituted phase is of intermediate polarity and is less retentive than silica gel but displays similar selectivity. It provides good selectivity for the separation of double bond isomers and ring compounds differing in either the position or number of double bonds [19]. With aqueous mobile phases, the alkyl nitrile-substituted stationary phases have been used for the separation of saccharides that are poorly retained on reversed-phase columns. The alkylamine-substituted phases provide a separation mechanism complementary to either silica gel or alkyl nitrile-substituted phases. The amino function imparts strong hydrogen bonding properties, as well as acid or base properties, to the stationary phase depending on the nature of the solute. The aminoalkyl-substituted stationary phase has been used for the class separation of polycyclic aromatic hydrocarbons [20,21]. Retention is based primarily on charge transfer interactions between the aromatic π-electrons of the polycyclic aromatic hydrocarbons and the polar amino groups of the stationary phase. Samples are separated into peaks containing those components with the same number of rings. Retention increases incrementally with increasing ring number, but is scarcely influenced by the presence of alkyl ring substituents. In contrast, reversed-phase separations show poor separation between alkyl-substituted polycyclic aromatic hydrocarbons and polycyclic aromatic hydrocarbons of a higher ring number.

The diol- and diethyl amino-substituted stationary phases are used mainly in size exclusion and ion-exchange chromatography, respectively. The practice of normal-phase chromatography is similar to that described for adsorption chromatography. A polar solvent modifier, such as isopropanol at the 0.5–1.0% v/v level, is used in nonpolar solvents to improve peak symmetry and retention time reproducibility. It is believed that the polar modifier solvates the polar groups of the stationary phase, leading to an improvement in mass-transfer properties. For the separation of carboxylic acids or phenols, either glacial acetic acid or phosphoric acid is used at low levels as a tailing inhibitor. Likewise, propylamine is a suitable modifier for the separation of bases.

Certain specific problems arise with the use of alkylamine-substituted stationary phases. Since amines are readily oxidized, degassing the mobile phase and avoiding solvents that may contain peroxides, e.g., diethyl ether and tetrahydrofuran, are recommended. Samples or impurities in the mobile phase containing ketone or aldehyde groups may react chemically with the amine group of the stationary phase, forming a Schiff's base complex [19]. This reaction will alter the separation properties of the column. The column may be regenerated by flushing with a large volume of acidified water [22].

15.7.3   Reversed-phase chromatography

Reversed-phase chromatography is performed on columns where the stationary phase surface is less polar than the mobile phase. The most commonly used column packings for reversed-phase separations have a ligand such as octadecyl (C-18), octyl (C-8), phenyl, or cyanopropyl chemically bonded to microporous silica particles. The silica particles can be either spherical or irregularly shaped, containing unreacted silanol groups. The unreacted silanol groups can be end-capped by silanization with a small silanizing reagent such as trimethylchlorosilane (TMCS).

It is estimated that over 65% (possibly up to 90%) of all HPLC separations are carried out in the reversed-phase mode. The reasons for this include the simplicity, versatility, and scope of the reversed-phase method [23]. The hydrocarbon-like stationary phases equilibrate rapidly with changes in mobile-phase composition and are therefore eminently suitable for use with gradient elution.

Retention in reversed-phase liquid chromatography (RPLC) occurs by nonspecific hydrophobic interactions of the solute with the stationary phase. The near-universal application of reversed-phase chromatography stems from the fact that virtually all organic molecules have hydrophobic regions in their structures and are capable of interacting with the stationary phase. Reversed-phase chromatography is thus ideally suited to separating the components of oligomers or homologues. Within a homologous series, the logarithm of the capacity factor is generally a linear function of the carbon number. Branched-chain compounds are generally retained to a lesser extent than their straight-chain analogues and unsaturated analogues. Since the mobile phase in reversed-phase chromatography is polar and generally contains water, the method is ideally suited to the separation of polar molecules that are either insoluble in organic solvents or bind too strongly to solid adsorbents for normal elution. Many samples with a biological origin fall into this category.

It is believed that retention in reversed-phase chromatography is a function of sample hydrophobicity, whereas the selectivity of the separation results almost entirely from specific interactions of the solute with the mobile phase [24]. Generally, the selectivity may be conveniently adjusted by changing the type of organic modifier in the mobile phase. For ionic or ionizable solutes, appropriate pH is used to suppress ionization, or ion-pairing reagents are used to increase lipophilicity to assist the degree of solute transfer to the stationary phase and thus control selectivity [25]. Metal-ligand complexes and chiral reagents can be added to the mobile phase to separate optically active isomers.

The details of the mechanism governing retention in reversed-phase chromatography, using chemically bonded hydrocarbonaceous phases, are not completely understood [26]. Solute retention in RPLC can proceed either via partitioning between the hydrocarbonaceous surface layer of the nonpolar stationary phase and the mobile phase, or by adsorption of the solute to the nonpolar portion of the stationary phase. In this context, the partitioning mechanism seems less likely since the hydrocarbonaceous layer is only a monolayer thick and lacks the favorable properties of a bulk liquid for solubilizing solutes. However, the less polar solvent components of the mobile phase could accumulate near the apolar surface of the stationary phase, forming an essentially stagnant layer of mobile phase rich in the less polar solvent [27]. As a result, the solute could partition between this layer and the bulk mobile phase without directly interacting with the stationary phase proper. The balance of evidence favors the adsorption mechanism either with the stationary phase surface itself or by interaction with the ordered solvent molecule layer at the stationary phase surface [28].

Retention of a solute because it is forced by the solvent to go to the hydrocarbonaceous layer is called solvophobic effect. To provide a simple view of solvophobic theory, we will assume that solute retention occurs by adsorption of the solute to the stationary phase, without defining the stationary phase. The solvophobic theory assumes that aqueous mobile phases are highly structured because of the tendency of water molecules to self-associate by hydrogen bonding, and this structuring is perturbed by the presence of nonpolar solute molecules. As a result of this very high cohesive energy of the solvent, the less polar solutes are literally forced out of the mobile phases and are bound to the hydrocarbon portion of the stationary phase. Therefore, the driving force for solute retention is not because of the favorable affinity of the solute for the stationary phase, but because of the solvent's forcing the solute to the hydrocarbonaceous layer.

The most commonly used solvents for RPLC are methanol, acetonitrile, and tetrahydrofuran, used in binary, ternary, or quaternary combinations with water. The effect of solvent strengths can be seen in

S. Ahuja

Table 15.6. A significant difference in separation selectivity can be achieved by replacing a given solvent with a different solvent.

Changes in pH can change the separation selectivity for ionized or ionizable solutes, since charged molecules are distributed preferentially into the aqueous or more polar phase (for further discussion, see Chapter 6, reference 3.) In general, separation conditions that are used to vary α values of various peaks are summarized in Table 15.7.

TABLE 15.6

Solvent strength in reversed-phase HPLC [2]

Solvent	P'	$k'*$
Water	10.2	—
Dimethyl sulfoxide	7.2	1.5-fold
Ethylene glycol	6.9	1.5
Acetonitrile	5.8	2.0
Methanol	5.1	2.0
Acetone	5.1	2.2
Ethanol	4.3	2.3
Tetrahydrofuran	4.0	2.8
i-Propanol	3.9	3.0

* Decrease in k' for each 10% addition of solvent to water

TABLE 15.7

Separation conditions used to improve α values

Variable	Impact
Stationary phase	Choice of C-18, C-8, phenyl, cyano or trimethyl has varying impact, depending on type of sample
Organic solvent	Change from methanol to acetonitrile or THF commonly produces large change in α values
pH	Change in pH can have a major effect on α values of acidic or basic compounds
Solvent strength	Changes in % organic solvent often provides significant changes in values
Additives	Ion-pair reagents have great impact on α values. Other additives such as amine modifiers, buffers, and salts, including complexing agents, can be used to produce various effects
Temperature	Vary between 0–70°C to control α values

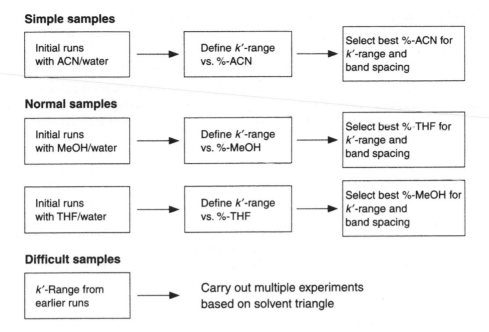

Fig. 15.8. A recommended approach for developing HPLC methods.

Frequently the choice of the mobile phase equates to methanol/water or acetonitrile mixtures in various proportions. The next step is to optimize the concentration of the organic solvent. Following that, low concentrations of tetrahydrofuran are explored to further improve separations (see Fig. 15.8).

Resolution can be mapped as a function of various proportions of acetonitrile, methanol, and THF in the mobile phase. Usually the k' range or run time is held constant during the process by varying the amount of water in the mobile-phase mixture so as to compensate for small differences in the strength of the three pure organic solvents. If further improvement in separations is needed, the additives given in Table 15.8 should be considered.

15.7.4 Ion-pair chromatography

Ionic or partially ionic compounds can be chromatographed on reversed-phase columns through the use of ion-pairing reagents. These reagents are typically long-chain alkyl anions or cations that, in dilute concentrations, can increase the retention of analyte ions. For cationic compounds, C5 to C10 alkyl sulfonates are commonly used; combinations may also be used

TABLE 15.8

Additives for RPLC [29]

Nature of Sample	Example	Additive concentration
Basic compounds	Amines	Phosphate buffer, triethylamine (buffered to pH 3.0)
Acidic compounds	Carboxylic	Phosphate buffer, 1% acetic acid (buffered to pH 3.0)
Mixture of acids and bases	Various	Phosphate buffer, 1% acetic acid (buffered to pH 3.0)
Cationic salts	Tetraalkyl quaternary ammonium compounds	Triethylamine, sodium nitrate
Anionic salts	Alkyl sulfonates	1% acetic acid, sodium nitrate

for difficult separations. In the case of anionic solutes, tetraalkyl ammonium salts (tetramethyl-, tetrabutyl-, etc. ammonium salts or (triethyl-, C5 to C8 alkyl ammonium salts) are generally used.

The following important observations can be made about ion-pair separations:

- In general, an increase in concentration of pairing reagent increases k' (capacity factor). Ion-pairing reagent concentrations range from 0.001 to 0.05 M; however higher concentrations have been recommended by some investigators.
- Increase in chain length increases k'. However, when k' is plotted versus surface concentration, different chain lengths show about the same increase in retention. Therefore, most changes of k' observed with increasing chain length can be reproduced with concentration changes of a single reagent [30].
- The k' changes little for neutral species with increases in concentration of the ion-pairing reagent [30].
- The k' for solutes having the same charge as the pairing reagent decreases with increases in concentration or chain length of pairing reagents [30].
- Removal of pairing reagents from the column by washing is more difficult as the chain length increases. This suggests that use of long-chain pairing reagents can change the nature of the column.

15.7.5 Ion-exchange chromatography

Ion-exchange chromatography utilizes the dynamic interactions between charged solute ions and stationary phases that possess oppositely charged groups. In separations of this type, sample ions and ions of like charge in the mobile phase, compete for sites (X) on the stationary phase:

$$A^- + X^+B^- \rightarrow B^- + X^+A^- \text{ (anion exchange)}$$
$$C^+ + X^-B^+ \rightarrow B^+ + X^-C^+ \text{ (anion exchange)}$$

The extent to which the ions compete with B for the charged sites (X) will determine their retention. In general, this type of chromatography may be used to separate ionic species, such as organic acids or bases, which can be ionized under certain pH conditions. Besides the reaction with ionic sites on the stationary phase, retention may also be affected by the partitioning of solutes between the mobile and stationary phases, as in reversed-phase chromatography. Thus, even nonionized solutes may be retained on ion-exchange columns.

As discussed before, ion-exchange chromatography is a flexible technique used mainly for the separation of ionic or easily ionizable species. The stationary phase is characterized by the presence of charged centers bearing exchangeable counterions. Both cations and anions can be separated by selection of the appropriate ion-exchange medium [31]. Ion exchange finds application in virtually all branches of chemistry. In clinical laboratories, it is used routinely to profile biological fluids and for diagnosing various metabolic disorders [32].

Ion exchange has long been used as the separation mechanism in automated amino acid and carbohydrate analyzers. More recently, ion-exchange chromatography has been used to separate a wide range of biological macromolecules using special wide-pore low-capacity packings designed for this purpose [33–35]. Ion exchange may be used to separate neutral molecules as their charged bisulfite or borate complexes and certain cations as their negatively charged complexes, e.g., $FeCl_4$. In the case of the borate complexes, carbohydrate compounds having vicinal diol groups can form stable charged adducts that can be resolved by anion-exchange chromatography [36]. Ligand-exchange chromatography has been used with cation exchangers (in the nickel or copper form) for the separation of amino acids and other bases. Ion-exchange packings may also be used to separate neutral and charged species by mechanisms not involving ion exchange.

Oligosaccharides and related materials can be separated by a partition mechanism on ion-exchange columns where water/alcohol mobile phases are employed [37]. Ion-exclusion may be used to separate charged species from uncharged species and also charged species from one another on the basis of their degree of Donnan exclusion from the resin pore volume. An ion-exchange packing having the same charge as the sample ions is selected for this purpose. Retention is dependent on the degree of sample access to the packing pore volume. An example of this mechanism is the separation of organic acids with a cation-exchange packing in the hydrogen form [33,38]. Strong acids are completely excluded and elute early; weak acids are only partially excluded and have intermediate retention values; and neutral molecules are not influenced by the Donnan membrane potential and can explore the total pore volume.

The packings for ion-exchange chromatography are characterized by the presence of charge-bearing functional groups. As mentioned before, sample retention can be envisioned as a simple exchange between the sample ions and those counterions that are originally attached to the charge-bearing functional groups. However, this simple picture is a poor representation of the actual retention process. Retention in ion-exchange chromatography is known to depend on factors other than coulombic interactions. For organic ions, hydrophobic interactions between the sample and the nonionic regions of the support are important. Since the mobile phase in ion-exchange chromatography is often of high ionic strengths, hydrophobic interactions are favored because of the "salting-out" effect. From a qualitative standpoint, the retention of organic ions probably proceeds by a hydrophobic reversed-phase interaction with the support, followed by diffusion, of the sample ion to the fixed charge center where an ionic interaction occurs.

Both sample retention and column efficiency are influenced by diffusion-controlled processes, of which the following steps are considered important [39]:

- Diffusion of ions through the liquid film surrounding the resin bead
- Diffusion of ions within the resin particle to the exchange sites
- The actual exchange of one counterion for another
- Diffusion of the exchanged ions to the surface of the resin bead
- Diffusion of the exchanged ions through the liquid film surrounding the resin bead into the bulk solution.

Slow diffusion of the sample ions within the resin beads contributes significantly to poor column performance. Reducing the particle size to less than 10 μm in diameter compensates for the poor mass-transfer kinetics exhibited with conventional resin beads, by reducing the length of intraparticulate channels.

Because the column packings used in ion exchange contain charged functional groups, an equal distribution of mobile-phase ions inside and outside the resin bead develops in accordance with the Donnan membrane effects. The ion-exchange bead behaves like a concentrated electrolyte solution in which the resin charges are fixed, whereas the counterions are free to move. The contact surface between the resin bead and the mobile phase can be envisioned as a semipermeable membrane. When equilibration occurs between the external and internal solution, and one side of the membrane contains a nondiffusible ion, then a combination of the Donnan membrane effect and the need to preserve overall electrical neutrality results in a greater concentration of free electrolytes within the bead. Therefore, diffusion of sample ions and counterions across the Donnan membrane barrier is often the rate-controlling process in ion-exchange chromatography.

Selectivity series have been established for many counterions: $Li^+ < H^+ < Na^+ < NH4^+ < K^+ < Cs^+ < Ag^+ < Cu^{2+} < Cd^{2+} < Ni^{2+} < Ca^{2+} < Sr^{2+} < Pb^{2+} < Ba^{2+} < F^- < OH^- < CH3COO^- < HCOO^- < Cl^- < SCN^- < Br^- < I^- < NO_3^+ < SO_4^{2-}$

The absolute order depends on the individual ion exchanger, but deviations from the above order are usually only slight for different cation and anion exchangers. For weak-acid resins, H^+ is preferred over any common cation, while weak-base resins prefer OH^- over any of the common anions.

Once a selection of the column type has been made, sample resolution is optimized by adjusting the ionic strength, pH, temperature, flow rate, and concentration of buffer or organic modifier in the mobile phase [39a]. The influence of these parameters on solute retention is summarized in Table 15.9.

The temperature at which separations are performed is another variable that can markedly affect separations. Temperatures up to 50 or 60°C often result in improved separations due to decreased viscosity and better mass transfer. Solute stability at these elevated temperatures should be determined prior to use.

15.7.6 Ion chromatography

Ion chromatography has found widespread application for the analysis of inorganic and organic ions with pK_a values less than 7. It combines the techniques of ion-exchange separation, eluent suppression, and

TABLE 15.9

Factors influencing retention in ion-exchange chromatography [39]

Mobile-phase parameter	Influence on mobile-phase properties	Effect on sample retention
Ionic strength	Solvent strength	Solvent strength generally increases with an increase in ionic strength. Selectivity is little affected by ionic strength except for samples containing solutes with different valence charges. The nature of mobile-phase counterion controls the strength of the interaction with the stationary phase.
pH	Solvent strength	Retention increases in cation-exchange and decreases in anion-exchange chromatography with an increase in pH.
	Solvent selectivity	Small changes in pH can have a large influence on separation selectivity.
Temperature	Efficiency	Elevated temperatures increase the rate of solute exchange between the stationary and the mobile phases and also lower the viscosity of the mobile phase.
Flow rate	Efficiency	Flow rates may be slightly lower than in other HPLC methods to maximize resolution and improve mass-transfer kinetics.
Buffer salt	Solvent strength and selectivity	Solvent strength and selectivity are influenced by the nature of the counterion. A change in buffer salt may also change the mobile-phase pH.
Organic modifier	Solvent strength	Solvent strength generally increases with the volume percent of organic modifier. Its effect is most important when hydrophobic mechanisms contribute significantly to retention. In this case, changing

(continued)

TABLE 15.9 (*continued*)

Mobile-phase parameter	Influence on mobile-phase properties	Effect on sample retention
Efficiency		the organic modifier can be used to adjust solvent selectivity as normally practiced in reversed-phase chromatography. Lowers mobile-phase viscosity and improves solute mass-transfer kinetics.

conductivity detection for the quantitative determination of a variety of ions such as mono- and divalent cations and anions, alkylamines, organic acids, etc. [40–45]. Its growth is due in part to the difficulty of determining these ions by other methods. Examples of ion chromatographic separations include common anions and the alkali earth elements [46].

The column packings are styrene-divinyl benzene bead-type resins that are surface functionalized to give low ion-exchange capacities of 0.001–0.05 M equiv/g [46,47]. These resins have good structural rigidity, allowing operation at high flow rates with only moderate back pressure. The mechanical strength of the column packings used limits pressures to about 2000 ψ. These resin beads are stable over the pH range of 1–14. The limited hydrolytic stability of silica-based packings curtails their use in ion chromatography compared to their dominant position in the modern practice of HPLC. For anion separations, a special packing that has an outer layer of fine (0.1–0.5 µm), aminated latex particles agglomerated to the surface of a surface-sulfonated resin bead is frequently used [48,49]. The latex layer is strongly attached to the surface-sulfonated core by a combination of electrostatic and van der Waals forces. The thinness of the exchange layer and the Donnan exclusion effect of the intermediate sulfonated layer provide excellent sample mass-transfer properties by ensuring that single penetration is confined to the outer latex layer.

In general, the efficiency of the columns used in ion chromatography is limited by the large-sized particles and broad particle size distributions of the resin packings. Resin beads are currently available in the ranges of 20–30, 37–74, and 44–57 µm.

The principal problem in the progress of ion chromatography was development of a suitable on-line detection system. Most common ions cannot be detected photometrically, and a method was needed that could detect the separated ions in the background of a highly conducting eluent. Since conductivity is a universal property of ions in solution and can be simply related to ion concentration, it was considered a desirable detection method provided that the contribution from the eluent background could be eliminated. The introduction of eluent suppressor columns for this purpose led to the general acceptance of ion chromatography [50].

The principal disadvantages of suppressor columns are the need to periodically replace or regenerate the suppressor column; the varying elution times for weak acid anions or weak base cations because of ion-exclusion effects in the unexhausted portion of the suppressor column; the apparent reaction of some ions such as nitrite with the unexhausted portion of the suppressor column, resulting in a varying response depending on the percentage exhaustion of the suppressor column; and interference in the baseline of the chromatogram by a negative peak characteristic of the eluent, which varies in elution time with the degree of exhaustion of the suppressor column [39]. Finally, there is some band spreading in the suppressor column that diminishes the efficiency of the separator column and reduces detection sensitivity.

Many of the problems encountered with conventional suppressor columns can be eliminated by using bundles of empty or packed ion-exchange hollow fibers for eluent suppression [51–53]. The sample anions in the column eluent do not permeate the fiber wall because of Donnan exclusion forces. The primary advantage of the hollow fiber ion-exchange suppressor is that it allows continuous operation of ion chromatography without varying interferences from baseline discontinuities, ion-exclusion effects, or chemical reactions. The main disadvantage is that the hollow fiber ion-exchange suppressors have approximately 2 to 5 times the void volume of conventional suppressor columns. This obviously leads to some loss of separation efficiency, particularly for ions eluting early in the chromatogram. Other limitations include restrictions on usable column flow rates to ensure complete suppression and the need for an excess of exchangeable counterions in the regeneration solution.

Common eluents in suppressor ion chromatography are dilute solutions of mineral acids or phenylenediamine salts for cation separations and sodium bicarbonate/sodium carbonate buffers for anion separations. These eluents are too highly conducting to be used without a suppressor column or conductivity detection. Fritz et al. [54–56] have shown that if separator columns of very low capacity are used in conjunction with an eluent of high affinity of the separator resin but of low conductivity, a suppressor column is not required.

15.8   SEPARATION MECHANISM IN REVERSED-PHASE HPLC

Since reversed-phase HPLC is the most commonly used technique in HPLC, it would be a good idea to develop a better understanding of separation mechanisms. The reader would benefit from the basic review on physicochemical basis of retention in HPLC (see Chapter 2, reference 5).

As mentioned before, selectivity in HPLC involves both the mobile phase and the stationary phase; this distinguishes it from GLC where the stationary phase is primarily responsible for observed separation. It is also important to recognize that the stationary phase can be influenced by its environment. The reason for the erroneous assumption to the contrary arises from experiences in GLC where the stationary phase is minimally influenced by the gaseous mobile phase. In the early 1960s, it was demonstrated by Ahuja *et al.* [4] that even in GLC, the stationary phase can be affected by the composition of the mobile phase. Compounds such as water, when present in the carrier gas, can be retained by the stationary phase and then participate in the partition process when a sample is injected. These investigations showed that the stationary phase is a dynamic system that is prone to change with the composition of the mobile phase. It is much more important to remember this fact in HPLC, where multiple components are constantly bathing the stationary phase. Discussed below are various considerations that would lead to better understanding of separation mechanisms in reversed-phase HPLC.

15.8.1   Stationary phase effects

It is important to recognize that the following three distribution equilibria contribute to retention:

(1)   Partition between the bulk phases
(2)   Adsorption on the solid surface
(3)   Adsorption on the liquid/liquid interface.

In a liquid/liquid chromatographic system, the interface between the phases can be involved in the distribution process. So, it is necessary to consider the distribution at the

- interface between solid support and the stationary liquid phase,
- stationary liquid phase,
- interface between the stationary liquid phase and the mobile phase,
- mobile liquid phase.

Reversed-phase (RP) silicas are extremely useful in solving many separation problems in modern LC. Nevertheless there are still problems with the reproducibility of chromatographic results obtained with these materials. One important obstacle to the improvement of RP silicas, with respect to constant properties and quality, might be the lack of reliable analytical methods giving detailed information about the silica matrix itself and about the ligand attached to the surface. Items of interest are the structure of the ligand and its purity, its degree of substitution, and its distribution on the surface of a mono-, di-, and trifunctional bound alkyl groups, and mono- and polymeric substitution [57]. Other important parameters are the proportion of unchanged silanol groups and the degree of end capping.

The interaction of amines with the chromatographic support is complex. The curvature in the k' versus amine concentration plot illustrates that more than one mechanism is operating. The behavior of amines can be compared to that of alcohols, phenols, and protonated thiols. The latter groups are readily chromatographed by silica-based RPLC. The fact that all these groups are poor hydrogen-bond acceptors but good donors implies that silanol groups are strong hydrogen-bond donors and poor acceptors, as would be expected from their Brønsted acidity. Thus, depending on the pK_a of the silanol groups and the mobile phase pH and ionic strength, both hydrogen bonding and ionic interactions can occur with the amine functional groups.

Since selectivity in HPLC involves both the stationary and mobile phases [5–9,58–60], it is important to note that the solvent strength of the mobile phase, as compared to the stationary phase, (composed of mobile-phase components reversibly retained by the bonded phase and silica support) determines the elution order or k' of the retained components. Unfortunately, the columns with the same stationary phase can exhibit significant variabilities from one manufacturer to another and even from the same manufacturer [5–8]. Based on discussions heard at various scientific meetings, this situation has not changed much. Variabilities can occur in the packing process even where all other conditions are supposedly constant. These factors have to be considered prior to developing an understanding as to how separations occur in HPLC.

The strategy employed for choosing a particular chromatographic separation generally entails maximizing the column efficiency and, at the same time, choosing a stationary and/or mobile phase that is thought to be suitable for the sample at hand. There is a real lack of quantitative guidelines for optimizing selectivity decisions made pertaining to appropriate experimental conditions, and the techniques to achieve this end have generally been based primarily on the experience of the analyst as guided by literature [9]. An alternative approach to optimizing the system selectivity, one more rational than that of trial-and-error, would be understanding the physicochemical basis of separations, followed by modeling of the system at hand, and then optimizing the system and conditions within the boundaries of the required resolution. Several theories have been proposed to explain physicochemical interactions that take place between the solute and the mobile and stationary phases as well as those that may occur between the mobile and stationary phases themselves [9,58].

It is generally acknowledged that despite the enormous popularity of RPLC, which may be broadly classified as involving distribution of a nonpolar or moderately polar solute between a polar mobile phase and a relatively nonpolar stationary phase, there is a lack of understanding of details of solute retention and selectivity. Such an understanding is crucial for informed control and manipulation of separations that ultimately require a sufficiently detailed description of how retention and selectivity depend on the various mobile- and stationary-phase variables [58]. The nature of the solute distribution process in RPLC, i.e., the retention mechanism, has also been a topic of much study, discussion, and speculation. The most rigorous treatment to date is based on the solvophobic theory, with primary focus on the role of the mobile phase. Solute distribution is modeled by invoking "solvophobic" interactions, i.e., exclusion of the less polar solute molecule from the polar mobile phase with subsequent sorption by the nonpolar stationary phase. This implies that the mobile phase "drives" the solute toward the stationary phase in lieu of the solute's being driven by any inherently strong attraction between the solute and the stationary phase. The basic premise of the theory is reasonable, and agreement with experimental data is generally good; however, the description is incomplete in that it does not provide a sufficiently detailed explanation of the dependence of solute retention and selectivity on the stationary-phase variables. Moreover, it has been reported that under certain chromatographic conditions (very polar mobile-phase solvent and relatively long n-alkyl chains), solute distribution in RPLC appears to approach that of partitioning between two bulk liquid phases, suggesting quasi-liquidlike behavior of unswollen n-alkyl chains.

527

With just a few exceptions, there is a dearth of published information providing systematic studies of retention volumes as a function of composition of the eluent over the whole composition range of binary solvents. To rectify this situation, a general equation for HPLC binary solvent retention behavior has been proposed [59] that should help generate a chromatographic retention model to fit Eq. (15.20):

$$\frac{1}{V_{AB}} = \frac{\varphi A}{V_A} + \frac{b\varphi_B}{1 + b'\varphi_B} + \frac{\varphi_B}{V_B} \qquad (15.20)$$

where

V_{AB} = the corrected elution volume of the solute (experimental elution volume-void volume) in the binary eluent $A+B$, with volume fractions φA and φB, respectively.
V_A = the corrected elution volume for the solute with pure A as eluent.
V_B = the corrected elution volume for the solute with pure B as eluent. Both b and b' are constants, and the term that includes them is reminiscent of the Langmuir adsorption isotherm.

15.8.2 Relationship of RPLC to partition in octanol

It is known that the RPLC retention parameters are often strongly correlated to the analyte's distribution coefficient in organic solvent/water. Generally, the relationship between liquid/liquid (LL) distribution and RPLC retention are of the form of the dimensionless Collander-type equations, e.g., see Eq. (15.21)

$$\log K_d = a_1 + b_1 \log k' \qquad (15.21)$$

where

K_d = the solute's organic solvent:water distribution coefficient.
k' = chromatographic capacity ratio ($k' = t_r - t_0/t_0$, t_r and t_0 being solute retention time and mobile phase "holdup" time, respectively).
a and b = coefficients whose magnitudes depend on the LL distribution and RPLC systems.

The general validity of this type of relationships depends on

- the physicochemical state (temperature, degree of ionization) of solutes in both LL distribution and RPLC;

- the RPLC system used, with emphasis on the mobile phase, particularly on the type, and also to some extent on the concentration of organic cosolvents (modifiers) used; and
- The type of organic solvent used in the distribution system modeling and RPLC system.

It has been shown, e.g., that RPLC capacity factors of unionized solutes obtained using aqueous methanol mobile phases can be generally correlated with octanol/water distribution coefficients for neutral, acidic, and basic compounds, whereas no such overall correlation can be made with alkane/water distribution parameters [61,62]. HPLC can be used for the measurement of octanol/water partition coefficients even when a small amount of sample is available. Solutes are equilibrated between the two phases by using the conventional shake-flask method, which entails shaking them together and then analyzing by HPLC sampled portions of both phases [63]. The results thus obtained with a variety of substances having partition coefficients in the range of 10^{-3} to 10^4 showed excellent agreement with literature data. The technique is rapid and has the advantage that small samples suffice, the substances need not be pure, and the exact volume of the phases need not be known. Furthermore, it could readily be developed into a micromethod for measurements with submicrogram quantities. Table 15.10 shows a comparison of data obtained by HPLC with the reported literature values by classical method.

Data collected with a simple RPLC procedure has been found to be in good agreement with 1-octanol shake-flask partition or distribution coefficients over a 3.5 log range [64]. A chemically bonded octadecylsilane support is coated with 1-octanol. With 1-octanol-saturated buffers as mobile phases, a stable baseline (compared to 1-octanol absorbed on silica) is

TABLE 15.10

Comparison of $\log P$ values by HPLC with literature [63]

Name	HPLC	Literature
tert-Butylbenzene	4.07	4.11
Propylbenzene	3.44	3.62
Toluene	2.68	2.58
Acetophenone	1.58	1.66
Resorcinol	0.88	0.80
Hydroquinone	0.54	0.56
Caffeine	−0.05	−0.07

obtained rapidly, and the log relative retention times are highly correlated with unit slope to log distribution or partition coefficients obtained from the classical shake-flask procedures. Only relatively basic, unhindered pyridines deviate, probably because of binding with residual silanol sites.

In addition, if the apparent pK_a of an ionizable compound lies within the pH operating range of the column support, the apparent pK_a usually can be determined simultaneously with $\log P$ by measuring the log distribution coefficient at several pH values (Table 15.11). The main advantages of the procedure are that it gives rapid results, requires little material, and can tolerate impurities.

Using an unmodified commercial octadecylsilane column and a mobile phase consisting of methanol and an aqueous buffer, a linear relationship has been established between the literature $\log P$ values of 68 compounds and the logarithms of their k' values [65]. For the determination of the partition coefficients of unknowns, standards are used to calibrate the system, the choice being made on the basis of the hydrogen-bonding character of the compounds being evaluated. The overall method is shown to be rapid and widely acceptable and to give $\log P$ data that is comparable to results obtained by classical and other correlation methods.

The studies with barbiturates revealed that the logarithm of the retention time is linearly related to the octanol/water partition coefficients [66,67]. It has been observed that the retention index of the drug is linearly related to the octanol/water partition coefficient ($\log P$), and that results are very close to that of the 2-keto alkane standard (solid line in Fig. 15.9).

TABLE 15.11

Comparison of $\log P$ and apparent pK_a determined by HPLC with literature values [64]

Compound	HPLC		Literature	
	pK_a	$\log P$	pK_a	$\log P$
Naproxen	4.28	3.21	4.53	3.18
	4.21	3.20	4.39	—
Benzoic acid	4.33	1.78	4.20	1.87
	4.38	1.77	4.18	
Salicylic acid	3.52	2.00	3.00	2.23
	3.29	2.18	—	—
p-Toluic acid	4.30	2.22	4.37	2.27
	4.41	2.26	—	—

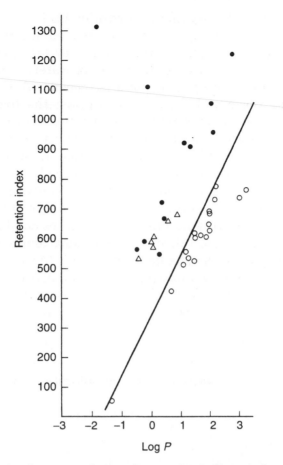

Fig. 15.9. Correlation between observed retention indices and octanol partition coefficients [66].

The retention data of catecholamine derivatives in reversed-phase chromatography with octadecyl-silica stationary phase and aqueous eluent has been analyzed. Good agreement is observed between the observed and predicted k' values [68]. Data obtained with different C-18 stationary phases at various temperatures suggest that quantitative structure-retention relationships can be transformed from one reversed-phase system to another as long as the eluent composition is the same. Kalizan [69] observed that C-18 columns without any additional coating give results that are precise enough for quantitative structure activity relationship (QSAR) purposes and are convenient over a wide range of operational conditions.

Tomlinson [70] suggests that a "hydrophobic bond" is formed when two or more nonpolar groups in an aqueous medium come into contact, thus decreasing the extent of interaction with the surrounding water molecules, and resulting in the liberation of water originally bound by the molecules. The hydrophobic bond is recognized as complex in nature, involving polar and apolar interactions; the hydrophobic bond concept has been useful in explaining association of organic and biologic molecules in aqueous solution. In QSAR models, the ability of a compound to partition between a relatively nonpolar solvent and water is normally used as a measure of its hydrophobic character.

Martin was the first to point out that a substituent changes the partition coefficient of a substance by a given factor that depends on the nature of the substituent and the two phases employed, but not on the rest of the molecule. Martin's treatment assumes that for any stated solvent system, the change in retention (ΔR_{min} TLC) caused by the introduction of group X into a parent structure is of constant value, providing that its substitution into the parent structure does not result in any intramolecular interactions with other functions in the structure. On the other hand, it can be appreciated that if the introduction of a group into a structure causes a breakdown in the additivity principle, then intra- or intermolecular effects are likely to be more significant within the substituted structure. These effects are as follows:

- Steric effects (including ortho effect)
- Intramolecular hydrogen bonding
- Electronic effects
- Intramolecular hydrophobic bonding
- Chain branching.

As mentioned before, the hydrophobicity and the partition coefficient can be related to the solubility of the solutes in water (also see Section 15.8.4). The partition coefficient (P) between octanol and water can be described as the π-constant of Hansch or the $\log P$ values of Rekker. The $\log P$ values calculated from the fragmental constant are then used for the optimization of RPLC. However, this method is not adequate to develop an optimization system for the mixtures of different types of compounds [71].

15.8.3 Impact of other physicochemical parameters on retention

Discussed below is the impact of various other physicochemical parameters excluding solubility (see Section 15.8.4) on retention in HPLC.

The van der Waals volume can be related to the hydrophobicity of the solutes, and retention of molecular compounds can be predicted from their van der Waals volumes, π-energy, and hydrogen-bonding energy effects [72–74]. It should be noted that the isomeric effect of substituents cannot be predicted with good precision because this is not simply related to Hammett's σ or Taft's σ^* constants. On the other hand, the hydrophobicity is related to enthalpy [75]. Retention times of non-ionizable compounds were measured in 70 and 80% acetonitrile/water mixtures on an octadecyl-bonded silica gel at 25–60°C and the enthalpy values obtained from these measurements.

Retention volumes of monosubstituted benzenes, benzoic acid, phenols, and anilines have been measured in RPLC [76]. Buffered acetonitrile/water and tetrahydrofuran/water eluents were used with an octadecylsilica adsorbent. From the net retention volumes, a substituent interaction effect was calculated and described with the linear free energy relationship developed by Taft. The data was interpreted in terms of hydrogen bonding between the solutes and the eluent.

Enthalpy–entropy compensation has been investigated in reversed-phase HPLC with octylsilica stationary phase [77]. The compensation temperatures were determined for this system, and the results show that their change with the composition of the mobile phase is almost similar to that with octadecylsilica stationary phase. It can be concluded that the retention mechanisms of the separation of alkyl benzenes is the same in both systems with the mobile phase exceeding 20% water content.

The separation of substituted benzene derivatives on a reversed-phase C-18 column has been examined [78]. The correlations between the logarithm of the capacity factor and several descriptors for the molecular size and shape and the physical properties of a solute were determined. The results indicated that hydrophobicity is the dominant factor to control the retention of substituted benzenes. Their retention in reversed-phase HPLC can be predicted with the help of the equations derived by multicombination of the parameters.

As mentioned before, retention in RPLC has been found to be related to the van der Waals volume, π-energy, and hydrogen-bonding energy effects. However, higher-molecular-weight compounds are retained more strongly than expected [79]. In order to investigate this effect more fully, the retention times of phenols were measured on an octadecyl-bonded silica gel in acidic acetonitrile/water mixtures at different temperatures. The enthalpies of phenols were then calculated from their $\log k'$ values. The magnitude of the enthalpy effect increases

S. Ahuja

with increasing molecular size, but the polarity of the molecule was the predominant factor in the enthalpy effect.

An increase in the number of methylene units in alkyl benzene did not significantly affect the π-energy effect on their retention, but the enthalpy effect increased dramatically [80]. This means that a hydrophobic compound can be adsorbed directly onto an octadecyl-bonded silica gel. The value of enthalpy effect of a methylene unit in alkyl benzene was calculated to be 500 cal/mol.

Schoenmakers *et al.* [81] investigated the relationship between solute retention and mobile-phase composition in RPLC over the entire range of composition, with emphasis on mobile phases with a high water content. It appears that a quadratic relationship between the logarithm of the capacity factor and the volume fraction of organic modifier is generally valid for mobile phases containing less than 90% water. When more water is added to the mobile phase, the quadratic equation turns out to be insufficient. An experimental study of ten solutes and three organic modifiers is used to show that an extension of the quadratic equation by a term proportional to the square root of the volume fraction leads to a description of the experimental retention data within approximately 10%.

The polarity values of binary acetonitrile/water and methanol/water mobile phases used in RPLC were measured and compared with methylene selectivity (α_{CH_2}) for both traditional siliceous bonded phases and for a polystyrene-divinylbenzene resin reversed-phase material [82]. The variation in methylene selectivity for both was found to correlate best with percent organic solvent in methanol/water mixtures, whereas the polarity value provided the best correlation in acetonitrile/water mixtures. The polymeric resin column was found to provide higher methylene selectivity than the siliceous-bonded phase at all concentrations of organic solvent.

The retention indices, measured on the alkyl aryl ketone scale, of a set of column test compounds (toluene, nitrobenzene, *p*-cresol, 2-phenyl ethanol, and *N*-methylaniline) were used to determine the changes in selectivity of a series of ternary eluents prepared from methanol/0.02 M phosphate buffer pH 7 (60:40), acetonitrile/0.02 M phosphate buffer pH 7 (50:50) and tetrahydrofuran/0.02 M phosphate buffer pH 7 (25:65). The analyses were carried out on a Spherisorb ODS reversed-phase column. The selectivity changes were often nonlinear between the binary composition [83].

Direct measurement of solute sorption-desorption kinetics in chromatographic systems provides some useful insights into the mechanism of the sorption process and a sensitive means of measuring slight differences in those stationary phase–solvent interactions that are responsible for determining the chemical contributions of the stationary phase.

Variations in retention and selectivity have been studied in cyano, phenyl, and octyl reversed bonded phase HPLC columns. The retention of toluene, phenol, aniline, and nitrobenzene in these columns has been measured using binary mixtures of water and methanol, acetonitrile, or tetrahydrofuran mobile phases in order to determine the relative contributions of proton donor–proton acceptor and dipole–dipole interactions in the retention process. Retention and selectivity in these columns were correlated with polar group selectivities of mobile-phase organic modifiers and the polarity of the bonded stationary phases. In spite of the prominent role of bonded phase volume and residual silanols in the retention process, each column exhibited some unique selectivities when used with different organic modifiers [84].

The physicochemical framework has been examined by comparing the predictions of two models for the combined effects of the composition of the hydro-organic mobile phase and the column temperature on the retention of n-alkyl benzenes on hydrocarbonaceous bonded stationary phases. The "well-mixed" model leads to expressions for the dependence of retention on three factors that are equivalent to those derived previously from linear extra-thermodynamic relationships. The "diachoric" model stems from the assumption that the mobile phase is identical to the retention model most widely used in chromatography with polar sorbents and less polar solvents. Over limited ranges of mobile phase composition and temperature, each model describes retention behavior. However, only the well-mixed model describes retention well over the entire range of mobile-phase composition and temperature studied here. The success of the well-mixed model and the limits of the model give insights into the role of organic solvents in determining the magnitude of chromatographic retention on a nonpolar stationary phase with hydro-organic eluents [85].

It has been shown that when the intracolumn effect of mass transfer and diffusion is the main factor controlling band broadening, the column efficiency decreases with the increase of the viscosity of the methanol/water mixture; on the other hand, when the extra-column effect is the main factor, an increase in viscosity of the eluents will help in improving column efficiency. Column efficiency is also related to the properties of the sample [86].

15.8.4 Solubility and retention in HPLC

The solubility parameter concept was established in the 1930s by the work of Hildebrand and Scatchard. The original concept covers regular solutions, i.e., solutions that do not show an excess entropy effect on mixing. The solubility parameter concept offers the following interesting features:

- The concept is based on the assumption that the properties of mixtures can be described by the properties of pure components. As a result, the arithmetic expressions involved (regular mixing rule) are relatively simple.
- The solubility parameter concept relates to compounds rather than to molecules. Because it is a macroscopic approach, it relates to practical data more conveniently than a molecular statistical approach does.

In earlier work on the applicability of the solubility parameter theory to HPLC, attention was focused on quantitation; for this work, the model did not prove to be successful. Schoenmakers *et al*. [87] believe that the potential of the solubility parameter can aid in designing a genuine framework for retention behavior in LC. Based on their work, the following conclusions have been drawn:

- Reasonable retention times are obtained if the polarity of the solute is roughly intermediate between the polarities of the mobile and stationary phases.
- For higher members of a homologous series with approximately the same polarity, the logarithm of the capacity factor varies linearly with the molar volume or the carbon number.
- The absolute difference in polarity between the mobile phase and the stationary phase may be defined as the general selectivity of an HPLC system.
- There are two commonly used ways to elute a given compound in HPLC: the normal-phase mode ($\delta s > \delta m$) and the reversed-phase mode ($\delta m > \delta s$). Reversed-phase systems offer superior general selectivity. Solutes are eluted in ascending order of polarity in normal-phase systems and in descending order of polarity in reversed-phase systems.
- Although stationary phases of intermediate polarity (alumina, silica, carbon) provide only moderate general selectivity, they are potentially most powerful for very polar solutes when operated in the reversed-phase mode.

- Perfluorinated stationary phases offer superior selectivity in comparison to the current hydrocarbon bonded-stationary phases.
- Specific separation effects can be understood from the multicomponent solubility parameter theory. Specific effects for nonpolar compounds are predictable with perfluorinated and graphitized carbon black stationary phases. Specific selectivity for polar compounds in reversed-phase HPLC can be realized with polar additives to the mobile phase.
- Previously formulated transfer rules for binary mobile phases in reversed-phase HPLC can be explained by solubility parameter expressions.

15.9   MOLECULAR PROBES/RETENTION INDEX

A variety of compounds have been used as molecular probes to evaluate HPLC columns and characterize them.

- Nonpolar compounds, e.g., benzene, naphthalene
- Polar compounds, e.g., hydroquinone or steroids
- Chelating compounds, e.g., acetyl acetone
- Quaternary compounds, e.g., quaternary ammonium compounds
- Basic compounds, e.g., amines
- Acidic compounds, e.g., toluic acid.

   Unfortunately, none of the commonly used molecular probes is adequate to evaluate column-to-column variabilities [88]. The absolute prediction of retention of any compound involves the use of a rather complex equation [89,90] that necessitates the knowledge of various parameters for both the solute and the solvent [91]. The relative prediction of retention is based on the existence of a calibration line describing the linearity between $\log k^*$ and interaction index. This second approach, although less general than the first, is simpler to use in practice, and it often gives more accurate results than the first. With a proper choice of calibration solutes, it is possible to take into account subtle mobile phase effects that cannot be included in the theoretical treatment.

   However, certain conditions must be verified prior to using a prediction model based on a calibration of the chromatographic system. First, it is necessary to limit the number of calibration solutes. It is clear that the use of a large number of calibration standards can give a high degree of accuracy, but this is too time-consuming. Five to six solutes appear to offer a good compromise between accuracy and convenience. Second, one must be able to use these calibration compounds in a rather large range of solvent composition. Finally, calibration compounds must be "simple" chemicals, stable, and easily available in any chemical laboratory.

Using steroids as solutes and 2-keto alkanes as reference compounds, simple linear equations have been developed for accurately predicting reversed-phase HPLC retention indices and resolution [92]. These equations have practical applications for predicting whether given pairs of compounds can be separated under given conditions, or for predicting the conditions that will separate mixtures of compounds in a minimum amount of time. The technique may be used to optimize isocratic or gradient separations of compound mixtures. The results of this study show that the Snyder-solvent-selectivity triangle concept for characterizing mobile-phase liquids fails to consistently group solvents according to selectivity for separating steroids. Contrary to theory, experimental separations often differ markedly within a given solvent group. Selctivity differences between solvents in the same group sometimes exceed those between solvents in different groups.

A study was conducted to utilize the 2-keto alkane system for calculating HPLC retention indices of a series of steroids, and then to utilize the retention index data to predict retention and resolution of the compounds as a function of solvent strength and selectivity. The main goal was to utilize this capability to predict resolution for optimizing the separation of steroid mixtures or pairs of individual steroids. A secondary objective was to study the selectivity characteristics of solvents grouped according to the solvent-selectivity triangle concept for their ability to separate given pairs of steroids.

These results contradict the theory of the solvent-selectivity triangle concept, which states that solvents in the same group should result in similar selectivity, while those in different groups should yield different selectivities [93,94]. A literature search showed a widespread usage of the solvent-selectivity triangle as a rationale for solvent selection. However, with the exception of a publication by Lewis et al. [95], definitive studies on the accuracy of the solvent groupings in the selectivity triangle appear to be lacking. They studied the separation of polystyrene oligomers using a total of 17 solvents representing all eight of the selectivity groups and concluded that the solvent triangle did not accurately predict selectivity for the separation being studied. These authors reported that the degree of the solute solubility in the pure mobile phase solvents was a better predictor of selectivity than were the groupings of the solvent triangle.

The studies with steroids and polystyrene oligomers have demonstrated that solvents classified within the same Snyder-solvent-selectivity group do not necessarily result in similar selectivity. This discrepancy might be due to the underlying assumptions of the solvent triangle theory, which assumes that selectivity is largely governed by the ability of the solvent to engage in hydrogen bonding and dipolar interactions and that dispersion interactions play an unimportant role in selectivity for solutions of polar solvents. The solvent triangle concept also fails to consider the role of the stationary phase and the nature of the solutes themselves in affecting a given separation. This concept further assumes that only three test solutes are needed to establish the primary selectivity characteristics of various solvents.

A method has been offered to characterize variations in the retention properties of RPLC by column–eluent combinations by using retention indices of a set of reference compounds, toluene, nitrobenzene, p-cresol, and 2-naphthylethanol [96]. These compounds were selected by multivariate analysis to give optimum discrimination between eluents and columns.

An interesting pair of compounds is caffeine and theophylline [97]; these compounds are relatively polar compounds with different functional groups (tertiary and secondary amine). In a few cases, more appropriate comparisons have been made such as between androstenedione/testosterone and methyl benzoate/anisole; these compounds are expected to be different in Snyder interaction groups.

A number of researchers have also used a proposed ASTM test mixture, benzaldehyde, acetophenone, methyl benzoate, dimethyl terphthalate, benzyl alcohol, and benzene to demonstrate separation on a column [98]. However the first four compounds are from the same interaction group and should behave in the same way on changing conditions. The first three have almost constant indices (respectively 760, 800, and 890) so that in effect they create an "index scale" with constant differences against which the last two compounds can be compared [96,99].

In order to determine the applicability of retention indices, based on the alkyl arylketone scale, as the basis of a reproducible method of reporting retentions, the separation of 10 barbiturates and a set of column test compounds were examined on an octadecylsilyl bonded silica (ODS-Hypersil) column with methanol-buffer of pH 8.5 as eluent [100]. The effects on the capacity factors and retention indices, on changing the eluent composition, pH, ionic strengthened temperature, showed that the retention indices of the barbiturates were much less susceptible to minor changes in the eluent than the capacity factors.

For nonionized compounds, the retention indices were virtually independent of the experimental conditions.

The silanophilic character of 16 reversed-phase high-performance liquid chromatographic columns was evaluated with dimethyl diphenycyclam, a cyclic tetraza macrocycle [101]. The method is rapid, does not require the removal of packing material, and uses a water-miscible solvent. The results demonstrate two points: first, cyclic tetraza macrocycles offer substantial benefits over currently used silanophilic agents; second, the method can easily differentiate the performance of various columns in terms of their relative hydrophobic and silanophilic contributions to absolute retention.

A mixture of acetyl acetone, 1-nitronaphthalene, and naphthalene has been proposed for evaluating reversed-phase packing material [102]. This reveals the usual optimum kinetic chromatographic parameters (the naphthalene peak), the degree of activity or end-capping status of the column (the ratio of the 1-nitronaphthalene and naphthalene retention times) and trace metal activity (the shape and intensity of the acetylacetone peak).

15.10   STRATEGY OF SEPARATION

A strategy to design a successful HPLC separation may involve the following steps:

a. Select the method most suitable for your sample based on solubility of sample and other relevant physical properties. For example, the preliminary choice of a method would include one of the following methods:
    • Normal phase
    • Reversed phase
    • Ion exchange
    • Size exclusion.
b. Select a suitable column based on the above-mentioned method for selection. For example, in reversed-phase HPLC, the choice may be to simply entail selecting C-8 or C-18.
c. Choose simple mobile phase for initial evaluations. For example, methanol: water (50:50) may be a good starting point in the case of RPLC.
d. Make desirable changes in the mobile phase in terms of proportions of solvent or vary solvents or include suitable additives.
e. Utilize specialized techniques such as gradient elution or derivatization to enhance detectability and/or improve separations.

15.10.1   Improving resolution

The primary goal of any separation process is to achieve optimum resolution of the components. Resolution can be improved by varying the three terms α, N, or k' in the resolution equation:

$$Rs = \frac{1}{4}\left(\alpha - \frac{1}{\alpha}\right)N^{1/2}\left(\frac{k'_2}{1+k'_2}\right) \qquad (15.22)$$

The effect on sample resolution with changes in k', N or α values is shown in Fig. 15.10. For example, an increase in separation factor α results in a displacement of one band center relative to the other, and a

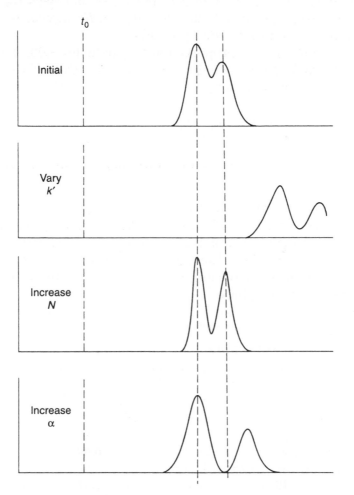

Fig. 15.10. Effect of α, k', or N on resolution.

rapid increase in Rs. The peak height or retention time is not significantly affected for a moderate change in α.

The increase in plate number N results in narrowing of the two bands and an increase in band height; the retention time is not affected if there is no change in weight ratio of the sample. The most dramatic effect on separation is caused by change in k' values. If k' for the initial separation falls within the range $0.5 < k' < 2$, a decrease in k' can make a separation look bad. On the other hand, an increase in k' can provide a significant increase in resolution. However, as k' is increased, band heights rapidly decrease and separation time increases.

When it is necessary to improve Rs, k' should be increased into the optimum range $1 < k' < 10$. When k' is <1, Rs increases rapidly with increase in k'. For values of k' greater than 5, Rs increases very little with further increase in k' (see Table 15.12).

If k' were infinity, its proportionality would be 1.00. This means there is very small improvement over that seen with k' of 10 or 5 in the table. These observations suggest that the optimum k' is between 1 and 10.

It is important to note here that no other change in separation condition provides as large an increase in Rs value for very little effort.

The k' values in HPLC can be controlled by means of solvent strength. When it is necessary to increase k' value, a weaker solvent is used. For example, in reversed-phase separations, solvent strength is greater for pure methanol than for pure water. The right proportionality of these solvents has to be found to get optimum separation.

When k' is already within the optimum range of values and the resolution is still marginal, the best solution is to increase N. This is generally achieved by increasing column length or by decreasing the flow rate of the mobile phase.

TABLE 15.12

Impact of k' on Rs value*

k'	$K'/1+k'$
0	0
1	0.5
2	0.67
5	0.83
10	0.91

* See Eq. (15.20)

When Rs is still small even though k' is optimum, an increase in N would not be very helpful because it unusually prolongs separation time. In this case, increasing α value would be more desirable. However, it should be recognized that predicting the right conditions for the necessary change in α is not as simple procedure (for more details, see Section 15.8).

15.11   MOBILE-PHASE SELECTION AND OPTIMIZATION

The conventional approaches to mobile-phase selection and optimization are discussed here. The primary focus is on compounds with molecular weight less than 2000. More detailed information including coverage of macromolecules may be found in some basic texts [2,5,39,103]. As discussed earlier, various modes of chromatography utilized to separate these compounds can be classified as follows:

1.   Adsorption chromatography
2.   Normal-phase chromatography
3.   Reversed-phase chromatography
4.   Ion-pair chromatography
5.   Ion-exchange chromatography
6.   Ion chromatography.

Ion-pair chromatography is frequently performed in the reversed-phase mode and is therefore discussed in Section 15.7. Since ion chromatography is an offshoot of ion-exchange chromatography, it has been discussed right after ion-exchange chromatography (Section 15.7.6).

15.11.1   General considerations

Various means have been used to optimize separations for each chromatographic technique (see Chapters 5 through 8 in reference 5); the discussion here is limited to conventional approaches used to select the mobile phase.

These approaches are frequently based on intuitive judgment and know-how of the chromatographer. For the latter, it is important to emphasize that a knowledge of physicochemical basis of retention and a basic understanding of separation mechanism in HPLC, discussed above, will go a long way in helping to select the right mobile phase quickly and then to optimize it by the usual experimentation. These experiments can be logically conducted even when the operator is not present, by letting a computer select mobile-phase combinations based on certain preset

S. Ahuja

requirements. This aspect of method development is better appreciated once the conventional approaches have been mastered.

15.11.1.1 Properties of sample

The selection of an HPLC method should be made primarily from the properties of the sample (alternate terms are solute, analyte, or eluite) once it has been established that it has a sufficiently low molecular weight, i.e. <2000, to justify use of the techniques mentioned above. The decision could be based on the solubility of the sample, i.e., whether it is soluble in polar or nonpolar solvents (Chart 1).

From Chart 1, it is clear that compounds with molecular weight >2000 are better separated by using gel permeation chromatography or size exclusion chromatography (SEC). Of course SEC can be used for molecular weights below 2000 just as other modes of chromatography can be used for compounds with higher molecular weight. The discussion here is concerned primarily with compounds of molecular weight <2000.

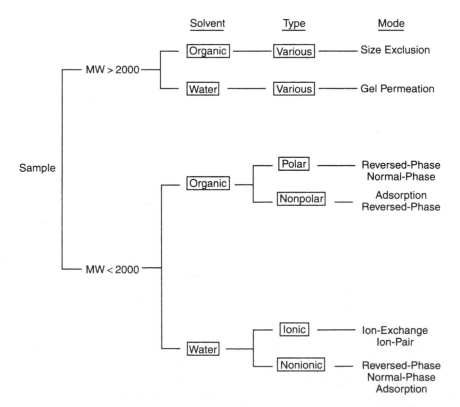

Chart 1. Selection of chromatographic method based on solubility.

544

Information other than solubility that can help select the suitable mode of chromatography, whether or not the sample is ionic. In this regard, the dissociation constant of the compound is of great value because with appropriate adjustment of pH, one can select a desirable percent ionization of the compound of interest, remembering when $pH = pK_a$, the compound is 50% ionized.

15.11.1.2 Column selection

The choice of column should be made after careful consideration of mode of chromatography, column-to-column variability, and a number of other considerations [3–5]. A short discussion on columns and column packings is given below. The column packings may be classified according to the following features [2]:

1. Rigid solids or hard gels
2. Porous or pellicular and superficially porous particles
3. Spherical or irregular particles
4. Particle size (dp).

Rigid solids based on a silica matrix are most commonly used as HPLC packings. Such packings can withstand the relatively high pressures (10,000–15,000 psi). The silica particles can be obtained in a variety of sizes, shapes, and varying degrees of porosity. Furthermore, various functional groups or polymeric layers can readily be attached to the silica surface, thus extending the utility of these particles for applications to any individual HPLC method.

Hard gels are generally based on porous particles of polystyrene cross-linked with divinyl benzene. Depending on how they are prepared, the resulting particles may vary in rigidity and porosity over fairly wide limits. They still find use in ion exchange and SEC; however, rigid solids are gradually replacing hard gels.

Packings for HPLC can be further described as either pellicular or porous. Pellicular particles are made from spherical glass beads, which are then coated with a thin layer of stationary phase. For example, a porous layer can be deposited onto the glass bead to produce a porous layer or a superficially porous particle. The porous layer can in turn be coated with liquid stationary phase or reacted to give a bonded stationary phase. Pellicular particles are generally less efficient than the porous layer of superficially porous particles.

As the particle size (dp) decreases, column plate height decreases and the column becomes less permeable. As a result, for small values of dp, a column of some fixed length generates higher plate count; i.e., higher efficiency, but greater pressure drop across the column is required for a given value of linear flow. This suggests that columns of small particles are more efficient, but require higher operating pressures.

For columns of similar efficiency, the maximum sample size allowed is generally smaller for pellicular particles than for porous particles. The reason is that a less stationary phase is available per unit volume of column. Roughly five times as much sample can be charged to a porous column before there is a significant decrease in k'. Since larger samples can be injected onto a porous-particle column, the resulting bands are larger and more easily detected. The columns of small porous particles give good detection sensitivities and are preferred for ultratrace analysis [6].

For comparison of similar columns, it is important that experimental conditions for the test chromatogram are faithfully reproduced and sufficient time is allowed for column equilibration before starting the test. The expected changes in column performance parameters that are due to changes in the experimental conditions are summarized in Table 15.13.

The columns commonly used in HPLC can be classified based on mode of separation, selected backbone, particle size, and functionalities (Table 15.14).

15.11.1.3   Column evaluations
As mentioned earlier, columns should be thoroughly evaluated prior to use [5]. Some of the desirable properties of test solutes are given below.

TABLE 15.13

HPLC parameters affecting column efficiency

Parameter	Change in efficiency (N)
Flow rate	Low flow rate generally gives high value of N
Particle size	Small particle size gives high value of N
Column length, L	N is proportional to L
Mobile-phase viscosity	Low value gives high value of N
Temperature, T	High values reduce viscosity and give high values of N
Capacity factor, k'	Low k' (<2) give low values of N; for high k' values (>2), N is influenced
Dead volume	N is decreased due to band-broadening contributions to peak width
Sample of size	Large amounts (mg) or large volumes decrease N

TABLE 15.14

HPLC columns

Mode	Material	Particle size (μM)	Treatment
Adsorption	Silica, irregular	2–20	Unreacted
	Silica, spherical	5–10	Unreacted
	Alumina, irregular	3–12	Unreacted
	Alumina	5–20	Unreacted
Reversed phase	Silica with long C chain	3–15	C-18
	Silica with intermediate C chain	5–10	C-8
	Silica with short C chain	5–10	C-1,C-3
Normal phase	Silica(weak)	5–15	Ester, ether, diester
	Silica(medium)	5–15	NO_2, CN
	Silica(high)	5–15	Alkylamino, amino
Ion exchange	Silica(anion)	5–15	NMe^{+3}, NR^{+3}, NH_2
	Resin(anion)	7–20	NMe^{+3}, $-NH^{+3}$
	Silica(cation)	5–10	$_-SO^3(H^+)$, $_-SO^3_-$, $(NH_4)^+$
	Resin	5–20	$_-SO^3_-$

1. Test solutes should be of low-molecular weight to ensure rapid diffusion and easy access to the packing pore structure.
2. The solute should include components that
   - characterize the column in terms of both kinetic and thermo-dynamic performance;
   - determine the column dead volume, i.e., of the test mixture as an unretained solute; and
   - differentiate retention with k' values between 2 and 10.
3. Test solutes should have strong absorbance, preferably at 254 nm.

15.11.1.4   Mobile phase selection in HPLC

Retention in HPLC depends on the strength of the solute's interaction with both the mobile and stationary phases as opposed to GC, where the mobile phase does not contribute to the selectivity. An intelligent selection of the type of stationary phase for the separation is made and selectivity is adjusted by modifying the mobile phase. The selection of the mobile phase for a particular separation is thus a very important consideration in HPLC.

For HPLC, some fairly broad generalizations can be made about the selection of certain preferred solvents from the large number available. A suitable solvent will preferably have low viscosity, be compatible with the detection system, be readily available in pure form, and if possible have low flammability and toxicity. In selecting organic solvents for use in mobile phases, several physical and chemical properties of the solvent should be considered. From the standpoint of detection, the refractive index or UV cutoff values are also important.

The term polarity refers to the ability of a sample or solvent molecule to interact by combination of dispersion, dipole, hydrogen bonding, and dielectric interactions (see Chapter 2 in reference 5). The combination of these four intermolecular attractive forces constitutes the solvent polarity, which is a measure of the strength of the solvent. Solvent strength increases with polarity in normal phase, and adsorption HPLC decreases with polarity in reversed-phase HPLC. Thus, polar solvents preferentially attract and dissolve polar solute molecules.

Common HPLC solvents with adequate purity are commercially available. Halogenated solvents may contain traces of acidic impurities that can react with stainless steel components of the HPLC system. Mixtures of halogenated solvents with water should not be stored for long periods, as they are likely to decompose. Mixtures of halogenated solvents with various ethers, e.g., diethyl ether, react to form products that are particularly corrosive to stainless steel. Halogenated solvents such as methylene chloride react with other organic solvents such as acetonitrile and, on standing, form crystalline products.

15.11.1.5 Solvent selection
Snyder (for more details, see Chapter 2, reference 5) has described a scheme for classifying common solvents according to their polarity or chromatographic strength (P' values) based on their selectivity or relative ability to engage in hydrogen bonding or dipole interactions. Various common solvents classified into eight groups (I–VIII) showing significantly different selectivities (Fig. 15.11).

Group	Common Solvents
I	Aliphatic ethers, tetramethylguanidine
II	Aliphatic alcohols
III	Tetrahydrofuran, pyridine derivatives, glycol ethers, sulfoxides
IV	Acetic acid, formamide, benzyl alcohol, glycols
V	Methylene chloride, ethylene chloride
VIa	Aliphatic ketones and esters, dioxane, tricresyl phosphate
VIb	Sulfones, nitriles
VII	Aromatic hydrocarbons, halo-substituted aromatic hydrocarbons, nitro compounds, aromatic ethers
VIII	Water, m-cresol, fluoroalcohols, chloroform

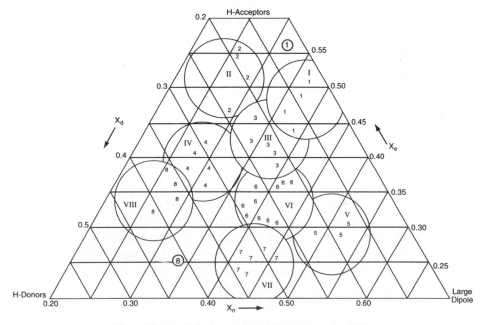

Fig. 15.11. Solvent-selectivity triangle [5].

The P' (polarity index) values and selectivity group classifications for some solvents commonly used in LC are given in Table 15.15.

The values of P' and selectivity factors are calculated from the experimentally derived solute polarity distribution coefficient for the test solutes ethanol, dioxane, and nitromethane. The solute distribution

TABLE 15.15

Polarity of some common solvents [5]

Solvent	δ	P'	ε
n-Hexane	7.3	0.1	0.00
Ethyl ether	7.4	2.8	0.43
Triethylamine	7.5	1.9	—
Cyclohexane	8.2	0.2	0.03
Carbon tetrachloride	8.6	1.6	0.11
Ethyl acetate	8.9	4.4	0.48
Tetrahydrofuran	9.1	4.0	0.53
Chloroform	9.3	4.1	0.26
Methylene chloride	9.6	3.1	0.30
Acetone	9.7	5.1	0.53
Dioxane	10.1	4.8	0.51
Dimethyl formamide	11.8	6.4	—
Isopropanol	12.0	3.9	0.60
Acetonitrile	12.1	5.8	0.52
Ethanol	12.7	4.3	—
Methanol	14.5	5.1	0.70
Formamide	19.2	9.6	—
Water	23.4	10.2	—

δ = Hildebrand solubility parameter
P' = Polarity index
ε = Solvent strength for silica adsorbent

coefficients are corrected for effects that are due to solute molecular size, solute/solvent dispersion interactions, and solute/solvent induction due to solvent polarizability. The resultant parameters P' and solvent selectivity should reflect only the selective interaction properties of the solvent. The test solutes ethanol, dioxane, and nitromethane are used to measure the strengths of solvent proton acceptor, proton donor, and strong dipole interactions, respectively.

Changes in the mobile phase can result in significant selectivity changes for various sample analytes. The greatest change in mobile phase selectivity can be obtained when the relative importance of the various intermolecular interactions between solvent and solute molecules is markedly changed. The changes in selectivity can be affected by making use of the following solvent properties:

- Proton donors: alcohols, carboxylic acids, phenols, and chloroform
- Proton acceptors: amines, ethers, sulfoxides, amides, esters, and alcohols

- Large dipole solvents: methylene chloride, nitrites, sulfoxides, and ketones.

Substitution of methanol by another alcohol such as propanol would not be expected to radically change selectivity because in both cases a proton donor solvent is present. However, a greater change in selectivity can be expected by using ethyl ether (proton acceptor) or methylene chloride (large dipole moment).

The solvent classification scheme is helpful in identifying solvents with different chromatographic selectivities. It is generally preferable to use mixtures of solvents rather than a single pure solvent as the mobile phase. For binary solvents, mixing a strength-adjusting solvent with various volume fractions of a strong solvent enables the complete polarity or solvent strength range between the extremes represented by the pure solvents themselves to be covered. The strength-adjusting solvent is usually a nonselective solvent, such as water for reversed-phase chromatography, and hexane for normal-phase applications. The solvent strength of a binary solvent mixture is the arithmetic average of the solvent strength weighting factors adjusted according to the volume fraction of each solvent. For normal-phase chromatography, the solvent strength weighting factor, S_i, is the same as the polarity index, P'. In reversed-phase chromatography, a different set of experimentally weighting factors is used [104].

The solvent strength for any solvent mixture can be calculated from this equation:

$$S_T = \sum_i S_i \theta_i \qquad (15.23)$$

where

S_T = total solvent strength of the mixture
S_i = solvent strength weighting factor
θ_i = volume fraction of solvent in the mixture

Binary solvent mixtures provide a simple means of controlling solvent strength but limited opportunities for controlling solvent selectivity. With ternary and quaternary solvent mixtures, it is possible to fine-tune solvent selectivity while maintaining a constant solvent strength [105–107]. In addition, there are only a small number of organic modifiers that can be used as binary mixtures with water.

The Snyder-solvent-selectivity triangle concept can be combined with a mixture-design statistical technique to define the optimum mobile-phase composition for a particular separation A feature of this mixture-design technique is that it leads to the selection of a quaternary mobile-phase system for most separations. The selection process can be controlled by a microprocessor in an interactive way if the solvent delivery system can pump four solvents simultaneously (for more details, see Chapter 11, reference 5).

15.11.1.6 *Mobile-phase additives*

At times it is necessary to add reagents such as buffers, ion-pairing reagents, or other modifiers such as triethylamine to the mobile phase to improve reproducibility, selectivity, or peak shape.

Buffers are used mainly to control the pH and the acid-base equilibrium of the solute in the mobile phase. They can also be used to influence the retention times of ionizable compounds. The buffer capacity should be maximum and should be uniform in the pH range of 2–8 commonly used in HPLC. The buffers should be soluble, stable, and compatible with the detector employed, e.g., citrates are known to react with certain HPLC hardware components.

Addition of compounds such as long-chain alkyl compounds in reversed-phase separations will alter the retention of ionic compounds but will have no effect on nonionic compounds unless the concentration is high enough to form micelles (for additional information, please refer to Chapter 8, reference 5).

Competing amines such as triethylamine and di-*n*-butylamine have been added to the mobile phase in reversed-phase separations of basic compounds. Acetic acid can serve a similar purpose for acidic compounds. These modifiers, by competing with the analyte for residual active sites, cause retention time and peak tailing to be reduced. Other examples are the addition of silver ions to separate geometric isomers and the inclusion of metal ions with chelating agents to separate racemic mixtures.

15.12 APPLICATIONS

HPLC can be used in virtually all fields. Listed below are the types of compounds that can be resolved by HPLC [7].

- Amino acids
- Dyes
- Explosives
- Nucleic acids
- Pharmaceuticals, excluding impurities and metabolites

- Pharmaceutical impurities
- Pharmaceutical metabolites
- Plant active ingredients
- Plant pigments
- Polar lipids
- Polysaccharides
- Proteins and peptides
- Recombinant products
- Surfactants
- Synthetic polymers.

It is clearly not possible to cover all these applications within the scope of this chapter. Detailed applications can be found in the literature [1–3,5,7]. Reproduced below is a case study to show how one may select a mode of separation in HPLC and at the same time develop a better understanding of separation mechanism.

15.12.1 A case study

Ionizable compounds such as baclofen (I), 4-amino-3-(p-chlorophenyl)butyric acid, can be chromatographed by several modes in high-pressure liquid chromatography (HPLC). It can be chromatographed as a cation or an anion on cationic or anionic exchange columns respectively, or as an ion-pair with the oppositely charged counterion by RPLC. Ionization suppression techniques may also be used in RPLC. The separation mechanisms involved in chromatography of baclofen (I) from its transformation product were investigated by comparison of ion-exchange and ion-pair reversed-phase HPLC. These investigations were aimed primarily at the amino group, which was protonated for the cation-exchange chromatography or ion-paired with pentane sulfonated anion for reversed-phase chromatography. Model compounds were used to evaluate selectivity for ionic and nonionic compounds. Table 15.16 ows selectivity of the ion-exchange method. These experiments clearly show that retention in ion-exchange chromatography can be considerably influenced by the nonionized portion of the molecule, e.g., see retention data of Compounds I, III, and VI.

A combination of different modes of chromatography can provide excellent resolution for those components that chromatograph poorly with a single mode [108,109]. The chromatographic separation of baclofen (I) from its potential transformation product (II) with dual mode chromatography entailing ion-pair reversed-phase chromatography and

TABLE 15.16

Selectivity of ion-exchange method

Compound	Structure	Retention volume
Lactam	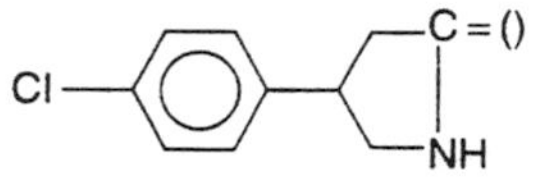	1.97
4-(p-Chlorophenyl) glutarimid	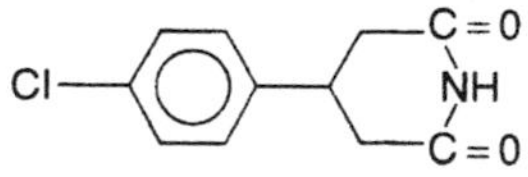	1.97
o-Chlorohydrocinnamic acid	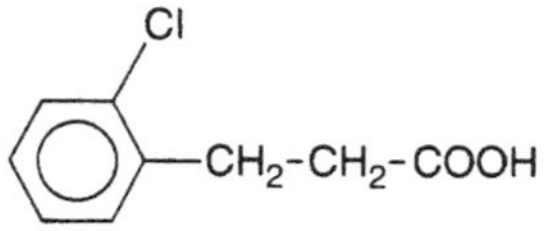	1.97
Baclofen	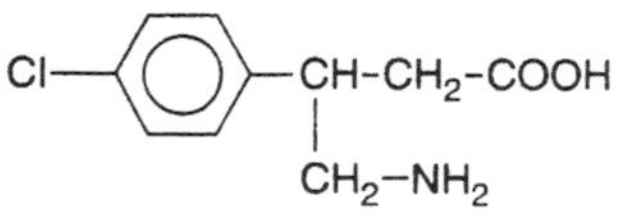	5.91
Benzylamine	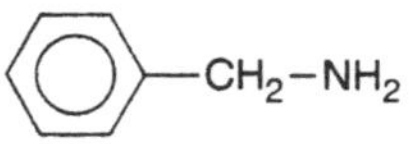	9.84
dl-α-Methylbenzylamine	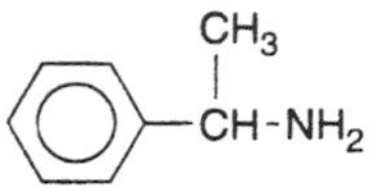	18.6

TABLE 15.17

Effect of sulfonic acid on retention time of baclofen

	Retention time (min)	
Concentration of C-5 reagent	Baclofen (I)	Lactam (II)
0.0 m	5.6 (asymmetrical peak)	22.0
0.007 m	10.1 (symmetrical peak)	24.0

reversed-phase chromatography, respectively, demonstrates this point clearly (Table 15.17).

Ion pairing between the amino group of baclofen and pentane sulfonic acid is primarily responsible for the chromatographic behavior of baclofen on a reversed-phase octadecylsilane column. However, the transformation

TABLE 15.18

Effect of increasing concentration of C-5 sulfonic acid on retention

Concentration of C-5 reagent	t_R Baclofen (min)	Calc. $\Delta\, t_R$ /mM
0.0	5.97	—
0.22	6.04	0.32
2.2	9.25	1.49
5.4	10.87	0.91
10.8	12.08	0.57
16.2	12.90	0.43
21.6	13.43	0.34
32.3	14.23	0.26
43.1	15.06	0.21

product (II) does not form an ion pair with pentane sulfonic acid and is, therefore, separated primarily by a reversed-phase partition process. It was noted that peak symmetry and analysis time of (I) can be significantly influenced by the concentration of C-5 sulfonic acid (Table 15.18). These experiments show the importance of investigations on stationary phase dynamics in separations by ion-pair RPLC.

CONCLUSIONS

The phenomenal growth in chromatography is largely due to the introduction of the technique called HPLC. It allows separations of a large variety of compounds by offering some major improvements over the classical column chromatography, thin-layer chromatography, GC, and it presents some significant advantages over more recent techniques such as SFC, CE, and electrokinetic chromatography. New developments in HPLC include utilization of smaller particle size and ultrahigh pressures or miniaturization to combine nanoLC with MS.

REFERENCES

1 S. Ahuja, *Chromatography and Separation Science*, Academic Press, San Diego, CA, 2003.
2 L.R. Snyder and J.J. Kirkland, *Introduction to Modern Liquid Chromatography*, Wiley, NY, 1979.
3 S. Ahuja and M. Dong, *Handbook of Pharmaceutical Analysis by HPLC*, Elsevier, Amsterdam, 2005.

S. Ahuja

4   S. Ahuja, G.D. Chase, and J.G. Nikelly, Pittsburgh Conference on Analytical Chemistry and Spectroscopy, Pittsburgh, PA, March 2, 1964; *Analytical Chemistry*, 37, 840, 1965.
5   L. R. Snyder, *Selectivity and Detectability Optimizations in HPLC*, In: S. Ahuja (Ed.), Wiley, NY, 1989, pp. 15–35.
6   S. Ahuja, "Recent Developments in High Performance Liquid Chromatography", Metrochem '80, South Fallsburg, NY, October 3, 1980.
7   S. Ahuja, *Trace and Ultratrace Analysis by HPLC*, Wiley, NY, 1992.
8   S. Ahuja, *Proceedings of Ninth Australian Symposium on Analytical Chemistry*, Sydney, April 27, 1987.
9   R.J. Laub, *Chromatography and Separation Chemistry,* S. Ahuja, Ed., Vol. 297, ACS Symposium Series, p.1, 1986.
10 H. Engelhardt, *Chromatography and Separation Chemistry*, 15 (1977) 380.
11 J.P. Thomas, A.P. Burnard and J.P. Bounine, *J. Chromatogr.*, 172 (1979) 107.
12 H. Engelhardt, *High Performance Liquid Chromatography*, Springer, Berlin, 1979.
13 H. Engelhardt and W. Boehme, *J. Chromatogr.*, 133 (1977) 67.
14 R.A. Bredeweg, L.D. Rothman and C.D. Pfeiffer, *Anal. Chem.*, 51 (1979) 2061.
15 J. Crommen, *J. Chromatogr.*, 186 (1979) 705.
16 B.A. Bidlingmeyer, J.K. Del Rios and J. Korpi, *Anal. Chem.*, 54 (1980) 442.
17 R.L. Smith and D.G. Pietrzyk, *Anal. Chem.*, 56 (1984) 610.
18 R.E. Majors. In: C. Horvath (Ed.), *High Performance Liquid Chromatography, Advances and Perspectives*, Vol. 1. Academic Press, 1980, p. 75.
19 R.D. Rassmussen, W.H. Yokoyama, S.G. Blumenthal, D.E. Bergstrom and B.H. Ruebner, *Anal. Biochem.*, 101 (1980) 66.
20 S.A. Wise, S.N. Chesler, H.S. Hertz, L.P. Hilpert and W.E. May, *Anal. Chem.*, 49 (1977) 2306.
21 J. Chmielowiec and A.E. George, *Anal. Chem.*, 52 (1980) 1154.
22 D. Karlesky, D.C. Shelley and I. Warner, *Anal. Chem.*, 53 (1981) 2146.
23 A.M. Krstulovic and P.R. Brown, *Reversed-Phase High Performance Liquid Chromatography: Theory, Practice and Biomedical Applications*, Wiley, New York, 1982.
24 P. Jandera, H. Colin and G. Guiochon, *Anal. Chem.*, 54 (1982) 435.
25 M. Otto and W. Wegscheider, *J. Chromatogr.*, 258 (1983) 11.
26 W.R. Melander and C. Horvath, *High Performance Liquid Chromatography, Advances and Perspectives*, Vol. 2, Academic Press, New York, 1980 P. 114.
27 R.P.W. Scott and P. Kucera, *J. Chromatogr.*, 142 (1977) 213.
28 H. Colin and G. Guiochon, *J. Chromatogr.*, 158 (1978) 183.

29 L.R. Snyder, J.L. Glajch and J.J. Kirkland, *Practical HPLC Method Development*, Wiley, NY, 1988.
30 J.A. Knox and R.A. Harwick, *J. Chromatogr.*, 204 (1981) 3.
31 S. Elchuk and R.M. Cassidy, *Anal. Chem.*, 51 (1979) 1434.
32 H. Miyagi, J. Miura, Y. Takata, S. Kamitake, S. Ganno and Y. Yamagata, *J. Chromatogr.*, 239 (1982) 733.
33 F.E. Regnier, K.M. Gooding and S.-H. Chang, *Contemporary Topics in Analytical Clinical Chemistry*, Vol. 1, Plenum, New York, 1977, p. 1.
34 F.E. Regnier and K.M. Gooding, *Anal. Biochem.*, 103 (1980) 1.
35 D.N. Vacik and E.C. Toren, *J. Chromatogr.*, 228 (1982) 1.
36 W. Voelter and H. Bauer, *J. Chromatogr.*, 126 (1976) 693.
37 O. Samuelson, *Adv. Chromatogr.*, 16 (1978) 113.
38 K. Tanaka and T. Shizuka, *J. Chromatogr.*, 174 (1979) 157.
39 C.F. Poole and S.A. Schuette, *Contemporary Practice of Chromatography*, Elsevier, New York, 1984.
39a F.M. Rabel, *Adv. Chromatogr.*, 17 (1979) 53 through reference 39.
40 E. Sawicki, J.D. Mulik and E. Wattgenstein, *Ion Chromatographic Analysis of Environmental Pollutants*, Vol. 1, Ann Arbor Science, Ann Arbor, MI, 1978.
41 J.D. Mulik and E. Sawicki, *Ion Chromatographic Analysis of Environmental Pollutants*, 2 (1979).
42 F.C. Smith and R.C. Change, *CRC Crit. Revs. Anal. Chem.*, 9 (1980) 197.
43 H. Small. In: J.F. Lawrence (Ed.), *Trace Analysis*, Vol. 1. Academic Press, New York, p. 269, 1982.
44 J.S. Fritz, D.T. Gjerde and C. Pohlandt, *Ion Chromatography*, Huthig, Heidelberg, 1982.
45 H. Small, *Anal. Chem.*, 55 (1983) 235A.
46 C.A. Pohl and E.L. Johnson, *J. Chromatogr. Sci.*, 18 (1980) 442.
47 D.J. Gjerde and S.J. Fritz, *J. Chromatogr.*, 176 (1979) 199.
48 T.S. Stevens and H. Small, *J. Liq. Chromatogr.*, 1 (1978) 123.
49 T.S. Stevens and M.A. Langhorst, *Anal. Chem.*, 54 (1982) 950.
50 R. Small, T.S. Stevens and W.C. Bauman, *Anal. Chem.*, 47 (1981) 1801.
51 T.S. Sevents, J.C. Davis and H. Small, *Anal. Chem.*, 53 (1981) 1488.
52 Y. Hanaoki, T. Murayama, S. Muramoto, T. Matsura and A. Nanba, *J. Chromatogr.*, 239 (1982) 537.
53 T.S. Stevens, G.L. Jewett and R.A. Bredeweg, *Anal. Chem.*, 54 (1982) 1206.
54 D.T. Gjerde, J.S. Fritz and G.S. Schmuckler, *J. Chromatogr.*, 186 (1979) 509.
55 D.T. Gjerde, J.S. Fritz and G.S. Schmuckler, *J. Chromatogr.*, 187 (1980) 35.
56 S. Matsushita, Y. Tada, N. Baba and K. Hosako, *J. Chromatogr.*, 259 (1983) 459.

57 C.L. Ulmann, H.-G. Genieser and B. Jastorff, *J. Chromatogr.*, 354 (1986) 434.
58 D. Martire and R.E. Boehm, *J. Phys. Chem.*, 87 (1983) 1045.
59 M. McCann, S. Madden, J.H. Purnell and C.A. Wellington, *J. Chromatogr.*, 294 (1984) 349.
60 T.L. Hafkensheid, *J. Chromatogr. Sci.*, 24 (1986) 307.
61 T. Braumann, G. Weber and L.H. Grimme, *J. Chromatogr.*, 261 (1983) 329.
62 T.L. Hafkensheid and E. Tomlinson, *Int. J. Pharm.*, 16 (1983) 225.
63 A. Nahum and C. Horvath, *J. Chromatogr.*, 192 (1980) 315.
64 S.H. Unger, J.R. Cook and J.S. Hollenberg, *J. Pharm. Sci.*, 67 (1978) 1364.
65 J.E. Hoky and A. Michael Young, *J. Liq. Chromatogr.*, 7 (1984) 675.
66 J.K. Baker, *Anal. Chem.*, 51 (1979) 1693.
67 J.K. Baker, R.E. Skelton and C.Y. Ma, *J. Chromatogr.*, 168 (1979) 417.
68 B.-K. Chen and C. Horvath, *J. Chromatogr.*, 171 (1979) 15.
69 R. Kalizan, *J. Chromatogr.*, 220 (1981) 71.
70 E. Tomlinson, *J. Chromatogr.*, 113 (1975) 1.
71 T. Hanai and A. Jukurogi, *Chromatography*, 19 (1984) 266.
72 T. Hanai and J. Hubert, *J. Chromatogr.*, 290 (1984) 197.
73 T. Hanai and J. Hubert, *J. Chromatogr.*, 291 (1984) 81.
74 T. Hanai and J. Hubert, *J. Chromatogr.*, 302 (1984) 89.
75 W. Melander, D.E. Campbell and C. Horvath, *J. Chromatogr.*, 158 (1978) 215.
76 M.C. Spanjer and C.L. Deligny, *Chromatography*, 20 (1985) 120.
77 K. Jinno and N. Ozaki, *J. Liq. Chromatogr.*, 7 (1984) 877.
78 K. Jinno and K. Kawasaki, *Chromatography*, 18 (1984) 90.
79 Y. Arai, M. Hirukawa and T. Hanai, *J. Chromatogr.*, 384 (1987) 279.
80 T. Hanai, A. Jakurogi and J. Hubert, *Chromatography*, 19 (1984) 266.
81 P.J. Schoenmakers, H.A.H. Billiet and L.D. Galan, *Chromatography*, 282 (1983) 107.
82 B.P. Johnson, M.G. Khaledi and J.G. Dorsey, *Chromatography*, 384 (1987) 221.
83 R.M. Smith, *Chromatography*, 324 (1985) 243.
84 W.T. Cooper and L.Y. Lin, *Chromatography*, 21 (1986) 335.
85 W.R. Melander and C. Horvath, *Chromatography*, 18 (1984) 353.
86 J.D. Wang, J.Z. Li, M.S. Bao and P.C. Lu, *Chromatography*, 17 (1983) 244.
87 P.J. Schoenmakers, H.A.H. Billiet and L. De Galan, *Chromatography*, 15 (1982) 205.
88 S. Ahuja, *Selectivity and Detectability Optimization in RPLC*, Academy of Science USSR Meeting, September 5–7, 1984.
89 P. Jandera, H. Colin and G. Guiochon, *Anal. Chem.*, 54 (1982) 435.
90 H. Colin, G. Guiochon and P. Jandera, through ref. 47.

91 H. Colin, G. Guiochon and P. Jandera, *Chromatography*, 17 (1983) 93.
92 S.D. West, *J. Chromatogr. Sci.*, 25 (1987) 122.
93 L.R. Snyder and J.J. Kirkland, *Introduction to Modern Liquid Chroma-tography*, Wiley, New York, 1979.
94 L.R. Snyder, *J. Chromatogr. Sci.*, 16 (1974) 223.
95 J.J. Lewis, L.B. Rogers and R.E. Pauls, *J. Chromatogr.*, 264 (1983) 339.
96 R.M. Smith, *Anal. Chem.*, 56 (1984) 256.
97 A.P. Goldberg, *Anal. Chem.*, 54 (1982) 342.
98 J.L. DiCesare and M.W. Doag, *Chromatogr. Newsl.*, 10 (1982) 12–18.
99 R.M. Smith, *J. Chromatogr.*, 236 (1982) 313.
100 R.M. Smith, T.G. Hurley, R. Gill and A.C. Moffat, *Chromatography*, 19 (1984) 401.
101 P.C. Sadak and W. Carr, *J. Chromatogr.*, 21 (1983) 314.
102 M. Verzele and C. Dewaele, *Chromatography*, 18 (1985) 84.
103 S. Ahuja, *Ultratrace Analysis of Pharmaceuticals and Other Compounds of Interest*, Wiley, New York, 1986.
104 L.R. Snyder, J.W. Dolan and J.R. Grant, *J. Chromatogr.*, L65 (1979) 3.
105 C.F. Simpson, *Techniques in Liquid Chromatography*, Wiley, New York, 1982.
106 P.A. Bristow and J.F. Knox, *Chromatographia*, 10 (1977) 279.
107 J.C. Chen and S.G. Weber, *J. Chromatogr.*, 248 (1982) 434.
108 S. Ahuja, *Retention Mechanism Investigations on Ion-pair Reversed Phase Chromatography*, American Chemical Society meeting, Miami, April 28, 1985.
109 S. Ahuja, *Probing Separation Mechanism of Ionic Compounds in HPLC*, Metrochem '85, Pocono, October 11, 1985.

REVIEW QUESTIONS

1. Discuss various modes of chromatography.
2. Which mode of chromatography is most commonly used in HPLC?
3. How can you vary α values in RPLC?
4. List some of the common additives for RPLC.
5. Describe various approaches to solvent optimization in RPLC.

Chapter 16

Supercritical fluid chromatography

Mary Ellen P. McNally

16.1   INTRODUCTION

What is a supercritical fluid?

The discovery of supercritical fluids occurred in 1879, when Thomas Andrews actually described the supercritical state and used the term critical point. A supercritical fluid is a material above its critical point. It is not a gas, or a liquid, although it is sometimes referred to as a dense gas. It is a separate state of matter defined as all matter by both its temperature and pressure. Designation of common states in liquids, solids and gases, assume standard pressure and temperature conditions, or STP, which is atmospheric pressure and 0°C. Supercritical fluids generally exist at conditions above atmospheric pressure and at an elevated temperature. Figure 16.1 shows the typical phase diagram for carbon dioxide, the most commonly used supercritical fluid [1].

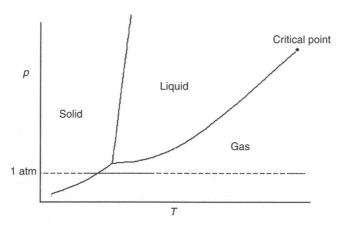

Fig. 16.1. Phase diagram for carbon dioxide critical temperature 31.3°C critical pressure 72.9 atm.

Comprehensive Analytical Chemistry 47
S. Ahuja and N. Jespersen (Eds)
Volume 47 ISSN: 0166-526X DOI: 10.1016/S0166-526X(06)47016-1

TABLE 16.1

Comparison of physical properties of liquids, gases and supercritical fluids

Phase type	Density (g/cm^3)	Diffusion (cm^2/s)	Viscosity (g/cm s)
Gas at 1 atm, 21°C	10^{-3}	10^{-1}	10^{-4}
Supercritical fluid	0.3–0.8	10^{-3}–10^{-4}	10^{-4}–10^{-3}
Liquid	1	$< 10^{-5}$	10^{-2}

Source: Van Wassen et al. [2]

The critical point of a material is the temperature and pressure conditions at which the liqiud state ceases to exist. As a liquid is heated, it becomes less dense and starts to form a vapor phase. The vapors being formed becomes more dense, with continued heating the liquid and vapor densities become closer to each other until the critical temperature point is reached. At this same point, the liquid-line or phase boundary disappears. This critical point was first discovered and reported in 1822 by Baron Charles Cagniard de la Tour.

As a fluid, the supercritical state generally exhibits properities that are intermediate to the properties of either a gas or a liqiud. In Table 16.1, the physical properties of liquids, gases and supercritical fluids are compared [2]. Examination of the values in Table 16.1 makes the intermediate nature of a supercritical fluid more obvious. The density of a supercritical fluid approaches the levels of a liquid as does its diffusivity, while its viscosity is similar to a typical gas. These properties offer rapid movement (equilibration) as in a gas, but solvation or solubilization as found in a liquid. The best of both worlds from a chromatographic viewpoint. This is because as a solute travels through a chromatographic column the number of equilibration points it reaches, as defined by the Van Dempter equation, defines how effective the separation will be, either through the number of theoretical plates, N, the resolution, R_s, or the alpha value, α, also known as a separation factor. For a detailed description of the theory of chromatography, the reader is refered to Snyder and Kirkland's text *Modern Practice of Liquid Chromatography* [3] or Chapters 12, 14, 15 of this text.

The ease of solubilizing the analyte is a key factor also. This is because the easier it is to get the solute to these potential equilibration points inside the chromatographic column when the fluid is moving with high diffusivity, the faster the equilibrations or separation can take place. Compared to a gas where the solubilities of the analytes of interest is almost non-existent, the liquid-like density of a supercritical fluid offers enhanced solubilization.

In supercritical fluid chromatography, fluids above their critical point are used as mobile phases. This chapter discusses the principles of operation, mobile phase considerations, parameters that can be adjusted in method development as well as an overview of instrumentation required and a few pertinent examples from current literature. Not everything can be illustrated, but the advantages of this diverse technology will be highlighted.

16.2   HISTORY OF SFC

In terms of chromatography, the first individual credited with the use of supercritical fluids as the mobile phase is Ernst Klesper when in 1962 he reported on the separation of metal porphyrins using dense-gas chromatography (GC) or SFC [3]. But it was not until the 1980s that the analytical community took hold of the abilities and advantages of the technique with the advent of several commercial instrumentation ventures.

Two approaches were taken during this time, one by gas chromatographers and the other by liquid chromatographers. Practicing gas chromatographers who experimented with supercritical fluid chromatography for its enhanced solvating powers pursued the first approach. They coupled supercritical fluids, in general pure carbon dioxide; with small narrow bore capillary columns, similar to leading GC columns of the time. With that coupling they were able to get enhanced resolution of compounds that are too difficult to analyze by GC, because they were not soluble in the nitrogen and helium mobile phases commonly used in GC.

The second approach was taken by practicing liquid chromatographers. They routinely dealt with thermally labile, highly polar molecules and frequently sacrificed resolution, and speed in their separations because of the aqueous mobile phases that were required. With the enhanced diffusion and decreased viscosity of supercritical fluids over liquids, chromatographic run-time and resolution could be improved when supercritical fluids were used. But solubility in pure carbon dioxide mobile phases, which has the solvating powers from hexane to methylene chloride under normal density ranges, was a problem for these polar molecules. To compensate for this, experimentalists started working with mixed mobile phases. These mixed phases were based on the addition of polar modifiers or additives to the carbon dioxide mobile phase. Similar to the first approach, this second group of scientists continued to use the columns with similar dimensions to liquid chromatography (LC) columns prevalent in the 1980s that they were already using, i.e., 25 cm by 4.6 mm; the columns were also functionalized with the common C-18, C-8 and phenyl phases as found in LC.

With both these approaches rapidly being pursued by a wide variety of academic, industrial and instrument company laboratories, the growth and publication rate in supercritical fluid chromatography during the mid-late 1980s and the early 1990s was exponential. A wide variety of applications are available from that time period as well as several references that explain the advantages of optimizing a supercritical fluid separation using all the parameters available in SFC compared to GC or LC [4–8].

16.3   BASIC PRINCIPLES IN SFC

The most common and widely used supercritical fluid in SFC is carbon dioxide. It is inert, in that it is non-toxic and non-flammable, it also has mild critical parameters, a low critical temperature of 31.3°C and a critical pressure of 72.8 atm [1]. Using pure, supercritical carbon dioxide eliminates organic solvent waste and with it waste disposal costs and concerns. This is extremely practical advantage in the industrial environment where the generation of waste requires special handling and significant cost.

Beyond this practical consideration, the advantages of SFC from a technical perspective are in the wider range of parameters that can be optimized to achieve the best separation [4]. Taken from the same two approaches described above, the parameter that is most commonly adjusted to change resolution and retention time in GC is the temperature, after that the only choice an experimentalist might have is to change the column used for separation. However, from the liquid chromatographers perspective, the mobile phase composition is the principal component that can be changed to effect a better separation, the mobile phase components are generally the second choice and then finally the choice of chromatographic column. Using supercritical fluid chromatography gives all the parameters optimized in LC as well as the ability to see drastic retention changes due to a temperature gradient, as in GC. In addition, SFC has the added parameters of pressure and/or density that can be selected to achieve the best separation conditions.

From an overall perspective, both the temperature and pressure of the mobile phase control the density or elution power. A change in either temperature or pressure changes the mobile phase density and will alter the chromatographic elution. It should be noted that increasing the temperature in SFC increases the retention time, generally an undesirable effect and the opposite effect that is exhibited in GC. Because of this, reverse temperature gradients, from a high temperature to a low temperature, are utilized in SFC to decrease retention. With decreasing temperature, density of the fluid increases; these higher density mobile phases solubilize the analytes of interest and results in earlier elution of the compounds of interest.

The solvating ability of carbon dioxide in SFC is significant when compared to gases used in GC. This is because as a pure component, a change in density of carbon dioxide causes a change in the material's Hildebrand solubility parameter. In carbon dioxide, this range of solubilities can be considered equivalent to the solubilities seen from hexane to methylene chloride. However, in terms of solvating moderately polar or highly polar molecules, this solubility parameter range is not sufficient and modifiers or additives to carbon dioxide must be included. Once a modifier has been added to the mobile phase, the critical parameters of the original solvent are no longer valid and a two-phase system can exist. As an example, a 10% methanol in carbon dioxide solution has a critical temperature of 51.5°C with a critical pressure of 74.2 atm. The critical parameters of pure methanol are 240.5°C, temperature, and 78.9 atm, pressure. From these values it can easily be determined that there is no linear relationship for critical point change when the components of a two-phase system are mixed together. What should be noted though is that above the critical point of a mixture, the mobile phase is one phase. This is extremely important in chromatography, an equilibrium-based process, where by definition the movement of the analyte is between two phases. If one of the phases is a not a pure component, equilibrium is difficult to reproduce precisely, the analyte is not transferring back and forth between the stationary phase and the mobile phase, but instead between three phases, the stationary phase and a two-component mobile phase. This discussion emphasizes that, from a practical standpoint, one of the key considerations, often neglected by initial practitioners, is the need to maintain the mobile phase above its critical point at all times during the chromatographic separation. This may be difficult with the wide range of parameters that are adjusted to optimize a separation.

16.3.1 Parameter optimization

Being able to change the density, *via* either changes in pressure or temperature, is the key difference in SFC over GC and LC separations. Typical density ranges are from 0.3 to 0.8 g/ml for pure carbon dioxide. Table 16.2 shows data obtained from ISCO's SF-Solver Program for the calculation of density (g/ml), Hildebrand Solubility Parameter and a relative equivalent solvent for pure carbon dioxide at a constant pressure of 6000 psi, approximately 408 atm.

But the real advantage in SFC is not just the ability to adjust the density of the mobile phase but the ability to adjust it from two different directions. Density changes achieved by a change in pressure can yield different separation factors then density changes achieved by adjusting the temperature. This offers an advantage in method development by SFC not available in GC or LC.

Beyond the density changes that can be used to control method modifications in SFC, the mobile phase composition can also be adjusted. Typical LC solvents are the first choice, most likely because of their availability, but also because of their compatibility with analytical detectors. The most common mobile phase modifiers, which have been used, are methanol, acetonitrile and tetrahydrofuran (THF). Additives, defined as solutes added to the mobile phase in addition to the modifier to counteract any specific analyte–column interactions, are frequently included also to overcome the low polarity of the carbon dioxide mobile phase. Amines are among the most common additives.

TABLE 16.2

Calculation of density, Hildebrand solubility parameter using ISCO's SF solver program at constant pressure of 6000 psi (408 atm)

Temperature (°C)	Density (g/mL)	Hildebrand solubility	Equivalent solvent	
40	0.967	8.248	C6H12	Cyclohexane
49	0.937	7.902	C2H4	Ethylene
55	0.917	7.814	CF4	Carbon tetrafluoride
64	0.886	7.554	C8H16	*n*-Octane
70	0.865	7.375	C6H16	*n*-Hexane
73	0.855	7.287	O2	Oxygen
79	0.834	7.113	C4H10	*n*-Butane
85	0.815	6.944	C8H18	2-Trimethyl pentane
94	0.785	6.694	F2	Fluorine

Modifiers and gradient compositions in LC are used broadly.

A gradient that runs with 30–80% methanol or acetonitrile is not uncommon. This amount of modifier is generally not needed in supercritical fluid chromatography to affect the same separation. Typical modifier composition in SFC is 1.0–10% and would achieve higher Hildebrand Solubility Parameter adjustment overall than the broader gradients found in LC.

16.3.2   Instrument requirements

The major difference in supercritical fluid chromatography and conventional LC equipment is the pumping systems as well as the safety features installed to maintain higher pressure. Unique SFC equipment differences are:

1. Carbon dioxide tank for mobile phase supply
   a. Equipped with a pressure relief value and rupture disk
2. High-pressure pump
   a. Chiller to maintain mobile phase in liquid state
3. High-speed injector
4. Pressure restrictor
   a. High-pressure tubing
5. High-pressure flow cell for UV detection
6. Solvent collection device with ability to vent to a laboratory hood or elephant trunk.

Carbon dioxide is usually purchased in a tank, inside the tank the mobile phase exists as a liquid. Typically, the tank does not come with a pressure gauge but is hooked up to a pressure relief valve and rupture disk, which are set above the tank pressure should a tank leak occur.

High-pressure pumps used in SFC can come in a variety of types; most of them are modifications of pumps used in LC. Piston, diaphragm and syringe pumps are used.

The pumps deliver the most accurate flow of the carbon dioxide if it is pumped in the liquid state. Since that is the case, if there is a great distance between the tank and the pump then a chiller is usually placed so that the tubing containing the carbon dioxide from the tank can be cooled maintaining the carbon dioxide in the liquid state. Most commercially available pumps operate at high enough pressures that the pump heads or pump bodies do not need ancillary cooling, but older models frequently required a chiller that provided a circulation of cold

567

fluids to the parts of the pump where the carbon dioxide has a potential to vaporize out of the liquid state.

A high-speed injector is required in supercritical fluid chromatography. This is to prevent loss of pressure during the injection process.

There are a variety of restrictor types that control the pressure of the fluid during the chromatographic process. Each manufacturer has patented their individual restriction devices. It is beyond the scope of this text to go into the detail about all of these restrictors. However, by way of explanation, most of the mechanical restrictors that are used in commercial equipments operate under the principle of decreasing a volume or a space that the mobile phase must pass through by some mechanical means. This decrease in volume, if metered accurately, increases the pressure in the system in a controlled manner. As a safety precaution, for any equipment where elevated pressures are used, tubing should be rated with a safety factor of at least 1.5 times the maximum pressure the SFC system can achieve.

Because the density is a controlling factor in achieving the separation, maintaining the elevated pressure used in the separation is necessary up to the point of detection. For UV measurements, this requires a high-pressure cell also rated to the maximum pressure the SFC system can achieve with a desired safety factor of 1.5 times. For other detection methods, i.e., flame ionization detection and mass spectrometry, where the sample is nebulized into the detector, the output of the restrictor is generally right at the nebulization point. This positioning of the column outlet eliminates any peak merging that could occur under low- or no- pressure movement.

16.4   CURRENT EXAMPLES

The examples illustrated herein are not all-inclusive but should give a representation of the advantages of the technique and its most common uses.

16.4.1   Chiral separations

The use of supercritical fluids to separate enantiomers is one of the most important tasks in several areas of research, especially pharmaceuticals and agrochemicals. This is because it is well known that the two enantiomeric forms of a molecule can display dramatically different biological activity. The use of supercritical fluids to separate, with higher efficiencies and shorter retention times, enantiomers is extremely advantageous in this difficult work. In their article entitled "Chiral separation of some triazole pesticides by supercritical fluid chromatography, Toribio and coworkers, separated six triazole pesticides and showed the effects of different organic modifiers on the resolution and retention *via* k' value [9]. The modifiers they examined were methanol, ethanol and 2-propanol. The addition of additives was also examined. Table 16.3 shows some of their modifier results. In general, separations could be conducted in less than 10 min and the best modifier for the individual separations was compound dependent. This

TABLE 16.3

Values of capacity factors and resolutions obtained for diniconazole, tetraconazole, hexaconazole and tebuconazole using different modifiers [9]

Compound	k_1'	k_2'	R_s	Compound	k_1'	k_2'	R_s
Hexaconazole				*Tebuconazole*			
Methanol (%, v/v)				Methanol (%, v/v)			
5	5.35	5.68	0.67	5	13.23	14.74	1.28
10	2.33	2.59	1.02	10	5.26	5.82	1.17
15	1.43	1.61	0.87	15	3.11	3.42	1.03
Ethanol (%, v/v)				Ethanol (%, v/v)			
5	5.14	6.23	**2.41**	5	7.16	7.16	0
10	2.01	2.28	1.14	10	5.04	5.04	0
15	1.19	1.32	0.65	15	2.69	2.69	0
2-Propanol (%, v/v)				2-Propanol (%, v/v)			
5	4.38	5.01	0.54	5	20.36	22.53	0.98
10	2.71	2.90	0.48	10	7.61	8.35	1.2
15	1.45	1.45	0	15	3.54	3.91	1.15
Tetraconazole				*Diniconazole*			
Methanol (%, v/v)				Methanol (%, v/v)			
5	4.30	5.23	**1.55**	5	7.29	12.44	**6.08**
10	2.05	2.41	1.27	10	2.81	5.69	**6.90**
15	1.06	1.83	1.02	15	1.63	3.43	**5.87**
Ethanol (%, v/v)				Ethanol (%, v/v)			
5	4.26	5.29	**3.05**	5	6.97	10.63	**4.55**
10	2.24	2.84	**2.49**	10	2.53	3.87	**3.77**
15	1.68	1.98	**2.06**	15	1.37	2.08	**3.11**
2-Propanol (%, v/v)				2-Propanol (%, v/v)			
5	5.86	8.75	**4.98**	5	12.46	21.06	1.23
10	3.28	5.46	**4.01**	10	4.09	4.09	0
15	2.40	3.69	**3.43**	15	2.04	2.04	0

is consistent with the understanding of modifier interactions with the solutes and their effect on the overall equilibrium between the stationary and mobile phases in SFC as previously described. Recalling, acceptable resolution, R_s, values for quantitation purposes are greater than 1.5, values that are greater than 1.5 for resolution are highlighted in Table 16.3. Figure 16.2 shows an example chromatogram of a typical separation. These separations were conducted on a Hewlett-Packard 1205A supercritical fluid chromatograph equipped with a photodiode array detector; detection was at 220 nm. A Chiralpak AD 25 cm × 4.6 mm column packed with the 3, 5-dimethylphenylcarbamate derivative of amylose, coated on a 10 μm silica support was used. The mobile phase was carbon dioxide modified as outlined in Table 16.3 for each of the individual separations. Other chromatographic conditions were 35°C, 2 ml/min flow rate and a pressure of 200 bar (2960 atm) (Fig. 16.3). (*Note*: Because of the low temperature of the separations, the mobile phase was not likely in the supercritical state for all of the modifier concentrations examined in this report.)

16.4.2 Polymer separations

The advantages of supercritical fluid chromatography for polymer separations have been illustrated in the literature for many years. A recent example is the separation of long-chain polyprenols using SFC with matrix-assisted laser-desorption ionization TOF mass spectrometry [10]. The generic name for 1,4-polyprenyl alcohols is polyprenol; these compounds generally have smaller polymerization chains of less

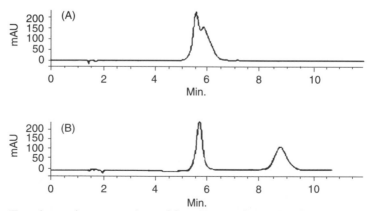

Fig. 16.2. Enatiomeric separation of hexaconazole at 200 bar, 35°C, 2 ml/min and 10 v/v 2-propanol. (A) Without using additives; (B) using 0.1% (v/v) triethylamine and 0.1% (v/v) trifluoroacetic acid [9].

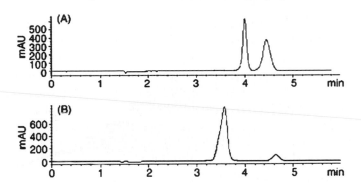

Fig. 16.3. Enatiomeric separation at 200 bar, 35°C, 2 ml/min. (A) Tetraconazole with 4%(v/v) ethanol; (B) diniconazole 15% (v/v) ethanol [9].

than 30 monomers. Chromatographic separation of these chains was conducted using a Jasco Super-201 chromatograph. The system has two separate pumps, one delivers the carbon dioxide, and the other delivered the THF modifier. The column temperature was 80°C, the pressure, controlled by a backpressure regulator was controlled at 19.6 MPa. An Inertsil Ph-3 (25 cm × 4.6 mm), 5 µm column was used. A flow gradient was used to introduce the THF modifier, this is not typical, but it is possible with the two pumps on this instrument. The consistent flow rate of the carbon dioxide is 3.0 ml/min; the THF flow rate started at 0.8 ml/min and was adjusted over 30 min- to 2.0 mL/min and then held constant. Figure 16.4 illustrates the resultant chromatogram obtained for *Eucimmia ulmoides* leaves, a plant producing fibrous rubber; over 100-mer (M_W: 6818) components were completely separated using the conditions described above.

16.4.3   High throughput screening of pharmaceuticals

At Abbott Laboratories, Hochlowski and coworkers used preparatory scale SFC to screen the output of high throughput organic synthesis (HTOS) for purity levels and classify chemical structures into libraries of similar analogs [11]. These scientists used a Berger Instruments SFC self-modified for this particular use. Figure 16.5 shows the schematic of this modified system. In this application, the authors reported adding SFC capability to their repertoire of instrumentation to solve their analytical challenges. Figure. 16.6 shows the comparison of a typical reaction mixture analyzed by both high performance liquid chromatography (HPLC) and SFC.

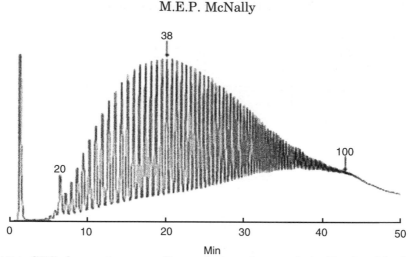

Fig. 16.4. SFC chromatogram of long-chain polyprenols in *E. ulmoides* leaves. The sample had M_n 3.99 × 10³ (calibrated against cis-1,4-polyisoprene standards) and M_w/M_n 1.41. The numbers in the chromatogram represent degrees of polymerization for polyprenol homologues [10].

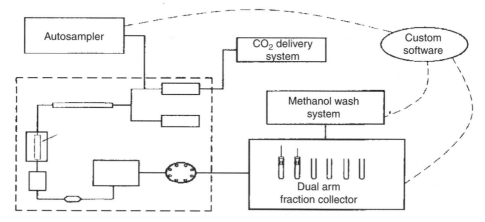

Fig. 16.5. Preparative supercritical fluid chromatographic system, customized for high throughput organic synthesis (HTOS) screening [11].

Actual operating conditions can be found in the figure caption. A given limitation of SFC, relative to HPLC, as described is the ability to dissolve samples in a solvent system compatible with the methanol/carbon dioxide mobile phase. For this particular mobile phase, other compatible sample diluents that worked effectively are pure methanol,

Supercritical fluid chromatography

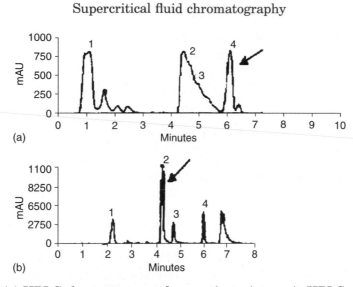

Fig. 16.6. (a) HPLC chromatograms for reaction mixture A. HPLC conditions: C18 (20 × 100 mm Nova–Pak TM), gradient 5–95% acetonitrile/water, with 0.1% TFA; (b) SFC chromatogram for reaction mixture A. SFC conditions: Diol column (21.2 × 150 mm, Berger Instruments), gradient 5–60% Methanol with 0.5% dimethylethylamine in carbon dioxide [11].

or mixtures of either methanol and acetonitrile or methanol and di-chloromethane. Dimethyl sulfoxide, DMSO, on the other hand was not compatible with the carbon dioxide mobile phase used.

16.5   CONCLUSIONS

Supercritical fluids offer properties intermediate to gas and liquids. As such, supercritical fluid chromatography is an alternative technique to both LC not liquid and or GC. The distinct advantages of supercritical fluids, because of gas like densities yields faster chromatographic elution and therefore shorter overall run times than in LC. Greater solvating power than gases, makes it more applicable to a wider variety of analytes that can typically be analyzed by GC. Ability to optimize a wider variety of parameters than GC or LC is also advantageous; density, temperature and pressure, mobile phase composition, gradient elution and the addition of additives to the mobile phase are all available. But, the challenge to the analytical chemist or technical operator is that there is more to understand to utilize the technique correctly, work in the optimized range of a phase diagram and deal with the higher-pressure equipment and safety precautions that are necessary. Because of this complication, SFC has not replaced LC or GC in the mainstream analytical laboratory although it has the capability to do so.

REFERENCES

1   *Supercritical Fluid Extraction: Principles and Practice*, M. A. McHugh and V. Krukonis, 2nd Edition, Butterworths-Heinemann, 1994, 608. ISBN: 0750692448
2   U. Van Wassen, I. Swaid and G.M. Schneider, *Angew. Chemie Int. Ed. Engl.*, 19 (1980) 575.
3   *Introduction to Modern Liquid Chromatography*, L.R. Snyder and J.J. Kirkland, 2nd Edition, Wiley Interscience, 1979
4   *Supercritical Fluids in Chromatography and Extraction*, R. M. Smith and S.B. Hawthorne, Elsevier, 1997, 414, ISBN: 0-444-82869-9
5   J.A. Cros and J.P. Foley, *Anal. Chem.*, 62 (1990) 378–386.
6   M.E. McNally and J.R. Wheeler, *J. of Chromatogr*, 477 (1988) 53–63.
7   M.E. McNally and J.R. Wheeler, *LC/GC Magazine*, 6(9) (1988).
8   M.E. McNally, *Anal. Chem.*, 67(9) (1995) 308A–314A.
9   L. Toribio, M.J. del Nozal, J.L. Bernal, J.J. Jimenez and C. Alonso, *J. Chromatogr A.*, 1046 (2004) 249–253.
10 T. Bamba, E. Fukusaki, Y. Nakazawa, H. Sato, K. Ute T. Kitayama and A. Kobayashi, *J. Chromatogr. A.*, 995 (2003) 203–207.
11 J. Hoxhlowski, J. Olson, J. Pan, D. Sauer, P. Searle and T. Sowin, *J. Liq. Chromatgr. & Rel. Tech.*, 26(3) (2003) 3333–3354.

REVIEW QUESTIONS

1. Describe the parameters that are available to the practicing chromatographer in LC, GC and SFC to conduct method development and obtain the best resolution of the components of a mixture.
2. What is the principal requirement in SFC instrumentation to control the pressure?
3. Define the critical point of a substance.
4. Describe the change that occurs in the critical point of a substance if a second component is added.
5. What is the main advantage of SFC over LC and GC? How is this obtained?

Chapter 17

Electromigration methods: origins, principles, and applications

M. Jimidar

17.1   INTRODUCTION

Electrophoresis is a separation technique that is based on the differential migration of charged compounds in a semi-conductive medium under the influence of an electric field [1]. It started with O. Lodge (1886) who studied the migration of hydrogenium ions (H^+) in a phenolphthalein gel [2] and F. Kohlraush (1897) who described the migration of ions in saline solutions [3]. During the following period, several authors applied these concepts in their methods. However, it was not until 1937, when Arne Tiselius [4] reported the separation of different serum proteins, that the technique was recognized as a potential analytical technique. The method described was called moving boundary electrophoresis and Tiselius was rewarded with the Nobel Price of chemistry for his effort [5].

   The efficiency of the technique was enhanced further, which resulted in well-established techniques as paper electrophoresis and gel electrophoresis. During the 1950s, paper electrophoresis became quite successful, but it is now considered to be obsolete. Gel electrophoresis, on the contrary, is still being applied on a routine basis especially in biochemistry, for the determination of proteins and nucleic acids. Although the method as such is useful, as is obvious from the vast number of applications, it still suffers from some important drawbacks. First of all one has to deal with the effect of 'Joule heating', which gives rise to heat production in the system. This heat cannot be dissipated efficiently and leads to temperature gradients that result in convection and viscosity gradients and finally in an increase of dispersion and thus band broadening. Therefore, Joule heating induces a decrease in the separation

Comprehensive Analytical Chemistry 47
S. Ahuja and N. Jespersen (Eds)
Volume 47 ISSN: 0166-526X DOI: 10.1016/S0166-526X(06)47017-3

efficiency. The reproducibility also deteriorates owing to these effects as it is strongly affected by the viscosity gradients.

The power production (W) is directly proportional to the applied potential for the generation of the electric field (Eq. 17.1)

$$W = V \times I = I^2 \times R \tag{17.1}$$

where V is the applied voltage (in volts), I the current flow (in A) and R the resistance (in ohm) of the medium. For this reason, the possibility of applying high voltages is limited to a maximum of about 10 kV. As a result, the analysis times are long, usually in the order of some hours. Another drawback is that the detection method is very tedious, labor intensive and involves a lot of chemistry. The fact that it is usually not possible to develop a fully quantitative assay also reduces the applicability of the technique. The best that can be achieved in conventional electrophoresis is a semi-quantitative analysis.

Better heat-dissipating systems were found in the use of narrow bore tubes. The small volume of the tube has a large surface of the internal wall to dissipate the produced heat. The lower the ratio of the volume to surface of a tube (Eq. 17.2), the better the heat dissipation and thus higher the separation efficiency.

$$\text{Ratio} = \frac{\text{volume of the tube}}{\text{surface of internal wall}} \tag{17.2}$$

In 1967, Hjerten [6] first applied this technology using tubes made of glass with an internal diameter (I.D.) of $\pm$ 3 mm. Seven years later Virtanen [7] reported the separation of three cations in a 215 μm I.D. glass tube. In 1979, Mikkers et al. [8] reported the application of this technique with 200 μm I.D. Teflon tubes and obtained separations with plate heights less than 10 μm. Simultaneous with this application, they also provided a theoretical basis for migration dispersion in free-zone electrophoresis [9]. Although the potential of such an approach was clearly demonstrated and also theoretically well described, it was only in the early 1980s that the interest in the technique increased. Probably this was due to the development of fused silica tubing coated with polyimide to give it flexibility, a technology first developed for the communication via optical fibers [7] and used in the 1970s for gas chromatography (GC) columns.

Jorgenson and Lukacs [10–12] introduced capillary electrophoresis (CE) in 1981, as they first applied these fused silica capillaries with diameters smaller than 100 μm. By the use of 75-μm I.D. capillaries, they were able to apply voltages up to 30 kV and obtain separations with plate heights of less than $1\,\mu m$ for proteins and for dansylated amino acids. In 1984, Terabe *et al.* [13] reported the next milestone in CE. They described the method of Micellar electrokinetic capillary chromatography (MECC). It was now possible to separate both charged and neutral compounds simultaneously. From this point on, the technique developed rapidly. The advantages of CE compared to conventional electrophoresis and other chromatographic techniques include the very high-separation efficiencies that can be obtained, the low sample volume consumption, on-column detection, the low organic solvent consumption, generally fast analyzes and lower operational costs. Table 17.1 summarizes these and other characteristics of CE and compares them with conventional electrophoresis and high performance liquid chromatography (HPLC) for the separation of biomolecules [14].

Owing to its potential of performing extremely high-efficiency separations, robustness of the equipment, automation, ease of use and flexibility, electromigration methods, i.e. CE have widely been applied to different problems in analytical chemistry. It is considered to be a complementary or even an alternative technique to established chromatographic techniques such as HPLC, GC and others.

17.2 FUNDAMENTALS OF CE

The basic requirements of electrophoresis are a semi-conducting medium and an electric field. The semi-conducting medium can be a paper soaked in an electrolyte or a gel placed in an electrolyte. In the case of CE, it is a capillary filled with an electrolyte or a gel. The electric field is generated by applying a voltage difference over the capillary.

17.2.1 Electrophoretic mobility (μ_i)

A charged compound in an electric field undergoes the influence of two important forces (Fig. 17.1), namely, the electric force ($\vec{F}_{el}$) and the frictional force ($\vec{F}_{s}$) approximated by Stokes law. At steady state, these two forces (in $kg \cdot cm \cdot s^{-2}$) are in equilibrium:

$$\vec{F}_{el} \rightleftharpoons \vec{F}_{s} \tag{17.3}$$

$$Q_i \times E = 6\pi\eta r_i \vec{v}i \tag{17.4}$$

where Q_i (in C) is the net ionic charge of a compound, E (in $volts \cdot cm^{-1}$) the electric field strength, η (in $g \cdot cm^{-1} \cdot s^{-1}$) the viscosity of the medium,

TABLE 17.1

Characteristics of CE in comparison with other separation techniques for the case of biomolecular separations

Characteristics	CE	Gel electrophoresis	HPLC
Separation mechanism	Charge based (polarity)	Size and charge	Polarity partitioning; size, ion exchange
Analysis times	5–30 min	2 hrs to several days	10–120 min
Efficiency	Very good to excellent	Very good	Good
Detection	On-column optical/ electrochemical/ conductivity/MS etc.	Visible and fluorescent stains and dyes/ autoradiography	Optical/ electrochemical/ conductivity/MS etc.
Sample volume	5–50 nL	1100 μL	10–100 μL
Precision	Good to excellent	Fair	Good to excellent
Accuracy	Good to excellent	Fair to poor	Good to excellent
Absolute detection limit	Zepto* to nanomoles	Nano to micro-moles	Nano to micro-moles
Relative detection limit	ppm	ppm	ppb
Organic solvent consumption	Milliliters	—	Liters
Equipment cost	High	Moderate	Moderate to high
Cost of supplies	Low	Moderate	Moderate to high
Manual labor	Low	High	Low
Automation	High	Limited	High
Preparative	No	Yes	Yes

*10^{-21}.

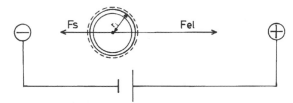

Fig. 17.1. Representation of the forces that influence a negatively charged compound in an electric field.

r_i (in cm) the hydrodynamic radius of the charged compound and $\vec{v}i$ (in $\mathrm{cm \cdot s^{-1}}$) the migration velocity of the compound in the system. From Eq. 17.4 we can derive an expression for the migration velocity in function of the system and component parameters

$$\vec{v}i = \frac{Q_i}{6\pi\eta r_i} \times E \qquad (17.5)$$

From Eq. 17.5 it can immediately be observed that the migration velocity is directly proportional to the ionic charge of a compound and the applied electric field strength. It is inversely proportional to the viscosity of the medium and the radius of the compound. The electrical field strength is determined by the applied voltage difference (V, in volts) and the length of the capillary (L, in cm) as described in Eq. 17.6. Hence, the length of the capillary and the applied voltage make the migration velocity a characteristic for one specific system. Therefore, comparisons between runs in capillaries of different lengths or with different applied voltages are impossible.

$$E = \frac{V}{L} \qquad (17.6)$$

For this reason, one utilizes a normalized velocity characteristic that is called the electrophoretic mobility (μ_i, in $\mathrm{cm^2 \cdot V^{-1} \cdot s^{-1}}$)

$$\mu_i = \frac{Q_i}{6\pi\eta r_i} \qquad (17.7)$$

In order to influence a migration it is obvious that one can alter the charge of the compounds, the viscosity of the medium and the dynamic radius of the compounds. According to Eq. 17.5, the electrophoretic mobility is the proportionality factor in the linear relationship of the migration velocity and the electric field strength

$$\vec{v}i = \mu_i \times E \qquad (17.8)$$

The charge of a compound is the major parameter of selectivity in CE. It can be adjusted by affecting the ionization of a compound depending on the degree of dissociation (α_i). For this reason, the migration velocity is also directly proportional to α_i. The proportionality factor in the relationship of the migration velocity and the electric field strength in such a case is called the effective electrophoretic mobility (μ_{eff}) and the migration velocity as the effective migration velocity (v_{eff})

$$\vec{v}_{\mathrm{eff}} = \alpha_i \times \mu_i \times E = \mu_{\mathrm{eff}} \times E \qquad (17.9)$$

From the migration velocity of a compound and the migration distance one can derive its migration time (t_i). The migration distance (l) in CE is equal to the length of the capillary from the inlet (injection site) to the point of detection.

$$t_i = \frac{l}{v_{\text{eff}}} = \frac{l}{\mu_{\text{eff}} \times E} = \frac{l \times L}{\mu_{\text{eff}} \times V} \tag{17.10}$$

The migration time increases with increasing capillary length and reduces with increasing mobilities of the compounds and applied voltages.

17.2.2 Electroosmotic mobility (μ_{eof})

Electroosmosis or electroendosmosis is the bulk movement of the solvent (electrolyte solution) in the capillary caused by the zeta (ζ) potential at the wall/water interface of the capillary. Any solid–liquid interface is surrounded by solvent and solute constituents that are oriented differently compared to the bulk solution. Figure 17.2 illustrates a model of the wall–solution interface of the widely applied capillaries. Owing to the nature of the surface functional groups, in silica capillaries the silanol groups, the solid surface has an excess of negative

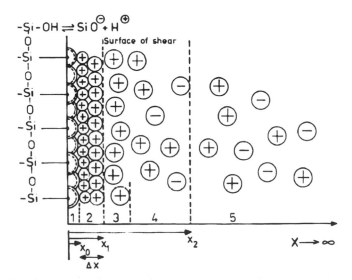

Fig. 17.2. The distribution of charges at the internal wall of a silica capillary. x is the length in cm from the center of charge of the negative wall to a defined distance, 1 = the capillary wall, 2 = the Stern layer or the inner Helmholtz plane, 3 = the outer Helmholtz plane, 4 = the diffuse layer and 5 = the bulk charge distribution within the capillary.

charge [15]. The anionic wall attracts counter ions from the bulk solution in the capillary. This results in a different distribution of charged
compounds in the neighborhood of the wall when compared to the bulk
solution. The distribution is such that a double layer of counter ions (in
this case cations) is formed adjacent to the wall. A fraction of the cations adsorb strongly to the wall, resulting in an immobilized compact
layer of tightly bound cations, the solvation sheath has been stripped
off. This layer is also called the Stern layer or inner Helmholtz plane.
Other cations arrange themselves in a mobile, loosely held layer of
solvated ions. This layer is known as the diffuse layer. The initial part
of the diffuse layer, at the side of the Stern layer, is known as the outer
Helmholtz plane [16,17]. Because of the Coulomb forces, this distribution of charges at the interface forms a potential field as is shown in
Fig. 17.3. The potential at the surface of the wall is given by [18]

$$\psi_0 = \frac{Q_w}{\varepsilon \times x_o} \tag{17.11}$$

where Q_w represents charge of the wall, ε is the dielectric constant of
the medium (in Farad $\cdot$ m^{-1} or A $\cdot$ s $\cdot$ V^{-1} $\cdot$ cm^{-1}) and x_o the distance from
the surface of the wall to the center of the charge. This surface potential

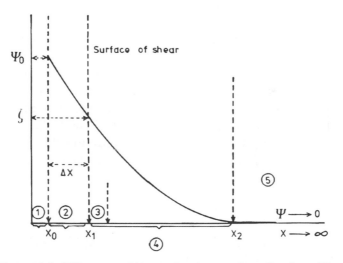

Fig. 17.3. Potential difference (ψ) at the internal wall of a silica capillary,
because of the distribution of charges. x is the length in cm from the center of
charge of the negative wall to a defined distance, ζ the zeta potential, $1 =$ the
capillary wall, $2 =$ the Stern layer or the inner Helmholtz plane, $3 =$ the outer
Helmholtz plane, $4 =$ the diffuse layer and $5 =$ the bulk charge distribution
within the capillary.

decreases with the distance in the double layer to reach a zero value at the bulk distribution of ions in the capillary.

The potential drop is exponential and can be expressed according to Gouy–Chapman as

$$\psi_x = \psi_0 \exp(-\kappa x) \tag{17.12}$$

where ψ_x represents the potential at a distance x and κ is the De-bye–Huckel constant

$$\kappa = \frac{8\pi e^2 n_o z^2}{\varepsilon k T} \tag{17.13}$$

Here e is the electronic charge, n_o the bulk concentration of ionic species (ionic strength), z the valence of the electrolyte, k the Boltzmann constant and T the absolute temperature.

The potential at the surface of shear at a distance $x_1 = x_o + \Delta x$ from the wall, Δx being the statistical average of the thickness of the hydration (solvation) shell, is called the electrokinetic or ζ potential [18,19]. The surface of shear coincides with the outer Helmholtz plane at the boundary of the Stern layer. The ζ potential governs the movement of ions in the diffuse layer when an electric field is applied. The irregular distribution of ions in the capillary results in an accumulation of cations close to the wall, compared to the bulk distribution of ions. The cations in the diffuse layer are set into motion in the direction of the cathode. Because of their solvation, they also carry the liquid along with them, resulting in the bulk movement of the solvent through the capillary, known as the electroosmotic flow (EOF). The velocity of the EOF rises from a zero value at the surface of shear (outer Helmholtz plane) to a limiting value (electroosmotic velocity $\vec{v}_{eof}$) at an infinitive distance from the wall where the distribution of the cations equals to that of the bulk.

According to the Helmholtz–Von Smoluchowski [2,16–18,20,21] equation, the electroosmotic velocity, $\vec{v}_{eof}$, is related to the ζ potential in the following way:

$$\vec{v}_{eof} = \frac{\varepsilon \zeta}{4\pi \eta} \times E \tag{17.14}$$

The electroosmotic velocity is also characterized by a mobility, namely, the electroosmotic mobility (μ_{eof}).

$$\mu_{eof} = \frac{\varepsilon \zeta}{4\pi \eta} \tag{17.15}$$

From Eq. 17.15 it can be derived that μ_{eof} can be adjusted by changing the dielectric constant and/or the viscosity of the medium, but also the ζ

potential is mainly influenced by the distribution of charges at the capillary wall. Every factor that affects this distribution can be utilized to adjust the velocity of the EOF. The pH for instance has an influence on the ionization of the surface silica groups. As ζ is directly dependent on the charge density of the wall, the relationship of μ_{eof} in function of the pH follows the behavior of the dissociation of the silanol groups. A sigmoid curve resembling a titration curve is observed and different capillary materials result in different profiles of the electroosmotic mobility in function of the pH (owing to differences in ζ [19,22].

From Eqs. 17.12 and 17.13 it can be derived that the ζ potential and the thickness of the double layer, which is referred to as $1/\kappa$, decreases rapidly with an increase in ionic strength or the valence of the electrolytes in the capillary. Therefore, the ionic strength and the nature of the ions of the electrolyte solution are also very important. Careful control of the ionic strength leads to a better reproducibility of the EOF [21]. The direction of the EOF can be reversed when the charge of the wall is changed. With a silica capillary this is easily achieved by adding a cationic surfactant, such as cetyltrimethylammonium bromide (CTAB), to the buffer electrolyte. By the addition of specific modifiers, it is also possible to reduce the EOF or even inhibit it. These effects are also obtained by modifying the internal wall of the capillary by derivatization or by applying a radial external field [2].

An important feature of the EOF is its flat flow profile, compared to the parabolic profile of hydrodynamic flows (Fig. 17.4). The reason for this characteristic is that the EOF originates almost at the wall of the capillary, owing to the extremely small size of the double layer

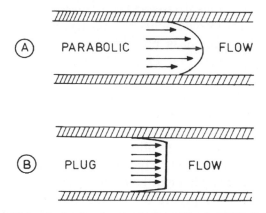

Fig. 17.4. Flow profiles of a hydrodynamic flow like in HPLC (A) and of electroosmosis like in CE (B).

(thickness ranges from 3 to 300 nm). Flat profiles in capillaries are expected when the radius of the capillary is greater than seven times the double layer thickness [14,23]. The flat profile is favorable in terms of peak dispersion and has a major contribution to the high-separation efficiency of CE.

17.2.3 Electrophoretic migration

The migration in CE is obviously influenced by both the effective and the electroosmotic mobility. Therefore, the proportionality factor in the relationship of the migration velocity and the electric field strength in such a case is called the apparent electrophoretic mobility (μ_{app}) and the migration velocity the apparent migration velocity ($\vec{v}_{app}$). The μ_{app} is equal to the sum of μ_{eff} and μ_{eof}. For this reason, the apparent migration velocity is expressed as

$$\vec{v}_{app} = \mu_{app} \times E = (\mu_{eff} + \mu_{eof}) \times E \tag{17.16}$$

The migration time is also determined by the apparent migration velocity

$$t_i = \frac{1}{\vec{v}_{app}} = \frac{l}{(\mu_{eff} + \mu_{eof}) \times E} = \frac{l \times L}{(\mu_{eff} + \mu_{eof}) \times V} \tag{17.17}$$

As the EOF in a silica capillary is directed toward the cathode, the apparent migration velocity of cations undergoes a positive effect of the EOF, while the migration of anions is negatively affected. The EOF also transports the neutral compounds through the capillary toward the cathode. In conditions where the magnitude of μ_{eof} is greater than μ_{eff}, anions that originally migrate toward the anode, are still carried toward the cathode due to a positive apparent velocity. When the velocity of the EOF is sufficiently high, it is possible to separate both cations and anions in one single run (Fig. 17.5). The neutral compounds migrate unresolved under one peak in the electropherogram, between the cations and anions with the velocity of the EOF.

The selectivity of a separation is determined by the effective mobility because the effect of the electroosmotic mobility is equal for all the sample constituents. In order to obtain μ_{eff} from μ_{app}, knowledge of the magnitude of μ_{eof} is required. Therefore, it is necessary to measure the velocity of the EOF. This can be done by several methods [2]; however, the procedure of applying a neutral marker is commonly used. The neutral marker is a neutral compound and thus migrates only because of the EOF. Its migration velocity represents the velocity of the

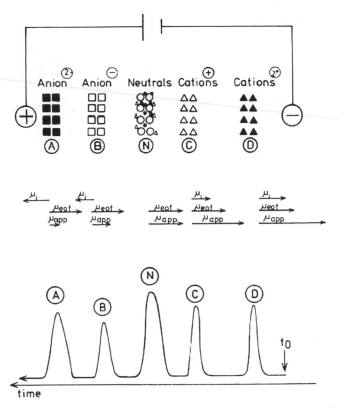

Fig. 17.5. The migration of cations, anions and neutral compounds in CZE in an ordinary fused silica capillary.

EOF when no disturbing processes such as wall adsorption occurs. The practically obtained migration time of the neutral marker (T_{eof}) can be utilized to calculate the velocity of the EOF:

$$\vec{v}_{eof} = \frac{l}{T_{eof}} = \frac{l \times L}{\mu_{eof} \times V} \qquad (17.18)$$

This procedure can be risky as it does not consider the problem of a possible adsorption of the neutral marker on the capillary wall. Nevertheless, it is widely used to determine the mobility of the EOF. The effective mobility of a compound can now be calculated from the practically obtained migration time (T_m) of that compound and the migration time of the neutral marker:

$$\mu_{eff} = \frac{l \times L}{V} \times \left(\frac{1}{T_m} - \frac{1}{T_{eof}} \right) \qquad (17.19)$$

585

17.2.4 Efficiency

A general goal in chromatography and related separation techniques such as electrophoresis is to obtain narrow bands in the final chromatogram or electropherogram, in order to achieve a good separation between two compounds with close mobilities. Whether the bands in the electropherogram of a CE separation will be narrow, is determined by dispersion of the migrating solute zones in the capillary. CE is a dynamic process, therefore dispersion effects are bound to occur. The extent of dispersion is determined by the efficiency of the separation system. This is usually expressed in terms of the number of theoretical plates (N) or the height equivalent to a theoretical plate (HETP). The number of theoretical plates may be defined as follows:

$$N = \frac{l^2}{\sigma^2} \tag{17.20}$$

where l is the migration distance of a zone (length of the capillary from injection to the point of detection) and σ the standard deviation of the zone distribution, which is assumed to be Gaussian. The HETP can be calculated from the number of theoretical plates as shown in Eq. 17.21 [3]. The number of theoretical plates is used more often in the literature to express the separation efficiency.

$$\text{HETP} = \frac{l}{N} \tag{17.21}$$

The standard deviation of the 'Gaussian' zones expresses the extent of dispersion and corresponds to the width of the peak at 0.607 of the maximum height [24,25]. The total system variance (σ_{tot}^2) is affected by several parameters that lead to dispersion (Eq. 17.22). According to Lauer and McManigill [26] these include: injection variance (σ_I^2), longitudinal (axial) diffusion variance (σ_L^2), radial thermal (temperature gradient) variance (σ_T^2), electroosmotic flow variance (σ_F^2), electrical field perturbation (electrodispersion) variance (σ_E^2) and wall-adsorption variance (σ_W^2). Several authors [9,24,27–30] have described and investigated these individual variances further and have even identified additional sources of variance, like detection variance (σ_D^2), and others (σ_O^2).

$$\sigma_{tot}^2 = \sigma_I^2 + \sigma_L^2 + \sigma_T^2 + \sigma_F^2 + \sigma_E^2 + \sigma_W^2 + \sigma_D^2 + \sigma_O^2 \tag{17.22}$$

It must be mentioned that the individual variances in Eq. 17.22 cannot be suppressed to a zero value as they are inherent to the principle of the technique. It is possible, however, to control the contributions of these sources of variance by proper instrumental design and selection of working conditions.

Injection variance. The injection variance is considered to be a component of the extra-column zone-broadening effects and is related to the width (w_{inj}) of the ideal rectangular sample plug (volume). The variance is given by

$$\sigma_I^2 = \frac{w_{inj}^2}{12} \tag{17.23}$$

The parameters that influence the injection plug width will be discussed in a further section of this chapter. In practice, it is necessary to keep the width of the injection plug smaller than 1% of the total length of the capillary, in order to avoid column overloading [27].

Detection variance. Because of on-column detection, the variance of the detection process is reduced to the variance related with the width of the detector cell (w_{det}). It is expressed similarly to injection as

$$\sigma_D^2 = \frac{w_{det}^2}{12} \tag{17.24}$$

According to the literature [29], the contribution of the detection variance to the total system variance is usually below 0.01%.

Axial diffusion variance. The variance due to longitudinal diffusion is expressed by the Einstein equation [24,27–30]

$$\sigma_L^2 = 2 \times D_i \times t_i \tag{17.25}$$

where D_i represents the diffusion coefficient of a solute (in $m^2 \cdot s^{-1}$) and t_i is its migration time. By substituting t_i in Eq. 17.25 with Eq. 17.17, one obtains

$$\sigma_L^2 = 2 \times D_i \times \frac{l \times L}{(\mu_{eff} + \mu_{eof}) \times V} \tag{17.26}$$

From Eq. 17.26 it is clear that the variance due to longitudinal diffusion is negatively influenced by the length of the capillary and the diffusion coefficient of the solute. However, it is positively affected by the applied potential and the apparent mobility of a solute. According to this equation, fast migrating zones will show less variance due to axial diffusion.

Radial thermal variance. When a potential difference is applied over an electrolyte solution, a current flow is generated between the electrodes in the solution. The current gives rise to Joule heating that leads to a parabolic zone deformation, owing to the formation of a radial temperature gradient [23]. In aqueous media, the viscosity shows a strong dependence on the temperature. A radial temperature gradient will result in a corresponding radial viscosity gradient. As discussed previously (Eq. 17.7), the electrophoretic mobility is a function of the viscosity of the medium. Therefore, in conditions of a radial viscosity gradient, a radial velocity gradient of the migrating solute is observed. It is established that the mobility of a solute increases by 2% when the temperature is increased by 1°C.

The temperature difference (ΔT) between the center of an open capillary and its wall is given by

$$\Delta T = \frac{E^2 \lambda c d_c^2}{16 \kappa_t} \tag{17.27}$$

where E corresponds to the electric field strength, λ is the equivalent conductivity of the electrolyte, c represents the molar concentration, d_c is the inner diameter of the capillary and κ_t represents the thermal conductivity of the medium. Eq. 17.27 clearly illustrates the dependence of the radial distribution of the temperature on the inner diameter of the capillary. It is therefore obvious that d_c has to be reduced ($< 100\,\mu$m) in order to apply the desired high potentials (E). In such a way, ΔT is reduced and radial zone distribution is controlled [22]. Nevertheless, an effective thermostating device is necessary to control the temperature difference between the wall of the capillary and the surrounding environment (Fig. 17.6). The difference in temperature at this interface can increase significantly and therefore requires a temperature-controlling device. Temperature control is performed by cooing

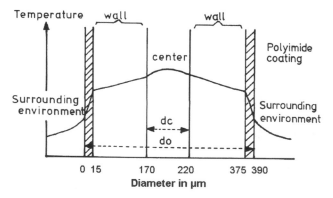

Fig. 17.6. The distribution of the temperature in the capillary and the surrounding environment, d_c and d_o represent the inner and outer diameter of the capillary, respectively. Modified with permission from Ref. [5].

with air or liquid circulation and is a must for obtaining reproducible peak areas and migration times [31].

Electroosmotic flow variance. As the flow profile of the EOF is flat, almost like a piston, its contribution to dispersion of the migrating zones is small. The EOF positively contributes to the axial diffusion variance, as can be derived from Eq. 17.26, when it moves in the similar direction as the analytes. However, when the wall of the capillary is not uniformly charged, local turbulence can occur and cause irreproducible dispersion [27].

Electrical field perturbation variance. It is well established in HPLC that when the injected sample solution has a different eluting strength than that of the mobile phase, peak deformation is bound to occur. A similar effect is observed in CE. Instead of eluting strengths, it concerns here differences in conductivity between the sample zone and the bulk electrolyte in the capillary [9,32]. The conductivity (γ, Ohm$^{-1}\cdot$m^{-1}) of a solution is given by the cumulative effect of the contributions of different ions:

$$\gamma = F \times \sum_i c_i \times \mu_i \times Z_i \tag{17.28}$$

where F is the Faraday constant and Z_i represents the charge of an ion. The conductivity of an injected sample may differ from that of the bulk electrolyte in the capillary. In order to maintain a constant current flow [9], the electric field strength in the sample zone and the bulk electrolyte is adjusted according to

$$I = \frac{\pi d_c^2}{4} \times E_i \times \gamma_i = \frac{\pi d_c^2}{4} \times E_b \times \gamma_b \tag{17.29}$$

where I represents the current (in A), E_i and γ_i correspond to the electric field strength and the conductivity in the sample zone, respectively, while E_b and γ_b represent the same for the bulk electrolyte. As can be seen in Eq. 17.28, the conductivity and thus the field strength (Eq. 17.29) is strongly dependent on the concentration (ionic strength), the electrophoretic mobility and the ionic charge of the ions in the capillary. Differences in field strength between the sample zone and the bulk electrolyte lead to band distortion, especially at the boundaries. For example, when the field strength in the sample zone is higher than in the bulk electrolyte, the ions in the sample zone have a higher migration velocity than those in the bulk electrolyte. The sample constituents migrate fast but are slowed again ('stacked') at the border of the sample zone, close to the bulk electrolyte. Therefore, the distribution of the analyte zone is skewed with a higher density at the side rather than at the top of the zone. Sometimes this effect is advantageously employed as a concentration step (this is discussed in a further section).

Three situations may occur regarding the migration velocities of the ions in the sample zone (v_s) and in the bulk electrolyte (v_b), as is shown in Fig. 17.7:

(a) $v_s < v_b$, the conductivity of the sample zone is higher than that of the buffer electrolyte. In other words, the electric field strength in the sample zone is smaller than in the buffer electrolyte. Under

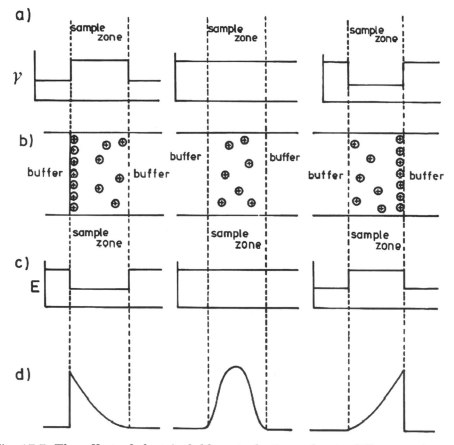

Fig. 17.7. The effect of electric field perturbations, due to differences in conductivity between the sample- and the buffer electrolyte zone on the shape of the peaks, (a) conductivity distribution, (b) sample ions distribution, (c) electric field strength perturbations, and (d) effect on the peak shapes.
Adapted with courtesy of Dr. D. Heiger of Agilent Technologies Inc. from Ref. [5].

these conditions, the top of the zone migrates faster than the back flank. A tailing peak with a steep front side is observed.

(b) $v_s = v_b$, the conductivity in the sample zone is equivalent to that of the buffer electrolyte. This occurs in the ideal situation when there are no electric field perturbations present. The migrating zones do not show electrodispersion and the peaks are symmetrical.

(c) $v_s > v_b$, the conductivity of the sample zone is lower than in the buffer electrolyte. The electric field strength in the sample zone is therefore higher than in the buffer electrolyte; in such a case, the front flank of the sample zones migrates faster than the top. A fronting peak with a steep backside is observed.

There are several ways to control electrodispersion. It can be avoided by increasing the bulk electrolyte concentration (limited by Joule heating) or by diluting the sample (limited by the sensitivity of the detection system). Mikkers et $al.$ [9] have demonstrated that when the sample concentration is smaller than the bulk electrolyte concentration with a factor of 100, electrodispersion due to concentration differences can be neglected. Besides differences in concentrations, the electrophoretic mobility and the ionic charge are also important. Proper selection of the bulk electrolyte consisting of ions with matching ionic charge and mobilities with the analytes is therefore crucial. The phenomenon of electrodispersion is sometimes advantageously used to enhance the detectability of analytes. By selecting appropriate conditions, it is possible to concentrate the analytes in a narrow band by using the stacking effect.

Wall-adsorption variance. Analyte–wall interactions are possible in fused silica capillaries depending on the characteristics of the wall and the analytes. Usually it concerns an ionic interaction mechanism, but sometimes hydrophobic interactions are also possible. Wall adsorption is frequently observed in the analysis of large molecules [26,33], but is also observed in separations of small ions [34]. The variance due to wall adsorption is given by the following expression [29]:

$$\sigma_W^2 = \frac{2(l - \theta)^2}{\tau} \times \mu \times l \times E \qquad (17.30)$$

where θ represents the fraction of free particles (not adsorbed) and τ the mean free lifetime of a particle. The effect of analyte–wall adsorption can be reduced by proper selection of the capillary material, the buffer pH, ionic strength, wall surface deactivation, the use of lower sample concentrations and capillaries of larger inner radii, because this reduces the surface area to volume ratio (inverse of Eq. 17.2). However, there is no single universal approach in finding a solution for wall adsorption [26,29,30].

Other sources of variance. The brand broadening of a zone can also be caused by other sources that are not yet identified or not considered to be important at this stage of knowledge. For example, it has been shown that the coiling of a capillary in a cassette or on a spool, also has an effect on zone broadening.

N and HETP. Under optimal working conditions [2,10,22,35], σ_T^2 (proper thermostating), σ_F^2 (proper capillary rinsing) and σ_E^2 (matching of sample and buffer electrolyte conductivity), σ_W^2 (proper capillary selection) are under control. When the influence of σ_O^2 can also be neglected, then σ_{tot}^2 is expressed as

$$\sigma_{tot}^2 = \sigma_I^2 + \sigma_D^2 + \sigma_L^2 \tag{17.31a}$$

$$= \frac{w_{inj}^2}{12} + \frac{w_{det}^2}{12} + 2 \times D_i \times \frac{l \times L}{(\mu_{eff} + \mu_{eof}) \times V} \tag{17.31b}$$

From Eqs. 17.31a and 17.31b it can be derived that when the amount injected is very small, the detection path length very narrow and the capillary diameter is also very small, only σ_L^2 influences the broadening of a migrating zone. In such a case, σ_{tot}^2 is approximated by σ_L^2 and Eq. 17.31b can be simplified to

$$\sigma_{tot}^2 = \sigma_L^2 = 2 \times D_i \times \frac{l \times L}{(\mu_{eff} + \mu_{eof}) \times V} \tag{17.32}$$

By substituting σ_{tot}^2 in Eq. 17.20 with Eq. 17.32, one obtains an expression for the number of theoretical plates:

$$N = \frac{l^2}{2 \times D_i \times \frac{l \times L}{(\mu_{eff} + \mu_{eof}) \times V}} \tag{17.33}$$

$$= \frac{(\mu_{eff} + \mu_{eof}) \times V \times l}{2 \times D_i \times L} \tag{17.34}$$

$$= \frac{(\mu_{eff} + \mu_{eof}) \times E \times l}{2 \times D_i} \tag{17.35}$$

From Eq. 17.35 it can be observed that N is directly proportional to the applied field strength, the migration distance from injection to detection and the apparent mobility of a solute. The diffusion coefficient and the length of the capillary are inversely related to the efficiency.

The height of a theoretical plate can also be calculated by substituting N in Eq. 17.21 with Eq. 17.34:

$$HETP = \frac{2D_iL}{(\mu_{\text{eff}} + \mu_{\text{eof}}) \times V}$$
(17.36)

17.2.5 Selectivity and resolution

The selectivity of separation is mainly affected by parameters of the bulk electrolyte in the capillary. These include type of anion and cation, pH, ionic strength, concentration, addition of modifiers such as complexing agents, organic solvents, surfactants, etc. It is expressed in terms of mobility differences $(\Delta\mu)$ or the mobility ratio's (α):

$$\Delta\mu = \mu 1 - \mu 2$$
(17.37)

$$\alpha = \frac{\mu 1}{\mu 2}$$
(17.38)

In general, the resolution (Rs) between two peaks in chromatographic and related separation techniques is determined by the efficiency of the column and the selectivity of the separation [10]:

$$Rs = \frac{1}{4} \times N^{1/2} \times \frac{\Delta v}{v_a}$$
(17.39)

where Δv and v_a represent the difference in velocities and the average velocity of the two migrating bands of interest, respectively. In CE this is expressed as

$$Rs = \frac{1}{4} \times N^{1/2} \times \frac{\Delta\mu}{(\mu_a + \mu_{\text{eof}})}$$
(17.40)

where $\Delta\mu$ and μ_a represent the difference in effective mobilities and the average effective mobility of the two migrating bands of interest, respectively. By substituting Eq. 17.35 for N in Eq. 17.40, one obtains the expression for the resolution

$$Rs = \frac{1}{4} \times \left[\frac{(\mu_{\text{eff}} + \mu_{\text{eof}}) \times E \times l}{2 \times D_i}\right]^{1/2} \times \frac{\Delta\mu}{(\mu_a + \mu_{\text{eof}})}$$
(17.41)

Eq. 17.41 is worked out in

$$Rs = 0.177 \times \Delta\mu \times \left[\frac{E \times l}{D_i \times (\mu_a + \mu_{\text{eof}})}\right]^{1/2}$$
(17.42)

Eq. 17.42 is the expression of the resolution for CE in electrophoretic terms. However, the application of this expression for the calculation of Rs in practice is limited because of D_i. The diffusion coefficient of different compounds in different media is not always available. Therefore, the resolution is frequently calculated with an expression that employs the width of the peaks obtained in an electropherogram. This way of working results in resolution values that are more realistic as all possible variances are considered (not only longitudinal diffusion in Eq. 17.42). Assuming that the migrating zones have a Gaussian distribution, the resolution can be expressed as follows:

$$Rs = 1.177 \times \left[\frac{T_{m2} - T_{m1}}{W_{1/2(1)} + W_{1/2(1)}} \right] \qquad (17.43)$$

where T_{m1}, T_{m2} and $W_{1/2(1)}$, $W_{1/2(2)}$ represent the migration times (in min) and the width of the peaks at half height (in min), respectively. The width of the peaks can be determined at different parts of the full height. In these cases the proportionality factor will change. For example, for the width at baseline, this value is equal to 2 instead of 1.177.

17.3 INSTRUMENTATION

In CE the semi-conducting medium consists of two electrolyte (buffer)-containing compartments connected with each other by a capillary (Fig. 17.8). The electric field in the capillary is generated by applying a voltage difference over the two electrolyte compartments by placing electrodes in the compartments and connecting them with a power supply. The detector is placed on a distance downstream from the inlet of the capillary and is usually connected with a data-processing system. The capillary, detection system (on-column), electrolyte compartments and sample vials (carousel) are housed in a thermostated compartment. This is the basic instrumental setup of CE.

The first commercial CE instruments were introduced in 1988 by Applied Biosystems (Foster City, CA) and Beckman Coulter (Fullerton, CA). The main challenge during these days was to find out how to overcome poor reproducibility and improve separation efficiency. In the early 1990s, many instrument-building companies introduced CE systems (e.g. Isco (Lincoln, NE), Bio-Rad (Hercules, CA), Waters (Milford, MA), Applied Biosystems, ThermoQuest (Santa Fe, NM) and Dionex (Sunnyvale, CA)), but at the end of the decade only Beckman Coulter

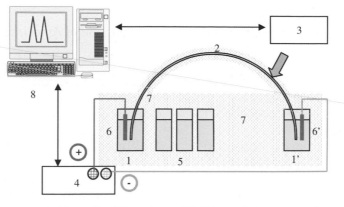

Fig. 17.8. Representation of a CE system. (1) Electrolyte compartments, (2) the capillary, (3) detector, (4) power supply, (5) sample carousel, (6) electrodes, (7) thermostatted areas and (8) data workstation.

and Agilent Technologies (Wilmington, DE) remained as major manufacturers of single-capillary instruments.

As is shown in Table 17.2, current CE systems can be categorized in three groups: (i) single-capillary-, (ii) capillary array- and (iii) chip-based instruments [36]. The single-capillary instruments are for general purpose applications, whereas the capillary array instruments are mainly applied for DNA sequencing because of the high-throughput capability. More recently, chip-based instruments are introduced for specific applications such as DNA sequencing. It is expected that the use of this type of instruments will grow in the future, covering other specific application methods.

17.3.1 Sample injection

Since capillaries have very small diameters, the injection volumes in CE are extremely small. Injection volumes in the order of 10–50 nL are commonly applied (a fog droplet is ± 10 nL). Several approaches have been applied for the injection of such small volumes of sample into the capillary. These included the use of rotary-, split- and micro-injectors, electrokinetic and hydrodynamic injection. Although all these injection techniques have shown to be quite appropriate, electrokinetic and hydrodynamic injection are mostly applied. All the recent commercial

TABLE 17.2

Instrumentation for CE (reproduced with permission of American Laboratory [36])

Manufacturer	Instrument	Application	Features
Single-capillary instruments			
Agilent Technologies	Agilent CE	General purpose	Diode array detection
Applied Biosystems	Prism 310 Genetic Analyzer	DNA chemistries	Single capillary with LIF detection
Beckman Coulter	MDQ	General purpose	96-well sampling, numerous detectors
CE Resources (Kent Ridge, Singapore)	CE P1	Portable	Potential gradient detection
	CEL1	General purpose	Many detectors available
Convergent Bioscience (Mississaqua, Toronto, Canada)	iCE280	Isoelectric focusing	Whole capillary imaging
Dycor (Edmonton, Canada)	VEGA	Research in bio/chem warfare	Epillumination fluorescence microscope
Micro-tek Scientific (Sunnyvale, CA)	Ultra-Plus II	CEC, μLC, CZE	Gradient elution
Prince Technologies (Emmen, The Netherlands)	Various models	General purpose	Modular system
Unimicro Technologies (Pleasanton, CA)	Trisep-100	CEC, μLC, CZE	Gradient elution, several detectors
Capillary array instrumentation			
Applied Biosystems	Prism 3100 Avant	Many DNA applications	4 capillaries, upgradeable
	Prism 3100 Genetic Analyzer	Many DNA applications	16 capillaries
	3700 Genetic Analyzer	Many DNA applications	96 capillaries

continued

TABLE 17.2 (*continued*)

Manufacturer	Instrument	Application	Features
Beckman	CEQ2000XL	DNA sequencing, fragments	8 capillaries
	Paragon	Serum proteins	7 capillaries
CombiSep (Ames, IA)	MCE 2000	Combinatorial screening amino acids p*K*a	96 capillaries, UV detector
Molecular Dynamics (Sunnyvale, CA)	MegaBACE 4000	Sequencing, genetic analysis	384 capillaries
	MegaBACE 1000	Sequencing, genetic analysis	96 capillaries
	MegaBACE 500	Sequencing, genetic analysis	46 capillaries
Spectrumedix (State College, PA)	SCE 9610	High-throughput, genetic analysis	96 capillaries
Chip-based instrumentation			
Agilent Technologies	Agilent 2100	DNA, RNA fragments	12 runs/chip SDS, proteins
Caliper Technologies (Mountain View, CA)	AMS 90 SE	DNA fragments	Reusable separation channel

instruments provide these two injection modes as standard methods [1,2,37].

17.3.1.1 Electrokinetic injection

Electrokinetic (also called electromigration) injection is performed by placing the inlet of the capillary and an electrode in the sample vial. Following this a voltage is applied during a defined period of time. The sample constituents are actively carried into the capillary, and when present, the EOF also passively carries them into the capillary. For this reason, neutral compounds are also injected. The active migration is due to the effective electrophoretic mobilities of the constituents. The amount (B), in units of concentration injected into the capillary is expressed by [2,38]

$$B = C \times \left[\frac{(\mu_{eff} + \mu_{eof})\pi r_c^2 V_{inj} t_{inj}}{L} \right] \tag{17.44}$$

where C is the concentration of the compound, r_c the radius of the capillary, V_{inj} the applied injection voltage and t_{inj} the time period of injection. By the accurate control of the injection voltage and the injection time, one can vary the injection amount and maintain precision of the injection. However, the electroosmotic mobility is usually difficult to control. As discussed previously, it is affected by several parameters, like the viscosity, ζ potential, dielectric constant, etc. This can lead to a high-injection variance, which can be corrected by applying the approach of internal standardization. Huang *et al.* [39], additionally, reported that electrokinetic injection is biased in two ways. Owing to mobility differences, the amount of a sample constituent injected does not represent the amount of that compound in the sample, unless μ_{eof} greatly exceeds μ_{eff}. The fastest moving ions are injected the most; therefore, the first peak will have a higher area under the peak compared to the rest of the peaks in the electropherogram. In order to avoid this problem, they suggested to calculate a bias factor (b), which is the migration times ratio of the second and the first peak. This ratio relates the apparent ratio of the injected amounts (given by the area ratio of first and second peak) to the ratio of the concentrations in the sample solution as follows:

$$b = \frac{T_{m2}}{T_{m1}} = \frac{\text{area first peak}}{\text{area second peak}} \tag{17.45}$$

Another kind of bias is related to the electrical resistance (conductivity) of the medium in which the sample constituents are dissolved. Owing to differences in conductivity, both the electrophoretic as well as the electroosmotic mobility is affected. Therefore, injections from samples with different conductivity are bound to show this bias [39].

17.3.1.2 Hydrodynamic injection

Hydrodynamic injections are performed by placing the capillary inlet in the sample vial and applying a pressure difference across the capillary. The pressure difference can be generated by applying either a positive or a negative pressure. The positive pressure is applied to the vial containing the sample, pushing it into the capillary. The negative pressure (vacuum) is applied on the opposite end of the capillary, drawing the sample from the sample vial into the capillary. The amount (B) and the volume (ϑ_{inj}) [40] injected can be calculated from the Poiseuille law and are given by Eqs. 17.46 and 17.47, respectively.

$$B = C \times \frac{\Delta P \pi r_c^4 t_{inj}}{8 \eta L} \tag{17.46}$$

$$\vartheta_{inj} = \frac{\Delta P \pi d_c^4 t_{inj}}{128 \eta L} \tag{17.47}$$

where ΔP (in $Nw \cdot m^{-2}$ or Pascal or $kg \cdot m^{-1} \cdot s^{-2}$) is the applied pressure difference. The volume and the amount injected are varied by changing the injection time and/or the pressure difference. The length of the injection time can be controlled accurately, but it is often difficult to fully control ΔP.

Hydrodynamic injection can also be performed by using gravity to generate ΔP [41]. This injection mode is also called hydrostatic injection. The inlet of the capillary is placed in the sample vial and this is then raised during a period of time, creating a difference in height (Δh, in cm) between the inlet and the outlet of the capillary. The sample enters the capillary by siphoning. The amount and volume injected are derived from Eqs. 17.46 and 17.47, respectively, after substitution of ΔP with Eq. 17.48.

$$\Delta P = \rho \times g \times \Delta h \tag{17.48}$$

where ρ represents the density of the liquid in the capillary (in $g \cdot ml^{-1}$) and g is the gravitational constant. Since it is possible to accurately control both the height and the injection time, the injection precision tends to be better compared to positive or negative pressure injections [37] and especially compared with electrokinetic injection [28].

It is important to note that with hydrodynamic injection a volume of sample is injected, which has representative amounts (concentration) of the sample constituents. This characteristic, together with a better precision, makes it the most widely used injection technique in CE. However, owing to the poor detectability in terms of concentration of the widely used detectors (see further), its application is rather limited to the analysis of high-concentration samples. In trace analysis, for example, electromigration is favored over hydrodynamic injection [42].

17.3.2 Detection

Detection in CE is performed on-column and frequently with optical detection systems. On-column detection minimizes zone broadening as it avoids connection devices or fittings. The detection systems in CE were copied from established separation techniques (especially liquid chromatographic techniques) and slightly modified for adjustment. On-column detection with UV–VIS- and fluorescence detectors has an important implication on the path length of the detection cell. The dimensions of the detection cell are limited by the inner diameter of the capillary (less than $100\,\mu m$). This, together with the small-injection volume (some nl), affects the sensitivity of the detection in terms of concentration (molar). In terms of mass detection, the sensitivity is very good (Table 17.3) [5].

Several kinds of detection systems have been applied to CE [1,2,43]. Based on their specificity, they can be divided into 'bulk property' and 'specific property' detectors [43]. Bulk-property detectors measure the difference in a physical property of a solute relative to the background. Examples of such detectors are conductivity, refractive index, indirect methods, etc. The specific-property detectors measure a physico-chemical property, which is inherent to the solutes, e.g. UV absorption, fluorescence emission, mass spectrum, electrochemical, etc. These detectors usually minimize background signals, have wider linear ranges and are more sensitive. In Table 17.3, a general overview is given of the detection methods that are employed in CE with their detection limits (absolute and relative).

The most applied detectors are based on UV or fluorescence detection, both direct and indirect. For this purpose, a window has to be made on the polyimide coating of the fused silica capillary, simply by

TABLE 17.3

Detection systems applied in CE and their limits (for approximately 10 nl injection volumes). Adapted from [5,15]

Method	Concentration detection limit (molar)	Mass detection limit (moles)
UV—VIS	10^{-5}–10^{-8}	10^{-13}–10^{-16}
Fluorescence	10^{-7}–10^{-9}	10^{-15}–10^{-17}
Amperometry	10^{-10}–10^{-11}	10^{-18}–10^{-19}
Indirect UV	10^{-3}–10^{-6}	10^{-11}–10^{-14}
Indirect fluorescence	10^{-5}–10^{-7}	10^{-13}–10^{-15}
Indirect amperometry	10^{-8}–10^{-9}	10^{-16}–10^{-17}
LIF	10^{-14}–10^{-17}	10^{-18}–10^{-21}
Conductivity	10^{-7}–10^{-8}	10^{-15}–10^{-16}
MS	10^{-8}–10^{-9}	10^{-16}–10^{-17}

burning off a small section of the polyimide coating (<1 cm). As the path length of the cell is already limited, it is advisable to work at wavelengths as low as possible. Usually one works in the neighborhood of 200 nm (cut off silica: 170 nm); therefore, fixed wavelengths detectors that produce a high-intensity light beam (for example, a zinc lamp at 214 nm) are preferred over variable monochromatic-based detectors.

17.3.2.1 Direct detection

Most compounds can be detected directly as they are able to produce a direct analytical signal. Photometric detection, especially UV (including diode array and multi-wavelength UV detection) is by far the most frequently applied detection technique. The application of mass spectrometry (MS) detection in CE is attractive as it can provide structural information [44]. Hologram-based refractive index detection [45] and electrochemical detection [46,47] were also reported. Conductivity [41,48–50] and amperometric [51,52] detection has shown to have advantages for the analysis of both organic and inorganic compounds.

Although coupling of a micro-column to a mass spectrograph is not as straightforward as coupling to optical detection techniques, recently CE–MS has advanced significantly [53,54]. CE–MS/MS is a very powerful and promising technique. Since the separation principle is different than in HPLC, a unique selectivity is obtained that provides novel opportunities to develop orthogonal methods. The CE–MS/MS approaches are very beneficial as an orthogonal chromatographic methodology during impurity profiling of drugs [55,56]. It is found that the applied instrumentation generally meets our expectation, i.e. the sensitivity, repeatability and reproducibility are excellent and allows detecting impurities at very low levels in pharmaceutical samples. These promising results are encouraging us to continue the application of CE–MS as an orthogonal technique besides HPLC for impurity profiling [56].

17.3.2.2 Indirect detection

As was mentioned previously, photometric detection is the most frequently applied detection technique. Most of the commercial CE-systems are equipped with at least a UV detector. Some compounds, such as low molecular weight organic and inorganic ions [57–60], do not produce a direct analytical signal. In such cases indirect detection, by indirect UV or fluorescence [59–64] is applied. Besides photometric detection, an application of indirect amperometric [65] detection was also reported. When the analytical signal results from a decrease in

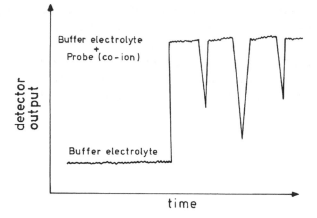

Fig. 17.9. Indirect detection of analytes by the replacement of the probe (co-ion or background electrolyte) ions that are responsible for a constant high-background analytical signal.

the constant background signal of a probe (background electrolyte or co-ion), owing to the replacement of this probe in the analyte zone (Fig. 17.9), one speaks of indirect detection. The probe is selected on the basis of its ability to produce a constant stable background signal and its mobility, which should be more or less equal to that of the analytes. This constant background signal results in the baseline of the electropherogram.

17.3.3 Sensitivity enhancement

Owing to the short path length, the sensitivity of photometric detection in CE is limited. For this reason, procedures are necessary to enhance the sensitivity. This can be achieved by sample pre-concentration, improvement of the optical design and/or alternative capillary geometries [66]. Sample pre-concentration can be done off- or on-column. The off-column procedures are well described in the literature and have been applied extensively in chromatography [67]. The on-column procedures are more specific to CE and are therefore briefly discussed.

An important pre-concentration method is the sample-stacking procedure [68,69]. When the sample is dissolved in a solvent with an electrical conductivity lower than that of the buffer electrolyte, sample stacking occurs because of the difference in electric field strength in the sample and the electrolyte medium. As the field strength in the sample zone is higher, the migration velocity in the sample zone is also higher. At the interface of the sample and the buffer electrolyte zone the components are slowed (stacked) again, causing a concentration of the components at the interface. This procedure can result in a 10-fold enhancement.

Another procedure was described by Nielen [69] and Burgi and Chien [70], known as field amplification. During this procedure, a large sample plug is injected hydrodynamically in a solvent of low conductivity. Following this the sample is focused in the opposite direction to the migration of the separation, by applying a voltage of the opposite polarity compared to the separation voltage. When a stable current is obtained, the polarity of the potential is reversed into the separation voltage. During the first voltage application, the zones are focused in highly concentrated bands due to a strong stacking effect. The result of this procedure is an enhancement with a factor 100. By applying the principles of isotachophoresis (ITP) (see further) prior to the separation in CE, it is possible to obtain an increase in sensitivity up to three orders of magnitude.

Pre-concentration methods using online trace enrichment by applying chromatographic principles are also reported [66,69]. As described by Guzman and Meyers [71,72], this can be achieved by incorporating e.g. a solid-phase CE-concentrator tip at the inlet of the capillary. Undesired sample components can be flushed out prior to the CE separation run, providing faster and more specific analyses. Especially in the field of bioanalysis, where sample clean up and pre-concentration of analytes is usually critical, this approach may be preferred.

Changing the geometry of the capillary in order to achieve a longer path length is another possibility to improve the sensitivity. As was described earlier, simply increasing the diameter of the capillary would result in an increased Joule heating, leading to an increased peak broadening and loss of resolution. Therefore, several alternatives for capillary geometry have been proposed, some with substantial success. Among others [66], these include the use of rectangular and bended capillaries in the shape of a Z (Z-cell) [69,73] or a bubble (bubble-cell) [73]. By altering a small part of the capillary only, e.g. at the detection window, the increase in peak broadening can be neglected.

Improving the optical design also leads to a higher sensitivity. For example, the application of high-intensity light sources are to be preferred. The use of laser light beams is becoming more and more common in CE-detection systems. Especially with fluorescence detection, lasers (argon) have shown their benefits. With laser-induced fluorescence (LIF) detection, one can obtain detection limits down to 10^{-16} M or 2×10^{-21} (zepto) moles or $\pm\,600$ molecules [66]. Sometimes derivatization is necessary to generate chromophores on the compounds in order to make them actively fluoresce. Of course, this is a major drawback of the method.

17.3.4 Time-corrected (normalized) areas

An important feature of the data obtained from CE photometric detectors is the change of the peak area in function of the migration time [28,74]. As can be observed in Fig. 17.10, the peak area is linearly ($r = 0.999$) correlated to the migration times, determined at different run voltages. From Eq. 17.17 it can be derived that an increase in migration velocity due to increasing voltages results in a decrease of the migration time. Although the concentration of the compounds is not changed, the peak area also decreases with the decreasing migration times. This is a characteristic for 'flow-dependent' detectors such as the UV-detector [75].

The area under the peaks obtained by these detectors is proportional to the number of molecules passing through the detector divided by the flow rate. The change in area is contributed to the fact that the area under the peak is proportional to the time (Δt) spent by a zone in the

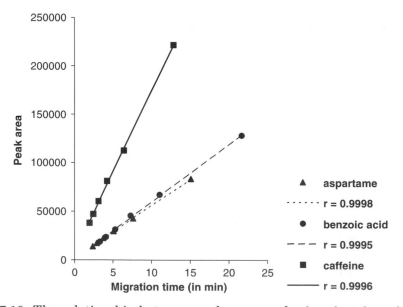

Fig. 17.10. The relationship between peak areas and migration times in CE. The obtained migration times were recorded by injection of three compounds at different voltages.

detecting light beam. Δt is determined by the migration velocity ($\vec{v}_{det}$) of a compound when it passes through the detector, which can be approximated by the overall migration velocity (Eqs. 17.16 and 17.17).

$$\Delta t = \frac{w_{det}}{\vec{v}_{det}} = \frac{w_{det} \times T_m}{l} \tag{17.49}$$

According to Beer's law, the peak area must be proportional to Δt and the absorbance (A):

$$\text{Area} = C \times A \times \Delta t \tag{17.50}$$

where C is a constant. By substituting Δt with Eq. 17.49 in Eq. 17.50, one obtains the expression of the area in function of the migration time:

$$\text{Area} = \frac{C \times A \times w_{det} \times T_m}{l} \tag{17.51}$$

From these expressions it can be seen that peak area is directly proportional to migration time (and inversely related to the potential). As the migration velocity in CE is different for different compounds, changes in peak area are bound to occur for compounds with similar concentration and molar absorptivities, but different mobilities. When the migration velocity (migration time) is not controllable, the resulting peak area will vary leading to poor reproducibility. Therefore, it is recommended to apply a time correction, by dividing the peak areas by the migration time. This should result in better reproducibility. However, when migration times are controlled properly, time-corrected areas may not be necessary. Sometimes the time correction can even lead to worse precision, as it assumes that the migration velocity of a zone when it passes through the detector ($\vec{v}_{det}$) is equal to the overall migration velocity ($\vec{v}_i$), determined by Eq. 17.16 [76].

17.3.5 Thermostatting

As was already addressed in Section 17.2.4, despite the use of narrow bore capillaries, the temperature difference between the wall of the capillary and the surrounding air/liquid can rise up to several degrees (exceeding 70°C) [31,77]. This was shown in Fig. 17.6 and is known as the self-heating of the capillary due to the power production within the capillary. An efficient thermostatting procedure for the capillary is one of the primary prerequisites in order to perform reproducible CE. The uncontrolled temperature augmentation obviously has effects on sample stability, buffer pH and viscosity. Especially, certain biomolecules are inherently unstable and therefore require efficient thermostatting during the electrophoretic separation.

The thermostatting systems that are applied maximize the Joule heat dissipation through cooling either air (forced air convection or with a Peltier device) or liquid (Peltier device) circulation. In most of the comparisons of temperature control by air and liquid circulation, it was shown that liquid-thermostatting systems perform best [31,77].

17.4   METHODS AND MODES IN CE

Electrophoresis has been applied for decades in analytical chemistry. It is still the method of choice for the determination of proteins and nucleic acids. The technique of CE is classified in three main groups of methods [1]:

(1)   Moving boundary CE
(2)   Steady-state CE
   – ITP
   – Isoelectric focusing (IEF)
(3)   Zone CE
   – Capillary gel electrophoresis (CGE)
   – Free-solution CE (FSCE)
       – Capillary zone electrophoresis (CZE)
       – MECC
       – Chiral CE (CCE)
(4)   Capillary electrochromatography (CEC)

The zone capillary electrophoretic methods are applied the most and are therefore discussed in more detail. The others are discussed briefly.

17.4.1   Moving boundary CE

Moving boundary electrophoresis was proposed by Tiselius [1,4]. The technique has become outdated, but it is now possible to utilize a recent CE system as described in Fig. 17.8. The sample is added as the only constituent in the electrolyte compartment at the injection side. In the other compartment, an electrolyte is used with a mobility that is higher than the mobilities of the sample constituents. The capillary also contains the high-mobility electrolyte and connects the two compartments. By applying an appropriate potential, the ions migrate in the generated electric field. The electrolyte in the capillary, having the highest mobility, migrates first along the capillary, followed by successively slower components of the sample. After a certain time, the fastest compound in the sample zone emerges from the bulk of the sample and is followed by a zone containing a mixture of the next slower compound and the fastest compound. This zone is again followed by a zone containing a mixture of the third slower compound and the first two fastest compounds, etc. In fact, it is only possible to obtain a pure zone of the fastest migrating ion of the sample. This and the problems connected with the detection make the method less attractive for systematic use [4,18].

17.4.2 Steady-state capillary electrophoresis

As the name of this group of methods suggests, a certain stage of steady state in terms of migration is achieved in these methods. ITP and IEF are the main representative modes. The experimental setup is similar to the one described in Fig. 17.8. In these modes a gradient is generated in order to obtain a separation. In ITP a potential gradient is created by using 'leading' and 'terminating' electrolyte systems. The compartment at the injection side contains a 'terminating electrolyte' that has a mobility much smaller than the mobility of the sample constituents. The capillary and the compartment at the detection side contain a 'leading electrolyte' that has a higher mobility than that of the sample constituents. The sample is injected in a small plug through the inlet of the capillary. When an appropriate voltage is applied, the ions migrate according to their mobility. First, the leading electrolyte, followed by the sample constituents with successive mobilities, and finally the terminating electrolyte. Owing to the differential migration rates, the zones tend to migrate away from each other. If this would occur, then parts of the capillary would be occupied with water only (with very low conductivity) and the current would be cut. In order to maintain the current flow through the capillary, the zones migrate in separate but adjacent zones. The potential gradient is constant within each zone, but increases in a stepwise fashion from zone to zone [1–3,37]. ITP results in very high-resolution separations, but is useful for the separation of ionic compounds only. It usually requires capillaries that are pretreated in order to avoid the occurrence of the EOF.

The gradient in IEF is a pH gradient that is generated by electrolyte systems consisting of ampholytes. The capillary wall is coated in order to avoid electroosmosis. The IEF mode is especially suitable for the separation of zwitter ionic compounds like proteins, peptides, amino acids, various drugs, etc. After injection of the sample and application of the potential, the sample constituents migrate to a pH in the capillary where their net-charge is equal to zero. At this point, the isoelectric point, the mobility of a compound is also equal to zero and therefore the migration of the compound is stopped. Band broadening is avoided because of the focusing effect of the pH gradient. In the next step, the zones are brought to the detector and removed from the capillary (for example, by applying a pressure difference) [1,2,37].

17.4.3 Zone capillary electrophoresis

In this group of methods the sample constituents can migrate differentially in zones through the capillary in a medium that can be either a gel (CGE) or an electrolyte (FSCE). Again, the experimental setup is similar to the one presented in Fig. 17.8.

17.4.3.1 Capillary gel electrophoresis
In this mode the capillary is filled with a gel (polyacrylamide or agarose) that forms a polymeric network [78] in the capillary. This network leads to molecular sieving. The separation mechanism is therefore especially based on differences in solute size (also determines the charge density) as the analytes migrate through the pores of the gel. Furthermore, the gel avoids zone broadening because of diffusion and prevents wall adsorption and electroosmosis. Separations of extremely high efficiency (30 million theoretical plates/m) have been obtained. CGE is especially useful for the separation of large biomolecules, like proteins, polynucleotides and DNA fragments. It has considerable advantages over conventional slab–gel electrophoresis: small sample requirement, automation, high sensitivity, high throughput, rapid and efficient molecular mass determination, trace quantitation and is applicable to a wide variety of large biomolecules [1–3,15,33,37,75,79–81]. Figure 17.11 shows a typical high-efficiency separation that is easily achieved in CGE with optimal single-base resolution of a polyU standard [82]. The application of capillary array instruments in this area is very typical for high throughput. CGE is the method of choice in the genomics field and thanks to the introduction of the high-throughput instruments, gene sequencing has advanced greatly today.

17.4.3.2 Free-solution capillary electrophoresis
In FSCE the capillary is filled with only an electrolyte, usually a buffer to maintain the pH. The experimental setup remains similar to the one used for the previous methods and presented in Fig. 17.8. FSCE is currently the most applied technique in CE. The separation is based on

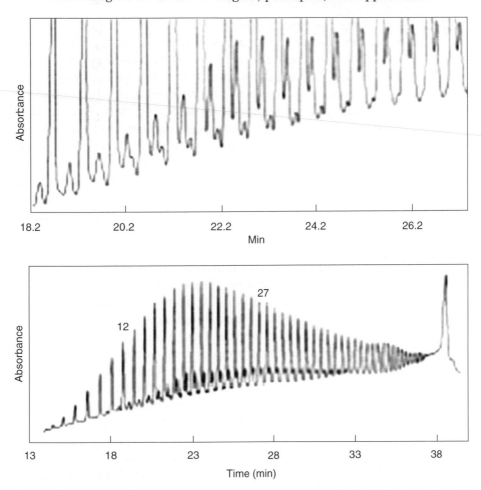

Fig. 17.11. Bottom: CGE separation of components of "poly U" (sigma) in 25% pluronic F127. Top: Note the resolution of two contaminants between each of the oligonucleotides from about 15 to 27 nucleotides long in this expanded section of the bottom electropherogram. Electrophoresis was performed in 25% pluronic F127 in tris–borate–EDTA buffer (90 mM tris, 90 mM boric acid, 2 mM Na EDTA, pH 8.3.) (25°C, 500 V cm^{-1}, effective column length 30 cm). Reprinted with permission from Ref. [82].

the difference in effective mobilities between compounds that result in differences in migration velocities. It can be applied for the separation of many compounds, including small and large ionic and neutral compounds, both organic and inorganic. FSCE is divided in three modes due to the addition of specific additives in the electrolyte: CZE, MECC and CCE. These modes are discussed in further sections.

17.4.3.2.1 Capillary zone electrophoresis

CZE is the basic technique of the free-solution mode in CE. A separation is achieved due to a differential migration caused by differences in effective mobilities. The capillary contains a buffer electrolyte that guarantees the current flow. Samples are injected and electrophoretically separated in different pure zones by applying an electrical field. The characteristics of the method have been discussed earlier in Sections 17.2 and 17.3. Because only ionized compounds can have a differential migration in CZE, neutral compounds are not separated. The neutral compounds are carried by the EOF to the detection site and therefore migrate with the mobility and velocity of the EOF. Several modifiers (complexing agents and organic solvents) are used to enhance the selectivity [1,2,81].

CZE has been applied for the determination of both organic and inorganic charged compounds. The analysis of low molecular weight inorganic and organic ions has developed into a distinct field of interest [83,84]. In order to distinguish this field from other areas of application in CE, it is referred to as 'Small Ion Capillary Electrophoresis (SICE)' [85]. A typical example for the determination of the bromide counter ion in galantamine hydrobromide drug substance is shown in Fig. 17.12. In such an anion separation, the capillary inner wall is coated with cationic surfactant, e.g. cetyl trimethylammonium bromide [85]. Simple CZE methods can be applied for the analysis of charged compounds ranging from small molecules to proteins and DNA fragments. Figure 17.13 shows the determination of Benzalkonium chloride in a nasal pharmaceutical formulation using a phosphate buffer at pH 2.3 (containing 75 mM phosphoric acid and 42.7 mM triethylamine), with an uncoated fused silica capillary. As can be seen, both benzalkonium derivates with chain lengths C12 and C14 present in the sample are easily separated from each other and from the active pharmaceutical ingredient [86].

17.4.3.2.2 Micellar electrokinetic capillary chromatography

In order to separate neutral compounds, Terabe et al. [13] added surfactants to the buffer electrolyte. Above their critical micellar concentration (cmc), these surfactants form micelles in the aqueous solution of the buffer electrolyte. The technique is then called Micellar electrokinetic capillary chromatography, abbreviated as MECC or MEKC. Micelles are dynamic structures consisting of aggregates of surfactant molecules. They are highly hydrophobic in their inner structure and hydrophilic at the outer part. The micelles are usually

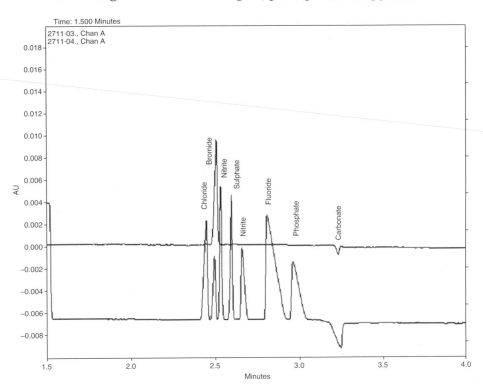

Fig. 17.12. The determination of typical anions using indirect UV detection. Conditions: 10 mM sodium chromate, 2.30 mM cetyltrimethylammonium bromide, 60 cm fused silica capillary (effective length 52 cm) × 75 µm I.D., injection 5 s at 35 mbar, 20°C, –15 kV (reversed polarity) resulting in a current of approximately 30 µA, detection UV 254 nm.

charged, positively or negatively, but can also be neutral, depending on the nature of the original surfactants. During electrophoresis these structures have a differential migration because of their effective mobility. The most applied surfactant in MECC is sodium dodecyl sulfate (SDS). Owing to their polarity, neutral compounds are solubilized by the micelles. When there is no affinity of the solutes for the micelles, they migrate with their own effective mobility. For example, a neutral compound with no affinity for the micelles migrates with the mobility of the EOF. If a solute is highly hydrophobic and therefore has a very strong affinity for the inner part of the micelles, then it will migrate with the mobility of the micelles. In other words, the solutes are partitioned between the micellar- and aqueous phase. Figure 17.14 shows a representation of the mechanisms in MECC. Note that not only neutral

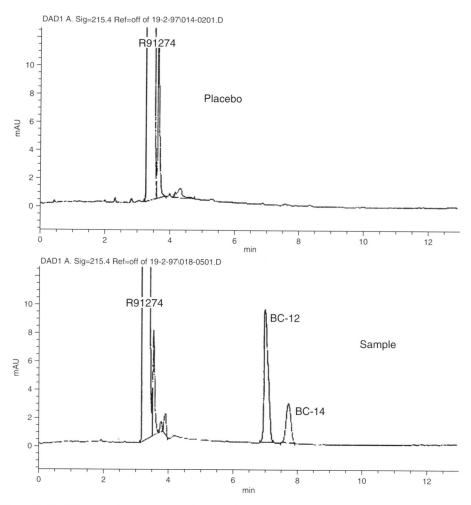

Fig. 17.13. Electropherograms obtained for the analysis of a nasal formulation for the determinations of benzalkonium chloride (BC) in the presence of active pharmaceutical ingredient (R91274) and other placebo ingredients. Conditions: 75 mM sodium phosphate buffer, pH = 2.3, 35 cm fused silica capillary (effective length 28.5 cm) × 75 μm I.D., injection 10 s at 35 mbar, 20°C, 15 kV (positive polarity) resulting in a current of approximately 95 μA, detection UV 215 nm.

compounds can be solubilized by the micelles, but also charged molecules as the partition is based on polarity. Therefore, the technique provides a way to resolve both neutral and charged compounds by CE. In fact, MECC combines electrophoretic and chromatographic principles in one technique [1–3,13,80,87–89].

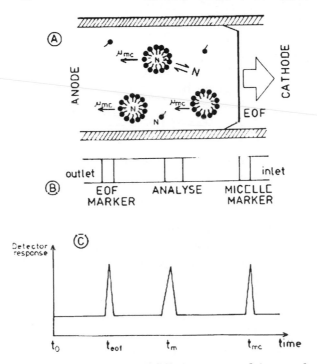

Fig. 17.14. Separation principle in MECC. A compound (neutral or charged) is partitioned between the micellar and aqueous phase. A fully solubilized neutral compound migrates with the velocity of the micelles. A neutral compound with no affinity for the micelles migrates with the velocity of the EOF. A neutral compound with an affinity for both the micellar and the aqueous phase migrates with an intermediate velocity. (A) Schematic overview of the partitioning of compound 'N', the EOF moves toward the cathode and the typical SDS micelles toward the anode. (B) Diagram of the zone distribution within the capillary. (C) Reconstructed typical electropherogram.

Owing to the interaction with the micelles, the migration time (T_m) of a neutral compound is situated in a 'migration time window', between the migration time of the EOF ($\pm 0\%$ solubilization) peak (T_{eof}) and that of the micellar ($\pm 100\%$ solubilization) peak (T_{mc}). The migration time of the micelles is experimentally obtained by injecting, e.g. Sudan III (highly hydrophobic dye), while T_{eof} can be obtained by injecting a neutral marker as methanol, mesithyloxide, formamide, etc. In chromatographic terms, the solubilization of compounds in the micelles can be interpreted as a retention of a solute by the micellar phase: the pseudostationary phase (since this stationary phase also moves). Similar to conventional chromatography, the extent of the retention can be expressed by a capacity factor, k', describing the ratio of the total

613

moles of a solute in the micellar and the aqueous phase:

$$k' = \frac{T_m - T_{eof}}{T_{eof} \times (1 - T_m/T_{mc})} \tag{17.52}$$

Resolution in MECC is given by

$$Rs = \frac{N^{1/2}}{4} \times \left(\frac{\alpha - 1}{\alpha}\right) \times \frac{k_2'}{(k_2' + 1)} \times \left(\frac{1 - T_{eof}/T_{mc}}{1 + k_1' \times (T_{eof}/T_{mc})}\right) \tag{17.53}$$

where α represents the separation factor given by k_2'/k_1' and k_1', k_2' are the capacity factors for peak 1 and 2, respectively. Eq. 17.53 is equal to the expression used in conventional chromatography with the exception of the last part.

As polarity plays an important role in the analysis of small organic compounds e.g. pharmaceuticals (most organic compounds can be neutral), the introduction of MECC resulted in a boom of applications in this area. In biomedical analysis with CE, MECC is the technique of choice for the analysis of drugs in pharmaceutical preparations or in body fluids. Several groups of drugs have been analyzed with considerable success [1,2,80,90]. A typical example of an MECC separation is shown in Fig. 17.15 for the determination of the organic purity of levocabastine drug substance [91]. For this product it was difficult to separate the ortho-, meta- and para-fluoro place isomers using HPLC. Therefore, an MECC method was developed, validated and filed to the regulatory authorities. Actually, this method is the first CE method to be published in a pharmacopoeia monograph.

17.4.3.2.3 Chiral capillary electrophoresis
Owing to the high-separation efficiency and the flexibility of CE, the separation of chiral compounds has been performed extensively by this technique [1,2,80,90,92]. By using diastereomeric complex formation with Cu^{2+}-aspartame, P. Gozel et al. [93] separated the enantiomers of some dansylated amino acids. Tran et al. [94] demonstrated that such a separation was also possible by derivatization of amino acids with L-Marfey's reagent. Nishi et al. [95] were able to separate some chiral pharmaceutical compounds by using bile salts as chiral selectors and as micellar surfactants. However, the utilization of cyclodextrins, first introduced by Fanali [96] as chiral selectors, resulted in the boom in the number of applications [2,90,96]. They are added to the buffer electrolyte during electrophoresis, and a chiral recognition is easily obtained because of the resulting hydrophobic inclusion, hydrogen bonding and steric repulsion mechanisms. Besides cyclodextrins, crown ethers have

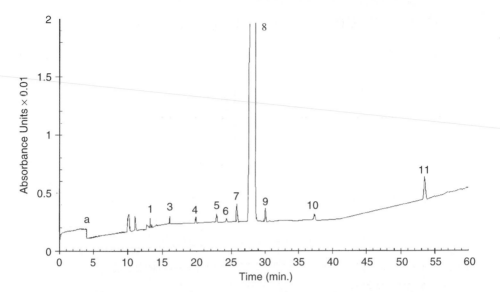

Fig. 17.15. Determination of the organic impurities (compounds 1–11) in levocabastine drug substance. Conditions: 225 mM borate buffer, 2.16% w/v SDS + 1.3% w/v hydroxypropyl-β-cyclodextrin + 10% v/v 2-propanol, pH = 9, 57 cm fused silica capillary (effective length 50 cm) × 75 μm I.D., injection 5 s at 35 mbar, 50°C, detection UV 214 nm, current program: 0–15 min: 75–130 μA, 15–40 min: 130 μA, 40–60 min: 130–200 μA. (a = auto zero).

also been applied for the separation of chiral compounds [80]. Owing to their spherical structure, the selectors (hosts) can spatially enclose another (guest) or part of another molecule. Therefore, this mechanism is also known as the 'host–guest' interaction. Chiral separation was also obtained by using proteins as chiral selectors. An example is the chiral separation of leucovorin with bovine serum albumin (BSA) as the chiral selector [97]. Other proteins like transferrin, ovalbumin, etc. can also be applied.

Cyclodextrins are the most frequently applied chiral selectors nowadays. Both the native as well as the derivatized cyclodextrins are massively applied in CCE. The major challenge, however, is finding the right cyclodextrin and at its optimal concentration. Up till now it has not been possible to predict e.g. on theoretical grounds which cyclodextrin will be the suitable chiral selector. The selection is based on trial and error approaches, leading to complex, laborious and time-consuming efforts. Strategies have been proposed to speed up and simplify this selection [98,99]. After selecting the right chiral selector, the test method has to be optimized for robustness and transferability. With respect to this, an optimization of the cyclodextrin concentration, the

615

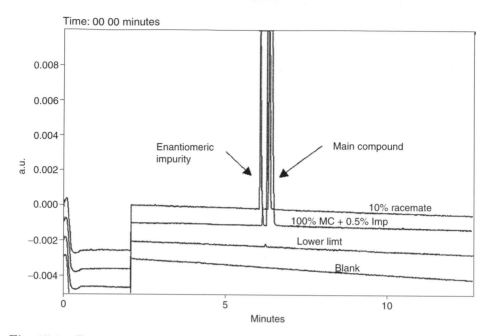

Fig. 17.16. Determination of the enantiomeric purity of an active pharmaceutical ingredient (main compound = MC). Conditions: 30 mM sodium phosphate buffer, pH = 2.3, 10 mM trimethyl β-cyclodextrin, 35 cm fused silica capillary (effective length 28.5 cm) $\times$ 75 μm I.D., injection 10 s at 35 mbar, 20°C, 20 kV (positive polarity) resulting in a current of approximately 65 μA, detection UV 215 nm.

buffer concentration, pH and the concentration of e.g. an organic modifier is very critical. In order to reduce the experimental work, design of experiments methodology is a preferred approach [100,101]. As a result, robust, reliable and transferable CE methods can be developed that are accepted in new drug applications to regulatory authorities [102]. Figure 17.16 shows a typical CCE separation for the determination of the chiral purity of a pharmaceutical ingredient. As can be seen, the unwanted enantiomeric impurity can be detected down to the necessary low levels in the samples, just by adding an optimized amount of cyclodextrin to a simple phosphate buffer.

17.4.3.3 Capillary electrochromatography

CEC is comparable to MECC, but with the major difference that the micelles are replaced by very small, i.e. less than 3 μm, solid or semi-solid particles in a packed or open column. The particles form a typical stationary phase as we know from ordinary HPLC. The mobile phase is obtained through the electrically driven flow resulting from electroosmosis: the EOF. Similar to MECC, separation of solutes is achieved either by differential migration resulting from chromatographic partitioning, or from electromigration, or from a combination of both. The particles can be sol–gel [103], molecularly imprinted polymers [104], continuous monolithic beds [105] and polymer/layered silicate nanocomposites [106] with reverse phase (e.g. ODS C18), normal phase, ion exchange (SCX) or size exclusion [107] properties in packed columns or open tubular mode [108]. In open tubular mode the stationary phase is coated on the inner surface of the capillary column. CEC is almost similar to HPLC, but results in a higher separation efficiency due to the flat profile of the EOF (mobile phase) and probably the stacking effect due to electrodispersion. Additionally, in CEC small particle sizes can be used since the EOF does not generate back pressure.

Typically CEC is performed at high pH to ensure a fast EOF, resulting in short analysis times. In combination with high-voltage applications and short capillary lengths, the analysis speed is increased further. CEC at lower pH is possible with ion-exchange stationary phases with sulfonic acid groups on e.g. an SCX phase. The sulfonic acid groups remain ionized at low pH and therefore generate a sufficiently high EOF. Lower pH may be needed for acidic compounds in order to reduce dissociation for better partitioning. Applications in CEC range from small organic neutral [109], basic [110] and acidic [111] drugs (including chiral compounds [112]) to peptides [113,114], proteins [114], DNA fragments [114], etc.

The main bottleneck in the further development of CEC is related with the state of the art of the column manufacturing processes and the robustness of the columns/instrumentation. Moreover, evidence to demonstrate reproducibility of separations from column to column still has to be established. The formation of bubbles in the capillaries due to the Joule heating and variations in EOF velocity on passing from the stationary phase through the frit and into the open tube is still very challenging in packed column CEC. A way to overcome this problem is to use monolithic columns or apply open tubular CEC [108]. Currently, many efforts are placed in improving column technology and in the development of chip-CEC [115] as an attractive option for 'lab-on-a-chip' separations.

17.5 CONCLUSIONS

The contribution of electromigration methods to analytical chemistry and biopharmaceutical science has been very significant over the decades. Especially in the biochemistry area for the analysis of proteins, amino acids and DNA fragments, electrophoresis is still the first choice method. By performing electrophoresis in a capillary, the technique has evolved into a high-performance instrumental method. The applications are wide spread to even small organic, inorganic, charged, neutral compounds and pharmaceuticals. Currently, CE is considered to be an established tool in the analytical chemist's hands to solve many analytical problems. The major application areas still are in the field of DNA sequencing and protein analysis, but also low molecular weight compounds (pharmaceuticals). Indeed, it was thanks to CE that the human genome project was completed many years earlier than initially planned.

In the beginning, electromigration and chromatographic methods developed as separate techniques over many decades. Nowadays, both methods have converged into a single approach: CEC. The approach is still under development, but has already demonstrated to be very promising. From the current findings and the overall CE expertise that was built up since the start of the last decade, it can be concluded that CE technology has grown to a certain maturity, which allows to develop and apply robust analysis methods. The technique is already described as general monographs in European Pharmacopoeia [116] and the USP [117]. However, a good understanding of CE-related issues and skilful CE analysts are needed to achieve this target. Applying general chromatographic know-how e.g. from HPLC to CE, is not sufficient. Usually, this is the major reason for failures in CE applications. Dedicated and trained CE analysts are necessary to be successful.

REFERENCES

1   R.A. Wallingford and A.G. Ewing, Advances in chromatography, *Biotechnological Applications and Methods*, Marcel Dekker Inc., New York, 1989.
2   S.F.Y. Li, *Capillary Electrophoresis; Principles, Practice and Applications*, Elsevier, Amsterdam, 1992.
3   E. Heftmann, Chromatography, fundamentals and applications of chromatography and related differential migration methods, *Part A: Fundamentals and Techniques*, 5th ed., Elsevier, Amsterdam, 1992.
4   A. Tiselius, *Trans. Faraday Soc.*, 33 (1937) 524.
5   D.N. Heiger, *High Performance Capillary Electrophoresis; An Introduction*, Hewlett-Packard GmbH, Walbronn Analytical Division, Waldbronn 2, Germany, 1992.

6 S. Hjerten, *Chromatogr. Rev.*, 9 (1967) 122.

7 R.L. Stevenson, *J. Cap. Elec. Microchip Technol.*, 1 (1994) 169.

8 F.E.P. Mikkers, F.M. Everaerts and Th.P.E.M. Verheggen, *J. Chromatogr.*, 169 (1979) 11.

9 F.E.P. Mikkers, F.M. Everaerts and Th.P.E.M. Verheggen, *J. Chromatogr.*, 169 (1979) 1.

10 J. Jorgenson and K.D. Lukacs, *Anal. Chem.*, 53 (1981) 1298.

11 J. Jorgenson and K.D. Lukacs, *J. Chromatogr.*, 218 (1981) 209.

12 J. Jorgenson and K.D. Lukacs, *Science*, 222 (1983) 266.

13 S. Terabe, K. Otsuka, K. Ichikawa, A. Tsuchiya and T. Ando, *Anal. Chem.*, 56 (1984) 111.

14 I. ISCO, *Isco Appl. Bull. 65R* (1991).

15 A.G. Ewing, R.A. Wallingford and T.M. Olefirowicz, *Anal. Chem.*, 61 (1989) 293A.

16 C. Schwer and E. Kenndler, *Anal. Chem.*, 63 (1991) 1801.

17 M.A. Hayes and A.G. Ewing, *Anal. Chem.*, 64 (1992) 512.

18 B.L. Karger, L.R. Snyder and C. Horvath, *An Introduction to Separation Science*, Wiley, Canada, 1973.

19 B.J. Kirby and E.F. Hasselbrink Jr., *Electrophoresis*, 25 (2004) 187–202.

20 Y. Xu, *Anal. Chem.*, 65 (1993) 425R.

21 B.B. VanOrman, G.G. Liversidge, G.L. McIntire, T.M. Olefirowicz and A.G. Ewing, *J. Microcol. Sep.*, 2 (1990) 176.

22 K.D. Lukacs and J.W. Jorgenson, *J. High Resolut. Chromatogr.*, 8 (1985) 407.

23 C. Schwer and E. Kenndler, *Chromatographia*, 30 (1990) 546.

24 X. Huang, W.F. Coleman and R.N. Zare, *J. Chromatogr.*, 480 (1989) 95.

25 P.J. Schoenmakers, *Optimization of Chromatographic Selectivity—A Guide to Method Development*, Elsevier, Amsterdam, 1996.

26 H.H. Lauer and D. McManigill, *TrAC*, 5 (1986) 11.

27 F. Foret, M. Deml and P. Bocek, *J. Chromatogr.*, 452 (1988) 101.

28 S.E. Moring, J.C. Colburn, P.D. Grossman and H.H. Lauer, *LC–GC INT.*, 3 (1991) 46.

29 S.L. Petersen and N.E. Ballou, *Anal. Chem.*, 64 (1992) 1676.

30 Y.J. Yao and S.F.Y. Li, *J. Chromatogr. Sci.*, 32 (1994) 117.

31 Y. Kurosu, K. Hibi, T. Sasaki and S. M, *J. High Resolut. Chromatogr.*, 14 (1991) 200.

32 J. Crommen, *Euro Training Course*, Monpellier, September 14–17, France, 1993.

33 W.G. Kuhr, *Anal. Chem.*, 62 (1990) 403A.

34 M. Jimidar, T. Hamoir, W. Degezelle, S. Soykenc, P. Van de Winkel and D.L. Massart, *Anal. Chim. Acta*, 284 (1993) 217.

35 S.L. Delinger and J.M. Davis, *Anal. Chem.*, 64 (1992) 1947.

36 R. Weiburger, *American Laboratory*, May (2002) 32–40.

37 J.D. Olechno, J.M.Y. Tso, J. Thayer and A. Wainright, *Int. Lab.* (1991) 42.
38 D.J. Rose and J.W. Jorgenson, *Anal. Chem.*, 60 (1988) 642.
39 X. Huang, M.J. Gordon and R.N. Zare, *Anal. Chem.*, 60 (1992) 375.
40 S.E. Moring, R.T. Reel and R.E.J. van Soest, *Anal. Chem.*, 65 (1993) 3454.
41 N. Avedalovic, C.A. Pohl, R.D. Rocklin and J.R. Stillian, *Anal. Chem.*, 65 (1993) 1470.
42 P.E. Jackson and P.R. Haddad, *TrAC*, 12 (1993) 231.
43 W.R.G. Baeyens, B. Lin Ling and K. Imai, *Euro Training Course*, Monpellier, September 14–17, France, 1993.
44 T.G. Huggins and J.D. Henion, *Electrophoresis*, 14 (1993) 531.
45 B. Krattiger, G.J.M. Bruin and A.E. Bruno, *Anal. Chem.*, 66 (1994) 1.
46 W. Lu, R.M. Cassidy and A.S. Baranski, *J. Chromatogr.*, 640 (1993) 433.
47 A.G. Ewing, J.M. Mesaros and P.F. Gavin, *Anal. Chem.*, 66 (1994) 527A.
48 X. Huang, J.A. Lucky, M.J. Gordon and R.N. Zare, *Anal. Chem.*, 61 (1989) 766.
49 W.R. Jones, C. Haber and J. Reineck, Paper Presented at the International Ion Chromatography Symposium 1994, Turin, Italy, 1994, Paper no. 56.
50 M.T. Ackermans, F.M. Everaerts and J.L. Beckers, *J. Chromatogr.*, 549 (1991) 345.
51 X. Huang, R.N. Zare, S. Sloss and A.G. Ewing, *Anal. Chem.*, 63 (1991) 189.
52 S. Sloss and A.G. Ewing, *Anal. Chem.*, 65 (1993) 57.
53 A. von Brocke, G. Nicholson and E. Bayer, *Electrophoresis*, 22 (2001) 1251.
54 C.W. Klampfl, *J. Chromatogr. A*, 1044 (2004) 131–141.
55 M. Jimidar, M. De Smet, R. Sneyers, W. Van Ael, W. Janssens, D. Redlich and P. Cockaerts, *J. Cap. Elec. Microchip Technol.*, 8 (2003) 45–52.
56 D. Visky, M. Jimidar, W. Van Ael, D. Redlich and M. De Smet, *Beckman P/ACE—Setter*, 8 (2004) 1–4.
57 P. Jandik and W.R. Jones, *J. Chromatogr.*, 546 (1991) 431.
58 W.R. Jones and P. Jandik, *J. Chromatogr.*, 546 (1991) 445.
59 P. Jandik, W.R. Jones, A. Weston and P.R. Brown, *LC– GC INT.*, 5 (1992) 20.
60 F.B. Erim, X. Xu and J.C. Kraak, *J. Chromatogr. A*, 694 (1995) 471.
61 W.G. Kuhr and E.S. Yeung, *Anal. Chem.*, 60 (1998) 2642.
62 W. Beck and H. Engelhardt, *Chromatographia*, 33 (1992) 313.
63 W. Buchberger, S.M. Cousins and P.R. Haddad, *Trends Anal. Chem.*, 13 (1994) 313.
64 S.M. Cousins, W. Buchberger and P.R. Haddad, *J. Chromatogr. A*, 671 (1994) 397.
65 T.M. Olefirowicz and A.G. Ewing, *J. Chromatogr.*, 499 (1990) 713.

66 M. Albin, P.D. Grossman and S.E. Moring, *Anal. Chem.*, 65 (1993) 489A.
67 R.D. McDowell, *J. Chromatogr.*, 492 (1989) 3.
68 J.L. Beckers and M.T. Ackermans, *J. Chromatogr.*, 629 (1993) 371.
69 M.W. Nielen, *Trends Anal. Chem.*, 12 (1993) 345.
70 R.-L. Chien and D.S. Burgi, *Anal. Chem.*, 64 (1992) 489A.
71 N.A. Guzman and R.E. Meyers, *LC–GC North America*, 19 (2001) 14–30.
72 N.A. Guzman, *J. Chromatogr. B*, 749 (2000) 197–213.
73 S.E. Moring, R.T. Reel and R.E.J. van Soest, *Anal. Chem.*, 65 (1993) 3454.
74 A. Pluym, W. Van Ael and M. De Smet, *Trends Anal. Chem.*, 11 (1992) 27.
75 S. Bay, H. Starke, J.Z. Zhang, J.F. Elliott, L.D. Coulson and N.J. Dovichi, *J. Cap. Elec.*, 1 (1994) 121.
76 M. Jimidar, T.P. Hamoir, A. Foriers and D.L. Massart, *J. Chromatogr.*, 636 (1993) 179.
77 J.P. Landers, R.P. Oda, B. Madden, T.P. Sismelich and T.C. Spelsberg, *J. High Resolut. Chromatogr.*, 15 (1992) 517.
78 B. Chu and D. Liang, *J. Chromatogr. A*, 966 (2002) 1–13.
79 E. Kenndler and H. Poppe, *J. Cap. Elec.*, 1 (1994) 144.
80 C.A. Monnig and R.T. Kennedy, *Anal. Chem.*, 66 (1994) 280R.
81 L.R. Gurley, J.S. Buchanan, J.E. London, D.M. Stavert and B.E. Lehnert, *J. Chromatogr. A*, 559 (2001) 411.
82 R.L. Rill, Y. Liu, D.H. Van Winkle and B.R. Locke, *J. Chromatogr. A*, 817 (1998) 287–295.
83 P. Jandik and G.K. Bonn, *Capillary Electrophoresis of Small Molecules and Ions*, VCH Publishers, New York, 1993.
84 M. Jimidar, C. Hartmann, N. Cousement and D.L. Massart, *J. Chromatogr. A*, 706 (1995) 479.
85 M. Jimidar, Q. Yang, J. Smeyers-Verbeke and D.L. Massart, *Trends Anal. Chem. (TrAC)*, 15 (1996) 91.
86 M. Jimidar, I. Beyns, R. Rome, R. Peeters and G. Musch, *Biomed. Chromatogr.*, 12 (1998) 128–130.
87 M.G. Khaledi, S.C. Smith and J.K. Strasters, *Anal. Chem.*, 63 (1991) 1820.
88 J.K. Strasters and M.G. Khaledi, *Anal. Chem.*, 63 (1991) 2503.
89 K. Ghowsi, J.P. Foley and R.J. Gale, *Anal. Chem.*, 62 (1990) 2714.
90 N.W. Smith and M.B. Evans, *J. Pharm. Biomed. Anal.*, 12 (1994) 579.
91 Monograph, *European Pharmacopoeia 4*, (01) [1484] (2001) 1458–1460.
92 S.G. Penn, E.T. Bergstrom and D.M. Goodall, *Anal. Chem.*, 66 (1994) 2866.
93 P. Gozel, E. Gassmann, H. Michelsen and R. Zare, *Anal. Chem.*, 59 (1987) 44.
94 P.A.D. Tran, T. Blanc and E.J. Leopold, *J. Chromatogr.*, 516 (1990) 241.
95 H. Nishi, T. Fukuyama, M. Matsuo and S. Terabe, *J. Chromatogr.*, 515 (1990) 233.

96   S. Fanali, *J. Chromatogr.*, 474 (1989) 441.

97   G.E. Barker, P. Russo and R.A. Hartwick, *Anal. Chem.*, 64 (1992) 3024.

98   M. Jimidar, W. Van Ael, R. Shah, D. Redlich and M. De Smet, *J. Cap. Elec. Microchip Technol.*, 8 (2003) 101–110.

99   M. Jimidar, W. Van Ael, P. Van Nyen, M. Peeters, D. Redlich and M. De Smet, *Electrophoresis*, 25 (2004) 2772–2785.

100 M. Jimidar, W. Van Ael, D. Redlich and M. De Smet, *J. Cap. Elec. Microchip Technol.*, 9 (2004) 13–21.

101 M. Jimidar, T. Vennekens, W. Van Ael, D. Redlich and M. De Smet, *Electrophoresis*, 25 (2004) 2876–2884.

102 M. Jimidar, W. Van Ael, M. De Smet and P. Cockaerts, *LC–GC Europe*, 15 (2002) 230–242.

103 W. Li, D.P. Fries and A. Malik, *J. Chromatogr. A*, 1044 (2004) 23–52.

104 J. Nilsson, P. Spegel and S. Nilsson, *J. Chromatogr. B*, 804 (2004) 3–12.

105 A. Maruska and O. Kornysova, *J. Biochem. Biophys. Methods*, 59 (2004) 1–48.

106 S. Sinha Ray and M. Okamoto, *Prog. Polym. Sci.*, 28 (2003) 1539–1641.

107 W.T. Kok, *J. Chromatogr. A*, 1044 (2004) 145–151.

108 E. Guihen and J.D. Glennon, *J. Chromatogr. A*, 1044 (2004) 67–81.

109 R. Dadoo, R.N. Zare, C. Yan and D.S. Anex, *Anal. Chem.*, 70 (1998) 4787.

110 I.S. Lurie, R.P. Meyers and T.S. Conver, *Anal. Chem.*, 70 (1998) 3255.

111 K.D. Altria, N.W. Smith and C.H. Turnbull, *J. Chromatogr. B: Biomed. Sci. Appl.*, 717 (1998) 341–353.

112 G. Blaschke and B. Chankvetadze, *J. Chromatogr. A*, 875 (2000) 3–25.

113 K. Walhagen, K.K. Unger and T.W. Hearn, *J. Chromatogr. A*, 887 (2000) 165–185.

114 I.S. Krull, A. Sebag and R. Stevenson, *J. Chromatogr. A*, 887 (2000) 137–163.

115 I.B. Stachowiak, F. Svec and J.M.J. Fréchet, *J. Chromatogr. A*, 1044 (2004) 97–111.

116 European Pharmacopoeia, Chapter 2. 2. 47, *Capillary Electrophoresis*, 5.0 (2004) 74–79.

117 USP, Chapter <727>, *Capillary Electrophoresis*, 27 (2004) 2315–2320.

REVIEW QUESTIONS

1. Which are the two most important parameters that cause migration in CE?
    The migration in CE is determined by both the effective and the electroosmotic mobility.

2. What determines the selectivity of a separation in CZE?

The selectivity of a separation is determined by the effective mobility because the effect of the electroosmotic mobility is equal for all the sample constituents.

3. Name two approaches that allow to separate neutral compounds in CE.

Neutral compounds can be separated either by applying MECC or by applying CEC.

4. What is meant by sample-stacking effect?

When the phenomenon of electrodispersion is advantageously used to enhance the detectability of analytes. This can be achieved by selecting appropriate conditions that allow to concentrate the analytes in narrow bands. For example, when the sample is dissolved in a solvent with an electrical conductivity lower than that of the buffer electrolyte, sample stacking occurs due to the difference in electric field strength in the sample and the electrolyte medium. As the field strength in the sample zone is higher, the migration velocity in the sample zone is also higher. At the interface of the sample and the buffer electrolyte zone the components are slowed (stacked) again, causing a concentration of the components at the interface.

Chapter 18a

Potentiometry

Martin Telting-Diaz and Yu Qin

18a.1   INTRODUCTION

Potentiometry is one of the most frequently used analytical methods in chemical analysis. In potentiometry, the voltage difference between two electrodes is measured while the electric current between the electrodes is maintained under a nearly zero-current condition. In the most common forms of potentiometry, the potential of a so-called indicator electrode varies depending on the concentration of the analyte, while the potential arising from a second reference electrode is ideally a constant. Most widely used potentiometric methods utilize an ion-selective electrode (ISE) membrane whose electrical potential to a given measuring ion, either in solution or in the gas phase, provides an analytical response that is highly specific. A multiplicity of ISE designs ranging from centimeter-long probes to miniaturized micrometer self-contained solid-state chemical sensor arrays [1] constitute the basis of modern potentiometric measurements that have become increasingly important in biomedical, industrial and environmental application fields. Indeed, a multitude of measurement applications as diverse as batch determination of medically relevant electrolytes or even polyions in undiluted whole-blood, the *in vivo* real-time assessment of blood gases and electrolytes with implantable catheters [2–4], the multi-ion monitoring in industrial process control [5] via micrometer solid-state chemical-field effect transistors (CHEMFETs) and the recent determination of heavy metals in environmental samples for low detection limit applications at trace levels [6] constitute just a few examples at the center of modern potentiometric methods.

Comprehensive Analytical Chemistry 47
S. Ahuja and N. Jespersen (Eds)
Volume 47 ISSN: 0166-526X DOI: 10.1016/S0166-526X(06)47027-6

18a.2   GENERAL PRINCIPLES

Potentiometry deals with the electromotive force (EMF) generated in a galvanic cell where a spontaneous chemical reaction is taking place. In practice, potentiometry employs the EMF response of a galvanostatic cell that is based on the measurement of an electrochemical cell potential under zero-current conditions to determine the concentration of analytes in measuring samples. Because an electrode potential generated on the metal electrode surface,

$$M^{z+}_{(aq)} + ze \rightleftharpoons zM_{(s)} \tag{18a.1}$$

depends on the activity of the metal ion in the solution and the charge of the metal ion, the Nernst equation can be written as

$$E = E^0 + \frac{RT}{zF} \ln a_I(aq) \tag{18a.2}$$

where E is the electrode potential, E^0 a constant, R, T and F are the gas constant ($8.314\,\mathrm{JK^{-1}mol^{-1}}$), absolute temperature (K) and Faraday constant (9.648×10^4 coulombs per mole of electrons), respectively, z is the charge of the analyte and $a_I(aq)$ the activity of the analyte in the aqueous sample. Ideally, membrane potentials show a linear response to the change of the logarithm of the sample activity.

As illustrated in Fig. 18a.1, all potentiometric measurements need an indicator electrode, a reference electrode and a voltmeter. As we see later in this chapter, the indicator electrode is most commonly an ISE.

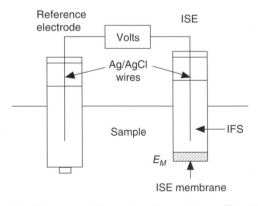

Fig. 18a.1. Schematic diagram of a potentiometric cell with an ion-selective electrode (ISE) as the indicator electrode. E_M is the electrical potential of the sensing membrane and IFS the internal filling solution.

18a.3   REFERENCE ELECTRODES

The electrochemical potential in a potentiometric cell is inevitably measured with respect to a standard electrode. Several types of electrodes are commonly used. The standard hydrogen electrode (SHE) is a hydrogen half-cell in which the cell reaction is as follows:

$$2H^+ + 2e^- \rightleftharpoons H_{2(g)} \tag{18a.3}$$

As an anode the hydrogen electrode can be represented schematically as

$$Pt,\ H_2(p = 1.00\ atm)/([H^+] = 1\ M)$$

where all components are at their standard states, with the hydrogen gas specified as having a partial pressure of 1 atm and the concentration of hydrogen ions in the solution is 1 M. This is not a convenient device to handle, so more often a secondary standard reference electrode is used, traditionally an Ag/AgCl electrode or a calomel electrode. These latter have redox potentials with respect to the standard H_2 electrode, which are well characterized. The redox potential for an electroactive species is usually assumed to be the value with respect to the standard reference electrode (the hydrogen electrode). If a value is measured with respect to a secondary standard, the secondary reference electrode potential with respect to the standard is added to this value. Among the main requirements of reference electrodes is the need for an accurately known and constant potential, which means that the electrode has to be insensitive to sample, temperature and other changes. Additionally, reference electrodes should be robust, easy to assemble and exhibit a nonvariable potential in the presence of a small current.

18a.3.1   Silver/silver chloride reference electrode

The most common and simplest reference electrode is the silver/silver chloride single-junction reference electrode. As shown in Fig. 18a.2, this type of electrode consists of a silver wire coated with silver chloride, which is in contact with an aqueous solution containing a concentrated salt such as 4 M KCl and saturated with AgCl. A cylindrical glass tube is used as the electrode body. The aqueous solution stays in contact with the sample solution by means of a ceramic or glass frit, a Teflon sleeve or a capillary to form the liquid junction. Since there is generally a large chloride concentration gradient across the junction, the reference potential will gradually change when used due to the slow diffusion of the chloride ions. In electrochemical terms, the half-cell can be represented

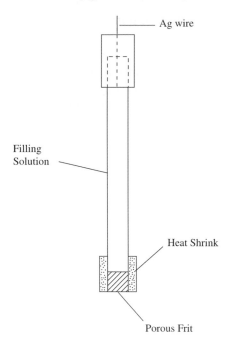

Fig. 18a.2. Schematic of a single-junction Ag/AgCl reference electrode.

by

$$Ag \ / \ AgCl_{(Satd)}, \ KCl_{(Satd)}$$

and the electrode reaction is

$$AgCl(s) + e^- \rightleftharpoons Ag_{(s)} + Cl^- \tag{18a.4}$$

The potential of a silver/silver chloride electrode with its corresponding saturated chloride activity against the SHE is 0.199 V at 25°C. Silver/silver chloride electrodes are available in a wide range of designs that include various shapes and sizes. Double-junction silver/silver chloride electrodes are commonly preferred since they minimize contact between the analyte sample and the KCl electrolyte in the electrode solution. The liquid junction of electrodes with porous restrains is often prone to clogging and thus requires careful attention to avoid erratic or unstable electrical signals. A free-flow capillary electrode design illustrated in Fig. 18a.3, and having a typical flow rate in the range of 1–10 µl/h, may be more desirable and particularly useful when assessing biological media or samples with elevated solid content.

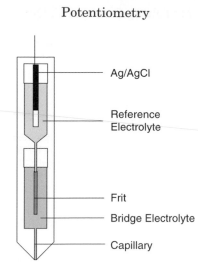

Fig. 18a.3. Reference electrode with a free-flow capillary design.

18a.3.2   Calomel electrode

Saturated calomel electrode (SCE) is another type of reference system of widespread use. The redox process for this electrode is

$$Hg_2Cl_2 + 2e^- \rightleftharpoons 2Hg + 2Cl^- \tag{18a.5}$$

The commercial SCE depicted in Fig. 18a.4 is generally an H-cell. One arm contains mercury covered by a layer of mercury(II) chloride (calomel). This is in contact with a saturated solution of potassium chloride; a porous frit is used for the junction between the reference electrode solution and the sample solution at the end of the other arm. Similar to the silver/silver chloride reference system, a calomel electrode also warrants precautionary measures to maintain the chloride concentration in the reference electrode.

In some applications, silver/silver chloride or calomel electrodes are considered cumbersome to use and maintain. More importantly, they are extremely difficult to miniaturize particularly with regard to their combined use with potentiometric membrane electrodes (see Section 18a.4.5.4) that have been fabricated into highly miniaturized and compact screen-printed sensor arrays for clinical use. Thus, several reference electrodes are manufactured with the same polymeric materials that are needed to design the responsive ion-selective membranes [7]. Incorporation of suitable active agents into such membranes leads to potentiometric responses that are ideally independent of the sample

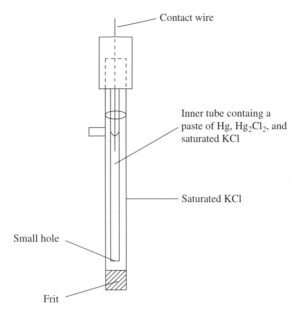

Fig. 18a.4. Diagram of a calomel reference electrode.

electrolyte. However, given that polymeric reference electrodes exhibit a common response behavior toward small sample ions, their operation is typically application specific. Their main advantage is clearly their ease of fabrication and integration into highly miniaturized and homogeneous measuring sensing arrays [7,8].

18a.3.3 Liquid junction potential

Since the composition of the reference electrode solution (i.e., high chloride ion concentration) is generally different from the composition of the sample solution, a potential difference, so-called liquid junction potential, across the interface of the two solutions is generated. The liquid junction potential (E_J) is the potential created at the sample/bridge electrolyte interface and it is caused by a separation of charge that is formed at the interface of the salt bridge of the external reference electrode and the sample solution. The charge separation is due to ions of different mobilities that migrate at different rates across the interface of two solutions of different concentrations. Unfortunately, this liquid junction potential cannot be measured. The influence of different sample compositions on the liquid junction potential can, however, be kept small and calculated to some extent on the basis of the

Henderson or Plank formalisms. Mobilities for several ions can be seen in Table 18a.1. Liquid junction potentials can become more problematic with voltammetric or amperometric measurements. For example, the redox potentials of a given analyte measured in different solvent systems cannot be directly compared, since the liquid junction potential will be different for each solvent system. However, the junction potential E_J can be constant and reproducible. It can also be very small (about 2–3 mV) if the anion and cation of the salt bridge have similar mobilities. As a result, for most practical measurements the liquid junction potential can be neglected [9].

18a.4   INDICATOR ELECTRODES

Two main groups of indicator electrodes are considered here. In one case, metal indicator electrodes that exhibit a potential difference as a consequence of a redox process occurring at the metal surface are examined. Later, ISEs that can respond to ionic species based on the principles of ion extraction across an active sensing membrane will be studied in detail.

18a.4.1   Metal electrodes

A metallic electrode consisting of a pure metal in contact with an analyte solution develops an electric potential in response to a redox reaction occurring at its metal surface. Common metal electrodes such as platinum, gold, palladium or carbon are known as *inert metal* electrodes whose sole function is to transfer electrons to or from species in solution. Metal electrodes corresponding to the *first kind* are pure metal electrodes such as Ag, Hg and others that respond directly to a change in activity of the metal cation in the solution. For example, for the reaction

$$Ag^+ + e^- \rightleftharpoons Ag_0 \tag{18a.6}$$

a silver metal electrode will be useful in determining the concentration of Ag^+ by expressing the Nernstian response as

$$E = E_0 - 0.0592\,V \times \log(1/[Ag^+]) \tag{18a.7}$$

It follows that metal electrodes of the *second kind* not only serve as indicator electrodes for their own cations but also react to changes in metal ion activity resulting from precipitation or complexation reactions. For example, a Cl^- ion electrode responds to the activity of the

TABLE 18a.1

Relative mobility of ions

	Full ion name	Valency	Relative mobility
Chol	Choline	1	0.51
Cs	Cesium	1	1.050
K	Potassium	1	1.000
Li	Lithium	1	0.525
NH_4	Ammonium	1	1.000
Na	Sodium	1	0.682
Rb	Rubidium	1	1.059
TEA	Tetraethylammonium	1	0.444
TMA	Tetramethylammonium	1	0.611
Acet	Acetate	−1	0.556
Benz	Benzoate	−1	0.441
Br	Bromide	−1	1.063
Cl	Chloride	−1	1.0388
ClO_4	Perchlorate	−1	0.916
F	Fluoride	−1	0.753
H_2PO_4	H_2PO_4	−1	0.450
HCO_3	HCO_3	−1	0.605
I	Iodide	−1	1.0450
NO_3	Nitrate	−1	0.972
Picr	Picrate	−1	0.411
Prop	Propionate	−1	0.487
SCN	Thiocyanate	−1	0.900
Ba	Barium	2	0.433
Ca	Calcium	2	0.4048
Co	Cobalt	2	0.370
Mg	Magnesium	2	0.361
Sr	Strontium	2	0.404
Zn	Zinc	2	0.359
HPO_4	HPO_4	−2	0.390
SO_4	Sulfate	−2	0.544
Ag	Silver	+1	0.842
H	Hydrogen	+1	4.763, 4.757
Cd	Cadmium	+2	0.37
Cu	Copper	+2	0.385, 0.365
Fe	Iron	+2	0.36, 0.37
Hg	Mercury	+2	0.433
Mn	Manganese	+2	0.364
Ni	Nickel	+2	0.340, 0.337

continued

Table 18a.1 (*continued*)

	Full ion name	Valency	Relative mobility
Pb	Lead	+2	0.48
Gd	Gadolinium	+3	0.306, 0.305
Fe	Iron	+3	0.313, 0.308
La	Lanthanum	+3	0.316

anion forming a sparingly soluble precipitate according to

$$AgCl_{(s)} + e^- \rightleftharpoons Ag_{(s)} + Cl^- \qquad (18a.8)$$

with its Nernstian response being

$$E = E^0_{(AgCl)} - 0.0592\,V \times \log[Cl^-] = 0.222\,V + 0.0592\,V \times pCl \qquad (18a.9)$$

Potentiometric metal electrodes are rather simple but lack major analytical relevancy due to a number of considerable disadvantages. In particular, they are not very selective, some metals are easily oxidized and still others (Zn, Cd) can dissolve in acidic solutions. Further, certain harder metals such as iron, cobalt and nickel do not provide reproducible potentials.

18a.4.2 Glass membrane pH electrodes

The glass membrane electrode was the first type of ISE and is still widely used for pH measurements. Its impressive selectivity and performance characteristics are almost unrivaled. The response of these electrodes is based on fixed SiO^- groups in the glass membrane that act as ion exchangers. The solid silicate matrix contains mobile alkaline metal ions inside the membrane and the electrode response is based on an ion-exchange mechanism [10]. When the membrane is in contact with an aqueous solution, the glass surface becomes hydrated and the alkali cations from the glass can be exchanged for other ions in the solution, preferably H^+, thus creating an electrical potential across the membrane. Doping the glass membrane with different proportions of aluminum oxide and other metal oxides can produce ion-selective glass membrane electrodes selective for other metallic ions such as Li^+, Na^+, K^+, Ag^+ or NH_4^+.

Illustrated in Fig. 18a.5 is a typical combination ion-selective glass electrode that houses both the reference and indicator electrode and is used to measure pH. Such glass electrodes have two main limitations.

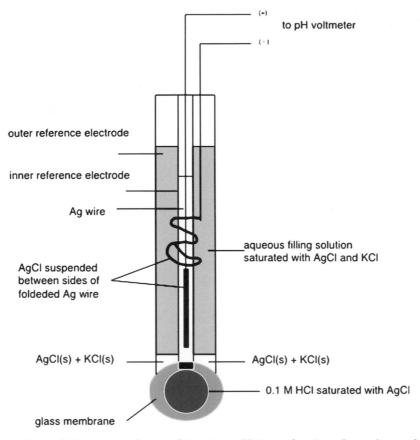

to pH voltmeter

outer reference electrode

inner reference electrode

Ag wire

aqueous filling solution
saturated with AgCl and KCl

AgCl suspended
between sides of
foldeded Ag wire

AgCl(s) + KCl(s)

AgCl(s) + KCl(s)

0.1 M HCl saturated with AgCl

glass membrane

Fig. 18a.5. Schematic of a combination pH ion-selective glass electrode.

First, measuring solutions containing particulate can damage the glass membrane and second, the glass membrane is fragile and can be easily broken. The high resistivity of glass also limits the degree of miniaturization and leads to relatively long response times for microelectrodes. Additionally, integration in highly compact and miniaturized arrays or its use in applications such as point-of-care testing are precluded. Many alternatives to the glass membrane do exist, though some are rarely used due to their own limitations. Reduced pH ranges and long response times are frequently cited. The antimony electrode is used as an alternative, mainly in solutions containing HF. A thin oxide layer formed on the surface of the antimony is sensitive to pH [11].

Various metal oxides including PtO_2, IrO_2, RuO_2, OsO_2, etc., have been used to prepare solid-state pH electrodes. In particular, the iridium oxide (mainly IrO_2) pH electrodes exhibit important advantages over conventional glass electrodes and other metal oxide electrodes including good stability over a wide pH range, high working temperatures up to 250°C [12], resistance to both high pressures [13] and aggressive environments [14,15], as well as fast response even in nonaqueous solutions [16]. The RuO_2 pH electrodes also exhibit a near-Nernstian response over a wide pH range, but drifts for both of the potential/pH slope and the apparent standard electrode potential ($E^{0'}$) are considered problematic [17,18].

18a.4.3   Ion-selective field effect transistors

Solid-state pH devices based on silicon microchips and known as ion-sensitive field effect transistors (ISFETs) constitute an important alternative to pH measurements with glass membranes. Their rapid expansion has been largely due to the fact that they can be made very small with current planar IC (integrated circuits) technology and have the advantage of fast response times. Clearly, they can also be fabricated massively at low costs. Important to their rapid acceptance is the use of field effect transistors (FETs). The FET measures the conductance of a semiconductor as a function of an electrical field perpendicular to the gate oxide surface. It contains a silicon dioxide insulating layer on top of which the electrolyte solution layer in contact with the reference electrode is placed, and then the SiO_2 gate oxide is placed directly in an aqueous electrolyte solution.

In ISFETs, the electric current flows from the source to the drain via the channel. The channel resistance depends on the electric field perpendicular to the direction of the current. It also depends on the potential difference over the gate oxide. These devices can transduce an amount of charge present on the surface of the gate insulator into a corresponding drain current. When SiO_2 is used as the insulator, the chemical nature of the interface oxide is reflected in the measured source–drain current. The surface of the gate oxide contains OH^- functionalities, which are in electrochemical equilibrium with ions in the sample solutions (H^+ and OH^-). The hydroxyl groups at the gate oxide surface can be protonated and deprotonated and thus, when the gate oxide contacts an aqueous solution, a change of pH will change the SiO_2 surface potential. A site-dissociation model describes the signal transduction as a function of the state of ionization of the amphoteric surface SiOH groups [19,20]. Typical pH sensitivities measured with SiO_2 ISFETs are 37–40 mV/pH unit [20]. The selectivity and chemical sensitivity of the ISFET are completely controlled by the properties of the electrolyte/insulator interface.

The ISFET can be modified with a hydrophobic ion-selective membrane that contains a highly selective ionophore [21]. When the gate oxide is covered with the selective membrane (see Section 18a.4.5.4), ions can be extracted into this layer, giving rise to a membrane potential. This potential is detected by the FET structure to enable the measurement of ion activity in the sample. Generally, the fabrication procedure involves solvent casting of poly(vinyl) chloride (PVC) membranes with incorporated plasticizer and ionophore on the top of the ISFET gate oxide. Alternately, FETs can be chemically modified and the attachment of the membrane can be improved by mechanical [22] or chemical [23,24] anchoring to the surface of the gate. For chemical attachment of polymer films, the gate oxide surface is silylated with 3-(trimethoxysilyl)propyl methacrylate. The methacrylate-modified surface can subsequently react with vinyl or methacryl monomers or prepolymers. The use of UV-photopolymerizable materials is advantageous for desired photolithographic mass production of CHEMFETs. Other polymeric materials such as polyurethane, silicone rubber, polystyrene, polyamide and several polyacrylates [23,25,26] have been used for ISFET fabrication to prevent leakage of plasticizer and improve the adhesion of the sensing membrane.

Miniaturized reference electrodes can also be made with the IC-compatible technology (reference field effect transistor—REFET) [27]. This modern technology provides a possibility to design multi-ion chemical sensors integrated with the reference cell (REFET). These sensors are very small, long-living and use only very small amounts of ion-sensing components [8,28].

18a.4.4 Solid-state electrodes

Insoluble inorganic materials with fixed sites can be used to fabricate ISEs. For example, Ag_2S, CuS, CdS, PbS, LaF_3, $AgCl$, $AgBr$, AgI and $AgSCN$ have all been tested as cation exchange sensitive membranes [29]. These solid materials can be incorporated into an electrode body in the form of single crystal, compressed powder discs or dispersed in suitable polymers. Copper sulfide, tungsten oxide and others are used for copper-selective electrodes [30] and triiodomercurate and chalcogenide have been found selective for mercury(II) ions [30]. These electrodes can respond to specific ions due to the mobility of the ions inside the ionic conductor. They are based on the low solubility of the solid materials. In solution, the membrane gives rise to an electrochemical potential between the internal reference and the sample solution, which depends on the activity of the target ions in the sample. A commercially available fluoride ISE, composed of a LaF_3 crystal doped with Eu^{2+} ions and shown in Fig. 18a.6, is one of the most well-known electrodes of this class. One of the principal applications of the fluoride electrode is in monitoring the levels of fluoridation of water supplies [31]. The membrane conducts F^- when immersed in solution and responds almost specifically to the activity of fluoride ions in the sample. Interference by hydroxyl ions is to be expected (selectivity, $K_{F^-,OH^-} = 0.1$). These materials are ionic conductors, though the conductivity is extremely small and mainly takes place through the migration of point defects in the lattice [31]. The response time of these membranes can be increased by incorporating ions such as Eu^{2+} into the lattice [31]. A series of electrodes for the detection of Ag^+, Cu^{2+}, Cd^{2+}, Pb^{2+}, S^{2-}, F^-, Br^-, I^-, SCN^- and CN^- ions can be constructed

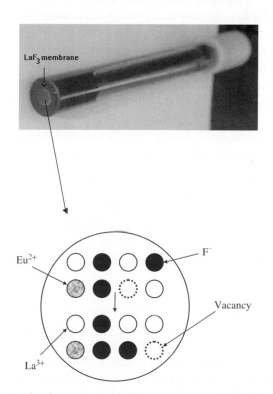

Fig. 18a.6. Photograph of a commercial La_3F selective electrode with expanded schematic view of the solid-state sensing membrane.

from such membranes [32,33]. The sensitivity of these electrodes arises from the dissolvation equilibria at the membrane surface and the Nernstian electrode function lies in the range of $1-10^{-6}$ M [33]. Interference effects are frequently encountered [33].

18a.4.5 Ion-selective polymer membranes

Ion-selective electrodes based on solvent polymeric membranes are called liquid membrane electrodes given the incorporation of a highly viscous solvent mediator in the organic polymer matrix. This type of ion-selective membrane is robust and has good mechanical properties. More importantly, the potentiometric membrane can be tailored to make selective electrodes for different analytes including cations, anions, polyionic macromolecules and to a lesser extent some acidic and basic gases. The most widely used membrane material is PVC doped with a key component, a complexing agent capable of reversibly binding ions. This agent is usually called ionophore or also ion carrier since it assists ion transport across a hydrophobic membrane. The essential part of a carrier-based ISE is the ion-sensitive membrane which separates two aqueous phases, i.e., the sample and the internal reference electrolyte solution. Additionally, the membrane contains a lipophilic salt as ion-exchanger. Carrier-based ISEs are by far the most popular due to their notably high selectivity and cover a rather large sensitivity range, typically $1-10^{-6}$ M. Among their chief advantages is their ability to directly assess analyte activity in colored turbid samples without requiring sample pretreatment.

18a.4.5.1 Response mechanism

The EMF across the entire potentiometric cell, shown in Fig. 18a.1, is the sum of the individual potentials that include the reference electrode potential and other sample-independent potentials (E_{const}), the liquid junction potential (E_J) and the membrane potential (E_M):

$$\text{EMF} = E_1 + E_2 + E_3 + E_4 + E_5 + E_J + E_M = E_0 + E_J + E_M \qquad (18a.10)$$

where E_0 is the constant that encompasses the various potential contributions E_1 through E_5 and the liquid junction potential (E_J) is the potential created at the sample/bridge electrolyte interface. The membrane potential (E_M) becomes the most important part of the signal. The sensing membrane is usually interposed between the sample and the inner filling solution; therefore, the membrane potential can be divided into three potential contributions, namely the phase boundary potentials at both interfaces and the diffusion potential within the membrane as shown in Eq. (18a.11).

$$E_M = E_{diff.} + E_{PB'} + E_{PB''} \qquad (18a.11)$$

where $E_{diff.}$ is the diffusion potential in the membrane, $E_{PB'}$ the phase boundary potential between the membrane and inner filling solution and $E_{PB''}$ the phase boundary potential between the membrane and the sample. A number of experiments have shown that the diffusion potential within the membrane is negligible in most practical experiments if the membrane is well conditioned and shows a Nernstian response. The membrane potential is then given by the following equation:

$$E_M = E_{const} + E_{PB'} + E_{PB''} \qquad (18a.12)$$

The only potential that varies significantly is the phase boundary potential at the membrane/sample interface $E_{PB''}$. This potential arises from an unequal equilibrium distribution of ions between the aqueous sample and organic membrane phases. The phase transfer equilibrium reaction at the interface is very rapid relative to the diffusion of ions across the aqueous sample and organic membrane phases. A separation of charge occurs at the interface where the ions partition between the two phases, which results in a buildup of potential at the sample/membrane interface that can be described thermodynamically in terms of the electrochemical potential. At interfacial equilibrium, the electrochemical potentials in the two phases are equal. The phase boundary potential is a result of an equilibrium distribution of ions between phases. The phase boundary potentials can be described by the following equation:

$$E_{PB} = k_I + \frac{RT}{zF} \ln \frac{a_I(aq)}{a_I(org)} \qquad (18a.13)$$

where k_I is the combined standard chemical potential

$$k_I = \exp\left(\{\mu^0(aq) - \mu^0(org)\}/RT\right) \qquad (18a.14)$$

μ^0 the standard chemical potential, R, T and F are the gas constant, absolute temperature and Faraday constant, respectively and a_I is the activity of the target ion. Eq. (18a.13) has been derived from basic thermodynamic theory by assuming that the equilibrium holds at all of the interfaces, and the electrochemical potentials for both the aqueous phase and organic phase are equal [9]. Therefore, the phase boundary potentials are proportional to the logarithm of the ratio of the activities of the free target ion in the aqueous phase and the organic phase.

Under the condition that $a_I(\text{org})$ remains constant, which is the case for well-conditioned carrier-based membranes, Eq. (18a.13) can be reduced to the well-known Nernst equation, i.e., Eq. (18a.15) (see also Eq. (18a.2))

$$E_M = E^0 + \frac{RT}{zF} \ln a_I(\text{aq}) \qquad (18a.15)$$

where E^0 includes all of the constants in the equation. Naturally, the membrane responds to the activity of the target ion. However, in order to get a constant composition in the membrane, the membrane must have permselectivity which is provided by the ion-exchange properties and hydrophobicity so that the coextraction of counterions is prohibited. Figure 18a.7 shows a typical calibration curve of an ISE for measuring monovalent cations. In the linear range of the calibration curve the electrode has a Nernstian response with a theoretical slope of 59.1 mV/decade for monovalent analytes.

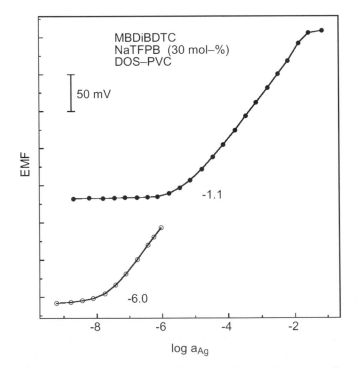

Fig. 18a.7. Typical calibration curve of a potentiometric sensor for measuring monovalent cations. From Ref. [70] with permission.

18a.4.5.2   Measurement and source of errors

Potentiometric measurements with ISEs can be approached by direct potentiometry, standard addition and titrations. The determination of an ionic species by direct potentiometry is rapid and simple since it only requires pretreatment and electrode calibration. Here, the ion-selective and reference electrodes are placed in the sample solution and the change in the cell potential is plotted against the activity of the target ion. This method requires that the matrix of the calibration solutions and sample solutions be well matched so that the only changing parameter allowed is the activity of the target ion.

A second approach is the standard addition method, which is commonly employed when the sample is unknown. The potential of the electrode is measured before and after addition of a small volume of a standard to the known volume of the sample. The small volume is used to minimize the dilution effect. The change in the response is related only to the change in the activity of the primary ion. This method is based on the assumptions that the addition does not alter the ionic strength and the activity coefficient of the analyte. It also assumes that the added standard does not significantly change the junction potential.

Often, ISEs can be used as potentiometric equivalence point indicators during a titration. In this case, a known volume of titrant is added to the sample during the titration procedure. The change in the electrode potential gives a measure of the change in concentration of a particular species. The potentiometric endpoint is widely applicable and provides accurate data. It is particularly useful for titration of colored or turbid species in solution and for detecting the presence of unsuspected species in the sample. By far, the most important application of this method is the routine determination of pH using pH autotitrators. In automated acid–base titrations, the potential change of a glass electrode is used to identify the equivalence point of the titration. In some cases, the location of the equivalence point can be easily identified from the pH profile of the titration when an excess of titrant is added to the initial solution. However, for weak acids or bases, and in particular polyprotic species, the titration profiles show nonvisible equivalence points. Here, the first or second derivative of the pH trace can be used to accurately identify the precise volume of the titrant.

Several sources of errors may be present during measurements. The response of an ISE is related to activity rather than to concentration. However, given that analysts are usually interested in concentration, the activity coefficient (γ) of the ionic species in the solution is required. In many situations, the activity coefficients are unavailable because the ionic strength of the solution is unknown or too high to be obtained from the Debye-Hückel equation. Therefore, the assumption that the activity and concentration are identical can lead to serious errors, especially when dealing with polyvalent ions. Also, the activity of the target ion must be in the dynamic measuring range of the ISE. The minimum activity that can be accurately determined depends on the type and activities of the interfering ions in the sample and the selectivity of the ISE. In addition, the formation of complexes between target ions and ligands in the sample can result in the underestimation of the total ion concentration as a result of the smaller free-ion activity being observed.

Interfering ions and sample effects may also contribute to errors in potentiometry. The electrode signal arising from interfering ions in the sample should be much smaller than that of the target ions, so that it can be made negligible in most of the situations. Practically, in order to minimize the variations in ionic strength, complexation and junction potential, it is important to match the background composition of the samples and the calibrants as closely as possible.

Errors in ISE measurements may also come from the sample composition. Samples may contain species other than the target ion and the interfering ions. For example, proteins or lipids may create suspensions or particle gels that can deteriorate the sensing membrane or dissolve the membrane components. In these situations, the normal performance of the electrode is affected and the working lifetime of the electrode may be reduced drastically. Furthermore, proteins and long-chain lipids may cover the membrane and thus block the sensing surface on the electrode. Chelates and other complexing agents may also develop side reactions with the target ions and interfere with the proper measurements. In addition, the sample composition and the ionic strength can change the junction potential of the reference electrode and lead to considerable errors.

18a.4.5.3 Membrane materials and performance optimization

Generally, the ion-selective membrane is made of plasticized PVC matrix, a highly lipophilic ionophore or ion-carrier and lipophilic ionic sites. Each component is critical to the performance of the electrodes. Adjusting the membrane composition is key to obtaining enhanced analytical responses and superior electrode selectivities. Additionally, the membrane components must have adequate lipophilicity so that each component is retained in the membrane phase for sufficiently long times. Immobilization of ionophores and ionic sites in polymer membranes can completely eliminate leaching effects [34]. In this respect, considerable efforts have been undertaken to ensure electrode systems with longer lifetimes [35,36].

Ionophores

The ionophore is undoubtedly a crucial component of polymer-based membrane electrodes. The complexation of the analyte and ionophore with specific stoichiometric ratios constitutes the driving force for analyte extraction into the organic phase. The strong complexation ensures that the free-ion concentration in the organic phase is kept constant. The complex formation constant is critical for the selectivity of the electrodes because high selectivity comes as a result of a strong complexation of the ionophore with a primary ion and a weak or no interaction with interfering ions (see Section 18a.4.5.5). Figure 18a.8 shows well-known ionophores for cations which are usually neutral macromolecules with oxygen or nitrogen atoms, while Fig. 18a.9 illustrates some common ionophores for anions which are often organic transition metal complexes and are, therefore, positively charged [37].

Ionic site additives

In ISE membranes, lipophilic salts are usually added as anion or cation exchanger. It is well known that for neutral carrier-based membranes to maintain their permselectivity characteristics (Donnan exclusion), counterions must be present in the membrane, otherwise, the analyte will be coextracted into the membrane phase (the so-called Donnan failure). A PVC membrane without any ionic sites may still exhibit a cationic response due to the presence of anionic impurities [38–40]. Detailed studies on ISEs demonstrate that adding the ionic sites also lowers the membrane resistance and reduces the interference from anions. Furthermore, the selectivity of the ISE can be adjusted by changing the ratio of concentrations of the ionophore and ionic sites. Even for electrically charged ionophores the ionic sites can still influence the selectivity of the electrode although the ionophore itself can induce Donnan exclusion. Clearly in this case, the ionic site has the same charge as the analyte. As illustrated in Fig. 18a.10, two types of ionic sites can be used. Cationic sites used in anion-selective electrodes are commonly lipophilic tetraammonium salts while tetraphenylborate derivatives are employed for cation-selective systems. Unfortunately, tetraphenylborate salts have long been known to suffer from important drawbacks, namely low lipophilicity, hydrolysis in acid and photosensitivity [41]. Until recently, such anionic sites were the only type of anionic sites available. Recently, halogenated carboranes displaying

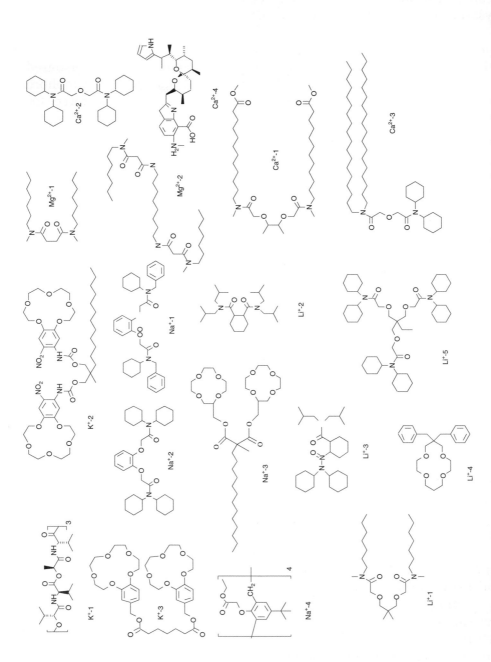

Fig. 18a.8. Chemical structures of some commonly known cation ionophores used in the design of ion-selective electrodes.

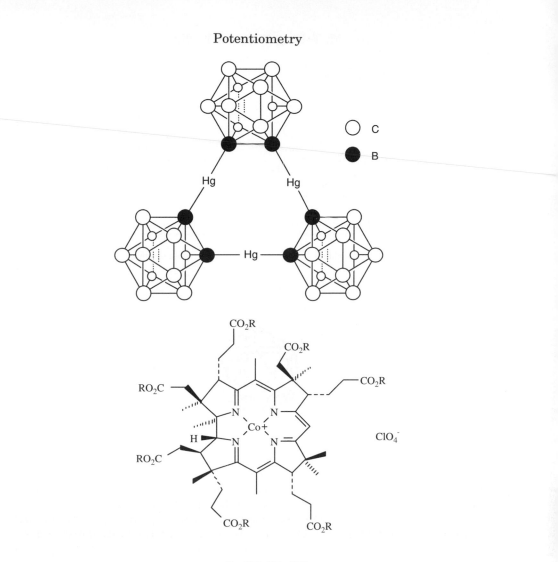

R: -CH$_2$-CH$_2$-C$_6$H$_5$

Fig. 18a.9. Anion-selective ionophores. Top: neutral ionophore with selectivity toward Cl$^-$ ions; bottom: charged ionophore selective for NO$_2^-$ ions.

improved lipophilicity and acid resistance have been documented and used successfully as anionic sites in potentiometric and optical-chemical sensors [42,43].

Polymer matrices and plasticizers
A membrane matrix made with 33% (w/w) of PVC and 66% (w/w) of plasticizer is most useful for ISEs. Adequate polymer matrices must

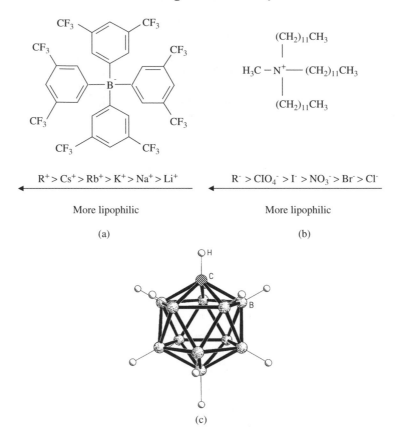

$$R^+ > Cs^+ > Rb^+ > K^+ > Na^+ > Li^+$$

More lipophilic

(a)

$$R^- > ClO_4^- > I^- > NO_3^- > Br^- > Cl^-$$

More lipophilic

(b)

(c)

Fig. 18a.10. Chemical structures of ion exchangers used in polymeric membranes of ion-selective electrodes with preference toward ions as indicated. (a) Tetraphenyl borates anion exchanger, (b) tridodecyl methyl ammonium chloride cation exchanger and (c) newly introduced carborane cation exchanger.

have a glass transition temperature (T_g) lower than room temperature [44]. Given that PVC exhibits a T_g of approximately 80°C, the inclusion of a low T_g plasticizer is needed to render the matrix mechanically robust and elastic. The incorporation of a plasticizer also provides a much-needed lipophilic environment in which other membrane components dissolve. A number of commonly used plasticizers are illustrated in Fig. 18a.11. These plasticizers have different structures, dielectric constants and lipophilicity characteristics and are often found to exert a strong influence on the electrode's response and its selectivity. The reasons ascribed for such effects are either the lipophilicity of the plasticizer that influences carrier solubility or a direct chemical effect over the ionophore–ion complex.

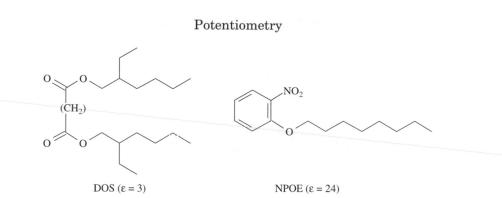

DOS (ε = 3)   NPOE (ε = 24)

Fig. 18a.11. Chemical structures of commonly used plasticizers in membranes of ion-selective electrodes. DOS, dioctyl sebacate; NPOE, nitrophenyl octyl ether.

Plasticized PVC membranes are by no means without limitations. A serious drawback is the leaching of the plasticizer from the membrane, which hampers the application of ISEs in samples of high lipophilic content (e.g., biological media) and shortens the lifetime of the electrodes [45]. The PVC matrix also exhibits poor adhesion to several surfaces, which poses significant problems for the fabrication of miniaturized electrode formats. Fortunately, a wide variety of alternative matrices are available although each with its own benefits and limitations. Suitable matrices for ISEs on the basis of derivatives of polyurethanes, silicone rubber, polysiloxanes and others have been thoroughly studied and used in many applications [46–48]. Methacrylate copolymers and methacrylate–acrylate copolymer are very attractive as they can be prepared by one-step polymerization and are completely plasticizer-free [49–53]. By using ionophores with polymerizable groups, such matrices permit one-step synthetic approaches resulting in electrodes with reduced leaching and prolonged lifetimes [54–57].

18a.4.5.4 Ion-selective electrodes with solid contact
The solid-contact electrode design shown in Fig. 18a.12 eliminates the internal electrolyte solution so that potentiometric sensor devices of various shapes and sizes can be manufactured. In this approach, the membrane is placed directly on a solid and electronically conducting surface. A recognized problem, however, is the ill-defined charge transfer at the interface between the ion-selective membrane that is ion-conductive, and the electronically conducting metal substrate [58]. Thus, a long-standing challenge for these electrodes has been how

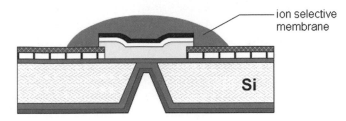

Fig. 18a.12. Schematic diagram of ion-selective electrode chips with solid contact.

to best define the reversibility and stability of the ion-to-electron transduction process at this interface. In a common approach, the proper charge transfer between the membrane and the conducting substrate rests on the incorporation of a conducting polymer. The conducting polymer is able to transduce an ionic signal into an electrical signal (mixed ionic and electronic conductivity), therefore, the phase boundary is thermodynamically well-defined without the intermediate internal solution. Conducting polymers such as poly(aniline), poly(3-octylthiophene) or poly(pyrrole) added in just a few mass percentage to the membrane provide a significant improvement of charge transfer with enhanced stability of the electrode signal [59]. Studies on poly(3, 4-ethylenedioxythiophene) as solid-contact material have shown that the potential stability of solid-contact-based ISEs is directly related to the redox capacitance of the conducting polymer [60–62]. Among those electrically conductive polymers of interest, polyaniline shows practical advantages since it deposits from aqueous acid solutions, adheres strongly to supports and exhibits good environmental stability [63]. Conducting polymers are deposited by electropolymerization or solution casting directly on the metal conductor. Alternately, they can be added directly in membranes of PVC or Siloprene to obtain mixed conductivity. Solid-contact ISEs fabricated on the basis of conducting polymers adopt a variety of formats similar to coated wire electrode, glassy carbon disc electrodes, screen-printing electrodes or planar microelectrode chips [1,8,30,61,63–65].

18a.4.5.5 *Selectivity and detection limit characteristics*
Selectivity
The Nernst equation (Eq. (18a.2)) describes the response of an ISE under ideal conditions in the absence of interfering ions in solution. However, real samples contain competing ions of the same charge as

the primary ion that may be extracted into the membrane. The extent to which interfering ions are extracted is described by the selectivity of the electrode for the primary ion over the interfering ion. Selectivity is one of the most important features of an ISE and is defined as the degree of discrimination of an electrode to ions in solution other than the primary ion [66–68]. The selective complexation of the target by the ionophore ensures that the membrane responds selectively to the analyte ion in an otherwise complex sample matrix.

Selectivity determination constitutes an important challenge because of the many variables that may affect the results. This section describes the theoretical foundations of selectivity and current methods employed to define this important characteristic of ISEs.

The traditional equation to describe the selectivity of ISEs is the Nicolskii-Eisenman equation:

$$E = E_I^0 + \frac{RT}{z_I F} \ln(a_I(I)) = E_I^0 + \frac{RT}{z_I F} \ln\left(a_I(IJ) + K_{IJ}^{pot} a_J(IJ)^{z_I/z_J}\right) \qquad (18a.16)$$

where $a_I(I)$ is the primary-ion activity in a solution without interfering ions and $a_I(IJ)$ and $a_J(IJ)$ are the activities of primary ion I and interfering ion J in the mixed sample. The Nicolskii coefficient, K_{IJ}^{pot}, is the potentiometric selectivity coefficient. If there is more than one interfering ion in solution, the sum of the selectivity constants $(\sum_j K_{IJ}^{pot})$ is used. If interference by J occurs, the potential for a solution with a certain activity I $(a_I(IJ))$ will be higher than it would be for the same activity of I in a solution containing no J $(a_I(I))$. The smaller the coefficient, the better the selectivity of the electrode. If K_{IJ}^{pot} is very small and $a_I(IJ)$ approaches the primary-ion activity $a_I(I)$ in a solution without interfering ions according to the following equation:

$$a_I(I) = a_I(IJ) + K_{IJ}^{pot} a_J(IJ) \qquad (18a.17)$$

it means that there is no interference. If interference is observed, a lower activity $a_I(IJ)$ of the mixed sample will give the same response as the activity $a_I(I)$ of a solution containing no interfering ions. The Nicolskii-Eisenman equation has been used to describe EMF responses of mixed ions in solution and is a very accurate formalism as long as the interfering ions are of the same charge as the primary ion. For cases where $z_I \neq z_J$ the Nicolskii-Eisenman equation is not valid in the activity range where the primary and the interfering ion significantly contribute to the potential. The Nicolskii-Eisenman equation will give different analytical expressions depending on which ion is treated as the primary ion and which is the interfering ion. The calculated EMF values E_{IJ} and E_{JI} will not be equal in such cases and therefore the Nicolskii-Eisenman equation can lead to substantial errors when used to determine selectivity coefficients and describe EMF responses to solutions of mixed ions of different charge. The Nicolskii-Eisenman equation is only valid in the linear ranges of the response curve, i.e., in the limiting cases of $a_I \ll K_{IJ}^{pot} a_J^{z_I/z_J}$ and $a_I \gg K_{IJ}^{pot} a_J^{z_I/z_J}$. Despite this deficiency of the Nicolskii-Eisenman equation, the potentiometric selectivity factor, K_{IJ}^{pot}, is still the best possible measure to quantify interference because it corresponds to the ion-exchange selectivity of the membrane and is constant for a specific electrode.

For carrier-based ISEs, the selectivity is determined by the composition of the membrane. Eq. (18a.18) describes the relationship between the ion selectivity and the membrane composition of cation-selective membranes by using charge balance and mass balance considerations [69].

$$K_{IJ}^{pot} = K_{IJ} \frac{(\beta_{JL_{n_J}})^{z_I/z_J}}{\beta_{IL_{n_I}}} \frac{R_T^-}{z_I[L_T - n_I(R_T^-/z_I)]^{n_I}} \left(\frac{z_J[L_T - n_J(R_T^-/z_J)]}{R_T^-} \right)^{z_I/z_J}$$

(18a.18)

where K_{IJ} is the equilibrium constant for the ion exchange between the uncomplexed primary and interfering ions between the sample and the membrane phase, L_T and R_T are the total concentrations of carrier and anionic site, z_I and z_J the charge of the primary ion and interfering ion, respectively, while β is the complex formation constant of the ion–ligand complex and n the complex stoichiometry. As shown in Eq. (18a.18), the ratio of the stability constants of the ionophore with primary ion and interfering ion is critical to the selectivity of the electrode. To be valid, this formalism assumes that each ion must be strongly complexed by the ionophore, the complex is of one fixed stoichiometry and that the effects of ion-pairing in the membrane are negligible.

Separate solution method

A commonly adopted approach to measure the selectivity of an ISE is the separate solution method, which is illustrated in Fig. 18a.13. Here, two separate solutions, which contain the primary ion and interfering ion individually, are measured separately. The selectivity is calculated from

$$\log K_{IJ}^{pot} = \frac{z_I F \{E(J) - E(I)\}}{2.303 RT} + \log \left(\frac{a_I(I)}{a_J(J)^{z_I/z_J}} \right)$$

(18a.19)

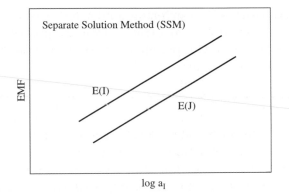

Fig. 18a.13. Schematic representation of ion-selective electrode selectivity as determined by the separation solution method (SSM). $E(I)$ is the potential of the electrode in primary-ion solution and $E(J)$ the electrode potential in the interfering ion solution. Both primary and interfering ions show Nernstian response.

where $E(J)$ and $E(I)$ are the recorded potentials in separate solutions for primary ion I and interfering ion J and $a_I(I)$ and $a_J(J)$ the activities of I and J in separate solutions. The selectivity coefficients obtained using the separate solution method are equal to the thermodynamically defined ones as long as the response slopes of all of the ions are Nernstian.

Mixed interference method
In this method, an entire calibration curve is measured for the primary ion in a constant background of interfering ion. $a_J(BG)$ is the activity of the constant interfering ion in the background. $a_I(DL)$ is the low detection limit (LDL) of the Nernstian response curve of the electrode as a function of the primary-ion activity. In the mixed interference method the selectivity is calculated from the following equation:

$$\log K_{IJ}^{pot} = \log\left(\frac{a_I(DL)}{a_J(BG)^{z_I/z_J}}\right) \tag{18a.20}$$

For both methods, a Nernstian response of both interfering ion and primary ion is required [9,67,68].

Matched potential method
A third relevant method consists in adding a specific amount of primary ions to a reference solution and the membrane potential is measured.

In a separate experiment, interfering ions are successively added to an identical reference solution until the membrane potential matches the one obtained before with the primary ion. The selectivity is calculated from the following equation:

$$K_{IJ}^{MPM} = \frac{\Delta a_I}{\Delta a_J} \tag{18a.21}$$

The advantage of this approach is that it can be used when the response of the electrode is not Nernstian or even linear. However, the drawback is that the selectivity coefficient may change under different experimental conditions, particularly the concentrations at which the measurements were made. In most situations, the selectivity obtained from matched potential method cannot be directly compared to the values obtained from other methods.

Traditionally, the membrane is conditioned in a solution containing a relatively high concentration of primary ion to ensure a stable and reproducible electrode behavior. This procedure however does not contemplate the fact that the response of the electrode is not always Nernstian, which is clearly the case if primary ions cannot be completely exchanged by the highly discriminated interfering ions [70]. In most situations, the slopes for the interfering ions are sub-Nernstian and the selectivity is biased [70]. A newly introduced method to determine unbiased selectivity contemplates conditioning the electrode in interfering ion solutions instead of primary-ion solutions and measuring the primary ions at the end of the measurements. Because the primary ions are preferred, they exchange easily with the interfering ions and the membrane shows Nernstian response again. In this manner, a Nernstian response can be obtained for both the interfering ions and primary ion as depicted in Fig. 18a.14.

Detection limits

According to IUPAC the detection limit is defined as the concentration for which the cell EMF deviates from the average EMF in a region by a multiple of the standard error of a single measurement of the EMF in this region. As shown in Fig. 18a.15, the lower detection limit corresponds to the primary-ion activity at the point of intersection of the extrapolated linear Nernstian segment and the final low concentration level segment of the calibration plot. On the other hand and as illustrated in Fig. 18a.15 the upper detection limit (UDL) is defined as the primary-ion activity at the point of intersection of the extrapolated linear Nernstian segment and the limiting high activity response [9].

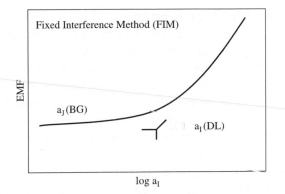

Fig. 18a.14. Illustration of selectivity determination by the fixed interference method (FIM). a_J(BG) is the activity of the interfering ion J in the background and a_I(DL) is the detection limit of the primary ion in the solution containing J as background.

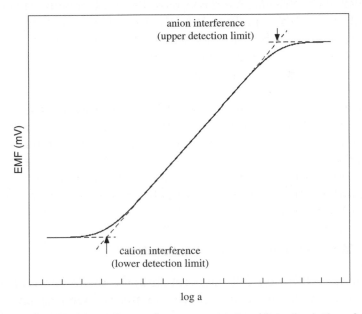

Fig. 18a.15. Ion-selective electrode response function depicting lower and upper detection limit zones.

Ideally, the LDL is the value of primary-ion activity where the activity of the primary ion equals the summation term for interfering ions in the Nicolskii-Eisenman equation (Eq. (18a.16)). When the activity of the primary ion is lower than the detection limit, the electrode responds to interfering ions and gives no response toward primary ions.

For a long time, the main limitations of ISEs were their insufficient detection limit and selectivity. Most of the potentiometric sensors had detection limits only in the micromolar range. Recently, the theory involving the LDL of ion-selective polymer membranes has become better understood [6]. Indeed, the main reasons dictating the lower detection limit of these electrodes is that the ion-selective membrane itself contains a certain amount of analyte because of the preconditioning of the membrane in relatively high concentration solutions. These analytes will leach out of the membrane when used in low concentration samples. The aqueous phase boundary is contaminated and the local concentration is higher than the sample concentration, which determines the response of the electrode. Very recently, the LDL of carrier-based ISEs could be improved down to 10^{-10} M by changing the composition of the inner filling solution and minimizing the ion flux in the membrane phase [71]. Monitoring of heavy metals in environmental samples with ISEs has been compared to routine analytical methods such as ICP-MS [72] (see Fig. 18a.16). In contrast to the complicated instrumentation of other methods, the dramatic improvement in the LDL is achieved by detailed understanding of the chemical principles and eliminating experimental biases.

Conversely, the fundamentals for the UDL lie on the coextraction of counterions into the membrane; therefore, the membrane is no longer permselective (Donnan failure) [9]. Ideally, when the ionophores are saturated by ions, the ion–ionophore complex functions as an ion-exchanger and the membrane shows an anion Nernstian response. The UDL can be estimated from the membrane composition, formation constant and coextraction coefficients obtained from the so-called sandwich membrane method [73].

18a.4.6 Modified ion-selective electrodes

Potentiometric ISEs do not respond directly to nonionic species, but different strategies to modify them permit broadening the range of measured analytes. Indeed, various electrode arrangements make use of traditional ISEs surrounded by an appropriate membrane film that confines the analyte species to the vicinity of the electrode surface. For example, the pH glass electrode can be covered with a hydrophobic gas-permeable membrane and a given buffer solution then fills the gap between the glass electrode and the membrane. This type of modified glass electrode is referred to as a potentiometric gas-sensing electrode. Carbon dioxide is often determined in blood samples by Severinghaus-type

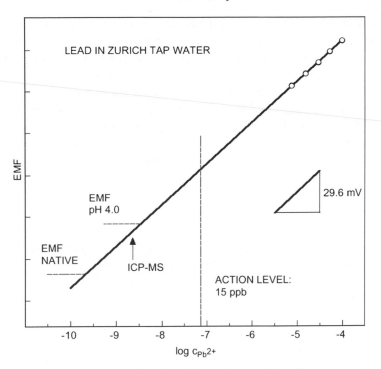

Fig. 18a.16. EMF of an ion-selective electrode used in the nanomolar determination of Pb_2^+ ions in tap water at pH 4. Measurement of the same sample via ICP-MS is indicated by an arrow. From Ref. [72] with permission.

potentiometric sensors which are based on the permeation of CO_2 through a gas-permeable membrane into an internal hydrogen carbonate solution and indirect determination of the gas occurs by measuring the pH of the buffer solution. Some acidic or basic gaseous species, namely, H_2S, NH_3, SO_2, organic amines, etc., dissolve in solution and diffuse through the membrane, causing a pH change that can be detected by a glass pH electrode [74]. Such an approach can be used to measure gases in solution or in the gaseous phase.

The use of additional membranes, which selectively convert nonionic analytes into ionic species that can be determined via ISEs is another common approach. An abundance of ingenious designs make use of biocatalysts for the development of potentiometric biosensors. Much of the earlier designs have made use of enzymes as the molecular recognition element. The products that are associated with such enzyme-catalyzed reactions can be readily monitored with the potentiometric transducer by coating the traditional electrodes with the enzyme.

The method of enzyme immobilization constitutes a key factor in the construction of these systems as it is the biocatalytic membrane that largely determines sensitivity, stability and response-time characteristics of the biosensor.

Potentiometric enzyme-based electrodes have found application in clinical, pharmaceutical, food and biochemical analyses to enable the selective determination of a wide range of important enzyme substrates, including amino acids, esters, amides, acylcholines, β-lactam antibiotics, sugars, enantioselective drugs and many others [74].

REFERENCES

1   J.E. Zachara, R. Toczylowska, R. Pokrop, M. Zagorska, A. Dybko and W. Wroblewski, *Sens. Actuators B*, 101 (2004) 207.
2   M.M. Reynolds, M.C. Frost and M.E. Meyerhoff, *Free Rad. Biol. Med.*, 37 (2004) 926.
3   M.C. Frost and M.E. Meyerhoff, *Curr. Opin. Chem. Biol.*, 6 (2002) 633.
4   M. Telting-Diaz, M.E. Collison and M.E. Meyerhoff, *Anal. Chem.*, 66 (1994) 576.
5   S. Ito, Y. Asano and H. Wada, *Talanta*, 44 (1997) 697.
6   E. Bakker and E. Pretsch, *Anal. Chem.*, 74 (2002) 420A.
7   M. Ciobanu, J.P. Wilburn and D.A. Lowy, *Electroanalysis*, 16 (2004) 1351.
8   W. Wróblewski, A. Dybko, E. Malinowska and Z. Brzózka, *Talanta*, 63 (2004) 33.
9   E. Bakker, P. Buehlmann and E. Pretsch, *Chem. Rev.*, 97 (1997) 3083.
10 A.I. Vogel and G.H. Jeffery, *Vogel's Textbook of Quantitative Chemical Analysis*, 5th ed., Longman Scientific & Technical, Harlow (Essex, England), 1989.
11 J.T. Stock, W.C. Purdy and L.M. Garcia, *Chem. Rev.*, 58 (1958) 611.
12 J.V. Dobson, P.R. Snodin and H.R. Thirsk, *Electrochim. Acta*, 21 (1976) 527.
13 T. Katsube, I. Lauks and J.N. Zemel, *Sens. Actuators B*, 2 (1982) 399.
14 I. Lauks, M.F. Yuen and T. Dietz, *Sens. Actuators B*, 4 (1983) 375.
15 M.L. Hitchman and S. Ramanathan, *Analyst*, 116 (1991) 1131.
16 K. Izutsu and H. Yananoto, *Anal. Sci.*, 12 (1996) 905.
17 K. Pa'sztor, A. Sekiguchi, N. Shimo, N. Kitamura and H. Masuhara, *Sens. Actuators B*, 13–14 (1993) 561.
18 H.N. McMurray, P. Douglas and D. Abbot, *Sens. Actuators B*, 28 (1995) 9.
19 L. Bousse and P. Bergveld, *Sens. Actuators B*, 6 (1984) 65.
20 A. van den Berg, P. Bergveld, D.N. Reinhoudt and E.J.R. Sudholter, *Sens. Actuators B*, 8 (1985) 129.
21 S.D. Moss, J. Janata and C.C. Johnson, *Anal. Chem.*, 47 (1975) 2238.
22 G.F. Blackburn and J. Janata, *J. Electrochem. Soc.*, 129 (1982) 2580.

23 P. van der Wal, M. Skowroñska-Ptasiñska, A. van den Berg, P. Bergveld, E.J.R. Sudholter and D.N. Reinhoudt, *Anal. Chim. Acta*, 231 (1990) 41.

24 D.J. Harrison, A. Teclemariam and L. Cunningham, *Anal. Chem.*, 61 (1989) 246.

25 M. Mascini and F. Palozzi, *Anal. Chim. Acta*, 73 (1974) 375.

26 U. Fiedler and J. Ruzicka, *Anal. Chim. Acta*, 67 (1973) 179.

27 E.S. Hochmair and O. Prohaska, *Implantable Sensors for Closed-Loop Prosthetic Systems*, Futura, Mountkisco, NY, 1985.

28 J. Janata, *Electroanalysis*, 16 (2004) 1831.

29 J.J. Dvorak and L. Kavan, *Principles of Electrochemistry*, 2nd ed., Wiley, NY, 1993.

30 M.J. Gismera, D. Hueso, J.R. Procopio and M.T. Sevilla, *Anal. Chim. Acta*, 524 (2004) 347.

31 J. Komljenovic, S. Krka and N. Radic, *Anal. Chem.*, 58 (1986) 2893.

32 M. Ghosh, M.R. Dhaneshwar, R.G. Dhaneshwar and B.B. Ghosh, *Analyst*, 103 (1978) 768.

33 Y.G. Vlasov and Y.E. Ermolenko, *Potential Dynamics of Crystalline Ion Selective Electrode*, Ion-Sel. Electrodes, 5, Proc. Symp., 5th (1989), 611.

34 T. Rosatzin, P. Holy, K. Seiler, B. Rusterholz and W. Simon, *Anal. Chem.*, 64 (1992) 2029.

35 M. Puntener, M. Fibbioli, E. Bakker and E. Pretsch, *Electroanalysis*, 14 (2002) 1329.

36 D.N. Reinhoudt, J.F.J. Engbersen, Z. Bròzka, H.H. van den Vlekkert, G.W.N. Honig, H.A.J. Holterman and U.H. Verkerk, *Anal. Chem.*, 66 (1994) 3618.

37 P. Buehlmann, E. Pretsch and E. Bakker, *Chem. Rev.*, 98 (1998) 1593.

38 M. Nägele and E. Pretsch, *Mikrochim. Acta*, 121 (1995) 269.

39 F.M. Karpfen and J.E.B. Randles, *J. Chem. Soc. Faraday Trans.*, 49 (1953) 823.

40 Y. Qin and E. Bakker, *Anal. Chem.*, 73 (2001) 4262.

41 E. Bakker and E. Pretsch, *Anal. Chim. Acta*, 309 (1995) 7.

42 S. Peper, M. Telting-Diaz, P. Almond, T. Albrecht-Schmitt and E. Bakker, *Anal. Chem.*, 74 (2002) 1327.

43 S. Peper, Y. Qin, P. Almond, M. McKee, M. Telting-Diaz, T. Albrecht-Schmitt and E. Bakker, *Anal. Chem.*, 75 (2003) 2131.

44 M.A. Simon and R.P. Kusy, *Polymer*, 34 (1993) 5106.

45 E. Lindner, V.V. Cosofret, S. Ufer, R.P. Buck, W.J. Kao, M.R. Neuman and J.M. Anderson, *J. Biomed. Mater. Res.*, 28 (1994) 591.

46 G.S. Cha, D. Liu, M.E. Meyerhoff, H.C. Cantor, A.R. Midgley, H.D. Goldberg and R.B. Brown, *Anal. Chem.*, 63 (1991) 11666.

47 E.J. Wang and M.E. Meyerhoff, *Anal. Chim. Acta*, 283 (1993) 673.

48 G. Hogg, O. Lutze and K. Cammann, *Anal. Chim. Acta*, 335 (1996) 103.

49 L.Y. Heng and E.A.H. Hall, *Anal. Chim. Acta*, 403 (2000) 77.

50 L.Y. Heng and E.A.H. Hall, *Anal. Chem.*, 72 (2000) 42.

51 L.Y. Heng and E.A.H. Hall, *Electroanalysis*, 12 (2000) 187.

52 L.Y. Heng and E.A.H. Hall, *Anal. Chim. Acta*, 443 (2001) 25.

53 Y. Qin, S. Peper and E. Bakker, *Electroanalysis*, 14 (2002) 1375.

54 E. Malinowaaks, L. Gawart, P. Parzuchowski, G. Rokicki and Z. Brzozka, *Anal. Chim. Acta*, 421 (2000) 93.

55 L.Y. Heng and E.A.H. Hall, *Electroanalysis*, 12 (2000) 178.

56 Y. Qin, S. Peper, A. Radu, A. Ceresa and E. Bakker, *Anal. Chem.*, 75 (2003) 3038.

57 Y. Qin and E. Bakker, *Anal. Chem.*, 76 (2004) 4379.

58 J. Bobacka, *Anal. Chem.*, 71 (1999) 4932.

59 J. Bobacka, A. Ivaska and A. Lewenstam, *Electroanalysis*, 15 (2003) 366.

60 A. Michalska, M. Ocypa and K. Maksymiuk, *Electroanalysis*, 17 (2005) 327.

61 M. Vázquez, P. Danielsson, J. Bobacka, A. Lewensta and A. Ivaska, *Sens. Actuators B*, 97 (2004) 182.

62 J. Sutter, A. Radu, S. Peper, E. Bakker and E. Pretsch, *Anal. Chim. Acta*, 523 (2004) 53.

63 H. Karami and M.F. Mousavi, *Talanta*, 63 (2004) 743.

64 A. Michalska and K. Maksymiuk, *Anal. Chim. Acta*, 523 (2004) 97.

65 R. Toczyowska, R. Pokrop, A. Dybko and W. Wr'oblewski, *Anal. Chim. Acta*, 540 (2005) 167.

66 G.G. Guilbault, R.A. Durst, M.S. Frant, H. Freiser, E.H. Hansen, T.S. Light, E. Pungor, G. Rechnitz, N.M. Rice, T.J. Rohm, W. Simon and J.D.R. Thomas, *Pure Appl. Chem.*, 48 (1976) 127.

67 Y. Umezawa, *Handbook of Ion-Selective Electrodes: Selectivity Coefficients*, CRC Press, Boca Raton, Ann Arbor, Boston, 1990.

68 Y. Umezawa, K. Umezawa and H. Sato, *Pure Appl. Chem.*, 67 (1995) 507.

69 E. Bakker, R.K. Meruva, E. Pretsch and M.E. Meyerhoff, *Anal. Chem.*, 66 (1994) 3021.

70 E. Bakker, *Anal. Chem.*, 69 (1997) 1061.

71 T. Sokalski, A. Ceresa, T. Zwickl and E. Pretsch, *J. Am. Chem. Soc.*, 119 (1997) 11347.

72 A. Ceresa, E. Bakker, B. Hattendorf, D. Guenther and E. Pretsch, *Anal. Chem.*, 73 (2001) 343.

73 Y. Mi and E. Bakker, *Anal. Chem.*, 71 (1999) 5279.

74 D.M. Pranitis, M. Telting-Diaz and M.E. Meyerhoff, *Crit. Rev. Anal. Chem.*, 23 (1992) 163.

REVIEW QUESTIONS

1. What are the essential parts of an electrochemical potentiometric cell and what is the main operational characteristic of a potentiometric measurement?

2. What are the main requirements of a reference electrode and explain what do you understand by liquid junction potential?

3. Explain the main mechanistic differences between a glass membrane electrode and an ion-sensitive field effect transistor (ISFET).

4. What are the main material components needed in the design of a polymer-based ion-selective membrane? What is the role of each of these components in the membrane?

5. What do you understand by selectivity? Explain the difference between at least two methods used in the selectivity characterization of an ion-selective electrode.

6. Explain how in recent years it has been possible to dramatically lower the detection limit of ion-selective polymer membranes from micromolar to picomolar levels.

Chapter 18b

Voltammetry: Dynamic electrochemical techniques

Abul Hussam

18b.1 INTRODUCTION

Voltammetry is a part of the repertoire of dynamic electrochemical techniques for the study of redox (reduction–oxidation) reactions through current–voltage relationships. Experimentally, the current response (i, the signal) is obtained by the applied voltage (E, the excitation) in a suitable electrochemical cell. Polarography is a special form of voltammetry where redox reactions are studied with a dropping mercury electrode (DME). Polarography was the first dynamic electrochemical technique developed by J. Heyrovsky in 1922. He was awarded the Nobel Prize in Chemistry for this discovery.

The characteristic shape of i–E curve depends on the nature of the redox couple in the condensed phase, its thermodynamics, kinetics, mass transfer, and on the voltage–time profile (E–t). In this section we will discuss various voltammetric techniques and their applications in modern chemistry.

With the introduction of modern electronics, inexpensive computers, and new materials there is a resurgence of voltammetric techniques in various branches of science as evident in hundreds of new publications. Now, voltammetry can be performed with a nano-electrode for the detection of single molecular events [1], similar electrodes can be used to monitor the activity of neurotransmitter in a single living cell in sub-nanoliter volume electrochemical cell [2], measurement of fast electron transfer kinetics, trace metal analysis, etc. Voltammetric sensors are now commonplace in gas sensors (home CO sensor), biomedical sensors (blood glucose meter), and detectors for liquid chromatography. Voltammetric sensors appear to be an ideal candidate for miniaturization and mass production. This is evident in the development of lab-on-chip

Comprehensive Analytical Chemistry 47
S. Ahuja and N. Jespersen (Eds)
Volume 47 ISSN: 0166-526X DOI: 10.1016/S0166-526X(06)47028-8

technologies with applications ranging from capillary electrophoresis chips with integrated electrochemical detection [3], chemical plume tracking [4], micro-electrophoresis system for explosive analysis [5], and clinical diagnostic devices [6–8]. There are hundreds of chemical and biochemical electroactive species, which are amenable to the voltammetric detection system. Table 18b.1 shows selected applications in three categories encompassing analytical, environmental, and biomedical sensing applications. In biosensing, the p-PAP probe for immunoassay is used for clinical diagnostic testing [9]. The redox active DNA tags could be potentially used for genotyping with significant clinical potential [10].

18b.2   TOOLS OF THE TRADE

All electrochemical reactions are carried out in a suitable cell with electrodes connected to a programmable voltage or current source. For analytical work the dimension of the electrodes are in the range of micrometers to millimeters. Almost in all electrochemical studies, one also needs an inert supporting electrolyte to carry most of the charges. The electrodes, their properties, and the working principle of the potentiostat is described below.

18b.2.1   Cells and electrodes

Any container or a flow system with three electrodes closely placed can be used for electrochemical studies. Some electrochemical cells are shown in Fig. 18b.1. Most electrochemical cells contain three electrodes. These are the working electrode (W), counter electrode (C), and the reference electrode (R). Table 18b.2 shows the materials and properties of W, R, and C.

The working electrode (W) is the substrate on which the redox reaction takes place. Generally, working electrodes are made of platinum, gold, mercury, and carbon. Solid working electrodes come in two most common shapes—button as planar electrodes and wire as cylindrical electrodes. Metal and carbon fibers are also used to make dot-shaped ultramicroelectrodes with few micrometers in diameter. Mercury is the classical electrode for polarography. In polarography, a glass capillary is used to deliver the liquid mercury in drops known as the DME. This is the only electrode where the surface of the electrode is renewed with each new drop. Hanging mercury drop electrode (HMDE) and static mercury drop electrode (SMDE) are also used but are less common.

TABLE 18b.1

Analytical applications of selected reactions studied by voltammetric techniques

Analyte	Redox reactions	$E°$ (V vs. ref)	Applications and comments
Probe molecules for general cell characterization (CA and CV, SWV)			$Fe(CN)_6^{3-} + e^- \Leftrightarrow Fe(CN)_6^{4-}$
0.200 V vs. Ag/AgCl	$Fe(CN)_6^{3-}$ Ferricyanide Reversible redox couple to test cell performance		
R_2Cp_2Fe Ferrocene $R = -H, -COOH$	$R_2Cp_2Fe \Leftrightarrow R_2Cp_2Fe^+ + e^-$ Cp: Cyclopentadiene	0.200 mV vs. Ag/AgCl	Reversible test candidate in non-aqueous solvents and a redox mediator
Toxic metals (ASV, SWASV)			
As	$H_3AsO_3 + 3H^+ + 3e^- \Leftrightarrow As(s) + 3H_2O$	0.247 V vs. SHE	Trace inorganic arsenic species in groundwater and biological samples
Pb	$Pb\ (II) + 2e^- \Leftrightarrow Pb(s)$	-0.126 V vs. SHE	Trace Pb (II) in water, blood, paint, etc.
Cd	$Cd\ (II) + 2e^- \Leftrightarrow Cd(s)$	-0.402 V vs. SHE	Trace Cd (II) in water and biological samples
Hg	$Hg\ (II) + 2e^- \Leftrightarrow Hg(s)$	0.852 V vs. SHE	Ultratrace Hg (II) in the environment (sub-ppb)
Gas sensing (CA)			
H_2 (g)	$2H^+ + 2e^- \Leftrightarrow H_2$ (g)	0.0	Hydrogen gas sensor
AsH_3 (g)	$AsH_3 + 4H_2O \Leftrightarrow H_3AsO_4 + 8H^+ + 8e^-$	0.247 V vs. SHE	Inorganic arsenic in the environment and biological sample through hydride generation

continued

Table 18b.1 (*continued*)

Analyte	Redox reactions	E^o (V vs. ref)	Applications and comments
CO (g)	$CO(g) + H_2O \Leftrightarrow CO_2 + 2H^+ + 2e^-$	0.0 vs. Ag/AgCl	Basis of CO gas sensor
Molecules for bio-sensing (CA, SWV)			
$H_2NC_6H_4OH$ p-aminophenol p-PAP	$H_2NC_6H_4OH \Leftrightarrow HN = C_6H_4O + 2H^+ + 2e^-$	0.25 V vs. Ag/AgCl	Probe molecule for ELISA immunoassay sensor for antigen and antibody
Glucose	Glucose + $O_2 \Leftrightarrow$ Gluconic acid + H_2O_2 (needs glucose oxidase as catalyst) $H_2O_2 \Leftrightarrow O_2 + 2H^+ + 2e^-$	0.60 V vs. Ag/AgCl	Blood glucose monitor. H_2O_2 is the redox species. O_2 is often replaced by a mediator molecule e.g., Ferrocene
$M(bpy)_x^{2+}$ M = Fe, Co, Ru	$M(bpy)_x^{2+} \Leftrightarrow M(bpy)_x^{3+} + e^-$ X = 2, 3, 5, 6	0.3–0.7 V vs. Ag/AgCl	Reversible molecular tag for DNA detection

CA—chronoamperometry, CV—cyclic voltammetry, SWV—square wave voltammetry, ASV—anodic stripping voltammetry, SHE—standard hydrogen electrode.

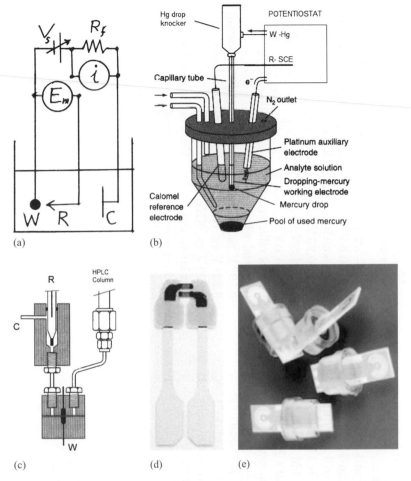

Fig. 18b.1. Electrochemical cells and representative cell configurations. (a) Schematic diagram of a cell–potentiostat system. (b) Typical laboratory cell with Hg-drop electrode and drop knocker. (c) Voltammetric cell as detector at the end of a high-performance liquid chromatographic column. (d) A two-electrode (graphite) chip cell for biosensor development. (e) Three-electrode chip cells on a ceramic substrate for bioanalytical work.

Owing to the toxicity of mercury and its disposal problem, solid electrodes are now very popular. In particular, electrodes made of carbon such as glassy carbon, graphite, carbon paste, and carbon fibers have gained popularity. Mercury, gold, bismuth, and other metals can be deposited as thin metal films on carbon and serves as thin metal film electrodes (TMFE) with excellent analytical advantages in trace metal analysis. The choice of working electrode is determined by the redox

TABLE 18b.2
Electrode materials and properties

Type of electrode	Material	Properties	Comments
W: Working or indicator electrode	Pt—Platinum in 1 M sulfuric acid	−0.25–1.2 V	Ultramicroelectrodes are made of metal or carbon fibers
	Pt—Platinum in 1M NaOH	−1.0–0.6 V	
	Hg—Mercury in 1 M sulfuric acid	−1.2–0.50 V	
	Hg—Mercury in 1 M KCl or NaOH	−2.0–0.20 V	
	C—Carbon in 1 M HClO$_4$	−0.2–1.5 V	
	C—Carbon in 0.1 M KCl	−1.5–1.0 V	
	(Glassy carbon, carbon paste, graphite, diamond film).	(shows effective potential window)	
R: Reference electrode	Ag/AgCl, satd. KCl (SSC)	0.197 V vs. NHE	At 25 °C. NHE—normal hydrogen electrode
	Hg (l)/Hg$_2$Cl$_2$ (s), satd. KCl (SCE)	0.241 V vs. NHE	
	Ag wire (Pseudo reference)	0.200 V vs. NHE	
C: Counter or auxiliary electrode	Platinum, carbon or steel	Similar to the working electrode except for steel	Must be larger than W electrode

potential of the analyte and the potential window within which the solvent and the supporting electrolyte remains electrochemically inert. Table 18b.2 shows the effective potential window for most common working electrodes.

Most common reference electrodes are silver–silver chloride (SSC), and saturated calomel electrode (SSC, which contains mercury). The reference electrode should be placed near the working electrode so that the W-potential is accurately referred to the reference electrode. These reference electrodes contain concentrated NaCl or KCl solution as the inner electrolyte to maintain a constant composition. Errors in electrode potentials are due to the loss of electrolytes or the plugging of the porous junction at the tip of the reference electrode. Most problems in practical voltammetry arise from poor reference electrodes. To work with non-aqueous solvents such as acetonitrile, dimethylsulfoxide, propylene carbonate, etc., the half-cell, Ag (s)/AgClO$_4$ (0.1 M) in solvent//, is used. There are situations where a conventional reference electrode is not usable, then a silver wire can be used as a pseudo-reference electrode.

The counter electrode is the current carrying electrode and it must be inert and larger in dimension. Platinum wire or foil is the most common counter electrode. For work with micro- or ultramicroelectrode where the maximum current demand is of the order of few microamperes, the counter electrode is not necessary. At very low current, a two-electrode system with the reference electrode can function as the current-carrying electrode with very little change in the composition of the reference electrode. Many commercial glucose sensors and on-chip microcells have such electrode configuration.

18b.2.2   Instrumentation: Potentiostats

The instrumentation for voltammetry is relatively simple. With the advent of analog operational amplifiers, personal computers, and inexpensive data acquisition-control system, many computer-controlled electrochemical systems are commercially available or custom made. Programming complex excitation waveforms and fast data acquisition have become a matter of software writing.

The function of a simple three-electrode system can be understood from Fig. 18b.2a. The variable voltage source, V_s, is placed between the working and the counter electrodes so that the electrochemical reaction can take place on both electrodes at the applied excitation potential. The current response flowing during a redox reaction is monitored by the voltmeter across a standard resistor, R_f. Since we are only interested in

A. Hussam

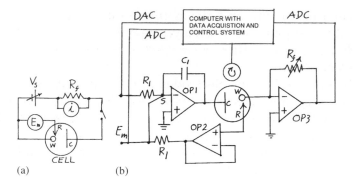

Fig. 18b.2. (a) A simple three-electrode system with a variable voltage source. (b) A potentiostat made of three operational amplifiers.

the redox reactions taking place on the working electrode, a reference electrode is placed near the working electrode through a high-input impedance voltmeter so that accurate electrode potentials can be measured. Note that all current is flowing between W–C, and none through R.

Figure 18b.2b shows a modern potentiostat made of three operational amplifiers. Here, the excitation voltage from a computer-controlled digital-to-analog converter is fed into OP1 through a resistor. The output of OP1 is connected to the counter electrode. The reference electrode is connected through a follower OP2 in a feedback configuration so that it does not draw any current at the summing point, S. The working electrode is connected to OP3, which is a simple current-to-voltage converter. The output voltage, E_o, of OP3 is related to cell current, $i = E_o/R_f$, through a feedback resistor, R_f. Since the working electrode is at virtual ground, the reference electrode tracks the W-potential at all times. The current through the cell is controlled by OP1 so that R is always at the applied potential, $-E_a$. Because W is grounded, E_m (vs. R) equals to E_a despite any change in cell resistance. A data acquisition system controlled by a computer can be used to measure E_m, i, and time. Since modern high-performance operational amplifiers are inexpensive the circuit above can be built with couple of dollars. Such a system can be used to perform any voltammetric experiments with software programmed excitation and current sampling.

18b.3   ELECTRODE PROCESSES

There are three major processes responsible for the response of a working electrode due to voltage excitation. These are electron transfer due to redox reaction, charging of the double-layer (an omnipresent capacitor on the electrode), and mass transfer dynamics. The electron transfer signifies the redox reaction, the rate of which is controlled by the excitation potential. This is the primary source of total current in the cell and is known as the faradaic current, i_f. The charging and discharging of the double layer due to voltage excitation demand some current flow, which is known as the charging current, i_c. Once the electrode reaction starts, the analyte is transported to the electrode surface through diffusion, convection, and migration. Diffusion is the transport of species in a concentration gradient created by the electron transfer reaction. Convection is the forced movement of fluid by mechanical means such as stirring, pumped flow, ultrasound, rotating electrode, etc. When charged species move under an electric field it is called migration. For most electroanalytical work, conditions are maintained to minimize the migration current by using a large concentration (tenth of a molar or higher) of supporting electrolytes such as KCl, $KClO_4$, strong acid or base, tetraethylammonium perchlorate (in nonaqueous solvents), etc. The supporting electrolyte carries all the charges through ion transport in the electric field. Thus the electroactive species present in small concentrations (mM or less) is not affected by migration effect. Therefore, the total current flow during an electrochemical reaction is a linear combination of all the contributions: $i_{Total} = i_d + i_c$, where i_d is the diffusion current. This relation assumes that the faradaic current is controlled by diffusion because the rate of electron transfer is much faster compared to diffusion. Let us look at the diffusion current, the faradaic current, and the charging current in a quantitative way.

18b.3.1 Diffusion current and mass transfer

Consider the fast redox reaction on the electrode surface (i.e., fast heterogeneous charge transfer kinetics):

$$O + ne^- = R \tag{18b.1}$$

(e.g., $O = [Fe(CN)_6]^{3-}$ and $R = [Fe(CN)_6]^{4-}$, $E^{o'} = 200\,mV$ vs. SSC)

For a reversible, fast reaction the net reaction rate, V_{rxn}, is controlled by the rate at which analyte, O, reaches the surface by mass transfer, V_{mt}.

Therefore,

$$V_{\mathrm{rxn}} = V_{\mathrm{mt}} = \frac{i}{(nFA)} \quad (\text{unit}: \text{mol s}^{-1} \text{cm}^{-2}) \tag{18b.2}$$

where i is the measured current (Ampere $= \text{C s}^{-1}$), n the number of electrons involved in the redox reaction, A the area of the electrode (cm^2), and F the Faraday constant (96,500 C mol^{-1}). The rate is the same as that of the flux, J, of species on the electrode surface. Let us now consider that analyte, O, at mM concentration is reduced on a small working electrode in a solution containing large concentration of a supporting electrolyte (e.g., 1.0 M KCl). The electrochemical cell also has a stirrer or the working electrode can be rotated mechanically as in a rotating disc electrode. The function of the supporting electrolyte is to minimize the migration current and the stirrer produces a constant diffusion layer thickness for O and R. We assume that only O is present in the beginning. When a potential is applied and O is reduced to R on the electrode surface, a concentration gradient is established near the working electrode. This is shown in Fig. 18b.3. Knowing that the rate of mass transfer is proportional to the concentration gradient (slope of line 1) on the electrode surface and assuming a linear concentration gradient, one can write

$$V_{\mathrm{mt}} = \frac{i}{(nFA)} = D_{\mathrm{o}} \frac{\left[C_{\mathrm{o}}^{*} - C_{\mathrm{o}}(X = 0) \right]}{\delta} \tag{18b.3}$$

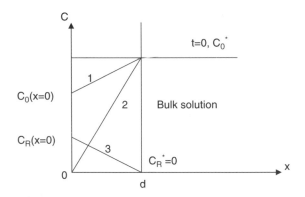

Fig. 18b.3. Ideal concentration gradients of O and R during a redox reaction on the surface of the working electrode. The x-axis shows the distance from the electrode surface. Bulk solution is stirred to maintain a constant diffusion layer thickness, d.

where D_o is the proportionality constant which is same as the diffusion coefficient (cm^2/s) of O, C_o^* the bulk concentration of O (mol/cm^3), C_o $(x = 0)$ is the surface concentration of O on the electrode surface, and δ the diffusion layer thickness (cm). Assuming δ is constant at a constant stirring rate, the term D_o/δ is a constant mass transfer coefficient, $m_o = D_o/\delta$, in cm/s, which is same as that of the first-order heterogeneous electron transfer rate constant. A similar equation can also be written for the concentration gradient for R (slope of line 3)

$$\frac{i}{(nFA)} = D_R \frac{[C_R(X = 0) - C_R^*]}{\delta} \tag{18b.4}$$

when the electrode potential reaches a value such that $C_o(x = 0) \ll C_o^*$, the concentration gradient can be shown by line 2. Therefore,

$$i_1 = nFAm_oC_o^* \tag{18b.5}$$

This is the limiting current where the mass transfer rate reaches a maximum value. The limiting current is achieved at a potential when O is reduced to R as fast as O reaches the electrode surface. Assuming there is no R in the beginning $(C_R^* = 0)$, and combining Eqs. (18b.3), (18b.4), and (18b.5), the surface concentrations of O and R can be derived as

$$C_o(X = 0) = \frac{i_1 - i}{(nFAm_o)} \tag{18b.6}$$

$$C_R(X = 0) = \frac{i}{(nFAm_R)} \tag{18b.7}$$

For fast electron transfer kinetics, the surface concentrations of O and R are at dynamic equilibrium and assumed to obey the Nernst law

$$E = E_o' + \left(\frac{RT}{nF}\right) \ln\left[\frac{C_o(X = 0)}{C_R(X = 0)}\right] \tag{18b.8}$$

Substituting Eqs. (18b.6) and (18b.7) in Eq. (18b.8), one gets the relation between current and voltage known as the steady-state voltammogram (Fig. 18b.4a)

$$E = E_o' + \left(\frac{RT}{nF}\right) \ln\left(\frac{m_R}{m_o}\right) + \left(\frac{RT}{nF}\right) \ln\left(\frac{i_1}{i - 1}\right) \tag{18b.9}$$

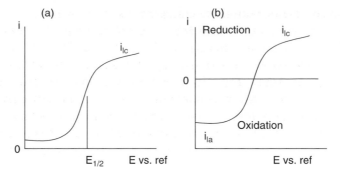

Fig. 18b.4. Steady-state voltammograms when initially (a) O is present and (b) both O and R are present.

In Eq. (18b.9), when $i = i_l/2$, the third term becomes zero and E becomes independent of concentrations. The corresponding potential,

$$E = E_{1/2} = E'_o + \left(\frac{RT}{nF}\right) \ln\left(\frac{m_R}{m_o}\right) \tag{18b.10a}$$

known as the half-wave potential is characteristic of the system. For small molecules the mass transfer coefficients (m_o, m_R) are similar and $E_{1/2} = E'_o$, the formal potential. The limiting or plateau current of a steady-state voltammogram is directly proportional to the bulk concentration of O as shown in Eq. (18b.5). Experimentally, a steady-state voltammogram can be obtained by scanning the potential starting from the nonfaradaic region (region of no electron transfer reaction) to several hundred millivolts more negative past the formal potential. Steady-state voltammogram can also be obtained under experimental conditions where the diffusion layer thickness changes very little over the time of the experiment i.e., when mass transfer coefficients remain almost unchanged. Such conditions can be achieved at a very slow scan rate or with ultramicroelectrode, in some pulse techniques, and in flow systems.

When both O and R are present as soluble electroactive species, Eq. (18b.9) can be rewritten as

$$E = E'_o + \left(\frac{RT}{nF}\right) \ln\left(\frac{m_R}{m_o}\right) + \left(\frac{RT}{nF}\right) \ln\left[\frac{(i_{lc} - i)}{(i - i_{la})}\right] \tag{18b.10b}$$

Here, we distinguish cathodic and anodic limiting currents as i_{lc} and i_{la}, respectively. The cathodic current being positive and the anodic current

672

being negative are conventions. Figure 18b.4b shows i–E curve with limiting currents and the location of the formal potential at zero.

18b.3.2 Faradaic current: Electron transfer kinetics

Consider the same general redox reaction,

$$O + e^- = R$$

is taking place on an electrode at a potential E. Assuming this reaction is entirely controlled by the kinetics of electron transfer, the net current flow is the difference between cathodic and anodic currents

$$i = i_f - i_a = nFA \left[k_f C_o(X = 0, t) - k_b C_R(x = 0, t) \right] \qquad (18b.11)$$

where, k_f and k_b are the forward and reverse rate constants (unit: cm/s).

Based on activation energies for anodic and cathodic processes and when the bulk concentrations of O and R are equal, one can write

$$k_f = k^\circ \exp(-\alpha f(E - E^{o'})) \; and \; k_b = k^\circ \exp((1 - \alpha)f(E - E^{o'})) \qquad (18b.12)$$

Combining Eqs. (18b.11) and (18b.12) one can write the familiar Butler–Volmer relation between current, potential, and the kinetic parameters

$$i = FAk^\circ \Big[C_o(X = 0, t) \exp(-\alpha f(E - E^{o'}))$$

$$- C_R(X = 0, t) \, \exp((1 - \alpha)f(E - E^{o'})) \Big] \qquad (18b.13)$$

where, k° is the standard heterogeneous electron transfer rate constant, $C_o(x = 0, t)$ and $C_R(x = 0, t)$ are the surface concentrations of O and R at any time t, α is the transfer coefficient which is a measure of the symmetry of the activation energy barrier, and $f = F/RT$. Eq. (18b.13) shows that both thermodynamic and kinetic information can be obtained from a voltammogram. A large k° indicates the system comes to equilibrium faster and a small k° indicates a slow equilibration. The value of k° can range 10–10^{-9} cm/s. It is found that for a steady-state experiment in an unstirred solution $k^\circ > 0.02$ cm/s leads to a reversible i–E curve and $k^\circ < 0.005$ cm/s to an irreversible i–E curve. Eq. (18b.12) shows that k_f and k_b can be made very large by changing the potential of the working electrode even if k° is small. When the system reaches equilibrium $(i = 0)$, Eq. (18b.13) reduces to the Nernst equation (Eq. (18b.8)). Note that for fast reaction the current is controlled by the nature of the mass transport and diffusion. However, in most real situations both kinetics and diffusion have to be considered together.

18b.3.3 Charging current: Effect of double layer

The electrode solution interface for an ideal polarizable electrode (IPE) has been shown to behave like a capacitor. The IPE is the metal solution interface where no charge transfer can occur when a potential is applied across the interface. Generally, most working electrode materials work like an IPE within a couple of voltage range. Figure 18b.5a shows the model of such a capacitor. At a fixed potential the metal side will assume a charge, which requires the assembly of opposite charges in the solution side. During this process there flows a current called the charging current. The transient current flows whenever the potential across this capacitor is changed. The origin of the capacitor is the compact inner layer of the water dipole as the dielectric medium separating an outer diffused charged layer. This model is called the double layer. Typical capacity of such a double-layer capacitor, C_d, is in the range of 10–$100\,\mu F/cm^2$.

Figure 18b.5b shows the equivalent circuit of the metal solution interface composed of C_d and the solution resistance R_s. When a voltage pulse, E, is applied across such a R_C circuit, the transient current flow

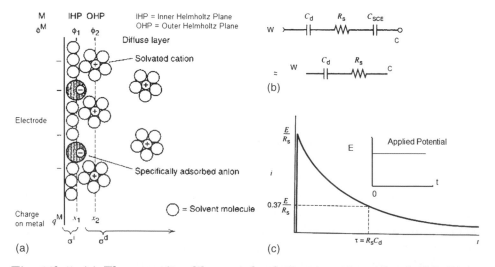

Fig. 18b.5. (a) The capacitor-like metal solution interface, the double layer. (b) The equivalent circuit with solution resistance and overall double-layer capacitor. (c) Charging current transient resulting from a step-potential at an IPE.

due to the charging of the capacitor is given by

$$i_{\mathrm{c}} = \left(\frac{E}{R_{\mathrm{s}}}\right) \exp\left(\frac{-t}{R_{\mathrm{s}}C_{\mathrm{d}}}\right) \qquad (18b.14)$$

The charging current decays exponentially with a time constant $\tau = R_{\mathrm{s}}C_{\mathrm{d}}$. Therefore, it is necessary to reduce $R_{\mathrm{s}}C_{\mathrm{d}}$ as fast as possible especially for fast voltammetry. At 3τ, the charging current decays 95% of its initial value. For example, if $R_{\mathrm{s}} = 10\,\Omega$ and $C_{\mathrm{d}} = 20\,\mu\mathrm{F/cm}^2$, $\tau = 200\,\mu\mathrm{S}$ i.e., the double-layer capacitor is discharged by 95% at $600\,\mu\mathrm{S}$. The time constant can be reduced by increasing the supporting electrolyte concentration. C_{d} is also directly proportional to the area of the electrode, thus by using an ultramicroelectrode (diameter $50\,\mu$ or less), one can drastically reduce the capacitance background and increase the response time of the cell. This is one of the most important advantages of ultramicroelectrode. However, some i_{c} is always present in a voltammetric experiment and regarded as the source of background signal. Since the measured current in a voltammetric experiment is the sum of faradaic and charging current, in order to increase the sensitivity of the technique the ratio of faradaic/charging current has to be increased. This is precisely the goal of various pulse voltammetric techniques. Charging current is also a weak function of electrode potential that can be reduced by differential current sampling at nearly the same potential.

18b.4   TECHNIQUES AND APPLICATIONS

The following section describes various voltammetric techniques and applications. In most cases we assume a reversible fast reaction and diffusion as the rate-controlling step.

18b.4.1   Chronoamperometry: The simplest experiment

The understanding of the nature of transient current after the imposition of a potential pulse is fundamental to the development of voltammetry and its analytical applications. Consider the same reaction, $O + ne^- = R$, taking place in a quiet solution at a potential such that the reaction is diffusion controlled. Figure 18b.6a shows the pulse and Fig. 18b.6b shows the concentration gradient O as a function of time and distance from the electrode surface.

Since the solution is quiet, and the surface concentration is C_{o} $(x = 0) = 0$, the reduction reaction can only continue through expansion

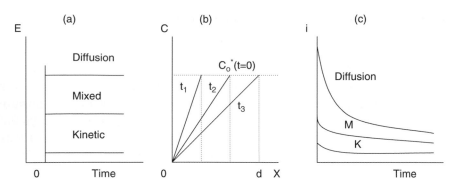

Fig. 18b.6. (a) Shape of the voltage pulses for diffusion control, mixed diffusion-kinetic control, and kinetic control, (b) concentration gradient of O showing expansion of the diffusion layer with time for complete diffusion controlled reaction, and (c) current transients show diffusion controlled, mixed kinetics and diffusion control, and complete kinetics controlled reactions corresponding to voltage pulses shown in (a). Note that the equations are derived only for the diffusion controlled case.

of the diffusion layer thickness, δ with time. For a planar electrode with linear diffusion,

$$\delta = (\pi D t)^{1/2} \tag{18b.15}$$

Combining Eqs. (18b.3) and (18b.15) one gets

$$i_t = \frac{(nFAD_o^{1/2}C_o^*)}{(\pi t)^{1/2}} \tag{18b.16}$$

This is known as the Cottrell equation. It shows that the faradaic transient current, i_t, decays $t^{-1/2}$. In contrast, the capacitance current decays exponentially and much faster. According to Eq. (18b.16) a plot of i_t vs. $t^{-1/2}$ is a straight line, the slope of which can be used to calculate the D of the analyte if the area of the electrode is known. Eq. (18b.16) is also used to measure the active area of an electrode by using species with known D. At a spherical electrode (such as HMDE) of radius, r, the Cottrell equation has an added spherical term

$$i_t = (nFAD_oC_o^*)\left[\frac{1}{(D\pi t)^{1/2}} + \frac{1}{r}\right] \tag{18b.17}$$

At short t the first term dominates and at longer t, i_t becomes independent of t. This is the limiting current. Also, when $r < 50\,\mu$, it

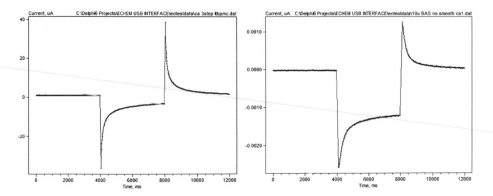

Fig. 18b.7. (a) Chronoamperogram showing the response due to a triple pulse 500–0–500 with a 3 mm diameter glassy carbon working electrode in 2.0 mM Potassium Ferricyanide in 0.1 M KCl. No current was recorded for the initial potential, 500 mV, where no faradaic reduction took place. (b) The same solution, except with a 10 µm diameter Pt working electrode. Current was recorded for the initial potential at 500 mV for 0–4000 ms where no faradaic reduction took place. Note the magnitude of current scale.

becomes independent of t. This happens with ultramicroelectrodes where the steady-state current dominates. Figure 18b.7a shows the chronoamperogram for the reduction (shown as the negative current) of 2.0 mM potassium ferricyanide in 0.1 M KCl after application of pulse 500–0 mV vs. SSC. The positive going current is the oxidation of the ferrocyanide to ferricyanide due to a reverse pulse 0–500 mV. Both show a high-current (200 µA) spike at shorter time and the limiting $i_t = 15$ µA at longer time. With a 3 mm diameter glassy carbon working electrode, the rate of analyte diffusion is slow enough to catch the reduced species to be oxidized by the reverse pulse.

Figure 18b.7b shows the response for the same solution, except with a 10 µm diameter Pt working electrode. Here, the pulse is 500–0–500 mV vs. SSC. The electrode shows a steady-state response of 1500 pA for the reduction of ferricyanide and near-zero response for the oxidation of ferrocyanide after the application of the reverse pulse. The loss of reduced species from the vicinity of an ultramicroelectrode is significant due to the spherical nature of diffusion. The current during the reverse pulse can therefore be regarded as the charging current of the double layer. Limiting and steady-state current measurement is the basis for electrochemical detection of species in flow systems such as flow-injection analysis, electrochemical detector for liquid chromatography, capillary zone electrophoresis (CZE), gas sensors, blood glucose measurements, etc.

A. Hussam

18b.4.2 Linear and cyclic voltammetry

Linear and cyclic voltage sweep techniques are the most popular voltammetric techniques due to their ease of use for diagnosis of various redox reactions and their chemical complications. In a linear scan voltammetry (LSV), the electrode potential is changed linearly at a fixed scan rate from an initial potential, E_1, where no faradaic reaction takes place to a potential, E_v, where the reaction is diffusion controlled in a quiet solution. In cyclic voltammetry (CV), the potential is scanned back to a final potential (E_f), which is often the initial potential from E_v. Figure 18b.8 shows the process and the resulting voltammogram. Since modern instruments use digital-to-analog converters the scan ramp is composed of small voltage steps (2–5 mV) to approximate a linear ramp. Current is sampled at fixed intervals on the step-pulse to construct the voltammogram. Generally, current sampling is delayed to the end of the step to minimize the charging current and increase the faradaic/charging ratio. This is not possible with conventional analog instruments.

The basic shape of LSV and CV, a peak-shaped voltammogram can be explained as follows. For a reduction reaction, $O + ne^- = R$, when the voltage is made more negative the surface concentration of O starts to decrease and thus increases the concentration gradient and the reduction current starts to rise. Eventually, at more negative voltage the surface concentration reaches zero and diffusion cannot deliver O to the surface at the same rate. This results in decreasing current and a

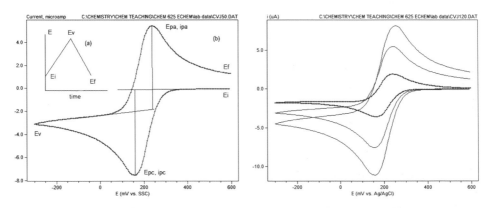

Fig. 18b.8. (a) Potential–time profile for CV, (b) Definition of parameters in a CV, (c) CV of a $[Fe(CN)_6]^{3-}/[Fe(CN)_6]^{4-}$ redox couple at 10, 50, and 100 mV/s scan rate. $E_i = 600$ mV, $E_v = -300$ mV, and $E_f = 600$ mV vs. SSC. Note that reduction is shown as negative current.

peak-shaped i–E curve. In a quiet solution, only at a very slow scan rate (e.g., $<5\,\text{mV/s}$), the concentration gradient remains low and unchanged which results in a steady-state current plateau. With a reversible system, when the scan is reversed, R accumulated near the electrode is oxidized back to O. The current on the reverse scan shows a peaked response with opposite sign. Note that the charge associated with the anodic process (oxidation current) is lower than the forward reduction process, due to the loss of accumulated R from the electrode surface by diffusion. Figure 18b.8c also shows that peak current increases with increase in scan rate. This is due to the faster rise in concentration gradient in a shorter time.

For a reversible system the peak current at $25\,^\circ\text{C}$ is given by the Randles–Sevcik equation

$$i_{\text{pc}} = -(2.69 \times 10^5)\, n^{3/2} A\, D_{\text{o}}^{1/2} C_{\text{o}}^* \, s^{1/2} \tag{18b.18}$$

where, i_{pc} cathodic (reduction) peak current and s the scan rate (V/s). All other terms are defined earlier. Eq. (18b.18) shows that the peak current is directly proportional to the bulk concentration of analyte and increases with the square-root of scan rate, s. Table 18b.3 lists the diagnostic criteria for reversible quasi-reversible and irreversible reactions. These criteria should be tested with CV performed over a wide range of scan rates (at least 10 to 1000 mV/s). One should realize that as the scan rate increases the charging current starts to dominate over the faradaic current and distorts the CV from which no meaningful measurements can be made. This limitation can be overcome by using ultramicroelectrodes where, at high scan rate, the faradaic current increases more than the charging current. Scan rates excess of 100 V/s were used to study fast electron transfer reactions and reactive intermediates with ultramicroelectrodes.

An irreversible reaction shows no oxidation (anodic peak) and the kinetic parameters ($\alpha_c n'$ and k_{o}) can be obtained from the shift in peak potentials as a function of scan rate

$$E_{\text{pc}} = E^{\text{o}'} - \left(\frac{RT}{(\alpha_c n'F)}\right)\left[0.780 + \ln\left(\frac{D_{\text{o}}^{1/2}}{k_{\text{o}}}\right) + 0.5\ln\left[\frac{(\alpha_c n'Fs)}{(RT)}\right]\right] \tag{18b.19}$$

In practice, the majority of redox reactions behave more like a quasi-reversible system. It is also common that a reaction that behaves reversibly at low scan rate becomes irreversible at high scan rate passing through a quasi-reversible region.

TABLE 18b.3

Diagnostic tests and quantitative criteria for cyclic voltammograms of reversible and irreversible redox reactions at 25°C

Reversible	Quasi-reversible	Irreversible
$\Delta E_p = \|E_{pa} - E_{pc}\| = 59/n$ mV	$\Delta E_p > 59/n$ mV and increases with s	No reverse peak
$\|i_{pa}/i_{pc}\| = 1$	$\|i_{pa}/i_{pc}\| = 1$ when transfer coefficient $= 0.5$	$i_{pc} = -(2.99 \times 10^5)\, n\, (\alpha_c n')^{1/2}\, A\, D_o^{1/2}\, C_o^*\, s^{1/2}$
		i_p vs. $s^{1/2}$ is a straight line
i_p vs. $s^{1/2}$ is a straight line	i_p vs. $s^{1/2}$ is a straight line	
E_{pc} and E_{pa}- independent of s	E_{pc} shifts negatively with increasing s	E_{pc} shifted by $-30(\alpha_c n_a)$ mV for each 10-fold increase in s
$i \propto t^{-1/2}$ beyond E_p O and R must be stable and kinetic of electron transfer is fast		O and R must be stable with no follow-up reactions

Definition of symbols: ΔE_p = peak potential difference, E_{pa} = peak potential at cathodic peak current, E_{pc} = peak potential at anodic peak current, i_{pa} = anodic peak current, i_{pc} = cathodic peak current, s = scan rate, t = time after peak (the Cottrell region), n = number of electrons involved in redox reaction. Rate parameters $(\alpha_c n')$ and heterogeneous rate constant can be found from irreversible wave.

CV is extensively used for the study of multi-electron transfer reactions, adsorbed species on the electrode surface, coupled chemical reactions, catalysis, etc. Figure 18b.9 shows some of the examples.

18b.4.3 Polarography

Polarography is the classical name for LSV with a DME. With DME as the working electrode, the surface area increases until the drop falls off. This process produces an oscillating current synchronized with the growth of the Hg-drop. A typical polarogram is shown in Fig. 18b.10a. The plateau current (limiting diffusion current as discussed earlier) is given by the Ilkovic equation

$$i_d = 708\, n\, D_o^{1/2}\, m^{2/3}\, t^{1/6}\, C_o^* \tag{18b.20}$$

where, m is the flow rate of Hg-drop (mg/s) and t is the drop time (s). All other parameters have their usual meanings. While the faradaic

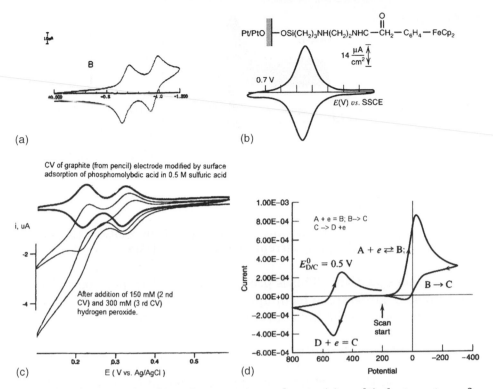

(a)

(b)

CV of graphite (from pencil) electrode modified by surface adsorption of phosphomolybdic acid in 0.5 M sulfuric acid

After addition of 150 mM (2 nd CV) and 300 mM (3 rd CV) hydrogen peroxide.

(c)

(d)

Fig. 18b.9. Example cyclic voltammograms due to (a) multi-electron transfer redox reaction: two-step reduction of methyl viologen $MV^{2+} + e^- = MV^+ + e^- = MV$. (b) ferrocene confined as covalently attached surface-modified electroactive species—peaks show no diffusion tail, (c) follow-up chemical reaction: A and C are electroactive, C is produced from B through irreversible chemical conversion of B, and (d) electrocatalysis of hydrogen peroxide decomposition by phosphomolybdic acid adsorbed on a graphite electrode.

current increases with the area of the DME, the charging current increases with rate of increase in area (dA/dt) that decreases as the drop-size increases. Therefore, charging current decays with drop growth time, t.

$$i_{c,t} = 0.00567 \, (E_z - E) \, m^{2/3} \, t^{-1/3} \, C_i \qquad (18b.21)$$

where, E_z is potential of zero charge for Hg (a constant value in a specific medium), E is the applied potential and C_i is the integral capacity of the double layer (in $\mu F/cm^2$). Figure 18b.10b shows the faradaic and charging current transients for the life of a drop. It shows that current sampling near the end of drop-life can maximize faradaic/charging thus

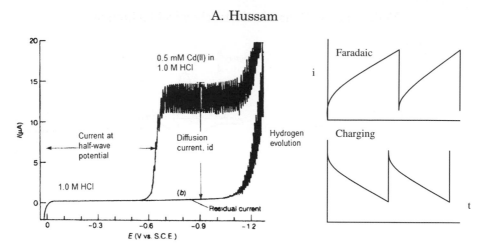

Fig. 18b.10. (a) Polarogram showing reduction Cd(II) in solution, (b) comparison of faradaic and charging current transients at a DME.

the detection sensitivity. This is precisely done in Tast polarography and other pulse polarographic techniques. Note that, like other techniques, i_d is proportional to the bulk concentration of analyte.

18b.4.4 Pulse voltammetry

The realization that current sampling on a step pulse can increase the detection sensitivity by increasing the faradaic/charging ratio is the basis for the development of various pulse voltammetric (or polarographic) techniques. Also, the pulses can be applied when it is necessary and can reduce the effect of diffusion on the analyte. Figure 18b.11 shows the waveform and response for three commonly used pulse voltammetric techniques: normal pulse voltammetry (NPV), differential pulse voltammetry (DPV), and square-wave voltammetry (SWV).

In NPV, the electrode is held at an initial potential, E_i, where no faradaic current flows. After a fixed delay, t_d (100–5000 ms), a potential pulse, ΔE (2–20 mV) is applied for pulse duration, t_p (20–50 ms). The faradaic current flows only during this short time. The current is then sampled near the end of this pulse for few milliseconds and averaged. The pulse returns to the initial potential and the next pulse is increased by few millivolts to increase ΔE. This process continues until ΔE reaches the potential where the reaction is diffusion controlled. The result is a sigmoid-shaped current response. In NPV, the pulse timing is such that the concentration gradient is large and the current is sampled on the pulse when the capacitance background decreases (within 1–10 ms) but not eliminated. After each current sampling the initial

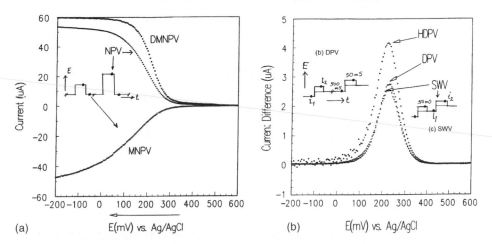

(a)   E(mV) vs. Ag/AgCl   (b)   E(mV) vs. Ag/AgCl

Fig. 18b.11. Figures show the pulse waveform and response for three techniques: (a) normal pulse voltammetry (NPV), (b) differential pulse voltammetry (DPV), and (c) square-wave voltammetry (SWV).

condition of the electrode is re-established by returning to the initial potential for a longer delay, t_d. In NPV, the maximum diffusion current is given by the equation

$$i_{d,max} = \frac{(nFA\,D_o^{1/2}C_o^*)}{(\pi t_s)^{1/2}}$$

(18b.22a)

where t_s is the sampling time after the application of the pulse. Clearly, the increase in sensitivity is due to the short pulse duration ($t_s = 20$–50 ms) compared to seconds in classical polarographic or voltammetric experiments. NPV can be easily used to measure 10^{-5} M analyte with a detection limit of 10–6 M. For a reversible redox couple, one can easily record the oxidation current of the reduced species by sampling the current after the pulse returns to the base potential. This is a form of reverse pulse voltammetry (RPV).

Detection sensitivity better than that in NPV can be achieved by differential current sampling with small pulses. Figure 18b.11b shows this scheme known as differential pulse voltammetry (DPV). Here, the potential is changed from an initial potential in small steps (2–5 mV) and a voltage pulse of a short duration (50 ms) is superimposed at the end of a long step (500–5000 ms). The current is sampled before the beginning of the pulse and near the end of the pulse as shown in Fig. 18b.12b. In differential pulse polarography (DPP), this is near the

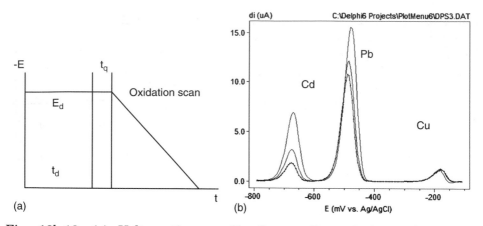

(a)　　　　　　　　　　　　　　　　(b)

Fig. 18b.12. (a) Voltage–time profile for anodic stripping voltammetry (ASV) and (b) ASV of an unknown solution with two aliquot additions of 100 ppb each of Cd and Pb in the final solution. The peak at $-190\,\text{mV}$ is that of Cu present in the unknown. Experimental conditions: Initial deposition potential, $E_d = -800\,\text{mV}$, final potential $= 0$, deposition time, $t_d = 120\,\text{s}$, quite time, $t_q = 30\,\text{s}$, step potential $= 5\,\text{mV}$, pulse height $= 20\,\text{mV}$, pulse delay $= 100\,\text{ms}$, sampling width $= 17\,\text{ms}$, and sampling frequency $= 6000\,\text{Hz}$.

end of the drop-time. The signal is the difference between these two current values, which is recorded as a function of the increased step potentials. The current difference shows a peak-shaped voltammogram, which is easy to recognize as a signal peak. The peak appears only in the potential region (near formal potential) because a small potential difference (pulse height) makes a significant change in current flow. Thus, the sensitivity of DPV can be increased by increasing the voltage pulse, however, with some decrease in resolution. The peak differential current and the peak potential is given by

$$\Delta i_{\text{peak}} = \frac{nFAD_o^{1/2}C_o^*Z}{(\pi t_s)^{1/2}} \tag{18b.22b}$$

$$Z = \frac{\left[1 - \exp\left(\frac{nFE_{\text{pulse}}}{2RT}\right)\right]}{\left[1 + \exp\left(\frac{nFE_{\text{pulse}}}{2RT}\right)\right]} \tag{18b.23}$$

$$E_{\text{peak}} = E_{1/2} - \frac{E_{\text{pulse}}}{2} \tag{18b.24}$$

where, E_{pulse} is the pulse height. As E_{pulse} increases Z approaches unity and Δi_{peak} reaches the $i_{\text{d,max}}$ of NPV. The advantage of DPV is the reduction in capacitance current realized by taking the current difference over a small potential change, while in NPV the charging current increases as E_{pulse} increases. The increase in faradaic/charging ratio is manifested in a lower detection limit of 10–8 M of redox species for a reversible couple in a suitable supporting electrolyte.

Square wave voltammetry (SWV) is a variation of DPV, where a pure square wave is superimposed on a staircase ramp as shown in Fig. 18b.11c. Like DPV, here the current is sampled at the end of the forward and reverse pulses and subtracted. The current difference plotted as a function of increased ramp potential yields the peak-shaped square wave voltammogram. In SWV, the square wave frequency can be increased to an effective scan rate of 1 V/s (200 Hz at 5 mV increment) compared to 50 mV/s in DPV or NPV, yet retaining similar advantages as DPV. Many slow reactions such as the irreversible reduction of dissolved oxygen produce very little faradaic response with fast SWV scan. Thus, the removal of oxygen is not necessary to obtain good response for many species in the reduction region. This is a clear advantage for the development of field analytical methods where solution de-aeration is difficult. Due to the fast response, SWV was used to measure the entire voltammogram of species on-the-fly in a flow system such as LC electrochemical detectors.

Variations of the three-pulse techniques were developed by choosing current sampling points to further minimize the effect of capacitance background and to deal with irreversible reactions. These can be found in modern electrochemical literatures.

18b.4.5   Stripping voltammetry

Stripping voltammetry or stripping analysis has a special place in electrochemistry because of its extensive application in trace metal analysis. Stripping voltammetry (SV) is a two-step process as shown schematically in Fig. 18b.12. In the first step, the metal ion is reduced to metal on a mercury electrode (thin mercury film on glassy carbon or a HMDE) as amalgam.

$$M^{n+} + ne^- \Leftrightarrow M(\mathrm{Hg}) \tag{18b.25a}$$

The deposition is performed at a potential where the current is diffusion controlled at a steady state. The steady-state diffusion is maintained for a deposition time, t_{d}, by stirring the solution, usually with a magnetic stirrer. At the end of the deposition time, the stirrer is turned off for a quiet time, t_{q} (about 30 s) while the deposition potential is held.

After t_q is passed, the second step starts by scanning the potential from E_d to a potential when all the deposited metals are re-oxidized (the reverse of reaction 25). The oxidation current recorded as a function of potential is the anodic stripping voltammogram (ASV). A typical ASV of three metals (Cd, Pb, and Cu) deposited on a mercury film electrode is shown in Fig. 18b.12b. The sensitivity of ASV can be improved by increasing the deposition time and by using the pulse technique to record the oxidation current. ASV in Fig. 18b.12b was obtained by using the square wave voltammetry. In most cases a simple linear or step ramp is sufficient to measure sub-ppm level of metals in aqueous solution. The peak current of a linear scan ASV performed on a thin mercury film electrode is given by

$$|i_{peak}| = \frac{(n^2 F^2 slA C_m^*)}{(2.7 RT)} \tag{18b.25b}$$

where, s is the scan rate, l is the mercury film thickness (few microns), A is the area of the electrode, C_m^* is the concentration of reduced metal (mol/cm^3), and n is the number of electrons involved in the oxidation reaction. Eq. (18b.25b) shows that the peak current is proportional to the concentration of the deposited metal which in turn is proportional to the bulk concentration of the analyte metal ion when all other parameters remain unchanged. Amalgam-forming metals such as Pb, Cd, Cu, Zn, Bi, Sb, In, Tl, Ga, Au, As, Se, Te, etc., can be determined at parts-per-billion (ppb) concentration in aqueous solution. By using a gold-film electrode, ASV was used to measure ppb As(III) and As(V) in thousands of groundwater samples in one of the largest environmental applications [11]. Speciation information such as these is difficult to obtain with other analytical techniques. Hand-held instruments employing the ASV principle are now available for the routine analysis of Pb in blood, and Zn, Pb, Cd, and Cu in drinking water samples. Table 18b.4 shows a list of ASV methods validated and approved by various environmental and analytical organizations such as EPA, AOAC, and ASTM.

While ASV is used to measure metal cations, cathodic stripping voltammetry (CSV) can be used to measure some anions. Mercury is known to dissolve as mercurous ion (Hg_2^{2+}) at a potential more positive than $220\,mV$ vs. SSC. The Hg_2^{2+} forms insoluble salts with many anions such as halides, S^{2-}, CN^-, SCN^-, SH^-, etc.

$$2Hg(l) + 2A^- \Leftrightarrow Hg_2A_2(s) + 2e^- \tag{18b.26}$$

In CSV, the anions are deposited as the Hg_2A_2 (s) salt on a mercury electrode in the deposition step and then reduced to Hg (l) in

TABLE 18b.4

List of approved stripping voltammetric methods for the measurement of metal ions in environmental and test solutions

Approved stripping voltammetric electroanalytical methods

- American Society for Testing Materials: ASTM Method D3557-95: Cadmium in water
- American Society for Testing Materials: ASTM Method D3559-95: Cadmium in water
- EPA Method 7198: Hexavalent chromium in water (1986)
- EPA Method 1001: Lead in drinking water by ASV (1999)
- EPA Method 7063: Arsenic and selenium in sediment samples and extracts by ASV
- EPA Method 7412: Mercury in aqueous samples and extracts by ASV
- AOAC Method 986.15: Arsenic, cadmium, lead, selenium, and zinc in human and pet foods (1988)
- AOAC Method 982.23: Cadmium and lead in food (not for fats and oils) (1988)
- AOAC Method 974.13: Lead in evaporated milk (1976)
- AOAC Method 979.17: Lead in evaporated milk and fruit juice (1984)
- NIOSH Method 7701: Lead in air by ultrasound/ASV (1999)
- Method 3130: Metals by anodic stripping voltammetry (proposed by American Public Health Association, American Water Works Association, and the Water Environment Federation, 1995)

the stripping step. CSV is not limited to Hg electrode. Any electrode forming insoluble salt with analyte anion that can be reversibly reduced to the metal can be used in CSV.

18b.5   FUTURE OUTLOOK

Topics discussed above are some basic principles and techniques in voltammetry. Voltammetry in the frequency domain where i–E response is obtained at different frequencies from a single experiment known as AC voltammetry or impedance spectroscopy is well established. The use of ultramicroelectrodes in scanning electrochemical microscopy to scan surface redox sites is becoming useful in nano-research. There have been extensive efforts made to modify electrodes with enzymes for biosensor development. Wherever an analyte undergoes a redox reaction, voltammetry can be used as the primary sensing technique. Microsensor design and development has recently received significant attention due to the new capabilities that micro-fabrication technology has made possible for the scientific sensor community. In particular, electrochemical sensors have attracted considerable attention due to its low cost and less-complex fabrication process when compared to optical-based sensors. In a voltammetric sensor, both excitation and response signals are electrical and hence do not require any conversion devices. Whereas, an optical sensor requires an optical source, optical filters, optical detection, and conversion devices that ultimately add to the fabrication complexity and its cost. For example, in CZE, a microfabricated electrochemical cell (as detector) enables the lowest detection limit (10 pM) with the highest sensitivity and a large linear dynamic range (>1 million) compared to other detection techniques [12]. Electrochemical detection has also been demonstrated to be a viable method for identifying biomolecules in lab-on-chip platforms [13,14]. Electrochemical detection based on redox potential is selective, sensitive, and rapid, so the method can be used as a general analyte detection with minimal sample preparation.

REFERENCES

1   R. Schuster, V. Kirchner, X.H. Xia, A.M. Bittner and G. Ertl, Nanoscale electrochemistry, *Phys. Rev.*, 80(25) (1998) 5599–5602.
2   C.D.T. Bratten, P.H. Cobbold and J.H. Cooper, Micromachining sensors for electrochemical measurements in subnanoliter volumes, *Anal. Chem.*, 69 (1997) 253–256.
3   A.T. Wooley, K. Lao, A.N. Glazer and R.A. Mathias, Capillary electrophoresis chip with integrated electrochemical detection, *Anal. Chem.*, 70 (1998) 684–688.
4   T. Kikas, H. Ishida and J. Janata, Chemical plume tracking. 3. Ascorbic acid: A biologically relevant marker, *Anal. Chem.*, 74 (2002) 3605–3610.
5   A. Hilmi and J.H.T. Luong, Micromachined electrophoresis chip with electrochemical detectors for analysis of explosive compounds in soil and groundwater, *Environ. Sci. Technol.*, 34 (2000) 3046–3050.
6   M. Eggers and D.A. Ehrlich, A review of microfabricate devices for gene based diagnostics, *Hematol. Pathol.*, 9 (1995) 1–15.
7   P. Selvaganapathy, M.A. Burns, D.T. Burke and C.H. Mastrangelo, *Inline electrochemical detection for capillary electrophoresis*, IEEE, 451–454, 2001.
8   J. Lorraine, Which chip will be in your diagnostic device? R&D Magazine, pg. 33, December 2003.
9   T.T. Hua, C.E. Lunte, H.B. Halsall and W.H. Heineman, p-Aminophenyl phosphate: An improved substrate for electrochemical enzyme immunoassay, *Anal. Chim. Acta*, 214 (1988) 187–195.

10   M. Rodriguez and and A.J. Bard, Electrochemical studies of the interactions of metal chelates with DNA, *Anal. Chem.*, 62 (1990) 2658–2662.

11   S.B. Rasul, Z. Hossain, A.K.M. Munir, M. Alauddin, A.H. Khan and A. Hussam, Electrochemical measurement and speciation of inorganic arsenic in groundwater of Bangladesh, *Talanta*, 58(1) (2002) 33–43.

12   C. Vogt and G. L. Klunder, *Fresenius J. Anal. Chem.*, 370 (2001) 316.

13   J. Rossier and H. Girault, Enzyme linked immunoabsorbent assay on a microchip with electrochemical detection, *Lab on a Chip*, 1 (2001) 153–157.

14   A. Frey, M. Jenkner, M. Schienle, C. Paulus and B. Holtzapfl, *Design of an Integrated Potentiostat Circuit for CMOS Bio Sensor Chip*, IEEE, 2003, pp. 9–12.

Suggested Reading and Resources

A. J. Bard and L. R. Faulkner, *Electrochemical methods: Fundamentals and Applications*, 2nd ed., Wiley Sons & Inc., New York, NY, 2002.

B. B. Damaskin, *The principle of current methods for the study of electrochemical reactions*, McGraw-Hill, NY, 1967.

D. Pletcher, R. Greff, R. Peat, L.M. Peter and J. Robinson, *Instrumental methods in electrochemistry*, Horwood Publishing, England, 2001.

A.M. Bond, W.N. Duffy, X.J. Guo, X. Jang and D. Elton, Changing the look of voltammetry—Can FT revolutionize voltammetric technique as it did for FT NMR?, Anal. Chem., 186A, May 1, 2005.

REVIEW QUESTIONS

1.   Explain the conditions under which a steady-state voltammogram is obtained.

2.   Explain ways to reduce the charging current over the faradaic current.

3.   List the kind of information one can get from cyclic voltammetry.

4.   Explain what is done in anodic stripping voltammetry. Why is stripping the most sensitive voltammetric technique? What are its limitations?

Chapter 19

Hyphenated methods

Thomas R. Sharp and Brian L. Marquez

19.1 INTRODUCTION

The topic of hyphenated analytical methods is too broad a topic to be covered by a single chapter. The topic, rather, should be the subject of an entire book in and of itself. Refinement of the subject matter with the editors of this book narrowed the topic of this chapter to include hyphenated chromatography–spectroscopy techniques, and further still to include only the three areas of combined gas chromatography–mass spectrometry (GC–MS), liquid chromatography–mass spectrometry (LC–MS) and liquid chromatography–nuclear magnetic resonance spectroscopy (LC–NMR). Niche areas where less conventional chromatographic methods have been coupled with these two spectroscopic methods—e.g., capillary electrophoresis–mass spectrometry (CE–MS), [1] supercritical fluid chromatography–mass spectrometry (SFC–MS) [1]—will not be covered here, but their absence should not diminish their importance. Tandem mass spectrometry (MS–MS) has been presented as a hyphenated technique and as a replacement for chromatography–mass spectrometry—the first stage of mass analysis assuming the role of the chromatographic separation. While aspects of this proposal are true, it is not a general-purpose replacement for chromatography–mass spectrometry. We will not cover MS–MS here, except where it specifically contributes to our discussion of the chromatography–spectroscopy methods. Discussions of MS–MS should of course start with the classical book by McLafferty, [2] but more recent treatises can be found readily in the literature [3]. Coverage of chromatographic coupling to the optical spectroscopies (UV-visible and -infrared) is relegated to another time and place.

As organic structure elucidation tools, these three techniques are logically grouped with each other. Together, MS and NMR spectroscopy

Comprehensive Analytical Chemistry 47
S. Ahuja and N. Jespersen (Eds)
Volume 47 ISSN: 0166-526X DOI: 10.1016/S0166-526X(06)47019-7
© 2006 Elsevier B.V. All rights reserved.

provide nearly all of the data required to "prove"[1] the structure of an organic molecule. The deficiency experienced by both MS and NMR—in requiring a sample with a high degree of purity—is well compensated by the ability to couple them to a chromatographic separation method. The relative certainty that one is examining a "chromatographically pure" compound by introducing a chromatographic effluent, adequately compensates for this deficiency.

GC–MS should be considered a relatively mature technique. One does not see major changes in the technique and practices of this method appearing in the current literature. LC–MS is an almost mature technique. One occasionally sees a new innovation concerning how this experiment is done. By distinct contrast LC–NMR is a technique still transitioning between infancy and adolescence.

In this chapter, we attempt to present the principles of these techniques, and some examples of how these techniques have been useful. We hope that the information will permit the reader to evaluate whether these techniques will be useful in his or her situation, and to pick and choose, approach(s) that are appropriate and useful for a given circumstance.

19.2   MASS SPECTROMETRY

Understanding the powers of the combined chromatography–mass spectrometry techniques that are present here presupposes that one understands the information that MS in its various forms and nuances can provide. Discussions of combined GC–MS and LC–MS will be all the more meaningful, given an understanding of some mass spectrometric principles. Discussions of some of these principles in the following will reinforce previous awareness (e.g., see Chapter 11 on mass spectrometry earlier in this book), and ensure that the discussions on GC–MS and LC–MS will be meaningful. Should the reader's curiosity stray beyond these bounds, we invite him or her to consult the following. A recent reasonably comprehensive textbook on the general topic of mass spectrometry has been published by Jürgen Gross [4]. Volume 101, issue 2 of *Chemical Reviews* covers many of the active areas of development in mass spectrometry. Grayson [5] has compiled an excellent history covering all aspects of the field of mass spectrometry for the 50th

---

[1]We are using the word "prove" here guardedly. In reality, one does not and cannot prove a structure, only disprove it.

Fig. 19.1. Sir J.J. Thomson (seated, 4th from left) and students, Cambridge University, 1897 [7].

anniversary of the annual meetings of the American Society for mass spectrometry.

19.2.1   It all started with J.J.

Most practitioners of the precise art and subtle science of mass spectrometry acknowledge the field to have originated with the work of J.J. Thomson (Fig. 19.1) and associates, [6] published in 1910–1912, using the parabola mass spectrograph. Seminal discoveries that he and his co-workers made include the fact that the elements could be polyisotopic, by discovering the isotopes of neon. Thomson was awarded the Nobel Prize in physics of 1906[2] for his work on "investigations on the conduction of electricity by gases."

19.2.2   Molecular weight vs. relative-molecular mass

"Molecular weight" is an incorrect scientific term! The numerical value being discussed, rather, is correctly known as a relative molecular mass. It is a mass, measuring a quantity of matter, not by the influence of an external gravitational field on that matter. It is a relative mass, not an

---

[2]http://nobelprize.org/physics/laureates/1906/index.html

absolute mass measured in grams, because it is expressed relative to the standard atomic mass of 12.0000 atomic mass units (daltons), defined in reference to the ^{12}C isotope of carbon. It is a relative molecular mass because it is a sum of all the relative atomic masses of the atoms contained within the molecule. The distinction between the incorrect term atomic weight and the correct term relative atomic mass follow the same logic, only applied to the atomic masses of the elements on the periodic table [8].

The standard atomic mass unit (dalton) is defined as being 1/12 of the mass of an atom of the most abundant naturally occurring isotope of carbon, ^{12}C. The ^{12}C standard was adopted by the International Union of Pure and Applied Chemists (IUPAC) in 1960 [5]. The previous standard had been assigned to be 1/16 of the atomic mass of oxygen. Two scales developed around this ambiguously defined oxygen standard, depending on whether one used the chemically determined atomic mass of oxygen, or the mass of the most abundant stable isotope, ^{16}O. The chemically determined mass incorporated the natural abundances of the minor stable oxygen isotopes, ^{17}O (0.04%) and ^{18}O (0.2%). The other was based upon mass measurement specifically of ^{16}O. The two scales are offset by 0.0044 Da. Accurate masses reported prior to 1960 are referenced to the oxygen standard. Care should be taken when comparing these numbers to each other, and to more modern measurements. It should be noted that to which of the two oxygen standard the measurements are referenced. The extensive table of the accurate mass values, published in Beynon's book, [9] is based upon the oxygen standard. If those values are used, then mass differences in that table should be correct, regardless of the original standard. Absolute masses should be corrected for the current definition.

19.2.3 Monoisotopic vs. chemical (average) relative molecular mass

The relative molecular mass that is measured by the mass spectrometer is numerically different from the chemical (average) molecular mass, which is used to calculate molar concentrations or reaction stoichiometries. For the elements commonly found in organic molecules, the numbers are close and in many contexts interchangeable. However, when one discusses a mass spectrum, and an accurate mass measurement made by a mass spectrometer, for a molecule that contains significant numbers of atoms of polyisotopic elements such as chlorine or bromine, or when the mass of the molecule gets large, the numbers become significantly different. One must exercise care when discussing molecular masses and be explicit about the kind of molecular mass number one is discussing. For example, the normal hydrocarbon hexatriacontane, $C_{36}H_{74}$, has a nominal relative molecular mass of 506 Da, an exact monoisotopic molecular mass of 506.5791 Da, and a chemical molecular mass of 506.98 Da. All three numbers are correct when used in their proper contexts.

19.2.4   Polyisotopic elements and isotope patterns

Thomson's discovery and characterization of the isotopes of neon, [6] and subsequent work by Francis Aston [6][3] (a student of Thomson's) and others, measuring the accurate masses of the isotopes, elaborated the isotopic diversity of the periodic table. Elements that have more than one naturally occurring isotope are called polyisotopic elements. The most familiar polyisotopic element to organic chemists is carbon, with two common stable isotopes—^{12}C (98.9% natural abundance) and ^{13}C (1.1% natural abundance).[4] Next are chlorine and bromine, each with two very abundant isotopes separated by 2 Da. Although hydrogen, oxygen and nitrogen have well-known naturally occurring stable isotopes, they are of low natural abundance. In practice, hydrogen is treated as monoisotopic in small molecules, where the number of hydrogen atoms is relatively small. Deuterium is only 0.015% naturally abundant. The stable isotopes of oxygen and nitrogen are also of sufficiently low abundance—^{17}O, 0.038%; ^{18}O, 0.2%; and ^{15}N, 0.37%—that, unless a molecule contains significant numbers of these heteroatoms, these elements do not contribute significantly to an isotope pattern. Fluorine and phosphorus are monoisotopic.[5] Sulfur is also a common polyisotopic element appearing in organic molecules. DeBievre and Barnes [10] report the natural abundances of the isotopes of the elements. Abundances and accurate masses of the isotopes can also be found in the *Handbook of Chemistry and Physics*. Masses and abundances of selected isotopes are given in Table 19.1.

---

[3]Aston was awarded the Nobel Prize for chemistry in 1922. http://nobelprize.org/chemistry/laureates/1922/index.html

[4] ^{14}C is naturally occurring, but is a radioactive (unstable) isotope. Its natural abundance is low enough not to be of concern in mass spectrometry, unless one makes special efforts, for example, using accelerator mass spectrometry.

[5]Naturally occurring stable isotopes! We are not considering ^{18}F and ^{32}P here.

TABLE 19.1

Accurate masses and natural abundances of selected isotopes

Isotope	Accurate mass	Natural abundance
^{12}C	12.0000	98.9
^{13}C	13.00336	1.1
^{1}H	1.007825	99.99
^{2}H	2.0140	0.01
^{14}N	14.00307	99.6
^{15}N	15.0001	0.4
^{16}O	15.99491	99.8
^{17}O	16.9991	0.04
^{18}O	17.99916	0.2
^{19}F	18.9984	100.0
^{23}Na	22.9898	100.0
^{31}P	30.9738	100.0
^{32}S	31.9721	95.0
^{34}S	33.9679	4.2
^{35}Cl	34.9689	75.8
^{37}Cl	36.9659	24.2
^{79}Br	78.9184	50.7
^{81}Br	80.9163	49.3

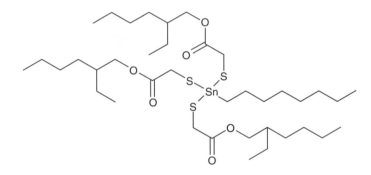

The most extensively polyisotopic element is tin, with 10 stable isotopes ranging in abundance from 0.4% to 33%, over a 12 Da range. Figure 19.2 illustrates a dramatic example of an isotope pattern for monooctyl tin ethylhexyl thioglycolate in the structure above. The upper panel of Fig. 19.2 expands the cationated molecular ion[6] region, $[M+Na]^{+}$, for this compound, taken from an LC–MS experiment

---

[6]The soft ionization methods, which will be discussed later, most often produce a molecular ion in which a charge-carrying species is attached to the neutral molecule. Typically, an H^{+} is the attaching species. Many structural classes of compounds, however, show the strong tendency to scavenge and attach monovalent cations—in this case, Na^{+}.

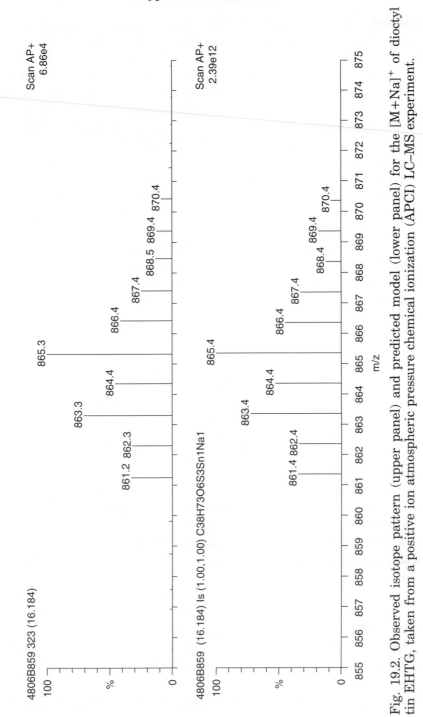

Fig. 19.2. Observed isotope pattern (upper panel) and predicted model (lower panel) for the [M+Na]⁺ of dioctyl tin EHTG, taken from a positive ion atmospheric pressure chemical ionization (APCI) LC–MS experiment.

conducted using atmospheric pressure chemical ionization (APCI). Which is the molecular ion? The answer is "all of them." The empirical formula for all of these species is $C_{38}H_{74}O_6S_3Sn_1Na$. While the chemical relative molecular mass for this compound is correctly calculated from the empirical formula to be 864.9 Da, the mass spectrometer reveals these 10 species, differing by isotopic content, but all represent the molecule under discussion. The 38 carbon atoms and the three sulfur atoms also contribute to this complex isotope pattern.

Revisiting the question, "which is the molecular ion species?" and calculating using isotopic masses, the measured values still bear further examination. For the most commonly encountered elements in organic molecules, the lowest mass isotope is the most abundant—e.g., carbon, oxygen, sulfur, chlorine. The calculated molecular mass using lowest mass isotopic masses in general corresponds to the most abundant species observed. Using ^{112}Sn, the species corresponding to this molecular ion should appear in Fig. 19.2 at m/z 857. Any signal at m/z 857 in the spectrum is insignificant. ^{112}Sn is only 1% abundant. The more appropriately designated molecular ion species here is the one at m/z 865, containing ^{120}Sn (the most abundant isotope of tin, at 33% natural abundance). The lower panel of Fig. 19.2 compares a theoretical prediction of the isotope pattern of this compound with the experimental observation. For the complex isotope patterns such as these, one can (and should) simulate the isotope pattern with isotope pattern modeling software, available in the better mass spectrometer data systems, and in several data system-independent programs. When discussing molecular mass results, especially mass spectrometric results for such polyisotope-containing molecules, one should specify how one is calculating results.

Further, the 865.3 Da assignment for this species in Fig. 19.2, and the position at which the peak is plotted, is correct and not a drafting artifact. The only isotope whose mass is truly an integer is ^{12}C, and only because it is arbitrarily assigned to be 12.0000. All other isotopic masses are either mass sufficient (slightly greater than the integral mass) or mass deficient (slightly less than the integral mass). The relative molecular mass of the m/z 865 species truly is 865.36 Da. The "sufficiency" results from the accumulated effect of the 74 hydrogen atoms,[7] offset by the deficiency contributed by all of the other mass-deficient elements (excluding carbon) in the empirical formula.

---

[7]64 hydrogen atoms cumulatively contribute 0.5 Da of additional mass above the integral mass.

19.2.5 High resolution measurements vs. accurate mass measurements

The accurate mass measurement technique is being actively re-discovered because of the recent improvements in instrumentation. John Beynon [9,11] established the validity of using high-precision mass measurements (to at least four decimal places of precision) to determine elemental compositions of ions with high certainty, capitalizing on the nonintegral nature of the precise atomic masses of the isotopes. Accurate mass measurements were the nearly exclusive province of magnetic sector instruments. Making such measurements presented difficulties, and required careful attention to detail. Reflectron time-of-flight instruments and Fourier transform (FT–MS) instruments have challenged magnetic sector dominance of this area and facilitated a somewhat easier route to making these measurements. Accurate mass measurement became accepted in the synthetic organic chemistry community as an appropriate measure of the "proof" of a structure. Acceptance criteria for including accurate mass measurements in the suite of data defending structure assignment of a newly synthesized compound are given in the instructions for authors of the synthetic and other journals,[8] even though Clayton Heathcock [12] (then editor-in-chief of the *Journal of Organic Chemistry*) acknowledges that combustion analysis, for which accurate mass measurements are often used as a substitute, is really a measure of purity. Accurate mass measurements are a measure of identity.

Biemann's discussion [13] lays out the circumstances and limitations very carefully where one should and should not use these kinds of measurements, and illustrates with examples applied to the elucidation of fragmentation pathways (the least demanding application), evidence supporting the confirmation of structural assignments and the determination of the structure of natural products (the most demanding application). Busch [14] discusses the utility of an accurate mass measurement in confirming the identification of a phthalate contaminant. Gilliam *et al.* [15,16] report on procedures and practices for making accurate mass measurements in a production type of environment, and the utility the measurements provide in a pharmaceutical discovery setting.

---

[8]Please consult, for example, recent instructions for authors for the *Journal of Organic Chemistry* or the *Journal of the American Society for Mass Spectrometry*.

Improvements in FT–MS instrumentation have permitted a re-evaluation of the value of accurate mass measurements to the analysis of very complex mixtures. Alan Marshall's group's work, for example, on vegetable oils [17] and on crude and refined petroleum fractions [18] has resurrected concepts which were developed in the early 1950s when MS was being vigorously applied in the petroleum industry. The very high resolving power, accessible on a high magnetic field FT–MS instrument emphasizes the potential elemental composition heterogeneity that can exist at a single mass in these complex mixtures, and demonstrates how much structural and compositional information can be extracted from a single mass spectrum. The mass deficiency of a single oxygen or sulfur atom imparts enough difference to distinguish compositions containing these elements from the more abundant, and more readily expected hydrocarbon compositions in a diesel oil sample (Fig. 19.3, top panel). A veritable forest of compositions appears in the negative ion electrospray ionization (ESI) spectrum of olive oil, due to a large number of oxygen atoms present in the compositions (Fig. 19.3, bottom panel).

A little recognized systematic error in the calculation of accurate masses of, for example, small radical cation molecular ions (as in electron ionization (EI)) or protonated molecular ions (as seen in the soft ionization methods) is the fact that the electron has a small, but finite mass. The accurate masses of radical cations, in which a valence electron has been removed, of anions that have been created by capture of an electron, and of protonated species produced by soft ionization processes, should take into consideration this small mass difference [19]. For example, there is a small difference between the relative atomic mass of a neutral hydrogen atom and a proton. The accepted accurate mass of $^1H^0$ is 1.007825 Da. The accurate mass of $^1H^+$ is 1.007276 Da. To be completely correct, expected accurate masses of protonated molecular ions, $[M+H]^+$, produced by electrospray should be calculated using the mass of one H^+, rather than all of neutral hydrogen atoms. Mamer and Lesimple do acknowledge, however, that, for large molecules, the error is of little consequence.

Notwithstanding the efforts of groups attempting to precisely define acceptable practices and requirements, [20] confusion still exists on the differences between definition of the terms "high resolution" and "accurate mass." [21] High resolution, as in any context, implies the ability to make confident measurements of small differences. In mass spectrometry, high resolution implies the ability to measure the small differences deriving from different elemental compositions, which have the

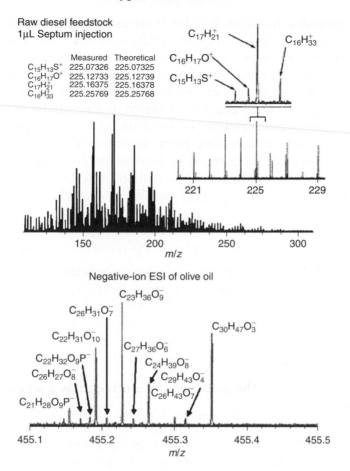

Fig. 19.3. Example of high-resolution experiments on complex mixtures, showing the multiple elemental compositions possible at a single nominal mass. Top panel: four elemental compositions at m/z 225 in the positive ion electrospray mass spectrum of a diesel oil feedstock (from Wu *et al.* [17]). Bottom panel: multiple compositions at m/z 455 in the negative ion electrospray spectrum of olive oil (from Marshall *et al.* [18]).

same nominal mass. Measurement can be made at sufficiently high enough resolution to see the heterogeneity of compositions at a given nominal mass without necessarily being able to assign the precise accurate masses of those compositions. Conversely, accurate mass measurement refers to the ability to confidently assign a precise relative-molecular mass (customarily four decimal places or more) to a signal. Accurate mass measurements can be made at low resolution, given a stable, reproducible scan by the instrument, appropriate accurate mass reference standards, and the confidence that one is observing a mass spectral peak composed of a single elemental composition. Indeed, Beynon's original demonstration [11] of the utility of calculating elemental compositions from accurate mass measurements was done on a single focusing magnetic sector instrument at a resolving power of only 250, while they were building an instrument intending to achieve a 2500 resolving power to enable them to make better measurements. Note these low-resolving powers, in contrast to the 5000 to 15,000 resolving powers used by Gilliam et al. [15,16] and the 350,000 resolving power used by Marshall et al. [18] and Wu et al. [17]

19.2.6 Ionization methods

Discovering and developing new ionization methods is one of the many topics that punctuates the history of mass spectrometry. All of the mass analyzers rely on the ability to use external force fields—magnetic fields, electric fields, radiofrequency fields—to steer electrostatically charged particles through space. Methods to convert neutral molecules to charged species, while preserving structural information are a perpetual topic of interest and development. Electric discharge in a gas produces ions, but is highly energetic, producing primarily atomic species. It was the ionization method used in Thomson's early parabola mass spectrograph studies, and in the early evaluations of isotopic composition of the elements. EI—interacting a beam of electrons with neutral molecule vapors—is the classical ionization method, producing radical cation molecular ions (when the molecule is stable enough to withstand dislodging of a valence electron) and extensive fragmentation. Development of softer ionization methods [9] is driven by the attempt to produce a molecular ion species without destroying structural integrity of the molecule. Chemical ionization (CI) using a variety of different reagent gases, field ionization (FI), field desorption ionization (FD), fast atom bombardment (FAB), APCI, ESI and matrix-assisted laser desorption ionization (MALDI) are more well known and prominent ionization methods. The order in which they are listed is an

---

[9]Hardness and softness of ionization refer to the energetics of forming an ion from a neutral molecule, and the extent of spontaneous fragmentation induced by the ionization process. Hard ionization techniques induce extensive fragmentation and have a high probability of completely destroying the molecular ion. Soft ionization techniques induce minimal or no fragmentation. The ideal soft ionization technique would produce a molecular ion species and nothing else.

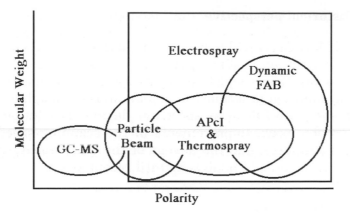

Fig. 19.4. Applicability of various combined chromatography–mass spectrometry techniques to the molecular domain.

approximate order from hardness to softness. The above list is also not comprehensive. Vestal [22] reviews these and a number of other minor specialty ionization methods.

Which method should one use for a particular application? The first factor to consider is the type of experiment to be used. Not all of these methods are compatible with GC–MS and LC–MS experiments. Figure 19.4 gives a first impression of where, within the universe of chemical compounds, GC–MS and LC–MS can be applied. The nature of GC–MS limits use to EI and CI, with some FI, and to compounds which are sufficiently volatile and thermally stable to pass successfully through a gas chromatography column. Attempts have been made to utilize nearly all of the major ionization methods for LC–MS, with varying degrees of success. Figure 19.4 again tries to capture the circumstances. FAB and MALDI are desorption ionization methods, and present particular engineering difficulties over how to present sample while preserving chromatographic separation. Workable solutions have been developed for FAB. While the general nature of pharmaceutically relevant compounds makes electrospray and APCI nearly ideal for this application area, they do not work well for all compounds. In short, there is no universally applicable ionization method.

19.3 GAS CHROMATOGRAPHY–MASS SPECTROMETRY

Of the three analytical techniques being discussed in this chapter, combined GC–MS can be considered as a mature technique—clearly the most mature of the three.

19.3.1 Historical perspective

Gohlke and McLafferty [23] review events in the early days of combined GC–MS. The first successful experiments using this technique were conducted in late 1955, coupled to a time-of-flight mass analyzer. Other analyzers available at that time could not scan fast enough to be able to record an adequate full mass spectrum during the elution of a gas chromatographic peak. Improvements in the scanning speed of magnetic sector analyzers, and the advent of the quadrupole, supplanted the original time-of-flight analyzers until recent years.

19.3.2 GC–MS interfaces

Both GC and MS are gas-phase experiments. However, coupling a gas chromatographic effluent to a mass spectrometer requires dealing with the problems associated with making the transition from a high pressure experiment to a low pressure experiment. Two older books on GC–MS by McFadden [24] and Message [25] describe the early days in the maturation of the GC–MS technique. The problems of coping with relatively high flows of carrier gas, in the case of packed column chromatography, and of enrichment of the analyte in a large excess of carrier gas have been dealt with straightforwardly.

Several types of carrier gas separators were developed to accommodate packed column gas chromatography—to accommodate the high gas flows and the necessity to enrich analytes. Most assume that helium is used as the carrier gas, and rely on the small molecular size of helium in relation to the larger size of analytes of interest. The Watson–Biemann effusion separator relied on a greater rate of passage of helium through the walls of a sintered glass tube than of larger analytes. Differentially permeable membranes of various types were employed, but showed some selectivity and discrimination concerning the types of molecules they would accommodate. The Ryhage double stage jet separator avoided these selectivity issues. It relied again on the molecular size differential between helium carrier gas and analytes of interest. Two jets are positioned in line with each other, as in Fig. 19.5. The gap between the ends of the jets is enclosed in an evacuated chamber. The effluent from the chromatographic column streams out of the input jet. Large molecules, having greater mass, and therefore greater forward momentum, tend to stream in a straight line and are more likely to flow into the opening of the opposing jet and be transported into the mass spectrometer. Helium carrier gas, being smaller and having less

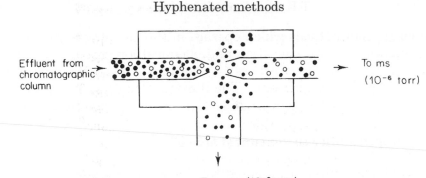

Fig. 19.5. Single stage jet separator.

forward momentum, is more likely to diverge and not enter the exit orifice. The analyte molecules are then enriched in the flow being transported on into the mass spectrometer. The original Ryhage design connected two such enrichment chambers in series. Subsequent experience demonstrated that a single stage was adequate for the purpose. Typical gas flows accommodated were 30 ml/min (or more) of helium carrier gas through a packed GC column, flowing into the separator, and 3–5 ml/min flowing on into the mass spectrometer.

Because of the substantially lower total gas flow through a fused silica capillary gas chromatography column, a separator is no longer needed. With flexible fused silica columns, the column end can be inserted directly into the mass spectrometer. Typical carrier gas flows through capillary columns of 1–2 ml/min can be readily accommodated by the pumping systems on modern mass spectrometers. The end of the capillary column can be positioned in close proximity to the ion source of the mass spectrometer, if not actually inserted into it. The sensitivity of detection is thus maximized, since this positioning delivers the entire sample directly into the ion source. Separator technology is not entirely obsolete, however. Megabore fused silica columns are often operated with carrier gas flows approaching those of packed columns. The higher flows need to be processed through a jet separator, with the requisite reduction in total gas flow and analyte enrichment.

19.3.3 Available ionization methods

Not all ionization methods are available for use in GC–MS experiments. Because the GC experiment is a gas-phase experiment, only those ionization methods which operate in the gas phase are appropriate. Standard electron ionization and chemical ionization methods are applicable. Recent years have seen a re-introduction of field ionization coupled to GC–MS experiments. The literature suggests spectra of quality comparable to that generated by conventional electron ionization.

19.3.4 Applicability to "chemical space"

The gas phase limitation imposed by the very nature of the GC–MS experiment limits its applicability to chemical space. The Venn diagram in Fig. 19.4 illustrates an approximate mapping of applicability of the combined chromatography–mass spectrometry techniques discussed here. Using only relative molecular mass and polarity as two dimensions that can map chemical space, the figure indicates that the applicability of GC–MS is limited to those relatively small, thermally stable, nonpolar compounds that can be readily volatized (or made to volatilize by derivatization [26]) to pass through a gas chromatographic column.

19.3.5 Identification of compounds

The primary qualitative application of combined gas chromatography–mass spectrometry is identification of compounds. The gas chromatographic retention time, while being characteristic of a particular compound, is insufficient by itself for positive identification of that compound. Application of Kovat's indices [27] and retention indices [28] assist by minimizing the effects of operational conditions on observed retention times. Retention indices work well for tracking compounds within defined groups, and are adequate to provide presumptive identifications. The combination of a retention time of a compound eluting from a gas chromatographic column coupled with the correct mass spectrum for that compound is much more certain evidence to present as identification. Even though MS by itself is not adept at differentiating isomeric structures—such as positional isomerism of groups around an aromatic ring, or branched structures—the gas chromatograph can differentiate isomers, and the combination of the two pieces of information is sufficient for agencies such as the Environmental Protection Agency to accept that GC–MS identifies a compound conclusively.

Computer programs have been developed that assist, at different levels, in the interpretation of mass spectra, the goal being identification of a molecular structure. Many early ones simply performed the book-keeping arithmetic necessary when one rationalizes a mass spectrum with a known structure—namely by systematically breaking a bond and computing the masses of the remaining fragments on either side of the broken bond. Mass spectral libraries and the search algorithms used with these libraries are another approach, which will be discussed more below. Two software systems that attempt to apply rational rules to the prediction of fragmentations from structures, exercising a degree of artificial intelligence, and which are commercially available are the efforts by ACD Laboratories, and the MassFrontier program, vended by Finnigan Instruments as an add-on to their data system. As with any predictive software, one should convince oneself how reliable the predictions are with known structures. These programs provide a convenient mechanism to remind us of the possibilities other than those upon which we have fixated.

19.3.6   Mass spectral libraries

The concept of the organic functional group—that a particular organic functional group will behave in essentially the same way, regardless of other functional groups attached to it—also holds true in the mass spectrometer. Mass spectra of compounds (at least EI spectra) are also reproducible. For these reasons, compilations of mass spectral fragmentation data for organic compounds are useful sources for help when interpreting mass spectra. Robust mass spectrometer operating conditions have been found, such that electron ionization mass spectra of compounds can be recorded on different instruments, different types of instruments and at different times and can be expected to be comparable. This situation holds true, primarily, for spectra recorded under 70 eV electron ionization conditions.

Two of the better-known libraries are the NIST/EPA/NIH/MSDC mass spectral database, maintained by the United States National Institute of Standards and Technology (NIST),[10] and the Wiley–McLafferty database, vended by Wiley Publishing Company. Heller [29] captures the early days in the development of the NIST library. These two libraries differ slightly in philosophy. The NIST collection originally contained only one spectrum representing a given compound, assuming the universality of the spectrum, independent of instrument and conditions. The 2005 version of this library contains 190,825

---

[10]http://www.nist.gov/srd/.

spectra of 163,198 compounds,[11] indicating that the NIST library is starting to embrace some concepts of the Wiley Library. The Wiley–McLafferty library includes multiple spectra for some compounds, intending to capture information on variability of spectra from observation to observation and instrument to instrument. There are a number of smaller and more specialized collections. The Thermodynamics Research Center at Texas A&M University still collects compounds and compiles data, [30,31], which eventually are incorporated into the NIST and Wiley–McLafferty libraries. Critical evaluations of the various algorithms for library searching, [32] and evaluation of the libraries themselves have been published [33].

Compilation of spectral libraries for the soft ionization techniques has not occurred to the same extent, for at least two reasons. First, the exact nature of a soft ionization spectrum is highly dependent upon instrument and sampling conditions. If one standardizes one's activities on a single set of operating conditions, it can be useful to assemble one's own library of reference spectra. Most of the instrument manufacturers provide the necessary software tools as a part of their mass spectometry data systems that will permit accomplishing this task. Independent software developers, including the NIST, also provide library building, maintenance and search tools. Second, for many compounds analyzed by the soft ionization techniques, the spectrum consists principally of a molecular ion. A compilation of soft ionization spectra, in its ideal form, would be a compilation of molecular masses. Movements have surfaced several times to promote the compilation of MS–MS product ion spectra. These, too, show a high degree of dependence upon exact operating conditions. An acceptable standard set of conditions has eluded efforts to make such a project happen. NIST, however, is also starting to tackle this problem.[11]

In addition to the searchable library compilations, several compendial books on the electron ionization fragmentation behavior of compounds have been published [34–36]. They are dated, but nevertheless effectively capture the collective fragmentation information prior to their publication. All of these information sources discuss electron ionization spectra. EI fragmentation rules, however, can be of limited assistance in interpreting soft ionization and MS–MS product ion spectra.

---

[11]The 2005 version introduces spectral duplication: 190,825 spectra of 163,198 compounds. The library also now contains a collection of gas chromatographic retention indices for 25,728 compounds on nonpolar chromatographic columns, and a collection of 5191 MS–MS spectra of 1920 ions. The latter multiplicity of MS–MS spectra reflects a dependence of the spectrum on the conditions under which it is recorded.

One small collection of CI information has been published [37]. Fragmentation behavior of molecules under the soft ionization conditions—chemical ionization, FAB, API—still remains largely scattered throughout the chemical literature.

19.3.7 The GC–MS data model

In order to compensate for the possible separatory shortcomings of the chromatography, a number of data processing algorithms have been developed to make use of the fact that repetitive scan GC–MS data are really a three-dimensional matrix of data. The x-axis describes the time dependence of the separation. The z-axis describes the mass spectrum of the GC effluent at any "instant" in time, and the y-axis describes the intensity of the signal. Imagine then a cube, with the lower left corner of the front face assigned to be the origin of our data space. Along the bottom edge of the front face is the time axis (the x-axis). The left edge of the front face is the intensity axis (the y-axis), and the axis from front to back is the m/z axis (the z-axis). In general, this model applies to any repetitive scan chromatography–spectroscopy experiment.

The total ion current (TIC) chromatogram is a summation of all the mass intensities for each individual mass spectrum, plotted as a function of time. The TIC chromatogram can be visualized as a projection of all the intensities of the mass spectra onto the front surface of the data cube. (Strictly speaking, the projection of data onto the front surface of a transparent cube would actually be the base peak intensity (BPI) chromatogram, in which the BPI of each spectrum is plotted as a function of time. Summing the intensities of all peaks in each spectrum, and plotting the sum as a function of time, is the TIC chromatogram.) Slicing the cube in a plane parallel to the front face and at a particular point on the m/z axis would expose a selected ion chromatogram of the time-dependent intensity of that particular m/z value. Slicing the cube orthogonal to the front face at any particular point in time would present on the exposed surface a mass spectrum of the GC effluent at that particular point in time.

19.3.8 Full (repetitive) scan data collection vs. selected ion recording

Full scan experiments set the mass spectrometer in a cycle of repeatedly scanning a selected range of m/z values. The experiment is done primarily for structure elucidation, and the richness of structural information the fragmentation (induced or spontaneous) provides. It is relatively insensitive. Observing a chromatographic peak relies on sufficient accumulated signal in the mass spectrum of a component to be discernable above background. If background in a given experiment is high, components can be easily missed. Selected ion recording (SIR or SIM) is used for quantitation. The experiment shows greater sensitivity because one maximizes the time that one is looking at the signal that tells one about what one wants to know, and discards everything else. Consider that, in a repetitive scan experiment, a finite amount of time is required for the mass spectrometer to scan across a range of m/z values, typically 1000 Da. One would envision spending equal amounts of time observing each of those nominal 1000 channels. Only a few (or even one) of those channels contain signals of interest—the molecular ion and/or fragments indicating the compound of interest. The remaining 99.9% of the time, one is looking elsewhere. By performing an SIR experiment, one is watching the information channel(s) containing the signal of interest nearly 100% of the time. One would expect to see sensitivity increases approaching three orders of magnitude. The price, of course, is that the maximum information is being measured on the targeted signal, but all other information about the composition of the original sample is being discarded.

19.3.9 GC–MS example—residual solvents analysis

Determination of residual solvents is a necessary test for quality of manufactured material, especially in a manufactured pharmaceutical drug substance. Analysis of a test mixture of residual solvents, typically used in the release testing of pharmaceuticals, is shown here as an example GC–MS application. The target solvents are those frequently encountered in the synthesis of drug substances. The solvents in this mixture are (in retention time order) methanol, ethanol, acetone, isopropanol, acetonitrile, dichloromethane, ethyl acetate, tetrahydrofuran and toluene. The analysis was conducted on a DB-624 bonded phase wall-coated open tubular fused silica capillary gas chromatography column. The DB-624 liquid phase was designed for use in EPA method 624 for analysis of volatile organic compounds. Chromatograms from analyses of this mixture are shown, detected by flame ionization detection (FID) on a stand-alone gas chromatograph (Fig. 19.6), and detected by a quadrupole mass spectrometer attached to a different gas chromatograph (Fig. 19.7). The columns used in each instance are nominally equivalent DB–624 columns. Aliquots of the same mixture were

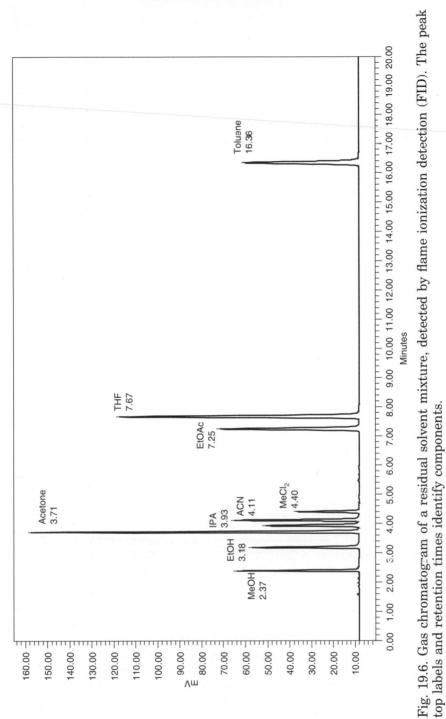

Fig. 19.6. Gas chromatogram of a residual solvent mixture, detected by flame ionization detection (FID). The peak top labels and retention times identify components.

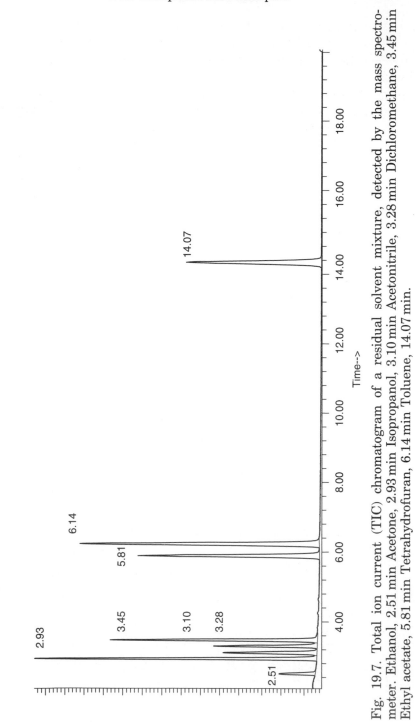

Fig. 19.7. Total ion current (TIC) chromatogram of a residual solvent mixture, detected by the mass spectrometer. Ethanol, 2.51 min Acetone, 2.93 min Isopropanol, 3.10 min Acetonitrile, 3.28 min Dichloromethane, 3.45 min Ethyl acetate, 5.81 min Tetrahydrofuran, 6.14 min Toluene, 14.07 min.

injected in each instance. (Data acquisition by the mass spectrometer was not started until after the methanol chromatographic peak had eluted. Methanol is missing from the chromatogram in Fig. 19.7). Absolute retention times are different because of the differences in instrumentation and columns. Because this is a well-known, well-behaved analysis of these solvents, the retention order and relative spacing of the chromatographic peaks are well known, and comparable between the two determinations. Retention index mapping would demonstrate this obvious fact, but is unnecessary here.

Electron ionization is used to generate the mass spectra. A single scan spectrum of toluene, at 14 min, is shown in the upper panel of Fig. 19.8. The spectrum is easily recognized by the characteristic abundant radical cation molecular ion, $M^{+\bullet}$, at m/z 92 and two fragments—the tropylium ion, $C_7H_7^+$, at m/z 91 and the cyclopentadienyl cation, $C_5H_5^+$, at m/z 65. Library search against the NIST library of electron ionization spectra identifies this spectrum as that of toluene, and gives a quality-of-fit factor of greater than 90%.

Figure 19.9 compares the observed and library spectra for dichloromethane (retention time 3.45 min in the GC–MS run). The prominent chlorine isotope pattern for the two chlorine atoms in this spectrum makes it readily identifiable. The primary fragmentation is loss of a chlorine atom, producing the m/z 49 fragment. While this fragment clearly manifests a chlorine isotope pattern still, it reflects the fact that only one chlorine atom remains. Library search identifies this spectrum as dichloromethane with quality-of-fit measures of greater than 95%.

Spectra here are intense and largely free of instrument background interference. They are recognizable without any spectral enhancements, such as spectral averaging and background subtraction. Were the observed spectra of significantly lower signal intensity, or clouded by substantial amounts of instrument background, averaging and background subtraction would be necessary to help reveal what is significant and what is background. Library searching using modified goodness-of-fit calculations (such as reverse fit and mixture fit comparisons), data processing algorithms such as the Hites–Biemann and Biller–Biemann algorithms, factor analysis and a variety of others have appeared in the literature over the years, and have achieved some levels of success in helping to extract information from noisy data.

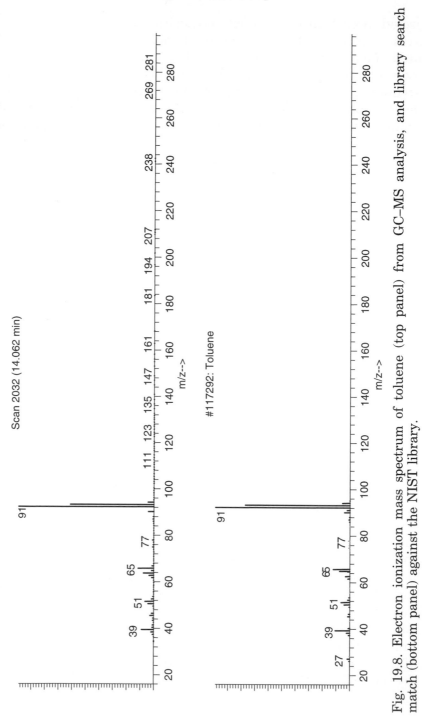

Fig. 19.8. Electron ionization mass spectrum of toluene (top panel) from GC–MS analysis, and library search match (bottom panel) against the NIST library.

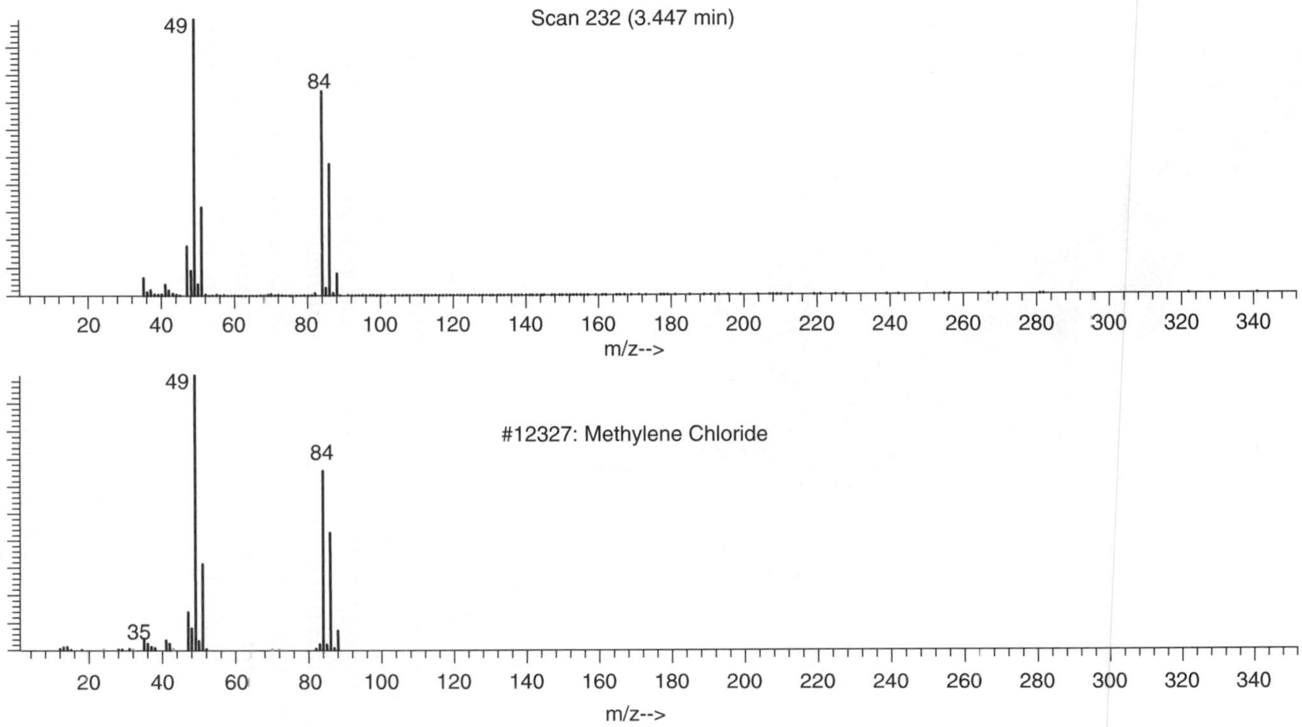

Fig. 19.9. Electron ionization mass spectrum of dichloromethane (top panel) from GC–MS analysis, and library search match (bottom panel) against the NIST library.

Fig. 19.10. Combining liquid chromatography with mass spectrometry involves a difficult courtship [38].

19.4   LIQUID CHROMATOGRAPHY–MASS SPECTROMETRY

Combined LC–MS is a rapidly maturing technique. Peter Arpino's cartoon (Fig. 19.10) captures well the essence of the difficulties associated with coupling the liquid chromatographic experiment with the mass spectrometric experiment. The historical perspective presented in the following is intended to illustrate the difficulties in achieving a useful LC–MS interface, and the variety of devices and the ingenuity of the people attempting, and eventually succeeding, to make it happen. The LC–MS experiment has only become an effective experiment since the early to mid-1990s.

19.4.1   Historical perspective

Attempts to couple condensed phase chromatographic methods with mass spectrometry start with the work reported by Victor Tal'Rose[12]

---

[12]Victor Tal'Rose recently passed away.

[39] and colleagues in 1968 [40]. By referring to condensed phase chromatographic methods here, we primarily imply high performance liquid chromatography (HPLC). Attempts, with varying degrees of success and acceptance, have been reported to interface other condensed phase separation methods, such as thin layer chromatography [41,42] (Chapter 13) and supercritical fluid chromatography [43,44]. Yergey *et al.* [45] summarized the progress on development of effective LC–MS interfaces, and compiled a review and comprehensive bibliography of interface development and applications up to 1990. They do not cover electrospray, as the explosion of efforts and applications of this technique only started in this timeframe.

Unlike combined GC–MS, LC–MS presents a significant incompatibility between the condensed phase liquid chromatographic experiment and the gas phase mass spectrometric experiment. The pumping system of the mass spectrometer must be capable of handling the gas load introduced by the chromatographic experiment. Typical gas loads generated by a gas chromatographic experiment are 3–5 ml/min for packed columns (after jet separation!), 0.5–1 ml/min for capillary columns. HPLC mobile phase must be vaporized. Typical mobile phase flow is 1–3 ml/min. Hexane, converted to gas phase, produces 180–540 ml/min of vapor. Chloroform produces 280–840 ml/min. Methanol produces 350–1650 ml/min. Water produces 1250–3720 ml/min. The design and engineering of an interface to connect a liquid chromatographic experiment with a mass spectrometric experiment must accommodate these large gas flows—approximately $1000x$ that of the gas chromatographic experiment—and still permit both the chromatographic experiment and the mass spectrometric experiment to work properly.

19.4.2   Evolution of LC–MS interfaces

19.4.2.1   *Direct liquid introduction*
Direct liquid introduction (DLI) is the simplest and most straightforward approach. It was first attempted and reported by Tal'Rose *et al.* [40] Baldwin and McLafferty [46] reported their attempts, utilizing chemical ionization. Chapman *et al.* [47] reported their experiences with interfacing to a magnetic sector instrument. A small portion of chromatographic effluent (10–20 µl/min) is sprayed into the ion source through a pinhole leak into the mass spectrometer. Chemical ionization is most frequently used, with the residual solvent molecules in the spray acting as the chemical ionization reagent.

19.4.2.2 Atmospheric pressure chemical ionization

Rather than leaking a portion of a chromatographic effluent into the vacuum system of the mass spectrometer, the effluent is introduced *via* a heated nebulizer into an API source. The source is outside the vacuum chamber, operating at or above atmospheric pressure. Ionization occurs in a corona discharge, struck in the plume of effluent sprayed from the heated probe. The solvent vapors provide a source of protonating reagent for the proton transfer chemical ionization reactions in the vapor. The corona discharge region is sampled through a sampling orifice by the mass spectrometer. The discharge was first struck by proximity with the ionizing radiation from a ^{60}Co radioactive needle, and later by a needle with an applied high voltage [48]. The ionization is soft, as the majority of compounds produce abundant protonated molecular ions.

19.4.2.3 Continuous flow-FAB interfaces

Fast atom bombardment (FAB) ionization revolutionized the application of mass spectrometry to large, thermally fragile, nonvolatile molecules. This desorption technique was described by Barber and colleagues in 1981 [49,50]. Dependence of the technique on a dispersing matrix and its inherent incompatibility with flowing sample introduction into the mass spectrometer prevented its immediate application to chromatographic separations. However, a mobile phase adulterated with 10% (by volume) of glycerol, the classical FAB matrix, was infused through a glass frit, the frit being the FAB target. Low-flow chromatographic effluent (*ca.* 2 µl/min) could be analyzed by FAB ionization [51]. Caprioli *et al.* [52] showed that the frit was not necessary, effluent being introduced directly onto the target through a hole in the center. The fine art of obtaining useful results from either of these approaches involved balancing the rate of evaporation of effluent from the FAB target with replenishment of fluid by chromatographic-flow. Adjusting solvent composition, glycerol content in the mobile phase and externally applied heat to the target were some of the operational variables.

19.4.2.4 The moving belt interface

Scott *et al.* [53] and McFadden *et al.* [54] first described this mechanical interface in 1974 and 1976. A diagram of a commercialized moving belt interface is shown in Fig. 19.11. The interface first consisted of a spool of wire, which was unrolled off one spool and onto another. As the wire was wound from spool to spool, the effluent from a liquid chromatographic separation was applied to the wire. As the wire was transported through

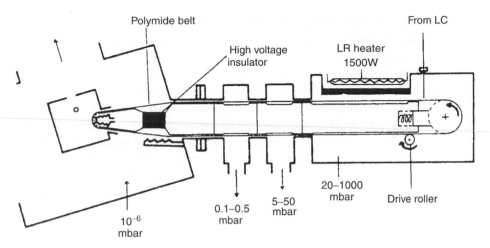

Fig. 19.11. Schematic diagram of a moving belt interface. Courtesy of VG Analytical.

the ion source of a mass spectrometer, it was heated to desorb analytes that had been applied to it. The desorbed materials were ionized and mass analyzed in the mass spectrometer. Later, a recycling continuous loop of wire, or a belt of an appropriate material, was used, with added mechanisms for "cleaning" the belt to avoid chromatographic confusion. The moving belt is a versatile interface, offering applicability in electron ionization, CI and FAB [55] ionization modes. However, it is mechanically the most complex, both in operation and optimization of the interface, and in its installation into a mass spectrometer.

19.4.2.5 Thermospray
The term thermospray refers to both a vaporization LC–MS interface device and a soft ionization method. Vestal and co-workers [56] developed thermospray, beginning in 1976. Plasmaspray is a variation on the thermospray concept, implemented by VG Instruments on their line of magnetic sector and quadrupole instruments. The technique and device were designed as an alternate means of rapidly vaporizing solvent and nonvolatile analytes by heating the tip of a stainless steel tube through which the LC effluent flowed. The resulting spray of vapor and droplets was directed against a heated probe to complete droplet vaporization. An electron beam from a hot filament bombarded this vapor, producing chemical ionization, and the vapor was sampled with a sampling orifice into the mass spectrometer. The accidental observation that a filament and electron beam was not needed was made when,

719

unknown to the operators, the filament had burned out and as many ions were being produced and detected without the filament as with the filament. Mass spectra produced by thermospray are soft-ionization spectra. The molecular species is typically a protonated molecular ion or an adduct—often an NH_4^+ cationated species, because thermospray works best from ammoniated mobile phases. Thoughts on the mechanism(s) involved in this process were reviewed by Vestal [57]. The Yergey *et al.* book [45] contains an extensive treatise on this technique. Empirical observations show that ionization is quite efficient when the LC separations are carried out in a polar solvent, which includes ionic solutes such as ammonium acetate. Separations carried out in a nonpolar solvent, without salts, require an assisting electron beam from a filament, if the application works at all.

19.4.2.6 MAGIC

The MAGIC acronym stands for "monodisperse aerosol generation interface for chromatography." This LC–MS interfacing technique was first described by Willoughby and Browner, [58] and evolved into the particle beam interface. Various instrument manufacturers implemented it under a variety of names. It contained no moving parts. Volatile solvents work best. Small portions of split aqueous flows could also be accommodated. Desolvation was done by evaporation and impinging with a "nebulizing gas." Only sufficient external heat was supplied to counteract the cooling caused by expansion and evaporation of the solvent. Reduction of net flow into the mass spectrometer is done by using the Ryhage jet separator technique (discussed above, developed for packed column gas chromatography–mass spectrometry). This interface presents the possibility of conducting electron ionization on samples introduced into the mass spectrometer, EI spectra being the "richest" in fragmentation and structural information. EI spectra could be obtained of sufficiently high quality to permit spectral library searching against, e.g., the NIST library. The particle beam technique occupies a niche in applications space. Compounds, which can produce fragmentation-rich electron ionization mass spectra (e.g., *via* a probe distillation experiment), but that will not pass well through a gas chromatographic column, work in liquid chromatographic experiments interfaced to the mass spectrometer through a particle beam interface.

Semivolatile and nonvolatile compounds of interest in environmental sciences, [59] such as the larger polycyclic aromatic hydrocarbons and pesticides, [60] have been effectively addressed. It is theoretically possible to introduce reagent gases into the ion source of the mass spectrometer and do CI on materials introduced through a particle beam interface. Extrel Corporation also developed and commercialized an ion source that included a fast atom source and a target for doing FAB ionization. The mobile phase was adulterated with a low percentage of glycerol. The low percentage, in general, did not interfere significantly with the chromatography. The glycerol acted as a continuously renewing matrix from which to do FAB ionization, the glycerol being sprayed onto the FAB target along with the analyte.

19.4.2.7 *Electrospray*

Abbe Jean-Antoine Nollet's 1750 experiments on electrostatic spraying of liquids[13] constitute the first experiments reported with this technique (Fig. 19.12[14]). Nollet demonstrated a number of ways in which electrostatically charged spraying devices could effectively spray fluids. John Fenn's [61] more recent elaboration of the electrospray phenomenon has led to a revolution in the application of mass spectrometry, first to biologically relevant large molecules, and subsequently in the application to polar molecules in general. Fenn was jointly awarded a Nobel Prize [62] in 2002, along with Koichi Tanaka and Kurt Wüthrich, for their development of methods for identification and structural analysis of biological molecules.[15]

Malcolm Dole [63] described his group's attempts to measure the relative molecular masses of oligomers of synthetic polymers based on ESI in 1968. Fenn's group elaborated on the Dole observations, and build the electrospray ion source shown in Fig. 19.13 [61]. In essence, liquid sample containing molecules of interest is flowed into the source through a small tube, on which a high-electrostatic voltage is placed. The liquid is sprayed from the tube tip, forming a Taylor cone and subsequently a stream of droplets, each electrostatically charged. A countercurrent flow of dry nitrogen gas opposes the spray, causing evaporation of solvent. As the droplets become smaller because of solvent loss, they reach the Rayleigh instability limit, where Coulombic charge repulsion overcomes solvent surface tensions and the droplets fragment further. Repetition of this process ultimately produces

---

[13]This interesting historical footnote contributed by Dr. Thomas Covey, MDS Sciex Instrument Corporation.

[14]Information available on the history of chemical instrumentation website maintained by Prof. Euginii Katz, Dept of Chemistry, The Hebrew University of Jerusalem. See http://chem.ch.huji.ac.il/~eugeniik/history/nollet.htm.

[15]http://nobelprize.org/chemistry/laureates/2002/index.html. Fenn's and Tanaka's work applied electrospray to the mass spectrometric examination of large molecules. Wüthrich's contributions were in application of NMR to this area.

Fig. 19.12. Abbe Jean-Antoine Nollet's experiments with the electrostatic spraying of liquids.

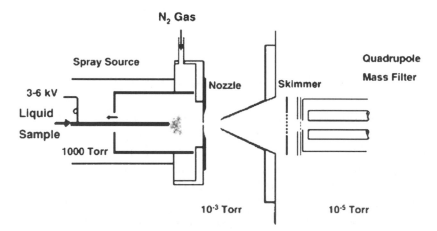

Fig. 19.13. Electrospray ion source. From Yamashita and Fenn [61].

solvent-free solute ions, which are subsequently swept into the sampling orifice of the mass spectrometer and detected.

The detailed mechanism of electrospray is still an active and lively topic of discussion at current scientific conferences. Numerous research laboratories are actively probing the subject. Volume 35, issue 7 (2000), of the *Journal of Mass Spectrometry* published a series of five papers [64–68] from research groups working in this area, followed by a lengthy printed discussion [69] in the subsequent issue.

Two models have been repeatedly discussed in the literature—ion evaporation, attributed to Iribarne and Thomson, [70] and coulombic repulsion, attributed to Dole. Both models rely on the Rayleigh instability of charged droplets to become smaller and smaller. They differ in that coulombic repulsion relies on the continuous repeating of the process until one is left with a single isolated desolvated ion. Ion evaporation invokes the idea that, before a droplet becomes small enough to contain a single solute molecule, the electrostatic field strength at the droplet surface becomes sufficiently intense to eject a surface ion from the droplet into the ambient gas. Kebarle reviewed the data available in 1991, [71] and updates and interprets [68] the summed results to propose that the data indicate that the mechanisms constitute a continuum, and that elements of both interplay, depending upon physical properties such as hydrophobicity or hydrophilicity, most notably the size of the molecule under consideration—small ions are produced by ion evaporation, large ions are produced by coulombic repulsion—and other environmental factors such as ion concentration in solution. Vestal [22] gives an additional concise summary of the current thoughts on how electrospray works, but more importantly summarizes a number of design variations of electrospray ion sources extant on modern instruments.

19.4.3   LC–MS compatible mobile phases

A point of concern, which repeatedly comes into play when planning LC–MS investigations, is that of a LC–MS compatible mobile phase. Phosphate buffers are a desirable reverse phase (RP) mobile phase buffering agent because these buffers can cover a wide pH range and are easily prepared gravimetrically from high, purity and stable inorganic components. They show UV-visible transparency well into the ultraviolet wavelength range—as low as 200 nm. This property makes them desirable because of the universality of UV-visible detection at low wavelength, the most common detector on an HPLC system being a UV-visible spectrophotometric detector.

However, phosphate salts are not volatile. We must constantly remember that mass spectrometry is a gas-phase experiment. Materials to be examined by mass spectrometry must ultimately be made gaseous. Figure 19.14 shows the atmospheric pressure ionization source chamber of a mass spectrometer after infusion of a 20 mM potassium phosphate-containing mobile phase into the instrument for a few hours. The accumulation of phosphate salts on the striker plate is evident. Visual evidence of salt accumulation is also apparent on the back wall of the source chamber, above the striker plate. The overall haziness of the image is not the result of poor photography, but rather due to the coating of dust on the inner walls of the chamber and all surfaces within.

We should be very clear, however, to emphasize that obtaining useful data from an LC–MS experiment using a nonvolatile salt-containing mobile phase (such as a phosphate system) is not impossible. The experiments being done when we had the opportunity to capture the

Fig. 19.14. Accumulation of phosphate salts in the atmospheric pressure ionization chamber. Note the large accumulation of salts on the striker plate. Additional accumulation of salts can be seen on the back walls of the chamber. The general "dustiness" in the chamber is salt accumulation. This source is a "z-spray" ionization source chamber of a Micromass Quattro Ultima mass spectrometer system.

photograph in Fig. 19.14 required our performing at least some work in a phosphate-containing mobile phase. The experiments were successful, but the consequences of doing the experiment had to be accepted. The complications ensuing from infusion of a phosphate-containing mobile phase can be managed. The instrument has to be cleaned more often, and additional care needs to be used to interpret the data obtained during this experiment.

Contrary to the general lore that ionic compounds are not amenable to mass spectrometry, we discuss the positive ion electrospray mass spectrum shown in Fig. 19.15. This spectrum represents the instrument background generated by the phosphate-containing mobile phase in the above-described experiment. The species appearing at m/z 213 corresponds to $K_3HPO_4^+$. The series of species appearing at 174 Da repeating intervals in the spectrum corresponds to clusters of the general empirical formula $(K_2HPO_4)_nK^+$. Ionic salt clusters were studied at length in the context of FAB ionization [72]. The science extends into the realm of ESI. Further, the details in the spectrum around the m/z 213 species and the clusters of increasing size reflect the fact that potassium is a polyisotopic element. Potassium exhibits two naturally occurring stable isotopes—^{39}K at 93.3% natural abundance and ^{41}K at 6.7% natural abundance.

If a phosphate buffer is not a good mobile phase for LC–MS, then what is? In general, buffer salts and mobile phase adulterants should be composed of volatile components. Both components of the buffering salt should be by themselves volatile components. The small organic acids are appropriate. At 0.1% (v/v) in aqueous solution, these acids will produce mobile phases at nominal pH values of 2, 3 and 4 for trifluoroacetic acid, formic acid and acetic acid, respectively. Adjustment upward in pH is done with ammonium hydroxide. Even though triethylamine and similar substituted amines are volatile, and are often used to adjust chromatographic peak shapes, they severely compromise mass-spectral sensitivity. Triethylamine is also very sticky. Residual traces of it persist in the mass spectrometer long after one has stopped using a buffer containing it. Figure 19.16 compares LC–MS TIC chromatograms of the same sample mixture injected onto the same chromatographic column using two different mobile-phase compositions. The mobile phases differed only by the buffer salts in the aqueous component. The top panel records the separation recorded by UV-visible detection. Separation was indistinguishable in both instances. The middle panel TIC chromatogram is from an aqueous triethylamine acetate buffer, the lower panel from an aqueous ammonium acetate buffer, at the same pH.

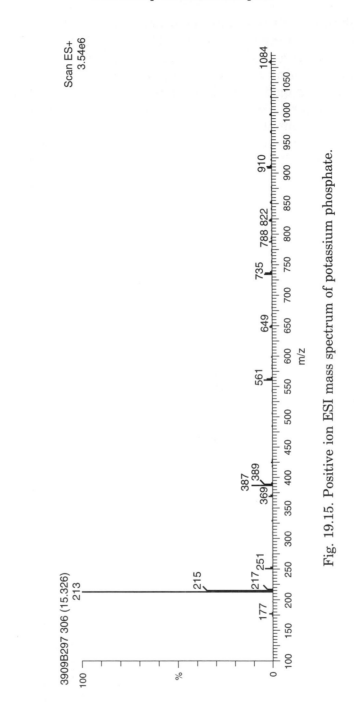

Fig. 19.15. Positive ion ESI mass spectrum of potassium phosphate.

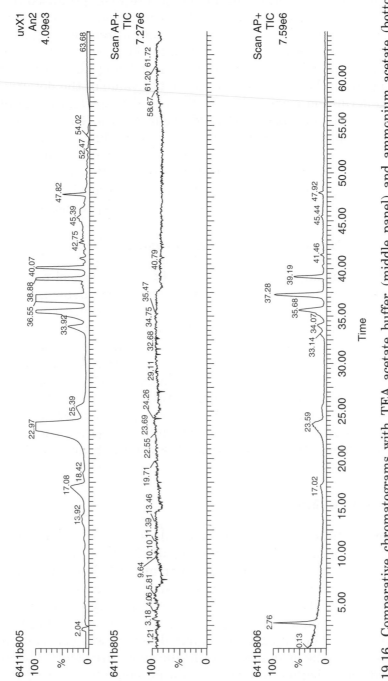

Fig. 19.16. Comparative chromatograms with TEA acetate buffer (middle panel) and ammonium acetate (bottom panel). The top panel shows chromatography detected by UV–Vis detection.

The presence of the triethylamine has severely compromised the ability of the mass spectrometer to see the components of interest in the mixture. The triethylamine analysis is not entirely useless, as the UV-visible chromatogram permits knowing where in the TIC chromatogram to look for features of interest. Not being able to use the TIC chromatogram, however, compromises the value of the determination.

In general, ion-pairing agents fit into the nonvolatile adulterants category. The small perfluorinated organic acids (e.g., trifluoroacetic, perfluoropropionic and perfluorobutyric acids) have been successfully used, even though trifluoroacetic acid is reported in the literature to compromise sensitivity [73]. Recently, HBF_4 [74] has been used as an LC–MS friendly mobile-phase adulterant, which acts like an ion-pairing agent in enhancing column retentiveness of analyte molecules.

19.4.4 LC–MS accurate mass measurements

With the maturing of a technique like LC–MS, a logical next step is to attempt to acquire accurate mass measurement data under LC–MS conditions. In theory, there is no reason not to be able to make accurate mass assignments, given the capability on the mass analyzer being used. The scan stability of modern time-of-flight mass analyzers permits reasonable measurements without included internal standards or lock-mass compounds. Technical refinements, such as mixing reference compounds into the flowing stream after chromatographic separation, but before introduction into the mass spectrometer, and building valving devices and software that present alternating sample effluent and a reference solution, are receiving attention and producing successful results.

19.4.5 An LC–MS example

Solutions to practical problems rarely depend upon a single technique or a single approach. The following example of an impurity identification in a pharmaceutical product illustrates the key role that LC–MS can play in such an investigation, but also illustrates the limitations of the technique. The identification of this impurity has been published elsewhere in complete detail [75]. The problem and solution is summarized here. The impurity, designated as H3, was observed at 0.15% in a bulk lot of the drug substance in the structure below. The impurity required identification before the bulk lot could be released for use in further studies.

An LC–MS examination revealed a 775 Da relative molecular mass, a mass almost large enough to implicate a dimer of the drug substance. Mass spectral fragmentation was similar to that of a known dimer-like impurity with a molecular mass of 790 Da, in which two molecules were coupled through an ether linkage involving the hydroxyisopropyl side-chain. The fragmentation of H3 was sufficiently anomalous to preclude a simple explanation. The numerology of the molecular masses also indicated a change in nitrogen atom count, according to nitrogen rule [76] considerations. The impurity was isolated by a combination of solid phase extraction (SPE) and preparative HPLC, taking care to keep the material in a neutral to alkaline environment because it was known that the drug substance and impurity were acid–labile.

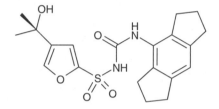

Analysis of NMR spectra showed an expected high degree of symmetry. Twenty-one unique carbon resonances were apparent, many of which appeared as paired or broadened peaks, providing clear evidence for a dimer-like type structure with a high degree of symmetry. The hexahydroindacene ring system remained intact, and there were indeed two such ring systems, although of very slightly different magnetic environments. Assuming the hexahydroindacene rings being furthest from the asymmetric coupling, working inward from these ends showed increasing amounts of magnetic inequivalency. Subtracting the molecular masses of each of the identified substructures from the 775 Da molecular mass left a 43 Da unit bridging the two halves of the molecule. Also, considering the number of nitrogen atoms required by the molecular mass, and the number accounted for by the above atomic bookkeeping, the bridge must contain a single nitrogen atom. The results pointed toward an amide linkage. The candidate structure is shown as the topmost molecule in Scheme 19.1.

We have seen materials change during the course of isolation and experimentation, such that the compound examined by NMR was not the compound of interest, its having decomposed or otherwise transformed during concentration or during the NMR observations. As a result, we routinely check for identity after the NMR experiment

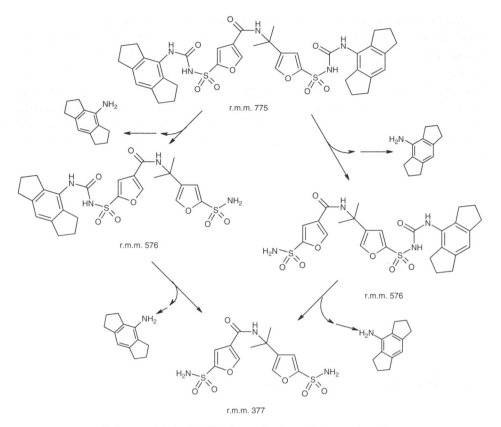

r.m.m. 775

r.m.m. 576

r.m.m. 576

r.m.m. 377

Scheme 19.1. Acidic degradation of impurity H3.

battery is complete, typically by re-injecting an aliquot into the LC–MS for confirmation of retention time and relative molecular mass. The NMR experiments on impurity H3 were conducted in alkaline D_2O. An aliquot from the NMR tube was injected directly into an acidic (pH 5.2) H_2O-containing HPLC mobile phase. We confirmed by retention time and mass spectrum that the desired molecule had survived the NMR experiments, and fully expected the compound to have had ample opportunity to exchange D for H during the chromatography. However, close examination revealed that a single deuterium had been retained in the molecule. We recalled the work appearing in the literature in recent years on the study of backbone amide hydrogen–deuterium exchange studies of protein folding, based upon the pH dependence of the rates of H–D exchange of amide protons [77]. Injecting the alkaline aliquot stolen from the NMR tube into the acidic mobile phase was in effect a pH quench experiment, suspending amide H–D exchange, and permitting us to further confirm the identity of the linkage. This se-rendipitous observation confirms the value that an inventory of ex-changeable hydrogen atoms has in an impurity identification, as advocated by Olsen *et al.* [78]. They report on the general utility of deuterium exchange LC–MS experiments for structure elucidation and impurity identification. However, one should not forget the importance of the chemical nature of the exchanging hydrogen. Exchange is not instantaneous and universal for exchangeable hydrogens, demon-strated in this example and emphasized by the amide hydrogen ex-change work.

Using the acid lability of these compounds provided further support for identification of the structure. Scheme 19.1 outlines the results of controlled acidic degradation of isolated impurity H3. HPLC examina-tion of the reaction mixture at short exposure times showed appearance of two chromatographically distinct components with identical 576 Da relative molecular masses and mass spectral fragmentation. At longer exposure times, both 576 Da species disappeared, and a new component with a 377 Da molecular mass appeared. Sequential acid-induced de-composition of the urea linkages on first one end of the impurity mole-cule, then the other, is mapped in Scheme 19.1. Efforts to synthesize the proposed impurity H3, examination of the synthetic route to the drug substance, and finding of the necessary intermediate impurities has provided final "proof of structure" of this impurity.

19.5 NUCLEAR MAGNETIC RESONANCE SPECTROSCOPY

19.5.1 Background

The field of Nuclear Magnetic Resonance (NMR) is a relatively young field of spectroscopy. Bloch *et al.* [79] and Purcell *et al.* [80] detected the first "NMR signal" in 1945, while exploring the physical behavior of condensed phases in a magnetic field. Subsequent to this discovery they were jointly awarded the 1952 Nobel Prize in physics [16]. Since these beginnings the field of NMR has advanced extremely rapidly, with some of the most notable advances being the observation of spin–spin coup-ling, two-dimensional correlation spectroscopy, inclusion of Fourier transformations, superconducting magnets, pulsed field gradients and the multitude of hardware advancements. All of these advances, and the many not mentioned, have lead to one of the most robust

---

[16]http://nobelprize.org/physics/laureates/1952/index.html.

techniques for exploring and determining molecular structure. While NMR is used to study everything from polymorphism in the solid state to determining the structures of fairly large proteins, its most significant impact is the elucidation of molecular structure in organic chemistry. The following will explore LC–NMR as it relates to the structure elucidation of small molecules (200–2000 Da) in solution. For a more comprehensive description of NMR please see Chapter 10 in this book.

This section of the "Combined Chromatography" chapter will deal with hyphenation techniques involving NMR spectrometers as detectors. The primary components of a hyphenated LC–NMR system are the chromatographic entity and the NMR spectrometer. However, the limits on a hyphenated NMR system are not limited to just LC–NMR. For example, LC–NMR–MS is a field that is demonstrating significant utility as a tool for structural analysis. One of the most important aspects of performing LC–NMR is the LC component. In this hyphenated mode it must be realized that the most crucial facet, of obtaining high-quality NMR data, is the separation itself, and hence, sample preparation. As mentioned above, the NMR spectrometer is utilized strictly as a detector in this mode. Therefore, if sufficient separations are not achieved the results will be poor-quality NMR data.

The field of LC–NMR is a field that started over 25 years ago with the first literature reference appearing in 1978. Since this first paper, there have been many examples throughout the literature describing both its limitations and successes. To this end, a quite comprehensive book detailing many of these aspects has been recently published [81]. The following sections will present a high-level perspective of LC–NMR and its various derivatives.

19.6 LIQUID CHROMATOGRAPHY–NUCLEAR MAGNETIC RESONANCE

19.6.1 Historical perspective

Watanabe *et al.* published the first paper to appear in the literature dealing with the coupling of LC and NMR in 1978 [82]. This early exploration of LC–NMR led to the modification of a standard NMR probe to include a "flow cell" comprised of a thin-wall Teflon tube with an inner diameter of 1.4 mm. The dimensions of this flow-cell were 1 cm in length and a total volume of 15 μl. This modification not only made the NMR spectrometer amenable to a "flow" system, but also overcame some of the inherent sensitivity issues associated with NMR as an LC detector. In addition, at this time in the world of NMR spectroscopy, very few advances had been made in the area of solvent suppression techniques. These techniques are required when using fully protonated solvent systems for the front-end separation technologies to ensure the signals arising from protonated solvents do not saturate the NMR receiver, allowing detection of the intended eluent. Very shortly after the appearance of the Watanabe manuscript, Bayer *et al.* carried out stop-flow experiments with a different flow-probe design [83]. In addition, in the early 1980s, several papers were published dealing with the analysis of jet fuel mixtures and separations using chloroform and Freon [84]. Few additional papers appeared in the literature throughout the 1980s. It was not until the early 1990s that many of the needed advances were made. These advances include pulse sequence development and hardware engineering improvements.

In its infancy, LC–NMR experiments were performed under chromatographic conditions containing fully protonated solvents. Considering the mole percent of the solvent relative to the solute it can be imagined that solvent resonances were the dominant signals observed. Due to the dynamic range of the receiver the solvent resonances made LC–NMR a prohibitive technique for "routine" analyses. The advent of eloquent solvent suppression schemes is arguably one of the most important breakthroughs in the field of LC–NMR. Water suppression enhanced through T1 effects (WET), published by Smallcombe *et al.*, significantly improved the quality of spectra [85]. Additional pulse sequence elements, including the double-pulsed field gradient spin echo (DPFGSE), published by Shaka *et al.*, provide amazing results for eliminating one to several solvent signals [86]. Significant hardware advances surrounding the LC and NMR interface have also led to LC–NMR becoming a more routine experimental technique. These include interfaces to control the transfer of eluent from the LC to the flow-probe and incorporation of instruments to store and trap eluting peaks for NMR data acquisition.

Of particular importance with the use of LC–NMR as an experimental technique is that it is suited for only a limited number of applications in reference to structure elucidation. As will be discussed in greater detail, the sensitivity issues that arise between the amount of compound one is able to load onto a particular chromatographic stationary phase, and hence elute into the flow-cell of an LC–NMR probe, limit what type of structural analysis that can be performed. It is this author's current opinion that most complete structure elucidations of unknown molecular entities are not amenable to LC–NMR. In these cases a more traditional approach of off-line isolation and subsequent tube-based NMR is more appropriate. However, analysis of compounds in which many of the structural features are known, or where the system being studied is well understood, or where 1D- and 2D-homonuclear correlation experiments will answer the questions at hand, are well suited for LC–NMR. We will also discuss further the advent of cryogenically cooled LC probes and the importance it has had on the utility of LC–NMR. These probes should provide the sensitivity gains needed for full structural analysis at the sub μg range.

19.6.2   Basic instrumentation

The basic components of an LC–NMR system are some form of chromatographic instrument and an NMR spectrometer equipped with a flow-probe, as shown in Fig. 19.17. In terms of the chromatography of choice, there are many examples in the literature of a wide array of separation instruments employed, from SFC to capillary electrophoresis (CE) [87,88]. By far the most common method (not necessarily the best choice from a separation point of view) of achieving the desired separation is through HPLC. There are many commercial

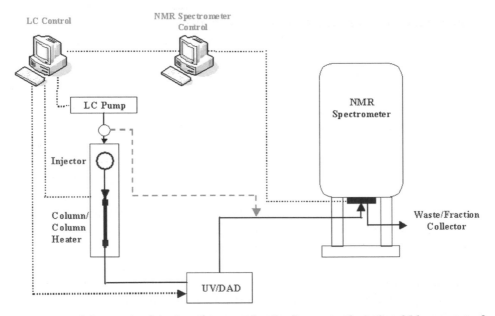

Fig. 19.17. Schematic showing the most basic elements that should be a part of an LC–NMR system. The dotted arrows represent electronic controls and the solid lines represent the flow path of chromatographic eluent.

suppliers of chromatography instruments, all of which can be employed for an LC–NMR system. The typical mode of separation utilized by HPLC is RP chromatography. The main reason for this is the separation methods performed in most pharmaceutical environments are performed under these types of conditions due to a fairly universal compatibility of drug-like molecules dissolution under these conditions. The typical solvent systems utilize combinations of methanol, acetonitrile and water (and additional modifiers). This is also convenient for the practicing LC–NMR Spectroscopist, as all three of these solvents are relatively reasonably priced in their fully deuterated form. This allows the researcher the ability to inherently reduce the amount of signal arising from protonated solvents, dealing instead with the residual protonated solvent "impurities" in the deuterated solvents themselves. The general mode of detection in LC systems is done through the use of UV. Typical modes of UV detection are the single wavelength, variable wavelength and diode array detectors (DAD). The DAD allows the most versatile options allowing data to be acquired over a specified UV range. Solvent systems and modifiers used to take advantage of the physical properties of the solute in its chromatographic environment must also be optimized for the NMR experiments themselves. Many of the common buffers used for analytical LC analysis utilize salts that contain moieties that contribute significant ^{1}H resonances and can hinder interpretation of the NMR data of the compound of interest. As discussed above, the LC–MS technique also requires certain considerations when determining an appropriate solvent system. This becomes applicable when we discuss the coupling of an MS instrument to an LC–NMR spectrometer later in this chapter.

The second basic component of an LC–NMR system is the NMR instrument itself. For a detailed description of the NMR spectrometer itself please refer to Chapter 10. The basic components of an NMR spectrometer consist of a superconducting magnet, a console containing all of the communication electronics, temperature regulation, RF electronics and very important to LC–NMR is the pulsed field gradient amplifier (used for gradient selected experiments and solvent suppression schemes). The NMR component of the LC–NMR system constitutes the majority of the physical footprint required for the LC–NMR instrumentation. In particular, the stray field of an unshielded instrument (e.g., unshielded 500 MHz has a five gauss line extending ~4 meters radially) dictates how close the LC component of an LC–NMR can be collocated. Significant improvements and innovation around superconducting magnets have allowed researchers to more easily collocate these two instruments in reasonable proximity. Use of an "UltraShield Plus" cryostat allows one to place the LC within <1 m of the magnet without affecting performance of either of the instrument [89]. This will become increasingly more important as we talk about LC–NMR–MS later in this chapter. The most important feature of an NMR instrument intended for LC–NMR experiments is a flow-probe.

Flow-probes are probes designed to allow a flow of eluent from the LC, through the probe (and hence the magnetic field of the spectrometer) to waste or some form of collection device. There are many types of flow-probes available to the researcher depending on the type of work to be done and the elaborateness of your system. A basic schematic is shown in Fig. 19.18. It can be that there is an inlet running up the body of the probe filling a flow-cell and exiting the probe from the top of the cell. The flow-cell has many different volumes associated with it depending on the vendor of choice. Typical volumes include 240, 120 and 60 μL. The volumes indicated are the total volume of the flow-cell. It is typical to hear the term "active volume" when talking about

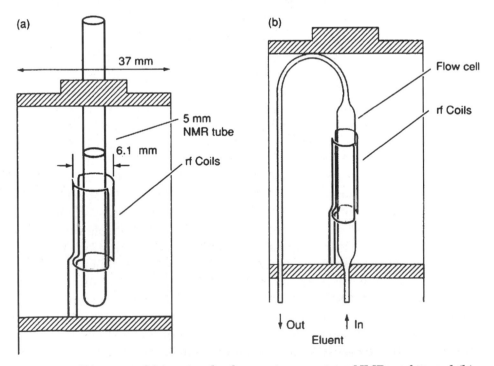

Fig. 19.18. Diagram of (a) a standard room temperature NMR probe, and (b) a flow NMR probe. Source: Albert, K. *On-Line LC–NMR and Related Techniques* p. 8.

flow-probes. The active volume is the volume of the flow-cell that is within the RF coils of the probe. For example, a 240 ml flow-cell has an approximate 120 µl active volume. Therefore, the spins that are observed are those contained within this active volume. For the purposes of this chapter only total volumes will be used Fig 19.18.

There are many new styles of flow-probes that have surfaced over the last few years that are providing much smaller cell volumes and capillary fluidics [90]. These probes will be discussed further in the section covering further hyphenation configurations. The NMR RF coil and pulsed field gradient coils (for probes fitted with RF gradient coils) are positioned around the cell providing increased fill factor in relation to a static tube-based NMR probe. The increased fill factor provides an advantage in signal to noise (S/N) values obtained in flow-mode as opposed to traditional tube-based NMR. These probes come in several configurations in terms of the nuclei that can be detected. Flow-probes are almost exclusively optimized to be "indirect" detection probes, with the ^{1}H coil being the closest in proximity to the flow-cell providing it the greatest sensitivity. This is a key factor when analyzing the typically small amount of material in these types of conditions. However, probes are also designed to take advantage of ^{19}F sensitivity and prove very useful in the analysis of fluorinated compounds. The reasons for this configuration will become clearer as we talk about the limitations of LC–NMR later in the chapter.

It is appropriate at this time to discuss some of the limitations associated with LC–NMR. It is more accurate to say the limitations of the NMR spectrometer in an LC–NMR instrument. As compared to MS, NMR is an extremely insensitive technique in terms of mass sensitivity. This is the key feature that limits NMR in its ability to analyze very small quantities of material. The key limiting factor in obtaining NMR data is the amount of material that one is able to elute into an active volume of an NMR flow-probe. The quantity of material transferred from the LC to the NMR flow-cell is dependant on several features. The first being the amount of material one is able to load on an LC column and retain the resolution needed to achieve the desired separation. The second is the volume of the peak of interest. The peak volume of your analyte must be reasonably matched to the volume of the flow-cell. An example would be a separation flowing at 1 ml/min with the peak of interest that elutes for 30 s. This corresponds to a peak volume of 500 µl, which clearly exceeds the volume of the typical flow-cell. This is the crux of the problem in LC–NMR. There is a balance that must be struck between the amount of compound needed to detect a signal in an

NMR experiment, and the load and volume of the peak of interest on the HPLC column.

A different issue is one that is quite common in the Pharmaceutical industry. A relatively frequent situation that arises is the need to identify a 0.1% impurity from a reaction mixture or metabolism sample. These samples are often quite convoluted in terms of the amount of compounds present as well as the general complexity of the separation, akin to a natural products extract, as can be seen in Fig. 19.19. However, to simplify this scenario to just a two-component mixture is appropriate for this section. Under common LC–NMR systems, it is typically required to have at least 50 µg of material for a complete structure elucidation (to enable the collection of long-range heteronuclear correlation data, HMBC). Therefore, one must be able to load 50 mg of the mixture on the column. Keep in mind, that if a 1D ^{1}H spectrum is all that is needed (in the case of a regiochemical issue in an aromatic system) this task becomes more amenable. The point trying to be made is that LC–NMR is a fantastic technique, but it must be used in

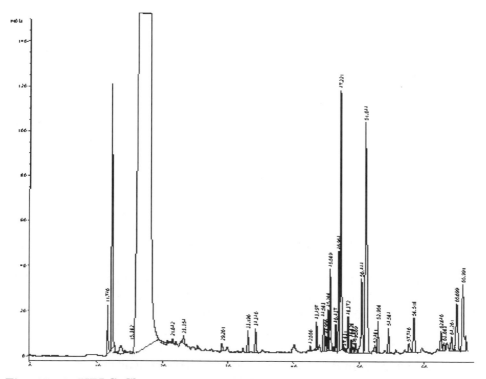

Fig. 19.19. HPLC Chromatogram (DAD detection) demonstrating the complexity of some compound mixtures.

the appropriate situation. This problem is being addressed in a multitude of ways through many instrumental modifications and the types of separations employed to afford the compound of interest.

19.6.3 Modes of operation

There are four main modes of operation of LC–NMR instruments, in terms of how the compound/fraction of interest is dealt with post chromatographic separation. These methods are continuous-flow, stop-flow, peak parking and peak trapping. There are several different variations within each of these modes (e.g. peak-slicing). Each one of these modes will be described in turn and advantages and disadvantages will be discussed. See Fig. 19.20.

19.6.4 Continuous flow

Historically the most common approach has been the utilization of continuous-flow methods. The basic schematic for an instrument set up for this mode is shown in Fig. 19.17. The premise of this methodology is that 1D ^{1}H data is acquired continuously, while the eluent is flowing from the HPLC. The residence time of the compounds of interest within the flow-cell is one of the most important features of this type of analysis. Again, using a 1 ml/min flow-rate as an example, and a peak volume encompassing 30 s and a flow-cell size of 240 μL the number of transients that can be collected on an analyte of interest is limited to approximately 10. Therefore, one must have enough sample contained within the peak that enough S/N to be acquired to allow interpretable data. As can be imagined, this limits the analysis to around 20 μg of material. This is a fairly effective method for looking at samples that have enough chromatographic resolution and are amenable to a single injection to provide this type of sample load. These features are also the disadvantage of this technique. Assuming enough separation under the chromatographic conditions and enough on-column sample load, one can collect only 1D ^{1}H data, which in many structural investigations is not nearly enough information. This type of approach works well for profiling fairly simple solutions allowing quick acquisition of data and potentially very valuable data. Examples would be the analysis of crude synthetic reaction mixtures, combinatorial chemistry mixtures and some metabolism samples.

Binary or Quaternary
LC Pump

Degasser

UV Detector   Manual Injection

RF Gradients

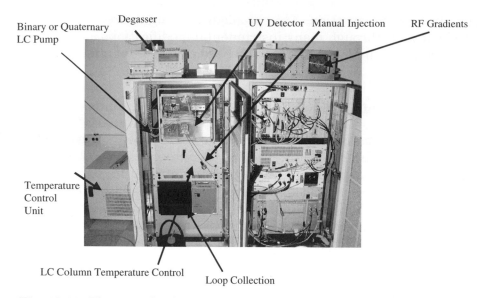

Temperature
Control
Unit

LC Column Temperature Control

Loop Collection

Fig. 19.20. Photograph of a standard 600 MHz Bruker LC–NMR console.

19.6.5   Stop-flow

Stop-flow mode is when the eluent flows directly into the NMR probe. When a peak of interest is observed (typically utilizing UV detection) the chromatography is stopped to allow data collection of that peak. This mode of collection can be incorporated into all three-instrument schemes shown in Fig. 19.17, Fig. 19.21 and Fig. 19.22. To properly employ stop-flow experiments, the flow and delay timings between the UV detector and the NMR flow-cell must be accurately determined. These are straightforward calibrations to be done. In this mode of data collection a larger array of experiments can be performed. These include many different types of 1D selective and 2D homonuclear correlation experiments. Examples include the 1D selective Correlation SpectroscopY (COSY) [91] and 2D Total Correlation SpectroscopY (TOCSY) [92], both of which are scalar coupling experiments and the Nuclear Overhauser Effect SpectroscopY (NOESY) [93] and the 2D Rotating-frame Overhauser Effect SpectroscopY (ROESY) [94], both dipolar coupling experiments. The added advantage of being able to perform these correlation experiments enables more detailed structural information to be gleaned and therefore solve a wider array of structural ambiguities. Some heteronuclear correlation experiments are possible to acquire. However, these experiments can take several days

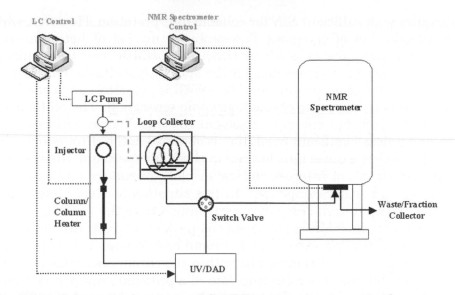

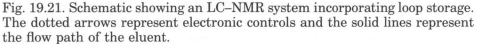

Fig. 19.21. Schematic showing an LC–NMR system incorporating loop storage. The dotted arrows represent electronic controls and the solid lines represent the flow path of the eluent.

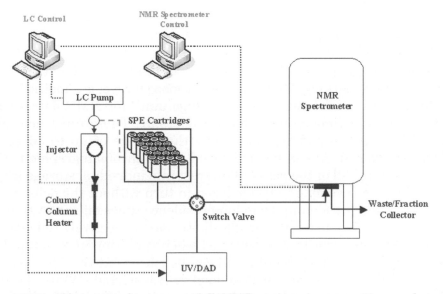

Fig. 19.22. Schematic showing an LC–NMR system incorporating peak trapping with SPE. The dotted arrows represent electronic controls and the solid lines represent the flow path of the eluent.

to acquire with sufficient *S/N* for reliable interpretation. However, with the introduction of cryogenic flow-probes a myriad of heteronuclear correlation experiments, such as HSQC and HMBC can be acquired, which can allow structural elucidation *via* NMR methods. One significant disadvantage to this mode of LC–NMR is potential carryover from major components in the chromatographic separation. The NMR flow-cell takes significant volumes of solvent to "wash" away residual analyte, and unless significant resolution is obtained in the chromatography the carryover will cause interference in the NMR spectrum. An example would be a peak of interest eluting in close proximity (for example, following) to a major component. In this situation it is very likely that you will see significant resonance contribution from the "tailing" major component in the NMR spectrum of the peak of interest. An additional problem that is often experienced is band broadening. This is best described as diffusion of compounds within the chromatographic stationary phase while you are collecting data of a preceding compound. Since data acquisition at times can be quite long (several hours), for 2D experiments on mass limited samples, many of the late eluting peaks can become very broad, making and collecting meaningful data very difficult. This problem has been virtually eliminated through the use of peak parking.

19.6.6 Peak parking

The band-broadening problem has been virtually eliminated through the use of peak parking. Peak parking, or loop storage, is a method used to divert peaks of interest into a "storage unit" for NMR data acquisition at a later time as pictured in Fig. 19.21. These units are commercially available and typically allow the storage of up to 36 peaks from a single run. As can be envisioned, a full chromatographic run is done "parking" all of the peaks of interest into individual storage loops. This allows each sample to be analyzed in turn without the deleterious effects of band broadening from long residence times on the stationary phase. In addition, using this mode allows one to effectively remove the problem of carry over that is observed when operating in the stop-flow mode. In between runs the flow-cell can be rinsed out through a series of valve switches before the next loop is transferred. While the disadvantages mentioned in the above modes are significantly diminished, the fact remains that you are still limited by the volume of the eluting peak and the dimension of the flow-cell.

19.6.7 Peak trapping

The most recent development in sample handling is the ability to peak trap. This is typically done through the use of SPE cartridges. This is a commercially available unit that can be put in-line. In addition, researches at Eli Lily have been successful in adapting a "homebuilt" peak-trapping method that is very effective [95]. The premise behind this method is quite simple in principle, but many hardware advances were made to incorporate it under software control within an existing LC–NMR system. As peaks are detected they are diverted to a unit that contains SPE cartridges. On the path to the cartridges (after separation and detection) the water content is increased to take advantage of the retention properties of the stationary phases in the SPE cartridge, thereby "trapping" the analyte. An additional feature of this technique is that one can do multiple chromatography runs and successively trap the peaks of interest on the same cartridges. Four successive trappings have been the loading limit of the SPE cartridges that are commercially supplied. Four trappings can, in effect, increase the total amount of analyte four-fold, assuming no breakthrough of material from the cartridge. Therefore, peak trapping has the potential to alleviate some of the mass sensitivity issues with LC–NMR. In this mode, the use of fully protonated solvents can be used for the separations. Once the analyte is trapped it is typically blown dry under a stream of nitrogen and eluted with deuterated solvent. This significantly decreases the financial overhead for doing LC–NMR including separations in fully deuterated solvents. The compounds are eluted with 100% organic solvents, which results in a very narrow band of compound eluted into the NMR flow-cell. This feature allows the cell dimensions to be reduced significantly (e.g. 30 µl) allowing the concentration to increase and thereby increasing the types of experiments that can be successfully completed.

LC–NMR techniques have been employed to solve a wide variety of structural ambiguities through the methods and modes of the operation described above.[17] The coupling of an LC to an NMR is just the beginning of this wave of NMR hyphenation. As can be read in the previous section on LC–MS, many crucial structural features of unknown compounds in question can be gleaned through careful analysis of the resulting MS data. In the field of structure elucidation there is no single technique that can provide all structural details with the arguable exception of X-ray diffraction studies. However, most mass limited

---

[17]Many examples are included in a special issue of *Magn. Reson. Chem.*, 2005, *43*.

samples are not amenable to diffraction quality crystals due in large part to sample quantity. This being said, the combination of LC retention time, NMR spectroscopy and MS provide much of the key physical data needed to stitch together the chemical structures of most unknown chemical entities. For this reason the complete melding of these three techniques has been implemented and commercialized.

19.6.8 Liquid chromatography–nuclear magnetic resonance–mass spectroscopy

The LC–NMR–MS systems are elaborately integrated instruments that can perform seamlessly under the right set of circumstances. The initial reports in the literature reported great successes [96]. Some of the most significant hurdles that needed to be overcome were the interface by which these systems would operate as well as how to deal with the differences in mass sensitivity. Commercially, interfaces were developed that allowed the LC–MS component to operate under a single piece of software that enabled the triggering of the NMR experiment once the peak of interest was observed. This can be done using a UV/DAD detector or the MS itself. Using MS as the discriminating feature of the separation, and hence NMR data collection gives the flexibility of specifically targeting a particular mass or more importantly a mass fragment induced through fragmentation (see above for the description of MS). The differences in mass sensitivity are addressed through various software controlled splitting mechanisms. This is typically done by a 95:5 split, with 5% going to the MS instrument and 95% going to the NMR flow-cell for data acquisition. This split mechanism also allows the manipulation of transfer times to ensure that you are able to collect both NMR data and MS data of the same peak. In the typical configuration of an LC–NMR–MS instrument an ion-trap spectrometer is utilized. An ion trap is quite ideal for the collection of MS^n data and therefore, is rich in the structural information it provides. The most common ionization source used is electrospray. However, APCI and APPI are becoming commonplace. In addition, recent reports have interfaced TOF instruments allowing accurate mass measurements (no MS^n experiments) and FTMS instruments. An FTMS instrument allows one to get MS^n data and with proper calibration, the elemental composition of the parent ion and subsequent fragmented daughter ions by accurate mass measurement.

Early implementations of LC–NMR–MS techniques used fully deuterated solvents (as mentioned above to reduce the abundance of

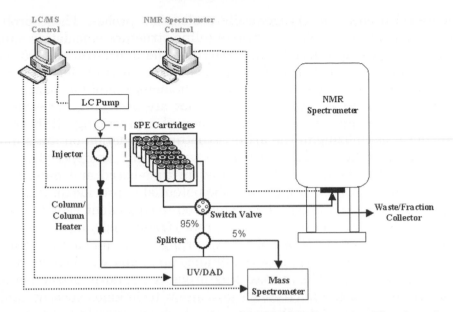

Fig. 19.23. Schematic showing an LC–SPE–NMR/MS system. The dotted arrows represent electronic controls and the solid lines represent the flow path of the eluent.

protonated solvents). The use of deuterated solvents for MS data acquisition can provide very useful information about the number of chemically labile hydrogens in the molecule under study. However, one must carry out MS experiments under protonated conditions to ensure the exchanged hydrogen atoms do not give rise to an exaggerated molecular mass. The analysis of MS data run under deuterium exchange conditions has been discussed previously in this chapter. The advent of the LC–SPE–NMR–MS systems has removed the issue of MS data collection under deuterium exchange conditions [97]. As mentioned above the initial chromatography is done under fully protonated conditions, allowing MS data collection under these conditions. After concentration and drying the peak is eluted from the SPE cartridge with fully deuterated 100% organic to the NMR probe for data collection Fig. 19.23 (see previous section on peak trapping).

19.6.9   Recent Advances

Hardware engineers are continually making elaborate advances in new technology. This can be exemplified in the advances that have been made within the last several years. In particular, the development of micro-coil probes and cryogenically cooled flow-probes. These probes have the potential to allow complete inline structure elucidation capabilities. Micro-coil probes first appeared in the literature in 1995 [90] with significant skepticism from the NMR community for their extremely small sizes, and therefore difficult sample handling issues. Despite these initial reservations the probes are becoming increasingly important and invaluable in the toolbox of a practicing NMR Spectroscopist specializing in the art of structure elucidation on mass limited samples. These probes utilize a "saddle coil" as opposed to the traditional Hemholtz coil that is used in tube-based and most flow-NMR probes. This coil design have the potential to provide an inherent ~2.5-fold increase in S/N (when compared to equivalent geometries of standard flow-probe designs). These probes traditionally have a volume of 5 µL with an active volume of 1.5 µL (a range of cell and active volume sizes are now available). All of the fluidics up to and within the probes are capillary tubes. This makes this probe particularly attractive for utilizing some of the micro-scale separation techniques such as capillary LC, capillary electrophoresis, ultra high-pressure LC and others. Using these types of separation mechanisms provide increased (with the exception of GC) theoretical plates and resolution, relative to traditional LC, to achieve a desired separation. Aside from the obvious advantages of a reduced cell volume and therefore increased concentration, the flow-cell size reduces the moles of residual protonated solvent inherent in the deuterated solvents. This reduction in protonated solvent impurities reduces the need for solvent suppression schemes even on samples in the 100s of nanograms ranges.

The introduction of the cryogenic flow-probe is one of the most important breakthroughs in the area of hyphenated NMR technologies. When comparing a room temperature and cryogenically cooled probe of equivalent geometries, an approximate four-fold increase in S/N can be expected. This gain in S/N comes from removing a great deal of the thermal noise generated from the electronics of the system. This is achieved by cryogenically cooling the probes RF electronics, including the preamplifiers, while keeping the sample at user defined temperatures around room temperature. In addition, these probes have the added feature of interchangeable flow-cells that allow one to change the size of the flow-cell within the probe as opposed to having several probes of differing cell sizes (should be pointed out that one of the commercial manufacturers of NMR probes has interchangeable flow-cells in their room temperature flow-probes). The smaller of these is fitted with capillary fluidics allowing one to take advantage of the chromatographic techniques mentioned above for the micro-coil probe. It should be noted that cryogenically cooled NMR probes suffer from losses as a result of "salty" solutions. These probes are best utilized with the column trapping features to allow the most benefit from the concentration, desalting and narrow band elution.

While the early days of LC–NMR and LC–NMR–MS were plagued by the poor sensitivity of the NMR spectrometer, the recent probe design advances have provided a means to potentially overcome this hurdle. As reported in the literature, it is possible to get both 1D and 2D homonuclear and heteronuclear correlation data on sub micrograms of materials in quite complex mixtures utilizing cryogenic flow-probes in tandem with SPE peak trappings [98]. While these technologies are still in their infancy, they have the potential to revolutionize LC–NMR as a structure elucidation technique.

19.7 CONCLUSIONS

In principle, these novel systems appear to have solved many of the problems associated with structure elucidation of low-level impurities. This however is not exclusively the case. It must be emphasized that many features come into play that still make the tried and trued method of off-line isolation and subsequent LC–MS and tube-based NMR the most reasonable approach. It is commonplace that some compounds behave poorly, spectroscopically speaking, when solvated in a particular solvent/solvent-buffer combinations. The manifestation of this poor behavior can be multiple stable conformations, and/or equilibration isomers in the intermediate slow-exchange paradigm resulting in very broad peaks on the NMR time-scale. Therefore, it can be envisioned that the flexibility of having an isolated sample in-hand can be advantageous. There are additional issues that can arise, such as solubility changes, the need to indirectly observe heteroatoms other than ^{13}C (e.g. triple resonance cryogenic probe), as well as many others. It must be stated that LC–NMR and its various derivatives is ideally suited for looking at relatively simple regiochemical issues in reasonably complex systems. In some cases full structure elucidation of unknown compounds can be completed. Although this will become more routine with some of the recent advances such as peak trapping in combination with cryogenic flow-probes. There have been many elegant examples in the literature of successful applications of LC–NMR and LC–NMR–MS on very complex systems such as natural products, drug metabolism and combinatorial synthesis. However, one could argue that it would have been quite effective to take a more traditional approach.

The LC–NMR instrument is a very attractive analytical tool in that it has the potential to provide a great deal of data detailing many structural features in an inline mode. The principle advantage to using hyphenated NMR technology is that in most cases one can collect NMR and MS data on the same sample reducing the possibility of decomposition during the isolation and sample preparation process.

It is the hope of these authors to convey that LC–NMR is a very powerful technique that has immense potential in the field of structure elucidation. However, a researcher should take care to not solely rely on this technique to solve all structure elucidation questions.

REFERENCES

1 K.B. Tomer, Separations combined with mass spectrometry, *Chem. Rev.*, 101 (2001) 297–328.
2 F.W. McLafferty, *Tandem Mass Spectrometry (MS/MS)*, Wiley, New York, 1983.
3 S.A. McLuckey and J.M. Wells, Mass analysis at the advent of the 21st century, *Chem. Rev.*, 101 (2001) 571–606.
4 J.H. Gross, *Mass Spectrometry: A Textbook*, Springer, Berlin, 2004, pp. 518.
5 M.A. Grayson, *Measuring Mass: From Positive Rays to Proteins*, Chemical Heritage Press, Philadelphia, 2002, pp. 149.
6 I.W. Griffiths, J.J. Thomson—the centenary of his discovery of the electron and of his invention of mass spectrometry, *Rapid Commn. Mass Spectrom.*, 11 (1997) 2–16.
7 J. Am. Soc. Mass Spectrom., 16 (8S) cover, (2005).
8 K.L. Busch, Masses in mass spectrometry: Balancing the analytical scales, *Spectroscopy*, 19(11) (2004) 32–34.
9 J.H. Beynon, *Mass Spectrometry and its Applications to Organic Chemistry*, Elsevier, Amsterdam, 1960.
10 P. DeBievre and I.L. Barnes, Table of the isotopic composition of the elements as determined by mass spectrometry, *Int. J. Mass Spectrom. Ion Processes*, 65 (1985) 211–230.
11 J.H. Beynon, Qualitative analysis of organic compounds by mass spectrometry, *Nature*, 174 (1954) 735–737.
12 C.H. Heathcock, Editorial, *J. Org. Chem.*, 55 (1990) 8A.
13 K. Biemann, Utility of exact mass measurements, *Methods in Enyzmology*, 193 (1990) 295–305.

14 K.L. Busch, Using exact mass measurements, *Spectroscopy*, 9(7) (1994) 21–22.

15 J.M. Gilliam, P.W. Landis and J.L. Occolowitz, Accurate mass measurement in fast atom bombardment mass spectrometry, *Anal. Chem.*, 55 (1983) 1531–1533.

16 J.M. Gilliam, P.W. Landis and J.L. Occolowitz, On-line accurate mass measurement in fast atom bombardment mass spectrometry, *Anal. Chem.*, 56 (1984) 2285–2288.

17 Z. Wu, R.P. Rodgers and A.G. Marshall, Characterization of vegetable oils: Detailed compositional fingerprints derived from electrospray ionization Fourier transform ion cyclotron resonance mass spectrometry, *J. Agric. Food Chem.*, 52 (2004) 5322–5328.

18 A.G. Marshall and R.P. Rodgers, Petroleomics: The next grand challenge for chemical analysis, *Acc. Chem. Res.*, 37 (2004) 53–59.

19 O.A. Mamer and A. Lesimple, Letter to the editor, *J. Am. Soc. Mass Spectrom*, 15 (2004) 626.

20 M.L. Gross, Accurate masses for structure confirmation, *J. Am. Soc. Mass Spectrom.*, 5 (1994) 57.

21 R. Kondrat, High resolution mass spectrometry, *J. Am. Soc. Mass Spectrom.*, 10 (1999) 661.

22 M.L. Vestal, Methods of ion generation, *Chem. Rev.*, 101 (2001) 361–375.

23 R.S. Gohlke and F.W. McLafferty, Early gas chromatography/mass spectrometry, *J. Am. Soc. Mass Spectrom.*, 4 (1993) 367–371.

24 W. McFadden, *Techniques of Combined Gas Chromatography–Mass Spectrometry: Applications in Organic Analysis*, Wiley, New York, 1973.

25 G.M. Message, *Practical Aspects of Gas Chromatography/Mass Spectrometry*, Wiley, New York, 1984.

26 D.R. Knapp, *Handbook of Analytical Derivatization Reactions*, Wiley, New York, 1979, pp. 741.

27 E. Kovats, Gas-chromatographische charakterisierung organischer verbindungen, *Helv. Chim. Acta*, 41 (1958) 1915–1932.

28 H. van den Dool and P.D. Kratz, A generalization of the retention index system including linear temperature programmed gas–liquid partition chromatography, *J. Chrom.*, 11 (1963) 463–471.

29 S.R. Heller, The history of the NIST/EPA/NIH mass spectral database, *Today's Chemist at Work*, 8(2) (1999) 45–50.

30 T.R. Sharp, H. Lee, A. Ferguson, K.N. Marsh and R.G. Harvey, Electron impact mass spectra of polycyclic aromatic hydrocarbons: A reference collection. Presented at the 36th ASMS Conference on Mass Spectrometry and Allied Topics, San Francisco, CA, 5–10 June 1988.

31 T.R. Sharp, M. Sutton, A. Ferguson and K.N. Marsh, Electron impact mass spectra of steroids: Androstanes and derivatives. Presented at the 37th ASMS Conference on Mass Spectrometry and Allied Topics, Miami Beach, FL, 21–26 May 1989.

32 F.W. McLafferty, M.-Y. Zhang, D.B. Stauffer and S.Y. Loh, Comparison of algorithms and databases for matching unknown mass spectra, *J. Am. Soc. Mass Spectrom.*, 9 (1998) 92–95.

33 P. Ausloos, C.L. Clifton, S.G. Lias, A.I. Mikaya, S.E. Stein, D.V. Tchekhovskoi, O.D. Sparkman, V. Zaikin and D. Zhu, The critical evaluation of a comprehensive mass spectral library, *J. Am. Soc. Mass Spectrom.*, 10 (1999) 287–299.

34 H. Budzikiewicz, C. Djerassi and D.H. Williams, *Mass Spectrometry of Organic Compounds*, Holden-Day, Inc., San Francisco, CA, 1967.

35 Q.N. Porter, *Mass Spectrometry of Heterocyclic Compounds*, 2nd Ed, Wiley, New York, 1985.

36 F.W. McLafferty and R. Venkataraghavan, *Mass Spectral Correlations*, 2nd Ed, American Chemical Society Advances in Chemistry Series #40, Washington, DC, 1982, pp. 124.

37 A.G. Harrison, *Chemical Ionization Mass Spectrometry*, 2nd ed, CRC Press, Boca Raton, FL, 1992.

38 P.J. Arpino, On-line liquid chromatography/mass spectrometry? An odd couple, *Trends Anal. Chem.*, 1 (1982) 154.

39 M.A. Baldwin, A.L. Burlingame and E. Nikolaev, *J. Am. Soc. Mass Spectrom.*, 15 (2004) 1517–1519.

40 V.L. Tal'Rose, G.V. Karpov, I.G. Grdetskii and V.E. Skurat, Capillary system for the introduction of liquid mixtures into an analytical mass spectrometer, *Russian J. Phys. Chem.*, 42 (1968) 1658–1664.

41 I.D. Wilson, The state of the art in thin-layer chromatography–mass spectrometry: a critical appraisal, *J. Chromatogr. A*, 856 (1999) 429–442.

42 K.L. Busch, Planar separations and mass spectrometric detection, *J. Planar Chromatogr.*, 17 (2004) 398–403.

43 L.G. Randall and A.L. Wahrhaftig, Direct coupling of a dense (supercritical) gas chromatograph to a mass spectrometer using a supersonic molecular beam interface, *Rev. Sci. Instrum.*, 52 (1981) 1283–1295.

44 R.D. Smith and H.R. Udseth, Mass spectrometry with direct supercritical fluid injection, *Anal. Chem.*, 55 (1983) 2266–2272.

45 A.L. Yergey, C.G. Edmonds, I.A.S. Lewis and M.L. Vestal, *Liquid Chromatography/Mass Spectrometry: Techniques and Applications*, Plenum Publishers, New York, 1990, pp. 306.

46 M.A. Baldwin and F.W. McLafferty, Liquid chromatography–mass spectrometry interface. I: The direct introduction of liquid solutions into a chemical ionization mass spectrometer, *Org. Mass Spectrom.*, 7 (1973) 1111–1112.

47 J.R. Chapman, E.H. Harden, S. Evans and L.E. Moore, LC–MS interfacing in sector mass spectrometers, *Int. J. Mass Spectrom. Ion Phys.*, 46 (1983) 201–204.

48 D.I. Carroll, I. Dzidic, R.N. Stillwell, K.D. Haegele and E.C. Horning, Atmospheric pressure ionization mass spectrometry: Corona discharge

ion source for use in liquid chromatograph-mass spectrometer-computer analytical system, *Anal. Chem.*, 47 (1975) 2369–2373.

49   M. Barber, R.S. Bordoli, R.D. Sedgwick and A.N. Tyler, Fast atom bombardment of solids (F.A.B): A new ion source for mass spectrometry, *Chem. Commun.*, (1981) 325–326.

50   D.J. Surman and J.C. Vickerman, Fast atom bombardment quadrupole mass spectrometry, *Chem. Commun.*, (1981) 324–325.

51   Y. Ito, T. Takeuchi, D. Ishii, M. Goto and T. Mizuno, Direct coupling of micro high performance liquid chromatography with fast atom bombardment mass spectrometry. II: Application to gradient elution of bile acids, *J. Chromatogr.*, 385 (1986) 201–209.

52   R.M. Caprioli, T. Fan and J.S. Cottrell, Continuous-flow sample prove for fast atom bombardment mass spectrometry, *Anal. Chem.*, 58 (1986) 2949–2953.

53   R.P.W. Scott, C.G. Scott, M. Munroe and J. Hess Jr., Interface for on-line liquid chromatography-mass spectroscopy analysis, *J. Chromatogr.*, 99 (1974) 395–405.

54   W.H. McFadden, H.L. Schwartz and D.C. Bradford, Direct analysis of liquid chromatographic effluents, *J. Chromatogr.*, 122 (1976) 389–396.

55   J.G. Stroh, J.C. Cook, R.M. Milberg, L. Brayton, T. Kihara, Z. Huang, K.L. Rinehart Jr. and I.A.S. Lewis, On-line liquid chromatography fast atom bombardment mass spectrometry, *Anal. Chem.*, 57 (1985) 985–991.

56   C.R. Blakley, J.C. Carmody and M.L. Vestal, Liquid chromatograph–mass spectrometer for analysis of nonvolatile samples, *Anal. Chem.*, 52 (1980) 1636–1641.

57   M.L. Vestal, Ionization techniques for nonvolatile molecules, *Mass Spectrom. Rev.*, 2 (1983) 447–480.

58   R.C. Willoughby and R.F. Browner, Monodisperse aerosol generation interface for combining liquid chromatography with mass spectrometry, *Anal. Chem.*, 56 (1984) 2625–2631.

59   T.A. Bellar, T.D. Behymer and W.L. Budde, Investigation of enhanced ion abundances from a carrier process in high-performance liquid chromatography particle beam mass spectrometry, *J. Am. Soc. Mass Spectrom.*, 1 (1990) 92–98.

60   C.J. Miles, D.R. Doerge and S. Bajic, Particle beam liquid chromatography–mass spectrometry of national pesticide survey analytes, *Archives Environmental Contamination Toxicol.*, 22 (1992) 247–251.

61   M. Yamashita and J.B. Fenn, Electrospray ion source: Another variation on the free-jet theme, *J. Phys. Chem.*, 88 (1984) 4452–4459.

62   J.B. Fenn, Electrospray wings for molecular elephants (Nobel Lecture), *Angew. Chem. (Int.)*, 42 (2003) 3871–3894.

63   M. Dole, L.L. Mack, R.L. Hines, R.C. Mobley, L.D. Ferguson and M.B. Alice, Molecular beams of macroions, *J. Chem. Phys.*, 49 (1968) 2240–2249.

64 R.B. Cole, Some tenets pertaining to electrospray ionization mass spectrometry, *J. Mass Spectrom.*, 35 (2000) 763–772.

65 G.J. Van Berkel, Electrolytic deposition of metals on to the high-voltage contact in an electrospray emitter: Implications for gas-phase ion formation, *J. Mass Spectrom.*, 35 (2000) 773–783.

66 M.H. Amad, N.B. Cech, G.S. Jackson and C.G. Enke, Importance of gasphase proton affinities in determining the electrospray ionization response for analytes and solvents, *J. Mass Spectrom.*, 35 (2000) 784–789.

67 M. Gamero-Castaño and J. Fernandez de la Mora, Kinetics of small ion evaporation from the charg and mass distribution of multiply charged clusters in electrospray, *J. Mass Spectrom.*, 35 (2000) 790–803.

68 P. Kebarle, A brief overview of the present status of the mechanisms involved in electrospray mass spectrometry, *J. Mass Spectrom.*, 35 (2000) 804–817.

69 J. Fernandes de la Mora, G.J. Van Berkel, C.G. Enke, R.B. Cole, M. Martinez-Sanchez and J.B. Fenn, Electrochemical processes in electrospray ionization mass spectrometry, *J. Mass Spectrom.*, 35 (2000) 939–952.

70 J.V. Iribarne and B.A. Thomson, On the evaporation of small ions from charged droplets, *J. Chem. Phy.*, 64 (1976) 2287–2294.

71 A.T. Blades, M.G. Ikonomou and P. Kebarle, Mechanism of electrospray mass spectrometry. Electrospray as an electrochemical cell, *Anal. Chem.*, 63 (1991) 2109–2114.

72 J.E. Campana, Cluster ions. I: Methods, *Mass Spectrom. Rev.*, 6 (1987) 395–442.

73 F.E. Kuhlmann, A. Apffel, S.M. Fischer, G. Goldberg and P.C. Goodley, Signal enhancement for gradient reverse-phase high-performance liquid chromatography–electrospray ionization mass spectrometry analysis with trifluoroacetic and other strong acid modifiers by postcolumn addition of propionic acid and isopropanol, *J. Am. Soc. Mass Spectrom.*, 6 (1995) 1221–1225.

74 J.M. Roberts, A.R. Diaz, D.T. Fortin, J.M. Friedle and S.D. Piper, Influence of the Hofmeister series on the retention of amines in reversed-phase liquid chromatography, *Anal. Chem.*, 74 (2002) 4927–4932.

75 K.M. Alsante, P. Boutros, M.A. Couturier, R.C. Friedmann, J.W. Harwood, G.J. Horan, A.J. Jensen, Q. Liu, L.L. Lohr, R. Morris, J.W. Raggon, G.L. Reid, D.P. Santafianos, T.R. Sharp, J.L. Tucker and G.E. Wilcox, Pharmaceutical impurity identification: A case study using a multidisciplinary approach, *J. Pharmaceutical Sci.*, 93 (2004) 2296–2309.

76 F.W. McLafferty, *Interpretation of Mass Spectra*, 3rd ed., University Science Books, Mill Valley, CA, 1980, p. 33.

77 Y. Bai, J.S. Milne, L. Mayne and S.W. Englander, Primary structure effects on peptide group hydrogen exchange, *Proteins: Struct. Function Genet.*, 17 (1993) 75–86.

78   M.A. Olsen, P.G. Cummings, S. Kennedy-Gabb, B.M. Wagner, G.R. Nicol and B. Munson, The use of deuterium oxide as a mobile phase for structural elucidation by HPLC/UV/ESI/MS, *Anal. Chem.*, 72 (2000) 5070–5078.

79   F. Bloch and W. Hansen, *Phys. Rev.*, 69 (1946) 127.

80   E.M. Purcell, *Phys. Rev.*, 69 (1946) 37.

81   K. Albert (Ed.) *On-line LC–NMR and Related Techniques*, Wiley: West Sussex, England, 2002.

82   N. Watanabe and E. Niki, *Proc. Jpn. Acad., Ser. B*, 54 (1978) 194.

83   E. Bayer, K. Albert, M. Nieder, E. Grom and T. Keller, *J. Chromatogr.*, 186 (1979) 197.

84   (a) J.F. Haw, T.E. Glass and H.C. Dorn, *Anal. Chem.*, 53 (1981) 2327;
     (b) J.F. Haw, T.E. Glass and H.C. Dorn, *Anal. Chem.*, 53 (1981) 2332.

85   S.H. Smallcombe, S.L. Patt and P.A. Keifer, *J. Magn. Reson. A*, 17 (1995) 295.

86   K. Stott, J. Stonehouse, J. Keeler, T.-L. Hwang and A.J. Shaka, *J. Am. Chem. Soc.*, 117 (1995) 4199.

87   K. Albert, *J. Chromatogr.*, 785 (1997) 65 References within.

88   P. Gforer, L.-H. Tseng, E. Rapp, K. Albert and E. Bayer, *Anal. Chem.*, 73 (2001) 3234.

89   Bruker-Biospin Inc. 15 Fortune Drive, Billerica, MA, 01821. http://www.bruker-biospin.com.

90   (a) D.L. Olson, M.E. Lacey and J.V. Sweedler, *Science*, 270 (1995) 1967–1968;
     (b) M.E. Lacey, R. Subramanian, D.L. Olsen, A.G. Webb and J.V. Sweedler, *Chem. Rev.*, 99 (1999) 3133–3152.

91   W.P. Aue, E. Bartholdi and R.R. Ernst, *J. Chem. Phys.*, 64 (1976) 2229.

92   L. Braunschweiler and R.R. Ernst, *J. Magn. Reson.*, 53 (1983) 521.

93   J. Jeener, B.H. Meier, P. Bachmann and R.R. Ernst, *J. Chem. Phys.*, 71 (1979) 4546.

94   (a) A. Bax and D.G. Davis, *J. Magn. Res.*, 63 (1985) 207;
     (b) A.A. Bothner-By, R.L. Stephens, J.M. Lee, C.D. Warren and R.W. Jeanloz, *J. Am. Chem. Soc.*, 106 (1984) 811.

95   A. Kaerner and D.A. Jackson, *43rd Experimental Nuclear Magnetic Resonance Conference*, Montery, CA, 2002.

96   J.C. Lindon and J.K. Nicholson. In: K. Albert (Ed.), *On-line LC–NMR and Related Techniques*, Wiley, West Sussex, England, 2002, p. 45.

97   L. Griffiths and R. Horton, *Magn. Reson. Chem.*, 36 (1998) 104.

98   V. Exarchou, M. Godejohann, T.A. Van Beek, I.P. Gerothanassis and J. Vervoort, *Anal. Chem.*, 75 (2003) 6288.

REVIEW QUESTIONS

Mass Spectrometry

1. Discuss the terms "molecular weight" and "relative molecular mass."
2. How is the atomic mass unit defined? How has this definition evolved?
3. Differentiate among the values for average (chemical), nominal (monoisotopic) and accurate mass when applied to a compound. What is the origin of the differences in these values? What are the various contexts where each value is appropriate and inappropriate to use?
4. What is a polyisotopic element? Give examples of monoisotopic and of polyisotopic elements.
5. Discuss the difference in meanings of the terms "high resolution" and "accurate mass".

Gas Chromatography–Mass Spectrography

1. What are the factors limiting the applicability of GC–MS to analysis of a particular compound or group of compounds?

Liquid Chromatography–Mass Spectrography

1. Discuss the concept of LC–MS–friendly mobile phases.

Liquid Chromatography–Nuclear Magnetic Resonance

1. Describe the features of a chromatographic separation that can limit the successful acquisition of NMR data in a hyphenated LC–NMR instrument.
2. What advantages are there for utilizing capillary-scale LC–NMR?

Chapter 20a

Optical sensors

Enju Wang

20a.1 INTRODUCTION

Chemical sensors are simple devices based on a specific chemical recognition mechanism which enables the direct determination of either the activities or concentrations of selected ions and electrically neutral species, without pretreatment of the sample. Clearly, eliminating the need for sample pretreatment is the most intriguing advantages of chemical sensors over other analytical methods. Optical chemical sensors (optodes or optrodes) use chemical sensing elements and optical transduction for the signal processing, thus these sensors offer further advantages such as freedom from electrical noise and ease of miniaturization, as well as the possibility of remote sensing. Since its introduction in 1975, this field has experienced rapid growth and has resulted in optodes for a multitude of analytes [1–6].

20a.2 SENSOR DESIGN

Figure 20a.1 shows a general schematic diagram of an optical sensor and detection instrumentation utilizing optical fibers. Optical fibers can guide the light from the light source to a remote distance without losing much of light intensity. In most optical sensors, dual optical fibers are used; one is to guide the light from the source to the sensing site and the other to collect the detecting light back to the detector. Single fibers are also used in many designs. By using optical fiber bundles, remote and multiple sensing is possible.

An optical sensor can be designed by utilizing the intrinsic property of certain analytes that interact with light (see scheme in Fig. 20a.2a). However, very often selectivity has to be improved by using a sensing mediator (the sensing layer) that reacts selectively with the analyte and generates an optical signal change (scheme 2b). The mediator is

Comprehensive Analytical Chemistry 47
S. Ahuja and N. Jespersen (Eds)
Volume 47 ISSN: 0166-526X DOI: 10.1016/S0166-526X(06)47029-X

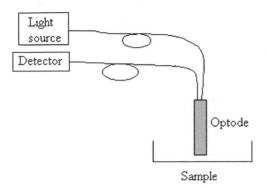

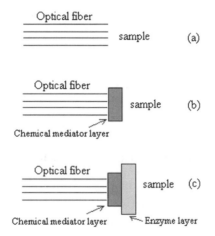

Fig. 20a.1. Schematic of an optical sensor using optical fibers as waveguide.

Fig. 20a.2. Optical sensor configurations: (a) intrinsic sensors, (b) mediator assistant chemical sensor, and (c) chemical sensor-based biosensor.

immobilized at the end of the fibers or on a solid support that is attached to the fiber at the sensing site. The third design is by combining an enzymatic reaction and an optical sensor according to scheme 2b, and the enzymatic reaction generates a species that can interact with the sensing element (scheme 2c) [3].

20a.3   DETECTION METHODS

The most often used detection method for the optical sensors are based on absorption, luminescence, reflectance, and Raman scattering measurements. The basic theory and instrumentation of most of these methods is described in earlier chapters. In this section, only the principle equations and the unique designs are presented.

In the case of *absorption* [3–8] measurements, the concentration (c) of the absorbing species is related to the absorbance (A) as stated in Beer's law:

$$A = \varepsilon bc \hspace{3cm} (20a.1)$$

where ε is the molar absorptivity of the absorbing species and b the light path. Both UV/Vis and IR light sources have been used. The remaining light intensity of source light is determined after passing the sensing element or sample . Generally, the absorbing species has a very high-absorption molar absorptivity ε for the sake to low detection limit and high sensitivity. Direct sensing of light absorbing species can be achieved with a single measurement. The sensitivity can be enhanced by increasing the length of the sample cell to increase b. For indirect sensing according to schemes 2b and 2c, chemical or biochemical reactions that alter the concentration of the absorbing species c, suitable fast and reversible reactions are utilized. The mediator has to be stable in the immobilized media to ensure constant concentration c. The sensor layer is generally thin to achieve complete reaction equilibrium within a limited time for fast response. Many pH indicators and other chromophores are used successfully for developing optical sensors for pH and metal ions.

The instrumentation for sensors based on absorption measurements can be designed on the traditional spectrophotometers by using a flow-through cell for automatic sampling with the sensors mounted inside the flow-through cell shown in Fig. 20a.3. For remote optical sensing using optical fibers, the chromophores can be immobilized in reflective

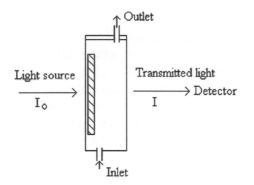

Fig. 20a.3. Schematic of optical sensing film mounted in a flow-through cell for absorbance measurements.

membranes, the remaining light after contacting with the chromophores are reflected by the reflective membrane and send to the detector
through a second fiber.

Luminescence [3] based optical sensors have a few different detection
modes. In these sensors, the energy of an excited state species is monitored. The species can be excited by photons, which includes fluorescence and phosphorescence sensors, or it can be excited using a
chemical reaction, i.e. chemiluminescence. By far fluorescence-based
sensors are most popular due to the larger selection of available fluorophores that respond to UV/Vis radiation. Detection of luminescence
signal can be the light intensity (I) or the fluorescence lifetime (τ). For
fluorescence intensity-based sensors, the light intensity is related to the
light source intensity (I_0), and the concentration (C) of the fluorophore
is as shown in the following equation:

$$I = 2.3 I_0 \O \varepsilon d k C \tag{20a.2}$$

With Ø the quantum yield, ε the molar absorption coefficient, d the
optical depth, and k a constant account for the fact that only a small
fraction of emission can be observed.

Lifetime [3,9–11] based sensors rely on the determination of decay
time of the fluorescence or phosphorescence. Typically, the fluorescence
lifetime is 2–20 µs and phosphorescence lifetime is 1 µs to 10 s. Lifetime-
based sensors utilize the fact that analytes influence the lifetime of the
fluorophore. Thus all dynamic quenchers of luminescence or suitable
quenchers can be assayed this way. The relationship between lifetimes
in the absence (τ_0) and presence (τ) of a quencher is given by Stern and
Volmer:

$$\tau_0/\tau = 1 + ksv[Q] \tag{20a.3}$$

There is a linear relationship between τ_0/τ and the concentration of
the quencher. In these sensors, the static fluorescence quenching has
no effect on the lifetimes.

Energy transfer [3,12] between molecules has also been used in the
design of optical sensors. Here, an excited molecule (donor) can transfer
its electronic energy to another species (acceptor). This process occurs
without the appearance of a photon and results from dipole–dipole interaction between the donor–acceptor molecules. The rate depends on
the fluorescence quantum yield of the donor, the overlap of the emission spectra of the donor with the absorption of the spectrum of the
acceptor, and their relative orientation and distance. It is the overlap of

the spectra and the relative distance that have been studied for optical sensors.

Chemiluminescence sensors utilize chemicals that generate photons, e.g.

$$A + B \ \rightarrow \ C^* \ \rightarrow \ C + h\upsilon \qquad\qquad (\text{rxn } 20a.1)$$

Here A and B are non-luminescence molecules. The C^* is the excited state of the product C. Often these reactions involve oxidation reactions and the presence of a catalyst. Both chemical and biochemical reactions could generate the photon. The intensity of the photons are collected through optical fibers and measured with a photon detector. The most successful chemiluminescence sensor for the detection of the hydrogen peroxide [13] is based on luminol using ferricyanide as catalyst

$$2H_2O_2 + \text{luminol} + OH^- \rightarrow \text{3-aminophthalate} + N_2$$
$$+ 2H_2O + h\upsilon \qquad\qquad (\text{rxn } 20a.2)$$

Many bioluminescence sensors were investigated using enzymes for H_2O_2, NADH, ethanol, glucose, and various amino acids.

20a.3.1 IR spectroscopy-based optical sensors

Many compounds exhibit near-IR and mid-IR absorption. By using IR transparent optical fibers, detection of an absorption band in the IR region is possible for optical sensing. Both direct sensing using the absorption property of the analyte or indicator sensing are widely exploited. Most mid-IR sensing schemes are based on the principles of internal reflection spectroscopy, or the attenuated total reflection (ATR) [3,14–21].

The technique of attenuated reflection involves prism coupling into fully guiding modes. In the optical waveguide (prism), total reflection at the interface between an optical dense medium and an optically rare one (low refraction index) occurs, when the angle of refraction is larger than a critical angle Ø. However, light is not instantaneously reflected when it reaches the interface. Light penetrates to some extent into the optically rare phase. More precisely, the amplitude of the electrical field does not drop abruptly to zero at the interface but has a tail that decreases exponentially in the direction of an outward normal boundary. This phenomenon is referred as an evanescent wave (see Fig. 20a.4a). If the optically rare medium, i.e. sample, contains an analyte that absorbs the evanescent wave, the reflected light intensity will be reduced. This results in the ATR.

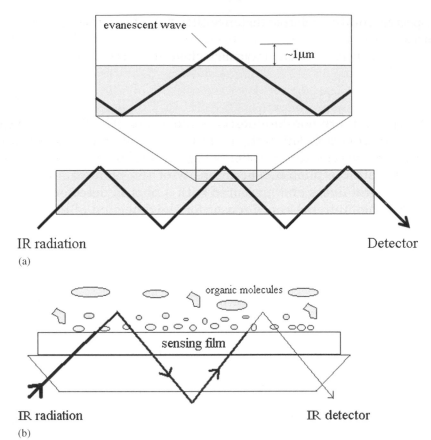

Fig. 20a.4. Schematic of light pathway of IR-evanescent wave (a) and IR-ATR optical sensor.

ATR is particularly useful when sampling is difficult, or the response time is too long due to a thick sensing element, in conventional spectroscopy. By using ATR, the light does not pass through the entire sample, but only up to a micrometer into the sample media, the thickness and the homogeneity of the sample is not critical for fast and accurate analysis. Using this method, the small signal of thin films can then be multiplied with multiple reflections passing through the sample. Figure 20a.4b illustrates a sensor based on this principle.

Raman spectroscopy is an emission technique involving the scatter of absorbed light often in the visible region. Raman bands arise from changes in polarizability in molecules during a vibration. Raman spectroscopy is widely used to monitor compounds that have highly polarizable bonds. Typical Raman spectrometers are used by coupling with visible and near-IR light sources and visible and NIR light transparent fiber optics. Sensitive sensing is often achieved by using metal films such as Cu, Ag, and Au that absorb molecules and plasma resonance to enhance the Raman scattering [22–30].

Surface-enhanced Raman scattering (SERS) has emerged as a powerful technique for studying species adsorbed on metal films, colloidal dispersions, and working electrodes. SERS occurs when molecules are adsorbed on certain metal surfaces, where Raman intensity enhancements of ca. 10^5–10^6 may be observed. The enhancement is primarily due to plasmon excitation at the metal surface, thus the effect is limited to Cu, Ag, and Au, and a few other metals for which surface plasmons are excited by visible radiation.

Although chemisorption is not essential, when it does occur there may be further enhancement of the Raman signal, since the formation of new chemical bonds and the consequent perturbation of adsorbate electronic energy levels can lead to a surface-induced RR effect. The combination of surface and resonance enhancement (SERRS) can occur when adsorbates have intense electronic absorption bands in the same spectral region as the metal surface plasmon resonance, yielding an overall enhancement as large as 10^{10}–10^{12}.

20a.4 SENSING PRINCIPLE OF SAMPLE OPTICAL SENSORS

The intrinsic sensors are based on the direct recognition of the chemicals by its intrinsic optical activity, such as absorption or fluorescence in the UV/Vis/IR region. In these cases, no extra chemical is needed to generate the analytical signal. The detection can be a traditional spectrometer or coupled with fiber optics in those regions. Sensors have been developed for the detection of CO, CO_2 NO_x, SO_2, H_2S, NH_3, nonsaturated hydrocarbons, as well as solvent vapors in air using IR or NIR absorptions, or for the detection of indicator concentrations in the UV/Vis region and fluorophores such as quinine, fluorescein, etc.

While there are only a few examples that can be used for direct detection of desired analytes, many simple molecules and ions do not have optical activity under regular conditions, a chemical reaction is needed to generate an optically active species. The reactions can be acid–base, ion pairing, complexation reactions, or quenching of fluorescence by O_2, paramagnetic molecules, etc. Optical sensors for a few analyte or group analytes are summarized below.

20a.4.1 Oxygen (O_2) gas sensors [3,31–35]

Oxygen is a good ligand and fluorescence quencher to many fluorophores. Many oxygen sensors have been designed based on the fluorescence quenching of polycyclic aromatic hydrocarbons such as pyrene, prenebutyric acid, fluoroanthene, decacylene, diphenylanthracene, etc. The sensors are designed by immobilizing these organic compounds in PVC membranes. Polar dyes such as perylene dibutyrate, fluorescent yellow, trypaflavin, and porphyrins were absorbed on silica gels for oxygen sensing. Other highly selective oxygen optical sensors are developed by using the quenching of Ru–Pt complexes.

$$Ru - TPP + O_2 \rightleftharpoons Ru - TPP : O_2$$

fluorescence active   non − fluorescent   (rxn20a.3)

Figure 20a.5 is a schematic of an O_2 sensor based on fluorescence Ru complexes doped in sol-gel. A corresponding response curve of this sensor to oxygen is shown in Fig. 20a.6.

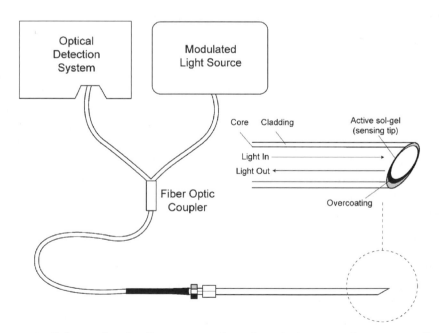

Fig. 20a.5. Schematic of a fluorescence-based optical sensor for oxygen. Courtesy of Ocean Optics, Inc.

Optical sensors

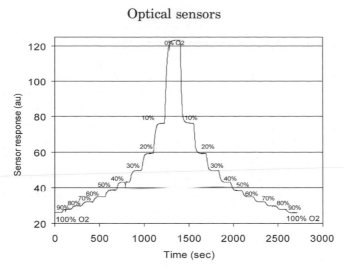

Fig. 20a.6. Response curve of a Ru-fluorescence complex based oxygen optical sensor to oxygen in gas mixture. Courtesy of Ocean Optics, Inc.

20a.4.2 pH sensors [3,4,36– 40]

The detection of proton is based on the acid–base reaction of a chromophore or fluorophore that are weak acid or base.

$$AQ : 2C_{(m)} + H^+ \text{ b } HC^+_{(m)} \rightleftharpoons C^- + H^+ \text{ b } HC_{(m)} \qquad \text{(rxn 20a.4)}$$

Here $C_{(m)}$ and $C^-_{(m)}$ are the neutral or anionic chromophore immobilized onto a gel or an organic polymer membrane. Upon protonation, positively charged HC^+ or neutral HC species are formed in the sensing element. Typically, the deprotonated form C (C^-) has a different optical activity than the protonated form HC or HC^+. If C_b is to represent the base (deprotonated) form of the indicator and C_t the total concentration of the indicator, then the pH and the concentration of the base form of the indicator is related in the Hendersen–Hassenbach equation:

$$pH = pKa + \log C_b/(C_t - C_b) \qquad \text{(20a.4)}$$

Thus as in the titration curves in the solution, the response range of the pH indicators for a signal protonation process is about 2–3 pH units for absorbance-based measurements. Long-range pH sensing has to be achieved by using mixed dyes doped in solid support. While fluorescence sensors have a wider linearity because it is only related to one form of the indicator.

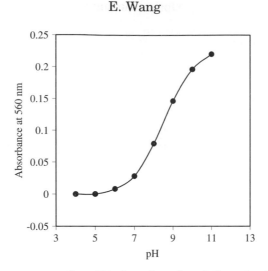

Fig. 20a.7. Response curve for pH of a phenol red doped sol-gel in universal buffers.

Many pH sensors have been developed using pH indicators active in the IR to UV region with absorbance or fluorescence spectroscopy, covering pH ranges of 0–14 using dyes of different pKa. The pH indicators can be immobilized on any solid support, such as silica, cellulose, ion-exchange resin, porous glass, sol-gel, or entrapped in polymer membranes depending on their polarity and lipophilicity. Figure 20a.7 is a pH response curve of a pH sensor with phenol red doped sol-gel using the sensing device shown in Fig. 20a.3 for absorbance measurement [39].

20a.4.3 Acidic or basic gas sensors using pH-sensitive dyes [41,42]

Acidic gas such as CO_2 and SO_2 can produce proton when hydrated in water. Using CO_2 as an example, the reaction between water and CO_2 is shown below:

$$CO_2 + H_2O \rightleftharpoons H_2CO_3 \text{ (hydration)} \qquad \text{(rxn 20a.5)}$$

$$H_2CO_3 \rightleftharpoons H^+ + HCO_3^- \text{ (first dissociation)} \qquad \text{(rxn 20a.6)}$$

$$HCO_3^- \rightleftharpoons H^+ + CO_3^{2-} \text{ (second dissociation)} \qquad \text{(rxn 20a.7)}$$

Thus pH responsive dyes or fluorophores can be used to design optical sensors for CO_2. To prevent pH interference, a hydrophobic membrane with high CO_2 permeability is used to selectively allow CO_2

Optical sensors

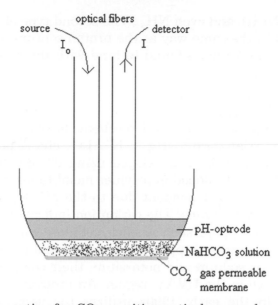

Fig. 20a.8. Schematic of a CO_2-sensitive optical sensor based on pH optical membrane. For more details see Ref. [39].

diffuse into the membrane, and hydrated in an internal solution between the CO_2 permeable membrane and the pH-sensitive membrane. Upon hydration CO_2 dissociated, the proton produced changes the pH of the internal solution, thus change the absorbance or fluorescence intensity of the pH sensor. A schematic of this type of sensor is shown in Fig. 20a.8.

Ammonia sensor can be designed in a similar method. Only here, the basic gas can directly interact with the indicator (reaction 8), thus no internal solution is needed, the reaction can be a single step.

$$NH_3 + HC \rightleftharpoons NH_4^+ + C^- \qquad \text{(rxn 20a.8)}$$

pH indicators that have pKa below 7 are very sensitive to NH_3 gas.

20a.4.4 Cation sensors [3,4,43–51]

Some transition metal cations have color in the visible or near IR region due to their unfilled d-orbitals, intrinsic sensing of these ions are possible. Such as the detection of Cu(II) in plating baths.

Yet the majority of cation sensors are mediator-based sensing. One type of optical sensor is based on the fluorescence quenching of fluorophore Rhodamine 6G by transition metals such as Co(II), Cr(III),

E. Wang

Fe(III), Cu(II), Ni(II), and even NH_4^+. The second type of optical sensor can be designed in the same way as the proton sensors if a fluorogenic or chromogenic compound is found to bind with the cation selectively and reversibly, e.g.

$$M^{n+} + mHIn \rightleftharpoons mH^+ + MI_m^{\,n-m} \qquad \text{(rxn 20a.9)}$$

Sensors based on the above reaction scheme have been developed for Al^{3+}, Zn^{2+}, Cu^{2+}, Ca^{2+}, Pb^{2+}, Hg^{2+}, K^+, Li^+, etc. A polycation, protamine sensor has also been developed using $2'7'$-dichlorofluorescein octadecyl ester (DCFOE) doped in polymer membranes. However, most of these sensors are pH dependent due to the pH dependence of the cation complexation reactions. The cation ion indicators can be immobilized on any solid support, such as silica, cellulose, ion-exchange resin, porous glass, sol-gel, or entrapped in polymer membranes.

For ions such as sodium and potassium, their complexation agents are usually inactive in the UV/Vis region. An innovative design by Dr. Simon's group in the early 1990s utilizes the same ion-extraction chemistries used in the potentiometric membrane electrodes. In this design, the ionophores are doped together with an appropriate pH indicator dyes, into thin ($< 10\,\mu m$) organic membranes, and the response is based on ion exchange for cation. When a membrane doped with neutral C and a cation carrier ligand L is bathed in a sample solution, exchange of H^+ and M^+ ions occur between the two phases, resulting in a change of the optical absorbance of the sensing film.

$$M_{(aq)}^+ + HC_{(m)}^+ + I_{(m)} + R_{(m)}^- \rightleftharpoons MI_{(m)}^+ + C_{(m)} + R_{(m)}^- + H_{(aq)}^+ \qquad \text{(rxn 20a.10)}$$

Similarly, an anionic chromophore HC can also be used for designing optical sensors for cations, in this case, no extra anionic site (R^-) is needed to maintain charge neutrality during the sensing.

$$M_{(aq)}^+ + HC_{(m)} + I_{(m)}(m) \rightleftharpoons MI_{(m)}^+ + C_{(m)}^- + H_{(aq)}^+ \qquad \text{(rxn 20a.11)}$$

The equilibrium constant for this reaction depends on the stability constants of the ionophore-M^+ complexes and on the distribution of ions in aqueous test solution and organic membrane phases. For a membrane of fixed composition exposed to a test solution of a given pH, the optical absorption of the membrane depends on the ratio of the protonated and deprotonated indicator which is controlled by the activity of M^+ in the test solution ($H_{(aq)}^+$ is fixed by buffer). By using α to represent the fraction of total indicator (C_T) in the deprotonated form ([C]), α can be related to the absorbance values at a given wavelength as follows:

$$\alpha = \frac{[C^-]}{[C_T]} = \frac{A - A_0}{A_1 - A_0} \tag{20a.5}$$

where A_0 and A_1 correspond to the absorbance of the completely protonated and deprotonated pH indicator, respectively. The dependence of relative optical absorption α on the activities of monovalent cations ($^aM^+$) and protons ($^aH^+$) in the test solution can be expressed using Eq. (20a.2):

$$\alpha^2/(1-\alpha)(L_T/C_T - \alpha) = K_{eq}{}^aH^+/(^aM^+ + K_{ij}^{opt}aj) \tag{20a.6}$$

where L_T is the total concentration of cation ionophore in the membrane, and K_{ij}^{opt} the selectivity coefficient defined by analogy to that for ion-selective electrodes. Using these designs, various optical sensors for Na^+, K^+, Ca^{2+}, Mg^{2+}, Pb^{2+}, and NH_4^+ have been designed. Figure 20a.9 shows an absorption spectra of a calcium responsive optical sensor with lipophilic anionic DCFOE as a chromophore doped together a calcium organophosphate salt in polymeric film. The membrane responds to calcium in the range of 1.0×10^{-6}–$0.2\,M$ at pH 7.4.

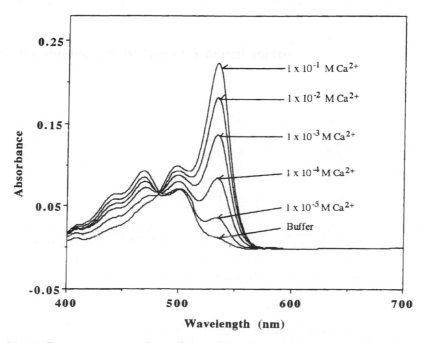

Fig. 20a.9. Response curve for calcium(II) of a calcium-organophosphate and dichlorofluorescein octadecyl ester doped in polymer films at pH 7.4. For more details see Ref. [49].

The fourth type of mediator-based cation optical sensing is using potential sensitive dye and a cation selective ionophore doped in polymer membrane. Strong fluorophores, e.g. Rhodamine-B C-18 ester exhibits differences in fluorescence intensity because of the concentration redistribution in membranes. PVC membranes doped with a potassium ionophore, can selectively extract potassium into the membrane, and therefore produce a potential at the membrane/solution interface. This potential will cause the fluorescent dye to redistribute within the membrane and therefore changes its fluorescence intensity. Here, the ionophore and the fluorescence have no interaction, therefore it can be applied to develop other cation sensors with a selective neutral ionophore.

20a.4.5   Anion sensing [3,4,52–57]

A few examples were reported for the anion sensing by principle of quenching of fluorophores with anions such as I^-, Br^-, and Cl^- as shown below:

$$X^- + \text{Alizarin} \rightleftharpoons \text{Alizarin-X} \qquad \text{(rxn 20a.12)}$$

Here X^- represents the halide anions.

Certain porphyrins have shown optical absorbance change when exposed to anions such as NO_2^-, Cl^-, I^-, and SCN^-, and optical sensors can be obtained accordingly.

An analogous system to the cation sensors is the use of optically non-active anion ionophores together with a pH-sensitive chromophore for the anion sensing. In this system, simultaneous extraction of H^+ and anion X^- from the aqueous solution into an organic phase results in protonation of the chromophore anion C^-, e.g.

$$L^+\text{-}C^-_{(m)} + H^+_{(aq)} + A^-_{(aq)} \rightleftharpoons L\text{-}A_{(m)} + HC_{(m)} \qquad \text{(rxn 20a.13)}$$

The equilibrium constant for reaction 5 depends on the complex formation constant, the association constant of C^- in the membrane and on the distribution coefficients of H^+, and ions between the organic membrane phase and aqueous sample solution, e.g.

$$K_{extr} = [L\text{-}A][HC]/\{[L^+][C][H^+][A^-] \qquad \text{(20a.7)}$$

With α as described as in Eq. (20a.5), the concentration of the anion is related to the α described in Eq. (20a.8).

$$[A^-] = (1-\alpha)[C_T]/\{\alpha^2(L_T/{}^aH^+K_{extr}\} \qquad \text{(20a.8)}$$

The influence of solution pH on the sensor's response can be eliminated by using an appropriate buffer solution. Thus, the selectivity of the optical sensors at a given pH depends only upon the interaction between the anion and its carriers, which gives the same selectivity sequence as that of ion-selective electrodes.

Thin plasticized polymer films, polyvinyl chloride (PVC) doped with a specific ion pairing quaternary ammonium compound, tridodecylmethylammonium chloride (TDMAC), and a lipophilic pH indicator, 3-hydroxy-4-(4-nitrophenylazo)phenyl octadeconate (ETH 2412) are shown to exhibit significant and analytically useful optical response toward lipophilic anions such as salicylate and perchlorate anions as well as macromolecular heparin. The response mechanism is based on favorable extraction of these into the bulk organic film, owing to their high lipophilicity of salicylate and perchlorate ions. For heparin response, it is due to the specific ion-pairing complexation reaction between the quaternary ammonium species and the polyanion. A simultaneous coextraction of hydrogen ions results in protonation of the pH chromophore, and hence a change in the optical absorbance of the polymeric film.

The anion optical sensor can also be fabricated with metalloporphyrins. For example, polymeric membranes doped with indium porphyrins and a lipophilic dichlorofluorescein derivative were shown to be very selective to chloride and acetate anions. The response mechanism is based on extraction of anions into the bulk organic film by indium porphyrins and a simultaneous coextraction of hydrogen ions. This results in protonation of the pH chromophore, and hence a change in the optical absorbance of the polymeric film.

$$In(P)\text{-}C_{(m)} + H^+_{(aq)} + A^-_{(aq)} \rightleftharpoons In(P)\text{-}A_{(m)} + HC_{(m)} \qquad \text{(rxn 20a.14)}$$

The type of metal ion as well as the porphyrin used greatly influences the selectivity and performance of the optical sensor, e.g. the selectivity sequence for a picket-fence porphyrin doped with DCFOE at 530 nm is $SCN^- \sim I^- > Sal^- \sim OAc^- > Br^- > Cl^- \sim NO_2^- \gg F^- > SO_4^{2-} > ClO_4^- > NO_3^- > H_2PO_4^-$, while the selectivity sequence of the membrane doped with (octaethylporphyrinato)indium at 520 nm is: $SCN^- \sim I^- > Sal^- > Br^- > Cl^- > NO_2^- > OAc^- > F^- > SO_4^{2-} > ClO_4^- > NO_3^- > H_2PO_4^-$. In the former, optical sensor can be used for acetate or acetic acid sensing, and the later is suitable for chloride sensing when ions such as $SCN^- \sim I^- > Sal^-$ are not present in an insignificant amount comparing to acetate or chloride ions [56].

769

And an Al(III) porphyrin shows very high selectivity to fluoride ion with sensitivity in the submicromolar range [57].

20a.4.6   Biosensors [58–60]

Optical biosensors can be designed using scheme 2c as shown in Fig. 20a.2 when a selective and fast bioreaction that produces chemical species which can be determined by an optical sensor. Like the electrochemical sensors, enzymatic reactions that produces oxygen, ammonia, hydrogen peroxide, or proton, can be utilized to fabricate optical sensors. For example, optical urea biosensor can be developed by immobilizing a urease enzyme layer on a thin ammonium-selective polymer membrane. The ammonium optical membrane utilized DCFOE as anionic chromophore and nonactin as neutral ionophore. The urease layer was coated on the top of the ammonium layer by gelatin entrapment combined with cross-linking with glutaraldehyde. Hydrolysis of urea catalyzed by urease produced ammonium ion that was extracted into the polymer film to form complexes with nonactin. A proton was released that resulted in a color change of the optical membrane due to the charge neutrality principle (see Fig. 20a.10). The biosensor showed response to urea in the range of 10^{-5}–0.1 M in 0.1 M Tris buffer at pH 7.4. The response time is 3–8 min and it is completely reversible. The biosensors were found to be reliable in diluted serum samples for urea detection from 0.1 to 10 mM, and good correlation was obtained with a potentiometric urea sensor.

Like the electrochemical sensors, optical biosensors for glucose are designed with an oxygen-sensitive optical sensor.

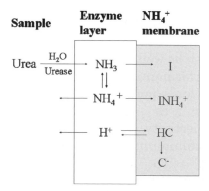

Fig. 20a.10. Schematic of the sensing principle of a urea optical sensor based on an ammonium-sensitive membrane employing anionic dye and neutral carrier.

REFERENCES

1 J. Janata, Chemical sensors, *Anal. Chem.*, 62(12) (1990) 33R–44R.

2 M.A. Arnold, Fiber optic chemical sensors, *Anal. Chem.*, 64(21) (1992) 1015A.

3 O.S. Wolfbeis (Ed.), *Fiber Optic Chemical Sensors and Biosensors*, Vols. 1 and 2, CRC Press, Boca Raton, FL, 1991.

4 O.S. Wolfbeis and R. Narayanaswamy, Optical Sensor; Industrial, Environmental and Diagnostic Applications, Springer Series on Chemical Sensors and Biosensors, 2004.

5 U.E. Spichiger, *Chemical Sensors and Biosensors for Medical and Biological Applications*, Wiley-VCH, Weinheim, 1998.

6 O.S. Wolfbeis, Fiber-optic chemical sensors and biosensors, *Anal. Chem.*, 76(12) (2004) 3269–3284.

7 L. Polerecky, C.S. Burke and B.D. MacCraith, Optimization of absorption-based optical chemical sensors that employ a single-reflection configuration, *Appl. Opt.*, 41(15) (2002) 2879–2887.

8 W.R. Seitz, Chemical sensors based on immobilized indicators and fiber optics, *Crit. Rev. Anal. Chem.*, 19(2) (1988) 135–173.

9 G.H. Victkers, R.M. Miller and G.M. Hieftje, Time-resolved fluorescence with an optical fiber probe, *Anal. Chim. Acta*, 192 (1987) 145.

10 T. Mayr, I. Klimant, O.S. Wolfbeis and T. Werner, Dual lifetime referenced optical sensor membrane for the determination of copper(II) ions, *Anal. Chim. Acta*, 462(1) (2002) 1–10.

11 G. Liebsch, I. Klimant, B. Frank, G. Holst and O.S. Wolfbeis, Luminescence lifetime imaging of oxygen, pH, and carbon dioxide distribution using optical sensors, *Appl. Spectrosc.*, 54(4) (2000) 548–559.

12 A. Sherman and O.S. Wolfbeis, Fiber optic fluorosensor for sulfur dioxide based on energy transfer and exciplex quenching, *Proc. SPIE*, 990 (1989) 116.

13 T.M. Freeman and W.R. Seitz, Chemiluminescence fiber optic probe for hydrogen peroxide based on the luminol reaction, *Anal. Chem.*, 50(9) (1978) 1242–1246.

14 B.D. Gupta and D.K. Sharma, Evanescent wave absorption based fiber optic pH sensor prepared by dye doped sol-gel immobilization technique, *Opt. Commun.*, 140(1–3) (1997) 32–35.

15 B.D. MacCraith, Enhanced evanescent wave sensors based on sol-gel-derived porous glass coatings, *Sensor Actuat. B-Chem.*, 11(1–3) (1993) 29–34.

16 V. Ruddy, S. McCabe and B.D. MacCraith, Detection of propane by IR-ATR in a Teflon-clad fluoride glass optical fibre, *Appl. Spectrosc.*, 44(9) (1990) 1461.

17 J.E. Walsh, B.D. MacCraith, M. Meaney and J.G. Vos, Mid-infrared fiber sensor for the *in-situ* monitoring of chlorinated hydrocarbons, *Sensor Appl. VII, IOP*, (1995), 75–82.

18 B.D. MacCraith, V. Ruddy and S. McCabe, Remote flammable gas sensing using a fluoride fiber evanescent probe, *Proc. SPIE*, 1267 (1990) 43.

19 U.E. Spichiger, D. Citterio and M. Bott, Analyte-selective membranes and optical evaluation techniques. Characterization of response behaviour by ATR measurements, *SPIE*, 2508 (1995) 179–189.

20 F. Regan, M. Meaney, J.G. Vos, B.D. MacCraith and J.E. Walsh, Determination of pesticides in water using ATR–FTIR spectroscopy on PVC/chloroparaffin coatings, *Anal. Chim. Acta*, 334(1-2) (1996) 85–92.

21 T.M. Butler, B.D. MacCraith and C.M. McDonagh, Development of an extended-range fibre optic pH sensor using evanescent wave absorption of sol-gel-entrapped pH indicators, *Proc. SPIE*, 2508 (1995) 168–178.

22 F. Akbarian, B.S. Dunn and J.I. Zink, Porous sol-gel silicates containing gold particles as matrices for surface-enhanced Raman spectroscopy, *J. Raman Spectrosc.*, 27 (1996) 775.

23 T. Vo-Dinh, SERS chemical sensors and biosensors: new tools for environmental and biological analysis, *Sensor Actuat. B-Chem.*, 29(1-3) (1995) 183–189.

24 C.E. Talley, L. Jusinski, C.W. Hollars, S.M. Lane and T. Huser, Intracellular pH sensors based on surface-enhanced Raman scattering, *Anal. Chem.*, 76(23) (2004) 7064–7068.

25 A. Tao, F. Kim, C. Hess, J. Goldberger, R. He, Y. Sun, Y. Xia and P. Yang, Langmuir–Blodgett silver nanowire monolayers for molecular sensing using surface-enhanced Raman spectroscopy, *Nano Lett.*, 3(9) (2003) 1229–1233.

26 A. Ruperez and J.J. Laserna, Surface-enhanced Raman sensor, *Analysis*, 23(2) (1995) 91–93.

27 J.M. Bello, V.A. Narayanan, D.L. Stokes and T. Vo-Dinh, Fiber-optic remote sensor for *in situ* surface-enhanced Raman scattering analysis, *Anal. Chem.*, 62(22) (1990) 2437–2441.

28 J.M. Bello and T. Vo-Dinh, Surface-enhanced Raman scattering fiber-optic sensor, *Appl. Spectrosc.*, 44(1) (1990) 63–69.

29 C.R. Yonzon, C.L. Haynes, X. Zhang, J.T. Walsh Jr. and R.P. Van Duyne, A glucose biosensor based on surface-enhanced Raman scattering: improved partition layer, temporal stability, reversibility, and resistance to serum protein interference, *Anal. Chem.*, 76(1) (2004) 78–85.

30 H. Schmidt, B.H. Nguyen, P. Jens, A. Hans and K. Heinz-Detlef, Kowalewska Grazyna detection of PAHs in seawater using surface-enhanced Raman scattering (SERS), *Mar. Pollut. Bull.*, 49(3) (2004) 229–234.

31 A.K. McEvoy, C.M. McDonagh and B.D. MacCraith, Dissolved oxygen sensor based on fluorescence quenching of oxygen-sensitive ruthenium complexes immobilized in sol-gel-derived porous silica coatings, *Analyst*, 121(6) (1996) 785–788.

32 B.J. Basu, A. Thirumurugan, A.R. Dinesh, C. Anandan and K.S. Rajam, Optical oxygen sensor coating based on the fluorescence quenching of a new pyrene derivative, *Sensor Actuat. B-Chem.*, 104(1) (2005) 15–22.

33 F.G. Gao, A.S. Jeevarajan and M.M. Anderson, Long-term continuous monitoring of dissolved oxygen in cell culture medium for perfused bioreactors using optical oxygen sensors, *Biotechnol. Bioeng.*, 86(4) (2004) 425–433.

34 S.M. Borisov and V.V. Vasil'ev, New optical sensors for oxygen based on phosphorescent cationic water-soluble Pd(II), Pt(II), and Rh(III) porphyrins, *J. Anal. Chem. (Translation of Zhurnal Analiticheskoi Khimii)*, 59(2) (2004) 155–159.

35 M. Ahmad, N. Mohammad and J. Abdullah, Sensing material for oxygen gas prepared by doping sol-gel film with tris(2,2'-bipyridyl)dichlororuthenium complex, *J. Non-Crystal. Solids*, 290(1) (2001) 86–91.

36 M.J.P. Leiner and P. Hartmann, Theory and practice in optical pH sensing, *Sensor Actuat. B-Chem.*, 11(1-3) (1993) 281–289.

37 J. Lin, Recent development and applications of optical and fiber-optic pH sensors, *TrAC Trend Anal. Chem.*, 19(9) (2000) 541–552.

38 N.K. Sharma and B.D. Gupta, Fabrication and characterization of a fiber-optic pH sensor for the pH range 2 to 13, *Fiber Integr. Opt.*, 23(4) (2004) 327–335.

39 E. Wang, K.-F. Chio, V. Kwan, T. Chin, C. Wong and A. Bocarsly, Fast and long term optical sensors for pH based on sol-gels, *Anal. Chim. Acta*, 495 (2003) 45–50.

40 Y. Kowada, T. Ozeki and T. Minami, Preparation of silica-gel film with pH indicators by the sol-gel method, *J.Sol-Gel Sci.Technol.*, 33(2) (2005) 175–185.

41 D.A. Nivens, M.V. Schiza and S.M. Angel, Multilayer sol-gel membranes for optical sensing applications: single layer pH and dual layer CO_2 and NH_3 sensors, *Talanta*, 58(3) (2002) 543–550.

42 O.S. Wolfbeis and H.E. Posch, Fiber-optic fluorescing sensor for ammonia, *Anal. Chim. Acta*, 185 (1986) 321–327.

43 E. Wang, G. Wang, L. Ma, C.M. Stivanello, S. Lam and H. Patel, Optical films for protamine detection with lipophilic dichlorofluorescein derivatives, *Anal. Chim. Acta*, 334 (1996) 139–147.

44 W.E. Morf, K. Seiler, B. Rusterholz and W. Simon, Design of a calcium-selective optode membrane based on neutral ionophores, *Anal. Chem.*, 62 (1990) 738–742.

45 K. Seiler and W. Simon, Theoretical aspects of bulk optode membranes, *Anal. Chim. Acta*, 266 (1992) 73–87.

46 E. Bakker and W. Simon, Selectivity of ion-sensitive bulk optodes, *Anal. Chem.*, 64 (1992) 1805–1812.

47 E. Bakker, P. Buehlmann and E. Pretsch, Carrier-based ion-selective electrodes and bulk optodes. 1. General characteristics, *Chem. Rev. (Washington, DC)*, 97(8) (1997) 3083–3132.

48 P. Buehlmann, E. Pretsch and E. Bakker, Carrier-based ion-selective electrodes and bulk optodes. 2. Ionophores for potentiometric and optical sensors, *Chem. Rev.(Washington, DC)*, 98(4) (1998) 1593–1687.

49 K. Watanabe, E. Nakagawa, H. Yamada, H. Hisamoto and K. Suzuki, Lithium ion selective optical fiber sensor based on a novel neutral ionophore and a lipophilic anionic dye, *Anal. Chem.*, 65(19) (1993) 2704–2710.

50 E. Wang, L. Ma, L. Zhu and C.M. Stivanello, Calcium optical sensors based on lipophilic dichlorofluorescein anionic dye and calcium organophosphate ionophore or neutral carriers, *Anal. Lett.*, 30(1) (1997) 33–44.

51 E. Wang, L. Zhu, L. Ma and H. Patel, Optical sensors for sodium, potassium and ammonium ions based on lipophilic fluorescein anionic dye and neutral carriers, *Anal. Chim. Acta*, 357 (1997) 85–90.

52 E. Wang and M.E. Meyerhoff, Anion selective optical sensing with metalloporphyrin-doped polymeric films, *Anal. Chim. Acta*, 283(2) (1993) 673–682.

53 E. Wang, M.E. Meyerhoff and V.C. Yang, Optical detection of macromolecular heparin via selective coextraction into thin polymeric films, *Anal. Chem.*, 67 (1995) 522–527.

54 I.H. Badr, R.D. Johnson, M. Diaz, M.F. Hawthorne and L.G. Bachas, A selective optical sensor based on [9]mercuracarborand-3, a new type of ionophore with a chloride complexing cavity, *Anal. Chem.*, 72(18) (2000) 4249–4254.

55 E. Wang, C. Romero, D. Santiago and V. Syntilas, Optical anion sensing characteristics of indium-porphyrin and lipophilic dichlorofluorescein doped polymer films, *Anal. Chim. Acta*, 433 (2001) 89–95.

56 C. Huber, C. Krause, T. Werner and O.S. Wolfbeis, Serum chloride optical sensors based on dynamic quenching of the fluorescence of photo-immobilized lucigenin, *Microchim. Acta*, 142(4) (2003) 245–253.

57 I.H.A. Badr and M.E. Meyerhoff, Highly selective optical fluoride ion sensor with submicromolar detection limit based on aluminum(III) octaethylporphyrin in thin polymeric film, *J. Am. Chem. Soc.*, 127(15) (2005) 5318–5319.

58 H. Chen and E. Wang, Urea optical sensors based on ammonium ion selective polymer membranes, *Anal. Lett.*, 33(6) (2000) 997–1011.

59 W. Trettnak, M.J. Leiner and O.S. Wolfbeis, Optical sensors. Part 34. Fibre optic glucose biosensor with an oxygen optrode as the transducer, *Analyst*, 113(10) (1988) 1519–1523.

60 M.C. Moreno-Bondi, O.S. Wolfbeis, M.J. Leiner and B.P. Schaffar, Oxygen optrode for use in a fiber-optic glucose biosensor, *Anal. Chem.*, 62(21) (1990) 2377–2380.

REVIEW QUESTIONS

1. What is an optical sensor? What is its advantages comparing to electrochemical sensors?
2. How can the optical sensors be designed for multiple sample and remote sensing?
3. What are the response principles for an absorbance-based optical sensor for hydrogen and other cations?
4. What is the advantage of a time-resolved fluorescence optical sensor?
5. Give an example of an anion optical sensor.
6. Use diagrams to illustrate how biosensors can be designed with O_2 or H^+ optical sensors?

Chapter 20b

Bioactivity detectors

Sue M. Ford

20b.1 BACKGROUND

20b.1.1 Needs for bioactivity detection

Routine monitoring of food and water for the presence of pathogens, toxins, and spoilage-causing microbes is a major concern for public health departments and the food industry. In the case of disease-causing contamination, identification of the organism is critical to trace its source. The Environmental Protection Agency monitors drinking water, ambient water, and wastewater for the presence of organisms such as non-pathogenic coliform bacteria, which are indicators of pollution. Identification of a specific organism can aid in locating the source of the pollution by determining whether the organism is from humans, livestock, or wildlife [1]. Pharmaceutical companies monitor water-for-injection for bacterial toxins called pyrogens, which are not removed by filter-sterilization methods. There is a need for methods to quickly detect biothreat agents that may be dispersed as bacteria, spores, viruses, rickettsiae, or toxins such as the botulism toxin or ricin [2,3].

20b.2 METHODS

20b.2.1 Traditional bioactivity methods

The FDA, USDA, and EPA have established methods and standards for detecting food- and water-borne pathogens; these methods are available on the agencies' websites [4–8]. Many of the procedures for microbe identification rely on culturing the sample for subsequent identification by multiple tests such as colony characteristics, selective growth conditions, and biochemical assays for metabolites. Such traditional methods take upward of 24 h and remain the standard against which new

Comprehensive Analytical Chemistry 47
S. Ahuja and N. Jespersen (Eds)
Volume 47 ISSN: 0166-526X DOI: 10.1016/S0166-526X(06)47030-6

tests are evaluated. Recently developed procedures for microbial detection include biochemical test kits, genetic and molecular biology techniques, and miniaturized instruments. New assays for bacteria are very sensitive, requiring fewer organisms for detection. Although an incubation period is still generally needed to expand the population above detection limits, these more sensitive methods shorten the incubation period. Testing for toxins such as the botulism toxin relied on observing the effects of a test compound injected in mice or the use of colorimetric [9] and immunoassays.

20b.2.2 Methods and instruments based on molecular recognition

Immunology-based methods have been the most important technology for detection and identification of microbes, viruses, and toxins for many years [10]. Traditional immunoassays rely on antibodies produced in immunized animals or in hybridoma cultures. The interaction of the antigen of interest (pathogen, toxin) with the antibody is combined with a detection protocol which may involve labeling the bound antigen/antibody complex with a reporter molecule that allows detection. The methods include latex agglutination, immunodiffusion, enzyme-linked immunoassays, immunoprecipitation, and immunochromatography, which are rapid, sensitive, and specific [5,11]. The reagents can be incorporated into a disposable device similar to a home pregnancy test kit so that the presence or absence of the test organism is indicated by a color change or development of a line. The devices have quality control mechanisms incorporated to increase reliability; nonetheless, some products have been reported to have high false-positive response rates [12].

The model immunoassay is the enzyme-linked immunosorbent assay (ELISA) in which a non-specific capture antibody is bound to a surface, such as a multi-well plate or small tube [13]. In the basic form of ELISA, a second antibody tagged with an enzyme interacts specifically with the analyte. The enzyme assay produces a colored product that is read with a spectrophotometer. There are many variations on the basic immunoassay format that serve to increase sensitivity, specificity, linear range, and speed. Many commercial instruments have been developed to take advantage of various technologies for reporter molecules. The immunoassay may be coupled to an electronic sensor and transducer, such as a surface acoustical wave (SAW) sensor. Electrochemiluminescence (ECL) is a method in which the detector antibody is tagged with a ruthenium-containing chelate [13–15]. When the tag is electrically stimulated, it undergoes oxidation–reduction cycling with tripropylamine in the mixture. The excitation of the tag results in the emission of light. ECL has been used for detection of Staphylococcus enterotoxin B, ricin, *Yersinia pestis* antigen, anthrax PA antigen, and equine encephalitis virus [11]. ECL is particularly useful for detection of toxins that are not conveniently detected by molecular biology-based assays. BioVeris (Gaithersburg, MD) BV M-SERIES® instruments use ECL technology.

A developing alternative to antibody-based molecular recognition is phage-display technology. Filamentous phages are thread-shaped viruses that attack bacteria [10,16]. Phage-display libraries are mixtures of phages containing foreign peptides encoded into their surface coat proteins. The number of phages containing different peptides in a population can exceed 10^6, providing a vast selection of possible antigen-binding domains. From such populations, phages can be selected for affinity for a particular antigen. The selected phage is then replicated in host bacteria and harvested. An advantage of phage display over traditional immunological methods is that these phage "antibodies" can be prepared against compounds such as botulism toxin or organisms that would kill the animals that would be used to produce antibodies, and for molecules with low antigenicity. A problem with phage display is the potential for false positives due to unpredictable non-specific interactions in a complex mixture [10].

20b.2.3   Methods and instruments based on nucleotide analysis

Many modern instruments for bioactivity detection are based on analysis of nucleic acids subsequent to multiplication of the amount of unknown DNA of interest by the polymerase chain reaction (PCR). The basic process of PCR involves adding the unknown DNA to a reaction mixture which contains (1) nucleotides for DNA strand synthesis, (2) the heat-stable Taq DNA polymerase to catalyze the process, and (3) nucleotide primers chosen to initiate polymerase activity along specific regions of target DNA. In the instrument, the temperature is increased to 90°C and the test DNA denatures into two single strands. After denaturation of DNA, the reaction is cooled to 55°C to allow the primer to bind (anneal) to the corresponding region on the strands of test DNA and then the temperature is raised to 75°C, the optimum for polymerase activity. At the end of the elongation reaction, the amount of target DNA has doubled. This process is repeated numerous times in the thermocycler, increasing the DNA exponentially. The resulting

DNA is subsequently analyzed by various characterization techniques, such as molecular weight determination, electrophoresis, and hybridization to known probes. Using these methods, the presence of DNA unique to a particular species can be confirmed. Alternatively, with appropriate primers, a mixture of DNA fragments, which are uniquely characteristic of an individual or organism, will result in a fingerprint that can be used for identification.

The conventional steps for PCR may take hours to days to obtain results. Numerous improvements and variations in the process have enhanced the sensitivity and speed of PCR for detection and identification of bacteria, spores, and viruses. The amplicons or the oligonucleotide probes can be tagged with fluorescent dyes so that the reaction can be followed as it progresses, a process referred to as real-time PCR (RT-PCR). The amplification process and detection are carried out in the same closed vessel [17]. This eliminates the need to use electrophoresis, for example, after PCR to determine if a sample is positive for an amplicon of interest [11] and also prevents contamination by nucleic acids during sample handling in the laboratory. Instrumentation for temperature cycling has been improved such that 30 cycles can be accomplished in 30 min, a process referred to as rapid-cycle, real-time PCR [17,18]. The use of reporter tags, such as fluorescent dyes that bind to double-stranded DNA or sequence-specific probes can be added to the reaction to allow quantitation of amplicons at the end of each cycle. Fluorescence resonance energy transfer (FRET) RT-PCR involves the transfer of energy between two dyes when they are in proximity to one another. An excited reporter fluorophore excites a nearby quencher fluorophore, which then emits a photon. This phenomenon is used to construct PCR probes with reporter and quencher components. FRET RT-PCR is a powerful technique that allows several strains of smallpox viruses, for example, to be detected simultaneously [19].

The temperature at which 50% of the DNA double strands separate is the melting point (T_m) [10]. Melting point profiles can also be used to identify DNA inasmuch as a single-base substitution can change the melting curve for DNA. The LightCycler^{TM} by Roche Applied Science (Indianapolis, IN) is a temperature-controlled microvolume fluorimeter with RT-PCR module and software to facilitate quantitation of DNA, as well as identification and genotyping by fluorescence detection or melting profiles analysis. RT-PCR assays are very sensitive, with limits of detection for smallpox virus being around 12 gene copies [20].

PCR is now the front-end process for other DNA profiling techniques such as pulse-field gel electrophoresis (PFGE), terminal restriction fragment length polymorphism (T-RFLP), dentaturing-gradient gel electrophoresis (DGGE), and others [1].

20b.2.4 Methods and instruments based on chemical and physical analysis

Even the simplest bacterium is composed of innumerable lipids, proteins, and polysaccharides; however, few of these molecules are specific to a particular species. Similarly, the fluorescence of small metabolites such as ATP and nucleic acids may confirm the presence of biological organisms in aerosols or water [12], but cannot identify the agent nor indicate its virulence. A major exception is dipicolinic acid (DPA), which is found in the spores of bacteria, particularly the anthrax organism, *Bacillus anthracis*. Although non-pathogenic bacteria form endospores as well, DPA is characteristic enough to be a marker for anthrax spores. Coherent anti-Stokes Raman scattering (CARS) of DPA appears to be promising in detecting anthrax spore clumps less than 6 μm in diameter [21], which is in the range of bioweapons grade anthrax [22].

Protein toxins such as botulism, staphylococcal enterotoxin B, or ricin can be separated with gas or liquid chromatography, electrophoresis, or a combination. The μChemLab^{TM} (Sandia National Laboratories Albuquerque, NM) series of instruments includes a hand-held Bio Detector. Proteins in the sample are labeled with fluorescent tags, and nanoliter volumes of samples are separated by microchannels etched into a glass chip. The separation occurs as the sample moves through the channels and identification is based on retention times. The analyses can be completed within 10 min.

20b.2.4.1 Fingerprint methods

Analysis of individual microbial compounds is generally of limited use for distinguishing pathogens from benign organisms because of the lack of specificity. Instead, a new approach to identifying bacteria through chemical analysis is evolving. *Chemotaxonomy* refers to the process by which a whole organism is deconstructed into a chemical signature or fingerprint. This concept shows promise for rapid detection of microbes of interest. For example, various species of bacteria have distinct profiles of lipid composition, which may be used for identification [23], although this method may have limitations if the spectra are altered by the nutrients available during microbial growth.

Proteins and other large biomolecules can be analyzed using MALDI (matrix-assisted laser desorption and ionization) with TOF (time-of-flight) mass spectrometry (see Chapter 11). The sample is mixed with a matrix of organic molecules, which absorbs energy from a laser and transfers it to the protein molecules, forming ions typically with a single positive charge. Whole, intact cells can be mixed with such a matrix, dried, and analyzed directly to produce a spectral fingerprint that can distinguish among various bacterial and viral species. However, for MALDI, as for all other fingerprint methods, the ability to identify pathogens mixed with non-pathogens requires further research into mathematical procedures to interpret the results (NRC [24]).

Pyrolysis has been successfully used to prepare samples for lipid profiling by mass spectrometry [23]. Rather than a separate step in sample preparation, the methylation of fatty acids is done during the pyrolysis, a process referred to as *in situ* thermal hydrolysis methylation (THM). This reduces the sample preparation time to 30 s and the total analysis with MS is less than 10 min. The authors developed a portable ion-trap MS using this procedure which did not require pressurized gas; however, it required a large number of bacteria (10^7–10^8 cells) in order to distinguish among the four strains studied. Py-GC/MS has low resolving power and is susceptible to false positives; however, the portability and low cost make it attractive as a screening device [25]. Ion mobility spectroscopy (IMS) can also produce fingerprints to identify bacterial strains within 1 min [26]. Plasmagrams from microgram quantities of bacteria are sufficiently unique to distinguish among 200 strains or species of bacteria. The technique can be enhanced by stepped increases in the desorber temperature during desorption. Such programed temperature ramping improves peak detection and adds components to the fingerprints. This increases the complexity of the output, permitting greater discrimination among species. Pyrolysis can also be used with instruments combining short capillary GC columns with IMS for detection of bacterial biomarkers in suspect aerosols, data can be obtained within 4 min with a hand-held instrument [25]; interestingly, 3 min of that time is for collection of the aerosol.

Another type of sensor relies on semiselective polymer films, which are fabricated to interact with multiple chemicals. When a chemical is absorbed onto a polymer, which contains carbon black for electronic conduction, a change in size of the polymer—and therefore change in electrical conductivity—occurs, initiating a signal to the transducer. The magnitude of the signal varies depending on the chemical and the polymer. Greater discrimination power can be achieved by using multiple sensors with varying chemical sensitivities, an arrangement called a sensor array. Individual chemicals interact differently with each of the polymer surfaces in the array, creating fingerprint signals. For example, the Cyranose 320 unit, which evolved from NASA's E-Nose, has 32 sensors. The unit is able to identify certain classes of bacteria with 96–98% accuracy using combined statistical algorithms; however, the difficulty of this analysis [27] may preclude its use for identification. Measuring volatile metabolites with such sensors requires a high-enough concentration of bacteria to produce sufficient material for detection. Although the method is rapid, it is not sensitive enough to detect small numbers of bacteria for early detection of a biothreat. Microcalorimetric spectroscopy-based detection uses vibrational photo-thermal spectra to identify organic molecules. It can be applied to the detection and identification of microorganisms. The specificity of the method can be enhanced by using a chemically selective layer to enhance trapping of the organisms on the thermal detector. This method can detect amounts of bacteria in an order of magnitude smaller than conventional IR and FTIR [28].

20b.3   BIOWEAPONS

20b.3.1   Characteristics of bioweapons

Particles containing bacterial and spore clumps to be used as weapons must fall within a certain size range. They must be small enough to remain suspended in the air for maximum distribution, yet large enough to be retained in the pulmonary tract; thus the sizes of such particles generally fall in the range of 0.5–20 μm [12]. Thus the size and uniformity of particulate sizes in a cloud can indicate the presence of bioweapons. The analytical problems in developing bioweapons detectors are substantially different from monitoring food and water for inadvertent contamination. First, time is often an issue as the need is to treat or warn individuals, who have been exposed to microbes that have been chosen for their ability to cause disease or death in as many people as possible. Second, the quantities of biological agent to be detected are very small. Biological agents are more potent on a mass basis than chemical agents because microbes and viruses multiply within the host. Third, aerosols are diluted as the cloud drifts [3]. In order to concentrate the agent, samplers need to draw in large amounts of air. The samples therefore contain dust and other solid materials that interfere with particulate analyses. Fourth, there are many types of biological agents that might be used as weapons (see charts in reference [3] for an extensive listing), which are diverse in physical and chemical nature and not amenable to a common analytical procedure. Fifth, there is a naturally occurring background of non-pathogenic microorganisms and chemicals. Consequently, the detection methods need to distinguish microbes from the non-biological particulates as well as to distinguish pathogenic microorganisms from benign ones in a mixture. Methods that rely on quantitation of a macromolecule such as DNA would be able to recognize biological agents in an aerosol or dust, but would not be able to determine whether the agent is a pathogen or benign microbe. Such a distinction depends on identification of the microbe, which in turn requires DNA sequencing.

20b.3.2   Detect-to-warn, detect-to-treat

Rapid identification of a specific bioagent is critical for assessing the level of threat. Ideally, detectors for pathogens and toxins would produce results that are rapid, specific for individual organisms, and sensitive; however, current technical limitations require compromises, which depend on the use of the instrument. Detectors for bioweapons may be divided into several categories based on urgency of results: detect-to-warn (DTW), detect-to-treat, and detection to monitor decontamination. The purpose of DTW is, upon a positive result, to immediately set in motion actions to protect personnel and limit the spread of the agent. Thus, time is the critical feature. The goal for technology is to develop DTW systems that provide reliable alerts within 1 min [24], i.e., sample collection, preparation, identification, and data reporting should be completed within that interval. The other phases of biological agent warning systems have less urgency but have increased need for specificity and sensitivity. Depending on the organism, the time frame for detect-to-treat emergencies may range from hours to days, in order to identify those who were exposed before symptoms appear and start prophylactic treatment with antibiotics. This approach would be suitable for agents such as *Bacillus anthracis* for which there is a reasonable window for treatment (up to 9 days) after exposure [29]. In this case, the goal is accurate and specific identification of potential agents [3]. For decontamination, time is less of a critical factor than sensitivity in the ability to monitor diminishing contamination levels. In order to facilitate detection, technologies for bioactivity analysis are often optimized either for rapid detection of biological agents or for identification.

20b.3.3 Stand-off detectors, point detectors, light scattering

For practical reasons, terrorists would discharge bioweapons as aerosols or particles, which may be mixed with dispersion aids or additional toxic materials [24]. The agents may be released as suspended aerosolized masses (clouds) in external environments such as military facilities or public arenas, may be transmitted by a more contained mechanism such as the HVAC systems of buildings, or may be dispersed through contaminated items such as mail. When suspicious clouds are observed, a first response might be to probe from a distance for the presence of biological agents such as spores or live bacteria mixed with inert material. Remotely operated instruments, referred to as *stand-off detectors*, detect biological material in a cloud by evaluating the general physical characteristics of the particulates. Methods to remotely detect aerosols of particles or droplets include Doppler radio detection and LIDAR (light detection and ranging) [12]. Spores or microbes have physical properties (characteristic shape, uniformity of size) and chemical properties—such as fluorescence due to ATP, NADH, tryptophan, riboflavin—that can be detected by such light-scattering instruments.

In contrast to stand-off detectors, the sampling mechanisms for *point detectors* must directly take in the air to be tested so that the sample must be collected at or within the test area. Point detectors can be transported into large suspicious clouds on trucks or carried by personnel to the point of measurement. Stationary or hand-carried units are used for evaluating biological threats in interior areas or surfaces. For high-risk buildings or public gathering spots, the ideal for DTW would be to have continuously sampling sensors placed in the various places, which would activate an alarm when a positive result is found. Such devices are referred to as biological "smoke detectors" [24]. Another option is the Biological Aerosol Sentry and Information System (BASIS) developed by the Lawrence Livermore and Los Alamos National Laboratories. This is a detect-to-treat system in which individual air samplers are placed in high-risk areas. The units pull air through filters, which retain particulates for analysis at regular intervals. Mobile laboratory units use PCR methods to identify pathogens with a high degree of reliability.

One type of point detector, the Aerosol Particle Sizer (APS), inhales air with a high-speed sampler, then counts and sizes the particulates. In the APS the counting and sizing of particles is done with a flow cytometer, which uses a precision fluidics system and a laser to count individual particles. The light scattered by the particle is gathered by multiple detectors. Flow cytometers have also been reduced in size to bench-top instruments, which can be easily transported to the field. Other configurations for point detectors include hand-held mobile units that can be used to scan contaminated internal or external environments, and units with tiered detection. In the latter case, the system contains four components—a trigger (and/or cue) which detects particulates in near real time, the collector, a detector, and an identifier [3]. The *trigger* is the sentinel unit. When it senses an increase in particulates and, in some cases, the cueing device senses the fluorescence of biomolecules such as tryptophan, ATP, NADH, or riboflavin, the rest of the detector components are activated. The *collector* filters air and concentrates the particulates in water. The *detector* is a flow cytometer that further examines the material for biological material, which is then passed to the identifier. The *identifier*, an immunoassay device, makes the final determination based on a set of pre-selected possibilities. These point detection systems are fairly large and are mobilized on armored trucks to enter suspected aerosol clouds. From trigger to detection takes 4 min, and the final immunoassay reports in 20 min [22].

Chemical sensors offer great flexibility and possibilities as bioactivity detectors, which can be distributed throughout buildings or vulnerable areas. The two basic components of such sensors are the substrate with sites that recognize and bind the specific analyte and a transducer that produces an electric signal in response to the binding. The variety of sensors and transducers provide flexibility in producing detectors. Common sensors include those which rely on biological recognition mechanisms such as antibody–antigen interactions and DNA hybridization to a probe. Such sensors will have a high degree of specificity for the target structure; however, possible fouling of the sensitive surface will create problems for continuous monitoring devices. Single-use detectors or those with renewable surfaces may be a better option inasmuch as the harsh conditions needed to dissociate the analyte from the antibody or DNA hybridization fragment may limit its lifetime.

REFERENCES

1   C.L. Meays, K. Broersma, R. Nordin and A. Mazumder, Source tracking fecal bacteria in water: a critical review of current methods, *J. Environ. Manage.*, 73 (2004) 71–79.

2 T. Cieslak and E.M. Eitzen, Bioterrorism: agents of concern, *J. Pub. Health Manage. Prac.*, 6 (2000) 19–29.

3 A.A. Fatah, J.A. Barrett, Jr., R.D. Arcilesi, K.J. Ewing, C.H. Lattin and T.F. Moshier, An introduction to biological agent detection equipment for emergency first responders. *NIJ Guide*, (2001) 7–12.

4 US FDA, ITG subject: bacterial endotoxins/pyrogens. In: Office of Regulatory Affairs (Ed.), *Inspector's Technical Guide*, Washington, DC, 1987.

5 US FDA Center for Food Safety and Applied Nutrition, Bacteriological Analytical Manual Online, 2001.

6 U.S. Environmental Protection Agency, Water science: analytical methods. EPA 821-B-03-004, 2005.

7 USDA Food Safety and Inspection Service, Microbiology Laboratory Guidebook, 2004.

8 USDA Microbiological Data Program, SOPs for laboratory activities, 2005. http://www.ams.usda.gov/science/MPO/SOPs.htm

9 N. Bouaïcha, I. Maatouk, G. Vincent and Y. Levi, A colorimetric and fluorometric microplate assay for the detection of microcystin-LR in drinking water without preconcentration, *Food Chem. Toxicol.*, 40 (2002) 1677–1683.

10 S.S. Iqbal, M.W. Mayo, J.G. Bruno, B.V. Bronk, C.A. Batt and J.P. Chambers, A review of molecular recognition technologies for detection of biological threat agents, *Biosens. Bioelectron.*, 15 (2000) 549–578.

11 J.A. Higgins, M.S. Ibrahim, F.K. Knauert, G.V. Ludwig, T.M. Kijek, J.W. Ezzell, B.C. Courtney and E.A. Henchal, Sensitive and rapid identification of biological threat agents, *Ann. NY Acad. Sci.*, 894 (1999) 130–148.

12 M.E. Kosal, The basics of chemical and biological weapons detectors, 2003. http://cns.miis.edu/pubs/week/031124.htm

13 P.E. Andreotti, G.V. Ludwig, A.H. Peruski, J.J. Tuite, S.S. Morse and L.F. Peruski Jr., Immunoassay of infectious agents, *BioTechniques*, 35 (2003) 850–859.

14 J. Liu, D. Xing, X. Shen and D. Zhu, Electrochemiluminescence polymerase chain reaction detection of genetically modified organisms, *Anal. Chim. Acta*, 537 (2005) 119–123.

15 X.B. Yin, S. Dong and E. Wang, Analytical applications of the electrochemiluminescence of tris(2,2'-bipyridyl) ruthenium and its derivatives, *Trends Anal. Chem.*, 23 (2004) 432–441.

16 V.A. Petrenko and V.J. Vodyanoy, Phage display for detection of biological threat agents, *J. Microbiol. Meth.*, 53 (2003) 253–262.

17 J.R. Uhl, C.A. Bell, L.M. Sloan, M.J. Espy, T.F. Smith, J.E. Rosenblatt and F.R. Cockerill, Application of rapid-cycle real-time polymerase chain reaction for the detection of microbial pathogens: the Mayo–Roche rapid anthrax test, *Mayo Clin. Proc.*, 77 (2002) 673–680.

18 M.J. Espy, J.R. Uhl, L.M. Sloan, J.E. Rosenblatt, F.R. Cockerill and T.F. Smith, Detection of Vaccinia virus, Herpes Simplex virus, Varicella-Zoster

virus, and *Bacillus anthracis* DNA by lightcycler polymerase chain reaction after autoclaving: implications for biosafety of bioterrorism agents, *Mayo Clin. Proc.*, 77 (2002) 624–628.

19   M. Panning, M. Asper, S. Kramme, H. Schmitz and C. Drosten, Rapid detection and differentiation of human pathogenic orthopox viruses by a fluorescence resonance energy transfer real-time PCR assay, *Clin. Chem.*, 50 (2004) 702–708.

20   D.A. Kulesh, R.O. Baker, B.M. Loveless, D. Norwood, S.H. Zwiers, E. Mucker, C. Hartmann, R. Herrera, D. Miller, D. Christensen, L.P. Wasieloski Jr., J. Huggins and P.B. Jahrling, Smallpox and pan-orthopox virus detection by real-time 3′-minor groove binder TaqMan assays on the Roche lightcycler and the Cepheid smart cycler platforms, *J. Clin. Microbiol.*, 42 (2004) 601–609.

21   G. Beadie, J. Reintjes, M. Bashkansky, T. Opatrny and M.O. Scully, Towards a FAST-CARS anthrax detector: CARS generation in a DPA surrogate molecule, *J. Mod. Opt.*, 50 (2003) 2361–2368.

22   R.C. Spencer and N.F. Lightfoot, Preparedness and response to bioterrorism, *J. Infect.*, 43 (2001) 104–110.

23   F. Basile, M.B. Beverly, K.J. Voorhees and T.L. Hadfield, Pathogenic bacteria: their detection and differentiation by rapid lipid profiling with pyrolysis mass spectrometry, *Trends Anal. Chem.*, 17 (1998) 95–108.

24   National Research Council (NRC), *Sensor Systems for Biological Agent Attacks: Protecting Buildings and Military Bases*, National Academies Press, Washington, DC, 2005.

25   J.P. Dworzanski, W.H. McClennen, P.A. Cole, S.N. Thornton, H.L.C. Meuzelaar, N.S. Arnold and A.P. Snyder, Field-portable, automated pyrolysis-GC/IMS system for rapid biomarker detection in aerosols: A feasibility study, *Field Anal. Chem. Technol.*, 1 (1998) 295–305.

26   R.T. Vinopal, J.R. Jadamec, P. deFur, A.L. Demars, S. Jakubielski, C. Green, C.P. Anderson and J.E.D.R.F. Dugas, Fingerprinting bacterial strains using ion mobility spectrometry, *Anal. Chim. Acta*, 457 (2005) 83–95.

27   R. Dutta, E. Hines, J. Gardner and P. Boilot, Bacteria classification using Cyranose 320 electronic nose, *BioMed. Eng. OnLine*, 1 (2002) 4.

28   E.T. Arakawa, N.V. Lavrik and P.G. Datskos, Detection of anthrax stimulants with microcalorimetric spectroscopy: *Bacillus subtilis* and *Bacillus cereus* spores, *Appl. Opt.*, 42 (2003) 1757–1762.

29   J.M. Teich, M.M. Wagner, C.F. Mackenzie and K.O. Schafer, The informatics response in disaster, terrorism, and war, *J. Am. Med. Inform. Assoc.*, 9 (2002) 97–104.

REVIEW QUESTIONS

1.   Describe some instruments based on molecular recognition.
2.   Describe some methods to detect bioweapons.

Chapter 20c

Drug detectors

Sue M. Ford

20c.1 DRUG DETECTION IN THE FIELD

20c.1.1 Background considerations

Drug detection technologies are used in various law enforcement scenarios that present challenges for instrumentation development. Not
only do the drugs which must be detected vary, but the amounts to be
detected range from microgram quantities left as incidental contamination from drug activity to kilogram amounts being transported in
hidden caches. The locations of hidden drugs create difficulties in detection as well. Customs and border agents need drug detection technology to intercept drugs smuggled into the country in bulk cargo containers or carried by individuals. Correctional facility personnel search
for drugs being smuggled into the prison by mail (e.g. small amounts
under stamps) or by visitors, and furthermore need to monitor drug
possession inside the prison. Law enforcement representatives may use
detectors in schools to find the caches of student dealers. Other customers for drug detection technology include aviation and marine carriers, postal and courier services, private industry, and the military. In
all these cases, the problem is to reveal the relatively rare presence of
drugs in a population of individuals or items, most of which will be
negative. Consequently, for drug detection purposes the ability to find
illicit contraband is of greater importance than accurate quantitation.

20c.1.2 Desirable characteristics of field detectors

Drug detectors are used on-site to screen items in transit, such as cargo,
pieces of mail, or personal belongings of individuals. The units must be
mobile or portable. Inasmuch as innumerable objects need to be scru-
tinized rapidly in order to minimize delays and inconvenience, the

Comprehensive Analytical Chemistry 47
S. Ahuja and N. Jespersen (Eds)
Volume 47 ISSN: 0166-526X DOI: 10.1016/S0166-526X(06)47031-8

process of screening should ideally be non-invasive, non-destructive, and rapid, and should display results within seconds. Instrumentation for drug detection in the field must be simple enough to be used by trained non-scientific personnel and the output should be easily interpreted. Sample acquisition should be simple with little requirement of sample preparation. With the exception of the X-ray and CT scanners, which weigh 0.5–3 tons [1], instrumentation for drug detection in the field is easily transported, and includes mobile desktop units as well as hand-held instruments. On-site drug detection necessitates all phases of analysis to be completed in the field, including sample collection, processing for presentation to the analytical module, identification, and quantitation.

20c.1.3 Sampling

The ability to successfully discover trace amounts of contraband depends on the sampling methodology as well as the detector. Drug residue detection on individuals, luggage and personal effects, mail, currency, and other items such as newspaper may be assessed by particulate detection. The surfaces are swiped with a swab, pad, or paper disk (sample trap), or the particles are collected on a filter by moving a vacuum over the surface. For screening of multiple items, it may be sufficient to swipe several times at a time with individual analysis needed only if a positive result is obtained. The swab or trap may be extracted with solvent and the sample then introduced into the instrument. In some cases, samples collected on a Teflon^{TM} pad are heated in the instrument in order to volatilize the analyte. Most instruments operate either as vapor detectors or particle analyzers, although there are a few that can operate in both modes. Newer sampling devices, such as those employed by the US Transportation Security Administration (produced by Smiths Detection (London, England and Pine Brook, NJ) for explosives detection, blow puffs of air at passengers from head to foot. Particles which subsequently fall to the ground are collected by being sucked into vents for chemical analysis.

20c.1.4 Background contamination

Detection of drug residues that contaminate surfaces and vapors seeping from caches requires that the sensitivity of such methods is generally in the sub-microgram range. Individuals who have handled drugs may have residues on their hands in the range of 1–10 µg [2], which can remain at least 90 min after handling and handwashing [3]. Another application of trace detection is in screening currency, suspected of being used in illicit drug activity. Although the instrument should pick up the low quantities expected from handling in such situations, in some metropolitan areas most currency is contaminated with drugs and an instrument with the threshold set too low will generate many false positives.

20c.2 INSTRUMENTATION

20c.2.1 Bulk detection

Instrumentation for revealing the presence of bulk quantities of concealed drugs will differ from those developed to find evidence of minute quantities on surfaces. Bulk detection is concerned with amounts ranging from grams to kilograms [4]. Bulk detection is done by manual inspection, X-ray, CT scans, and acoustic inspection. X-ray or CT scanners used as bulk detectors have sensitivity of 2–10 g, and suspect items are subsequently confirmed by chemical analysis. Hand-held acoustic inspection instruments such as the Acoustic Inspection Device (AID) and the Ultrasonic Pulse Echo (UPE) developed by Pacific Northwest National Laboratories/Battelle, can be used for analysis of cargo liquids in sealed containers of various sizes within seconds [5]. The acoustical velocity and attenuation of multiple echoes returned to the instrument is evaluated by software which compares the data to the shipping manifest.

20c.2.2 Trace detection

20c.2.2.1 Wipe and spray
For evaluating a small number of samples, non-instrument-based kits are available which are rapid and do not require sample cleanup. Wipe and spray detection kits such as those from the Mistral Group (Bethesda, MD) can be purchased for single drugs, multiple drugs, and explosives. The surfaces to be tested are swiped with paper sample traps, which are then sprayed with a reagent which turns the color depending on the compound(s) present. The reagents have been modified to be stable in the kit, for example, a modified Fast Blue BB reagent for cannabinoids and a modified cobalt thiocyanate (Scott) reagent for cocaine. The sensitivity of these kits is in the 1–10 µg range.

20c.2.2.2 Vapor detection

Some contraband drugs have sufficient vapor pressure to be detected by vapor detectors ("sniffers," electronic noses), which draw in air around the test area and introduce the sample directly into the instrument. With the ZNose™ (Electronic Sensor Technology, LP, Newbury Park, CA), the components of the vapor are separated by a fast (10 s) gas chromatography system. The detector contains molecular recognition sensor arrays based on surface acoustic wave (SAW) technology. The individual sensors each contain a chemoselective polymer sorbent applied to the surface of acoustic wave transducers [6]. As each vapor component interacts with a particular polymer, the properties of the polymer change (mass, viscoelasticity, swelling), which in turn perturbs the velocity of the SAW. The resulting shift in resonance frequencies of multiple sensors contributes to the pattern of output. Electronic noses provide a distinctive visual output, similar to an inkblot, which can be used to identify complex vapors by the shape of the "blot" or print.

Detection of vapors seeping from closed containers may be sufficiently sensitive that opening the container may not be necessary. Vapor detectors work for drugs such as heroin and marijuana which have volatile components with sufficient vapor pressure for detection, whereas particle detection and swiping are more appropriate for other drugs. However, drugs without appreciable vapor pressure may be detectable in warmer temperatures or may have components that are distinctive enough to comprise an odor signature. For example, methyl benzoate is the component of street cocaine that dogs recognize and represents a signature molecule that can be detected with instrumentation.

20c.2.2.3 Biosensors

Biosensors represent another technique for simple and rapid analyses in the field (see Chapter 20a). Biosensors utilize proteins, including antibodies, enzymes, or receptors, which have affinities for specific chemicals. The proteins are incorporated into devices which produce a detectable response (e.g. color change) proportional to the amount of binding of analyte to the protein. Figure 20c.1 shows the general procedure for an enzyme-linked immunosorbant assay (ELISA). The proteins are chosen to bind specific compounds, such as the use of antibodies to detect cocaine or the enzyme acetylcholinesterase to detect organophosphate pesticides [7]. However, there is some degree of cross-reactivity. In some cases this may be advantageous in detecting structurally related derivatives such as metabolites [8]. Biosensors may

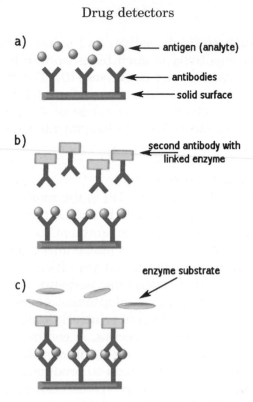

a) — antigen (analyte)
— antibodies
— solid surface

b) — second antibody with linked enzyme

c) — enzyme substrate

Fig. 20c.1. ELISA assay. (a) Antibodies to the drug of interest are secured to a solid substratum such as a test tube or micro-well plate. The sample containing the analyte antigen is added to the reaction surface. (b) After the analyte has bound to the antibody, the vessel is rinsed to remove unbound antibody. A second antibody to the analyte is added. This antibody has a bound enzyme which has been chosen because its reaction produces a colored product which can be detected spectrophotometrically. (c) After this second antibody has bound to the first antibody–antigen complex, the surface is again rinsed to remove unbound-antibody enzyme. The enzyme substrate is added in sufficient excess such that the *rate* of product formed is proportional to the amount of enzyme present. The enzyme-linked assays are very sensitive, since each enzyme can rapidly catalyze thousands of substrate to product reactions.

be incorporated into electronic equipment using piezoelectric transduction [7,9], in which case the binding event of antigen to antibody causes a change in the frequency of the signal from a piezoelectric transducer. A simpler, yet effective, use of immunoassay is illustrated by the single-use disposable devices such as the Drugwipe®. The Drugwipe® (Securetec Detektions-Systeme AG, Germany) is an immunoassay-based biosensor device that is available for opiates, cocaine, cannabis, and amphetamines. It can be used to assay surfaces as well as bodily fluids such as sweat, saliva, and urine. Limited testing suggests that the low sensitivity of such biosensors in biological fluids results in a high incidence of false negatives [10]. The method is about an order of magnitude more sensitive for surface detection [11]. The detection limits for cocaine are in the range of 34 ng/l (in water) for a piezoelectric immunosensor [7,9] to 50 ng/ml (in sweat, saliva) [11] for Drugwipe®.

20c.2.2.4 Ion mobility spectrometry

Ion mobility spectrometry (IMS) [3,12] is the most widely used instrument for drug detection. The sample is heated to vaporize the analyte, which is then ionized by atmospheric (ambient) pressure chemical ionization (APCI) [3]. The resulting gas-phase ions travel through a drift tube and are separated by their distinct velocities (mobilities) in a weak electrostatic field. IMS instruments use ambient air or nitrogen as the carrier gas, making it particularly adaptable to field applications.

The ionization process for IMS is relatively mild, such that little or no fragmentation of the analyte occurs. A common ionization source for APCI is ^{63}Ni, which is coated on a foil of gold or nickel [3]. The β-particle emitted reacts with nitrogen to produce a cascade of positive and negative ions, and secondary electrons. Collisions and subsequent reactions can result in proton transfer reactions and formation of clusters, dimers, and adducts. The drift gas for drug detection contains trace amounts of nicotinamide, which becomes protonated and can transfer the proton to the sample molecules [13]. ^{63}Ni as an ionization source does limit the linear range and selectivity; however, it is advantageous for field equipment because it does not require an external power supply, it is not prone to malfunction, and it has good specificity for its intended analytes. Other ionization methods, including photoionization, corona discharge, flame ionization, laser ionization, surface ionization, and electrospray ionization provide opportunities for extending the range and selectivity of IMS [3,14].

The linearity of IMS is not sufficient for accurate quantitation of analytes. However, detection of contraband essentially requires a positive/negative response, so that for this purpose linearity is not critical. The method is sensitive (sub-nanogram range) and has sufficient resolution ability [15,16]. The instrument can be adjusted to provide even better resolution for specific analytes of interest (e.g. drugs vs. explosives). These features, along with the reliability of the instrument, have made it the method of choice for field operations. IMS instrumentation has undergone numerous modifications that are promising for field detection technology, including miniaturization to palm-sized units, redesign of drift tubes to increase resolution [17–19], inclusion of a GC column or GS/MS as a pre-separator to reduce matrix effects [3,20], and improvements in sampling techniques [3,14]. Smiths Detection has recently introduced the Ionscan 500DT, which contains two IMS detectors and allows for simultaneous detection of narcotics and explosives in a single analysis.

20c.2.2.5 Mass spectrometry
Mass spectrometry (MS) is a reliable laboratory method for identification and analysis of organic compounds (see Chapter 11). Successive generations of instruments have increased the power of analysis by interfacing MS with separation techniques such as gas chromatography. The advantages of MS methodologies over IMS include versatility in sample analysis, improved linearity, and accurate quantitation when necessary. The size of mass spectrometers has been reduced such that it is now feasible to transport the instruments for analysis in the field [21,22]. Although the performance is generally inferior to laboratory-sited instruments, software can help minimize problems related to resolution. The limitations of MS compared to IMS are the bulk and complexity of instrumentation (e.g. vacuum, heating/cooling of GC column, power consumption) and more involved sample preparation. Additionally, the sampling and analysis times are longer, in the range of 15 min, compared to less than 20 s for IMS. MS technology is particularly valuable when specific identification of compounds is mandatory. The instrumentation size and the length of analysis preclude the use of MS for rapid screening of mail, luggage, and other high-throughput tasks.

20c.2.2.6 Raman spectroscopy
Raman spectroscopy is used to identify substances based on the inelastic scattering of light. When light is applied to a molecule, the energy change between incident and scattered light depends on the structure of the molecule. Drugs and other complex molecules have many possibilities for interaction with light, and the spectrum of scattered light contains a characteristic pattern of peaks ("fingerprint") that can be used to identify substances of interest [2]. The process is rapid, non-destructive, and requires minimal sample preparation [23]. The instruments can be configured with fiber optic sample probes to detect chemicals on the ground and other surfaces. One application is to detect drugs in fingerprints while preserving the prints for other

analyses [24,25]. It will also work through certain closed containers such as dark glass, plastic bags, and thin plastic containers. This, in particular, makes it more useful for field applications than other spectroscopic methods such as FT-IR. A database of spectra is needed to determine whether compounds of interest are present or absent. FT-Raman spectra libraries are commercially available, including for illicit drugs [23]. Portable instruments suitable for field use are available by InPhotonics (InPhotote™), GE StreetLab™ (GE Ion Trak, Plainville, CT), RSL-1 Raman (Spectrolab, Berkshire, England), and RA series (Renishaw plc, Gloucestershire, UK). Software is available to dissect the spectra of interfering materials which may be mixed with the target substances.

REFERENCES

1   D. Blackburn, Counterdrug technology development program. Identification, demonstration, and assessment of drug detection technology, 2003.

2   J.E. Parmeter, D.W. Murray and D.W. Hannum, Guide for the selection of drug detectors for law enforcement applications, *NIJ Guide*, 601-00 (2000) 1–53.

3   G.A. Eiceman and Z. Karpas., *Ion Mobility Spectrometry*, CRC Press, Boca Raton, 1994.

4   S. Wright and R.F. Butler, Technology takes on drug smugglers. Can drug detection technology stop drugs from entering prisons?, *Corrections Today*, 63 (2001) 66–68.

5   A. Diaz, Acoustic inspection and analysis of liquids in sealed containers, *Sensors*, 20 (2003) 14.

6   E.J. Houser, T.E. Mlsna, V.K. Nguyen, R. Chung, R.L. Mowery and A. McGill, Rational materials design of sorbent coatings for explosives: applications with chemical sensors, *Talanta*, 54 (2001) 469–485.

7   J. Halamek, A. Makower, K. Knosche, P. Skladal and F.W. Scheller, Piezoelectric affinity sensors for cocaine and cholinesterase inhibitors, *Talanta*, 65 (2005) 337–342.

8   P.J. Devine, N.A. Anis, J. Wright, S. Kim, A.T. Eldefrawi and M.E. Eldefrawi, A fiber-optic cocaine biosensor, *Anal. Biochem.*, 227 (1995) 216–224.

9   J. Halamek, A. Makower, P. Skladal and F.W. Scheller, Highly sensitive detection of cocaine using a piezoelectric immunosensor, *Biosens. Bioelectron.*, 17 (2002) 1045–1050.

10 P. Kintz, V. Cirimele and B. Ludes, Codeine testing in sweat and saliva with the Drugwipe, *Int. J. Legal Med.*, 111 (1998) 82–84.

11 R. McCullough, The use of the Drugwipe in a drug court setting. *National Law Enforcement and Corrections Technology Center Rocky Mountain Regional Center, 5th Annual Innovative Technologies for Community Corrections Conference*, Boston, MA, June 14–16, 2004.

12 T. Keller, A. Schneider, E. Tutsch-Bauer, J. Jaspers, R. Aderjan and G. Skopp, Ion mobility spectrometry for the detection of drugs in cases of forensic and criminalistic relevance, *Int. J. Ion Mobility Spectrom.*, 2 (1999) 22–34.

13 T. Keller, A. Miki, P. Regenscheit, R. Dirnhofer, A. Schneider and H. Tsuchihashi, Detection of designer drugs in human hair by ion mobility spectrometry (IMS), *Forensic Sci. Int.*, 94 (1998) 55–63.

14 F. Li, Z. Xie, H. Schmidt, S. Sielemann and J.I. Baumbach, Ion mobility spectrometer for online monitoring of trace compounds, *Spectrochim. Acta-B*, 57 (2002) 1563–1574.

15 L.M. Matz and H.H. Hill Jr., Separation of benzodiazepines by electrospray ionization ion mobility spectrometry-mass spectroscopy, *Anal. Chim. Acta*, 457 (2002) 235–245.

16 J.A. McLean, B.T. Ruotolo, K.J. Gillig and D.H. Russell, Ion mobility-mass spectrometry: a new paradigm for proteomics, *Int. J. Mass Spectrom.*, 240 (2005) 301–315.

17 I.A. Buryakov, Qualitative analysis of trace constituents by ion mobility increment spectrometer, *Talanta*, 61 (2003) 369–375.

18 I.A. Buryakov, Express analysis of explosives, chemical warfare agents and drugs with multicapillary column gas chromatography and ion mobility increment spectrometry, *J. Chromatogr. B*, 800 (2004) 75–82.

19 R.R. Kunz, W.F. Dinatale and P. Becotte-Haigh, Comparison of detection selectivity in ion mobility spectrometry: proton-attachment versus electron exchange ionization, *Int. J. Mass Spectrom.*, 226 (2003) 379–395.

20 A.M. DeTulleo, P.B. Galat and M.E. Gay, Detecting heroin in the presence of cocaine using ion mobility spectrometry, *Int. J. Ion Mobility Spectrom.*, 1 (2000) 38–42.

21 P.A. Smith, M.T. Sng, B.A. Eckenrode, S.Y. Leow, D. Koch, R.P. Erickson, C.R.J. Lepage and G.L. Hook, Towards smaller and faster gas chromatography–mass spectrometry systems for field chemical detection, *J. Chromatogr. A*, 1067 (2005) 285–294.

22 A.L. Makas and M.L. Troshkov, Field gas chromatography–mass spectrometry for fast analysis, *J. Chromatogr. B*, 800 (2004) 55–61.

23 H. Tsuchihashi, M. Katagi, M. Nishikawa, M. Tatsuno, A. Nishioka, A. Nara, E. Nishio and C. Petty, Determination of methamphetamine and its related compounds using Fourier Transform Raman spectroscopy, *Appl. Spectrosc.*, 51 (1997) 1796–1799.

24 J.S. Day, H.G.M. Edwards, S.A. Dobrowski and A.M. Voice, The detection of drugs of abuse in fingerprints using Raman spectroscopy II:

cyanoacrylate-fumed fingerprints, *Spectrochim. Acta Part A: Mol. Biomol. Spectrosc.*, 60 (2004) 1725–1730.

25 J.S. Day, H.G.M. Edwards, S.A. Dobrowski and A.M. Voice, The detection of drugs of abuse in fingerprints using Raman spectroscopy I: latent fingerprints, *Spectrochim. Acta Part A: Mol. Biomol. Spectros.*, 60 (2004) 563–568.

REVIEW QUESTIONS

1. Describe some of the desirable characteristics of field detectors.
2. Describe briefly operation of biosensors.

Chapter 21

Problem solving and guidelines for method selection

Alan H. Ullman and Douglas E. Raynie

Analytical chemistry is designed to answer the questions, "What is in this sample and how much is there?" It is an important part of how we perceive and understand the world and universe around us, so it has both very practical and also philosophical and ethical implications. In general, in this chapter we will be more concerned with the analytical approach rather than being primarily technique-oriented, even when techniques are discussed. It is expected that you will further develop your critical-thinking and problem-solving skills and, at the same time, gain more insight into an understanding and appreciation of the physical world surrounding you. In the previous chapters of this text, you learned the principles underlying the majority of measurement techniques used in analytical chemistry labs around the world. In this chapter, we introduce the kinds of thinking needed to solve real-world problems, including how to select the "best" technique for the problem. Let us begin the discussion with an example.

21.1 THE BULGING DRUMS [1]

As an analytical chemist for a company that formulates industrial cleaning products, such as detergents for automobile mechanic uniforms, you receive a call from Tom, one of your colleagues. He has just completed a routine walk-through of one of your company's warehouses and is extremely agitated about his findings. He informs you that many of the 55-gallon drums of ethoxylated alcohol (EO) are bulging; rather than "sitting" flat on the ground, their tops and bottoms are rounded and the approximately 400-lb drums can be caused to rock with a gentle

Comprehensive Analytical Chemistry 47
S. Ahuja and N. Jespersen (Eds)
Volume 47 ISSN: 0166-526X DOI: 10.1016/S0166-526X(06)47021-5

push. He is extremely worried because he used an explosimeter[1] to test the vapor in one of the drums and the test was positive! This is a critical problem because of the safety implications. Only secondary is the concern that the raw material might be unusable, delaying production. Tom asks for your assistance in understanding the cause of the bulging drums and the explosive vapor in them.

When you were less experienced as an analytical chemist with the company, your first instinct would have been to rush out and analyze something, but you have learned to stop and think, and plan your next steps. Since you have not worked on this particular detergent or its raw materials in a while, you pull up some information from company records on your computer. The company buys EO from two suppliers; both use the same synthesis, described in reaction 21.1.

21.1 Synthesis of ethoxylated alcohol (EO)

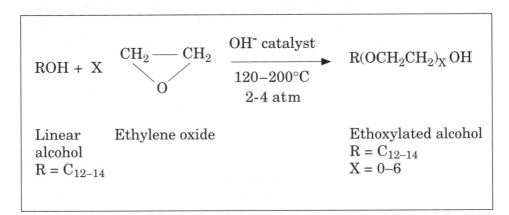

After studying this reaction, you wonder if this is a simple situation of excess ethylene oxide (ETO) dissolved in the ethoxylate. That would explain the bulging since Tom said the drums that bulged were in the warmer part of the warehouse. Maybe we just have gaseous ETO coming out of solution because of the higher temperature, building pressure, and causing the bulging. ETO would be explosive, so the observations can be explained by this hypothesis.

The simplest way to prove this hypothesis would be to find ETO in the vapor. If Tom can sample another drum with a gas-tight sampling device and get it to you, you can do IR or GC on the sample. GC on the

---

[1]Explosimeter: a device that determines the combustibility of a sample by drawing a sample of gas across a hot wire. Combustible gases ignite causing a measurable change in the resistance of the wire. The measurement is not specific for any compound.

vapor would be more sensitive and we have an instrument in Alan's lab that is already configured for such an analysis.

Unfortunately, analysis of the sample headspace (the gas above the liquid in the drum) by GC does not show any ETO. The easy hypothesis has been disproven! You need another hypothesis of what is happening—and time is of the essence!

Further research in the library and discussion with other chemists in the company leads you to a new mechanism: autooxidative degradation. Fatty chemicals are known to undergo such degradation with the formation of a series of compounds, some of which are volatile and potentially explosive (Table 21.1). These reactions would occur more readily at elevated temperatures and in the presence of trace metals, such as iron, cobalt, and nickel.

In this case, the fastest analytical results would be obtained using methods already in use in the lab. Since this proposed hypothesis would result in degradation of the bulk EO, analytical testing of the liquid to show such degradation of the EO or the presence of higher-than-normal levels of the degradation products would support the hypothesis. Tom sends samples of the EO from several drums for analysis. A series of wet chemistry and instrumental tests are used on the samples sent by Tom. In each case, the tests were selected because they were available in the lab and would indicate either degraded ethoxylate or the products of the oxidation reaction. Because of the urgency (safety!) of the matter, tests were performed simultaneously by several different analysts. Results for the tests are shown in Tables 21.2 and 21.3.

The hydroxyl value (or hydroxyl number) method measures the number of –OH groups on a molecule. The result is reported in milligram of potassium hydroxide titrated by a sample of 1 g size; thus it is a measure of both the chain length/molecular weight and purity of a

TABLE 21.1

Possible products of auto-oxidative degradation of EO

Ethoxylated alcohol—fewer EO units
Esters
Polyethylene glycols
Peroxide intermediates
Formaldehyde
Formic acid
Ethylene

Source: Reprinted/retyped with permission from *Analyt. Chem.*, **56**, 603A (1984). Copyright 1984 American Chemical Society.

TABLE 21.2

Wet chemical analyses to detect degradation products

Analytical method	Description	Result
Hydroxyl number	Titrimetric determination of alcohol content	174–177 mg KOH/g alcohol (molecular weight 317–322)
Peroxide value	General quantitative measure of peroxides or hydroperoxides produced by oxidative degradation	0
Karl Fisher moisture	Titrimetric determination of H_2O	1.71–2.63%
pH	Hydrogen ion concentration	9.5–10.5
Color	Visual examination typical of acceptable material	Colorless to yellow

Source: Reprinted/retyped with permission from *Analyt. Chem.*, **56**, 603A (1984). Copyright 1984 American Chemical Society.

TABLE 21.3

Summary of analytical results to detect degradation of ethoxylated alcohol

Analytical method	Result	
Wet chemical analyses	Degradation products not detected	
Gas chromatography	Suspect product qualitatively identical to "good" sample; average carbon chain = 12.6	
Atomic absorption	Fe	0.5 ppm
	Co	0.6 ppm
	Ni	< 3 ppm
Headspace gas chromatography	Gaseous components dissolved in "suspect" product not detectably different from those in "good" product	

Source: Reprinted/retyped with permission from *Analyt. Chem.*, 56, 603A (1984). Copyright 1984 American Chemical Society.

sample. It is a two-step titration shown by the chemistry in reactions 21.2a, 21.2b and 21.3.

21.2a. Preliminary reaction (hydroxyl groups are converted into methyl esters with formation of hydrochloric acid).

$$R-(OCH_2CH_2)_x-OH + CH_3COCl \longrightarrow$$

$$R-(OCH_2CH_2)_x-O\overset{\overset{\displaystyle O}{\|}}{C}-CH_3 + HCl + CH_3COCl$$

21.2b. Hydrolysis reaction (excess acetyl chloride is converted into acetic and hydrochloric acids).

$$CH_3COCl + H_2O \rightarrow CH_3COOH + HCl$$

21.3. Titration reaction (the acetic and hydrochloric acids are titrated with sodium hydroxide).

$$\left\{ \begin{matrix} CH_3COOH \\ HCl \end{matrix} \right\} + 2NaOH \rightarrow CH_3COONa + NaCl + 2H_2O$$

The peroxide value measurement is a general quantitative measure of peroxide/hydroperoxide intermediates produced during oxidative degradation.

21.4a. Preliminary reaction (peroxides oxidize iodide ion and become reduced).

$$\text{Peroxide Intermediate} + \underset{\text{excess}}{3I^-} \rightarrow I_3^-$$

$$+ \text{Reduced Peroxide Intermediates}$$

21.4b. Titration reaction (the iodine complex is titrated with thiosulfate to a starch endpoint).

$$I_3^- + 2S_2O_3^{2-} \rightarrow S_4O_6^{2-} + 3I^-$$

The Karl Fischer titration is a generally accepted method for the determination of moisture in a sample [2].

Results for the hydroxyl value determination were inconclusive; the precision of the method and the normal range of results for receipts of EO were not significantly different. Combined with the finding of "zero" for the peroxide value, the oxidation hypothesis seems less likely. Moisture content and pH results were more interesting. Normally, pH readings on the ethoxylate were above 10. The moisture values were somewhat above "typical" values. These two results do not really support or refute the hypothesis of EO oxidation, but when solving a problem, any unusual or unexpected results are noteworthy.

Two different instrumental measurements were used to test the oxidation hypothesis: GC and atomic absorption (AA). GC was used to determine the chain length distribution of the EO. From the GC results, an average chain length number was calculated. The gas chromatogram is shown in Fig. 21.1. In addition, headspace GC measurement of the gaseous compounds dissolved in the liquid ethoxylate was made and compared to typical or "good" product. Atomic absorption spectroscopy was used to determine the levels of several trace metals known to catalyze oxidation of fatty chemicals.

Results of the instrumental measurements did not support the degradation hypothesis. The chain length GC measurement of the ethoxylate was not appreciably different from "normal" product. The average carbon number of 12.6 was in the expected range of 12–13. Comparison of headspace gas chromatograms of the dissolved volatile components in the bulging drums and "normal" raw material did not show any appreciable differences. The levels of the three heavy metals were not so high as to be indicative of significant catalytic activity.

At this point, it is clear that no evidence supports the hypothesis of oxidative degradation of the raw material. None of the results was positive for the presumed degradation products, nor did the ethoxylate

Problem solving and guidelines for method selection

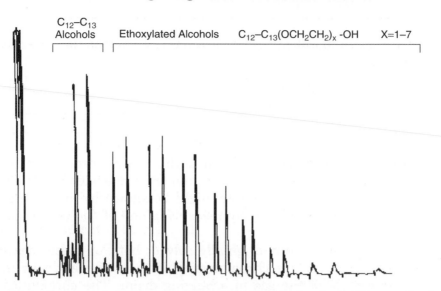

Fig. 21.1. Gas-liquid chromatogram of BSTFA-derivatized EO sample (column: 6-ft X 1/8-in o.d. stainless steel packed with 3% SP-2100 on 100/120 mesh Supelcoport; column programmed from 150°C to 350°C at 4°C/min; sample size: 2 μL). (Reprinted/redrawn with permission from *Analyt. Chem.*, **56**, 603A (1984). Copyright 1984 American Chemical Society.)

itself show any degradation. Thus, a significant amount of effort, in library research, conversations with colleagues, and in the lab has been expended and the bulging drums remain in the warehouse. While the warehouse staff has taken appropriate precautions, it is still imperative that the problem be understood so appropriate action can be taken.

Additional research in the company library, including both company reports and the external literature does not lead to any new information, but a lunchtime conversation with a colleague who recently returned from vacation does. Joanne remembered reading some sales literature from a company describing uses of sodium borohydride as an antioxidant, including its use in fatty chemicals. Could this be the culprit?

Additional library work, and a discussion with one of the company's organic chemists, leads to Hypothesis #3: sodium borohydride decomposing to form hydrogen gas. Chemically, this hypothesis may be written as reaction 21.5.

21.5 Sodium borohydride anti-oxidant decomposition reaction producing $H_2(g)$.

$$BH_4^- + 4H_2O \rightarrow B(OH)_4^- + 4H_2$$

as pH drops below 10

In order to prove or disprove this hypothesis, you need to find hydrogen in the bulging drum headspace and/or find evidence of sodium borohydride in the EO. The latter might be shown by measuring the element, boron, or by specifically measuring one of the boron species shown in reaction 21.6.

When this problem was originally solved, both of these measurements were made. A gas chromatographic measurement was made of another sample of the gas in a bulging drum. The chromatogram (Fig. 21.2) shows the presence of hydrogen—clearly an explosive gas! The amount of hydrogen gas in the drum was calculated to be 11.9%. This is the first data in support of any of the proposed hypotheses to date! To confirm the sodium borohydride mechanism, several additional experiments were performed. The concentration of boron in the ethoxylate was determined by atomic absorption spectroscopy and by mannitol titration. This titration is an acid–base titration that makes use of the ability of some polyhydroxy compounds, such as mannitol, to form a complex with certain weak acids. In this case, the mannitol complexes with boric acid. The complex titrates as if it was a stronger acid, thus allowing the endpoint to be more readily detected (reactions 21.6a–21.6c).

21.7 The mannitol titration for boric acid.

21.6a Preliminary reaction (sodium borohydride is reacted with acid to form boric acid).

$$H_3O^+ + BH_4^- \rightarrow 2H_2 + H_3BO_3$$

21.6b Complexation reaction (boric acid forms the mannitoboric acid complex).

$$H_3BO_3 + C_6H_{12}O_6 \rightarrow C_6H_{12}O_6 \bullet H_3BO_3$$

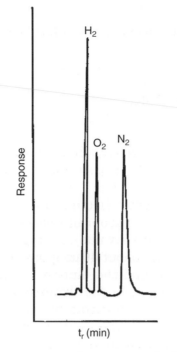

Fig. 21.2. Gas-solid chromatogram of vapor sample from bulging drum (column: 5-ft X 1/8 in o.d. aluminum packed with molecular sieve 5A 60–80 mesh; column temperature: ambient 25°C to 28°C; detector: thermal conductivity at 80°C; carrier gas: Ar at 15 cm³/min. (Reprinted/redrawn with permission from *Analyt. Chem.*, **56**, 603A (1984). Copyright 1984 American Chemical Society.)

21.6c Titration reaction (the boric acid is titrated with sodium hydroxide).

$$C_6H_{12}O_6 \cdot H_3BO_3 + NaOH \rightarrow C_6H_{12}O_6 \cdot H_2BO_3Na + H_2O$$

The two methods were in good agreement as shown in Table 21.4. These results allow calculation of the original sodium borohydride concentration (approximately 100 µg/g). Measurement of the remaining, unreacted sodium borohydride would permit calculation of how much borohydride had already reacted and how much hydrogen gas was evolved.

Remembering the basic gas laws of general chemistry, we can determine the remaining borohydride by reacting it quantitatively

TABLE 21.4

Summary of results from evaluation of NaBH4 breakdown

Analytical method	Result
Gas–solid chromatography (GSC)	11.9% H_2
Atomic absorption spectroscopy	28 µg/g B
Mannitol titration	30 µg/g B
Hydrogen evolution	60 µg/g $NaBH_4$

Source: Reprinted/retyped with permission from *Analyt. Chem.*, **56**, 603A (1984). Copyright 1984 American Chemical Society.

with dilute acid and carefully collecting the evolved hydrogen gas. The volume of hydrogen, after appropriate corrections for standard temperature and pressure, gave a value of 60 µg/g sodium borohydride in the drums. In other words, about 40 µg/g has reacted to form hydrogen gas. Based on this value and the amount of material in a drum, an estimate of ~14% hydrogen in the drum headspace was made. This agrees quite well with the result from the GSC.

Therefore, we have proven the sodium borohydride hypothesis. The different analytical methods have given us consistent results. The bulging is due to the release of hydrogen gas from the borohydride caused by excess water and slightly low pH.

These results must be shared (ASAP) with Tom in the warehouse. Together we conclude that careful venting of the drums and their prompt use would be safe. The raw material is perfectly acceptable— and has probably received more thorough analysis than any lots received in quite some time! Furthermore, an investigation, with the supplier of the ethoxylate, is undertaken to learn whether sodium borohydride has been used in our shipments in the past or if this was a one-time mistake. If it was not a mistake, can it be eliminated? In addition, the recommendation was made to change our raw material specification by reducing the amount of allowable moisture and raising the lower pH limit. Finally, a full report on this incident was written for the company files.

Thinking Questions:

1. Why did the first headspace GC not show the hydrogen, when the latter one did?
2. Was the Material Safety Data Sheet (MSDS) consulted? Would it have indicated the presence of sodium borohydride?

3. Could the supplier of the ethoxylate help? What issues need to be considered before consulting the company?

4. In Table 21.2 and on page 808, it is stated that the peroxide value was determined to be zero. Why is this incorrect?

5. Most laboratories doing organic analysis do not use AA spectroscopy. How would you have performed the trace metals (second hypothesis) and boron (third hypothesis) analyses if you did not have an AA spectrometer available in your laboratory?

6. With a safety-related problem of this magnitude, is it likely that you and Tom worked alone on this problem? If you need to recruit others from outside of your organization, how can you get them to place the same priority on this problem that you have?

7. Would you have approached this problem differently if only one or two drums were bulging instead of an entire warehouse full of drums? How?

21.2 STUDENT EXERCISE

Think about the Bulging Drum problem. Outline a scheme that describes the steps used to reach the solution. This can be a scheme that is very specific to the Bulging Drum Problem, at least as a start, but our preference is for you to think about the generic process and create a generic plan. It can take whatever form you like (list, chart, diagram, etc.). Please do this exercise before you turn the page and see Figs. 21.3 and 21.4. We want you to think about this, not just memorize it.

The problem solving flow charts in Figs. 21.3 and 21.4 represent different levels of detail. Figure 21.3 presents a general scheme in solving problems. We start by identifying the problem, which often requires us to gather more information and is often the most vital part of the process. For example, in the bulging drum, the problem was not "what is the headspace gas in the bulging drum," but rather, "what caused the drums to bulge?" Properly defining the problem leads us through the rest of the process. Even after we have defined the problem and gathered relevant information, we do not rush to the laboratory. We must first consider which analytical methods are appropriate in proving or disproving any hypothesis we may have formulated. Finally, an analysis is performed. If our interpretation of the analytical results confirms our hypothesis, the problem is solved (after communicating the results to the problem originator and appropriately documenting

A.H. Ullman and D.E. Raynie

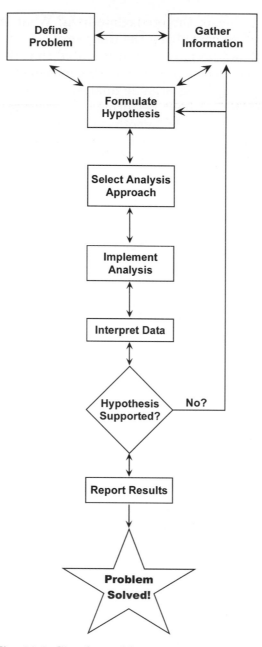

Fig. 21.3. Simple problem solving flow chart

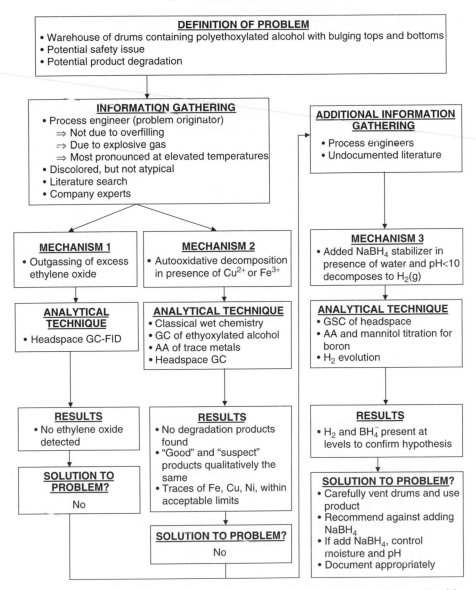

Fig. 21.4. Detailed problem solving flow chart for the Bulging Drum Problem

our findings so that we can prevent the problem from happening again).
Fig. 21.4 is very specific to the steps just discussed in the Bulging Drum
Problem, and follows a similar format. Note that both are flow charts
containing multiple branch points that go back to previous steps. In the

next section of this chapter, we will explain, in detail, each step of the Fig. 21.3 flow chart.

21.3 THE ANALYTICAL APPROACH

As you learned in the Bulging Drum Problem, the modern analytical laboratory works closely with its partners (customers) in the other parts of the company, research institute, etc. When one of these partners comes to the analytical lab with a sample, the request may be as straightforward as, "Please tell me what the hardness of this water sample is." Even when the "problem" is as (apparently) simple as that, the analytical chemist must dig deeper to understand the question (problem) as thoroughly as possible. Many times the submitter may ask for something that does not provide the information he or she really needs. Perhaps the question about water hardness was really an attempt to understand why a precipitate was forming in a pipe in the plant. The submitter has asked for a measurement that supports the submitter's apparent hypothesis. As an analytical chemist, you might help your customer with the problem by analyzing the actual precipitate and then, if necessary, determining the hardness of the water. Remember, the most important step in solving a problem is properly understanding and defining it!

21.3.1 Defining the problem and gathering information

21.3.1.1 Defining the problem
In the Bulging Drum Problem just discussed, the business partner was Tom, the engineer in the warehouse who discovered the bulging drums. It was in the discussion with him that you, the analytical chemist, began to understand the situation and define the analytical problem (note that we really have two problem definitions—the "real" problem (e.g., what caused the drums to bulge) and the "analytically defined" problems (e.g., what is the headspace gas, how is the product different from previous batches, etc.)). In this example, it is clear that the most critical issue is safety; the problem of the drums must be solved rapidly before anything untoward happens. In addition, there are business objectives to solving the problem. These include possible interruptions to production should the raw material be unusable or the quality of previously produced product be unacceptable. To the chemist, the goal of understanding the issue means determining things like the identity of the gas and changes in the composition of the bulk liquid. Business (the

TABLE 21.5

Defining the problem: Describe and define the chemical or physical system(s) from which information must be obtained in order to meet technical or organizational objectives

Technical objective	Organizational objective
Compositional determinations	Assure product quality
Assays	Correct product problem
Specific determinations	Understand competitor's products
Qualitative	Regulatory compliance
Quantitative	Develop new products
Understand the chemistry responsible for proper/improper product performance	
Provide technical leads for new developments	

organization's) objectives often limit the analytical chemists' scope in understanding a problem. Time and money often cause invocation of the 80–20 rule. ("You've already solved 80% of the problem, so we know what's going on well enough. Don't waste any more time.") In Table 21.5, we have summarized some of the different issues involved in defining the problem. The bottom line is that defining the problem is the most important step in solving the problem. If you do not get this right, none of the work you do in the lab will get the needed answer. Like a good newspaper reporter, you need to try to frame every problem with the classic questions: "What, When, Where, How, Why, and maybe even, Who?"

21.3.1.2 Gathering information

In defining the problem, more often than not, it is necessary to gather information to help understand the problem. This begins with your conversation with the individual who brings the problem to you—Tom in the Bulging Drum Problem. The originator is only the starting point, however. Generally there will be conversations with other people in the company; sometimes with another individual in your lab, frequently with someone more senior in your organization, often with someone in a plant, and most importantly, with an information manager in the company library. Of course, sometimes you can find the information you need yourself by searching the company reports or the public literature, but do not be too proud to make a few calls and solicit help. Your goal is to solve the problem for the company in the most efficient manner possible. Frequently, it is more efficient to make a few phone calls, than to spend hours trying to dig through old files, be they electronic or paper! In Table 21.6, we have listed some of the many resources you have in gathering the information you need to define the problem and lead you to the appropriate analytical method to use.

During the information gathering process, you may find yourself wondering if you have really defined the problem correctly. Very often the things you learn from your colleagues or that old organic text will cause you to rethink the whole problem. That is not only okay; it is

TABLE 21.6

Gathering information: Some of the sources of information that may help you in defining the problem and lead to the best analytical approach

Problem originator(s) (the person or people who brought the problem to your attention)

Internal
    Company reports and memoranda
    Laboratory notebooks
    Company analytical methods
    Colleagues (including those at other Company locations), such as managers, project engineers, other analytical chemists, etc.
External
    Published articles (The "Literature," Journals reporting original research)
    Handbooks, encyclopedias, and treatises
    Monographs and review articles
    Textbooks
    Standard analytical methods (e.g., ASTM, EPA, AAOAC, etc.)
    Internet searches (double check the validity of the information)
    Patent literature
    Consultants (University and others)*
    Contract research laboratories*
    Government agencies
    Instrument manufacturers and suppliers

Source: Reprinted/retyped with permission from Analytical Problem Solving: Selection of Analytical Methods, A.H. Ullman, T.M. Thorpe, and G.D. Boutilier in Encyclopedia of Analytical Chemistry, R.A. Meyers (Ed.), 2000. Copyright John Wiley & Sons Limited.
*These contacts will likely require both a confidentiality agreement and funding.

exactly what should happen. The problem solving approach is iterative! As you work through the steps, you will often have to go back to a previous step in the approach. Recall that in the Bulging Drum Problem, three different hypotheses were worked before the problem was solved. After finding that the data did not support the "overfilled with ethylene oxide" hypothesis, additional information led to the "oxidative degradation" hypothesis.

21.3.2 Formulating a hypothesis

After gathering appropriate information about the problem, you will be ready to develop a hypothesis that provides a reasonable explanation of the problem. The hypothesis should be something that you can test. However, keep in mind that, many times, it may be easier to disprove your hypothesis than to prove it. Only after developing a testable hypothesis are we ready to jump to the next step, which begins the truly "analytical" portion of the problem-solving process.

21.3.3 Selecting an analytical approach

When the problem has been defined and needed background information has been studied, it is time to consider which analytical methods will provide the data you need to solve the problem. In selecting techniques, you can refer back to the other chapters in this book. For example, if you want to measure the three heavy metals (Co, Fe, and Ni) that were suspect in the Bulging Drum Problem, you might immediately think of atomic absorption or inductively coupled plasma atomic emission spectroscopies and reread Chapter 8 of this book. How would you choose between them? Which would be more accurate? More precise? Does your lab have both instruments? Are they both in working order? What if you have neither of them? What sample preparation would be needed?

The methods you select, whether instrumental or wet chemistry, must provide data that helps you answer a question related to the hypothesis. Ideally, the methods go right to the heart of the problem, to detect or measure an analyte central to the hypothesis. Ideally, the problem has a "handle" you can try to measure. (We use the term "handle" to describe the problem's key attribute. If a sample is off-color, the color is the "handle" you want to use to solve the problem. If you have bulging drums, the gas causing the pressure is the "handle.")

The analytical methods you consider must meet criteria in the following areas:

Accuracy: The degree of accuracy needed is determined by the question being asked. In the Bulging Drum Problem, does it matter if the hydrogen gas was found to be 12.0% or 12.1% of the headspace? Which calibration method—external standards, internal standards, or standard addition—is appropriate for your analysis?

Precision: The reproducibility of the method likewise must match the needs of the problem. The hydroxyl value method used in the Bulging Drum Problem might not have been precise enough to detect any differences between the "good" and "bad" EOs.

Selectivity: The method you choose must be selective enough to measure the analyte of interest in what may be a complicated matrix. Frequently, not one method is selective enough, and a separation technique must be used before the determination step. Selectivity is a continuum from highly selective to completely non-specific for a given analyte. Different degrees of selectivity can be achieved in different ways (Table 21.7).

TABLE 21.7

Means of achieving selectivity

Chemical reaction	Inorganic qualitative analysis
	Organic functional group analysis
	Electrochemistry
	Radioimmunoassay
	Enzymatic analysis
Physical separation	Solvent extraction
	Chromatography
	Volatilization
	Precipitation
	Filtration
Spectroscopy	
Hyphenated-techniques	Chromatography+selective detectors (e.g., LC-IR, GC-MS)
Data reduction and manipulation	Factor analysis
	Pattern recognition

Limit of detection: The method you choose must be able to detect the analyte at a concentration relevant to the problem. If the Co level of interest to the Bulging Drums was between 1 and 10 parts per trillion, would flame atomic absorption spectroscopy be the best method to use? As you consider methods and published detection limits (LOD), remember that the LOD definition is the analyte concentration producing a signal that is three times the noise level of the blank, i.e., a S/N of 3. For real-world analysis, you will need to be at a level well above the LOD. Keep in mind that the LOD for the overall analytical method is often very different than the LOD for the instrumental analysis.

Sample size and matrix: Your choice of analytical method will also be dependent on the amount of sample you have, especially if the amount is limited and some of the methods under consideration are destructive to the sample. In the Bulging Drum Problem, sample size was not an issue. However, sampling the gas in the drum was challenging, since loss and contamination were quite likely. Getting the samples to the lab presented other challenges. Sample matrix is another important factor in method choice. As you know, some methods and instrumental techniques are not suitable for analysis of solids, without sample preparation. Table 21.8 lists some of the issues that must be considered for different sample matrices.

Interferences: The method must not detect components of a sample, other than the analyte, or substances chemically similar to the analyte.

TABLE 21.8

Sample matrix. The physical form of the material containing the analyte

Solid	Liquid	Gas
• Uniformly dispersed • Segregated in/with particles of various sizes, shapes, etc. • Segregated at the surface • Segregated at a single point	• Uniform solution • Colloidal suspension • Micelles	• Uniform mixture • Suspended particles • Aerosols

Source: Reprinted/retyped with permission from Analytical Problem Solving: Selection of Analytical Methods, A.H. Ullman, T.M. Thorpe, and G.D. Boutilier in Encyclopedia of Analytical Chemistry, R.A. Meyers (Ed.), 2000. Copyright John Wiley & Sons Limited.

See Selectivity, above, and Table 21.9. Industrial problems usually generate samples with complex matrices and many potential interferences. Selective analytical methods or sample preparation are normally required. Separation techniques are quite commonly used. In the average industrial analytical lab, the most numerous instruments are usually gas or liquid chromatographs because they combine separation with detection.

Interferences are the last of the "technical" considerations in selecting an analytical method. Now we will address a series of consideration that are equally important, but are not related to the science of the measurement.

Personnel: Choosing an appropriate technique sometimes comes down to having the right person on your staff to perform the technique. If you, or no one in the lab, had had experience with the hydroxyl value titration, it might not have been an option in the Bulging Drum Problem. Certainly, if the operator of your triple quadrupole mass spectrometer is on vacation and there is no "back-up" operator, another method, or sending the samples to another lab, may be necessary. Should the method need to be transferred to another lab, does that lab have the requisite people for the job?

Number of samples: In the Bulging Drum Problem, several, but not dozens or hundreds, of samples were analyzed for different analytes. That means the complexity of the methods was not a major issue. What if the problem had the potential to recur frequently, requiring daily measurement, or installation of the method in a production facility lab?

TABLE 21.9
Interferences

Sources	Minimizing interferences
Sample	Chemical suppression or Elimination
Reagents	Masking agents
Environment	Precipitation
	Separation
	Extraction
	Volatilization
	Chromatography
	Alternate analytical technique

Analysis time: The time involved, both hands-on and total or elapsed time, is another factor that must be considered. In the same way that the number of samples can be critical to your method selection thinking, if it takes two days to get a result, or ties up an instrument or analyst for hours on end, it may not be a satisfactory choice.

Equipment availability: Obviously, if your lab does not own the equipment needed for a given method, you either choose a different method, or find a lab that can help you out. This might be another lab in your company, a contract lab, or university that will do you a favor or take your money.

Cost: Another consideration is the cost of the method being considered. This could be the cost of consumable supplies (reagents, solvents, etc.) or capital equipment if purchasing an instrument is an option. The cost of using an outside lab must also be weighed. The cost of the measurement plus the number of samples (above) can be significant factors, especially for an on-going need.

Safety: While we list safety last, safety must always be at the top of the list of considerations. For example, you should not even consider a method requiring perchloric acid digestion for sample preparation unless your facility has the appropriate quality fume hood. In Table 21.10, we have listed some of the many safety issues you need to evaluate.

TABLE 21.10

Safety Considerations. Some of these factors can be controlled with appropriate precautions or equipment, but others may lead you to select a different approach to the analysis

Chemical	Physical
Flammability or explosivity	Electrical (high voltage or current)
Corrosivity	Light (intense sources, such as lasers, UV lamps)
Toxicity	Radioactivity/Radiation
Carcinogenicity or teratogenicity	
Breathability (noxious or poisonous fumes)	

Source: Reprinted/retyped with permission from Analytical Problem Solving: Selection of Analytical Methods, A.H. Ullman, T.M. Thorpe, and G.D. Boutilier in Encyclopedia of Analytical Chemistry, R.A. Meyers (Ed.), 2000. Copyright John Wiley & Sons Limited.

A.H. Ullman and D.E. Raynie

21.3.4 Implementing the analysis

Once an appropriate analytical technique has been selected, it is time to consider the actual steps of the measurement process. No, it is still not the time to run into the lab! Proper implementation of an analytical measurement requires the steps shown in Table 21.11.

The last steps in the Implementation are so important they merit further discussion.

21.3.5 Reducing the data and interpreting the results

Raw results from an analytical measurement are rarely useful without some data reduction. For example, what does the number of counts from a GC detector mean by itself? These data must be converted into concentrations and then into information relevant to the problem. The degree of data reduction varies with the kind of data and the needs of the experiment. In the Bulging Drum Problem, the raw counts (signal) from the atomic absorption spectrometric determination of the heavy metals had to be converted into concentrations using calibration curves. The raw GC signal from the measurement of the headspace composition showed the presence of hydrogen gas—the first information supporting a hypothesis. However, converting the signal to a

TABLE 21.11

The steps included in making an analytical measurement

Planning	Plan the experimental program.
Preparation	Assemble the chemicals, equipment, instrument, etc.
Sampling	Arrange to get samples of the material to be analyzed, or better yet, go and perform the sampling yourself.
Pretreatment	Perform the preliminary physical or chemical treatment needed for the samples, standards, and blanks.
Calibration/ standardization	Standardize the reagents; prepare the needed calibration (external, internal or standard addition) standards
Analysis	Make the measurements of the samples, blanks, standards, control samples
Data reduction	Analyze the data generated
Interpretation	Explanation of the results obtained

concentration provided information that was even more useful. Another important aspect of Data Reduction is the statistical validation of the data. Are the results valid? Do the results show a real difference or are they a statistical anomaly? Remember the three key statistical tests you will likely use are Student's t-test (to determine whether there is a significant difference between the mean of replicate measurements and the known, accepted, or certified value), the F-test (to compare different sets of measurements), and the Q-test (to determine outliers).

Nowadays, generating huge amounts of data is relatively simple. That means Data Reduction and Interpretation using multivariate statistical tools (chemometrics), such as pattern recognition, factor analysis, and principal components analysis, can be critically important to extracting useful information from the data. These subjects have been introduced in Chapters 5 and 6.

21.3.6 Problem solved? Not until you report the results!

The final step in the Problem Solving process is to report the results to the customer with whom you have been working so hard. However, it is not enough to "report just the results." The problem is not solved until appropriate recommendations have been made. In the Bulging Drum Problem, these included proper use/disposition of the raw material, changes to raw material specifications, and conversations with the supplier. The problem is not solved until the paperwork (i.e., Memo, Report, Recommendation, etc.) is completed!

21.3.7 Summary

Go back and review Figs. 21.3 and 21.4. Do you notice a logical flow in the sequence of the steps performed? One thing that separates the "skilled" problem-solver from the novice is knowledge of when to move on to the next step. In the Bulging Drum Problem, for example, at what point do you conclude that autooxidative degradation (second hypothesis) is not the solution and it is time to rethink your hypothesis? Keep in mind that solving "real" problems is different than identifying an unknown in undergraduate chemistry classes—they truly are unknown samples and IR spectroscopy is not the appropriate technique to use because you just finished the chapter on IR. When solving problems, there is no single "correct" approach. Anything that gets you the answer in an efficient manner is deemed to be acceptable. (There are, of course, "wrong" answers!!)

Now, let us shift our attention away from the process of solving problems to actually working problems. For the remainder of this chapter we will present you with actual examples of real-world problem-solving. The problems vary in complexity from fairly straightforward to very complex. Each of these really did happen and we present the manner in which it was solved at the time. As you read them, think about the process, as well as the chemistry involved in solving the problem. At each step, ask yourself if there is enough information or if a different analytical technique would also have helped solve the problem. Remember, there is no right way to solve a problem; there are many different ways to get to the solution.

21.4 EXAMPLE PROBLEMS

21.4.1 Problem 1: The cruddy crude coconut oil

For this problem, you are an analytical chemist with a company that processes crude coconut oil into fatty chemicals. The first step in the process is refining the crude oil to remove the few percent of free fatty acids from the oil. Today one of the engineers in the plant half way across the country calls you to explain the problem he is having. An unknown solid in the crude oil is clogging the refinery centrifuge requiring cleanout and causing excessive downtime. He is shipping samples of the oil and the material from the centrifuge to you.

Problem definition: Identify the material that clogs the centrifuge in order to recommend changes that will prevent or reduce future blockages and keep the process running. Lost production time is lost money!

Information gathering: Crude coconut oil consists primarily of triglycerides (the oil), free fatty acids and glycerine (from hydrolysis of the oil), and water. The process of refining the oil uses continuous centrifugation after washing with a dilute caustic solution. (The caustic reacts with the free fatty acids to form water-soluble soaps. The soap/water layer is removed by the centrifuge.) The sample clogging the centrifuge is a slimy, black solid. The crude coconut oil is a yellowish liquid with a slight haze. Previous experience with build-up in pipes and tanks in related parts of the process was sometimes caused by iron particles from worn or shattered pump bearings or damaged piping.

Formulating an hypothesis: The clogging of the centrifuge is due to iron particles mixed with coconut oil, soaps, and other "normal" ingredients of crude coconut oil.

Selecting an approach: The centrifuge sludge was rinsed with solvent to remove coconut oil. It was a fine, dark brown or black powder without metallic appearance. After the solvent evaporated, it was brought near a magnet. The residue was not attracted to the magnet; iron was not the culprit clogging the centrifuge.

Selecting an approach: The solid from the centrifuge might be identifiable by infrared spectroscopy.

Implementation: The IR spectrum is obtained as a smear on salt crystals. Not surprisingly, the spectrum shows soap and triglyceride, but is of poor quality and not suitable for identification of the other components. The dark color of the sample may indicate charring and is likely partly responsible for the quality of the spectrum.

Formulating a hypothesis: A different approach is needed. Identification of the haze in the crude coconut oil may lead to understanding the clogging.

Selecting an approach: The suspended solid material in the crude oil might be identifiable by infrared spectroscopy, but obtaining an identifiable spectrum of the small amount of solid material suspended in the oil requires separating it from the oil. One approach would be to filter and wash the oil out of the residue with a solvent such as hexane. Another approach would be to subtract the spectrum of the oil from the spectrum of the oil+haze.

Implementation: A spectrum of the filtered coconut oil (Fig. 21.5) is obtained and subtracted from the spectrum of the oil containing the haze (solid) (Fig. 21.6). The haze in the crude coconut oil is filtered out of the oil and washed with hexane to remove residual oil. An infrared spectrum of the residue is obtained (Fig. 21.7). The spectra are essentially equivalent (Fig. 21.8). Identification is obtained by search against

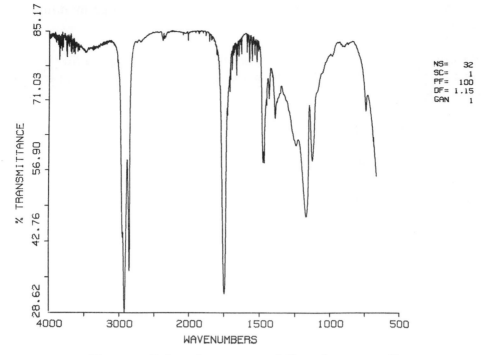

NS= 32
SC= 1
PF= 100
DF= 1.15
GAN 1

Fig. 21.5. Infrared spectrum of filtered coconut oil.

the Sadtler infrared spectral library. The list of matches includes several complex carbohydrates (e.g., guar gum, carob bean gum).

Formulating a hypothesis: It is hypothesized that copra (coconut meal), another complex natural material, is the solid found in the oil and in the centrifuge.

Selecting an approach: A sample of copra is needed so that a reference spectrum can be obtained and compared to the spectra of the material in the oil.

Implementation: A sample of copra is prepared as a reference standard fresh coconut. The spectrum of the standard is a good match for the spectrum of the unknown material from the centrifuge.

Interpretation/report: Coconut meal, not adequately removed during the oil extraction process, is the haze in the oil and is building up in the centrifuge.

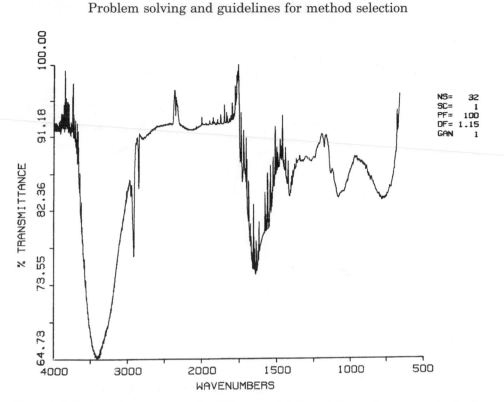

NS=   32
SC=   1
PF=   100
DF= 1.15
GAN   1

Fig. 21.6. Infrared spectrum of solid material (haze) from the coconut oil obtained by subtracting the spectrum of the oil from the hazy oil.

Solution: Recommend tighter specifications for solids in future coconut
oil purchases.

Commentary: The only hypothesis presented was that iron from some-
where in the process might be mixed with the oil and other expected
materials and clog the centrifuge. A simple test with a magnet refuted
this theory. More complex, time consuming, and expensive tests, such
as X-ray fluorescence or atomic absorption, could also have been used,
but would have been "overkill." The gross amount of iron needed to
cause the observed problem would have been detected with the simple
use of a magnet. Infrared spectroscopy is an excellent technique for
identification of organic (and some inorganic) compounds. Combined
with simple separation techniques, or spectral subtraction and over-
lays, the IR proved that higher than normal levels of coconut meal was
the cause of the centrifuge clogging.

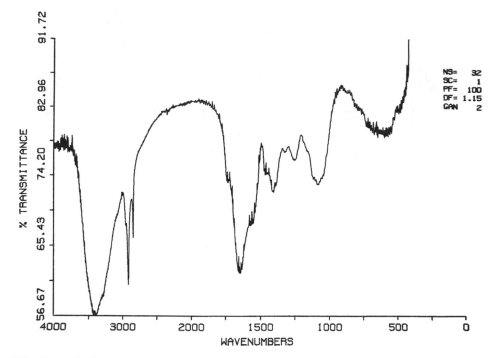

Fig. 21.7. Infrared spectrum of the filtered and hexane rinsed solid material (haze) from the coconut oil.

Thinking Questions and Exercises:

1. Prepare a problem-solving flow chart specific to this problem, similar to Fig. 21.4 for the Bulging Drum Problem.
2. How might you have approached this problem differently? (Experience led to initial suspicion of metal shavings from worn machine parts. What if you do not have the experience to develop this initial hypothesis? Would you have thought of using of a magnet, as opposed to a more sophisticated method like atomic absorption spectroscopy?)
3. The "culprit" in this problem is a complex mixture (copra) as opposed to a single element or compound, like the hydrogen or sodium borohydride in the Bulging Drum Problem. How might this change the way you approach the problem? (Knowledge of the material (in this case) is very helpful to the identification of properties of expected mixtures. Otherwise, statistical algorithms or separation methods might be necessary.)

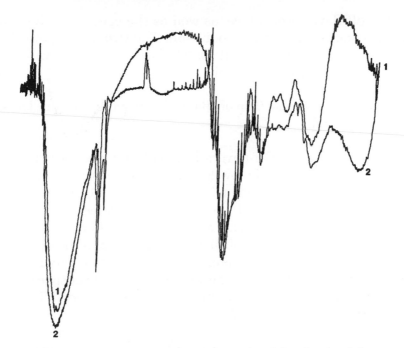

Fig. 21.8. Infrared spectra of the coconut oil solids obtained by extraction (filter and rinse), 1, and infrared subtraction, 2.

21.4.2   Problem 2: An off-flavor in cookies [3]

For this problem, you are an analytical chemist with a company that manufactured ready to eat cookies sold in supermarkets.

Problem definition: A colleague from the cookies product development department of your company has come to your office to describe what appears to be a significant business issue. Complaints of an off-flavor in your chocolate chip cookies, sold throughout the country, are being reported. Samples of the product are being shipped from the plant by overnight air. Production has been stopped. A recall of all product is under consideration! Identification of the off-flavor and its cause are now your top priority!

Information gathering: The off-flavor was noticed intermittently. No formulation changes have been made and no new ingredients have been used during the time at which the off-flavor was noticed. You ask your colleague to collect the lot numbers and batch codes for all ingredients and materials used in making the cookies during the time before and during the off-flavor occurrence, as well as the specific equipment and processing conditions used in production. The two of you pore through all this data and find an apparent correlation between particular lots of polypropylene packaging material.

Formulating a hypothesis: The off-flavor is caused by certain lots of the packaging material.

Selecting an approach: Off-flavors are typically due to volatile compounds present at extremely low levels. (Flavor is sensed more by the olfactory system than the tongue, which senses only 5 flavors, sweet, sour, bitter, salty, and umami). GC is ideal for detecting low levels of volatile components. In this case, headspace GC will allow you to treat the plastic directly. Since the off-flavor is suspected to be derived from the polypropylene packaging material, you decide to compare different samples ("good" vs. "bad") of the material using headspace GC with both a flame ionization detector (FID) and a "sniff port." These chromatograms are shown in Fig. 21.9.

Implementation: The chromatograms of the two samples look identical (Fig. 21.9), but an odor similar to the off-flavor is noted in one region of the chromatogram of the suspect sample.

Selecting an approach: Since the chromatograms appeared identical using the FID, a different technique is needed to allow detection of the off-flavor. GC–mass spectrometry (GC–MS) provides more information than GC–FID and, in selected ion monitoring mode, can detect lower concentrations.

Implementation: GC–MS in an ion-profiling mode was used to compare the two samples. At the retention time where sniff-port detection noted the off-flavor, differences were observed (Fig. 21.10). A compound with a mass-to-charge ratio (m/z) of 104 (the off-flavor suspect?) is co-eluting with another compound.

Selecting an approach: Physical separation of the two compounds will not be easy, but mass spectral subtraction techniques may allow you to obtain a spectrum of the peak of interest.

Implementation: Subtract the spectrum of the interfering peak from one in the region containing the 104^{+} ion so that identification becomes

Problem solving and guidelines for method selection

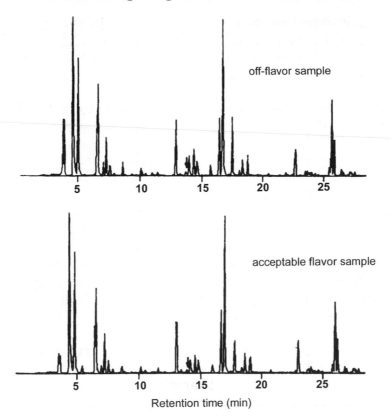

Fig. 21.9. Flame ionization gas chromatograms of the headspace of the acceptable and unacceptable flavor samples. (Reprinted/redrawn from *J. Chromatogr.*, **351**, R.A. Sanders, and T.R. Morsch, "Ion profiling approach to detailed mixture comparison. Application to a polypropylene off-odor problem, 525–531, Copyright (1986) with permission from Elsevier.)

possible. Figure 21.11 shows the spectrum of the unknown (after subtraction), as well as, the best match found in the mass spectral library (ethyl isopropyl sulfide). Absolute confirmation of the identity of the off-flavor is needed, so some ethyl isopropyl sulfide is obtained, and its mass spectrum compared with the unknown. An excellent match is obtained.

Interpretation/report: Ethyl isopropyl sulfide is the off-flavor compound and its source is the polypropylene packaging material.

Solution: Discuss this result with the packaging material supplier and with your purchasing department. Monitor new lots of the packaging

829

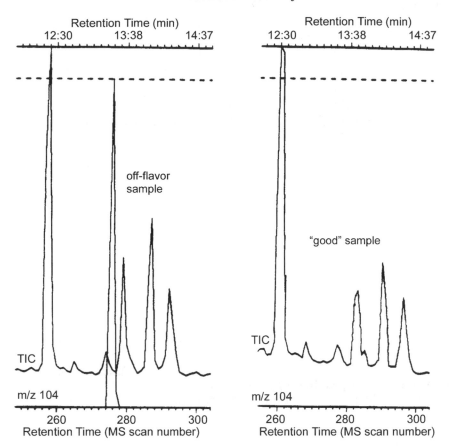

Fig. 21.10. Total ion (TIC) and selected ion monitoring (at m/z = 104) chromatograms of the headspace of the acceptable and unacceptable flavor samples. (Redrawn/redrawn from *J. Chromatogr.*, **351**, R.A. Sanders, and T.R. Morsch, "Ion profiling approach to detailed mixture comparison. Application to a polypropylene off-odor problem, 525–531, Copyright (1986) with permission from Elsevier.)

material for ethyl isopropyl sulfide and share the data with the supplier until the supplier can assure you that the compound is no longer detectable in the polypropylene. File a report in the company library.

Commentary: This problem, like many other off-flavor and off-odor problems, was difficult because the offending compound was present at a very low level and was difficult to separate from another compound present in the sample. The two keys to this problem were the use of the "handle" (the off-flavor), detected with the sniff port, and the mass

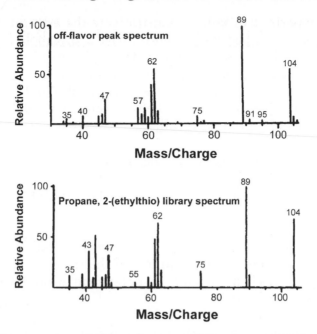

Fig. 21.11. Mass spectra of the unknown off-flavor compound after spectral subtraction from the co-eluting peak and the matching spectrum from the NIST library. (Redrawn/redrawn from *J. Chromatogr.*, **351**, R.A. Sanders, and T.R. Morsch, "Ion profiling approach to detailed mixture comparison. Application to a polypropylene off-odor problem, 525–531, Copyright (1986) with permission from Elsevier.)

spectral techniques that allowed the "hidden" peak to be uncovered and identified.

Thinking Questions and Exercises:

1. Prepare a problem-solving flow chart specific to this problem, similar to Fig. 21.4 for the Bulging Drum Problem.
2. What was the significance of this problem for the cookie company? Before solving the problem, you (the company) might be thinking, "Is it simply stale cookies? Is there a health or safety issue for our consumers? Is something wrong with our cookie recipe or the production process? If we don't solve this right away, how do we win back customers?"
3. How might you have approached the problem differently?

4. Was it important to know quantitatively the amount of ethyl iso-propyl sulfide, or just its presence above a threshold level?
5. Ultimately, whose responsibility is the problem of the off-flavor, the cookie company or the packaging supplier?
6. At one point, a product recall was considered. What considerations do you think a company makes before issuing a recall? What are the implications of a product recall?

21.4.3 Problem 3: The fish kill mystery [4]

For this problem, you are an analytical chemist with the state department of natural resources.

Problem definition: Joe, a water resource specialist calls to tell you that a huge fish kill has just occurred in one of the rivers in your state. The two of you will work as a team to determine what has killed over 300,000 fish.

Information gathering: Fish kills are often due to natural phenomena, such as algal blooms that can deplete the oxygen content of the water. This one was especially severe.

Formulating a hypothesis: Oxygen depletion is killing the fish.

Selecting an approach: The *in situ* dissolved oxygen monitors in the river near the area of the kill report normal dissolved oxygen levels.

Implementation: Confirmation of the *in situ* monitors' results is obtained when river water samples are brought to the lab and tested for dissolved oxygen using a lab dissolved oxygen probe (a polarographic electrode-based measurement) and the classic Winkler oxygen titration method.

Information gathering: Fish kills may also be caused by toxic industrial discharges or agricultural run-off that contains high levels of fertilizers, herbicides, or pesticides. Sometimes old barrels of industrial waste dumped in the river years ago begin leaking into the river. Such barrels occasionally come to the surface. Solutions of toxic heavy metals, such as chromium, are sometimes dumped surreptitiously into rivers. The Environmental Protection Agency (EPA) has published many methods for pesticides, herbicides, etc. in water.

Problem solving and guidelines for method selection

Formulating a hypothesis: River contamination with a toxic material is killing the fish.

Selecting an approach: A nearby lab specializes in mass spectrometric analysis and can perform the EPA screening method for pesticides and other toxic chemicals. Your own lab just bought an inductively coupled plasma emission spectrometer and can analyze the water for heavy metals.

Implementation: Samples of the river water, sediments, and even some of the dead fish are taken to the contract lab where their chemists use GC–MS, after appropriate sample preparation, to determine that there are no toxic compounds at significant levels; i.e., high enough to harm the fish. Your own results for heavy metals are also negative—nothing above safe limits was detected.

Information gathering: The town's sewage plant discharge is located near the section of the river; could the chlorine concentration from the treatment facility be too high? Even low levels can be toxic, not only to fish directly, but also to many organisms lower in the food chain that would ultimately kill the fish.

Formulating a hypothesis: Residual chlorine in treated sewage, discharged into the river, is killing the fish.

Selecting an approach: To check for chlorine in the river water, there are many potential techniques. The simplest one would be a total chlorine test strip, in which the chlorine in the water reacts with chemicals on the paper to change the color.

Implementation: On his next sampling trip on the river, Joe uses the chlorine test strips to check the water.

Interpretation/report: The chlorine results are inclusive so Joe proposes a more direct test.
Selecting an approach: Since fish are the ultimate detector, Joe proposes using fish to test the water.

Implementation: A cage of fish is lowered into the water at the point of the plant discharge.

Interpretation/report: The fish in the cage survive, eliminating chlorine and anything else in the sewage plant discharge as the culprit.

Information gathering: A company that manufactures outboard motors tests prototype motors on stationary boats in the river near the site of the fish kill. With every other hypothesis refuted, that has to be it! The engines burn gasoline, so hydrocarbons could be released/leaked into the water. The exhaust would contain predominantly carbon dioxide and water, but also would have some unburned gasoline, carbon monoxide (CO), and perhaps nitrogen oxides. Volatile organics and unburned gasoline were not at levels high enough to harm the fish. CO seems to be an unlikely culprit because, while it is poisonous, it is not very soluble in water. Nevertheless, it is the only suspect remaining on the list! CO is not easy to detect in water. CO monitors, such as the ones used in homes, are not applicable to water systems. Rather than trying to test the water, what about testing the fish? CO is poisonous because it reacts with hemoglobin forming a stronger bond than oxygen does. This prevents oxygen from being transported to the cells of the body by the blood. Carboxyhemoglobin is different from oxyhemoglobin in several ways and one of them is its UV–Vis spectrum.

Formulating a hypothesis: Though it seems far-fetched, CO poisoning is hypothesized to be the cause of the fish kill.

Selecting an approach: If available, a CO-Oximeter, an instrument used to test human blood by measuring its UV-Visible light spectrum, would be ideal. If unavailable, a standard UV–Vis will be used.

Implementation: The spectrum of samples of blood from several of the dead fish from the kill and some control fish blood are compared.

Interpretation/report: The spectra of the two different blood samples are different (Fig. 21.12). The Oximeter, though calibrated for human blood, indicates 60–70% of the hemoglobin in the fish from the fish kill is carboxyhemoglobin. The control fish showed less than 10%. Human fatalities have carboxyhemoglobin levels above 30%!

Information gathering: Additional techniques are needed to confirm these results.

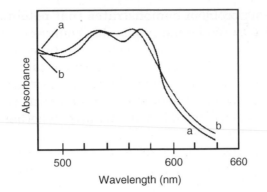

Fig. 21.12. Visible light absorbance spectra of fish blood saturated with oxygen (a) and CO (b). (Reprinted/redrawn with permission from *Chem. Matters*, 6–9 (October 1990). Copyright 1990 American Chemical Society.)

Selecting an approach: The palladium reduction method involves acid decomposition of the carboxyhemoglobin releasing CO. The CO is passed through a solution of a palladium chloride, reducing the Pd^{2+} to elemental Pd that is observed as a "mirror" on the surface of the liquid. $(CO+PdCl_2+H_2O \rightarrow Pd+CO_2+2HCl)$ Fish in a tank were exposed to CO bubbled into the tank. A cage of live fish was placed in the river near the motor testing site; another cage of fish was placed elsewhere in the river.

Interpretation/report: A palladium mirror formed, confirming the presence of carboxyhemoglobin in the fish. The live fish in the tank died quickly after the CO entered the water. After 8 h, the fish at the test site were either dead or dying, while the control fish were healthy. The blood of the dead fish showed CO saturation over 50%. CO from the outboard motors is dissolving in the river, killing the fish.

Solution: Outboard motors discharge their exhaust into the water to reduce noise. Though low in solubility, the propellers, churning at high velocity, were delivering the CO as small bubbles, helping it to dissolve. The company was building more powerful engines every year, increasing the CO. The company installed an exhaust venting system that carried engine exhaust high into the air where it dissipated. The company paid a fine to the state for the fish kill and funded additional studies on fish and the health of the river.

Commentary: This problem demonstrates that potential health issues receive high priority. While many possible sources of the fish kill were identified, they were likely approached simultaneously and the most obvious sources were eliminated from consideration first. While there may be some ethical concerns with the use of live fish as a "pollution detector," it is important to solve this problem in the most efficient manner possible, at least until it is determined whether this is a human health concern.

Thinking Questions and Exercises:

1. Prepare a problem-solving flow chart specific to this problem, similar to Fig. 21.4 for the Bulging Drum Problem. How might you have approached the problem differently?
2. There were many potential sources of the problem, such as algae bloom, municipal wastewater treatment, the outboard motor company, etc. How do you decide which to examine first? Do you approach these many possible avenues simultaneously?
3. To solve this problem, cages of live fish were used to confirm or refute a given hypothesis. How might you have approached this problem without sacrificing live fish?

21.4.4 Problem 4: The unknown pesticide [5]

For this problem, you are an analytical chemist at the local university. The on-campus museum received a donation of a locally produced insecticide from 1927. The museum curator calls you to identify the contents of the bottle.

Problem definition: When you question the curator further, you discover that what she really wants to know is, "Is the bottle of pesticide hazardous? Are there any special storage or handling considerations?" *Information gathering*: You start simply by observing the container. Reading the labels, shown in Fig. 21.13 provided several pieces of information. You remember from your undergraduate ecology class that the organophosphate and organochlorine pesticides used today were not in existence in 1927. You find two phrases on the label rather curious ... "100% Active Ingredients" and "Do not spray into or near open flame." Since it is a clear glass bottle, you can see that it contains a straw-colored liquid. After you take the cork off the bottle, you smell

RID-SECT
"Insecticides of Merit"
Copyright 1927 by Rid-Sect Insecticide Co.

MOTH-KNOX
(100% Active Ingredients)

Kills
Clothes Moths Mosquitoes
Fleas Flies Bed Bugs

Not to be used on Garden or House Plants

CAUTION - Do not spray into or near open flame.

Manufactured by
Rid-Sect Insecticide Co.
Brookings, S. Dak. USA

DIRECTIONS
For best results, use Rid-Sect Sprayer

CLOTHES MOTHS—Hang garments out of doors, brush thoroughly and spray surfaces lightly with *Moth-Knox*, especially all seams and folds. In clothes-packing, spray inside of chest or trunk before putting sprayed garments in. In closets or wardrobes, spray garments thoroughly every week and close tightly.

Moth-Knox Does Not Stain

BED BUGS—Take wooden beds apart. Spray *Moth-Knox* in all crevices and on all slats, springs and mattresses. In severe cases spray behind baseboards, moldings, door-frames, pictures and all loose wallpaper.

FLIES and MOSQUITOES—In rooms, close all doors and windows. Spray freely with *Moth-Knox* filling room thoroughly with vapor. Around garbage cans and other places where flies congregate, spray thoroughly.

FLEAS—In flea-infested rooms, spray floor cracks and crevices thoroughly. Do likewise with cat and dog cushions, rugs, kennels, etc.

PRICE 50 CENTS

Fig. 21.13. Front and back labels from a bottle of insecticide donated to your local museum.

the contents and recognize a petroleum-like odor, while your coworker detects a slight smell that she describes as "mothballs."

Formulate hypothesis: This leads to your hypothesis that the sample consists of naphthalene (i.e., mothballs) dissolved in a petroleum product.

Technique selection: Since gasoline, kerosene, diesel fuel, and other petroleum products are complex mixtures, you use high-resolution capillary GC–MS.

Implementation: Samples of gasoline, kerosene, and diesel fuel are each spiked with naphthalene and are characterized using the same conditions as the sample (Fig. 21.14).

Interpretation/report: The GC retention time of a naphthalene standard and the mass spectrum of this peak confirm its presence. Because of the complexity of the chromatograms of the petroleum products and the pesticide sample, you find it impossible to examine the chromatogram of each. However, a comparison of the GC "fingerprints" (i.e., the matching of chromatographic peaks and comparison of peak ratios) clearly shows that the sample consists of naphthalene dissolved in kerosene.

Information gathering: As you report your results to the museum curator, she decides that there is nothing terribly unique about "naphthalene dissolved in kerosene." She asks you to dispose of the sample according to your university protocols and she will display the empty bottle in the museum.

Solution: You pour the contents of the bottle into your organic waste container and return the bottle to the museum.

Commentary: Ultimately, solving this problem was rather straightforward once the characteristic odor of mothballs was noted. (Keep in mind the safety aspects of smelling unknown chemicals, especially pesticides.) During the information gathering stage, perhaps interviews with elderly residents or perusal of the local newspaper at the time may have provided additional information to guide the analysis.

Thinking Questions and Exercises:

1. Prepare a problem-solving flow chart specific to this problem, similar to Fig. 21.4 for the Bulging Drum Problem.
2. How might you have approached this problem differently?
3. If the museum curator wanted to display the bottle and its contents, would you have enough information to make a recommendation on display conditions?

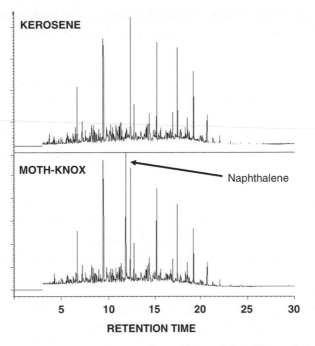

Fig. 21.14. Temperature-programmed capillary GC–MS total ion chromatograms for kerosene (upper trace) and Moth-Knox pesticide (lower trace). Note the similarity in the pattern of peaks, with the exception of the large peak in the pesticide sample (at a retention time of about 12.5 min). The mass spectrum and the retention time of this peak both corresponded to a standard of naphthalene.

4. This problem is a great example of our advice to "find a handle, then separate and identify." What was the handle in this problem?
5. Should you be concerned with the stability of the sample, or any potential degradation, over the past 80 years?

21.4.5 Problem 5: The mummy's make-up [6]

For this problem, you are the analytical chemist at an archaeological research institute.

Problem definition: During the excavation of a Roman temple dating to the middle of the second century AD, a fully intact, small tin container containing a whitish cream was discovered. Such creams are extremely rare and you are asked to characterize and identify the cream.

839

Information gathering: Very little is known about such creams from the Roman age, so you decide on a preliminary analysis.

Formulate hypothesis: Due to the creamy nature of the sample contained in a tin container, you hypothesize that this is a container of a medicinal or cosmetic cream.

Technique selection: Organic elemental analysis (i.e., "CHN analysis") and other simple tests can provide a starting point to refine your approach.

Implementation: Elemental analysis shows that the organic portion of the sample is about 50% carbon, 8% hydrogen, and has no detectable nitrogen or sulfur. About 40 wt% of the sample is soluble in an organic solvent (2:1 chloroform:methanol). GC analysis of the organic-soluble portion is found to be fatty acids (Fig. 21.15).

Interpretation/results: The lack of organic nitrogen and sulfur tells you that there are no significant levels of proteins (e.g., gelatin). The high

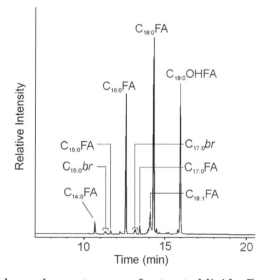

Fig. 21.15. Partial gas chromatogram of extracted lipids. Peaks: $C_{14:0}$ to $C_{18:0}$, saturated straight fatty chain acids (FA) with 14 to 18 carbon atoms; $C_{15}br$ and $C_{17}br$, *iso-* and *anteiso-* branched-chain fatty acids; $C_{18:1}$ FA, mono-unsaturated; $C_{18:0}$OH-FA, hydroxylated form derived by bacterial hydration of the original oleic acid. (Reprinted/redrawn from *Nature*, **432**, 35–36, Copyright 2004, Nature Publishing Group, with permission.)

proportion of oleic acid found in the GC analysis indicates that the fat is of animal origin.

Technique selection: Carbon-13 ratios of the extracted fatty acids may provide further information on the source of the animal fat.

Interpretation/results: A GC-isotope ratio MS technique is used on known animal fats and the ancient samples. (In this technique, each peak eluting from the GC is combusted to CO_2 and its ^{12}C-^{13}C ratio is measured by a mass spectrometer [7].) The ^{13}C ratio of the C-16:0 and C-18:0 fatty acids are plotted. The knowns are concentrated in specific areas on the plot, shown as ellipses in Fig. 21.16. The position of the cream sample points on this "pattern recognition" plot indicates that the fatty portion of the cream is from ruminant adipose tissue.

Technique selection: Organic analysis methods, including GC–MS and FTIR, can further characterize the sample.

Implementation: The GC–MS of the sample headspace finds no perfume compounds. The cream is found to be greater than 80-wt% organic matter. Pyrolysis-GC–MS identified significant levels of glucose polymers, which were confirmed by FTIR to be either cellulose or starch. The iodine test revealed that the glucose polymer was starch. Further GC–MS analysis did not find cholesterol, but did find trace levels of a cholesterol degradation product.

Interpretation/results: The absence of cholesterol, combined with the presence of one of its degradation products, $\Delta^{3,5}$-cholestadiene, indicates that the animal fat had been heated, perhaps as a means of bleaching. Given the mass balance of organic compounds, a significant level of inorganics must be present.

Technique selection: X-ray fluorescence spectroscopy and X-ray diffraction can be used to identify inorganic compounds and their crystal structure.

Implementation: A dry-ashing procedure removed the organic portion of the cream (85%) and X-ray fluorescence indicated large amounts of tin. X-ray diffraction of the cream provided a scattering pattern that was determined to be tin (IV) oxide (Fig. 21.17).

A.H. Ullman and D.E. Raynie

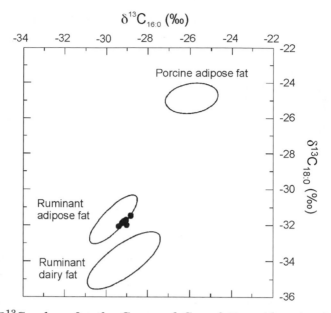

Fig. 21.16. $\delta^{13}C$ values for the $C_{16:0}$ and $C_{18:0}$ fatty acids extracted from the Roman cream, compared with confidence ellipses (1σ) corresponding to those from modern cow, sheep, and pig adipose fat and sheep and cow butter fat (reference $\delta^{13}C$ values are adjusted for post-Industrial Revolution effects of fossil-fuel burning; analytical precision $\pm 0.3\%$). (Reprinted/redrawn from *Nature*, **432**, 35–36, Copyright 2004, Nature Publishing Group, with permission.)

Information Gathering: Research institute historians find accounts that Roman women preferred fair complexions and used whitening creams. Furthermore, Greco-Roman literature indicates that tin compounds had little medicinal value.

Formulate hypothesis: The Roman cream sample is a cosmetic.

Technique selection: Recreation of the sample following the "formula" you uncovered can provide some insight and bolster the hypothesis that this was a beauty cream.

Interpretation/results: Your white cream had a texture similar to the Roman sample. The fat acts as an emollient and the starch reduces the greasiness of the cream. The SnO_2 provides a white appearance when

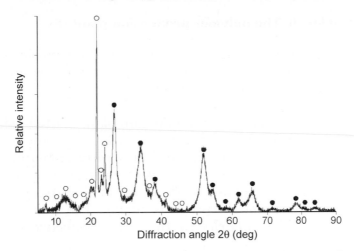

Fig. 21.17. X-ray diffraction pattern of the cream, showing that SnO_2 is present (filled circles). The crystalline fatty acids (open circles) probably formed by deterioration of the animal fat component during burial. Starch is not evident and so must be in a non-crystalline form, consistent with mode of preparation. (Reprinted/redrawn from *Nature*, **432**, 35–36, Copyright 2004, Nature Publishing Group, with permission.)

the cream is applied to skin. All of these ingredients were readily available in Roman times.

Problem solution: The Roman cream is a cosmetic, perhaps some kind of whitening foundation.

Commentary: Since this problem is solved by an academic research institute, they were likely to have greater access to cutting-edge analytical tools than are considered standard. Unlike a problem such as the Bulging Drum, time was not an important consideration in this problem. Re-creating the ancient cream, based on the analytical results, was important to postulating its end-use.

Thinking Questions and Exercises:

1. Prepare a problem-solving flow chart specific to this problem, similar to Fig. 21.4 for the Bulging Drum Problem.
2. How might you have approached this problem differently?
3. How did the limited amount of sample change your approach?
4. Was the problem definitively solved?

21.4.6   Problem 6: The polymer production plant [8]

For this problem, you are a fairly inexperienced (3 years since graduate school) mass spectroscopist working in the corporate research and development organization of a large polymer producer.

Problem definition: Ever since a production plant started using a new supplier of a carbon decolorizing agent, consistently acceptable polymer has not been obtained. The plant's technical director recently read an article about the power of new mass spectrometric techniques, so she sends you an urgent e-mail and sends a sample of the offending carbon and a retained sample of the previous supply. The technical director asks you to find the impurity in the new, ineffective decolorizing carbon.

Information gathering: Polyimide (PI) is made by reacting an aromatic dianhydride (DAN) with an aromatic diamino ether (DA), Fig. 21.18, or more simply,

DAN + DA → PI

The resulting PI is very novel and useful. It is an important product for your company. The polymer and its production process are relatively new and suffer from several problems, including variability in the quality of the DA monomer. To produce DA monomer, it is crystallized from alcohol using a decolorizing carbon. Last month, your supplier of the carbon-decolorizing agent went out of business, so your company

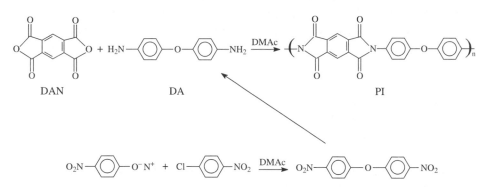

Fig. 21.18. Synthesis of a polyimide from an aromatic dianhydride (DAN) and an aromatic diaminoether (DA). The DA is synthesized by the second (lower) two-step reaction. All reactions were performed in the solvent, demethylacetamide (DMAC).

was forced to switch suppliers. That is when the problems were observed! While it is tempting to analyze the carbon samples sent by the technical director, you know that it is important to define the problem before analyzing the samples. The e-mail you received did not have a lot of information, or at least it left you with many questions. You suspect that an engineer closer to the problem will be able to provide more information. A phone call to the plant puts you in touch with an engineer who explains the significance of the problem—the plant has been shut down and 200 employees have been furloughed!! The engineer confirms that the only change in the process is the new supplier of carbon decolorizing agent.

Hypothesis: The new source of carbon decolorizing agent leads to inconsistent production of PI.

Technique selection: Since you are a mass spectrometrist, you decide a liquid–solid extraction of the solid followed by MS analysis can identify impurities in the carbon.

Interpretation/results: You find no difference between the carbon samples.

Information gathering: While you were running MS on the carbon samples, the plant engineer has been doing some investigation. He finds that the plant making the DA monomer—it is made at a separate plant, 200 miles away—reports that their DA production frequently fails a color test. Their DA can be as much as 6% different from the pure DA standard. The engineer's experiments determine that any difference in the color test greater than 1% leads to poor polymerization. You ask the engineer to send samples of "good" and "bad" DA monomer.

Technique selection: The DA monomer is not volatile enough to characterize with GC, so you choose LC with an IR detector to provide structural information.

Implementation: You extract the "good" and "bad" samples, run them by LC–IR, and compare the results.

Interpretation/Results: The colored component of the "bad" DA monomer is identified by LC–IR as N,N-dimethyl paraphenylene diamine.

Your knowledge of polymer chemistry is weak, so you talk to a polymer chemist down the hall. He informs you that this compound is known to cause chain termination. However, he runs some experiments in his lab and finds that this is not significant for this particular process. Further inspection of the LC–IR data finds another impurity in both the "good" and "bad" samples. You use MS to confirm its identity as a trifunctional derivative of DA. There is no correlation between this impurity and the color test, but the polymer chemist confirms that the presence of this impurity leads to poor polymerization. You call the technical director and inform her that if the impurity levels are minimized, the polymerization process will work.

Problem definition: You are shocked when the plant technical director tells you that her question was not "what caused the failures at the plant," but rather "how do we ensure consistent output from the plant."

Formulate hypothesis: You go back to the original hypothesis that the change of sources of carbon led to the erratic production results.

Information gathering: You review the chemistry and the lack of impurity in the suspect carbon. However, you note that the new carbon source is much more basic than the original source of carbon.

Problem solution: You document the carbon basicity differences and submit the report to your report and the technical director.

Problem definition: The next day the technical director expresses her clear aggravation with you when she tells you that the DA plant finds your answer incorrect.

Information gathering: You decide to contact the DA plant engineers directly. You find that the polymerization takes place in dry solvent. The engineer suspects that plant operators may take shortcuts in drying the solvent.

Implementation: You and the polymer chemist conduct polymerization experiments using DA with varying levels of moisture.

Interpretation/results: It becomes obvious that the solvent moisture during the DA production leads to the inconsistent polymer production results. You share these experimental results with the engineers at both the PI and DA plants.

Problem solution: Both engineers reproduce your results and agree that the moisture level is the controlling factor in polymerization success. The threshold moisture level is determined and documented in the operating procedures for both plants.

Commentary: To solve this problem, the mass spectrometrist needed to work outside of his graduate school specialty (his "comfort zone"). This required him to call upon other resources, like the polymer chemist. The vital importance of communication (between the analyst, the technical director, the PI and DA plant engineers, and the polymer chemist) was essential to defining the problem, and ultimately solving it.

REFERENCES

1 T.M. Thorpe, *Analyt. Chem.*, 56 (1984) 603A.
2 H.A. Laitinen and W.E. Harris, *Chemical analysis*, 2nd ed., McGraw-Hill, NY, 1975, pp. 361–363.
3 R.A. Sanders and T.R. Morsch, *J. Chromatogr.*, 351 (1986) 525–531.
4 H. Black, *Chem. Matters*, 6–9, 1990.
5 T. Froke, Department of Environmental Health and Safety, South Dakota State University, personal communication, 2001.
6 R.P. Evershed, R. Berstan, F. Grew, M.S. Cosley, A.J.H. Charmant, E. Barnam, H.R. Mottram and G. Brown, *Nature*, 432 (2004) 35–36.
7 R.P. Evershed, K.I. Arnot, J. Collister, G. Eglinton and S. Charters, *Analyst*, 119 (1994) 909–914.
8 P.J. Crean, E.I. DuPont deNemours, http://www.udel.edu/ccr/analyt/analyt12.html, January 27, 2005.

REVIEW QUESTIONS

1. Prepare a problem-solving flow chart specific to this problem, similar to Fig. 21.4 for the Bulging Drum Problem.
2. Comment on the role of interpersonal communications, developing a network of contacts, and teamwork in problem solving.
3. How might you have approached this problem differently?
4. Look through your local newspaper and find an article that describes an analytical problem. (This might take you a week or so.

A.H. Ullman and D.E. Raynie

The article could be about an oil spill, a product recall, a forensics case, a health concern, or something else.) In a general readership newspaper, the article probably is devoid of technical details. Think about each step of the problem solving flow-chart. What is the problem? If you were the analytical chemist working with this problem, what information would you need to formulate your hypothesis and where would you have obtained this information? The newspaper article probably only presented the hypothesis that proved to be "correct." What other possible hypotheses may have been considered? What analytical techniques were likely used in solving the problem? Does the fact that this problem was deemed newsworthy influence how you approach the problem? (If the problem was local, perhaps your professor could invite the chemist to talk to your class or you could take a field trip to their lab.) Remember the importance of good communication skills, both written and oral, in solving problems. You need to be able to gather information to define the problem, and you need to report the results and recommendations to complete the solution to the problem. Use Table 21.12 and the Flow Chart (Fig. 21.3) as reminders on problem-solving and method selection.

TABLE 21.12

Problem solving is a valuable and important skill

- Follow the steps of the problem solving flow chart!
- When gathering information, ask the question, "What did you do differently?"
- When beginning to consider analytical methods, ask yourself, "What's the handle to this problem?"
- Remember that real-world samples are frequently very complex and will require you to "separate and then identify or quantify!"

Subject Index

Subject Index

Subject Index

retention
indices 531
in HPLC 536
factors in ion-exchange
chromatography 522
parameters in HPLC 493
times 447, 494, 497
volume 495
reversed-phase chromatography 514,
543
reversed-phase HPLC 525
to partition in octanol 528
reversed-phase liquid chromatography
(RPLC) 514
reversible reaction 669, 675
R$_f$ determination 425
rheology 49
ricin 777, 779, 781
Ritter, Johann Wilhelm 111
rotational transitions 115
RSL-1 Raman 796
Ru complexes 762
rubidium 233

safety 819
salicylic acid 530
sample 16, 544
acquisition 790
direct exposure probe 322
direct insertion probe 322
heated reservoir 322
injection 595
introduction 234, 320–322, 325, 478,
489, 719
matrix 817
number 818–819
preparation 2, 15, 240, 817–819,
833
size 817
trap 790–791
types of 23
sampling 2, 15–16
challenges 16
error 17
rate 472
techniques 478
valid 19

saponification 171
scalar coupled experiments
COSY and TOCSY 280–282
INADEQUATE 282
scandium 233
Schroedinger equation 160
screening 790–791, 795
screw axis 183
sealed containers 791–792, 796
sector mass spectrometers 327, 345
selected ion monitoring 828, 830
selectivity 521, 526, 554, 593, 633, 635,
637–638, 642–643, 646, 648–652,
654, 816
for counter ions 521
self-absorption 229
semiconductors 165
sensitivity 149, 505
enhancement 602
sensor
cation 765, 768
design 755
luminescence 758
optical 11, 755–759, 761–762, 764,
766–770, 775
oxygen 762
pH 763–764
Raman-based 772
sensor, gas
acid or basic 764
separate solution method 650–651
separation 489, 540
factor 562
mechanism 525
process 489
techniques and mass spectrometry 364
theory 447
septum 457
SERS 761
SFC 8, 489
shims for NMR 267
sigma bonding electrons (σ) 118
signal averaging 145
signal-to-noise ratio 148
silicon 233
silver, See Ag
single beam 165
spectrometer 138